INTRODUCTION TO FORESTRY AND NATURAL RESOURCES

T0333737

INTRODUCTION TO FORESTRY AND NATURAL RESOURCES

SECOND EDITION

DONALD L. GREBNER, PhD
Professor
Department of Forestry
College of Forest Resources
Mississippi State University, Mississippi State
Mississippi
United States

PETE BETTINGER, PhD
Professor
School of Forestry and Natural Resources
University of Georgia, Athens
Georgia
United States

JACEK P. SIRY, PhD
Professor
School of Forestry and Natural Resources
University of Georgia, Athens
Georgia
United States

KEVIN BOSTON, PhD, JD
Adjunct Professor
Department of Forestry and Wildland Sciences
Humboldt State University, Arcata
California
United States

ELSEVIER

ACADEMIC PRESS
An imprint of Elsevier

Academic Press is an imprint of Elsevier
125 London Wall, London EC2Y 5AS, United Kingdom
525 B Street, Suite 1650, San Diego, CA 92101, United States
50 Hampshire Street, 5th Floor, Cambridge, MA 02139, United States
The Boulevard, Langford Lane, Kidlington, Oxford OX5 1GB, United Kingdom

Library of Congress Cataloging-in-Publication Data
A catalog record for this book is available from the Library of Congress

British Library Cataloguing-in-Publication Data
A catalogue record for this book is available from the British Library

ISBN: 978-0-12-819002-9

For information on all Academic Press publications visit our website at
https://www.elsevier.com/books-and-journals

Publisher: Charlotte Cockle
Acquisition Editor: Charlotte Cockle
Editorial Project Manager: Kelsey Connors
Production Project Manager: Beula Christopher
Cover Designer: Ryan Cook

Typeset by TNQ Technologies

Instructor review access can be requested here: https://educate.elsevier.com/9780128190029

Printed in the United States of America
Last digit is the print number: 9 8 7 6 5 4 3 2

Dedication

Donald L. Grebner: To my family members, Donald and Ann Grebner, Andrew Friedrich, Karl Friedrich, and Josef Grabner, who inspired my dreams, love of travel, and helped me develop a global perspective. To my mentors, William R. Bentley, Charles Hatch, Benjamin Hoffman, and Gregory S. Amacher, for not giving up on me and helping me grow as a forestry and natural resource professional. Finally, to my wife, Brenda, and children, Karl, Daniel, Amanda, Kristin, and Wayne, for their persistent and unwavering love and support.

Pete Bettinger: To Kelly Bettinger, for never once suggesting a career change was necessary. To Jack Chappell, and other field foresters we have known, for helping us understand.

Jacek P. Siry: To my children, Victoria and Maximilian.

Kevin Boston: To my parents for taking me to the forest early in my life and developing my deep interest in it.

Contents

17. Ethics

18. Forestry and natural resource management careers

Preface

Introduction to Forestry and Natural Resources arose through a desire to produce an introductory text on forestry and natural resource management for first-year college students. However, anyone with an interest in learning about these wide-ranging professions for the first time may find this book enjoyable. Our professional experiences in North America, South America, Europe, Asia, and Oceania provide us with regional, national, and global perspectives on forestry and natural resource management, and these experiences have guided us in the writing of this book. Throughout the book we have included numerous practical examples to help reinforce key concepts of the application of forestry and natural resource management in various locations around the world. One challenge encountered in writing a book of this scope is to adequately cover the vast array of biological, social, and economic factors affecting forests and their natural resources while creating it to be accessible for newly initiated and beginning practitioners. Along these lines, we made a concerted effort to balance the depth and scope of topics presented. We hope readers of this text will be inspired to delve more deeply into topics that interest or concern them, or into issues that they face in their daily lives. We apologize in advance for any omissions or errors found within the text. They are solely our own.

Introduction to Forestry and Natural Resources, Second Edition, is divided into 18 chapters, two appendices, and a glossary. In Chapter 1, we provide an introduction to humankind's interaction with forests and the natural resources commonly associated with forested areas. In Chapter 2, we provide a description of the varied forest regions around the world and then concentrate on the socioeconomic and political issues that have influenced forest development in selected countries. This discussion is segmented by continent, although, given the potential scope of this chapter, we could not provide a summary of every country. Our intent is not to downplay the importance of those countries not mentioned, as many of them play important roles in the development and use of forests and natural resources. In Chapter 3, we outline common objectives of forest landowners and the typical constraints that they face. The differences in management challenges by ownership type should be carefully considered, as these set the tone for forest management

activities. In Chapter 4, we describe the wide variety of timber- and nontimber-based products that can be derived from forests. In Chapter 5, we highlight wildlife—habitat relationships and discuss important issues such as edge effects and habitat management strategies. In Chapter 6, we introduce the concept of ecosystem services and provide a context for further discussion regarding ecological concerns. In Chapter 7, we focus first on the evolution of forest recreation and then describe the importance of recreational interests as a nontimber natural resource on the landscape. In Chapter 8, we provide a summary of the basic forest measurement techniques and the technology commonly employed in collecting information that will eventually help guide management decisions in forested landscapes. This discussion includes topics ranging from the measurement of tree crown sizes and tree ages to the use of global positioning systems and satellite imagery. In Chapter 9, we provide a synopsis of tree anatomy and physiology, covering topics such as tree cell composition, sap flow, and photosynthesis. In Chapter 10, we describe forest dynamics, which includes treatments of forest communities, succession, disturbances, gradients, and niches. In Chapter 11, we provide an outline of common forestry and natural resource practices for managing forested landscapes and provide a discussion of the diversity of practices employed in meeting forest objectives around the world. In Chapter 12, we introduce basic harvesting systems for forested landscapes and describe the various systems in brief so that resource managers will gain an understanding of the methods used to extract commodities from forested areas. In Chapter 13, we introduce the application of economic theory in forestry and natural resource management decision-making, since key components of many public and private decisions involve costs, revenues, and returns on investment. In Chapter 14, we discuss various types of forest disturbances, ranging from wind and volcanic activities to insects and diseases. In Chapter 15, we summarize the types of forest policies and external pressures that can influence the management of forested landscapes. Although each country will have a set of pertinent policies that guide forest management, to illustrate the complexity and scope of these, we provide only a few North American examples. In Chapter 16, we introduce the concept of urban forestry and the role it can play in the health and welfare of forests within towns, cities, and metropolitan areas. In Chapter 17, we describe a portion of the vast realm of conservation and professional ethics, emphasizing the importance of these to our profession. Finally, in Chapter 18, we highlight numerous potential career paths for forestry and natural resource graduates. Of course, the list we provide of careers mentioned is not exhaustive, yet it should provide readers with an indication of the type of responsibilities each professional group accepts.

Two appendices are included with this book. In Appendix A, we list the Latin and common names for the various flora and fauna mentioned in the book, organized by the typical regions of the world in which they are found. In Appendix B, we provide a list of important forestry and natural resource organizations around the world. These organizations are organized by continent. Finally, we provide a glossary of many of the technical terms mentioned throughout the book and, as with Appendix B, this list is not intended to be exhaustive.

Many of the figures provided in this book are from our personal digital libraries or from those of friends and colleagues. We would like to acknowledge and thank several individuals for providing us with photographs used in the second edition of our book. These generous individuals include Stephen C. Grado, Andrew J. Sánchez Meador, Kathy Freeman, A. Taylor Hall, Thomas O'Shea, Starling Childs, Samantha Langley, Florian Geyer, Robert Crook, David W. Wilkinson, Joshua P. Adams, Dawn M. Grebner, Kelly A. Bettinger, Francisco Vilella, Gustavo Peréz-Verdín, Zachary A. Parisa, Walter Sekot, Enoki Yoshio, Jodi Roberts, Jessica Taylor, Lynn Sheldon, J. Taylor Thomas, David Hobson, Hugh Bigsby, David Jones, Dirk Stevenson, Amy Castle Blaylock, Jon D. Prevost, Stacey Herrin, Tibor Pechan, Alexis Londo, Tor Schultz, Jack Chappell, Rich Reuse, Jean-Luc Peyron, Matt Elliott, Brenda F. Grebner, Erika Mavity, Catherine Kirk, Mike Strange, Rachel Reyna, Brian Reed, Joe Burnam, Jason Hall, Wes Sprinkle, Andrew McCarley, Michael Brown, Rachel Martin, Kimpton Cooper, Colleen Sloan, Todd Tietjen, Mark Ducey, Paul Doruska, Michael Torbett, Steve Hunt, Kathleen Knight, Travis Howard, Maralyn Renner, Pete Fulé, Evelyn Martin, Michael Starbuck, Zennure Ucar, and Travis McDonald. We would also like to acknowledge that additional digital imagery was used from sources such as Bugwood.org, Goodfreephotos.com, Mississippi State University, National Aeronautics and Space Administration, Pikrepo.com, PresentationMaps, University of Georgia, US Department of Interior, USDA Forest Service, US Geological Survey, US National Oceanic and Atmospheric Administration, USDA Natural Resources Conservation Service, USDA Technology Development Program, Wikimedia Commons, and Wikipedia. The photo credits for these images can be found in the figure captions throughout the text.

We would like to take a moment to acknowledge and thank a large group of people who made this effort successful throughout both editions of our book. First, we would like to thank Brenda F. Grebner for her tireless efforts in critically reviewing the text of the first edition. Second, we would like to thank A. Taylor Hall for his efforts in the first edition in proofreading, converting measurement units, developing appendices, and assisting with the development of the instructor's manual. Third, we would like to thank Krista Merry for developing several of our figures. We would also like to thank several individuals who provided important information for developing the manuscript and figures. These individuals include Armin Offer, David L. Evans, Jarrod H. Fogarty, Omkar Joshi, Andrew W. Ezell, Andrew J. Londo, Stephen C. Grado, Zhaofei Fan, Isabelle Fahimi, Jeff Hatten, Dave Godwin, Francesco Carbone, Federica Alisciani, Olga Pechan, Richard Campbell, Duffy Neubauer, Matt Elliott, Ikuko Fujisaki, Gabi Geyer, Scott Roberts, Robert K. Grala, Karen Brasher, Laurie Grace, Nancy Thomas, Susan Blanton, Hunter Miles, Sara Baldwin, and Matt Gaw. We would also like to apologize to anyone whom we may have inadvertently forgotten to mention.

The goal of this work was to provide readers with a global and comprehensive overview of the exciting fields of study that center on forests. Although the development process was difficult at times, we were and continue to be excited to have this opportunity to portray the challenging image of evolving forested landscapes to those interested in forestry and natural resource management. Environmental, social, and economic structures constantly change and continuously present us with a complex set of forest and natural resource problems. Fortunately, numerous approaches have been successful in addressing many of these issues, whether they are silvicultural, mathematical, or social in nature. We hope that our brief introduction to forestry and natural resource management will enhance your enthusiasm to delve more deeply into these fascinating fields.

D.L.G.
P.B.
J.P.S.
K.B.

1

A brief history of forestry and natural resource management

With the human population exceeding 7.7 billion individuals in 2019, there is an increasing concern for the world's forests and natural resources due to stresses placed upon them. Forest ecosystems are often at the center of contentious and global debates concerning carbon sequestration, climate change, and biodiversity (Gallardo et al., 2003). With a growing awareness of the role forests play in the global carbon cycle, including the possibility of reducing carbon emissions and increasing carbon sequestration through forest growth processes, a broad segment of human society has developed an interest in forest and natural resource management. Forests not only represent many of the most diverse ecosystems on Earth, they also serve as refuges for many species and provide sustenance and employment for a large portion of the human population (Food and Agriculture Organization of the United Nations, 2016). Forests also provide a number of inherent values (i.e., aesthetic, religious, and simply by their existence) that are important to a wide range of societies around the world. This introductory chapter was designed to provide an overview of some of the very broad themes in forestry and natural resource management in order to set the stage for more in-depth analyses

of forest uses and values later in the book. At the completion of this chapter, readers should understand:

- what forests are and why they are important to humans;
- how human use of forests has progressed from ancient to modern times;
- what differences might exist between the extractive perspective of forest resources and the renewal and managed perspective;
- how the practice of forestry and natural resource management is viewed in the current political and environmental context; and
- what issues affect forestry and natural resource management as they relate to human developmental pressures and the juxtaposition of wildlands and developed areas.

As we noted, this chapter sets the stage for more comprehensive examinations of many of the issues associated with the practice of forestry and natural resource management. The development of a broad understanding of these issues will be of great benefit to new and experienced professionals as they encounter and address complex management problems during their careers.

Introduction to Forestry and Natural Resources, Second Edition
https://doi.org/10.1016/B978-0-12-819002-9.00001-8

1.1 What is forestry and natural resource management?

Throughout history, forests have been revered and championed through poems, proverbs, stories, and fairy tales. Some societies once believed, and some may still, that forests are a gift from a higher power (Zyryanova et al., 2010). Many people will agree that the role of humans within the natural environment has changed over the last 50−100 years, and therefore an understanding of the history of forestry and natural resource management helps develop a context for deeper investigations into the associated environmental, social, and managerial issues. In addition, past experiences can help foresters and natural resource managers develop methods for addressing complex forest and natural resource problems. The problem posed in this section of the book is very basic and is perhaps the motivation for individuals to read a book such as this one. The field of forestry may appear to some to involve only the trees growing on an area of land, while the field of natural resource management may appear to involve everything else that is found in that area. In reality, the fields of forestry and natural resource management are intertwined, often involving multiple types of resources and a diverse suite of expected outcomes. The terms *forestry* and *natural resource management* have existed for a very long time; however, their definitions have evolved owing to shifts in human perspectives toward natural resources. Chiras et al. (2002) defined a *natural resource* as "any component of natural environment, such as soil, water, rangeland, forest, wildlife, and minerals that species depend on for their welfare." With the exception of minerals and petroleum resources, we address many of these components in this book. *Natural resource management* is thus the management of these resources for consumption, conservation, and preservation purposes. *Management* inherently involves forming decisions regarding the use of resources, including forming decisions to do nothing at all. According to the Society of American Foresters, *forestry* is defined as "the profession embracing the science, art, and practice of creating, managing, using, and conserving forests and associated resources for human benefit and in a sustainable manner to meet desired goals, needs, and values" (Helms, 1998). Forestry is, therefore, the art, science, and business of managing forests to achieve a diverse set of goals that range from timber production to ecosystem services. From these definitions, it seems that natural resource management is broader in scope than forest management (perhaps a valid conclusion), yet forests are natural resources, and forestry emphasizes the management of these to address conservation, preservation, and consumptive objectives defined by humans. These

objectives can involve providing wood products, promoting the development of wildlife habitat, maintaining or improving water quality, sequestering carbon dioxide (CO_2) from the atmosphere, promoting the production of meat through grazing, creating recreational opportunities, and facilitating spiritual experiences. In other words, the field of forestry can involve the management of natural resources other than trees. Both fields, if one concludes that they are different, still share a number of common characteristics. For example, each involves broad management principles that rely on economics and ecology, and each involves actions that can be applied at various spatial and temporal scales. Each field also involves the examination of trade-offs among management alternatives, which are inherent in planning, organization, policy development, and public relations activities.

In a broad sense, forestry and natural resource management can each be used to manage land to meet the various needs of humans and other species, and each can be viewed as fields of expertise in which professionals must implement the actions necessary to maintain or enhance biodiversity and ecosystem processes. While this seems all-encompassing, it is not necessarily so, since management practices vary from region to region based on differing social and economic conditions and on differing policies. Further, the objectives and constraints of managing forests and natural resources may change from one landowner (or land management organization or government) to the next. As a result, we are likely to find that certain resources are managed for a single purpose (e.g., to produce timber, to maintain wildlife habitat, or to provide recreational experiences), while other resources are managed for multiple purposes (e.g., to produce some level of income while also improving wildlife habitat quality and providing recreational opportunities; Anderson and Smith, 1976). While some may find the idea controversial, *practical forestry* involves both the use and the preservation of forests, as was noted by Pinchot (1905) over 100 years ago. Forestry is also usually considered different from *arboriculture*, as the former deals with collections of trees while the latter deals with individual trees (Pinchot, 1900). However, *urban forestry* concepts seem now to bridge the gap between the management of large collections of trees and individual trees.

The role and importance of forests to human society continue to evolve. At one point in recent human history, forestry in Europe was associated with the protection of Royal herds of deer. At other points in recent human history, forestry concentrated on the extraction of wood chiefly for the development of housing material and military equipment, such as ships. The use of forestry practices should not imply that forests will disappear; in fact, the Sihlwald Forest in Switzerland was managed for timber production for nearly 1000 years (Roth, 1902), yet within the last 25 it has become a forested, recreational park. It is clear that there are differences in opinion regarding how forests and natural resources should be used. At times, these opinions can be traced to differences in human political or socioeconomic conditions and, at other times, these opinions can be based on differences in educational backgrounds. For example, German forestry professionals perceive changes in nature as predictable and long lasting; thus, they tend to view nature in a manner differently than the general public (Storch, 2011). The following section of this book provides a basic historical background on human views of forests and natural resources, with examples from ancient to modern times. Human interaction with forests, and the associated perceptions of forests and natural resources, has led to a number of opportunities and challenges worldwide. These broad views on the value and use of forests have, in many ways, helped shape the development of modern forestry and natural resource management.

1.2 What are forests?

What are forests? may seem an odd question to ask, but it is one that must be addressed in order to inform the context and background of this book. The word *forest* is said to arise from the Latin words *foris* (outdoors) and *forestis* (unenclosed open ground or woods). Quite some time ago, Allen (1938) expressed the obvious to an early generation of natural resource managers—*forests are essential to human existence*. A forest is a source of food, fuel, shelter, and material, and, beyond that, it is a place within which people can travel and enjoy outdoor experiences, such as forest bathing (unwinding and connecting with nature). Many have attempted to categorize land as either forest or some other land use class. For example, the Society of American Foresters defines a forest as an ecosystem with a dense and extensive tree cover that varies by species composition, structure, age class, associated processes, and which commonly includes meadows, streams, fish, and wildlife (Helms, 1998). The Food and Agriculture Organization of the United Nations (2004), a leading international source of information on agricultural, forestry, fisheries, and rural development, defines a forest as an area of land covering at least 0.5 ha (1.2 ac) containing trees taller than 5 m (16.4 ft) and having more than 10% of the area covered by tree canopies. This definition also comprises land areas recently reforested or currently understocked with trees and areas that are expected to achieve these thresholds over time. Bamboo (e.g., *Bambusa* spp. and others), palms (e.g., *Elaeis* spp.), and other types of

plantations (e.g., rubberwood [*Hevea brasiliensis*] or cork oak [*Quercus suber*]) used primarily for fiber production or protection purposes are also included in this definition of forests, as long as the criteria (land area, tree heights, and canopy cover) are met. The same definition applies to windbreaks, shelterbelts, and corridors of trees, provided these are at least 20 m (66 ft) wide. For better or for worse, this definition excludes land uses that are designed predominantly for agricultural or urban purposes, such as fruit tree plantations or city parks; urban forests are discussed in greater detail in Chapter 16.

Forests can be established by humans or can be naturally regenerated after disturbances (e.g., hurricanes, fires, or logging operations). Tree seeds can also be dispersed by wind and animal action, and these may facilitate the establishment of forests in formerly open areas. In fact, forests are continually changing due to environmental factors, natural disturbances, and human interactions. In essence, forests are constantly changing systems of vegetation (at a slower pace in some areas and at a faster pace in others) that provide a wide variety of functions. A forest can therefore be described as a living community of trees (Fig. 1.1) within which plants and animals reside, reproduce, and forage, and from which humans derive numerous economic, environmental, and social values.

Forests can be located close to rivers or streams, and when these types of water bodies flow through them they may be considered *riparian forests* (Fig. 1.2), as opposed to *upland forests* that are found further away from water bodies. Often, a forested area of a few dozen hectares will usually have some type of stream (perennially flowing, intermittent, or ephemeral) flowing through it, thus it may have both riparian and upland forests. Forests can be found next to large bodies of water, such as lakes or oceans (e.g., *coastal forests*; Fig. 1.3),

FIGURE 1.1 Deciduous forest community in Indiana, United States. *Donald L. Grebner.*

FIGURE 1.2 Riparian and upland forests along the Sacramento River in northern California, United States. *Kelly A. Bettinger.*

FIGURE 1.3 Windswept coniferous forests at Cape Perpetua, Oregon, United States. *Kelly A. Bettinger.*

FIGURE 1.4 Alpine forests around Mount Baker, Washington, United States. *Kelly A. Bettinger.*

FIGURE 1.5 Forests located on the steep mountain slopes of Maui, Hawaii, United States. *Donald L. Grebner.*

FIGURE 1.6 Pine forests in the arid environment of New Mexico, United States. *Donald L. Grebner.*

or high in the mountains (*cloud forests*, *montane forests*, or *alpine forests*; Fig. 1.4). Forests can be found on flat lands (e.g., the flatwoods of Florida) or steep terrain (e.g., the Sierra Nevada or the mountains of Hawaii; Fig. 1.5). Forests can also be found in arid environments (e.g., the *xeric forests* of northern Mexico and the southwestern United States; Fig. 1.6) and in standing bodies of water (e.g., the *bottomland forests* in the swamps and bayous of the southern United States; Fig. 1.7). Some forests thrive in moist environments (e.g., forests along the southern coast of Chile) and humid environments (e.g., forests in the Congo and Indonesia), while others

FIGURE 1.7 Baldcypress (*Taxodium distichum*) forest in the Okefenokee National Wildlife Refuge, Georgia, United States. *Kelly A. Bettinger.*

FIGURE 1.8 Boreal forests of interior central Alaska, United States. *Kelly A. Bettinger.*

FIGURE 1.9 Landscape view from the Waldsteinburg (also known as the Red Castle), Fichtelgebirge, Oberfranken, Germany. *Donald L. Grebner.*

have adapted themselves to heavy snow accumulation during winter (e.g., forests in Siberia, eastern Asia, and northern North America; Fig. 1.8). Forests can currently be found in every continent on Earth except in Antarctica, although forests also existed there millions of years ago (O'Hanlon, 2004). The growth rate of forests varies considerably, from the fast-growing pines and eucalypts (*Eucalyptus* spp.) of the southeastern United States and South America, to the slower-growing spruce and fir forests of northern Canada, Scandinavia, and Russia. Forests sometimes grow next to cities and towns or are scattered across the countryside (Fig. 1.9) and intermixed with agricultural land. Even though the

Food and Agriculture Organization of the United Nations (2004) defines a forest as being as small as 0.5 ha, in developed areas forests might be considered as small as 2 ha (about 5 ac—imagine a rectangle containing approximately five American football or soccer fields), yet in undeveloped areas they might be as expansive as 400,000 ha (1 million ac) or more.

For simplicity's sake, many people think forests are composed primarily of trees. However, a forest is a collection of woody, shrubby, and herbaceous vegetation that can extend vertically from a few centimeters off the ground to as much as a 100 m (about 330 ft) or more vertically (e.g., the redwood [*Sequoia sempervirens*] forests of California). This collection of vegetation can be comprised of trees of various ages and shrub, vine, and herbaceous material distributed in either a random or systematic pattern over a defined land area. To grow, forest vegetation uses light energy from the sun and nutrients and water from the Earth. Light may be intercepted by the canopy (leaves and needles) of the trees but may also infiltrate through gaps in the canopy to the shrubby or herbaceous vegetation below. As a result, some interior forest areas are very dark and forbidding (a value statement) while others are very open and park-like (another value statement) owing to the presence or absence of gaps in the tree canopy. Forest vegetation absorbs carbon dioxide (CO_2) from, and emits oxygen (O_2) to, the atmosphere. Forest vegetation also holds soil in place and prevents it from washing away during severe storms containing heavy rainfall. A forest may contain the structure necessary to meet the nesting, roosting, and foraging requirements of numerous wildlife species. Some of the more common species dependent on forests include animals, both big and small (Table 1.1). The size and location of a forest can significantly influence the types of wildlife that reside there. Some animals, such as deer, prefer forests composed of

TABLE 1.1 Examples of wildlife found in forested areas.

Broad group	Specific example	
	Common name	Scientific name
Deer	White-tailed deer	*Odocoileus virginianus*
	Mule deer	*Odocoileus hemionus*
Elk	Rocky mountain elk	*Cervus canadensis*
	Red deer	*Cervus elaphus*
Bear	American black bear	*Ursus americanus*
	Brown bear	*Ursus arctos*
Moose (Fig. 1.10)	Eurasian elk	*Alces alces*
Beaver	North American beaver	*Castor canadensis*
	Eurasian beaver	*Castor fiber*
Otter	Northern river otter	*Lontra canadensis*
	Southern river otter	*Lontra provocax*
Bobcat	Bobcat	*Lynx rufus*
Fox	Red fox	*Vulpes vulpes*
	Gray fox	*Urocyon cinereoargenteus*
Wolf	Gray wolf	*Canis lupus*
Snake	Eastern diamondback rattlesnake	*Crotalus adamanteus*
Frog	European tree frog	*Hyla arborea*
Salamander	Red salamander	*Pseudotriton ruber*
Worm	Earthworm	*Arctiostrotus vancouverensis*
Rabbit	Forest rabbit	*Sylvilagus brasiliensis*
	Eastern cottontail	*Sylvilagus floridanus*
Chipmunk	Eastern chipmunk	*Tamias striatus*
	Siberian chipmunk	*Eutamias sibiricus*

patches of trees interwoven with agricultural fields and open areas. Other animals, such as chipmunks, have habitat requirements that are smaller in scale and perhaps linked to other resources (such as the character of the soil). Therefore, forests can be viewed as factories storing carbon, generating oxygen, and providing shelter and subsistence to a variety of animals, including humans.

Local climatic conditions can be affected by the amount and character of forest vegetation. For example, within a dense, closed-canopy, mature forest, air temperatures will generally be cooler in the summer and warmer in the winter than the surrounding areas. Humidity will also be higher there owing to lower evaporation rates, and soils will be less subject to frost owing to the insulating effect of forest litter and organic matter. The influence of trees on erosional processes is

significant, as the branches and leaves (or needles) of trees act to mitigate the physical impact of rain through interception processes. Once precipitation reaches the ground, forest litter and organic matter will absorb some of the water, mitigating surface runoff and facilitating seepage into the soil. Water will then move downward through forested soils at a slower rate than if it were moving laterally over bare ground. This downward movement of water through forest soils may allow suspended sediments to be filtered prior to the groundwater entering a stream system. In some areas of the world, the condition and character of forests can also help reduce the rate of snow melting, which may prolong the period of runoff and facilitate seepage of water into the soil system. Forests that provide shade to aquatic systems can also help mitigate stream temperatures and thus provide a higher quality habitat for certain fish species during the summer. Forests within riparian areas may also act as sources of coarse woody debris for aquatic systems, which can result in the development of both food sources and shelter (pools) for fish populations. Within developed areas, whether urban, suburban, or rural residential areas, forests can abate noise pollution generated by vehicles, neighbors, and nearby businesses. In arid regions, forests, planted as windbreaks, may be able to mitigate the drying effect of winds and reduce the detrimental effects of blown soil and snow, thus protecting crops, livestock, and homes. Further, the presence of forests can often enhance property values, and trees are frequently used as barriers designed to reduce glare and reflection from structures made of glass or concrete (Anderson and Smith, 1976). Therefore, while some simply view forests as collections of trees, these systems of vegetation provide a multitude of functions (water filtration, noise abatement, etc.) that humans and other species favor.

1.3 Why are forests important to humans?

As noted in the previous section, forests are areas where humans and other species of animals and insects can forage and take shelter (Fig. 1.10). Since the birth of humankind, forests have played an important role in the survival of our species. The forest canopy provides cover during hot days or rainstorms, and individual trees have traditionally been used as a source of building materials for shelters in which people can protect themselves. In addition, wood has always been a critical source of energy for heating and cooking. Even today, about half of the global wood harvest is consumed for energy purposes (Food and Agriculture Organization of the United Nations, 2010). The woody material of trees is amenable to the creation of many simple and

FIGURE 1.10 Moose in the Kenai National Wildlife Refuge, Alaska, United States. *Kelly A. Bettinger.*

useful tools, some of which can be fashioned or created with relatively little effort. For example, young trees were once the main source of material for the creation of primitive spears, bows, and arrows used for hunting animals (e.g., elk or deer) or for the creation of rudimentary baskets used to transport edible berries and roots. Among others, insects, birds, and fur-bearing animals found in forests provide protein for human consumption. The fur from some wildlife species was, and still is in some parts of the world, viewed as an important source of clothing, especially in the cold climates. Over time, humans became adept at using all parts of hunted wildlife species, either as consumable food or as tools used to enhance their lives. For instance, the stomach linings of some animals were once commonly used to make water jugs, small bones were once sharpened and used as needles, and large bones were once used as weapons for hunting or for self-defense. Uses such as these may continue in some areas of the world.

Forests have also provided humans with many types of plants used for medicinal or consumptive purposes. Two examples of medical drugs that are produced from forest plants are taxol and quinine. Taxol is a mitotic inhibitor used to treat ovarian cancer and is derived from the bark of the yew tree (*Taxus brevifolia*) of the Pacific Northwest of North America. Quinine is an alkaloid used to reduce fevers and to treat malaria, lupus, and arthritis. Natural sources of quinine are trees of the genus *Cinchona* of South America. The complete list of forest plants that can be used by humans for nutritional consumption is extensive and includes a large array of edible berries, nuts, leaves, shoots, roots, and mushrooms. For example, edible berries in North America include blackberry, blueberry (*Vaccinium* spp.), dewberry (*Rubus* spp.), huckleberry (*Gaylussacia* spp.), raspberry, and salmonberry, all of which are popular nutritional sources for humans and other animal species (e.g., bears).

Following the arrival of early colonists in North America, forests provided free forage or feed for animal stock. This forage included acorns, beechnuts, chestnuts, grasses, and roots and shoots of various plants (Williams, 1989). Settlers used wood as a material to create fences and homes (Fig. 1.11), as a fuel for cooking, and as a base for making potash and pearl ash (Williams, 1989). Potash could be used not only as a fertilizer but could also be used in manufacturing processes, such as in the production of bleached textiles, glass, gunpowder, and soaps. Further, during the colonial period in North America, the British Navy reserved large white pines (*Pinus strobus*) having a diameter at breast height of 61 cm (24 in) or greater for potential use as masts for their ships. This reserve became known as the Broad Arrow Policy and was extended to all of America by 1729 (Cubbage et al., 1993). During the early development

FIGURE 1.11   Cabins such as this, in the Watoga State Park in West Virginia, were built by colonial settlers of the United States in the early 20th century. *Kelly A. Bettinger.*

of North America, other products derived from forests included cordage, mask, resin, rosin, tar, and turpentine. Some of these were further used to make lubricants, paints, polishes, roofing materials, soaps, and varnishes (Smith, 1940; Williams, 1989).

Over the last 100 years, a wide variety of wood products have been produced from wood, including pulp and paper, books, newspapers, magazines, sticky notes, folders, personal hygiene products, and toilet paper. Construction materials produced from forests include lumber, plywood, oriented strand board, and other engineered woods. Recently, there has been significant interest in using wood to produce energy pellets and liquid fuels such as butanol, ethanol, and syngas (Perez-Verdin et al., 2009). Other forest products that are more important to us today than in earlier times (e.g., 100 years ago) include sequestered carbon, recreational opportunities, aesthetic values, wilderness experiences, and wildlife habitats. The types of products or services facilitated by forests are extensive, and this discussion has provided only a few examples of the broad range of important products derived from forests. In Chapter 4, we will delve further into the products and other inherent values derived from forests. It may be impossible to predict the pressures that will be applied to forests in the future, but it seems likely that nontimber benefits of forested areas will become increasingly valued (McIntosh, 1995).

1.4 A brief history of human interaction with forests

Forests provide many resources that are useful to humanity's survival, and the need for natural materials to support human life has led to the ongoing contraction and expansion of forested areas around the world. As the world's human population approaches 8 billion, forested areas face the difficult burden of meeting our increasing recreational (Fig. 1.12), conservation, and consumptive needs. Some may argue that our reliance on forest resources to meet these needs may be irrational and that changes in societal values (e.g., increasing funding to expand forested areas, lowering expectations, or finding other ways to meet our needs) may be necessary. Mather (1990) introduced a convenient conceptual model that can be used to evaluate the relationship of humans to forests and how this relationship may change over time. This model suggests that when nonindigenous humans interact with and control forests for the first time, they tend to view forest and natural resources as *unlimited.* This perspective eventually leads to a philosophical contradiction regarding the value of forested areas, since initially people have little regard to sustaining the resources over the long term. A society in general views the *unlimited resource* perspective in a favorable light because of the need for materials related to basic human survival (e.g., food, fuelwood, and building materials). Although warnings may arise, the potential problems of overconsumption may be overlooked given the abundance of the resource. Over time, as a human society begins to deplete the forest and natural resources under their control, they become more concerned with resource reduction and destruction and may then take action (e.g., conservation, reduced consumption, or importing goods) to address local issues. As conservation measures are put into place, at some point local forest growth will recover and forested areas will expand from their depleted state. A society's perception of this common trend (from consumption to conservation) may be viewed either positively or negatively. Although Mather's (1990) conceptual model is simple, it illustrates

FIGURE 1.12   Day hikers along a trail near Mount Rainier, Washington, United States. *Kelly A. Bettinger.*

how forest management perspectives may change over time in relation to changes in human needs and concerns. In addition, the evolution of human perspectives regarding the use of natural resources, as portrayed in this simple model, can influence the development and evolution of the forestry and natural resource professions. Ultimately, there are two main perspectives on the use of forests that shape societal views of the resource: the forest extraction perspective and the forest renewability and management perspective.

1.4.1 The forest extraction perspective

In a developing area, the initial depletion of forests and natural resources may be viewed by some as a natural process, since resources may be considered to have little value owing to their initial abundance. As human communities grow, forests and other natural areas may be converted to agricultural use or may be managed to provide better habitats for wildlife species used as food sources in those communities. For instance, as early European settlements were established, forests and natural areas were converted to agricultural fields to grow food from seeds acquired from other communities across the European continent (Williams, 1989). When early European explorers traveled across what is now the southern United States, forested landscapes were often described as having park-like characteristics, primarily due to the frequent fires initiated by aboriginal people to develop the forage necessary for potentially consumable wildlife populations (Walker, 1991). Over human history, numerous factors have contributed to changes in the character and condition of forests and natural areas, as well as the natural resources that rely on the vegetative structure and protection provided by forests. These factors include human population growth, influences of religious beliefs, iron ore smelting, political instability, poorly designed governmental policies, warfare, fluctuations in international trade, and a general lack of land tenure, along with changes in climatic conditions.

Humans have been affecting the landscape since as far back as 1.4 million years ago during the time of *Homo erectus* (Westoby, 1989). Archeological evidence indicates that humans used fire for cooking purposes and possibly for influencing the types of vegetation found on the landscape. Mesolithic peoples during the Bronze Age (3300−1200 BC) used stone axes and saws for forest clearing but also used forests to graze their domesticated animals, which led to the clearing of large areas of forests. During the Classical Era (18th century BC to 5th century AD), the Greek and Roman societies were heavily dependent on wood from forests as a source of fuel (Meiggs, 1982; Perlin, 1989; Westoby, 1989; Mather, 1990). Greek city-states harvested wood from local forests to make

charcoal, which was then used for home heating purposes as well as cooking and toolmaking (Mather, 1990). As city-states grew, more forest area was depleted, while animal grazing was allowed unchecked in the remnant forests. Greek city-states, such as Athens, eventually had to import wood and charcoal from locations as far away as Macedonia (i.e., at least 350 km or 217 mi). Wood was also important for the Greek navy's need to build a fleet of ships large enough to defeat the Persian navy during a series of military conflicts in the 5th century BC (Meiggs, 1982). These ships, called *triremes*, were a type of galley ship that used a series of wooden oars for propulsion (Fig. 1.13). Alexander the Great of Greece (4th century BC) required wood for building ships that were used to explore new and unknown territories and expand his empire (Meiggs, 1982).

Similar patterns of forest use were observed during the rise of the Roman Empire, as wherever Roman settlements existed, forests were cleared. Roman officials viewed the forested landscape north of the Alps as a limitless resource (Perlin, 1989). At one point, the Roman government implemented incentive programs giving landowners land tenure if they cleared 20 ha (about 50 ac) of forestland (Westoby, 1989), in part because the Roman navy needed wood to build ships in their pursuit of defeating the Carthaginian Empire (Meiggs, 1982). One location, known today as modern Spain, was largely deforested by the Roman Empire because of both their need to build ships and to support the metallurgy of gold, silver, and iron using fuelwood (Perlin, 1989). Another location, southern England, was once covered by small farms dedicated to the export of food to major Roman cities (Perlin, 1989; Westoby, 1989). Efforts

FIGURE 1.13 Greek trireme. *Source: Matthias Kabel, through Wikimedia Commons, from a model located at the Deutsches Museum, Munich, Germany.*
Image Link: https://commons.wikimedia.org/wiki/File:Model_of_a_greek_trireme.jpg
License Link: https://creativecommons.org/licenses/by-sa/3.0/deed.en

further afield led to the clearing of forested areas not only for producing agricultural and metallurgical products, but also for the elimination of hiding cover used by the Picts and Scots during border raids. Roman and other governments of this era also used large amounts of wood to build corduroy (wooden) roads (Fig. 1.14), in part to help defend settlements. During the Roman Era, forests were constantly depleted during times of war for producing weapons such as siege engines, for the smelting of swords and spears, and for shipbuilding purposes (Meiggs, 1982; Perlin, 1989). Both Greek and Roman governments were aware of the potential for resource depletion, and in some ways they implemented governmental control over resources in an attempt to minimize the rate of depletion and to maintain fiber supplies (Westoby, 1989). For example, the Roman government attempted to recycle items such as glass because of the diminished wood supplies needed to manufacture it (Perlin, 1989). In essence, early societies viewed forests as strategic assets, yet unfortunately the dry Mediterranean environment posed a difficult challenge for forest reestablishment via natural means. The resulting topsoil loss from erosional processes associated with these land uses still creates problems today (Butzer, 2005). In fact, the Fertile Crescent of the Middle East, an area composed of parts of Iraq, Israel, Jordan, Kuwait, Lebanon, and Syria, is still unable to respond well to deforestation because of existing climatic conditions (Diamond, 2005). Great Britain responded better to the effects of deforestation, in part due to the more favorable rainfall patterns in northern Europe (Westoby, 1989).

Forests in western and central Europe were largely converted to agricultural lands by the 12th and 13th centuries (Laarman and Sedjo, 1992). The remaining forests experienced intensive grazing, which subsequently

FIGURE 1.14 Excavation of an ancient corduroy road near Oranienburg, Germany. *Saxo, through Wikimedia Commons.*
Image Link: https://commons.wikimedia.org/wiki/File:Oranienburg-breite-strasse.jpg
License Link: Public Domain

degraded their inherent productive capacity. Through the 14th century, the expansion of ore smelting and glassmaking processes greatly depended on fuelwood as the main source of energy. The construction of basilicas (large public buildings) also played an important role in forest depletion from the 10th century to the 14th century (Westoby, 1989). By the 16th century, entire mountain ranges were deforested, and this depletion of forestlands led to greater levels of soil erosion, which fostered landslides and abnormal flooding events (Laarman and Sedjo, 1992). Deforestation spread to eastern Europe by the 17th century, primarily due to the spread of wheat production and the conversion of forests and other natural lands to support agricultural practices. Due to human use of wood, by the end of the 17th century, forests in England had declined to around 8%, from about 60% of the land area prior to the early Middle Ages (Holmes, 1975). Throughout this time, trees were often used in a discriminant manner, by the value that each species provided society (Filková et al., 2015). During the 18th century, the North American colonies, Brazil, and the Caribbean islands became the centers of wood supply (Perlin, 1989), partly due to the Napoleonic Wars in Europe. The British Navy was very dependent on imported wood from Baltic nations in eastern Europe (Perlin, 1989), but Napoleon eventually blockaded this resource. As a result, Britain had to look for alternative sources of wood fiber and, as we mentioned earlier, their quest for alternative wood supplies stretched from eastern North America to India, Australia, and other Pacific Islands. Even as recent as the late 19th century, some areas of Europe were entirely dedicated to charcoal, fuelwood, and lumber production (Ericsson et al., 2005).

When the first Australian state of New South Wales was founded in 1786, forests were viewed as obstacles to growth (Rule, 1967). From the first arrival of Europeans in Sydney Cove, forests were cleared for settlements and for agricultural purposes, and wood derived from these forests was used for a wide variety of purposes, such as cooking, home building, shipbuilding, and toolmaking. An important local tree species, the Australian red cedar (*Toona ciliata*), was also heavily exported. For the next 100 years, an extensive pattern of forest clearings and woodcutting extended throughout Australia, from Tasmania to Queensland and western Australia (Carron, 1985). New Zealand followed a similar pattern of forest development. New Zealand was the last large landmass settled by humans. Polynesian immigrants, later known as Māori, settled there about 800 years ago, and their activities led to the removal of 50% of the forested area and the extinction of several flightless bird species such as the Moa (Perry et al., 2014). When European immigrants began to settle in New Zealand in the late 18th and 19th centuries, large areas of the remaining native forests were felled to facilitate agricultural production as well as to provide wood commodities for export, to support gold mining enterprises, and to build homesteads (Owen, 1966). This extractive perspective of forests continued through the 19th century, but eventually, society became increasingly concerned with the status and condition of the diminishing resource.

Although the extractive perspective was prevalent in the North American colonies since before the 1800s (Perlin, 1989), it accelerated after about 1810. About this time, large sections of the American population began to emigrate from their settlements along the eastern seaboard in search of new homesteads west of the Appalachian Mountains (Williams, 1989). This led to a process of forest clearing for promoting small-scale agricultural production (Fig. 1.15). In some cases, it required a family's entire generation to clear a tract of land for agricultural purposes (Williams, 1989). Between 1810 and 1860, the United States transformed from an agricultural economy to a more commercial one, leading to widespread and commonplace forest clearing for domestic fuel and construction lumber use. Clearing methods involved both clearcutting and girdling of trees, and if the removal of stumps were necessary, this process could require nearly 5 person-weeks per ac using the technology of the time (Williams, 1982). As forests were being cleared for agricultural purposes during this period of growth, the United States entered the Industrial Revolution, which required enormous amounts of wood products for metal production and mining enterprises (Williams, 1989). Various technological innovations, such as the iron plow (or plough) and the steam engine (used both in trains and boats), facilitated the reduction in forested areas and an increase in the speed of wood removal from forests (Perlin, 1989). In the early 1900s, Green (1908) noted that the

supply of mature white pine was decreasing rapidly in the northern and Lake States and that most of the land of good quality was destined for agricultural uses. After the American Civil War and into the early 20th century, the need for wood in various industrial processes led to a wave of exploitation that began in the Lake States and then moved to the southern region and eventually to the West Coast (Williams, 1989) after the completion of transcontinental railroads and associated feeder lines (Hessburg and Agee, 2003).

By the late 19th century, vast forested areas were being utilized for human consumptive needs across the globe, and forest clearing became a major issue in countries such as Australia, eastern Asia, India, New Zealand, South Africa, and the United States (Laarman and Sedjo, 1992). The pattern of rapid declines in forested areas was similar in eastern Europe. For example, in Ukraine, total forested area declined by about 30% during a 100-year period, primarily as a result of economic activity in the latter half of the 19th century (Nijnik and van Kooten, 2000). In the mid-1900s, technological change brought gasoline-powered crawler-tractors and chainsaws to ground-based logging operations, which increased the efficiency of logging and thus enhanced productivity (Hessburg and Agee, 2003). With appropriate forethought, forest management practices can be applied to regenerate and reestablish new forests after logging operations. However, in some cases reestablishment does not occur (causing deforestation) and in other cases land uses change after trees are removed. Despite the growing public concern over forests and natural resources in many developed countries, the pattern of deforestation and land use changes continued into the 20th century in developing countries such as Brazil, Indonesia, and the Philippines (Kummer, 1991). In an estimate by the Food and Agriculture Organization of the United Nations (1997), approximately 180 million ha (445 million ac) of forests were lost between 1980 and 1995, which represents an area larger than Mexico or Indonesia. In the 1990s, 16 million ha (40 million ac) of forests were lost annually. Since 2000, the pace of global deforestation has slowed to about 13 million ha (32 million ac) of forests each year (Food and Agriculture Organization of the United Nations, 2010). While Brazil and Indonesia have managed to reduce their deforestation rates, Australia has lost vast swaths of forests to severe droughts and fires. At the same time, forest area gains through natural forest regeneration and large-scale forest restoration efforts in countries such as China, which has embarked on a massive forest plantation development effort, have reduced the net global forest area loss. More recently, between 2010 and 2015, an annual loss of forest area amounted to 7.6 million ha (18.3 million ac) and an annual gain to 4.3 million ha (10.4 million ac), resulting

FIGURE 1.15 Old farms and agricultural areas where deciduous forests once stood in West Virginia, United States. *Kelly A. Bettinger.*

in a net forest area loss of 3.3 million ha (nearly 8 million ac) (Food and Agriculture Organization of the United Nations, 2016). South America and Africa continue to account for the vast majority of global forest area loss.

1.4.2 The forest renewability and management perspective

In describing the history of Kielder Forest in northern England, McIntosh (1995) suggested that the role of forestry in England changed in the mid-20th century from one of managing strategic reserves of timber and facilitating employment for rural people to managing forests for a broader set of values (timber production, wildlife conservation, and recreational opportunities). Numerous examples such as this can be found at local, regional, and national levels in North America and other continents. According to Mather's conceptual model, as a society increasingly utilizes the resources of its forests, some members of that society become increasingly concerned by forest resource depletion and the loss of intrinsic values. Even though a few historical examples of this evolution in perspective were provided earlier, this concept is as relevant today as it was in earlier times (Perlin, 1989). Unfortunately, while forests are widely considered a renewable resource, the reestablishment of forests through natural or artificial means has not always been successful (Fig. 1.16). For example, tropical

Asian forests are part of a cultural landscape complex, and humans living there use traditional agriculture-based management of the land (shifting agriculture), which some view as an obstacle to effective forest conservation efforts (Ramakrishnan, 2007). With regards to the Mediterranean experience, some widespread natural reforestation did occur at various points of time, especially following the fall of the Roman Empire (Westoby, 1989). Unfortunately, the combination of arid or semiarid climatic conditions and the frequent misuse of the land led to extensive soil erosion, which resulted in many formerly forested sites becoming infertile and subsequently difficult to reestablish with a forest of any type. This combination of conditions is also evident within the Fertile Crescent in the Middle East (Diamond, 2005).

Other parts of the world clearly followed the trajectory of Mather's conceptual model. For example, three locations that experienced widespread forest renewal after the perceived depletion of the original resource became a concern were western and northern Europe, North America, and Oceania. The renewal of European forests is believed to a be result of several factors, including the migration of rural populations to urban areas, the afforestation of agricultural lands, the intensification of agriculture, and the conversion to nonwood energy (Laarman and Sedjo, 1992). In the United States, widespread forest renewal was due to several similar factors, and is also believed to include the abandonment of agricultural fields in the Northeast (Fig. 1.17), the afforestation of cotton and tobacco lands in the South,

FIGURE 1.16   Planted area in southern Georgia, United States, where reforestation success was relatively low. *U.S. Department of Agriculture, Natural Resources Conservation Service, 2019.*

FIGURE 1.17   Onset of the reversion of former agricultural lands to forest vegetation in New England, United States. *U.S. Department of Agriculture, Natural Resources Conservation Service, 2019.*

advances in fire suppression, and decreases in per capita consumption of timber products (Laarman and Sedjo, 1992).

In the United States, concern about forest renewal began about 200 years ago. In an early example from 1818, President James Madison expressed concern over the wasteful destruction of timber in rural America (Williams, 1989). By the mid-1800s, novels and poems written by American authors such as James Fenimore Cooper and William Cullen Bryant were published, and these helped romanticize the beauty, virtue, and wildness of the American forest (Williams, 1989). By the late 1800s, the need to train foresters to address the shortage of timber was being emphasized by Franklin B. Hough and, later, by Gifford Pinchot. By the mid-1900s, timber harvesting in the Pacific Northwest began to slow because most of the original old-growth forest on private land had been harvested and nontimber forest resource values were rising. The value of habitat for protecting the northern spotted owl (*Strix occidentalis caurina*), and endangered species, is one example.

In New Zealand, public concern for the loss of forest and natural resources led to many advances in the afforestation of old pasturelands. In conjunction with the perceived need to afforest vast areas, numerous planting trials of a variety of exotic tree species were conducted throughout the country to meet the growing need for forest products. Today, there still stands a magnificent redwood forest in Rotorua, on the North Island of New Zealand, a legacy of the early tree species trials from the late 1890s (Rotorua District Council, 2019). Between 1925 and 1935, nearly 400,000 ha (about 1 million ac) of land were planted with exotic tree species in New Zealand. Since then, the main emphasis has concentrated on afforesting pasturelands with radiata pine (*Pinus radiata*) originally from the West Coast of North America. Oddly enough, radiata pine was first planted on the plains near Christchurch on New Zealand's South Island where no trees had previously existed (Hegan, 1993) because, while deforestation was occurring elsewhere, this part of New Zealand was desperate for wood. In the latter half of the 20th century, governmental reform led to a distinct classification of forests for two purposes: wood production and the preservation of native forests (Birchfield and Grant, 1993).

The renewability perspective of forests combined with the greater value placed on products they generate has led to stable levels of forestland area in some countries, and even positive growth in countries located in the temperate regions of the world. Even though many tropical countries continue to experience the negative aspects of deforestation and expanding human populations, a significant amount of research and international collaboration has been focused on the afforestation and reforestation of denuded landscapes. In countries such as Brazil, forests created through afforestation and reforestation efforts are an increasingly important aspect of the national economy. Interestingly, the Canadian, United States, and Australian experiences with respect to these two philosophies are very similar. Although the policies, legislation, and terminology used are different in the development of these countries, similar stages of forest exploitation, wood resource protection, multiple-use management, and ecosystem management are evident, particularly with regard to public lands (Lane and McDonald, 2002).

1.5 Forests in the current world political and environmental context

Covering nearly one-third of the Earth's landmasses (Food and Agriculture Organization of the United Nations, 2016), forests are prominently featured in many national and international developmental and environmental policies. As described earlier, forests are essential for the conservation of biodiversity and water and soil resources. Forests are meeting our needs for wood and nonwood products and significantly contribute to carbon cycles. Forests are home to millions of people, whose livelihoods depend almost entirely on the services they provide, and a multitude of other animal species (Fig. 1.18). Yet, too often forests are still perceived by humans as an obstacle to development. It seems clear that a forest renewability perspective on the management of these areas could provide jobs in the logging and wood-processing industries, facilitate certain forms of food production, and help alleviate poverty. However, the chief threat to the health and existence of forests may be, interestingly, of human origin. Therefore, it may be necessary to first address human population problems before forests can be saved.

FIGURE 1.18 Golden-fronted woodpecker (*Melanerpes aurifrons*) located in a riparian forest along the Rio Grande in southern Texas, United States. *Kelly A. Bettinger.*

World leaders, concerned with the continued depletion of Earth's natural resources, gathered at the 1992 United Nations Conference on Environment and Development (or the Earth Summit) in Rio de Janeiro, Brazil. The theme of the meeting centered on the contribution of forests toward achieving sustainable development. One outcome of the meeting was a set of *Forest Principles* for protecting the world's forests. Following the Earth Summit, numerous countries developed criteria and indicators (C&I) to measure and monitor successes in achieving sustainable forest management (SFM) goals: criteria represent forest values that one desires to sustain while indicators measure the progress toward sustaining these values.

The Montréal Process, initiated by the government of Canada in 1993, is the largest of the resulting C&I initiatives, encompassing 60% of the world's forests, 35% of the population, and 45% of the trade in wood and wood products (Montréal Process Working Group, 2005). The Montréal Process criteria include (1) conservation of biological diversity; (2) maintenance of the productive capacity of forest ecosystems; (3) maintenance of forest ecosystem health and vitality; (4) conservation and maintenance of soil and water resources; (5) maintenance of forest contribution to global carbon cycles; (6) maintenance and enhancement of long-term multiple socioeconomic benefits; and (7) development and maintenance of legal, policy, and institutional frameworks for conservation and sustainable management. As is evident in these criteria, progress toward achieving SFM seems to be consistent in many ways with the conservation of biological diversity and other life-supporting ecosystem functions within forests.

Inspired by the Earth Summit, the United Nations Forum on Forests was established in 2000 and served to promote sustainable management of all forests and strengthen the political agreements developed to achieve this goal (United Nations Forum on Forests Secretariat, 2019). Goals of the United Nations Forum on Forests are to reverse the loss of forest cover, to enhance a wide range of forest-related benefits, and to increase the area of sustainably managed forests. The Forum is composed of all member countries of the United Nations. The United Nations General Assembly also declared 2011 as the *International Year of Forests* to promote forest conservation and sustainable management.

Since the Earth Summit, the world achieved some measures of success in reducing the deforestation rate, although deforestation remains alarmingly high in some countries. In China, massive tree planting efforts that began before the Earth Summit and have continued after it appeared successful and helped to reduce the net global deforestation rate (Li et al., 2012). However, only 36% of the world's forests can be classified as primary forests, which are composed of native species with no evident human disturbance (Food and Agriculture Organization of the United Nations, 2016). The area of primary forests continues to decline, primarily in the tropics. While nearly 13% of forests are formally reserved for the conservation of biological diversity, in many cases the effectiveness of these conservation efforts is unknown. Forest insect and disease problems, along with severe wildfires, complicate conservation efforts. For example, since 1990 the mountain pine beetle (*Dendroctonus ponderosae*) has affected 11 million ha (27 million ac) of forests in Canada (Fig. 1.19) and the western United States in a vast outbreak traced in part to milder-than-average winters (Food and Agriculture Organization of the United Nations, 2010).

Progress in the development of national forest programs has been achieved by numerous countries since 2000. Many of these programs are designed to enable comprehensive legal, policy, and institutional framework to support SFM and thus fulfill international forest protection commitments. A Food and Agriculture Organization of the United Nations (2010) report shows that nearly 75% of the world's forests are now covered by national forest programs. Throughout the world, forest policy for, and administration of, public land is primarily the domain of ministries or departments of agriculture. It is estimated that about 1.3 million people work for public forest institutions; however, this number has been declining in recent years. Another 20,000 people work for public forest research institutions. Encouragingly, the number of students worldwide graduating from university forestry programs is rising and exceeds 60,000 annually (Food and Agriculture Organization of the United Nations, 2010). Female graduates account for about one-third of all students, and this proportion

FIGURE 1.19   Mountain pine beetle damage south of Field, British Columbia, Canada. *Mark A. Wilson through Wikipedia.*
Image Link: https://commons.wikimedia.org/wiki/File:BarkBeetleDamageBC.jpg.
License Link: Public Domain

is rising as well. These trends should help forestry and natural resource management organizations maintain a professional forest workforce necessary to address increasingly complex forest management problems.

In recent years, the world's attention has turned to the role of forests in addressing climate change. As trees grow, they store carbon in biomass through the photosynthetic process. About one-quarter of a tree's green weight is composed of carbon. Due to their vast expanse, forests store more carbon in biomass, litter, and soils than does the entire Earth's atmosphere (Food and Agriculture Organization of the United Nations, 2010). When forests are lost through development or land use change processes, massive amounts of carbon are returned to the atmosphere. Carbon emissions resulting from forest loss account for about 13% of global greenhouse gas emissions, about the same as the emissions of the entire global transportation sector (14%) but less than electricity and heat generation (25%) (Herzog, 2009). For these and other reasons, the importance of managing forests in a sustainable manner must be emphasized.

To address these challenges, in 2008 the United Nations launched the Collaborative Initiative on Reducing Emissions from Deforestation and Forest Degradation Program, known formally as UN-REDD or informally as simply REDD (United Nations REDD Programme Secretariat, 2019). This effort is designed to create financial values for carbon stored in trees, with the goal of preventing further losses of forest area and the associated increases in carbon emissions. Since benefits of avoided carbon emissions are global, and forest protection costs are usually local, the underlying idea is that rich countries should reimburse poor countries for their preservation efforts. Saving forests from destruction, in addition to preventing carbon emissions, yields a range of cobenefits that include, among others, the conservation of biodiversity. An enhanced version of this program, REDD+, reaches beyond deforestation and degradation to encompass conservation and enhancement of forest carbon storage through improvements in forest management.

Broad-ranging and multifaceted conservation programs such as REDD encounter numerous challenges. For example, in these cases long-term program funding is uncertain, and when funds are available they flow to countries with the worst deforestation record. This, in turn, may create a moral hazard and induce other countries to accelerate deforestation processes in order to induce donors to pay them to cease these activities. Furthermore, many countries experiencing rapid deforestation suffer from endemic corruption, which raises questions regarding how effectively funds will be spent and the extent to which governments will honor their forest conservation commitments. However, if programs such as REDD succeed in making sustainable forestry worth more, then there is a good chance that these programs will achieve their broad range of goals.

1.6 Human developmental pressures on forests

As human populations continue to expand outward from urban population centers, the ecological, economic, and social pressures on outlying forests and natural areas will ultimately increase (Fig. 1.20) and perhaps come into conflict with the broad scope of a region's natural resource management objectives. The *rural-urban fringe* is a general term for human development near to, or abutting, forested or other natural areas. This area represents a transition zone between urban and rural land uses. Other names for this transition zone are *urban hinterland* and *wildland urban interface*. Some general factors that are important in defining the rural-urban fringe include the housing density, the human population density, the distance from homes to wildland vegetation, and the condition of the current and future wildland vegetation.

In Europe, the rural-urban fringe is generally described as landscapes where the majority of land is devoted to some sort of rural use, yet the presence of urban influences (e.g., shopping centers, manufacturing facilities, or dense human housing areas) is obvious. In the United States, three categories of land use have been defined in relation to the rural-urban fringe: interface communities, intermixed communities, and occluded communities (U.S. Forest Service, Bureau of Indian Affairs, Bureau of Land Management, Fish and Wildlife Service, and National Park Service, 2001). Population density, human structure density, and proximity to wildland fuels are generally used to determine whether land belongs to these groups.

FIGURE 1.20 Development activities in Charlotte, North Carolina, United States. *U.S. Department of Agriculture, Natural Resources Conservation Service, 2019.*

In the rural-urban fringe of North America, the main concerns are fire behavior potential, at-risk human values, and the infrastructure to support firefighting capabilities. Rural-urban fringe areas have become the central focus of a number of wildland fire policies, particularly in the United States (Stewart et al., 2007). Many resource management organizations are actively working with local communities to help increase the level of awareness of issues associated with the gradual encroachment of human populations into areas that may now or in the future be characterized as having a high fire risk. Fire behavior potential is a function of the condition of the landscape and the condition of wildland fuels to support large fires, the frequency of fires, and the effectiveness of firefighting agencies to control fires once they have begun. Therefore, the condition, extent, and proximity of fuel loads to homes and businesses (Fig. 1.21) may need to be addressed if land development is allowed to proceed into what were once rural areas. The values at risk include personal human possessions, property, companion animals, and the potential for residents or businesses within a community to incur economic losses. Ecological values, such as the potential for soil erosion or flooding after fires, may also be considered as a set of values at risk. Concerns about the infrastructure necessary to access water resources and control fires, and the road systems that provide firefighting equipment access, are also important. The location and condition of these resources may directly affect the ability of firefighting organizations to control fires within rural-urban fringe areas.

FIGURE 1.21 High-density housing area adjacent to the Jones State Forest in Texas, United States. *Douglas J. Marshall, from Marshall et al. (2008).*

1.7 What are the major challenges to forests in the future?

Given the importance of forests to humankind, a major challenge to their well-being and ability to provide critical ecosystems services to modern societies is the growing presence of climate change and its potential impact on the world's economy and terrestrial biomes. The uncertainties surrounding the impacts of climate change will make the management of forested landscapes more complex and difficult as changes in weather patterns and projected human population growth stress existing forested ecosystems. The potential impacts of climate change on forested landscapes are numerous and include species shifts, changes in growth rates of existing plant species, changes in the prevalence of water availability due to greater frequency of drought conditions as well as a decline in water quality, and increased vulnerability of forested landscapes to new forest pests and invasive species.

Climate change is expected to have an extensive impact on the management of water resources in many parts of the world. In particular, many climate models expect that rainfall patterns will change around the world which could not only have an important impact on growing conditions, but also limit the ability of forests to supply water to local communities. Numerous studies suggest that snow pack size and streamflow will diminish and the duration of and frequency of droughts will increase in some areas of the world (Ali et al., 2018; Azmat et al., 2018; Pérez-Palázon et al., 2018). This situation will be further challenged by the expected rise in global human population levels, as more people will demand more water from forested landscapes.

Another impact of climate change on forested landscapes will be the introduction and spread of invasive species. The introduction of any number of these plant, animal, and insect invaders can have a dramatic effect on the structure and function of forested landscapes and its natural resources. The spread of invasive grasses such as cogongrass (*Imperata cylindrica* L. [Beauv.]) into southern United States forested landscapes can outcompete natural vegetation which could displace the growing space from producing more traditional ecosystems services such as timber production or wildlife habitat, but it could also change the system by making it more dependent on fire. Other invasive species such as the emerald ash borer (EAB) (*Agrilus planipennis* Fairmaire), which kills over 95% of the ash trees it invades, greatly diminish the biodiversity of forest landscapes and the many ecosystems services it provides.

Climate change may also gravely impact forested landscapes through the potential escalation of extreme weather events. It is expected that extreme weather events such as hurricanes (typhoons), tornadoes, floods, droughts, and wildfires will drastically affect the structure and function of forested landscapes by destroying the canopy layer, altering the species composition, inhibiting regeneration through extended drought conditions, and salt desiccation from coastal flooding. The impact of extreme weather events associated with climate change will have the potential of diminishing the ecosystems services that forest landscapes provide to local and national communities and reduce the economic outputs dependent on these natural resources.

Summary

Historically, humans have enjoyed, feared, used, and conserved forests and natural resources. Forests and natural resources have played important roles in the development and history of human society, dating back to early hunting and gathering communities (Ramakrishnan, 2007). Forests have consistently provided humans with a variety of consumable foods and medicines, as well as wood for cooking and heating and materials for building shelters and homes. As human settlements have expanded, forests and their associated natural resources have increasingly been removed to provide land for agricultural activities, as well as to provide materials to support economies, defense activities, and trade. This phenomenon has occurred consistently around the world. Generally, forest areas in relatively wet climates that have experienced widespread deforestation and degradation recover well through natural regeneration processes or through active planting programs. Forest areas in drier climates, such as countries around the Mediterranean Sea (e.g., Italy, Greece, and Turkey), fare worse due to greater human population pressure, steep topography, and challenging reforestation issues.

For much of human history, forests and their associated natural resources have been viewed from an extractive perspective. There are many instances where natural resources such as timber and wildlife have been used with little regard to their long-term sustainability. In both ancient and modern times, as human settlements began to locally and regionally exhaust these resources, public consciousness toward sustainability grew. An increased concern for renewability generally leads to the development of policies and technologies aimed at expanding the supply of these resources. Current forests are a by-product of numerous landscape changes, either natural or anthropogenic. The manner in which human societies have valued forests and natural resources does in fact change. Today, our general perspective of the role of forests is complicated, in part, by increased international trade and greater interactions between dispersed communities through advances in communication technology. As a result, emerging conflicts between developing areas and developed countries regarding the use of forests and natural resources are real and continue to grow (Vogt et al., 2010).

As you may have gathered, the main focus of many sections of this book is North American or European forestry and natural resource management in countries where ample documentation of forest management activities is available. This limitation is recognized and, admittedly, we very likely have missed valuable knowledge arising from countries that have not yet played a significant role in the global wood products market. Further, valuable knowledge disseminated through non-English sources, journals of limited circulation, and unpublished reports has likely been overlooked. These types of omissions should not be seen as a dismissal of the significant role that forestry and natural resource management has played in countries such as Venezuela (Kammesheidt et al., 2001), for example. An individual pursuing a career in forestry and natural resource management is likely to encounter numerous interesting and challenging developments facing the conservation and use of forests and natural resources, and prior examples of these are likely be found in other regions of the world.

Questions

(1) *Your daily life.* Write a short essay describing how forests affect your daily life. Look around your home or school and describe things you use that come from a forest. Also describe those things you use that do not come from a forest. If you are unsure, place these in an "uncertain" category. The Internet may be helpful in determining the origin of some products. Of the three categories, which category contains more items? Why? Is this what you expected prior to beginning this exercise?

(2) *Daily news.* Examine either the paper copy or the Internet version of your local newspaper and identify articles that discuss forestry or natural resources. Perform this task for five business days, or Monday through Friday. Develop a short, professional PowerPoint presentation that illustrates the sources of

Continued

QUESTIONS   (cont'd)

these stories and the topics discussed. Describe how many articles you found, and whether the topics varied throughout the week.

(3) *Criteria and indicators.* Locate a land management organization that has developed a resource management plan based on C&I. Either arrange a meeting (in person or over the phone) with one of the land managers knowledgeable about the plan or obtain the plan itself. In a one-page memorandum format, develop a briefing that summarizes the use of the C&I in guiding the management of the land.

(4) *REDD programs.* Perform an Internet search and locate a description of a recently developed REDD project. In a short report, describe the stakeholders involved, the purpose of the project, the organization that funded

the project, and the location of the project. Once developed, share this with another person or group who has done the same, and compare and contrast the two projects.

(5) *Forest development.* For either the place where you live or a place of interest to you, develop a short report on the history of its forest development. Locate a person with intimate knowledge of the area and interview them. You might also rely on both published literature and anecdotal stories from other people with knowledge of the area. You should define a time frame and focus on issues relevant to that period in developing this overview. In addition, visit the location and take a few pictures of relevant resources. Finally, develop a 5 to 10 slide PowerPoint presentation on the forest.

References

Ali, S.A., Aadhar, S., Shah, H.L., Mishra, V., 2018. Projected increase in hydropower production in India under climate change. Scientific Reports 8 (1), 12450.

Allen, S.W., 1938. An Introduction to American Forestry, first ed. McGraw-Hill Book Co., Inc., New York.

Anderson, D.A., Smith, W.A., 1976. Forests and Forestry. The Interstate Printers & Publishers, Inc., Danville, IL.

Azmat, M., Qamar, M.U., Huggel, C., Hussain, E., 2018. Future climate and cryosphere impacts on the hydrology of a scarcely gauged catchment on the Jhelum river basin, Northern Pakistan. The Science of the Total Environment 639, 961–976.

Birchfield, R.J., Grant, I.F., 1993. Out of the Woods: The Restructuring and Sale of New Zealand's State Forest. GP Publications, Wellington, New Zealand.

Butzer, K.W., 2005. Environmental history in the Mediterranean world: cross-disciplinary investigation of cause-and-effect for degradation and soil erosion. Journal of Archaeological Science 32 (12), 1773–1800.

Carron, L.T., 1985. A History of Forestry in Australia. Australian National University Press, New South Wales, Australia.

Chiras, D.D., Reganold, J.P., Owen, O.S., 2002. Natural Resource Conservation, eighth ed. Prentice-Hall, Inc., Upper Saddle River, NJ.

Cubbage, F.W., O'Laughlin, J., Bullock, C.S., 1993. Forest Resource Policy. John Wiley & Sons, Inc., New York.

Diamond, J., 2005. Guns, Germs, and Steel: The Fates of Human Societies. W.W. Norton & Company, New York.

Ericsson, T.S., Berglund, H., Östlund, L., 2005. History and forest biodiversity of woodland key habitats in south boreal Sweden. Biological Conservation 122 (2), 289–303.

Filková, V., Kolár, T., Rybníček, M., Gryc, V., Vavrčík, H., Jurčík, J., 2015. Historical utilization of wood in southeastern Moravia (Czech Republic). iForest 8, 101–107.

Food and Agriculture Organization of the United Nations, 1997. State of the World's Forests, 1997. Food and Agriculture Organization of the United Nations, Rome, Italy. http://www.fao.org/3/w4345e/w4345e00.htm (accessed 13.08.19).

Food and Agriculture Organization of the United Nations, 2004. Global Forest Resources Assessment Update 2005, Terms and Definitions.

Food and Agriculture Organization of the United Nations. Forest Resources Assessment Programme, Rome, Italy. Working Paper 83/E.

Food and Agriculture Organization of the United Nations, 2010. Global Forest Resources Assessment 2010. Food and Agriculture Organization of the United Nations, Rome, Italy. FAO Forestry Paper 163.

Food and Agriculture Organization of the United Nations, 2016. Global Forest Resources Assessment 2015, second ed. Food and Agriculture Organization of the United Nations, Rome, Italy.

Gallardo, F., Fu, J., Jing, Z.P., Kirby, E.G., Cánovas, F.M., 2003. Genetic modification of amino acid metabolism in woody plants. Plant Physiology and Biochemistry 41 (6–7), 587–594.

Green, S.B., 1908. Principles of American Forestry. John Wiley & Sons, Inc., New York.

Hegan, C., 1993. Radiata—prince of pines. New Zealand Geographic 20, 88–114.

Helms, J.A. (Ed.), 1998. The Dictionary of Forestry. Society of American Foresters, Bethesda, MD.

Herzog, T., 2009. World Greenhouse Gas Emissions in 2005. World Resources Institute, Washington, D.C.

Hessburg, P.F., Agee, J.K., 2003. An environmental narrative of Inland Northwest United States forests, 1800–2000. Forest Ecology and Management 178 (1–2), 23–59.

Holmes, G.D., 1975. History of forestry and forest management. Philosophical Transactions of the Royal Society of London - B 271 (911), 69–80.

Kammesheidt, L., Lezama, A.T., Franco, W., Plonczak, M., 2001. History of logging and silvicultural treatments in the western Venezuelan plain forests and the prospect for sustainable forest management. Forest Ecology and Management 148 (1–3), 1–20.

Kummer, D.M., 1991. Deforestation in the Postwar Philippines. University of Chicago Press, Chicago, IL. University of Chicago Geography Research Paper No. 234.

Laarman, J.G., Sedjo, R.A., 1992. Global Forests: Issues for Six Billion People. McGraw-Hill, Inc., New York.

Lane, M.B., McDonald, G., 2002. Towards a general model of forest management through time: evidence from Australia, USA, and Canada. Land Use Policy 19 (3), 193–206.

Li, M.-M., Liu, A.-T., Zou, C.-J., Xu, W.-D., Shimizu, H., Wang, K.-Y., 2012. An overview of the "Three-North" Shelterbelt project in China. Forestry Studies in China 14 (1), 70–79.

Marshall, D.J., Wimberly, M., Bettinger, P., Stanturf, J., 2008. Synthesis of Knowledge of Hazardous Fuels Management in Loblolly Pine Forests. U.S. Department of Agriculture, Forest Service, Southern Research Station, Asheville, NC. General Technical Report SRS-110.

Mather, A.S., 1990. Global Forest Resources. Timber Press, Inc., Portland, OR.

McIntosh, R., 1995. The history and multi-purpose management of Kielder Forest. Forest Ecology and Management 79 (1–2), 1–11.

Meiggs, R., 1982. Trees and Timber in the Ancient Mediterranean World. Oxford University Press, Oxford, United Kingdom.

Montréal Process Working Group, 2005. The Montréal Process. Montréal Process Liaison Office, Tokyo. https://www.montrealprocess.org/ (accessed 13.08.19).

Nijnik, M., van Kooten, G.C., 2000. Forestry in the Ukraine: the road ahead? Forest Policy and Economics 1 (2), 139–151.

O'Hanlon, L., 2004. Antarctic Forests Reveal Ancient Trees. ABC Science, Sydney, New South Wales, Australia. https://www.abc.net.au/science/articles/2004/11/09/1239175.htm (accessed 13.08.19).

Owen, R.E., 1966. New Zealand Forestry. Government Printer, Wellington, New Zealand. New Zealand Forest Service Information Series No. 41.

Pérez-Palazón, M.J., Pimentel, R., Polo, M.J., 2018. Climate trends impact on the snowfall regime in Mediterranean mountain areas: future scenario assessment in Sierra Nevada (Spain). Water 10 (6), 720.

Perez-Verdin, G., Grebner, D.L., Sun, C., Munn, I.A., Schultz, E.B., Matney, T.G., 2009. Woody biomass availability for bioethanol conversion in Mississippi. Biomass and Bioenergy 33 (3), 492–503.

Perlin, J., 1989. A Forest Journey: The Role of Wood in the Development of Civilization. Harvard University Press, Cambridge, MA.

Perry, G.L.W., Wheeler, A.B., Wood, J.R., Wilmshurst, J.M., 2014. A high-precision chronology for the rapid extinction of New Zealand moa (Aves, Dinornithiformes). Quaternary Science Reviews 105, 126–135.

Pinchot, G., 1900. A Primer on Forestry, Part I—The Forest. U.S. Department of Agriculture, Bureau of Forestry, Washington, D.C. Bulletin No. 24.

Pinchot, G., 1905. A Primer on Forestry, Part II—Practical Forestry. U.S. Department of Agriculture, Bureau of Forestry, Washington, D.C. Bulletin No. 24.

Ramakrishnan, P.S., 2007. Traditional forest knowledge and sustainable forestry: a north-east India perspective. Forest Ecology and Management 249 (1–2), 91–99.

Roth, F., 1902. First Book of Forestry. Ginn & Company, Publishers, Boston, MA.

Rotorua District Council, 2019. About the Forest. Rotorua District Council, Rotorua, New Zealand. https://redwoods.co.nz/about/ (accessed 13.08.19).

Rule, A., 1967. Forests of Australia. Angus & Robertson, LTD, Sydney, Australia.

Smith, H.F., 1940. Primary Wood-Products Industries in the Lower South. U.S. Department of Agriculture, Forest Service, Southern Forest Experiment Station, New Orleans, LA. Forest Survey Release No. 51.

Stewart, S.I., Radeloff, V.C., Hammer, R.B., Hawbaker, T.J., 2007. Defining the wildland-urban interface. Journal of Forestry 105 (4), 201–207.

Storch, S., 2011. Forestry professionalism overrides gender: a case study of nature perception in Germany. Forest Policy and Economics 13 (3), 171–175.

United Nations Forum on Forests Secretariat, 2019. United Nations Forum on Forests. United Nations Forum on Forests Secretariat, New York. https://www.un.org/esa/forests/index.html (accessed 13.08.19).

United Nations REDD Programme Secretariat, 2019. UN-REDD Programme. United Nations REDD Programme Secretariat, Geneva, Switzerland. https://www.un-redd.org/ (accessed 13.08.19).

U.S. Department of Agriculture, Natural Resources Conservation Service, 2019. Web Soil Survey. U.S. Department of Agriculture, Natural Resources Conservation Service, Washington, D.C. https://websoilsurvey.sc.egov.usda.gov/App/HomePage.htm (accessed 12.09.19).

US Forest Service, Bureau of Indian Affairs, Bureau of Land Management, Fish and Wildlife Service, and National Park Service, 2001. Urban wildland interface communities within the vicinity of federal lands that are at high risk from wildfire. Federal Register 66 (3), 751–777.

Vogt, K.A., Patel-Weynand, T., Shelton, M., Vogt, D.J., Gordon, J.C., Mukumoto, C.T., Suntana, A.S., Roads, P.A., 2010. Sustainability Unpacked: Food, Energy and Water for Resilient Environments and Societies. Earthscan Ltd., London, United Kingdom.

Walker, L.C., 1991. The Southern Forest: A Chronicle. University of Texas Press, Austin, TX.

Westoby, J., 1989. Introduction to World Forestry. Basil Blackwell Ltd., Oxford, United Kingdom.

Williams, M., 1982. Clearing the United States forests: pivotal years 1810–1860. Journal of Historical Geography 8 (1), 12–28.

Williams, M., 1989. Americans & Their Forests. Cambridge University Press, Cambridge, United Kingdom.

Zyryanova, O.A., Terazawa, M., Koike, T., Zyryanov, V.I., 2010. White birch trees as resource species of Russia: their distribution, ecophysiological features, multiple utilizations. Eurasian Journal of Forest Research 13 (1), 25–40.

2

Forest regions of the world

A broad perspective on the distribution, composition, development, and use of forests around the world is a necessary foundation for fully understanding the challenges facing forestry and natural resource management. This is increasingly important today, given the continued expansion of human populations and the ominous threat of global climate change. The world's total amount of forest area has been estimated to be around four billion hectares (ha; nearly 10 billion acres [ac]). While the rate of deforestation may currently be high in some areas of the world, deforestation is slowing down on a global level due to large-scale afforestation programs (planting trees on previously nonforested land) and the natural expansion of forests. However, gains in forest areas have mainly been achieved in higher global latitudes (temperate and boreal forests), while losses continue to be incurred in tropical latitudes (Food and Agriculture Organization of the United Nations, 2010). In developed countries, domestically produced forest products and imported forest products from a wide variety of forest regions are used in the everyday lives of people. In developing countries, people directly rely on immediately accessible forests for their daily consumptive needs, to heat or cook with, and for generating income. It is indeed an understatement to note that in every country one can identify a need for the conservation and preservation of forest values. Some of the issues facing forestry and natural resource management are specific to a certain locale or culture. Ultimately, the regional, national, and international issues that affect forests and natural resources are rooted in the social, political, environmental, and economic fabric of society. The challenges that social, political, environmental, and economic issues pose for forests and natural resources is therefore highly pertinent to the conservation and use of forests and natural resources. Upon completion of this chapter, readers should be able to understand

- the historical aspects of forest use and development on different continents;
- some of the social, political, environmental, and economic forces that have shaped the current composition and distribution of forested areas in a number of nations; and
- the range of tree species that inhabit various forested areas and the timber and some of the nontimber forest products (NTFPs) that are available for internal and exportable uses.

The evolutionary development of natural forests around the world is a function of climatic, edaphic (a general term related to soil characteristics), and dynamic plant relationships, and these factors help explain why a certain type of forest is growing in a certain area (Barnes et al., 1998). Several common, basic characterizations of forests are used and very generally suggest that forests are composed of

- either softwoods (e.g., pines) or hardwoods (e.g., oaks) or both;
- either coniferous (e.g., pines, cedars, cypress) or deciduous trees (e.g., oaks, hickories, poplars) or some mixture of tree species;
- trees that have either needles or leaves, in either pure or mixed stands; and
- trees that are either gymnosperms (or *naked seed* plants, since their seeds are not enclosed during pollination), angiosperms (where the seeds are contained in fruits), or, again, some combination of these.

In describing forests of the world, we could use an ecological approach to address the broad floristic provinces that correspond to widespread climatic and physiographic conditions. For example, United States forests can be categorized into three broad classes: coniferous (west coast, Alaskan, and interior west forests), mixed coniferous and deciduous (southern and southeastern forests), and deciduous (northeastern, midwestern, and Lake States forests). In this case, the outward appearance (physiognomy) and structural condition of forests are widely used in a relatively broad sense to describe the type of forest found in a region.

Using an ecological approach, forests can also be classified by their major ecological community, or biome (Fig. 2.1). Three major, broad types of forest biomes are found on Earth: boreal, temperate, and tropical. While geographical latitude plays a significant role in defining where these biomes occur, elevation and climate also do. For example, the boreal forest (Fig. 2.2), or *taiga*, is generally found in areas with relatively short, warm summers, and long, cold winters. Boreal forests compose the largest of the three biomes and are situated in North America, Europe, and Asia, as well as in a small strip of South America along the Andes Mountains. Boreal forests are associated with (or are in proximity to) high and low Arctic tundra conditions as climates get colder (generally in higher latitudes), and a transition from northern to southern boreal forests occurs as climates get warmer (generally in lower latitudes). Tundra encompasses treeless areas near the polar deserts, where air temperatures are below freezing for about seven months of the year, and where continuous permafrost is present in the soil resource. Vegetation in a boreal forest is simple, when compared to the other forest types, and usually consists of hardy, coniferous tree species in the overstory, low-growing ericaceous shrubs (e.g., blueberries, cranberries, and heather), and a thick ground layer of lichens and mosses (Hicks, 2011). The length of a typical tree-growing season in a boreal forest is generally less than four months but varies depending on whether the forest is situated in the northern or southern boreal zone. Precipitation often arrives in solid

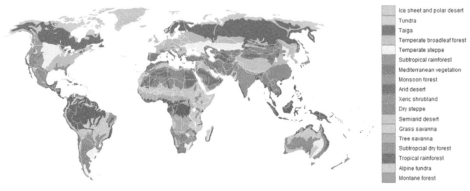

FIGURE 2.1 Major biomes of the Earth. *Ville Koistinen, through Wikimedia Commons.*
Image Link: https://commons.wikimedia.org/wiki/File:Vegetation.png
License Link: https://creativecommons.org/licenses/by-sa/3.0/deed.en

FIGURE 2.2 White spruce (*Picea glauca*) forest in central Alaska, United States. *L.B. Brubaker, U.S. National Oceanic and Atmospheric Administration, through Wikimedia Commons.*
Image Link: https://commons.wikimedia.org/wiki/File:Picea_glauca_taiga.jpg
License Link: Public Domain

form (snow or ice) in boreal forests, rather than liquid form. These forests generally have a continuous covering of snow in the winter, but there is no continuous permafrost present in the soil resource (Hicks, 2011). However, the soil resources found in this biome are generally shallow and of poorer quality than soil resources that might be found in other biomes. In addition, the decomposition rate of forest litter and coarse woody debris is fairly slow due to primarily low temperatures. Moisture is fairly abundant in boreal soils; thus, peat bogs (or muskegs) are common throughout the biome. While this biome can accommodate forests, because of harsh environmental conditions, tree growth is generally slower here than in other areas of the world.

The temperate forest biome (Fig. 2.3) covers approximately one-fifth of the available land areas in the mid to upper latitudes and is generally situated in areas with a well-defined, yet relatively mild, winter season. Eastern

North America, western and central Europe, eastern Asia, and parts of Australia, New Zealand, and Patagonia all host temperate forest biomes. Temperate biomes are productive and dynamic ecosystems that continuously change in response to climatic variation, diseases, and human alterations of the landscape (Willis, 2011). Seasonal variation is determined mainly by changes in air temperature, and air temperature extremes are greater in this biome than those that might be experienced in the tropical biome. Over the course of a year, air temperature extremes may range from −30 to 30°C (−22−86°F). During winter, forests in this biome enter dormancy. This phase of the year involves the cessation of tree growth, the loss of old leaves and needles through abscission (a total loss in the case of deciduous trees), and a series of biochemical changes that protect trees from frost damage, drought, and starvation (Willis, 2011). The soils of temperate forests are generally richer in nutrient content than boreal forest soils, and the decomposition rate of forest litter and coarse woody debris (CWD; which varies by tree species) is generally more rapid. Temperate deciduous and temperate coniferous forests are characterized as experiencing precipitation events that are spread relatively evenly throughout the year. Subdivisions of temperate forests are based not only on changes in topography, but also on differences in the intensity and timing of climatic events. For example, dry coniferous forests are generally found at higher elevations, where annual precipitation is low; moist coniferous or deciduous forests are found in areas lower in elevation, where there is a wet winter and a dry summer. Mediterranean forests have precipitation events mainly in the winter months, the extent of which may be limited.

The tropical biome contains a variety of forest types (Fig. 2.4), situated in terrestrial areas within the tropical

FIGURE 2.3 Temperate coniferous forests along the Columbia River in the western United States. *David W. Wilkinson.*

FIGURE 2.4 The Amazon rainforest, as seen from the Alto Madre de Dios River, in Peru. *Martin St. Amant, through Wikimedia Commons.*
Image Link: https://commons.wikimedia.org/wiki/File:7_-_Itahuania_-_Ao% C3%BBt_2008.JPG
License Link: https://creativecommons.org/licenses/by-sa/3.0/

zone of the planet, at lower latitudes around the equator. In terms of plant and animal life, tropical forests are some of the most complex and diverse areas on the planet. However, in comparison to boreal and temperate biome forests, the soil resources found in tropical forests may be poor and the decomposition rate of forest litter and CWD is fairly rapid. Thus, nutrients are continually recycled in these ecosystems. Interest in tropical forests is great because they contain an immense gene pool. Tropical forests have survived changing climates and geological upheavals and have expanded their ranges (e.g., across land bridges) when opportunities became available (Morley, 2011). Tropical forests generally have two seasons: rainy and dry. The availability of electromagnetic energy for photosynthesis (i.e., daylight) is very consistent, and year-round variations in air temperature are minimal. The average air temperature range is generally 20−25°C (68−77°F), rarely falling below 18°C, and these parts of the world are basically frost-free. Subdivisions of tropical rainforests are determined partly by the amount and duration of precipitation that occurs and partly by physical habitat conditions. For example, monsoon rainforests have a prolonged dry season and a short rainy season, whereas seasonal rainforests have a short dry period every year. In contrast, evergreen rainforests generally do not have a dry season. Semievergreen rainforests have a longer dry season than evergreen rainforests, and some of the canopy is occupied by deciduous tree species. Mangrove forests are a form of tidal forest situated in brackish wetlands between land and sea, in river deltas, and along sheltered coastlines. The tree species found in some of these tropical forests (e.g., mangroves) need to be very salt-tolerant, given their proximity to saltwater bodies. Tropical lowland evergreen forests are perhaps the richest and most luxurious of all plant communities in the world. These forests are found in tropical areas that are relatively wet all year round. They are composed of a dense, evergreen tree canopy and multiple layers of vegetation that form distinct vegetative strata. These forests have also been characterized through books and movies as the *Jungle* (Morley, 2011).

Other types of forests are found in the transition zones between the three main biomes. Cloud forests, for example, are often found in tropical or subtropical areas, where there is consistent low-level cloud cover that hangs in the sky at about the maximum height of the forest canopy. These forests are characterized as having a significant amount of mossy vegetation, and many are located in Central and South America, Africa, and Asia. Montane forests, on the other hand, are situated along mountain ranges, and can be characterized as boreal, temperate, or tropical, depending on the prevailing climate of the region. Savannas are grassy ecosystems that may contain a variable density forest of trees that are widely spaced; thus, the tree canopy

does not completely close. Savannas can be found in any biome and represent a transition between closed-canopy forests and prairies or deserts. Finally, steppe forests are generally transition zones between boreal and temperate biomes. Examples of these include the quaking aspen (*Populus tremuloides*) forests of Canadian prairie provinces and the birch (*Betula* spp.) and aspen groves of the Daurian forests between Siberia and Mongolia. Steppe forests contain not only parklike stands of trees but also a grassy understory.

An ecological approach is very useful for describing the natural forests of the world, and we allude to these biomes as individual countries are discussed in the next section of this chapter. We structured the remainder of the chapter in a geographic manner to provide a synthesis of forests in countries based on the continent with which they are typically associated. Five countries (Brazil, Canada, China, the Russian Federation, and the United States) contain more than half of the world's total forest area, while over 50 countries are less than 10% forested (Food and Agriculture Organization of the United Nations, 2015). This chapter focuses on only a few of the major and perhaps topically interesting (from a forestry perspective) countries located within each continent. Although in general this chapter provides a positive, objective view of forest regions of the world, the use of land and the management of forests have been associated with their fair share of recent political and environmental controversies. As an introduction to forestry and natural resource management, we allude to and briefly discuss some of the various conflicts, challenges, and national forest policies of select countries. Forest and natural resource policies arise from political, social, environmental, and economic problems. The following sections describe some of these issues; however, delving deeply into the vast array of socioeconomic and political controversies that face forest management throughout the world is beyond the scope of this book. For example, critical social and political issues and events have shaped the profession of forestry in Thailand (Usher, 2009) and, while these are intriguing, they require extensive treatment in order to adequately describe the actors, events, and issues related to the management of the environment. Although explorations of deeper issues in forestry and natural resource management are left for others to pursue, this chapter should broaden the reader's perspective on the use and care of forests in different regions of the world.

2.1 North America

The history of forests in many parts of North America has been shaped by the advances and retreats of ice sheets from the most recent ice age, the use of land by indigenous people, and the colonization of land by

Europeans and others. Approximately 20,000 years ago, an ice sheet 3—4 kilometers (km; 1.9—2.5 miles [mi]) thick stretched from the North Pole southward, covering all of Canada and most of the northern United States from Maine to Washington and as far south as Illinois, Indiana, Iowa, New Jersey, Ohio, and Pennsylvania. This last period of glaciation stripped the landscape of all vegetation and lowered sea levels. The period also led to the formation of the Bering land bridge, which was instrumental in connecting human populations in Asia with the unpopulated American continents. As the ice sheets receded, plant and wildlife species spread into the empty lands via natural processes such as animal defecation, water, and wind. In addition, the sea level rose, and the Bering land bridge disappeared, leaving the American continents with human occupants.

Among the different indigenous groups of North America were nomadic and agrarian-based tribes. The nomadic tribes, which were basically hunter-gathers, included the Cheyenne, the Sioux, and agrarian-based communities such as the Hopi and Navajo of the southwestern United States and the Iroquois of the northeastern United States. Indigenous populations depended to a great extent on the forests and the natural resources that they encountered. For instance, in the southwestern United States, indigenous communities were often concentrated near pinyon-juniper (*Pinus edulis, Juniperus* spp.) forests because pinyon seeds were an important part of their diet. Further, diaries from early European explorers indicate that large populations of indigenous people lived near the Mississippi River owing to the vast available areas of cultivable farmland. Southeastern indigenous groups were also known to burn large areas of the forested landscape in efforts to improve wildlife habitat quality (Lavender, 1958). Unfortunately, explorers and early colonists exposed vulnerable native populations to several infectious diseases (such as chicken pox, measles, and smallpox), which killed approximately 80% of the existing indigenous human population (Lewy, 2004).

When European settlers began to colonize land in what is now the United States and Canada, they believed they had found an endless sea of forests that could be used to build ships and heat homes, and land on which they could grow agricultural crops. With the decline of the indigenous populations, due both to infectious diseases and to intermittent wars with European settlers, many indigenous lands devoid of trees reverted back to a natural forested state, enhancing the image of an endless sea of forests. Colonists from Europe generally had a different perspective on how forests and natural resources should be managed and utilized than did indigenous people, who managed forests and natural resources under an informal communal model. European settlers brought with them a system of private property rights and more sophisticated agricultural and forest management technologies that transformed the

landscape much more quickly than indigenous methods did. At one time in the development of North America, trees from New England forests were removed not only to create agricultural land, but also to produce exports that would enhance the shipbuilding industry in England and the sugar industry in Barbados (Perlin, 1989). More recently, both developmental issues (aesthetics, housing, recreation, water, etc.) and external trade with other countries (e.g., China and Japan) have influenced the amount of pressure placed on North American forests.

A large portion of North America is too arid to support natural forest growth (Fig. 2.5). However, North American forests are diverse, ranging from the tropical forests of Mexico and the Caribbean, to the temperate forests of the eastern and western United States, and the boreal forests of Canada and Alaska. Canada and the United States each have over 900 million ha (over 2.2 billion ac) of land area, and each is currently about one-third forested (Table 2.1). The other main North American country, Mexico, is also about one-third forested. Table 2.1 provides data for areas that the Food and Agriculture Organization of the United Nations (2015, 2020) considers to be integrally tied to the continent of North America. In our geographical approach to describing the forested areas of the world, we discuss Central American countries separately in the next section. Broadly speaking, the Food and Agriculture Organization of the United Nations estimates

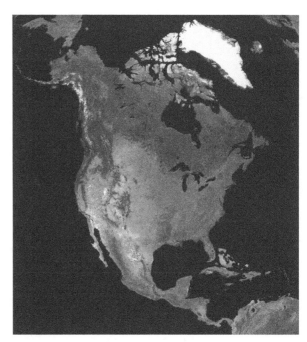

FIGURE 2.5 Satellite image of North and Central America. *National Aeronautics and Space Administration, Reto Stöckli, and Robert Simmon, through Wikimedia Commons.*
Image Link: https://commons.wikimedia.org/wiki/File:North_America_satellite_orthographic.jpg
License Link: Public Domain

TABLE 2.1 Demographic and physiographic data for North and Central America.[a]

Country	Human population 2018[1] (million)	Total land area[2] (1000 ha)	Total land area[2] (1000 ac)	Forest cover[2] (%)	Forest area annual change 2010–15[2] (%)	Per capita GDP (PPP) 17[1&b] (1000 US$)
Belize	<1	2281	5634	60	−0.4	5.1
Canada	37	909,351	2,246,097	38	–	45.0
Costa Rica	5	5106	12,612	54	1.1	11.7
Cuba	11	10,644	26,291	30	1.8	8.4
Dominican Republic	10	4832	11,935	41	1.8	7.1
Guatemala	17	10,716	26,469	33	−1.0	4.5
Honduras	9	11,189	27,637	41	−2.4	2.5
Mexico	126	194,395	480,156	34	−0.1	9.0
Nicaragua	6	12,034	29,724	26	–	2.2
Panama	4	7434	18,362	62	−0.4	15.1
United States of America	327	916,192	2,262,994	34	0.1	60.1
Other countries or territories (28)[c]	30	50,192	123,974	26	0.1	8.0
North and Central America (total)[b]	582	2,134,366	5,271,884	35	–	33.3

[a]That are more than 20% forested and have more than 3,000,000 ac (about 1,214,100 ha) of forest.
[b]Gross domestic product (GDP) expressed as purchasing power parity (PPP).
[c]Excludes Greenland and Saint Pierre and Miquelon.
[1]Food and Agriculture Organization of the United Nations 2020.
[2]FAO. Global Forest Resources Assessment 2015. Desk Reference. Rome, 2015.

that, as a whole, North and Central American countries experienced negligible forest area change over the five-year period from 2010 to 2015.

2.1.1 United States of America

The United States of America (Fig. 2.6) contains about 916 million ha of land (about 2.263 billion ac), 34% of which is forested (Table 2.1). In the early development of the country, fire was used extensively in forested areas by both indigenous people and early settlers to remove cover, provide firebreaks, facilitate hunting activities, and improve the condition of grassy vegetation for stock feed (Lavender, 1958). From the early 15th to the late 19th centuries, forests both assisted and hindered the settlement of the United States by nonindigenous people. Forests provided early settlers with fuel and building materials, clothing, food, and medicine, yet were also cleared for agricultural and grazing purposes (Anderson and Smith, 1976). As settlers moved west across the United States, early pioneers considered forests as an impediment to progress and, since agricultural uses of the land predominated, little thought was given to the conservation of natural resources (Muhn and Stuart, 1988). The clearing of land for agricultural uses increased from about 1700 to 1860 and then began to decline; however, fuelwood use peaked in the United States around 1875 (Birdsey et al., 2006). Forest harvesting activities began in the northeastern states and Lake States, then shifted to the southern states, and finally moved to the western states by the early 20th century. Eventually, segments of

the public grew concerned over the exploitation of forests in numerous regions of the country, and conservation of forests and natural resources became a significant issue. More recently, the rate of change in forested areas of the United States over the 2010–2015 period was slightly positive (Table 2.1).

Of the forested land in the United States, about 57% is privately owned (Food and Agriculture Organization of the United Nations, 2010), yet land ownership patterns vary across the country, from high percentages of private land in the eastern and southern states to relatively even percentages of state, federal, and private land in some of the western states. Private forest land in the United States is managed by company foresters, consultants, or the landowners themselves. Most of the larger companies have, or are seeking, forest certification through programs such as the *Sustainable Forestry Initiative*, which is described in Chapter 15. A large number of private landowners also belong to the *American Tree Farm System*. Aside from the requirement to comply with federal and state regulations, private landowners are relatively free to employ a wide variety of forest management systems. State lands are managed by representatives (foresters and natural resource managers) employed by each state. Each state also develops its own goals and objectives for state-owned forests, and these too are guided by certain aspects of applicable federal and state regulations as well as the goal of the state lands. Much of the state forests in the western United States are managed in a trust to generate revenues for the state schools. Federal lands (e.g., the Forest Service and Bureau of Land Management [BLM]) are managed by federal foresters and natural resource managers. A number of federal regulations guide the development of goals, objectives, and plans for federal forests in the United States. Other types of landowner groups can also be found in the United States, and these are described in more detail in Chapter 3.

In 1905, Gifford Pinchot became the first chief of the U.S. Forest Service under President Theodore Roosevelt. Pinchot was a great advocate of conservation for forests and their natural resources. He championed sustainable management of forests and coined the term *conservation ethic*. His perspective on the use of forests clashed with those of preservationists such as John Muir, but Pinchot was instrumental in the early development of the U.S. National Forest System. Since the early 20th century, the United States has developed a series of national policies and agencies to support the notion of sustainable forest management (SFM). For instance, the U.S. National Park Service was created in 1916 to manage national parks and other historically significant areas for conservation, education, and recreation purposes. Further, the BLM was formed in 1946 by the merger of the Grazing Service and General Land Office, two entities within the

FIGURE 2.6 Countries in North America. *Modified image from PresentationMaps.com.*

US Department of the Interior (Muhn and Stuart, 1988). This agency now administers about 109 million ha (270 million ac) of forest and rangeland in the United States. The US Forest Service administers about 78 million ha (about 193 million ac) of predominantly forested land. Finally, both the US Environmental Protection Agency and the Department of Energy, neither of which directly manages land, currently have programs that promote increasing carbon sequestration by forests (Birdsey et al., 2006).

Forests of the southern United States, which encompass a broad area ranging from Virginia to Texas, are naturally considered temperate broad-leaved (deciduous) forest areas and subtropical forest areas. Historically, naturally regenerated native coniferous forests could be found in areas where natural or anthropogenic events allowed successional stages to revert to states where conifers could capture resources and compete effectively with deciduous trees. In this regard, fire was one of the main tools used by early settlers and Native Americans. During the first part of the 20th century, fire was also the predominant tool used in the management of land in the southern United States and was widely used to reduce tree stocking and enhance range conditions for livestock (Birdsey et al., 2006). With the onset of fire prevention policies, the development of pine plantations, and the establishment of forests on marginal agricultural lands, the broad-scale use of prescribed fire is now limited. Forests of the southern United States now consist of a broad array of coniferous and deciduous forests, some managed very intensively and others very lightly. The pine plantation forests (Fig. 2.7) of this region are among the most productive in the world. Seventy-five percent of the pulpwood produced in the United States arises from the southern states (Johnson et al., 2011b). Some important tree species that are grown, harvested, and exported both to other regions of the United States and internationally include loblolly pine (*Pinus taeda*), slash pine (*Pinus elliottii*), longleaf pine (*Pinus palustris*), and shortleaf pine (*Pinus echinata*). In addition, southern forests contain numerous commercially important hardwood tree species (e.g., black walnut [*Juglans nigra*], maples, oaks, and yellow-poplar [*Liriodendron tulipifera*]) that support furniture industries in states such as Mississippi, North Carolina, and Tennessee.

Forests of the southern United States face several challenges in the future, including those related to climate change, human population growth, invasive species, and uncertainties in timber markets (Wear and Greis, 2011). Of the forestland in the southern United States, over 68% is owned by private individuals and families (perhaps as family farms), around 19% is owned by forest companies and corporate entities, and over 12% is owned and managed by various local, state, or federal public agencies. Each landowner can have its

FIGURE 2.7 Young and mature pine stands, as viewed from above, in southern Alabama, United States. *US Department of Agriculture, Natural Resources Conservation Service (2020).*

own distinct objectives and constraints, as we will see in Chapter 3. The larger landowners and the national forests in this region have usually developed long-term management plans, yet very few of the smaller landowners have developed these. Some states and local governments have developed policies and regulations that govern forestry activities but, for the most part, forest management activities are guided by state-level, voluntary *Best Management Practices* (BMPs) in the southern United States. The generation of commercial forest products from the southern United States is very high, making it one of the most competitive markets in the world (Table 2.2). Coniferous tree species are mainly used as pulpwood (for pulp and paper) and sawlogs (for lumber and other solid wood products) and, while deciduous tree species are also used mainly for pulpwood and sawlogs, these species are the main fuelwood sources for households equipped to utilize wood stoves.

In the northeastern and Lake States regions of the United States, forests are generally temperate broad-leaved forests with some boreal forests in the higher elevations and latitudes. Forests generally regenerate naturally on harvested or disturbed lands and on marginal agricultural lands. However, these forests (Fig. 2.8) are generally managed less intensively than forests in the southern states (Birdsey et al., 2006). In Maine and northern New England, one can find forests composed of tree species such as American beech (*Fagus grandifolia*), balsam fir (*Abies balsamea*), eastern white pine (*Pinus strobus*), paper birch (*Betula papyrifera*), sugar

TABLE 2.2 Roundwood output in the Southern United States, 2015, by product (1000 Cubic Meters).

Product	Softwood (coniferous)	Hardwood (deciduous)	Total
Sawlogs	1,836,576	689,748	2,526,325
Veneer logs or bolts	299,676	28,059	327,735
Pulpwood	2,708,699	834,637	3,543,336
Composite panels	185,350	7,580	192,929
Poles and posts	72,754	100	72,855
Other	596,720	198,162	794,882
Fuelwood	374,808	–	374,808
Total	6,074,584	1,758,286	7,832,870

USDA. https://www.fia.fs.fed.us/program-features/tpo/. Forest Inventory and Analysis National Program. Website. US Forest Service. 2015; Johnson, T.G., Bentley, J.W., Howell, M., 2011. The South's Timber Industry—An Assessment of Timber Product Output and Use, 2009. US Department of Agriculture, Forest Service, Southern Research Station, Asheville, NC. Resource Bulletin SRS-182. 44 p.

maple (*Acer saccharum*), white spruce (*Picea glauca*), and yellow birch (*Betula alleghaniensis*), grown in even-aged or uneven-aged, mixed species stands. In the Midwest and Lake States of the United States, one might find forests composed of American elm (*Ulmus americana*), black ash (*Fraxinus nigra*), black spruce (*Picea mariana*), jack pine (*Pinus banksiana*), paper birch, quaking aspen, red pine (*Pinus resinosa*), and sugar maple, tamarack (*Larix laricina*). When managed for commercial purposes, even-aged forests have relatively long rotation lengths (80–100 years) compared to forests of the southern United States (20–40 years). Forestland ownership varies by state, but private land ownership dominates the northern region of the country. As an example, in Maine, 93% of the forestland is privately owned, by individuals, corporations, timberland investment organizations, and real estate investment trusts (REITs;

FIGURE 2.8 Forested landscape during the fall season in Maine, United States. *Samantha Langley.*

McCaskill et al., 2011). In Minnesota, public forestland ownership (56%) is the highest of any eastern state, and this is nearly evenly divided among state, county, and federal ownership. Forty-four percent of forestland in Minnesota is privately owned, 37% by nonindustrial landowners, and about 7% by corporations. Corporations also own about 10% of the forestland in Wisconsin and 15% of the forestland in Michigan (Miles et al., 2011).

The central United States is composed of temperate broadleaved forest areas, and oak forests encompass over 50% of the timberland in the region (Chapman et al., 2006). As in other areas of North America, Native Americans influenced forest cover types through the use of fire to clear land for agricultural purposes and to drive wildlife game species during hunts (Drury and Runkle, 2006). Lumber production activity began in earnest in this region in the mid-19th century. In the late 19th century large expanses of forests in this region were cleared for charcoal production purposes and the production of pig-iron (Williams, 1982). Most of these areas regenerated naturally and, today, typical forests in the central states are composed of overstories of black cherry (*Prunus serotina*), black walnut, oaks and hickories, red maple (*Acer rubrum*), sugar maple, white ash (*Fraxinus americana*), and yellow-poplar. American beech, blackgum (*Nyssa sylvatica*), and maples can be found in the midstory of the forest canopy, and maples and sassafras (*Sassafras albidum*) may be prevalent in the understory (Chiang et al., 2008). As an overstory tree species, yellow-poplar prefers lower slopes and sheltered coves, while oaks are more abundant in unglaciated areas with drier conditions and thinner soils (Widmann et al., 2009). However, in some areas, younger forests are composed of overstories of maple tree species (Drury and Runkle, 2006). Northern hardwood forests also occur in some areas of the central

United States, and these may typically be represented by overstories of American beech and sugar maple. In both types of forests, maples could very likely be the dominant understory tree species (Widmann et al., 2009).

The influences of agriculture, expanding deer populations, fire, grazing, insect and disease outbreaks, and periodic seed crops can affect the future composition of forest ecosystems in the central states (Drury and Runkle, 2006). In fact, fire suppression in the last 80 years or so has favored fire-sensitive tree species over fire-adapted oaks, and in many areas oaks are being replaced by shade-tolerant, fire-intolerant tree species. Other important causes of forest disturbance include droughts, ice storms, insects and diseases (e.g., Dutch elm disease), land use changes, and logging, followed perhaps by abandonment of the land (Widmann et al., 2009). In some areas of the central states, recent losses of forestland to developmental land uses (e.g., urban areas) have been more than offset by gains in forestland from the reversion of abandoned farmland to forests (Widmann et al., 2009). As with other regions, the distribution of land ownership varies by state. For example, private family forest landowners own about 75% of the forestland in Indiana (Woodall et al., 2011) and about 85% of the forestland in Iowa (Nelson et al., 2011). In Ohio, 12% of the forestland is publicly owned; 73% is owned by private individuals and families (over 336,000); and 15% is owned by corporations, nonfamily partnerships, Native American tribes, nongovernmental organizations (NGOs), clubs, and other nonfamily private groups (Widmann et al., 2009).

The Ozark Mountains are situated in an area between the southern United States and Lake States (mainly Arkansas and Missouri), where temperate broadleaved forests naturally reside. Oak-hickory forests are the dominant vegetation type in this area. Throughout the early 20th century, this region had a long history of forest fires and unregulated grazing activities (Chapman et al., 2006). The understory and midstory vegetation of these forests usually consists of dogwood (*Cornus florida*), hickories, and oaks. The overstory vegetation may consist of black oak (*Quercus velutina*), white oak (*Quercus alba*), and perhaps shortleaf pine. However, since fire suppression activities began around 100 years ago, blackgum, dogwood, eastern hophornbeam (*Ostrya virginiana*), and red maple have become increasingly important and abundant tree species in these forests (Chapman et al., 2006). The influence of fire on forest composition has always been closely associated with the anthropogenic use of the land (Guldin, 2008), which changes as human settlement patterns change. For example, prior to 1820 the mean fire frequency was about once every 5 to 20 years. Yet between 1820 and 1920, the frequency was once every 2 to 5 years due to the migration of American Indians from the eastern

states to Oklahoma and, after 1920, the frequency has risen from once every 10 years to once every 80 years or more in some areas (Guyette et al., 2006). Forest composition continuously changes and, as upland oak-hickory forests in this area mature, large-scale oak decline and mortality has increased, a process closely associated with disease, drought, forest succession, stress, and other factors (Fan et al., 2011).

Wood utilization reached its peak in 1899 in the Ozark Mountains, which supported the industrialization of the area. By 1920, much of the original mature forest cover had been harvested for wood products and a period of forest recovery began (Guldin, 2008). Fire control was one of the main tools used in the recovery process, although the forest succession reaction is, as we noted a moment ago, complex. Since about the 1940s, the Missouri Ozark forests have experienced a demand for products and services that range from lumber to wildlife habitat and water. In the Ozarks and other areas of the United States, a number of state-level and federal policies have been implemented over the last 70 years to enhance the value of forests, physically protect some watersheds, and generally promote forest vigor and health (Guldin, 2008).

In the western United States, two general types of forests exist: wet forests and dry forests. The wet forests (Fig. 2.9) are those generally to the west of the Cascade and Sierra Nevada mountains or in the coast ranges. The dry forests (Fig. 2.10) are those generally to the east of these ranges, due to the orographic lifting of air masses over the mountains. As air from the Pacific Ocean travels east across the landscape and gains altitude, it cools and the air masses lose water content

FIGURE 2.9 Coniferous forests in Yamhill County, western Oregon, United States. *M.O. Stevens, through Wikimedia Commons. Image Link: https://commons.wikimedia.org/wiki/File:Northern_Oregon_ Coast_Range_mountains.JPG License Link: Public Domain*

FIGURE 2.10 Ponderosa pine (*Pinus ponderosa*) forests in south-central Washington, United States. *Pierre Nordique, through Wikimedia Commons. Image Link: https://commons.wikimedia.org/wiki/File:Pinus_ponderosa_Manashtash_Ridge.jpg License Link: https://creativecommons.org/licenses/by/2.0/deed.en*

through precipitation. These temperate, and relatively dry montane forests can be found in Arizona, Colorado, Idaho, Montana, New Mexico, Utah, and Wyoming, with minor amounts in South Dakota. In the drier southwestern and interior western United States, one may find grand fir (*Abies grandis*), subalpine fir or Rocky Mountain fir (*Abies lasiocarpa*), lodgepole pine (*Pinus contorta*), pinyon pine (*Pinus quadrifolia*), and ponderosa pine (*Pinus ponderosa*) forests and forests composed of other tree species that are adapted to arid climates, including an inland variant of Douglas-fir (*Pseudotsuga menziesii*). Wildfire and the suppression of fire have shaped the historical and current character of the forests of the interior western United States. In addition, recent outbreaks of the western spruce budworm (*Choristoneura occidentalis*) have caused massive defoliation and dramatically changed the structure and vegetation dynamics of vast areas of interior northwest forests.

In the Pacific Northwest states (Alaska, Oregon, and Washington), it is common to find Douglas-fir, red alder (*Alnus rubra*), Sitka spruce (*Picea sitchensis*), and western hemlock (*Tsuga heterophylla*) forests in the coastal areas, along with other tree species such as Pacific yew (*Taxus brevifolia*). These forests are also characterized as temperate, montane forests. Bracken fern (*Pteridium aquilinum*), salal (*Gaultheria shallon*), salmonberry (*Rubus spectabilis*), sword fern (*Polystichum munitum*), thistle (*Cirsium* spp.), and vine maple (*Acer circinatum*), among other plant species, are commonly found in the understory of

these forests (McIntosh et al., 2009). Douglas-fir is the dominant commercial tree species and is often found in plantations, and is managed in a manner similar to loblolly pine in the southern United States, although with a few notable exceptions with regard to site preparation (rarely mechanical), planting method (mainly by hand due to an area's steepness), the use of intermediate treatments such as precommercial thinning, and longer even-aged rotation lengths (35–55 years). The Pacific Northwest is a very productive timber-growing region, and timber harvest levels over the last 50 years reached a peak in the late 1960s and early 1970s but hit a low point in 2009. In 2010, 68% of the timber harvest in these two states was produced from private forestlands and, of the public land harvest, about two-thirds came from state rather than federal lands (Warren, 2011). The land ownership distribution is skewed more toward public ownership in this region of the country than in the eastern United States. For example, in Washington State, about 44% of the forests are controlled by the federal government and about 13% are owned by the state. The remaining 43% are owned by private individuals, corporations, Timberland Investment Management Organizations (TIMOs) and Real Estate Investment Trusts (REITs) (Campbell et al., 2010).

In northern California, one might find Douglas-fir, redwood (*Sequoia sempervirens*), and western red cedar forests in temperate coastal forests. Ferns, salmonberry, and thistle can be found in the understory of these lush forests. Douglas-fir, true firs, Jeffrey pine (*Pinus jeffreyi*), lodgepole pine, ponderosa pine, and sugar pine (*Pinus lambertiana*) forests are generally found in the inland northern forest areas of California. Southern California has a number of pine species unique to this area and is quite different and composed of Mediterranean vegetation and forests, where the climate is generally very dry during the summer months. These areas range from the Central Valley and Sierra Nevada foothills to the central coast ranges of California, and south to Mexico (Fenn et al., 2011). Southern California forests are diverse, and blue oak (*Quercus douglasii*), brittlebush (*Encelia* spp.), coastal sagebrush (*Artemisia californica*), manzanita (*Arctostaphylos* spp.), and other plant species are typically found in the understory. Chaparral and oak woodlands are widespread throughout the Mediterranean forest areas of California. These areas are composed of evergreen shrublands that contain, among those mentioned earlier and other plant species, big-leaf maple (*Acer macrophyllum*), boxelder (*Acer negundo*), ceanothus (*Ceanothus* spp.), and perhaps Oregon ash (*Fraxinus latifolia*) (Fenn et al., 2011). Sudden oak death, caused by the fungus *Phytophthora ramorum*, is of great concern for oak species in California and other states.

A large portion of the forested area in California can be described as being composed of mixed conifers, and over half is managed by the federal government. Public ownership of California's forests in 2005 was about 60%, nonindustrial private landowners owned about 26% of the forestland, and private corporations owned about 14%. Most of the merchantable wood volume is located in the northern part of the state. Over the last 50 years, timber harvest levels have been declining gradually on both federal and private land. Of the tree species that are harvested, most are conifers: Douglas fir (27%), true firs (21%), ponderosa pine (18%), and redwood (16%). Recently, it was estimated that forest growth in California substantially exceeded harvest rates (Christensen et al., 2008).

In Alaska, coniferous and mixed species boreal forests are generally located in the central part of the state, along with large expanses of tundra, and montane and boreal coniferous forests are located in the southeast part of the state. The boreal forest of Alaska occupies an area that spans from the Pacific Ocean to the Canadian border, and has perhaps the most severe climate conditions in the world, with air temperatures as low as −70°C and annual precipitation rates that rarely exceed 50 cm. As a result, few tree species can be found here; some of these include balsam poplar (*Populus balsamifera*), Alaska white birch (*Betula neoalaskana*), Kenai birch (*Betula kenaica*), quaking aspen, black spruce, white spruce, and tamarack (Anderson and Brubaker, 1994; Liang, 2010). Since about 1990, the boreal forests of the Kenai Peninsula and interior Alaska have experienced significant spruce beetle (*Dendroctonus rufipennis*) outbreaks (Boucher and Mead, 2006). Forests in southeast and south-central Alaska contain Alaska yellow-cedar (*Chamaecyparis nootkatensis*), Sitka spruce, western hemlock, and western red cedar (*Thuja plicata*) at low elevations; mountain hemlock (*Tsuga mertensiana*) at higher elevations; and some areas of aspen, birch, and Barclay's willow (*Salix barclayi*), black spruce, and white spruce (Barrett and Christensen, 2011). The economies and communities of southeast Alaska are diverse and affected by forest management direction, and while southeast Alaska has an abundant supply of forest resources, they are primarily controlled by the federal government (Crone, 2005). Most (about 75%; over US$100 million) of the forest product exports from Alaskan forests are to Asian countries, and most of these are in log form, from private forestlands (Roos et al., 2011).

The islands of Hawaii contain over 200,000 ha of potential commercial forestland, currently consisting of extensive coniferous, deciduous, and mixed forests; however, some of this land is also (or was recently) used for sugarcane production, and perhaps now lies fallow (Whitesell et al., 1992). Coastal forests are very diverse and may consist of alahe'e (*Psydrax odorata*), hala (*Pandanus tectorius*), and kopiko (*Psychotria hawaiiensis*), or numerous other native tree species and species introduced from Polynesia (Mascaro et al., 2008). Mesic forests on the windward slopes of mountains consist of species such as koa (*Acacia koa*), one of the few endemic tree species that dominate the montane forests of Hawaii (Baker and Scowcroft, 2005). Koa is also commonly found in wet forests along with 'ōhi'a lehua (*Metrosideros polymorpha*) and many other tropical tree species. Further, of the native tree species, koa is perhaps the most ecologically and economically important. Currently, forestry operations in Hawaii are limited, although there are areas managed as commercial forest plantations and, along with short-rotation fiber plantations, agroforestry is also being considered. High-density biomass plantations have been tested in Hawaii using eucalypt (*Eucalyptus* spp.) (Whitesell et al., 1992) and loblolly pine (Harms et al., 2000) tree species, and in some cases greater growth potential can be attained here than in these species' native environments.

2.1.2 Canada

The 10 provinces and 3 territories of Canada amount to a little over 909 million ha of land (2.246 billion ac), of which about 38% is forested (Table 2.1). The forests of Canada represent about 10% of the total world forest area, an area similar to that of the United States. Canada is said to be represented by at least 10 major forest regions (Drushka, 2003), including the extensive boreal forest that stretches the entire width of the country, and the Great Lakes forests of the southern and eastern half of the country. A large portion of the unforested land in Canada lies in the northern territories, where the climate is too harsh to support forest growth. Further, the prairies of the central provinces generally lack the moisture necessary to support natural forest growth. The dominant tree species of the boreal forests are black spruce, white spruce, and tamarack. The Acadian forests of New Brunswick, Nova Scotia, and Prince Edward Island (Fig. 2.11) contain a mixture of coniferous and deciduous trees, including aspen, eastern hemlock (*Tsuga canadensis*), white pine, red maple, sugar maple, jack pine, red pine, oaks, black spruce, red spruce (*Picea rubens*), white spruce, tamarack, yellow birch, and others. Forests of the Rocky Mountain region of Canada are composed primarily of conifers and aspen. In areas west of the Rocky Mountains, mainly the province of British Columbia, the montane and alpine forests are composed of Douglas-fir, grand fir, western hemlock, western redcedar, and other conifers (Drushka, 2003).

Prior to widespread development of the country, aboriginal people shaped the Canadian landscape by setting fires in order to develop meadows that would attract the wildlife necessary to meet basic consumptive needs. Once widespread development of the country

FIGURE 2.11 Forested coastline, Nova Scotia, Canada. *Dylan Kereluk, through Wikimedia Commons.*
Image Link: https://commons.wikimedia.org/wiki/File:Nova_Scotia_-_by_Dylan_Kereluk.jpg
License Link: https://creativecommons.org/licenses/by/2.0/deed.en

began around the 17th century, fires initiated by steam engines had a significant role in shaping the structure of the forests. Large volumes of wood were harvested in the early development of the country to support the buildings, docks, and warehouses needed to support the burgeoning fur and fishing industries. Early forest policies concerned the reservation of trees for their potential use for masts and spars of ships. The use of wood was prevalent in the early colonization of Canada, as settlers required about 30 cords of firewood each winter, and about 500 split rails per hectare to fence fields for livestock production. As with the development of the United States, the predominant view of forests by colonists was that they impeded the development of the landscape (Drushka, 2003).

Forest policies in Canada in the mid-19th century provided the mechanisms for the development of temporary leases and licenses for timber harvesting activities, with much of the land remaining in the control of the federal government or individual provinces. Some land was eventually transferred from the Crown to individuals to promote settlement. Currently, about 92% of the forested areas in Canada are publicly owned (Food and Agriculture Organization of the United Nations, 2010). The majority of commercial forestland in Canada that is controlled by federal or provincial governments is licensed to companies, and representatives of these companies develop management plans that are guided by provincial forest policies. The Canadian Forest Service is a research and policy organization of the government that is currently acting to promote both the sustainability of Canadian forests and the economic competitiveness of the Canadian forestry sector.

The development of railroads in the late 19th century further shaped the structure of the forests, as railroads required large volumes of wood for the development of bridges and the production of railroad ties. Nearly 2000 railroad ties were needed per kilometer of railroad track, and these needed to be replaced every three or four years prior to advancements made in the development of wood preservatives (Drushka, 2003). In the early 20th century, agricultural uses of the land, primarily in the eastern forests, reduced the amount of forest cover and served to fragment the forests. Petroleum exploration, well sites, and access roads have served to fragment some of the western forests. Forest conservation efforts began in the early 20th century but were temporarily hampered by World Wars I and II. Currently, over half of the country's forests are actively managed and, as in the western United States, firefighting policies have inadvertently contributed to the character of these forests and have raised forest health concerns (disease problems, high forest densities, high fuel loads, and insects). Canada has had the highest area of insect disturbance among world countries over the past 10 years, with extensive outbreaks of the mountain pine beetle (*Dendroctonus ponderosae*) and forest tent caterpillar (*Malacosoma disstria*) (Food and Agriculture Organization of the United Nations, 2010). SFM is now a pervasive theme in Canadian forestry, and by the end of the 20th century most forest companies sought certification through one of the main forest certification schemes (Drushka, 2003), which are described in Chapter 15.

2.1.3 Mexico

Mexico is located on the southern border of the United States (Fig. 2.6). It has a land base of about 194 million ha (480 million ac), with forests covering about 34% or about 66 million ha (about 163 million ac). The forest types found in Mexico vary from tropical rainforests to subtropical temperate forests. The tropical rainforests are found on slopes near the Gulf of Mexico and the Pacific Ocean, the Isthmus of Tehuantepec, and the southern Yucatán Peninsula. Recent changes in the area of forest cover have been slightly negative (Table 2.1), mainly due to wildfires and the conversion of forestland to agricultural uses. The forests of Mexico are very diverse; for example, across the country one can find 72 species, varieties, and forms of pines (Fig. 2.12). Other tree species found in more temperate areas of Mexico include Mexican beech or haya (*Fagus mexicana*), oaks, sweetgum (*Liquidambar styraciflua*), and various coniferous species of cypress, fir, juniper, and pine. In the tropical regions, broad-leaved and semi-deciduous trees are more common, and tree species found here include breadnut (*Brosimum alicastrum*),

FIGURE 2.12　Pine forests in Mexico. *Gustavo Perez-Verdin.*

balché tree (*Lonchocarpus violaceus*), caoba or mahogany (*Swietenia macrophylla*), cedro rojo (*Cedrela odorata*), guava (*Terminalia oblonga*), sapodilla (*Manilkara zapota*), and white olive (*Terminalia amazonia*). Forest plantations (eucalypts, pines, and teaks [*Tectona* spp.]) have also become common throughout the country.

About 90% of the forest production in Mexico arises from coniferous and deciduous forests located in the temperate climate zones, mainly in the states of Chiapas, Chihuahua, Durango, Guerrero, Jalisco, Michoacan, and Oaxaca. Tropical and subtropical forests, located in the states of Campeche, Chiapas, Oaxaca, Quintana Roo, Tabasco, and Yucatán, are about the same size, but account for only about 10% of forest production. Most of the softwood and hardwood timber produced in Mexico is consumed within the country (Food and Agriculture Organization of the United Nations, 2020). Forest policy at the federal level is currently guided by the Secretaría del Medio Ambiente y Recursos Naturales (Environment and Natural Resources Secretary), formed in 1995, and the Comisión Nacional Forestal (the Mexican National Forestry Commission), formed in 2001.

Although about a quarter of the forested areas in Mexico are privately owned (Food and Agriculture Organization of the United Nations, 2010), land tenure can also consist of *comuneros* or *ejidatarios*. *Comuneros* are members of ethnic groups who have acquired collective access to portions of land since colonial times, and *ejidatarios* are members of cooperatives called *ejidos* (Bocco et al., 2001). In 1917, in response to a strong agrarian reformist presence related to the Mexican Revolution that began in 1910, the common property ejido system of land tenure was established. This system applied to land in which the Mexican federal government once held title. A typical ejido has a small center village and is surrounded by agricultural land and forests. The agricultural lands are assigned to individuals,

while the forests are held in common trust. An ejido manages the forested area for forest products and services detailed in plans submitted for approval to the Environment and Natural Resources Secretary. The ejidatarios are generally male heads of households, and the right to become an ejidatario is passed down through patrilineal family lines. Selected agricultural lands within an ejido are given to individual ejidatarios for cultivation. In 1991, with the enactment of a new agrarian law and an amendment to the Mexican Constitution, ejidatarios can, with the approval of the ejido, lease or sell land to individuals or corporations (Kiernan, 2000).

2.2　Central America

The political and economic history of Central America (Fig. 2.13) is integrally linked to the developmental history of neighboring countries to the north and the evolution of various European countries who once acted to expand their empires across the Atlantic Ocean. The Food and Agriculture Organization of the United Nations (2015, 2020) groups countries in this area with those of North America, but because of its location it was important to discuss countries located here separately. Unlike Canada and the United States, this region of North America was not adversely affected by the last ice age period. When the first European explorers found these lands, they discovered a vast indigenous human population. The two largest groups were the Aztecs and the Mayans. Although the main center of those civilizations was located in Mexico, which we discussed in the previous section, the influence of these groups extended throughout the countries of Belize, Costa Rica, El Salvador, Guatemala, Honduras, Nicaragua, and Panama. These indigenous people formed mainly agrarian-based societies that had a similar impact on forests and natural resources as did the indigenous groups that were located in the United States and Canada prior to European settlement.

Prior to the arrival of the European settlers, approximately 90% of the land base (50 million ha or about 123.5 million ac) in Central America was still forested. With the arrival of the Spanish and British settlers, large areas of forestlands were converted to agriculture to support growing colonial civilizations. Later, many areas were converted to agricultural crops, such as bananas, coffee, and sugarcane that would be exported to the United States and Europe. In the 1980s, it was noted that forested areas had declined to about 20 million ha (about 49.4 million ac), partly due to the expansion of cattle ranches (Myers and Tucker, 1987). It is not uncommon today to drive along the Pan-American Highway and see large areas of pasture lands that are burned annually

FIGURE 2.13 Countries in Central America. *Modified image from PresentationMaps.com.*

in an effort to remove dead and decaying grassy vegetation. Since the end of the 20th century, expanding human populations have applied pressure to the existing forests and natural resources, despite attempts by several national governments to create forest protection areas.

The forests of Central America are diverse and, while most are broadly considered tropical rainforests, there are numerous ecoregions that span this isthmus, including the dry tropical forests, the moist Atlantic forests, the Chiapas highlands, the montane forests, the pine-oak forests, and the Sierra Madre. The dry tropical forests stretch from the Pacific coast of Mexico south to northwestern Costa Rica. The average rainfall in these areas is below 200 cm (78.7 in) per year. Tree species in these areas are mainly deciduous and cactus is common in some forested understories. Some of the common tree species found here include guanacaste tree (*Enterolobium cyclocarpum*), mata ratón (*Gliricidia sepium*), and pochote (*Pachira quinata*). In the Atlantic forests, the average annual rainfall is over 400 cm (157.5 in) per year, and there is no discernible change in season throughout the year. Some common tree species found in this ecoregion include rosewood (*Dalbergia retusa*) and Spanish elm or Ecuador laurel (*Cordia alliodora*). Montane forests include tree species such as Caribbean pine (*Pinus*

caribaea) or pino costanero. A complete history of each country in the region would be too extensive for this book. Therefore, our discussion of Central American countries focuses briefly on only five: Costa Rica, Guatemala, Honduras, Nicaragua, and Panama.

2.2.1 Costa Rica

Costa Rica (the *rich coast* in English) is a small country covering an area of about 5.1 million ha (about 12.6 million ac), wedged between Nicaragua to the north and Panama to the south (Fig. 2.13). The human population of the country is essentially an agrarian society focused primarily on the production of agricultural commodities such as bananas, coffee, pineapples, sugarcane, and beef. Local legend suggests that Christopher Columbus named the area the *rich coast* on his last voyage, despite the fact that few mineral resources could be found on these lands. Costa Rica was part of the Spanish Empire until 1821, the end of the Mexican War for Independence. It was then considered a province of the Federal Republic of Central America for almost two decades until it proclaimed itself a sovereign country in 1838.

About 45% of the land in Costa Rica is under control of the Costa Rican government, and about 55% is privately owned. Although 70% of the country's economy is based on agriculture, the landscape is ill suited for most forms of agriculture because of the mountainous terrain. Lowland areas on the Caribbean coast, Guanacaste, and the southern portion of Puntarenas have been sites of extensive cultivation of exported commodities (bananas, beef, and pineapples). Coffee is typically grown in the highlands around the central valley where the capital city of San José is located. Costa Rica's forests (Fig. 2.14) have declined from about 75% of the total land area in 1940 to about 39% in 2000, primarily due to demands for agricultural land and forest resources. The Food and Agriculture Organization of the United Nations (2015, 2020) estimates that forest cover now stands at about 54%, with forest area increasing by about 1% per year (Table 2.1).

Commercial wood products are derived from cedar, laurel, mahogany, oak, and other tree species found in Costa Rica. The wide variety of NTFPs include chicle (a gum from which Chiclets were named), ipecac (which induces vomiting and is prescribed after swallowing something poisonous), medicinal plants, and rubber. About two-thirds of the roundwood produced is used within the country for fuelwood purposes. Efforts to conserve forests outside of protected areas began in the 1970s (Pool et al., 2002). Natural forest management has been promoted in Costa Rica through programs of the US Agency for International Development (Pool et al., 2002) and, as with other Central American countries, NGOs and local associations have been important in this regard. Since the 1990s, Costa Rica has been

assertive in promoting reforestation programs and developing official forest protection areas. These forest protection areas have typically served as buffer zones for the numerous national parks found in the country. There is even one private preserve, Monteverde Cloud Forest Reserve, created in 1951 by 44 Quakers from Alabama, who settled the area (Monteverde Costa Rica Cloud Forest Nonprofit Organizations, 2006). The reserve is located on the Continental Divide and contains a cloud forest that is home to over 2000 plant species, hundreds of bird and mammal species, such as the resplendent quetzal (*Pharomachrus mocinno*), and amphibians, such as the golden toad (*Bufo periglenes*).

Costa Rica differs from other countries in the region by its advanced economic and social progress that has enhanced health care services, living standards, and literacy rates (Pool et al., 2002). Along with the creation of 26 national parks, there has been an extensive effort to promote nontimber products such as ecotourism. Recognition of the potential of ecotourism has allowed this area of work to surpass agriculture as the leading earner of foreign exchange (Pool et al., 2002). The national government has enacted numerous progressive policies (including a National Forest Policy in 2000) to promote conservation and sustainable ecotourism opportunities, particularly within the national park system (e.g., birdwatching, climbing volcanoes, and visiting tropical forests). Recent statistics suggest that over 1.7 million people travel to Costa Rica each year, generating US$1.7 billion in tourism revenue (Embassy of Costa Rica in Washington D.C., 2017).

2.2.2 Guatemala

Ancient Mayan civilizations were once prevalent throughout Guatemala. The Mayan lowlands were characterized by shifting agriculture and terraced cultivation and, although little of the forested areas were removed, the original cover was said to have been greatly altered by the 15th century (Myers and Tucker, 1987). European influence began with Spanish colonists arriving in the 16th century. As with other Central American countries, Guatemala gained independence in 1821 and was briefly a part of the Federal Republic of Central America. During a good portion of the 20th century, the Guatemalan government was embroiled in turmoil due to a number of coups and disputed elections. Guatemala has suffered from nearly four decades of internal conflict, partly due to inequalities in land ownership and income between the rural, indigenous population and the urban population. Natural resources in Guatemala are threatened by continued social and political problems, pressure from cattle and petroleum interests, and illegal logging (Pool et al., 2002). Large-scale deforestation began in

FIGURE 2.14 Natural deciduous forests in Limón Province, Costa Rica. *Donald L. Grebner.*

Guatemala in the 1970s as a result of a land colonization plan developed by the government. This accelerated in the 1980s as large numbers of people sought refuge in the rural areas during the Guatemalan Civil War (International Tropical Timber Organization, 2006).

Guatemala is nearly 11 million ha (about 26.8 million ac) in size and is currently about 33% forested (Table 2.1). About 52% of the forests are privately owned, but ownership rights are still obscure in some places in the aftermath of the civil war, which formally ended in 1996. About 42% of the forests are controlled by the government, and indigenous communal lands (*ejidales*) have a special status by law. Nearly 60% of the forests (Fig. 2.15) are located in the Petén region (a flat low-lying region that borders Mexico) and provide wood for the manufacture of cabinets and the development of chemical extracts, dyes, gums, and oils. However, most (86%) of the domestic roundwood production is burned for fuelwood and about 7% is used in the manufacture of charcoal (Pool et al., 2002). The forest area of Guatemala has been declining by about 1% per year over the last 5 years (Table 2.1). The main commercial tree species found in the Petén region are cypress, oaks, Mexican yellow pine (*Pinus oocarpa*), smooth-bark Mexican pine (*Pinus pseudostrobus*), thinleaf pine (*Pinus maximinoi*), and Spanish cedar (*Cedrela odorata*). Coniferous forests can also be found throughout Guatemala's highlands. Pines and teaks, along with *Gmelina arborea* (a white teak or beechwood), comprise most of the forest plantations in Guatemala. The Pacific plain of Guatemala was once covered by tropical moist forests but now has been developed into banana, rubber tree, and sugar plantations and cattle ranches (International Tropical Timber Organization, 2006).

The Forest Law of 1996 requires forest management plans to be developed by long-term forest resources users. A 1999 forest policy promoted the concept of productive management of natural forests in order to both conserve biodiversity and improve living conditions of forest-dependent communities. A formal community forestry program (*Proyecto Fortalecimiento Forestal Municipal y Comunal*) was begun in 2001 to provide employment to local municipalities and to implement reforestation measures with the assistance of the National Institute of Forests (International Tropical Timber Organization, 2006). Some community managed forest concessions have been awarded to local villages, and assistance has been provided by NGOs and the United States Agency for International Development to improve the technical capabilities of community forestry managers. However, as with other areas of the world, local organizational commitment is essential to increase participation in these programs and is dependent on sustained economic returns, secure land titles, and clear land tenure policies (Pool et al., 2002). Therefore, some local communities struggle with responsive leadership at a variety of levels, and the success of these programs may be problematic (Larson, 2008).

2.2.3 Honduras

Honduras is located between Guatemala and Nicaragua (Fig. 2.13) and is the second largest country in the region, with a land base of about 11 million ha (about 28 million ac). Forests cover about 41% of the country (Table 2.1). Currently, the main forest types in Honduras are cloud forests, coniferous forests, deciduous forests (Fig. 2.16), dry forests, and mangrove forests. The International Tropical Timber Organization (2006) indicates that tree species found below 700 meters (m;

FIGURE 2.15 Forests in the Petén region of Guatemala, near Tikal. *Nerdoguate, through Wikimedia Commons.*
Image Link: https://commons.wikimedia.org/wiki/File:TikalTemples2015_02.jpeg
License Link: https://creativecommons.org/licenses/by-sa/4.0/deed.en

FIGURE 2.16 Deciduous forest in the mountains of Sierra de Agalta, Honduras. *Dennis Garcia, through Wikimedia Commons.*
Image Link: https://commons.wikimedia.org/wiki/File:MontanasdelaSierradeAgalta_Honduras.jpg
License Link: https://creativecommons.org/licenses/by/3.0/deed.en

about 2300 feet [ft]) in elevation include Caribbean pine (or pino costanero) and Mexican yellow pine (or pino ocote), which are commonly referred to as Honduran yellow pines (Pool et al., 2002). Above 1500 m (about 4900 ft) and up to 1900 m (about 6200 ft) in elevation, mixtures of Caribbean pine, pino rojo (*Pinus tecunumanii*), and thinleaf pine can be found (International Tropical Timber, 2006). Above 2000 m (about 6600 ft), Hartweg's pine or pino de México (*Pinus hartwegii*), smooth-bark Mexican pine (or pinabete), Mexican white pine or pino blanco (*Pinus ayacahuite*), and various species of fir are commonly found in mixed stands (International Tropical Timber Organization, 2006).

With the exception of the influence of Mayan culture near the border with Guatemala, Honduran history is similar to that of other countries in Central America. Honduras was part of the Spanish Empire until 1821, and then was considered a province of the Federal Republic of Central America for almost two decades until it became independent in 1838. Honduras was historically a source of precious hardwoods (mahogany and Spanish cedar) that were exported for furniture production purposes to the United States and Europe in the early 1900s (Pool et al., 2002). Land tenure is a persistent concern in Honduras, owing to arson, cattle grazing, and squatting, and obscure property rights claims. One estimate of the distribution of land has about 36% in government control, about 28% in *ejidal* control (communal land administered by municipal governments), and about 36% in private ownership (Unidad de Reconstrucción Nacional, 1999). Unfortunately, forestland was lost at a rate of over 3% per year in the late 20th century due to the agricultural use of land suitable for forests, land clearance for cattle ranching purposes, and mismanagement that focused more attention on logging than on SFM (Merrill, 1995). The rate of forest loss for the last 5 years has been about 2.4% per year (Table 2.1). The establishment of banana plantations was once a major contributor to forest loss. Some agroforestry systems have been developed, but they rely on exotic species of legumes (e.g., mata ratón), English beechwood (e.g., *Gmelina arborea*), eucalypts, and teaks (e.g., *Tectona grandis*) (International Tropical Timber Organization, 2006).

The *Corporación Hondureña de Desarrollo Forestal* (National Corporation for Forestry Development) was established in 1974 with the intent of increasing the management of the forest sector and preventing exploitation by nondomestic companies (Merrill, 1995). However, inefficient management practices and debt created by military-dominated governments has limited the role of the National Corporation for Forestry Development, which was decentralized in the late 1980s but remains vital to wildfire response and protection (Merrill, 1995). In 2007, the new Forest Law was enacted, which established the National Institute of Forest Conservation (*Instituto Nacional de Conservación y Desarrollo Forestal, Áreas Protegidas y Vida Silvestre*) that replaced the National Corporation for Forestry Development for developing forest management standards (Brown et al., 2008).

Honduran agencies have supported the development of social agroforestry cooperatives in some communities with the intent that these communities would actively protect forests from problems associated with illegal logging, overgrazing, shifting agriculture, and wildfires. Most cooperatives operate within forests owned by the government or within *ejidal* forests (Jones, 2003). The Honduran government took steps to promote reforestation in the early 1990s by making large areas of government-owned land accessible to private investors (Merrill, 1995). This privatization of forests may help intensify the use of forest resources, yet restrictions on exports are affecting growth in some industries. Since 2000, sustainable development programs have been funded to increase the ecological, economic, and social benefits of forestry in Honduras. The hope is that these programs will modernize the agencies and policies that oversee forestry activities and bolster the role of the National Forestry Administration. The United States Agency for International Development's Forestry Development Project has assisted the Honduran forestry sector by providing training and thereby increasing technical skills and helping the country to improve harvesting practices (Pool et al., 2002). Collaborations such as these are complemented by activities funded through NGOs and entities such as Scandinavian governments.

2.2.4 Nicaragua

Nicaragua contains about 12 million ha of land (about 30 million ac) (Table 2.1) and is characterized by a warm tropical climate dominated by moist easterly trade winds. The topography of the country leads to three distinct regions: a warm Pacific coastal region (Fig. 2.17), a humid interior mountain region, and a Caribbean lowland area that has a warm and wet climate. Most of the forests in Nicaragua are considered tropical rainforests, although several other forest zones have been classified (deciduous hardwood, evergreen hardwood [largely inaccessible], lowland pine, and mountain pine). Compared to other Central American countries, a relatively small portion of the land is controlled by the government (about 11%), while the remainder is mainly privately owned. About 96% of the roundwood production from Nicaraguan forests is for domestic fuelwood uses (Merrill, 1994). In one study of the Masaya region of Nicaragua, virtually all of the fuelwood was derived from natural forests rather than forest plantations (McCrary et al., 2005). Transportation issues in the Caribbean lowlands are prominent and road systems are relatively

FIGURE 2.17 Cloud forests on the island of Ometepe, in Lake Nicaragua, Nicaragua. *Adrian Sampson, through Wikimedia Commons. Image Link: https://commons.wikimedia.org/wiki/File:Trees_on_Maderas.jpg License Link: https://creativecommons.org/licenses/by/2.0/deed.en*

undeveloped, preventing intensive use of the hardwood tree species found there. Large areas of pine stands in the northeastern part of the country support a small manufacturing industry (Merrill, 1994).

Nicaragua's history with regard to natural resource use is similar to other Central American countries. Internal struggles have also shaped the current management situation. Spanish conquistadors founded cities on the western side of Nicaragua in the 16th century and used forest resources for shipbuilding purposes. In the 17th century, the British occupied the eastern Miskito coast and utilized the big-leaf mahogany (*Swietenia macrophylla*) and Spanish cedar forest resources found throughout the region. From the 17th century until about 1950, large areas of forests in Nicaragua were converted to coffee plantations; from about 1950 to 1980, considerable areas of the western part of the country were converted to cotton plantations; and from 1970 to 1980 cattle production and small-scale agriculture influenced land conversion on the eastern (Caribbean) side of the country. Lately, tourism has been the fastest growing sector in the Nicaraguan economy (Weaver et al., 2003).

Before 1979, concessions provided by the Nicaraguan government for logging on national lands led to forestry practices that degraded the resource. With the Sandinista takeover, all concessions were revoked and the concession system was abolished. The Nicaraguan government then took control of the forestry sector, and the *Corporación Forestal del Pueblo* was developed to administer timber harvests on public lands. However, income from timber harvests on public land declined precipitously over the next decade or so. In 1990, after winning the national election, the Chamorro government began to promote an economic role in the forestry sector and created a forestry commission to oversee new

investments. Unfortunately, implementation of the process has been complicated by the unclear land tenure rights, particularly in the Caribbean region of Nicaragua (Castilleja, 1993; Weaver et al., 2003).

Deforestation in Nicaragua was slowed by internal conflicts during the 1980s but was said to have resumed again after these conflicts ended (Weaver et al., 2003). However, current estimates suggest that the forested area in Nicaragua is neither declining nor growing (Table 2.1). Several institutions are involved in the control and regulation of environmental activities within the country. From a forest management perspective, the Nicaraguan Forest Authority (*Instituto Nacional Forestal*) oversees and approves forest management plans. The *Ministerio Agropecuario y Forestal* addresses laws, policies, and regulations for forest management within Nicaragua and was instrumental in the development of the National Forest Law (Law No. 462) in 2003 (Brown et al., 2008).

2.2.5 Panama

Panama is a country comprising about 7.4 million ha of land (18.4 million ac), of which about 62% is forested (Table 2.1). Panama was once inhabited by indigenous populations of the Cuevas and Coclé tribes until Spain colonized the isthmus in the 16th century, controlling this country and others until 1821. Panama then became a department of Colombia for the next eight decades until it declared independence. In a treaty with the United States in 1903, Panama granted a zone of land about 16 km (10 mi) wide (thence called the *Canal Zone*) where the Panama Canal would be developed. This zone reverted to Panamanian control in 1999. The current rate of forest loss in the country is about 0.4% per year (Table 2.1), and deforestation was once said to be most pronounced along the Panama Canal, which may pose long-term water level issues (Meditz and Hanratty, 1989) due to sedimentation resulting from rainwater runoff. The Panamanian government has implemented a program promoting reforestation practices, but the current pace of forest depletion exceeds that of replanting. Recent forest losses have occurred due to cattle ranching, poor logging practices, shifting cultivation (rozas), and urbanization (International Tropical Timber Organization, 2006).

Forests of Panama are diverse (Fig. 2.18), and the most prevalent forest types in Panama are the semideciduous tropical moist forests, lowland submontane forests, and montane evergreen forests. The tropical moist forests contain cuipo (*Cavanillesia platanifolia*) and wild cashew or espavé (*Anacardium excelsum*), along with some species of palm. Oak forests can be found in the Talamanca Mountains, and cativo (*Prioria copaifera*) forests can be found along rivers on inundated areas. Caribbean pine and teak comprise over 80% of the forest plantations in

FIGURE 2.18 Deciduous forest canopy in Barro Colorado, Panama. *Christian Ziegler, through Wikimedia Commons.*
Image Link: https://commons.wikimedia.org/wiki/File:Forest_on_Barro_Colorado.png
License Link: https://creativecommons.org/licenses/by/2.5/deed.en

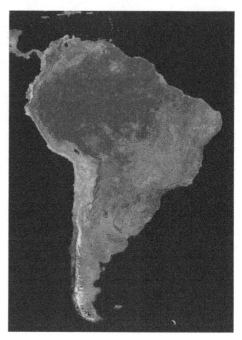

FIGURE 2.19 A satellite image of South America. *National Aeronautics and Space Administration, Dave Pape, and Reto Stöckli, through Wikimedia Commons.*
Image Link: https://commons.wikimedia.org/wiki/File:South_America_-_Blue_Marble_orthographic.jpg
License Link: Public Domain

Panama. Most of the logging practices involve some sort of selection system, and about 95% of the roundwood produced is used for fuelwood or charcoal. Planting trials of native tree species have provided some insight into their potential for establishing tree cover, stabilizing soils, and restoring forests (Wishnie et al., 2007).

Panama passed a Reforestation Incentive Law in 1992 to promote and increase reforestation efforts through tax incentives and other means. The Forest Law of 1994 provides a framework for SFM and classifies land into protection and production classes. The General Environmental Law of 1998 also promotes the sustainable use of natural resources. Decree Law No. 2 (2003) provides forest management guidelines for Panamanian forests and, while most of the land is owned by the government, the 1972 constitution recognizes collective landholding units known as *comarcas* (indigenous reserves) (International Tropical Timber Organization, 2006).

2.3 South America

As with North America, South America (Fig. 2.19) has been heavily influenced by indigenous peoples that crossed Beringia from Asia thousands of years ago. South America has also been heavily influenced by European settlers, primarily arriving from Spain and Portugal. Unlike North America, the last ice age had

little effect on this continent. The topography of South America ranges from the relatively steep Andes mountain range in the west, which stretches from Colombia in the north to the southern tip of Chile and Argentina, to the relatively flat coastal areas of Brazil and other countries bordering the Atlantic Ocean. The ecoregions of the continent vary from montane and temperate rainforests in the southwest, to grass savannas in the southeast, the rainforests of Amazonia in the north–central portion of the continent, and the coastal forests of Ecuador, Venezuela, and Guyana. The largest areas of forests designated for the conservation of biological diversity are found on this continent (Food and Agriculture Organization of the United Nations, 2010). Species richness is relatively high in South America, with over 40,000 plant species; hundreds of mammalian, amphibian, and bird species; and millions of insect species living on the continent (Da Silva et al., 2005). Another area of great species richness is the Galápagos Islands, which are part of Ecuador, located off the west coast of South America. Charles Darwin studied the large number of endemic species found here, which inspired his development of the theory of natural selection. One of the famous and unique species found in the Galápagos Islands is the marine iguana (*Amblyrhynchus cristatus*), which lives and forages in the ocean.

South America includes 12 independent countries (Fig. 2.20), as well as French Guiana (a department of

FIGURE 2.20 Countries in South America. *Modified image from PresentationMaps.com.*

France) and the Falkland Islands, South Georgia, and the South Sandwich Islands (all of which are associated with the United Kingdom). The land area covers approximately 1.8 billion ha (a little over 4.3 billion ac) and hosts a population of over 385 million people (Table 2.3). South America as a whole is about 53% forested, and the annual rate of change in forested areas is currently about −0.2% per year, a rate which is slower than what was observed in the early 2000s. However, the net loss of forest area in South America is estimated to be about 4 million ha (10 million ac) per year since 2000 (Food and Agriculture Organization of the United Nations, 2010). As with North America, the colonization of European settlements in South America led to changes in the land uses that were employed by indigenous people. As European colonies grew, pressure was applied to forested areas to facilitate agricultural operations, provide sustenance, and produce timber for

domestic and export uses. On a more current note, although deforestation in the Amazon rainforest is a continuing problem, afforestation of suitable forest sites in Chile, Brazil, and other countries has greatly expanded and challenges the timber production potential of other countries due to faster tree growth and favorable environmental conditions. Our discussion of South American countries briefly focuses on seven of the largest and heavily forested countries: Argentina, Bolivia, Brazil, Chile, Colombia, Peru, and Venezuela.

2.3.1 Argentina

Argentina is the second largest country in South America, about 274 million ha in size (about 676 million ac) yet is only about 10% forested (Table 2.3). It is bordered by the Atlantic Ocean to the east and the Andes mountain range to the west and is composed

TABLE 2.3 Demographic and physiographic data for South American countries.

Country	Human population 2018[1] (million)	Total land area[2] (1000 ha)	Total land area[2] (1000 ac)	Forest cover[2] (%)	Forest area annual change 2010–15[2] (%)	Per capita GDP (PPP) 2017[1&a] (1000 US$)
Argentina	44	273,669	675,962	10	−1.1	14.3
Bolivia	11	108,330	267,575	51	−0.5	3.3
Brazil	209	835,814	2,064,461	59	−0.2	9.8
Chile	19	74,353	183,652	24	1.8	15.3
Colombia	50	110,950	274,047	53	—	6.3
Ecuador	17	24,836	61,345	51	−0.6	6.2
Falkland Islands	<1	1217	3006	—	—	—
French Guiana	<1	8242	20,358	99	—	—
Guyana	<1	19,685	48,622	84	−0.1	4.6
Paraguay	7	39,730	98,133	39	−2.0	4.3
Peru	32	128,000	316,160	58	−0.2	6.6
Suriname	<1	16,066	39,683	95	—	6.8
Uruguay	3	17,502	43,230	11	1.3	17.1
Venezuela	29	88,205	217,866	53	−0.3	8.0
South America (total)	424	1,746,599	4,314,100	53	−0.2	9.4

[a]*Gross domestic product (GDP) expressed as purchasing power parity (PPP).*
[1]*Food and Agriculture Oraganization of the United Nations 2020.*
[2]*EAO. Global Forest Resources Assessment 2015. Desk Reference. Rome, 2015.*

of 23 states. The Inca Empire once ruled part of northern Argentina, and among the various indigenous people who inhabited the remainder of the country were the Guaraní, Mapuche, and Sanavirones. European explorers arrived to settle Argentina in the early 16th century, and the current composition of the country was once part of the *Viceroyalty of Peru*, a Spanish colony. Argentina declared independence from Spain in 1816 but, through various conflicts (internally, and with Brazil and Chile), did not enact a constitution until the mid-1850s. Since gaining independence, Argentina has had periods of unstable leadership and economic problems and has employed a variety of approaches for governance that range from neoliberal to protectionist (Gulezian, 2009).

There are four major biophysical regions in Argentina. The first, the Pampas, is an area of grasslands and plains in the center of the country, which forms the main agricultural area of the country. Over time, a good portion of the forested area that once stood in the central part of the country has been converted to agricultural use. Soy has become one of the major exports, and Argentina has an extensive livestock industry that is centered in the Pampas. The second major biophysical region, Patagonia, is a colder and drier place that is located in the southern part of the country and shared with Chile. Patagonia contains Tierra del Fuego, and the dominant vegetation in this area is characterized as shrubs and grasses. The third major biophysical region is an area of subtropical dry forests in the northern part of the country, where most of the commercial forestry activity is located. Deciduous and semideciduous forests are found throughout this region; however, carob (*Prosopis* spp.) and palm grow naturally here, and pine and eucalypt plantations are also prevalent in this region. The fourth major biophysical region contains areas adjacent to the Andes mountain range in the western part of the country (Fig. 2.21). Here, broad-leaved evergreen and semideciduous tree species are common in forested areas.

Nearly 80% of Argentine forests are privately owned (White and Martin, 2002). The current rate of forest change in Argentina is estimated to be about −1.1% per year (Table 2.3). About 70% of the roundwood produced from Argentine forests is consumed within the country, and the remaining 30% is exported as various wood products. Brazil, Chile, South Africa, Spain, and the United States are the most important export markets for wood. Forest policies in Argentina prohibit the development of natural forests yet promote the development of forestry sectors (Rubio, 2006). The 2007 Forest Law (Ley de Bosques) reinforced the country's desire to manage its natural forests in a sustainable manner (Gulezian, 2009).

FIGURE 2.21 Tucuman region of northwestern Argentina. *Susana Mutti, through Wikimedia Commons.*
Image Link: https://commons.wikimedia.org/wiki/File:Tafi_del_valle_(Tucuman-Argentina).jpg
License Link: https://creativecommons.org/licenses/by/2.0/deed.en

2.3.2 Bolivia

Prior to settlement of the country by people of European descent, Bolivia was part of the Inca Empire. Bolivia was colonized by Spain in the 16th century, and silver eventually became the main commodity derived from Bolivian mines. The country declared independence in 1809 but did not formally become an independent republic until 1825. At various times in its history Bolivia had direct access to the Pacific Ocean but, as a result of military struggles with Peru and Chile, it is now a landlocked country (Fig. 2.20). Bolivia has three biogeographic zones: a high altitude, treeless area in the Andes Mountains; forested valleys on both sides of the Andes Mountains that may also be associated with grass savannas; and tropical forest lowlands to the east that form part of the Amazon basin. With a total land area of a little more than 108 million ha (about 268 million ac), around 51% of Bolivia is currently considered forested (Table 2.3).

Most of the forested area in Bolivia is located in the eastern tropical forest lowlands (Fig. 2.22). This area contains a wide variety of broad-leaved evergreen and semideciduous forests, and tree species that include big-leaf mahogany, Brazil nut (*Bertholletia excelsa*), guanandi (*Calophyllum brasiliense*), rubber tree (*Hevea brasiliensis*), and sandbox tree (*Hura crepitans*). The forested valleys along the Andes mountain chain ultimately transition into alpine tundra, yet can include tree species such as curupay (*Piptadenia macrocarpa*), laurels (*Lauraceae* spp.), mahogany (*Meliaceae* spp.), urundel (*Astronium urundeuva*), walnut (*Juglans australis*), and *Andean alder* or aliso (*Alnus acuminata*). In areas where plantations can suitably be established, blue gum (*Eucalyptus globulus*) and patula pine (*Pinus*

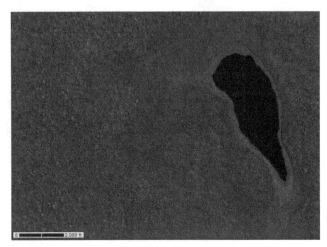

FIGURE 2.22   Forests and forested wetlands around a small lake in eastern Bolivia. *U.S. Department of Agriculture, Natural Resources Conservation Service (2020).*

patula) are the two tree species that comprise the majority of forest plantations. Brazil nut, cacao, and palm hearts are the main NTFPs derived from Bolivian forests (International Tropical Timber Organization, 2006).

Recent estimates of forest loss are in the order of 0.5% per year (Table 2.3) owing to planned and unplanned human settlement activities and agricultural activities. Large soybean plantations are prevalent in some areas, and smaller-scale farmers also cause forest loss through slash-and-burn agricultural practices (International Tropical Timber Organization, 2006). About 53% of the forestland in Bolivia is controlled by the government, about 31% is also publicly owned but has specific user rights, such as those allocated to indigenous people, and about 16% is privately owned. The 1996 Forest Law introduced the idea of sustainable use and development of the forest resources and helped clarify ancestral rights, although land tenure issues continue to be important obstacles to forest management in Bolivia (White and Martin, 2002).

2.3.3 Brazil

Brazil is the largest and most populous country in South America. It has a landmass of over 836 million ha (about 2.064 billion ac), of which forests cover about 59% (Table 2.3). Brazil was home to a number of indigenous tribes prior to colonization by Portugal at the beginning of the 16th century. During colonization, the important exports from Brazil were gold and sugar. The country became independent in 1822, while Portuguese continues to be the official language. Brazil is now composed of 26 states and one federal district, and across these are five broad forest types which include Amazon rainforest, Atlantic rainforest, central *cerrado* grass savanna, arid *caatinga*, and the Pantanal (International

Tropical Timber Organization, 2006). Deforestation has been an important issue across Brazil; however, since the late 2000s the rate of forest loss due to deforestation has been significantly reduced to about −0.2% per year, as compared to the activity of the 1990s (Food and Agriculture Organization of the United Nations, 2015). The Amazon region has received much attention owing to the rich diversity of plant and animal species found in these tropical rainforests. It is speculated that many plant and animal species that inhabit these lands have still not been identified. Since the late 20th century, much concern has been raised over the clearing of Amazon rainforests for agricultural and livestock uses (Morton et al., 2006). The broad-leaved evergreen and semideciduous forests of the tropical rainforests and the central plains of southern and eastern Brazil have been the most affected by deforestation.

In Brazil, there is a large and active forest management program that involves harvesting native and plantation forests. Some common native species harvested in Brazil include cambara or cedrinho (*Erisma uncinatum*), cow tree or amapa (*Brosimum utile*), faveira (*Parkia* spp.), guanandi, and jatobá (or guapinol; *Hymenaea courbaril*) (International Tropical Timber Organization, 2006). About one-third of the total amount of roundwood produced comes from plantation forests. Forest plantations, composed of eucalypts (Fig. 2.23), pines, or other species (e.g., yopo [*Anadenanthera peregrina*]; Fig. 2.24), are mainly located in southern Brazil (Fearnside, 1999),

FIGURE 2.23   Eucalypt (*Eucalyptus* spp.) plantation in Brazil. *Denis Rizzoli, through Wikimedia Commons.*
Image Link: https://commons.wikimedia.org/wiki/File:Eucaliptal_Aracruz.JPG
License Link: https://creativecommons.org/licenses/by-sa/3.0/deed.en

FIGURE 2.24 Planted yopo (*Anadenanthera peregrina*) forest in Brazil. *Zachary A. Parisa.*

where the biomes are classed as grass savannas or subtropical rainforests. In 2006, it was estimated that there were 2 million ha (4.8 million ac) of pine and 3.3 million ha (8.1 million ac) of eucalypt plantations in Brazil (International Tropical Timber Organization, 2006).

A Forestry Code developed in 1965 is still in effect in Brazil, although recent controversial reforms to the code have been proposed. Since independence was declared, most Brazilian forests have historically been controlled by the government, and property rights for some forested areas in Brazil have been characterized as ambiguous (Banerjee and Alavalapati, 2010). In 2009, a large portion of the Amazon basin was made available for privatization to the settlers located there, who some suggest have been illegally using the land. A number of NGOs are active in Brazil and apply political pressure for action on natural resource conservation issues (International Tropical Timber Organization, 2006).

2.3.4 Chile

Located on the southwestern Pacific coast of South America (Fig. 2.20), Chile is approximately 4300 km (2700 mi) long, and the Andes mountain range runs the length of the country along its eastern border with Argentina. Prior to Spanish colonization, Inca and Mapuche cultures shaped the Chilean landscape mainly through agricultural activities. In the early 16th century, the Spanish, after discovering what are now called the Straits of Magellan, eventually colonized the country after numerous confrontations with the indigenous population and with other countries seeking the gold resources that were thought to reside there. Independence came in 1818 after nearly a decade of attempts to become autonomous from Spain. The extent of Chile has varied since colonization, and today Chile has a total land area of about 74 million ha (about 184 million ac). Moist temperate and subtropical rainforests composed of broad-leaved or semideciduous tree species dominate the forested areas, although the country is currently only about 24% forested (Table 2.3). Of the original forest area

that existed before European settlers arrived, approximately 45% remains (Global Forest Watch, 2002). Chile, like other countries in South America, has experienced a period of extensive land conversion for agricultural and livestock production purposes (Donoso and Otero, 2005). However, the rate of change in forested area since 2010 is about +1.8% per year (Table 2.3).

The northern portion of the country contains the Puna grassland ecoregion, which is extremely arid and boasts the driest place on Earth. There is little vegetation in some of this region except for tola (*Proustia pyrifolia*) and grasses covering the mountain slopes. In the central region of the country, one can find assorted species of cactus as well as the monkey puzzle tree or Chilean pine (*Araucaria araucana*), which has edible seeds. The capital, Santiago, is located in the middle third of the country, which generally has a temperate climate. This is also where the majority of the Chilean population resides. The southern third of the country, which includes Patagonia, is where the moist temperate and subtropical rainforests can be found among numerous canals, fjords, islands, and peninsulas that stretch south to the Straits of Magellan. The forests in the southern part of the country (Fig. 2.25) are comprised of various coniferous species, laurels, magnolias, and southern beeches (*Nothofagus* spp.). Numerous wildlife species inhabit this area, including the Andean wolf (*Dasycyon hagenbecki*); Andean fox (*Lycalopex culpaeus*); the South American gray fox or gray zorro *Lycalopex griseus*; the guanaco, a camelid (*Lama guanicoe*); pudú, a species of deer (*Mapudungun pudú*); pumas; and various rodents, rabbits, birds, and lizards.

The predominant system of land tenure in Chile is private ownership (about 75%), and the remaining 25% is publicly owned. Although laws allow communal ownership of land, it represents a minor system. Conflict with indigenous communities with regard to land ownership continues today. The Chilean forest sector plays an important role in the economy, as one estimate suggested that 13% of the country's exports arose from

FIGURE 2.25 Forest along a river near Parque Nacional Vicente Pérez Rosales, Chile. *Donald L. Grebner.*

the forestry sector (Food and Agriculture Organization of the United Nations, 2003). Chile exports its products to Germany, Japan, and the United States, among others. Chile has been active in developing eucalyptus and radiata pine (*Pinus radiata*) plantations in the central region of the country for quite some time (Scott, 1954). In addition to traditional forest products, Chile has also been active in exporting nonwood natural resources such as boldo leaves (*Peumus boldus*), soap bark tree (*Quillaja saponaria*), sweetbriar rose (*Rosa eglanteria*), and wild mushrooms (Global Forest Watch, 2002). Forest policy in Chile is guided by the Comprehensive Environmental Law of 1994 and a Native Forest Law (2008). The latter prohibits the clearing of native forests for any reason, including the establishment of forest plantations. Some support for forest development has been provided by the German government in an effort to encourage the maintenance of farm families who use SFM to derive economic and environmental benefits and thus improve living conditions for this segment of the population (Food and Agriculture Organization of the United Nations, 2003).

2.3.5 Colombia

The area represented by present-day Colombia (Fig. 2.20) was once inhabited by a number of indigenous groups, including the Muisca and Tairona tribes. Spanish colonization began around the 16th century and ended in the early 19th century. Colombia has endured periods of unrest and instability over the last 200 years, and has been governed under several names, including the *Viceroyalty of New Granada*, the *Republic of Colombia*, the *Granadine Confederation*, and the *United States of Colombia*. Portions of Ecuador and Venezuela were once politically tied to Colombia, as was current-day Panama. Although Colombia is smaller today than it was immediately after the end of Spanish control, it is still the fourth largest country in South America and encompasses about 111 million ha of land (about 274 million ac), of which 53% is forested (Table 2.3). Colombia has two distinct biogeographic zones: the Andean mountain zone, which leads to alpine tundra; and the tropical plains that contain tropical rainforests and grass savannas. One of the three tropical plains contains a portion of the Amazon basin and the Orinoco basin in the southern portion of the country. Another tropical plain is located in the northern part of the country and faces the Caribbean Sea (Fig. 2.26). The third tropical plain is located in the western part of the country and faces the Pacific Ocean (International Tropical Timber Organization, 2006).

Until recently, the description and registration of land tenure and grants of land in Colombia, as in other countries of this region, was vague and this insecurity

FIGURE 2.26   Forests above the town of Taganga near Santa Marta, Magdalena, Colombia. *Juancplanb, through Wikimedia Commons. Image Link: https://commons.wikimedia.org/wiki/File:Taganga01.jpg License Link: https://creativecommons.org/licenses/by-sa/3.0/deed.en*

did not encourage people to take care of the land (van Bottenburg, 1952). Nearly two-thirds of the forestland in Colombia is privately owned and in the last five years the deforestation rate in Colombia has been negligible. However, deforestation may still be relatively high in the foothills of the Andes Mountains, where significant settlement and land conversion has recently occurred and where most of the coca (from which the stimulant cocaine is derived) is produced. Modern Colombian forest laws date back to the 1950s. The latest forest policy (*Política de Bosques*) stresses sustainable forestry practices and improvements in the quality of life for the Colombian people. Indigenous groups and Afro-Colombian communities can now register their rights to territories that they have historically occupied (White and Martin, 2002). Some of the broadleaved and semideciduous forests common to this country are managed under concessions or cooperatives between the government and the forest industry. A Green Plan (*Plan Verde*) was developed in 1998 to promote reforestation efforts in degraded areas. The reforestation program is now well developed, and forest plantations include Caribbean pine, Mexican yellow pine, patula pine, eucalypts, mangium (*Acacia mangium*), and white teak or beechwood. Most of the roundwood harvested within the country is consumed domestically (International Tropical Timber Organization, 2006), but some quantities of pulp, paper, and lumber are now exported (Mendell et al., 2006). Important NTFPs include palm fruits, rubber, and a neotropical species of bamboo (*Guadua angustifolia*) that is used as a building material. Colombia is one of the most biologically diverse areas in the world and the potential for ecotourism is very good, but security problems make the development of these opportunities problematic (International Tropical Timber Organization, 2006).

2.3.6 Peru

The Republic of Peru is about 128 million ha in size (about 316 million ac) and about 58% forested. Over the past 5 years, the rate of forestland loss has been about −0.2% per year (Table 2.3). The land that now represents Peru was once home to the Inca Empire and the Norte Chico civilization. As with other South American countries, in the 16th century Peru became part of the Spanish Empire and obtained independence in the early 1800s. The Amazon region of Peru harbors some of the last uncontacted people in the world, who have chosen to remain isolated and are referred to by anthropologists as *indigenous people of voluntary isolation* (Portilla and Eguren, 2007). The forests of Peru are home to about 2500 native tree species (Portilla and Eguren, 2007). The ecoregions include the dry steppe of the coastal plain (along the Pacific coast), the temperate Andes highlands (Fig. 2.27), and the extensive tropical forests in the Amazon basin. In the arid plains along the Pacific coast, cattle ranches and soybean farms have been developed. Along the coastal plain are also found hualtaco (*Lonopterygium huasango*), huarango or mesquite (*Prosopis pallida*), and mangroves. However, gold mining and petroleum exploration have affected the extent and composition of Peru's forests, and large areas of forests along the foothills of the Andes have been converted to coca plantations. The tropical forests of the Amazon basin include cloud forests and other areas considered to be highly biologically diverse. From a commercial perspective, some of the more important tree species are ishpingo (*Amburana cearensis*), mahogany, marupa (*Simarouba amara*), pumaquiro (*Aspidosperma macrocarpon*), Spanish cedar or cedro rojo, and *Virola* spp.

About one-third of the country's forests are privately owned and, of the remainder, about 1% is assigned as communal forests for indigenous people (White and Martin, 2002). Peru passed a forestry law in 1975, but the Forestry and Wildlife Law (Law No 27,308) passed in 2000 and the General Law of the Environment (Law No 28,611) passed in 2005 now provide (along with other laws) the primary legal bases for environmental compliance and management requirements for Peruvian forests (Portilla and Eguren, 2007). Although adherence varies for a number of reasons (Smith et al., 2006), under these laws, timber extraction rights on public lands are assigned through concessions to logging operators. Currently, revisions to the forest laws are being debated.

2.3.7 Venezuela

The history of Venezuela, from habitation by indigenous people to Spanish colonization, is similar to that of Colombia. Venezuela has had periods of political and economic instability in the 20th century, some of which is tied to the development of its petroleum industry (Rodríguez, 2000). Venezuela contains about 88 million ha of land (about 218 million ac) and is now about 53% forested (Table 2.3). Logging efforts beginning around 1950 were aimed at transitioning land use from forests to agricultural or cattle production. In the 1980s, the annual rate of deforestation was among the highest in Latin America (Kammesheidt et al., 2001). Forestland area has continued to decrease since 2010 at a rate of about −0.3% per year (Table 2.3). Over 90% of the forest area in Venezuela is owned by the government and concessions are provided to forestry companies for the rights to harvest trees on public land. As with other South American countries, a number of permanent forest estates have been established to protect native forests.

The ecoregions of Venezuela are numerous and complex. They range from mountain highlands to central plains (Llanos) and coastal areas that include deltas and islands located along or within the Caribbean Sea and Atlantic Ocean. In the western part of the country, in the area of Lake Maracaibo, the ecosystems range from arid to tropical over a relatively short distance and can be composed of physiographic groups that some might better recognize using the terms high plains, low plains, mesas, or Piedmont (Henri, 2001). In the Piedmont areas, many forests are composed of semideciduous tree species. South of the Orinoco River in the eastern portion of the country (Fig. 2.28) are large expanses of deciduous forests and plains that contain vegetation that may still be considered precolonial in character (Rodríguez, 2000).

Nearly all of the roundwood produced in Venezuela is consumed in the country, and most (about 70%) is used for fuelwood purposes. Cedar, mahogany, and saquisaqui (*Bombacopsis quinata*) have historically been the main tree species harvested from native forests to

FIGURE 2.27 Countryside surrounding Pampas, in the Huancavelica region of Peru. *Digary, through Wikimedia Commons.*
Image Link: https://commons.wikimedia.org/wiki/File:Anexo_de_Pamuri-3.JPG
License Link: https://creativecommons.org/licenses/by/3.0/deed.en

FIGURE 2.29 Satellite image of Europe. *National Aeronautics and Space Administration.*

FIGURE 2.28 Deciduous forests along the Orinoco River in Venezuela. *Pedro Gutiérrez, through Wikimedia Commons.*
Image Link: https://commons.wikimedia.org/wiki/File:Orinoco4.jpg
License Link: https://creativecommons.org/licenses/by/2.0/deed.en

produce lumber products. Cinnamon, cumin, ginger, nutmeg, palm nuts, resins, rubber, tonka beans, and other NTFPs are also important. In the dry tropical region of western Venezuela, forest plantations containing Caribbean pine, eucalypts, and teak have been developed (Henri, 2001). In terms of natural resource policy, Venezuela passed a Forest Law in 1966 (*Ley Forestal de Suelos y de Aguas*) that integrated the uses of forests, water resources, and soil resources. A 1999 Organic Environmental Law placed jurisdiction of forest management under the Ministry of the Environment and Natural Resources (International Tropical Timber Organization, 2006), and the 2013 Law of Forests further established the role of the ministry in forest management.

2.4 Europe

The history of modern forests in Europe, as in North America, is highly dependent on the role that the last ice age played across the European landscape (Fig. 2.29). As the ice sheets advanced south, they scoured the landscape and destroyed all of the terrestrial vegetation. The ice sheet covered all of Scandinavia, the northern half of central Europe, northern Russia, and the British Isles. As the ice sheet slowly retreated, vegetation recolonized the newly exposed landscape. The pioneering tree species were coniferous because of their greater ability to retain water. Later, broad-leaved species spread and recolonized the European landscape. Mesolithic peoples soon returned to these areas, but they had little impact on the forests (Westoby, 1989). Later, stone ax-wielding Neolithic peoples settled the land and cleared large areas of forestlands for agricultural purposes. This

clearing was further accelerated by the arrival of the Celts and, later, the expansion of the Roman Empire.

The spread and contraction of forests after this period have varied greatly due to ecological conditions, population growth, and plague epidemics, as well as social and political unrest. As human populations grew, people cleared more land for agriculture, but the lands reverted to forests when whole communities perished as a result of the black plague (Loude, n.d.). Numerous wars over the past several thousand years contributed greatly to the degradation of forests, which were used as a source of raw material for the development of weaponry and defenses. Some countries, such as the United Kingdom, were able to withstand these dramatic changes to their forested landscape because of their higher annual precipitation levels. Unfortunately, some Mediterranean countries were unable to withstand the dramatic changes in forest conditions due to their more arid climate and experienced high levels of soil erosion and subsequent losses of forested landscapes.

European forests are situated in the temperate and boreal biomes. The Iberian Peninsula hosts forests characterized as having Mediterranean vegetation. The Alps, Pyrenees, and other mountainous areas host montane forests. The higher latitudes of Scandinavian countries host boreal forests as biomes transition from temperate broad-leaved and coniferous to taiga and then to tundra. Some may argue that European forests are simpler, in terms of the diversity of tree species, than other parts of the world. The vegetation in Europe has been managed fairly intensively for several centuries longer than the vegetation in America. Nevertheless, their role in the world forestry sector is critical. Since 2010, Europe has experienced stability in forested area (Table 2.4) and, while rates of change vary, many countries with a significant forest area have had a positive rate of change since late 2010 (Table 2.4). There are too many countries contained in this region to explore in sufficient depth in this book. Therefore, we will

TABLE 2.4 Demographic and physiographic data for European countries.[a]

Country	Human population 2018[1] (million)	Total land area[2] (1000 ha)	Total land area[2] (1000 ac)	Forest cover[2] (%)	Forest area annual change 2010—15[2] (%)	Per capita GDP (PPP) 2017[1&b] (1000 US$)
Austria	9	8,244	20,363	47	–	47.7
Belarus	9	20,748	51,248	42	0.2	5.7
Bosnia and Herzegovina	3	5,100	12,597	43	–	5.2
Bulgaria	7	10,856	26,814	35	0.5	8.2
Croatia	4	5,596	13,822	34	–	13.2
Czech Republic	11	7,722	19,073	35	0.1	20.3
Estonia	1	4,239	10,470	53	–	19.8
Finland	6	30,390	75,063	73	–	45.7
France	65	54,766	135,272	31	0.7	38.4
Germany	83	34,861	86,107	33	–	45.0
Greece	11	12,890	31,838	32	0.8	18.2
Hungary	10	9,127	22,544	23	0.2	14.4
Italy	61	29,414	72,653	32	0.6	32.7
Latvia	2	6,220	15,363	54	–	15.6
Lithuania	3	6,268	15,482	35	0.1	16.4
Norway	5	30,427	75,155	40	–	75.3
Poland	38	30,622	75,636	31	0.2	13.8
Portugal	10	9,026	22,294	35	-0.4	18.9
Romania	20	23,002	56,815	30	1.0	10.8
Russian Federation[c]	146	1,637,687	4,045,087	50	–	11.0
Serbia	9	8,746	21,603	31	0.1	5.9
Slovakia	5	4,809	11,878	40	–	17.6
Slovenia	2	2,014	4,975	62	–	23.3
Spain	47	49,880	123,204	37	0.2	28.4
Sweden	10	41,034	101,354	68	–	54.0
Switzerland	9	4,000	9,880	31	0.8	80.1

Continued

TABLE 2.4 Demographic and physiographic data for European countries.[a]—cont'd

Country	Human population 2018[1] (million)	Total land area[2] (1000 ha)	Total land area[2] (1000 ac)	Forest cover[2] (%)	Forest area annual change 2010–15[2] (%)	Per capita GDP (PPP) 2017[1&b] (1000 US$)
Ukraine	44	57,938	143,107	17	0.2	2.5
United Kingdom	67	24,193	59,757	13	0.5	39.8
Other countries (23)	49	44,126	108,991	12	0.4	33.0
Europe (total)	746	2,213,945	5,468,444	46	—	27.2

[a]That have over 3,000,000 ac (about 1,214,100 ha) of forest.
[b]Gross domestic product (GDP) expressed as purchasing power parity (PPP).
[c]A large portion of the Russian Federation is located in Asia as well.
[1]Food and Agriculture Organization of the United Nations 2020.
[2]FAO. Global Forest Resources Assessment 2015. Desk Reference. Rome, 2015.

concentrate on a few (Austria, Belarus, Finland, Germany, Russia, Spain, and the United Kingdom) to provide an indication of the diversity of forest conditions and settings this region provides.

2.4.1 Austria

Austria is a land-locked, mountainous country in central Europe, bordered by the Czech Republic, Germany, Hungary, Italy, Liechtenstein, Slovakia, Slovenia, and Switzerland (Fig. 2.30). Of the 8.2 million ha (over 20 million ac) of land area in Austria, 47% is forested (Table 2.4). The history of Austria includes periods where the country was an integral part of the Roman Empire and the Holy Roman Empire, and periods where the country was associated with Hungary, forming the Austro-Hungarian Empire. After World War I, Austria became an independent republic. The country was annexed by Germany during World War II and was then occupied by the Allies for a few years until its

sovereignty was restored. Austria is currently a federal state with nine provinces and, given its location in the Alps, forests (Fig. 2.31) are characterized as cool temperate and boreal (montane or alpine), with variations of these according to the elevation of the land. The forests are composed of various species of European beech (*Fagus sylvatica*), European larch (*Larix decidua*), European silver fir (*Abies alba*), Norway spruce (*Picea abies*), alders, maples, oaks, and others.

Shortly after World War II, as with other countries, Austria's most important agricultural and forestry objectives were directed toward self-sufficiency of resources. Austria has since developed a wood products industry to the point that it now exports more products than it consumes for development or fuelwood purposes. The Forest Act of 1975 (amended in 2002) provides guidance for forest management activities. The Forest Act also allows people access to all forests for recreational purposes, with the exception of hunting, which requires a permit for a specific hunting district. Austria

FIGURE 2.30 Countries in Europe. *Modified image from PresentationMaps.com.*

FIGURE 2.31 Mountain forests of the eastern Alps south of Vienna in Lower Austria. *Walter Sekot.*

joined the European Union (EU) in 1995 and, as with other European nations, a number of EU regulations also directly or indirectly affects the management of forests (e.g., afforestation and protection activities) (Voitleithner, 2002). About 80% of Austrian forests are privately owned, and the remainder are publicly owned and managed. Small-scale forestry (managing less than 200 ha or about 500 ac) and the management goals of small forest holdings dominate forestry practices (Kvarda, 2004). The area of forests has expanded since World War II, due to natural regeneration on abandoned agricultural land and afforestation efforts. Although most of the flat, arable land in Austria is used for agricultural purposes, since the late 2010s the forest area in Austria has not really changed (Table 2.4).

2.4.2 Belarus

Belarus is a landlocked country bordered by Latvia, Lithuania, Poland, Russia, and Ukraine (Fig. 2.30) that was once inhabited by Baltic and Slavic tribes. Prior to the 16th century, portions of the country were governed by Poland and Lithuania. After World War I, Poland and Russia governed parts of modern-day Belarus. In 1939, the country became a republic within the Soviet Union, and in 1941 Belarus was embroiled in World War II. The country and its natural resources were among those most ravaged from activities related to the war, and several decades were required to rebuild the infrastructure. After the collapse of the Soviet Union (1991), it became an independent country, and of the nearly 21 million ha (a little over 51 million ac) that now comprise Belarus, temperate deciduous and coniferous forests cover about 42%. About two-thirds of the forests of Belarus are composed of coniferous tree species (e.g., Norway spruce and Scots pine [*Pinus sylvestris*]) and

the area of forests has been increasing at a rate of about 0.2% per year since 2010 (Table 2.4).

The forests of Belarus are controlled by the government, and the Ministry of Forestry is the agency assigned responsibility for the related activities on most of these lands. Roundwood is produced mainly from final harvests (54%) and thinnings (35%) (Gerasimov and Karjalainen, 2010). For the most part, thinnings and selection harvests are allowed in forests located along rivers and lakes, and around cities. Extraction of timber from forested areas near the Chernobyl nuclear accident in nearby Ukraine (Fig. 2.32) that occurred in 1986 is problematic due to high levels of radiation (Krott et al., 2000). These areas represent about one-quarter of the country's forests. Of the roundwood harvested in Belarus, about 43% is used for fuelwood and about 25% is exported. Over 80% of the harvested wood comes from coniferous tree species. NTFPs of importance include berries, birch sap, mushrooms, and turpentine (Food and Agriculture Organization of the United Nations, 2016). A number of national plans have been developed, such as the National Strategy for Sustainable Socioeconomic Development of Republic of Belarus, the Forest Code of 2000, and the National Forest Program (Gerasimov and Karjalainen, 2010). These were designed to promote a sustainable forest industry and to recognize important environmental and ecological aspects of natural resource management.

FIGURE 2.32 Radioactivity warning sign on a hill at the eastern end of Red Forest, Belarus, which received the highest levels of radiation following the Chernobyl nuclear power plant accident. *Timm Suess, through Wikimedia Commons.*
Image Link: https://commons.wikimedia.org/wiki/File:Red_Forest_Hill.jpg
License Link: https://creativecommons.org/licenses/by-sa/2.0/deed.en

2.4.3 Finland

Finland is located in the northeastern corner of Europe between Sweden and Russia on a strip of land it shares with Norway along the northern border (Fig. 2.30). Although ethnically a different language group, Finland historically was either a member of the Swedish kingdom or the Russian Empire until 1918, when it won independence. Human settlement spread widely into Finland in the 16th century with the advance of slash-and-burn cultivation practices (Parviainen et al., 2010). Currently, Finland is a sparsely populated nation with over 30 million ha of land (over 75 million ac), of which 73% is forested and most is considered to be taiga or boreal forests. A recent estimate of the rate of change in forestland area indicates that it is neither increasing nor decreasing (Table 2.4). Northern Finland, also known locally as Lapland, is a region containing tundra and sparsely distributed boreal forests. Southern Finland, relatively warmer but cold by United States standards, is heavily forested (Fig. 2.33) with numerous inland water bodies that rival Minnesota's 10,000 lakes. Finnish forests contain many common European tree species, including birches, larches, pines, and spruces. Finnish wildlife includes numerous species of birds and mammals such as the Eurasian eagle owl (*Bubo bubo*), Tengmalm's owl (*Aegolius funereus*), martens, whooper swans, elk, reindeer, wolves, and wolverines.

Over 75% of the forestland in Finland is privately owned, mainly by families or family groups. Forest practices are generally based on even-aged management principles that utilize a number of intermediate thinnings (Parviainen et al., 2010). Given their location, Finland's forests have played an important role in supplying forest products to European markets. This has allowed the country to develop a forest sector that produces valued-added wood products such as pulp and paper, furniture, wood-based panels, and other products for export. Finland and Sweden are the two largest producers of coniferous roundwood in the EU. Finland has also become a leading developer of forest harvesting machines, industrial paper-making machines, and innovative sawmilling technologies that are used throughout the world. Recreational use of forests in Finland is free under the Everyman's Right (right of public access), which allows use of land owned by others for bicycling, hiking, horseback riding, skiing (provided these activities do not cause damage), temporary camping, and extraction of some NTFPs such as berries and mushrooms (Parviainen et al., 2010). The concept of sustainable management of forests has been recognized in legislation since the Forest Act of 1886, and private forest legislation of 1928 and 1966 has also influenced how forests of Finland are managed (Siiskonen, 2007). The Forest Act of 1996 now guides and regulates silvicultural activities on all Finnish lands. The current law promotes ecological, economic, and social sustainability of forests and natural resources (Hirakuri, 2003).

2.4.4 Germany

Germany is located in central Europe (Fig. 2.30) and contains 16 states covering about 35 million ha (about 86 million ac). About 33% of the land area is currently forested (Table 2.4) and the recent rate of change in forest area has been negligible. Five land regions are recognized: the Bavarian Alps, Black Forest, the central highlands, the north German plain, and the southern German hills. Most of these areas support temperate broad-leaved or coniferous forests, although the higher elevations support montane forests. The state of Bavaria has the largest forested area (2.6 million ha or 6.3 million ac), Baden-Württemberg has the second largest forested area (1.4 million ha or 3.4 million ac), and Saarland has the smallest amount of forested area (102,634 ha or 253,614 ac) (German Federal Ministry of Food and Agriculture, 2012). Historically, German forests have been dominated by beech and oak forest communities, and currently 60% of German forest area is coniferous and 40% is broad-leaved. The broad-leaved forests are composed of tree species such as ash (*Fraxinus* spp.), English oak (*Quercus robur*), European beech, horse-chestnut or conker tree (*Aesculus hippocastanum*), and silver birch (*Betula pendula*). The coniferous forests are composed of spruce, fir, Douglas-fir, larch (*Larix* spp.), and Scots pine.

Germany provides a good example of the history of European forest use and its development. For example, one event of historic importance to forests suggests that Germanic tribes repelled three Roman legions in

FIGURE 2.33 Forests typical of southeastern Finland. *Thomas O'Shea.*

desolate forested areas during the rule of Caesar Augustus, thus preventing the expansion of the Roman Empire into parts of modern-day Germany. Another example includes the fuelwood famines of the 12th and 13th centuries, which shaped the character of European and German woodlands over time. Germany was also involved in several wars over the last 200 years that greatly impacted its natural resources. The current German landscape is therefore a by-product of peaceful and wartime manipulations by humans over many centuries (German Federal Ministry of Food and Agriculture, 2011). Land tenure has recently been undergoing a transition in some parts of the country. For example, some state-owned lands in eastern Germany are reverting to the previous ownership status (private or public) that existed prior to expropriation in the mid-20th century.

German concern for agricultural and forest sustainability has significantly altered the current landscape to the point where nearly all of the forested area has been influenced by humans' activities at some point in history (German Federal Ministry of Food and Agriculture, 2011). Increasing demands for forestlands and wood have led to the planting of coniferous trees, but efforts are under way to convert the character of current forests to one that represents the natural vegetation found historically on German land. German forests and those who work with them have been at the forefront of forestry development for a very long time and currently forests are managed for a variety of commercial and noncommercial uses (Fig. 2.34). A number of German foresters have had important influences on the forest sciences. In fact, the environmental concept called *sustainable management* was developed in 1713 by H C. von Carlowitz, a German forester (Niekisch, 1992).

2.4.5 Russian Federation

Forests in Russia compose over one-fifth of the world's total forest area and, although a good portion of Russian forests are technically located in Asia, we discuss the forests of the Russian Federation in this European section. The Russian Federation (Fig. 2.30) covers about 1.638 billion ha of land (about 4.045 billion ac), and about 50% of the total land area is covered by forests (Table 2.4). Forest areas in Russia consist of the northern boreal forest (containing Norway spruce, Scots pine, and Siberian pine [*Pinus sibirica*], among others) and the southern temperate deciduous and mixed forests (containing alders, aspen, birches, and oaks), although various ecoregions are found within each of these. Active forest management is concentrated mainly in the north European region of the country (Fig. 2.35) and the far eastern region. The majority of old-growth forests that remain are located in the boreal region of northern Russia. The rate of change in forest area has been negligible in the last few years (Table 2.4). Wood removals in Russia account for only 6% of the world's total production (Food and Agriculture Organization of the United Nations, 2010), yet Russia is the largest exporter of roundwood in the world, with European and Asian markets as the main customers (Solberg et al., 2010). NTFPs of importance include berries, mushrooms, pine nuts, turpentine, and a large number of medicinal plants.

The vast forests in the western part of the Russian Federation were initially inhabited by ancient Slavs, who were hunter-gatherers and fishermen, and Slavic and Byzantine cultures influenced what would eventually become Russian culture. The importance of forests to the region has been emphasized in many laws dating back over 1500 years. Early Russian laws, beginning in

FIGURE 2.34 Forest trail near the Waldsteinburg, also known as the Red Castle, Fichtelgebirge, Oberfranken, Germany. *Donald L. Grebner.*

FIGURE 2.35 Russian forest situated on the northwestern border with Finland. *Thomas O'Shea.*

about the 5th century, noted the importance of forests for hunting and obtaining honey. Ownership of land and property rights was introduced into legislation around the 12th century and was debated through other legislation for another 700 years. Since 1918, most forests are considered to be owned by the state and nearly all are administered by the Forest Fund (Teplyakov et al., 1998). Some forests continue to be privately held through an exemption to the 1997 Forest Code. During the period of Civil War (1917–23) and during World War II, Russian forests sustained heavy damage, especially along waterways. The 1997 Forest Code promoted the sustainable use of forests, and a new Forest Code, implemented in 2007, facilitates the development of long-term leases to forested areas for production purposes (Teplyakov et al., 1998). Given the size of the country and the vast resources contained within, illegal logging is a concern in some areas (Torniainen et al., 2006). Lately, wildfire has become a major issue in western Russia as unusually hot and dry weather precipitated a rash of fires in 2010.

2.4.6 Spain

Spain has a land base of about 50 million ha (about 123 million ac), of which about 37% is currently forested (Table 2.4). The country is located on the Iberian Peninsula of western Europe between France to the northeast and Portugal to the west (Fig. 2.30). The original people who inhabited the Iberian Peninsula were the Basques, Celts, and Iberians, and through its long course of history, this area of the world has been ruled by various Christian, Muslim, and Roman empires and kingdoms. As already mentioned, countries in the Mediterranean region have experienced periods of serious deforestation dating back to ancient times. In Spain, forestland was not only cleared to facilitate agricultural purposes but also to provide fuel for metallurgical operations as far back as during Roman rule. During the 15th and 16th centuries, Spain was expanding its empire around the world, and this expansion led to the clearing of large forested areas to provide timbers for the development of the Spanish Armada (Oosthoek, 1998). The expansion of the empire eventually led to conflicts with other nations. In 1588, Spain tried to invade England, but the Armada was destroyed, which heralded the decline of Spanish supremacy (Oosthoek, 1998). Currently, the forest area in Spain has been increasing in size (about 0.2% per year; Table 2.4) with the aid of various governmental assistance programs.

About two-thirds of the forestland in Spain is privately owned, while the remainder is publicly owned. A small portion of the public lands are considered community or municipal forests. Most of the Iberian Peninsula supports temperate, Mediterranean vegetation and coniferous forests. A small portion of land along

the Pyrenees supports montane forests, and a small portion of land along the Atlantic Ocean supports temperate broad-leaved deciduous forests. Many forests in Spain are composed of dwarf beech (*Fagus sylvatica*), eucalyptus or blue gum, Mesogean or maritime pine (*Pinus pinaster*), radiata pine, mixed forests of English oak and sessile oak (*Quercus petraea*), Pyrenean oak (*Quercus pyrenaica*) (Fig. 2.36), and European silver fir. European silver fir is commonly found in the Pyrenees bordering France. Stone pine (*Pinus pinea*) is also found in Spain and is commonly used to provide edible pine nuts used in salads and other culinary dishes. There are few forests in the southern portion of the Iberian Peninsula, which is characterized as having scrub and other low-lying Mediterranean vegetation. Although the Spanish forest products sector is significant and per capita consumption of wood is lower here than in other European countries, Spain is a net importer of wood products. NTFPs of importance in Spain include cork, fruits, medicinal plants, mushrooms, and nuts. Ecotourism is also becoming increasingly important, although wildfire is a particular concern in the southern portion of the country. The 2003 Forest Law (*Ley 43/2003*) provides the current regulatory framework for the management and conservation of forests in Spain.

2.4.7 United Kingdom

The history of the United Kingdom (Fig. 2.30) is complex but perhaps no more so than other countries in Europe. Inhabitants of this island nation included the original Celts, as well as descendants of Anglo-Saxon, Norman, and Roman invaders. The four countries that now comprise the United Kingdom (England, Northern Ireland, Scotland, and Wales) are the remnants of a much larger British Empire that, at its height in the early

FIGURE 2.36 Pyrenean oak (*Quercus pyrenaica*) forest in central Spain. *Luis Fernández García, through Wikimedia Commons.*
Image Link: https://commons.wikimedia.org/wiki/File:Quercus-pyrenaica-20071110-c.jpg
License Link: https://creativecommons.org/licenses/by-sa/2.5/es/deed.en

20th century, included nearly one-quarter of the land area in the world. Within these four countries, 90% or more of the natural forests were cleared by the start of the 15th century, as the demand for agricultural and rangeland increased, along with an expansion of the human population. England, Scotland, and Wales each had 10% of their land composed of forests at about this time. Plantations of European larch, Norway spruce, and Scots pine were begun in the 17th century, and later Douglas-fir and Sitka spruce were introduced from North America. The United Kingdom relied heavily on imported wood for its domestic use, but the shipping blockade associated with World War I underscored the country's vulnerability with regard to forest resources and a large-scale afforestation program was begun around 1919 with the passage of the Forestry Act of 1919. However, nearly two-thirds of the woodlands in the United Kingdom were felled to meet the needs of World War II. As a result, a renewed afforestation program was begun in 1948 to encourage private landowners to restock their woodlands (Richards, 2003). The recent rate of change in forest area has been positive, about 0.5% per year (Table 2.4).

The four countries of the United Kingdom contain a little over 24 million ha of land (almost 60 million ac), and forests now cover approximately 13% of the land area (Table 2.4), three-quarters of which are plantations. Agriculture dominates the various land uses of the United Kingdom, but the use of land for the extraction of other resources (e.g., coal, chalk, iron ore, and others) is important as well. The climate of the United Kingdom is temperate, and the prevailing winds are from the southwest; thus, the eastern portion of the island of Great Britain (containing England, Scotland, and Wales) is drier than the western portion. As it pertains to forests, the entire country can be said to support temperate broad-leaved and coniferous forests. Currently, about 80% of the rural land in the United Kingdom is privately owned and the remainder is in various public ownership classes. From a forest management perspective, the rural countryside of the United Kingdom is urbanizing, and the public in general now expects rural land to produce multiple goods and services (Fig. 2.37) (Munton, 2009).

Community forest programs were begun in the United Kingdom in 1989 to help restore derelict portions of the urban fringe and promote the creation of jobs (Richards, 2003). Interestingly, the area of deciduous forests planted annually in the United Kingdom now exceeds the area of coniferous forests planted (Richards, 2003). However, most of the roundwood produced in the United Kingdom comes from coniferous plantations. The production of roundwood represents about 2% of that produced in Europe, excluding the Russian Federation, and the United Kingdom is essentially a net importer of wood products (Eurostat, 2019). In 2007, *A Strategy for England's Trees, Woods and Forests* was

FIGURE 2.37   Bluebell (*Hyacinthoides non-scripta*) flowers under a deciduous forest canopy in Buckinghamshire, England, United Kingdom. *Keith Hulbert and Paul Zarucki, through Wikimedia Commons. Image Link: https://commons.wikimedia.org/wiki/File:Bluebells-2005-05-02-2p.jpg License Link: https://creativecommons.org/licenses/by-sa/3.0/deed.en*

developed, which outlines priorities for forest resources for the next 50 years in England. Similar policies have been developed in Northern Ireland and Scotland to promote sustainable use of the resource.

2.5 Asia

Asia (Fig. 2.38) is the largest continent on Earth and its history has gripped the human imagination for thousands of years, including periods influenced by (among others) the Ming Dynasty in China, the Mongolian hordes led by Genghis Khan, and the Japanese Samurai. The continent is the site of the Sphinx, Angor Wat, adventures of Marco Polo along the Silk Road, the Trans-Siberian rail line, Gandhi's silent protest (India), and oil production (the Middle East). Within this continent there are 51 countries that cover about 43.8 million square kilometers (km^2; 11.9 million square miles [mi^2]) of land and support over 4.1 billion people. Given the diverse climatic, developmental, and topographic history of the continental region, the extent and character of forested areas vary by country. For example, eastern Siberia, which is part of the Russian Federation, is still largely forested due to poor infrastructure and a low population density. Older civilizations such as China and India have each experienced similar patterns of forest use due to repeated expansions and contractions of human settlements over thousands of years. Burgeoning populations

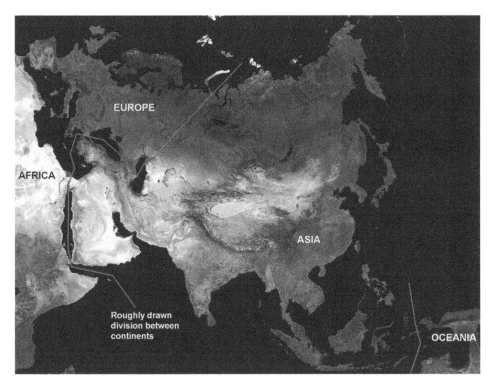

FIGURE 2.38 Satellite image of Asia. *National Aeronautics and Space Administration.*

in these countries over the past half century have accelerated the need for afforestation, restoration, regeneration programs, and environmental programs (such as those aimed to prevent or mitigate soil erosion) that promote increased self-sufficiency.

While the loss of forested area is a concern in some Asian countries, on the whole, the forested area has increased in China since 2010 due to massive afforestation efforts (Food and Agriculture Organization of the United Nations, 2015). The forest vegetation and associated wildlife species stretch across numerous ecoregions, ranging from tundra and taiga forests in Siberia to temperate forests in China, Japan, and Korea, to grasslands in Kazakhstan, Mongolia, and Uzbekistan, and tropical forests in India, Indonesia, the Philippines, Thailand, and Vietnam. Precipitation patterns range greatly from the more arid regions in Middle Eastern countries such as Israel, Pakistan, Saudi Arabia, and Turkey to the moist climatic environments found in eastern China, Indonesia, Japan, and Thailand.

In the Asian boreal forests, it is common to find tree species such as the Siberian fir (*Abies sibirica*). Siberian larch (*Larix sibirica*), and Siberian spruce (*Picea obovata*). Other plants include Siberian ginseng (*Eleutherococcus senticosus*) and numerous plant species associated with peat bogs. Wildlife species include brown bears, mink, moose, reindeer, the Siberian tree frog (*Rana amurensis*), and the Siberian tiger (*Panthera tigris tigris*). In more temperate forests, it is common to find tree

species such as Japanese cypress (*Chamaecyparis obtusa*), Korean pine (*Pinus koraiensis*), and Mongolian oak (*Quercus mongolica*). Wildlife species commonly found in temperate regions include the golden snub-nosed monkey (*Rhinopithecus roxellana*), the Manchurian hare (*Lepus mandschuricus*), and the giant panda (*Ailuropoda melanoleuca*), which is currently endangered. In tropical forests, numerous tree species exist such as kemenyan (*Styrax benzoin*), Philippine mahogany (*Shorea almon*), and sal (*Shorea robusta*). Other plants include the kitul palm tree (*Caryota urens*) and nutmeg (*Myristica fragrans*). Wildlife species may include the hog badger (*Arctonyx collaris*), Indian cobra (*Naja naja*), or the Sri Lankan elephant (*Elephas maximus maximus*). While a good portion of Asia includes the eastern part of the Russian Federation, which we discussed in the previous section, our discussion of Asian forests will center on China, Indonesia, Japan, and the Republic of Korea.

2.5.1 China

China (Fig. 2.39) has one of the world's oldest human civilizations, with nearly 4000 years of documented history. Shang, Zhou, Han, Wu, Tang, and Song dynasties ruled the various empires of China from about the 17th century BC to the 13th century AD. Chinese dynasties were the first to develop paper and books, and the first governmental bodies to use and distribute paper money. Other dynasties followed, but China effectively missed

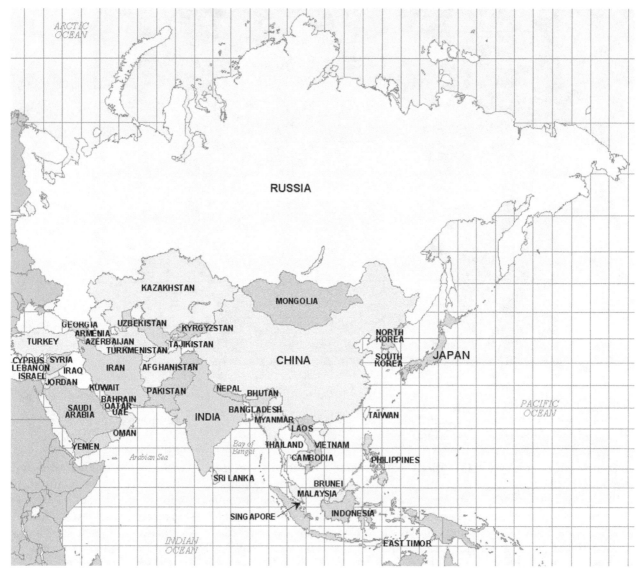

FIGURE 2.39   Countries in Asia. *Modified image from PresentationMaps.com.*

the Industrial Revolution, and intermittent wars and disorder led to economic decline. After the Chinese Civil War (1947–49), a socialist state developed and since the 1980s a number of economic reforms have helped the country to now become the second largest economy in the world. A land reform law was enacted right after the founding of the People's Republic of China in 1949, which allowed peasant ownership of land, but in 1956 this changed to collective ownership of land (Keliang et al., 2010). Today, the majority of land in China is owned by these collectives (townships and villages) and the remainder is controlled by the central or state governments. State forest management is more prevalent in the northeastern provinces. Southern China and north–central China are the major regions for collective forest management enterprises (Rozelle et al., 2003). Some forested lands that are collectively owned

have been distributed to people through a variety of contractual arrangements (Chen and Kurokawa, 2005), and a 1984 Forest Law, revised in 1998, legalized private ownership of trees (but not land) (Keliang et al., 2010). However, the 2007 Property Law affirms ownership of land to the collectives but does not address who actually has control over the land (whether the collectives or individuals given contract to part of the collectively owned land). The absence of a rural land registry affects management of the land and leaves unsolved the question of prior ownership. As a result, in some areas there is no system that can, by title, protect a landowner and their forest investment (Ho, 2006).

China is a vast and diverse country that encompasses over 942 million ha of land (almost 2.3 billion ac) (Table 2.5). The current forest area in China is estimated to be about 207 million ha (about 511 million ac), which

TABLE 2.5 Demographic and physiographic data for Asian countries.[a]

Country	Human population 2018[1] (million)	Total land area[2] (1000 ha)	Total land area[2] (1000 ac)	Forest cover[2] (%)	Forest area annual change 2010 −15[2] (%)	Per capita (PPP) 2017[1&b] (1000 US$)
Bangladesh	161	13,017	32,152	11	−0.2	1.5
Bhutan	<1	3812	9416	72	0.4	3.2
Cambodia	16	17,652	43,600	54	−1.3	1.4
China	1459	942,530	2,328,049	22	0.8	8.6
Democratic People's Republic of Korea	26	12,041	29,741	42	−2.3	0.7
Georgia	4	6949	17,164	41	—	3.9
India	1353	297,319	734,378	24	0.3	1.9
Indonesia	268	171,857	424,487	53	−0.7	3.8
Japan	127	36,450	90,032	69	—	38.2
Loa People's Democratic Republic	7	23,080	57,008	81	1.0	2.5
Malaysia	32	32,855	81,152	68	0.1	10.0
Myanmar	54	65,755	162,415	44	−1.8	1.3
Nepal	28	14,335	35,407	25	—	0.8
Philippines	107	29,817	73,648	27	3.3	3.0
Republic of Korea	51	9710	23,984	64	−0.1	30.0
Sri Lanka	21	6271	15,489	33	−0.3	4.2
Thailand	69	51,089	126,190	32	0.2	6.6
Turkey	82	76,963	190,099	15	0.9	10.5
Viet Nam	96	31,007	76,587	48	0.9	2.3
Other countries (30)	600	1,275,131	3,149,574	3	−0.2	7.0
Asia (total)	4561	3,117,640	7,700,571	19	0.1	6.5

[a]That are more than 10% forested and have over 3,000,000 ac (about 1,214,100 ha) of forest.
[b]Gross domestic product (GDP) expressed as purchasing power parity (PPP).
[1]Food and Agriculture Organization of the United Nations 2020 [2]FAO. Global Forest Resources Assessment 2015. Desk Reference. Rome, 2015.

represents about 22% of the total land area. The largest forest areas are located in the northeast (formerly called Manchuria, now called Dongbei), eight southern provinces (Fujian, Guangdong, Guangxi, Hubei, Hunan, Jiangxi, Sichuan, and Yunnan), and mid-China (Shaanxi) (Ho, 2006). Given its size, the physiographic zones within China are also diverse, ranging from arid deserts to dry humid forests. For example, the Xinjiang region of northwest China is mainly arid, but certain areas can support forest vegetation (Fig. 2.40). Northern forests are mostly temperate coniferous or mixed coniferous-deciduous, while some southern forests are lowland subtropical rainforests or monsoon forests (Ho, 2006). A broad expanse of temperate coniferous, semideciduous, and mixed forests currently resides in southern China. About 61.7 million ha (152.5 million ac) of Chinese forests are planted forests. Desertification is a serious environmental and social issue in the country and, although logging of native forests has been banned, the rate of forest loss has been, at times, high (Chen and Kurokawa, 2005). During China's Cultural Revolution (1966—76), large areas of forests were cleared for agricultural production and, by the onset of economic reforms (1978), a large portion of China's forests were depleted or in poor condition (Hyde et al., 2003). More recently, the Chinese government has placed a great deal of emphasis on wood supply and high-profile afforestation projects, such as the Upper/Middle Yangtze River Valley afforestation campaign and the Great Green Wall in Inner Mongolia (Rozelle et al., 2003). These large-scale afforestation programs have allowed the country to show a positive rate of change in forest area, estimated to be about a 0.8% increase per year since 2010 (Table 2.5). Recent reforms have stimulated forest management by providing farmers property rights to trees planted on contracted forestland (Petry et al., 2010).

FIGURE 2.40 Herdsman in the forest/range interface of the Xinjiang region of China. *Pikrepo.com.*

Prior to 1950, most of China's forests were naturally regenerated. Since then, reforestation of cleared areas has been accomplished using Chinese fir (*Cunninghamia lanceolata*), larch (*Larix* spp.), and poplar (*Populus* spp.) (Chen and Kurokawa, 2005). Northeast China has one of the most important forest areas in the country due to the size of the natural forests contained there. Forests are characterized as mixed broad-leaved Korean pine and larch forests. Korean aspen (*Populus davidiana*), Korean pine (which is similar to white pine), Mongolian oak, and Olgan larch (*Larix olgensis*) are among the many tree species found in this region (Chen, 2006). In the southwestern region of the country, the main tree species in these temperate forests are Chinese Douglas-fir (*Pseudotsuga sinensis*), camphor laurel (*Cinnamomum camphora*), dragon spruce (*Picea asperata*), nanmu (*Konishi nathaphoebe, Nantou litsea, Phoebe nanmu* and others), teak, Yunnan pine (*Pinus yunnanensis*), and zitan (*Pterocarpus santalinus*). As a country, China is among the top five producers of wood products, particularly fiberboard, particleboard, and plywood. About two-thirds of the roundwood produced is used for fuelwood purposes, and internal consumption of wood exceeds production; thus, the country is a net importer of wood products (Ho, 2006).

2.5.2 Indonesia

Indonesia consists of over 17,000 islands and forms a large archipelago in Asia (Fig. 2.39). The largest of the islands are Sulawesi, Sumatra, the western half of New Guinea, and the Indonesian portion of Borneo (Kalimantan). Indonesia has been a major trading region since around the 7th century, and Indonesian society has been influenced over time by Buddhist, Hindu, and Islamic cultures. The Dutch colonized portions of the archipelago for nearly 350 years, then Japan occupied the country during World War II, after which Indonesia became independent. Of the 172 million ha of land (nearly 424 million ac), about 53% is currently forested and the rate of change has been about −0.7% per year since 2010 (Table 2.5). Deforestation has been an important issue for Indonesia; however, as with Brazil, the rate of forest loss due to deforestation since the late 2000s has been significantly reduced compared to the 1990s (Food and Agriculture Organization of the United Nations, 2010). Most of the forestland in Indonesia is characterized as a tropical moist forest, consisting of broad-leaved evergreen or semideciduous forest trees. Subtypes of forests include monsoon forests, and montane forests, swamp forests, and tidal forests (Fig. 2.41) (International Tropical Timber Organization, 2006). The largest areas of tropical peat swamp forests are found in coastal Borneo and Sumatra, which contain valuable tree species such

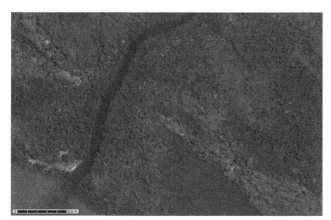

FIGURE 2.41 Tidal forests typical of Indonesia. *U.S. Department of Agriculture, Natural Resources Conservation Service (2020).*

as light red meranti (*Shorea albida*) and ramin (*Gonystylus bancanus*) (Brown, 1998). In forest plantations, acacias (*Acacia* spp.), eucalypts, Sumatran pine (*Pinus merkusii*), and teak comprise the majority of species grown. Bamboo, copal, rattan, and many medicinal plants are a few of the NTFPs derived from these tropical forests (International Tropical Timber Organization, 2006).

Indonesia experienced a significant expansion in its economy from the mid-1960s through to the end of the 20th century. The petroleum industry, along with advancements in manufacturing, helped improve the standard of living in the country. Nearly two-thirds of the roundwood produced is for fuelwood purposes, yet Indonesia is a net exporter of manufactured wood products (mainly to China, Japan, and Korea). Individual tree selection cutting is the main silvicultural technique due to the wide range of species found in the tropical forests. A 1999 Forest Law defines two types of ownership, one in which land is registered by title and the other in which land belongs to the state, although White and Martin (2002) suggest that nearly all of the forestland is controlled by the government. Community forest rights (*adata*) are also recognized, and the administration of forests for production and protection purposes has been decentralized to the provinces and districts within the country. Forest concessions are granted to companies or cooperatives and, in some cases, ownership is shared with villagers who have rights to the forest resources through cooperatives (Pirard and Irland, 2007). Given the widespread geography of the country and the decentralized administration of the forest resources, illegal logging is seen as a major problem. Another controversy associated with the rapid advancement of the Indonesian economy is the issue of land conversion. As production and exportation of wood products have increased, the area of palm oil and rubber plantations has also expanded (Gellert, 2005).

2.5.3 Japan

Japan is a nation comprised of over 3000 islands that form an archipelago in the Pacific (Fig. 2.39). The character of the country has evolved considerably from the original Jōmon culture (hunter-gatherer) that ended around 300 BC to shogunates, clans, and dynasties of recent centuries. Currently, Japan is a constitutional monarchy that has evolved over the last 100 years into the third largest economy in the world. Reforms in forestland ownership began at the turn of the 20th century and now about 31% of the land in Japan is considered national forest, 11% is owned by local governments, and 58% is privately owned. Japan covers over 36 million ha of land (about 90 million ac), of which about 69% is now forested (Table 2.5). Vast areas of forest were devastated during the conflicts of the mid-20th century; however, a large-scale reforestation effort has been very successful and the rate of change in forested area has been negligible since the late 2000s (Table 2.5). Japan has three broad forest regions. One is a subtropical forest zone that includes many species of oak and beech (also known as stone oak; *Lithocarpus* spp.) (Fig. 2.42) and other broad-leaved and coniferous evergreen tree species. The second is a temperate zone dominated by beech, fir, and pine tree species, along with several important Japanese conifers that include hinoki or Japanese cypress (*Chamaecyparis obtuse*), and sugi or Japanese cedar (*Cryptomeria japonica*). The third broad forest region is a cold zone that includes montane forest species such as fir, hemlock, and spruce.

Japan actually began developing forest plantations in the late 17th century. Currently, about 40% of the forest area in Japan is composed of plantations, and sugi and hinoki are two of the main tree species currently being grown for commercial purposes (Fujikake, 2007). Although significant efforts have been made to reforest

FIGURE 2.42 Deciduous forest in Japan. *Enoki Yoshio.*

large areas of the country, imports of wood into Japan began to rise significantly in the 1960s due to high silvicultural and logging costs, low import prices, and the need for a stable supply of wood with uniform characteristics (Iwamoto, 2002). As a result, about 80% of the wood product demand in Japan is met with imports, mainly from Canada and Russia (Tóth et al., 2006). Japan enacted a Basic Forestry Law in 1964 and amended it in 2016 (now the Forest and Forestry Basic Act), aimed to promote sustainable forestry and the development of multifunctional forests.

2.5.4 Republic of Korea (South Korea)

The Korean Peninsula (Fig. 2.39) has a long and rich human history dating back nearly 4000 years. The region was ruled by several dynasties until 1910, when Japan annexed the peninsula. After World War II, the Korean Peninsula was divided into two countries, North Korea and South Korea. Like many other countries around the world, South Korea has experienced periods of severe deforestation. However, unlike many countries that have experienced deforestation, South Korea has achieved remarkable success in restoring its forests. Deforestation was most rampant in the 20th century during the Japanese occupation (1910–45) and the Korean War (1950–53). Following the war, the impoverished nation relied heavily on its forests for food, fuelwood, and timber. As a result, vast expanses of forestland were denuded and, with heavy monsoon rains, soil erosion became a major problem. One measure of forest resource conditions, the average growing stock volume of forests, reached its nadir of only 6 m^3 per ha (about 86 ft^3 per ac) in 1952 (Chun, 2010), implying that vast areas of forestland were devoid of trees. The driving force behind forest degradation was a tremendous demand for fuelwood to meet more than 60% of the country's energy needs (Lee, 2010). In 1960, an estimated 42% of forestland area was still unstocked.

Land reform in the mid-20th century affected the land tenure system and currently most of the forestland in Korea is privately owned. The South Korean economy has grown significantly since the Korean War and is now an important economic power in Asia. With growing wealth, people could afford to switch to coal, slash-and-burn practices were abandoned, and rural residents moved to rapidly growing cities, all of which reduced pressure on the forests and natural resources. This growth has been overshadowed by an impressive reforestation effort and, by 2009, the average growing stock volume of forests had increased considerably to 103 m^3 per ha (about 1472 ft^3 per ac) as a result of these efforts. To illustrate the point, while South Korea is about 10 million ha in size (about 24 million ac)

(Table 2.5), between 1946 and 2000 nearly 100,000 ha of land (about 247,000 ac) was planted annually (1% of the land annually). During this period, nearly 5.3 million ha (13 million ac) was reforested.

Today, diverse forests (Fig. 2.43) cover about 64% of the land area of Korea and the recent rate of change is about −0.1% per year (Table 2.5). While most of the peninsula can be described as an area containing temperate deciduous forests, of the forested area about 30% is now comprised of plantations, some of which are composed of coniferous tree species. Originally, alder (*Alnus* spp.), black locust (*Robinia pseudoacacia*), pitch pine (*Pinus rigida*), and sawtooth oak (*Quercus acutissima*) plantations were created. Since then, chestnut, hybrid aspen, hybrid poplar, larch, and Korean pine plantations have also been established (Lee and Lee, 2005). While rapid economic growth since the late 20th century may be straining South Korea's native environmental resources, the Demilitarized Zone (DMZ) seems to be an unlikely exception. Running horizontally across the Korean Peninsula, the DMZ has separated North Korea and South Korea since the cease-fire armistice in 1953. The DMZ is roughly 4 km (2.5 mi) wide and 250 km (155 mi) long, crossing coastal areas, high mountains, plains, rivers, and wetlands. It serves as a buffer between the two countries to prevent further hostilities. This area has been closed to civilians since its inception. In addition, South Korea maintains the Civilian Control Zone (CCZ) that extends another 10–20 km (6–12 mi) and in which human access is extremely restricted. The combined area of DMZ and CCZ within South Korea amounts to nearly 98,000 km^2 (37,833 mi^2), 71% of which is forested (Korea

FIGURE 2.43 Coniferous forest in the Republic of Korea. *Jacek P. Siry.*

Forest Research Institute, 2010). While this area was heavily damaged during the war, since the armistice it has been virtually a no-man's land. As a result, it has evolved into a unique nature reserve teeming with a wide variety of plants, mammals, fishes, and birds, such as the extremely rare white-naped crane (*Grus vipio*) and red-crowned crane (*Grus japonensis*).

2.6 Africa

Africa (Fig. 2.44) is commonly referred to as the *cradle of mankind*. Although human ancestors slowly evolved from the area we now know as the Horn of Africa, Africa's development by today's standards has been relatively slow. Diamond (2005) asks a basic question: why did some nations advance faster than others? The answer largely depends on the initial set of natural resources available to each country. Therefore, the physical location of Africa influenced the rate of development on the continent. In contrast, countries located in Europe and Asia had a competitive advantage in that several species of animals and plants (e.g., wheat) were domesticated in the Fertile Crescent, the arch of land encompassing Iraq, Israel, Jordan, Lebanon, and Syria. In addition, since a large portion of the European and Asian continents are located along the same set of global latitudes and natural resources were abundant, the transfer of technologies across the landscape was easier.

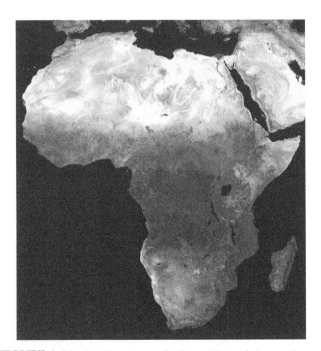

FIGURE 2.44 Satellite image of Africa. *National Aeronautics and Space Administration.*
Image Link: https://commons.wikimedia.org/wiki/File:Africa_(satellite_image).jpg
License Link: Public Domain

Unfortunately, locations further south, such as Africa, had climatic, managerial, topographic, and vegetative barriers that made the exchange of technologies more difficult. Early attempts to domesticate animals were largely unsuccessful in Africa, given the abundance of cheetahs, hyenas, lions, and other carnivores. This decreased the potential set of transferable technologies and slowed the initiation of development of the modern state of Africa until the period of colonization that began in the late 19th and early 20th centuries. During the colonization period, European nations created territories in Africa from land that they thought they could control, with little concern for traditional tribal or ethnic boundaries. Unfortunately, many regional and local wars have since occurred, which have acted to delay the development of some African societies. Further, high birth rates and political unrest have led to governmental policies that have not always favored sustainable management of forest and natural resources. These circumstances have resulted in extensive deforestation in some parts of Africa, as well as the exploitation and degradation of natural resources such as gorillas in the mountains of Rwanda. In addition to political and societal problems, some of the highest percentages of forest areas burned by wildfire since the late 2000s were associated with African countries (Food and Agriculture Organization of the United Nations, 2010), further complicating the natural resource management situation on this continent. In this section, we describe five African countries with important forest resources: the Democratic Republic of the Congo, Ghana, Nigeria, South Africa, and Tanzania.

2.6.1 Democratic Republic of the Congo

The Democratic Republic of the Congo is located in the center of Africa (Fig. 2.45) and is considered to be the most heavily forested country on the continent. It has a land area of about 227 million ha (about 560 million ac) and forest cover of 67% (Table 2.6). Former names of this country are Congo Free State, the Belgian Congo, and Zaire. Early villages consisted of people of the Imbonga, Ngovo, and Urewe cultures. Indigenous pygmy populations and Bantu-speaking cultures also were found in the area this country now occupies. As many as 250 ethnic groups have been distinguished and named in the country; some of the larger are the Anamongo, the Kongo, and the Luba. The Democratic Republic of the Congo was colonized by Belgium in 1908 and gained independence in 1960 (U.S. Central Intelligence Agency, 2019). It is believed that some of the forested areas in this country have continuously reproduced and survived since before the ice age nearly 12,000 years ago. Butler (2006) indicates that there are

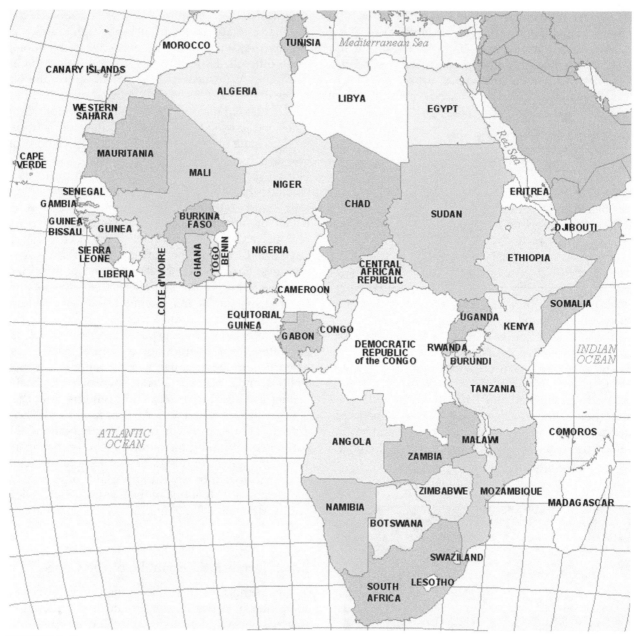

FIGURE 2.45   Countries in Africa. *Modified image from PresentationMaps.com.*

more than 11,000 plant species, 1150 bird species, 450 mammalian species, 300 reptile species, and 200 amphibian species that inhabit areas of the Democratic Republic of the Congo. In general, the forests are mostly composed of savannas and tropical rainforests. A closed-canopy, broad-leaved, tropical rainforest encompasses the Congo River basin and moist evergreen and semideciduous forests can be found in the west and center of the country (Fig. 2.46). Montane forests consisting of deciduous broad-leaved tree species can be found in the terraces and plateaus of the mountainous part of the country. Compared to other developing countries, deforestation was slow to arrive to this region because

of poor infrastructure, but some forested areas have now been degraded due to slash-and-burn agriculture.

The state is the sole owner of land in the Democratic Republic of the Congo, which was reinforced by the Forest Code of 2002 (Law 11/2002) (International Tropical Timber Organization, 2006). Although foreign logging companies have increasingly purchased concessions from the state, a serious threat to forests in this country are the periodic wars that have occurred in its recent history. These wars have led to the displacement of millions of people and resulted in increased degradation of plant and wildlife populations. Since the end of the last war in 2002, more effort has been expended to protect the

TABLE 2.6 Demographic and physiographic data for African countries. [a]

Country	Human population 2018[1] (million)	Total land area[2] (1000 ha)	Total land area[2] (1000 ac)	Forest cover[2] (%)	Forest area annual change 2010–15[2] (%)	Per capita GDP (PPP) 2017[1&b] (1000 US$)
Angola	31	124,670	307,935	46	−0.2	4.2
Benin	11	11,062	27,323	39	−1.1	0.8
Botswana	2	56,673	139,982	19	−0.9	7.6
Burkina Faso	20	27,360	67,579	20	−1.1	0.6
Cameroon	25	47,271	116,759	40	−1.1	1.5
Central African Republic	5	62,298	153,876	36	−0.1	0.4
Congo	5	34,150	84,351	65	−0.1	2.1
Côte d'Ivoire	25	31,800	78,546	33	–	1.6
Democratic Republic of the Congo	84	226,705	559,961	67	−0.2	0.5
Equatorial Guinea	1	2805	6928	56	−0.7	9.9
Eritrea	3	10,100	24,947	15	−0.3	1.1
Ethiopia	109	109,631	270,789	11	0.3	0.7
Gabon	2	25,767	63,644	89	0.9	7.2
Ghana	30	22,754	56,202	41	0.3	2.0
Guinea	12	24,572	60,693	26	−0.6	0.8
Guinea-Bissau	2	2812	6946	70	−0.5	0.7
Liberia	5	9632	23,791	43	−0.7	0.6
Madagascar	26	58,154	143,640	21	−0.1	0.5
Malawi	18	9428	23,287	33	−0.6	0.3
Morocco	36	44,630	110,236	13	−0.1	3.1
Mozambique	29	78,638	194,236	48	−0.5	0.4
Nigeria	196	91,077	224,960	8	−5.0	2.0
Senegal	16	19,253	47,555	43	−0.5	1.3
Sierra Leone	8	7162	17,690	43	2.2	0.5
Somalia	15	62,734	154,953	10	−1.2	0.1
Sudan	42	186,665	461,063	10	−0.9	3.0
Uganda	43	19,981	49,353	10	−5.5	0.6

Continued

TABLE 2.6 Demographic and physiographic data for African countries.[a]—cont'd

Country	Human population 2018[1] (million)	Total land area[2] (1000 ha)	Total land area[2] (1000 ac)	Forest cover[2] (%)	Forest area annual change 2010–15[2] (%)	Per capita GDP (PPP) 2017[1&b] (1000 US$)
United Republic of Tanzania	56	88,580	218,793	52	−0.8	0.9
Zambia	17	74,339	183,617	65	−0.3	1.5
Zimbabwe	14	38,685	95,552	36	−2.1	1.1
Other countries (28)	369	1,225,137	3,026,088	3	−0.4	5.0
Africa (total)	1276	2,986,544	7,376,764	21	−0.5	1.8

[a]That are more than 8% forested and have over 3,000,000 ac (about 1,214,100 ha) of forest.
[b]Gross domestic product (GDP) expressed as purchasing power parity (PPP).
[1]Food and Agriculture Organization of the United Nations 2020.
[2]FAO. Global Forest Resources Assessment 2015. Desk Reference. Rome, 2015.

FIGURE 2.46 Forest located along a river in the Democratic Republic of the Congo. *Kathy Freeman.*

country's forests. Unfortunately, increasing population growth rates are expected to generate greater pressure on forested areas over time. Since the late 2000s, the rate of forest loss has been about 0.2% per year (Table 2.6).

Some of the wide variety of tree species located in the Democratic Republic of the Congo include afrormosia (*Pericopsis elata*), bilinga (*Nauclea diderrichii*), bossé (*Guarea cedrata*), bubinga (*Guibourtia pellegriniana*), iroko (*Chlorophora excelsa*), kosipo (*Entandrophragma candollei*), longhi (*Gambeya africana*), sapelli (*Entandrophragma cylindricum*), sipo (*Entandrophragma utile*), tiama (*Entandrophragma angolense*), tola (*Gossweilerodendron balsamiferum*), and wengé (*Millettia laurentii*) (International Tropical Timber Organization, 2006). For many people of the country, forests are the main source of food, fuelwood, and medicine. Wood products from tola and wengé trees are the most valuable export products. Forest plantations are also being established and species such as cypress, dibétou or African walnut (*Lovoa trichilioides*), eucalypts, filao tree (*Casuarina equisetifolia*), Gold Coast bombax (*Bombax flammeum*), kapok (*Ceiba pentandra*), limba (*Terminalia superb*), sipo, and southern silky oak (*Grevillea robusta*) are commonly used to establish these plantations. In some cases, plantations were established in order to stabilize the soil and to prevent further erosion (International Tropical Timber Organization, 2006).

2.6.2 Ghana

Ghana is a republic that lies on the western coast of Africa (Fig. 2.45). Prior to initial contact by Portuguese and Dutch explorers in the 15th and 16th centuries, Ghana was inhabited by people of the Akan kingdoms. Formerly known as the Gold Coast, Ghana obtained independence in 1957. Ghana lies near the equator and is nearly 23 million ha in size (a little over 56 million ac). Currently, the landscape of Ghana is about 41% forested, although forest cover is increasing at a rate of over 0.3% per year (Table 2.6). Ghana contains tropical rainforests

(Fig. 2.47); moist and dry deciduous and evergreen forests, most of which are located in the southern part of the country along the Atlantic coast; and savannas in the central and northern portion of the country. Common tree species include danta (*Nesogordonia papaverifera*), ebony (*Diospyros* spp.), guarea (*Guarea cedrata*), Lagos mahogany (*Khaya ivorensis*), makore (*Tieghemella heckelii*), mangroves, mansonia (*Mansonia altissima*), palms, and wawa (*Triplochiton scleroxylon*) (International Tropical Timber Organization, 2006).

Illegal occupation, land use changes, and wildfire are all concerns in cutover forest areas. Human population growth and poverty, along with infrastructure (e.g., road construction) and economic development programs, contribute to the pressures placed on the country's forest and natural resources (International Tropical Timber Organization, 2006). Forest policy is codified in a number of acts, decrees, laws, regulations, and statements, and responsibility for developing these is centralized in the governmental structure (Wiggins et al., 2004). All of the land in Ghana is held by the government in trust for the people, although traditional lands that were previously under control by tribal chiefs continue to be used by those communities as long as the resource is not degraded (Owubah et al., 2001). While timber harvesting contracts rather than concessions are now the norm in the country, illegal harvesting is still a concern. Although 93% of the wood harvested in Ghana is for fuelwood purposes, fromager (*Ceiba pentandra*), teak, and wawa are common tree species harvested for industrial purposes. NTFPs that are integrally associated with Ghana's forests include fruits, honey, medicinal plants, tubers, wild meat, and seeds that facilitate the production of margarine, oils, and sweeteners (International Tropical Timber Organization, 2006).

FIGURE 2.47 Tropical forests along Lake Volta, Ghana. *Jurgen, through Wikimedia Commons.*
Image Link: https://commons.wikimedia.org/wiki/File:Lake_Volta_Ghana.jpg
License Link: https://creativecommons.org/licenses/by/2.0/deed.en

2.6.3 Nigeria

Nigeria is located in western Africa near the equator (Fig. 2.45), is a federal republic of 36 states and one federal capital territory (Abuja), and is the most populous country in Africa. Nigeria contains a little over 91 million ha of land (225 million ac), and about 8% of this area is forested (Table 2.6). Nigeria has the highest deforestation rate of African countries, as nearly 5% of forest has been lost per year since 2010. The area contained within Nigeria is thought to have been the homeland of the Bantu people who eventually migrated across the continent. Indigenous people included the Fulani, Igbo, Kano, Nok, and Yoruba, each of which have shaped the culture of the country. Colonization by the British Empire occurred early in the 20th century, although British, Portuguese, and Spanish explorers had visited the area earlier and developed trade with local tribes along the coast. In 1960, Nigeria gained independence from the United Kingdom. A number of internal conflicts have since troubled the country, and the rich oil reserves of the country have both positively and negatively influenced the course of Nigeria's natural resource development (Maconachie et al., 2009).

For the most part, the forests of Nigeria are located in the southern half of the country. About 70% of the forested area is considered savanna and the remainder along the coast is considered tropical rainforest, consisting of coastal forests (Fig. 2.48), lowland wet forests, mangrove forests, and swamp forests. These are mainly broad-leaved deciduous forests or broad-leaved evergreen forests. The savannas mainly contain riparian forests that consist of elolom or subaha (*Mitragyna ciliata*) and yeye (*Uapaca* spp.) tree species. African walnut (*Lovoa trichilioides*), agba (*Gossweilerodendron balsamiferum*), and Lagos mahogany are characteristic tree species of the rainforest areas, while iroko, obeche (*Triplochiton scleroxylon*), and otutu (*Nesogordonia papaverifera*) are characteristic tree species of the other coastal forests (International Tropical Timber Organization, 2006).

The forests of Nigeria are either locally owned as communal forests or owned by the state as forest reserves (or both in some cases), some of which were delineated through the Land Use Decree of 1978. Concessions for forest harvesting activities are awarded by the state, yet illegal logging activities are a concern. A new National Forestry Policy was adopted in 2008 to address problems related to the conservation of forest resources. Federal, state, and local governments administer various aspects of the forestry sector, and a number of NGOs have also assisted with the management of forest resources in the country. Over 300 tree species found in Nigeria could be utilized, yet only a handful are actually harvested. These include agba, iroko, ofun (*Mansonia altissima*), sapele (*Entandrophragma cylindricum*), teak, and English beech, the latter mainly from planted forests. While some of these tree species are used to develop lumber and veneer products, over 85% of the roundwood harvested from Nigeria's forests is used for domestic fuelwood purposes. Some of the important NTFPs derived from Nigeria's forests include bark, gum arabic (used in the food products industry as a stabilizer), leaves, mushrooms, nuts, rattan, resins, and other medicinal plants (International Tropical Timber Organization, 2006).

2.6.4 South Africa

Although it has less than 1.2 million ha (over 3 million ac) of forest and is less than 10% forested (thus not listed in Table 2.6), South Africa (Fig. 2.45) is important for forestry purposes because the development of forestry science here is perhaps the most advanced in Africa, and commercial forestry operations are important to the country's economy. The topography of South Africa is very diverse, yet the country generally has a temperate climate. The interior is mainly flat and arid and supports savannas and associated trees and grasses. In this area, a number of agricultural and game-based enterprises can be found (Shackleton et al., 2007). Forests along the eastern coast with the Indian Ocean are wet and lush, and forests in the interior east are composed of mixed tree species and occupy relatively dry sites. Mangrove and swamp forests also occur in small patches along

FIGURE 2.48 Kwa Falls along the Kwa River in Cross River State, Nigeria. *Shiraz Chakera, through Wikimedia Commons.*
Image Link: https://commons.wikimedia.org/wiki/File:Kwafalls.jpg
License Link: https://creativecommons.org/licenses/by-sa/2.0/deed.en

the low-lying areas of the eastern coast. There are virtually no forested areas along the western Atlantic Ocean coast. However, some Mediterranean-type forests can be found along the coast near Cape Town (Fig. 2.49).

Indigenous Bantu, Khoikhoi, San, Xhosa, and Zulu tribes once settled the area now known as South Africa. Some of these tribes were agriculturists and used iron tools to manipulate the ground, while others were hunter-gatherers. The Portuguese first reached the southern tip of Africa in the late 15th century. The Dutch established what would become Cape Town in the mid-17th century. Once gold and diamond resources were found, a struggle over control of the area prompted wars between Britain and Boer colonists (Dutch, French, German, and Portuguese) that lasted over 20 years in the late 19th century and early 20th century. In 1931, independence was granted from the United Kingdom. A transition to democracy began in 1991 and rising public awareness of environmental concerns has since occurred in the current, postapartheid period (Tewari, 2001). Currently, land in South Africa is about 40% privately owned and 60% publicly owned. A National Forests Act (Act No. 84 of 1998) provides measures for protecting and managing forests and prohibits cutting natural forests without a license.

Among the commercial tree species in natural forests, hard pear (*Olinia ventosa*), ironwood (*Olea capensis* spp.), stinkwood (*Ocotea bullata*), white alder (*Platylophus trifoliatus*), and yellowwood (*Podocarpus latifolius*) comprise the majority of roundwood harvested. Black wattle (*Acacia mearnsii*), pines, and eucalypts comprise over 95% of the tree species in forest plantations. In terms of annual value, forest product exports are important, but fall behind coal, gold, and other mineral ores (Grundy and Wynberg, 2001). The forest industry has grown steadily since World War II with active government support (Tewari, 2001). However, since the transition to democracy, concerns about the commercial side of forestry have arisen from a diverse network of stakeholders (Tewari, 2001). Natural limits on the productive range of forests and some uncertainty in ownership rights are hindrances to further expansion of the sector (Grundy and Wynberg, 2001).

2.6.5 United Republic of Tanzania

Tanzania is a country on the eastern coast of Africa (Fig. 2.45) that has nearly 89 million ha of land (nearly 219 million ac), of which about 52% are forested (Table 2.6). The forests of Tanzania transition from a thin strip of mangroves along the coast to the subtropical, dry forests called *coastal hinterland forests* that stretch about 200 km (124 mi) inland, and the upper montane forests (Fig. 2.50) that reach further west into the Tanzanian Eastern Arc Mountains. Much of interior Tanzania, however, is composed of grassy savannas that host broad-leaved evergreen or deciduous tree species. Mount Kilimanjaro and part of Lake Victoria are contained within this country's boundary. Tanzania has a rich and diverse, and perhaps the longest, human history. Early indigenous people were hunter-gatherers, and migrants from western Africa, India, and Persia (now Iran) have influenced Tanzanian culture. The coastal area of the country was controlled by the Portuguese from the early 16th century to the late 19th century, and afterward by the Omani Sultanate. Germany took over administration of the area in 1891, and this control lasted until the end of World War I. The country was then part of the British Empire until autonomy became effective in 1961.

Germans introduced state-administered forestry programs to Tanzania in the late 19th century, which provided

FIGURE 2.49 Assegai tree (*Curtisia dentata*) on the eastern slopes of Table Mountain, South Africa. *Abu Shawka, through Wikimedia Commons.* *Image Link: https://commons.wikimedia.org/wiki/File:Curtisia_dentata_-_Assegai_tree_-_Table_Mountain_slopes_5.JPG* *License Link: Public Domain*

FIGURE 2.50 Highland forests in the Usambara Mountains in eastern Tanzania. *Joachim Huber, through Wikimedia Commons.* *Image Link: https://commons.wikimedia.org/wiki/File:Usambara_Mountains.jpg* *License Link: https://creativecommons.org/licenses/by-sa/2.0/deed.en*

a model for the British programs that replaced them after World War I. However, due to a number of factors, including the relative weakness of the colonial state, a low commitment to forest conservation, unfamiliarity with African timber characteristics, and resistance from local citizens, scientific forestry has been a less than successful endeavor. African teak (*Milicia excelsa*) is one of the principal tree species found throughout central Africa that provides wood for charcoal, construction, furniture, and shipbuilding uses among others. Copal, a resin that resembles amber, was widely collected from *Hymenaea verrucosa* forests in east Africa in the 18th and 19th centuries and was highly prized as a source of varnish for carriages and furniture. Rubber-bearing shrubs, trees, and vines in coastal forests of east Africa also produce resinous material that was highly valued in the early 20th century. The coastal mangrove forests were also a source of bark and wood for construction, local fuelwood, and shipbuilding enterprises (Sunseri, 2009). For many local populations in Tanzania, forests currently provide commercial value, spiritual value, and refuge from political changes. In addition, charcoal derived from forests is the most important domestic energy source, with annual consumption estimated to be 1 million tons (Sander et al., 2010).

Almost all of the land in Tanzania is now owned by the state and, in accordance with the 1999 Village Land Act, about 75% is administered by village councils (Haugen, 2010). Land use rights and ownership status are often unclear or unknown to village officials and, as a result, few village councils proactively manage their forests (Sander et al., 2010). Although the development of Tanzanian forests has been influenced by British and German systems of forest management, some forest reserves in certain parts of east Africa are now managed by local communities through community forest management programs, with assistance from NGOs. One side effect of this policy is that local people are often confined to poorer quality lands between the boundaries of the reserves. As a result, the authority over the use of forests is constantly negotiated according to shifts in power structures and global events (Sunseri, 2009). The forest management situation in Tanzania continues to evolve; however, the recent rate of forest loss is about 0.8% per year (Table 2.6).

2.7 Oceania

Oceania (or *Australasia*) includes the countries of Australia, New Zealand, Papua New Guinea, and several other small island nations such as Fiji, the Solomons, and New Caledonia. During the last ice age, when the mean sea level was lower, there were two land formations that extended and connected the Asian and Australian landmasses: the Sunda and Sahul Shelves. The Sunda Shelf was basically a land bridge that extended from Asia through much of Indonesia, and the Sahul Shelf bridged Australia with Papua New Guinea. As in the case of Beringia, the sea level rose when the ice sheets receded, covering these two shelves, isolating numerous islands (Hanebuth et al., 2000), and stranding numerous species of plants, animals, and insects. Scientists have noted a distinct difference between species originating from Asia and those originating from Australia, which occurs along what is known as the Wallace line (Mayr, 1944). Diamond (2005) suggests that the two shelves facilitated the movement of humans from Asia to the Australian continental landmass during the last ice age or perhaps earlier.

The land area of Oceania (Fig. 2.51) is at least 850 million ha (2.099 billion ac) (Table 2.7) and is inhabited by about 33 million people (United Nations Department of Economic and Social Affairs, 2008). Stretched across this landmass lies approximately 187 million ha (462 million ac) of forestland. The ecosystems within this continent vary widely. For example, in New Zealand one might find tropical and temperate rainforests and highland grasslands on the South Island and subtropical forests on the North Island. In Australia, the interior of the nation is extremely arid; thus, forests are commonly found along the eastern, north—central, southern, and southwestern coasts. When European settlers arrived, they altered much of the forested landscape for many of the reasons mentioned in conjunction with the development of other continents. Given the island nature of the countries in this region, increased human population growth and the overuse of forests have prompted the development of active reforestation and management programs. Since 2010, Oceania has reported a net gain of forest (about 0.2% per year), despite climatic conditions and wildfires in Australia.

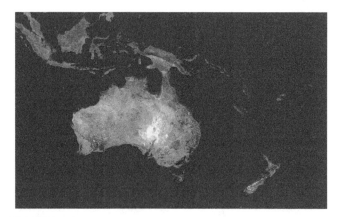

FIGURE 2.51 Satellite image of Oceania. *National Aeronautics and Space Administration.*
Image Link: https://commons.wikimedia.org/wiki/File:Oceania_satellite.jpg
License Link: Public Domain

TABLE 2.7 Demographic and physiographic data for Oceania.

Country	Human population 2018[1] (million)	Total land area[2] (1000 ha)	Total land area[2] (1000 ac)	Forest cover[2] (%)	Forest area annual change 2010 –15[2] (%)	Per capita (PPP) 2017[1&a] (1000 US$)
Australia	25	768,230	1,897,528	16	0.2	57.6
New Zealand	5	26,331	65,038	39	–	42.9
Papua New Guinea	9	46,312	114,391	73	–	2.7
Solomon Islands	<1	2799	6914	78	–0.3	2.0
Other countries or territories (21)	3	6007	14,837	48	0.2	7.0
Oceania (total)	42	849,679	2,098,707	20	0.2	41.0

[a]Gross domestic product (GDP) expressed as purchasing power parity (PPP).

[1]Food and Agriculture Organization of the United Nations 2020.

[2]FAO. Global Forest Resources Assessment 2015. Desk Reference. Rome, 2015.

2.7.1 Australia

Australia (Fig. 2.52) was originally inhabited by over 200 distinctly different groups of indigenous people. Australia was discovered by the Dutch in the early 17th century and was colonized by the British in the late 18th century. In 1901, the Commonwealth of Australia became a dominion within the British Empire in 1907, where it remained until most of the formal ties were severed in the late 1930s and is now a member of the Commonwealth. The largest landmass and country in Oceania contains over 768 million ha of land (about 1.898 billion ac), yet only about 16% is considered forested (Table 2.7). Of the forested area, about 98.6% is native forests and 1.4% is plantation forests (Australian Bureau of Agricultural and Resource Economics and Sciences, 2019a). Australian forests are primarily located along the coastlines in New South Wales, Queensland, South Australia, southwestern West Australia, Victoria, and throughout the island of Tasmania. These include tropical, subtropical, and temperate rainforests, with dry forests and savannas further inland. The interior of the country consists of xeric shrublands and arid deserts, and forests are therefore relatively rare in interior Australia owing to the arid environment. Some Mediterranean-type vegetation can be found along the southern coasts of the country. A severe drought and numerous forest fires exacerbated the loss of forest area in Australia from 2000 to 2013

(Australian Bureau of Agriculture and Resource Economics and Sciences, 2018). However, the current rate of forest change is about +0.2% per year (Table 2.7).

Australia is composed of both closed and open forest areas (Food and Agriculture Organization of the United Nations, 1979). Most (98%) of the forests are broadleaved, composed of about 1000 species of *Acacia*, 13 species of *Callitris*, 59 species of *Casuarina*, about 800 species of eucalypts, and numerous mangrove, *Melaleuca*, and rainforest tree species (Australian Bureau of Agriculture and Resource Economics and Sciences, 2018). The dominant species groupings are eucalypts (78%), acacias (7%), and melaleucas (5%). Of the plantation forests, approximately 50% are composed of exotic conifers such as radiata pine and the other 50% are composed of native hardwood species (Fig. 2.53).

About 70% of the land in Australia is publicly owned, about 20% is privately owned, and about 10% is controlled by community or indigenous groups (White and Martin, 2002). A National Forest Policy Statement that provides the vision for sustainable management of forests and associated goals and objectives was issued in 1992. The forest sector is small but regionally important. Australia is a net exporter of wood, mainly to China, Japan, and New Zealand (Australian Bureau of Agricultural and Resource Economics and Sciences, 2019b). Of the NTFPs derived from Australia's forests, a growing arts and crafts industry uses wood, bark, and chemicals from trees for carvings, dyes, weavings

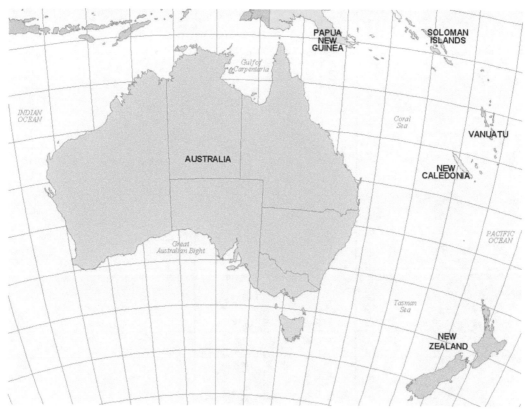

FIGURE 2.52 Countries in Oceania. *Modified image from PresentationMaps.com.*

FIGURE 2.53 Eucalyptus plantation near Mount Gambier, South Australia, Australia. *Donald L. Grebner.*

and other products. In addition, Australia's native forests are important to the apiary industry as a source of nectar for honeybees (Montreal Process Implementation Group for Australia, 2008).

2.7.2 New Zealand

New Zealand has a landmass of about 26 million ha (about 65 million ac) and is comprised of two large islands and several smaller ones (Fig. 2.52). Forests cover about 39% of the country (Table 2.7), and most are located in areas considered to be representative of subtropical or temperate rainforests or temperate steppes. Prior to the Māori settlements about 800 years ago, most of the land below the natural tree line was forested (New Zealand Ministry of Agriculture and Forestry, 2009). Pastoral agriculture now comprises approximately 50% of the land area, while indigenous forests occupy 23% and plantation forests (Fig. 2.54) account for 7%, or 1.8 million ha (about 4.4 million ac). Currently,

FIGURE 2.54 Coniferous plantations on the South Island, New Zealand. *Donald L. Grebner.*

most of the indigenous forests are owned by the government for conservation purposes, and harvesting operations are restricted or prohibited. These forests are found mostly in mountainous areas on the west coast of the South Island (New Zealand Ministry of Agriculture and Forestry, 2009). The type of forest, which New Zealanders commonly refer to as "bush," range from subtropical kauri forests on the North Island, to temperate rainforests on the west coast, and alpine forests in the Southern Alps, located in the interior of the South Island. Some of the native tree species found in these forests include cabbage tree (*Cordyline australis*), hinau (*Elaeocarpus dentatus*), horoeka or lancewood (*Pseudopanax crassifolium*), kapuka or broad-leaf (*Griselinia littoralis*), karaka or New Zealand laurel (*Corynocarpus laevigatus*), kauri (*Agathis australis*), manuka or tea tree (*Leptospermum scoparium*), puka (*Meryta sinclairii*), northern rātā (*Metrosideros robusta*), southern rātā (*Metrosideros umbellate*), rimu (*Dacrydium cupressinum*), tawa (*Beilschmiedia tawa*), and titoki (*Alectryon excelsus*). Most of the common names of these trees are Māori in origin.

The slow growth of native tree species and the general domestic demand for wood resources prompted New Zealand forest managers to search alternative tree species for afforestation purposes. A variety of species such as black walnut, redwood, and others were planted, but the most successful species was radiata pine. About 80% of the land is privately owned (the remainder is managed by the state), and forests considered to be industrially managed are to a large extent controlled by foreign companies. Douglas-fir has been planted at higher elevations on the South Island and eucalypts have also been planted there. However, currently radiata pine forests occupy 90% of the plantation area, while Douglas-fir forests occupy about 6% (New Zealand Ministry for Primary Industries, 2019). The remaining 4% is distributed among a variety of tree species, including eucalypts. As a result of reforestation and afforestation efforts, much of the country is composed of coniferous or mixed coniferous-deciduous forests. Currently, New Zealand is a net exporter of wood products, which are therefore important to the economy. Exports are primarily directed to Australia, Japan, and Korea (New Zealand Ministry of Agriculture and Forestry, 2009). Since the late 20th century, tourism has become a significant sector of economic importance to the country.

2.7.3 Papua New Guinea

Papua New Guinea is a nation in the South Pacific, located north of Australia, that occupies the eastern half of the island of New Guinea and a number of smaller islands to the east (Fig. 2.52). This nation has over 800 indigenous languages and is currently experiencing rapid population growth. Early inhabitants were Austronesian people. Since about the 16th century, a series of countries, including Australia, Germany, Portugal, and Spain, have claimed or administered parts of the current country. Independence was gained from Australia in 1975. The standard of living in Papua New Guinea is very low compared to developed countries and the major land use is subsistence-based agriculture. Forests are within the tropical biome and are generally considered to fall into four groups that include moist forests, montane forests, savannas or woodlands, and tropical rainforests (International Tropical Timber Organization, 2006). For the most part, forests are composed of broad-leaved evergreen or semideciduous tree species. Some forest areas are regularly flooded with either freshwater or saltwater.

Nearly 97% of the land is owned by communities or cooperatives, while the remaining 3% is owned by the government (White and Martin, 2002). Of the over 46 million ha of land area (almost 114 million ac; Table 2.7), some estimates suggest that there are only 7 million ha (about 17 million ac) of operable forest area, with 10 million ha (about 25 million ac) being unsuitable owing to mountainous or swampy terrain (Brunton, 1998). The amount of forest cover is around 73% (Table 2.7). Currently, 40% to 65% of the original forest cover still exists (United Nations Department of Economic and Social Affairs, 2008). A 1991 Forestry Act defined two objectives for the forests of Papua New Guinea: recognition of forests as renewable natural resources and utilization of these resources to obtain economic growth (Fig. 2.55). A number of subsequent regulations have stressed other aspects of forest management, including conservation, reforestation, and export of wood products. Individual

tree selection logging is the main silvicultural practice and, of the over 200 tree species found here, some of the commonly harvested species are eucalypts, kwila (*Intsia bijuga*), and taun (*Pometia pinnata*). Butterflies, rattan, and sandalwood oil are important NTFPs (International Tropical Timber Organization, 2006). It is believed that many of the plant species that exist in the interior region of the island of New Guinea have yet to be discovered and documented.

Summary

With the basic socioeconomic, land, and forest information for world regions presented in this chapter, an understanding of the relative availability of forest resources (forest area, forest cover, forest area change, and per capita forest area), demands placed on these resources by human populations, and resulting outcomes (forest area change) should be apparent. While it may have been mentioned in passing, little emphasis was placed on the current economic climate of different countries and the effect of a populace's standard of living on the character of today's forest resources. One measure of economic climate, the per capita gross domestic product (GDP), is a useful indicator of economic wealth or standard of living. The GDP represents the market value of final goods and services produced per person in a country and it is measured as either the sum of income or the sum of expenditures. In essence, it represents the income and expenditures of the average person in the country, and it is used by economists to compare the standard of living over time and across countries. For international comparisons, GDP is expressed at purchasing power parity (PPP) prices to account for the differences in the cost of living in different countries.

The relevance of using per capita GDP in considering the state of forest resources may be significant. In examining global occurrences of environmental problems, it could be argued that they are much more common in poorer countries that have lower per capita GDP. Nigeria, with per capita GDP of US$2,000, is an example of a very poor country experiencing rampant deforestation (Table 2.5). Poor people often have no choice but to rely on forests for their basic subsistence needs. In poor countries, forests are the main source of food, wood, and many other of life's necessities. With no other option, forests may be cleared to grow food. Human poverty levels, combined with expanding human populations, can thus influence the character of forests and natural resources. At the other end of the spectrum is the United States, with a per capita GDP of US$60,100 (Table 2.1). While not without its own environmental problems, it can be argued that both public and private forests in the United States have now been sustainably managed for decades (depending on your definition of *sustainability*), and

FIGURE 2.55 Hikers traversing the Kokoda Track, Papua New Guinea, which was a battlefield during World War II. *Luke Brindley, through Wikimedia Commons.*
Image Link: https://commons.wikimedia.org/wiki/File:Kokoda_track_Papua_New_Guinea.JPG
License Link: https://creativecommons.org/licenses/by-sa/4.0/

deforestation and forest decline are not currently an issue. On the contrary, the area of forests has been increasing in recent years. Because economic and social resources are available for adequate conservation and protection efforts, most forest resources are used efficiently in order to minimize waste. In addition, affordable alternatives to traditional forest products also exist. As a result, the overall sustainability of forests and natural resources is not as great a concern here as it may be in other countries.

In assessing the forest or natural resource character of a country, broad economic measures are not without limitations, and one limitation particularly relevant to forestry is that some economic measures, such as GDP, do not sufficiently account for environmental quality. Consider, for example, virgin tropical forests in Indonesia which, if left alone, hardly contribute to the production of goods and services of the country's GDP. However, tree harvesting and the eventual destruction of native forests to establish timber or agricultural (e.g., palm oil) plantations may actually result in an increase in the country's GDP. While in general economic growth is relatively good for a country mired in poverty, natural resources may be overconsumed to stimulate and maintain economic advances. China, for example, has enjoyed sustained high economic growth rates since the 1990s, but environmental costs have also been extremely high. Only in recent years has China moved decisively to protect the environment. The limitations of some economic measures for assessing the forest or natural resource character of a country have prompted the development of the field of *green accounting*, which allows the incorporation of environmental assets into GDP calculations.

Understanding the different forest regions of the world helps us understand how our personal lives might be interwoven with global resources and contemporary economic, environmental, and social issues. This understanding broadens perspectives, can lead to improved outcomes for complex forest and natural resource management problems in local and regional areas, and can facilitate greater career opportunities as time passes. As illustrated in this chapter, there are many different forest types (and associated tree species) around the world, and in many cases these have been used to create products we use every day. In effect, our demand for certain products can affect the character and condition of forests in other places where these products arise. As you may have gathered, in response to regionalized supply problems, some tree species have been used for afforestation purposes outside of their native range. These decisions (good or bad) were made by people with concerns and knowledge we may not fully understand. As a result, there may be times when we need to ponder an issue thoroughly before challenging the decisions made by others. In addition, natural forest type distributions may have been greatly affected by a diverse set of events in the last two or three centuries. Anthropogenic events (conflicts and economic growth) have arguably been the primary factor in these changes. Contemporary thought on climate change also suggests that the natural range of some tree species will change as temperatures and precipitation patterns change. The countries portrayed in this chapter are only meant to provide a sense of the forested areas and forest-related issues that are, or have been, posed on each continent. The short list of countries mentioned in this chapter is not intended to indicate that natural resource issues of other countries are irrelevant nor to diminish the importance of forestry and natural resources issues in the omitted countries. With sufficient time, one could write a substantive forest description for most of the other countries in the world (see Question 1).

Questions

(1) *Characterizing another country.* Select a country that was not highlighted in this chapter. Using your library, published papers, and Internet searches, develop an oral presentation that describes

(a) the country's developmental history

(b) the extent of forest resources within the country

(c) the types of forest regions within the country

(d) the tree species one might expect to find

(e) the uses of the country's forests and natural resources

(f) the ownership pattern of land within the country

(g) the policies that guide the uses of the forests and natural resources

(2) *Adventurous tree species.* Select a tree species you were unaware of prior to reading the country profiles. Write a short essay about this tree species. In this essay, discuss

(a) the native habitat where the species originated

(b) why the tree is located in a different part(s) of the world

(c) the climatic conditions the species needs to grow effectively

Continued

QUESTIONS *(cont'd)*

(d) the type of soils and topographic position in which the species will more commonly be found

(e) the uses of the bark, leaves, seeds, and wood of the tree species

(f) the nesting, roosting, or foraging opportunities the species provides for wildlife

(3) *Movies and threatened species.* Locate information about a movie that documents or emphasizes either an endangered or threatened plant or animal species, or a large forested area facing a crisis. Develop a short summary that describes the issues noted below and discuss your findings with a small group of others who have performed the same analysis of other species or forested areas. Include

(a) the resource of interest and the area of the world where it is situated

(b) the major threat to its existence

(c) whether any effort is under way to prevent a loss

(d) information that contradicts the accuracy of what the movie portrayed

(4) *Daily news.* Using a national newspaper such as the *New York Times, Wall Street Journal, Washington Post, USA Today* or one more geographically appropriate, identify news articles concerning a social conflict in a foreign country. Do this over a period of a week. In a short, one-page memorandum, describe

(a) the situation

(b) the opponents

(c) statistics that describe the characteristics and effects of the event

(d) the connections or consequences, direct or indirect, for forestry or natural resources

References

Anderson, P.M., Brubaker, L.B., 1994. Vegetation history of northcentral Alaska: a mapped summary of late-quaternary pollen data. Quaternary Science Reviews 13, 71—92.

Anderson, D.A., Smith, W.A., 1976. Forests and Forestry. The Interstate Printers & Publishers, Inc., Danville, IL, p. 432.

Australian Bureau of Agriculture and Resource Economics and Sciences, 2018. Australia's State of the Forest Report, 2018. Australian Bureau of Agricultural and Resource Economics and Sciences, Canberra.

Australian Bureau of Agricultural and Resource Economics and Sciences, 2019a. Australia's Forests at a Glance, 2019 with Data to 2017-18. Australian Bureau of Agricultural and Resource Economics and Sciences, Canberra.

Australian Bureau of Agricultural and Resource Economics and Sciences, 2019b. Snapshot of Australia's Forest Industry. Australian Bureau of Agricultural and Resource Economics and Sciences, Canberra. ABARES Insights Issue 5.

Baker, P.J., Scowcroft, P.G., 2005. Stocking guidelines for the endemic Hawaiian hardwood, *Acacia koa*. Journal of Tropical Forest Science 17, 610—624.

Banerjee, O., Alavalapati, J., 2010. Illicit exploitation of natural resources: the forest concessions in Brazil. Journal of Policy Modeling 32, 488—504.

Barnes, B.V., Zak, D.R., Denton, S.R., Spurr, S.H., 1998. Forest Ecology, fourth ed. John Wiley & Sons, Inc., New York, NY. 792 p.

Barrett, T.M., Christensen, G.A., 2011. Forests of Southeast and South-Central Alaska, 2004—2008. US Department of Agriculture, Forest Service, Pacific Northwest Research Station, Portland, OR. General Technical Report PNW-835. 155 p.

Birdsey, R., Pregitzer, K., Lucier, A., 2006. Forest carbon management in the United States: 1600—2100. Journal of Environmental Quality 35, 1461—1469.

Bocco, G., Rosete, F., Bettinger, P., Velázquez, A., 2001. Developing a GIS program in rural Mexico. Journal of Forestry 99 (6), 14—19.

Boucher, T.V., Mead, B.R., 2006. Vegetation change and forest regeneration on the Kenai Peninsula, Alaska following a spruce beetle outbreak, 1987—2000. Forest Ecology and Management 227, 233—246.

Brown, N., 1998. Out of control: fires and forestry in Indonesia. Trends in Ecology and Evolution 13, 41.

Brown, D., Schreckenberg, K., Bird, N., Cerutti, P., Del Gatto, F., Diaw, C., Fomété, T., Luttrell, C., Navarro, G., Oberndorf, R., Theil, H., Wells, A., 2008. Verification and Governance in the Forest Sector. Overseas Development Institute, London, UK.

Brunton, B.D., 1998. Forest loss in Papua New Guinea. In: Compendium of Discussion Papers in the Oceania Region. Pacific Bioweb, Nadi, Fiji.

Butler, R.A., 2006. Democratic Republic of Congo (Formerly Zaire). Mongabay.com, San Francisco, CA. http://rainforests.mongabay.com/20zaire.htm (accessed 19.03.20).

Campbell, S., Waddell, K., Gray, A., 2010. Washington's Forest Resources, 2002—2006. US Department of Agriculture, Forest Service, Pacific Northwest Research Station, Portland, OR. General Technical Report PNW-800. 189 p.

Castilleja, G., 1993. Changing Trends in Forest Policy in Latin America: Chile, Nicaragua and Mexico. Unasylva 44 (4). http://www.fao.org/3/v1500e/v1500e07.htm#changing%20trends%20in%20forest%20policy%20in%20latin%20america:%20chile,%20nicaragua%20and%20mexico (accessed 19.03.20).

Chapman, R.A., Heitzman, E., Shelton, M.G., 2006. Long-term changes in forest structure and species composition of an upland oak forest in Arkansas. Forest Ecology and Management 236, 85—92.

Chen, X., 2006. Carbon storage traits of main tree species in natural forests of northeast China. Journal of Sustainable Forestry 23, 67—84.

Chen, H., Kurokawa, Y., 2005. Studies on the relationship between desertification and forest policies in China. Journal of Forest Planning 11, 9—15.

Chiang, J.-M., McEwan, R.W., Yaussey, D.A., Brown, K.J., 2008. The effects of prescribed fire and silvicultural thinning on the aboveground carbon stocks and net primary production of overstory trees in an oak-hickory ecosystem in southern Ohio. Forest Ecology and Management 255, 1584—1594.

Christensen, G.A., Campbell, S.J., Fried, J.S. (Eds.), 2008. California's Forest Resources, 2001—2005: Five-Year Forest Inventory and Analysis Report, US Department of Agriculture, Forest Service, Pacific Northwest Research Station, Portland, OR. General Technical Report PNW-763. 183 p.

Chun, Y.W., 2010. The National Anthem and Tree Planting: A Story of Reforestation of South Korea. Forests and Culture Publishers, Seoul, Korea, 45 p.

Crone, L.K., 2005. Southeast Alaska economics: a resource-abundant region competing in a global marketplace. Landscape and Urban Planning 72, 215–233.

Da Silva, J.M.C., Rylands, A.B., Da Fonseca, G.A.B., 2005. The fate of the Amazonian areas of endemism. Conservation Biology 19, 689–694.

Diamond, J., 2005. Guns, Germs, and Steel: The Fates of Human Societies. W.W. Northon & Company, New York, p. 512.

Donoso, P.J., Otero, L.A., 2005. Hacia una definición de país forestal: ¿Dónde se sitúa Chile? (Towards a definition of forest country: where is Chile?). Bosque 26 (3), 5–18.

Drury, S.A., Runkle, J.R., 2006. Forest vegetation change in southeast Ohio: do older forests serve as useful models for predicting the successional trajectory of future forests? Forest Ecology and Management 223, 200–210.

Drushka, K., 2003. Canada's Forests, a History. McGill-Queen's University Press, Montreal, Canada, 97 p.

Embassy of Costa Rica in Washington D.C., 2017. About Costa Rica. http://www.costarica-embassy.org/index.php?q=node/19 (accessed 20.03.20).

Eurostat, 2019. Wood Products — Production and Trade. European Commission. https://ec.europa.eu/eurostat/statistics-explained/index.php/Wood_products_-_production_and_trade (accessed 21.03.20).

Fan, Z., Fan, X., Spetich, M.A., Shifley, S.R., Moser, W.K., Jensen, R.G., Kabrick, J.M., 2011. Developing a stand hazard index for oak decline in upland oak forests of the Ozark Highlands, Missouri. Northern Journal of Applied Forestry 28 (1), 19–26.

Fearnside, P.M., 1999. Plantation forestry in Brazil: the potential impacts of climate change. Biomass and Bioenergy 16, 91–102.

Fenn, M.E., Allen, A.B., Geiser, L.H., 2011. Mediterranean California. In: Pardo, L.H., Robin-Abbott, M.J., Driscoll, C.T. (Eds.), Assessment of Nitrogen Deposition Effects and Empirical Critical Loads of Nitrogen for Ecoregions of the United States. US Department of Agriculture, Forest Service, Northern Research Station, Newtown Square, PA, pp. 143–169. General Technical Report NRS-80.

Food and Agriculture Organization of the United Nations, 1979. Eucalypts for Planting. Food and Agriculture Organization of the United Nations. FAO Forestry Series No. 11, Rome, Italy, 677 p.

Food and Agriculture Organization of the United Nations, 2003. State of Forestry in the Latin American and Caribbean Region 2002. Food and Agriculture Organization of the United Nations. Regional Office for Latin America and the Caribbean, Santiago, Chile, 107 p.

Food and Agriculture Organization of the United Nations, 2010. Global Forest Resources Assessment 2010. Food and Agriculture Organization of the United Nations. FAO Forestry Paper 163, Rome, Italy, 340 p.

Food and Agriculture Organization of the United Nations, 2015. Global Forest Resources Assessment 2015. Desk Reference. Food and Agriculture Organization of the United Nations, Rome, Italy.

Food and Agriculture Organization of the United Nations, 2016. Forests and the Forestry Sector: Belarus. Food and Agriculture Organization of the United Nations, Rome, Italy. http://www.fao.org/forestry/country/57478/en/blr/ (accessed 19.03.20).

Food and Agriculture Organization of the United Nations, 2020. Global Forest Resource Assessment. FAOSTAT. Food and Agriculture Organization of the United Nations, Rome, Italy. http://www.fao.org/faostat/en/# data (accessed 17.02.20).

Fujikake, I., 2007. Selection of tree species for plantations in Japan. Forest Policy and Economics 9, 811–821.

Gellert, P.K., 2005. The shifting natures of "development": growth, crisis, and recovery of Indonesia's forests. World Development 33, 1345–1364.

Gerasimov, Y., Karjalainen, T., 2010. Atlas of the Forest Sector in Belarus. Working Papers of the Finnish Forest Research Institute, No. 170. Finnish Forest Research Institute, Vantaa, Finland.

German Federal Ministry of Food and Agriculture, 2011. German forests: nature and economic factor. Bundesministerium für Ernährung und Landwirtschaft (Federal Ministry of Food and Agriculture). Berlin, Germany. https://www.bmel.de/EN/Forests-Fisheries/Forests/forests_node.html (accessed 20.03.20).

German Federal Ministry of Food and Agriculture, 2012. Third National Forest Inventory. Bundesministerium für Ernährung und Landwirtschaft und Verbraucherschutz (Federal Ministry of Food and Agriculture). Berlin, Germany. https://bwi.info/inhalt1.aspx?lang=en (accessed 20.03.20).

Global Forest Watch, 2002. Chile's Frontier Forests: Conserving a Global Treasure. World Resources Institute, University Austral of Chile, Valdivia, Chile. http://pdf.wri.org/gfw_chile_full.pdf (accessed 20.03.20).

Grundy, I., Wynberg, R., 2001. Integration of biodiversity into national forest planning programmes, the case of South Africa. In: Paper Prepared for Integration of Biodiversity in National Forestry Planning Programme, Held in CIFOR Headquarters. Bogor, Indonesia, 13–16 August 2001. https://www.cbd.int/doc/nbsap/forestry/SouthAfrica.pdf (accessed 21.03.20).

Guldin, J.M., 2008. A history of forest management in the Ozark Mountains. In: Guldin, J.M., Iffrig, G.F., Flader, S.L. (Eds.), Pioneer Forest: A Half-Century of Sustainable Uneven-Aged Forest Management in the Missouri Ozarks. US Department of Agriculture, Forest Service, Southern Research Station, Asheville, NC, pp. 3–8. General Technical Report SRS-108.

Gulezian, S.E., 2009. Environmental politics in Argentina: The Ley de Bosques. University of Vermont, Burlington, VT. Honors thesis.

Guyette, R.P., Spetich, M.A., Stambaugh, M.C., 2006. Historic fire regime dynamics and forcing factors in the Boston Mountains, Arkansas. USA. Forest Ecology and Management 234, 293–304.

Hanebuth, T., Stattegger, K., Grootes, P.M., 2000. Rapid flooding of the Sunda Shelf: a late-glacial sea-level record. Science 288, 1033–1035.

Harms, W.R., Whitehsell, C.D., DeBell, D.S., 2000. Growth and development of loblolly pine in a spacing trial planted in Hawaii. Forest Ecology and Management 126, 13–24.

Haugen, H.M., 2010. Biofuel potential and FAO's estimates of available land: the case of Tanzania. Journal of Ecology and the Natural Environment 2 (3), 30–37.

Henri, C.J., 2001. Soil-site productivity of Gmelina arborea, Eucalyptus urophylla and Eucalyptus grandis forest plantations in western Venezuela. Forest Ecology and Management 144, 255–264.

Hicks, S., 2011. Boreal Forest, Tundra, and Peat Bogs. Encyclopedia of Life Support System, Earth System History and Natural Variability, vol. III. Eolss Publishers Co. Ltd., Isle of Man, UK.

Hirakuri, S.R., 2003. Can Law Save the Forests? Lessons from Finland and Brazil. Center for International Forestry Research, Jakarta, Indonesia.

Ho, P., 2006. Credibility of institutions: forestry, social conflict and titling in China. Land Use Policy 23, 588–603.

Hyde, W.F., Xu, J., Belcher, B., 2003. Introduction. In: Hyde, W.F., Belcher, B., Xu, J. (Eds.), China's Forests, Global Lessons from Market Reforms, Resources for the Future, Washington, D.C.

International Tropical Timber Organization, 2006. Status of Tropical Forest Management 2005. International Tropical Timber Organization, Yokohama, Japan. Technical Series No. 24. 302 p.

Iwamoto, J., 2002. The development of Japanese forestry. In: Iwai, Y. (Ed.), Forestry and the Forest Industry in Japan. UBC Press, Vancouver, Canada, pp. 3–9.

Johnson, T.G., Bentley, J.W., Howell, M., 2011a. The South's Timber Industry—An Assessment of Timber Product Output and Use, 2009. US Department of Agriculture, Forest Service, Southern Research Station, Asheville, NC. Resource Bulletin SRS-182. 44 p.

Johnson, T.G., Piva, R.J., Walters, B.F., Sorenson, C., Woodall, C.W., Morgan, T.A., 2011b. National Pulpwood Production, 2008. US Department of Agriculture, Forest Service, Southern Research Station, Asheville, NC. Resource Bulletin SRS-171. 57 p.

Jones, M.J., 2003. Evaluation of Honduran Forestry Cooperatives: Five Case studies. Master of Science thesis. Michigan Technical University, Houghton, MI.

Kammesheidt, L., Torres Lezama, A., Franco, W., Plonczak, M., 2001. History of logging and silvicultural treatments in the western Venezuelan plain forests and the prospect for sustainable forest management. Forest Ecology and Management 148, 1–20.

Keliang, Z., Vhugen, D., Hilgendorf, N., 2010. Who Owns Carbon in Rural China? an Analysis of the Legal Regime with Preliminary Policy Recommendations. Rights and Resources Initiative, Washington, D.C.

Kiernan, M.J., 2000. The Forest Ejidos of Quintana Roo, Mexico. US Agency for International Development, Biodiversity Support Program, Washington, D.C.

Korea Forest Research Institute, 2010. Forest Eco-Atlas of Korea. Research Note No. 384. Korea Forest Research Institute, Seoul, Korea.

Krott, M., Tikkanen, I., Petrov, A., Tunytsya, Y., Zheliba, B., Sasse, V., Rykounina, I., Tunytsya, T., 2000. Policies for Sustainable Forestry in Belarus, Russia, and Ukraine. European Forest Institute Research Report 9. Koninklijke Brill NV, Leiden, The Netherlands, 174 p.

Kvarda, M.E., 2004. "Non-agricultural forest owners" in Austria— a new type of forest ownership. Forest Policy and Economics 6, 459–467.

Larson, A.M., 2008. Indigenous peoples, representation and citizenship in Guatemalan forestry. Conservation and Society 6, 35–48.

Lavender, D., 1958. Land of Giants. The Drive to the Pacific Northwest 1750–1950. Doubleday & Company, Inc., Garden City, NY, 468 p.

Lee, D.K., 2010. Korean Forests: Lessons Learned from Stories of Success and Failure. Korea Forest Research Institute, Seoul, Korea, 74 p.

Lee, D.K., Lee, Y.K., 2005. Roles of Saemaul Undong in reforestation and NGO activities for sustainable forest management in Korea. Journal of Sustainable Forestry 20, 1–16.

Lewy, G., 2004. Were American Indians the victims of genocide? History News Network, Seattle, WA. http://hnn.us/articles/7302.html (accessed 19.03.20).

Liang, J., 2010. Dynamics and management of Alaska boreal forest: an all-aged multi-species matrix growth model. Forest Ecology and Management 260, 491–501.

Loude, D.G. (n.d.). The Black Death. http://www.castilles.adrianempire.org/pages/artsarchive/The_Black_Death.pdf (accessed 19.03.20).

Maconachie, R., Tanko, A., Zakariya, M., 2009. Descending the energy ladder? Oil price shocks and domestic fuel choices in Kano, Nigeria. Land Use Policy 26, 1090–1099.

Mascaro, J., Becklund, K.K., Hughes, R.F., Schnitzer, S.A., 2008. Limited native plant regeneration in novel, exotic-dominated forests on Hawai'i. Forest Ecology and Management 256, 593–606.

Mayr, E., 1944. Wallace's line in the light of recent zoogeographic studies. The Quarterly Review of Biology 19, 1–14.

McCaskill, G.L., McWilliams, W.H., Barnett, C.J., Butler, B.J., Hatfield, M.A., Kurtz, C.M., Morin, R.S., Moser, W.K., Perry, C.H., Woodall, C.W., 2011. Maine's Forests 2008. US Department of Agriculture, Forest Service, Northern Research Station, Newtown Square, PA. Resource Bulletin NRS-48. 62 p.

McCrary, J.K., Walsh, B., Hammett, A.L., 2005. Species, sources, seasonality, and sustainability of fuelwood commercialization in Masaya, Nicaragua. Forest Ecology and Management 205, 299–309.

McIntosh, A.C.S., Gray, A.N., Garman, S.L., 2009. Canopy Structure on Forest Lands in Western Oregon: Differences Among Forest Types and Stand Ages. US Department of Agriculture, Forest Service, Pacific Northwest Research Station, Portland, OR, p. 35. General Technical Report PNW-794.

Meditz, S.W., Hanratty, D.M. (Eds.), 1989. Panama: A Country Study. Federal Research Division, Library of Congress, Washington, D.C.

Mendell, B.C., De La Torre, R., Sydor, T., 2006. Timberland Investments in South America: A Profile of Colombia. Timber Mart-South. Market News, Athens, GA, pp. 13–14. Third Quarter 2006.

Merrill, T. (Ed.), 1994. Nicaragua: A Country Study. Federal Research Division, Library of Congress, Washington, D.C.

Merrill, T. (Ed.), 1995. Honduras: A Country Study. Federal Research Division, Library of Congress, Washington, D.C.

Miles, P.D., Heinzen, D., Mielke, M.E., Woodall, C.W., Butler, B.J., Piva, R.J., Meneguzzo, D.M., Perry, C.H., Gormanson, D.D., Barnett, C.J., 2011. Minnesota's Forests 2008. US Department of Agriculture, Forest Service, Northern Research Station, Newtown Square, PA. Resource Bulletin NRS-50.

Monteverde Costa Rica Cloud Forest Nonprofit Organizations, 2016. Monteverde Costa Rica. http://www.monteverde.org/ (accessed 19.03.20).

Montreal Process Implementation Group for Australia, 2008. Australia's State of the Forests Report 2008. Bureau of Rural Sciences, Canberra, Australia.

Morley, R.J., 2011. Tropical Rain Forests. Encyclopedia of Life Support System, Earth System History and Natural Variability, vol. III. Eolss Publishers Co. Ltd., Isle of Man, UK.

Morton, D.C., DeFries, R.S., Shimabukuro, Y.E., Anderson, L.O., Arai, E., del Bon Espirito-Santo, F., Freitas, R., Morisette, J., 2006. Cropland expansion changes deforestation dynamics in the southern Brazilian Amazon. Proceedings of the National Academy of Sciences United States of America 103, 14637–14641.

Muhn, J., Stuart, II.R., 1988. Opportunity and Challenge, the Story of the BLM. US Department of Interior, Bureau of Land Management, Washington, D.C., 303 p.

Munton, R., 2009. Rural land ownership in the United Kingdom: changing patterns and future possibilities for land use. Land Use Policy 26S, S54–S61.

Myers, N., Tucker, R., 1987. Deforestation in Central America: Spanish legacy and north American consumers. Environmental Review E 11, 55–71.

Nelson, M.D., Brewer, M., Woodall, C.W., Perry, C.H., Domke, G.M., Piva, R.J., Kurtz, C.M., Moser, W.K., Lister, T.W., Butler, B.J., Meneguzzo, D.M., Miles, P.D., Barnett, C.J., Gormanson, D., 2011. Iowa's Forests 2008. US Department of Agriculture, Forest Service, Northern Research Station, Newtown Square, PA. Resource Bulletin NRS-52. 48 p.

New Zealand Ministry of Agriculture and Forestry, 2009. A Forestry Sector Study. New Zealand Ministry of Agriculture and Forestry, Wellington, New Zealand. https://ndhadeliver.natlib.govt.nz/delivery/DeliveryManagerServlet?dps_pid=IE1092261 (accessed 20.03.20).

New Zealand Ministry for Primary Industries, 2019. National Exotic Forest Description as of 1 April 2019. New Zealand Ministry for Primary Industries, Wellington, New Zealand.

Niekisch, M., 1992. Nontimber forest products from the tropics: the European perspective. In: Plotkin, M., Famolare, L. (Eds.), Sustainable Harvest and Marketing of Rain Forest Products. Island Press, Washington, D.C., pp. 280–288

Oosthoek, K., 1998. The Role of Wood in World History. https://www.eh-resources.org/the-role-of-wood-in-world-history/ (accessed 19.03.20).

Owubah, C.E., Le Master, D.C., Bowker, J.M., Lee, J.G., 2001. Forest tenure systems and sustainable forest management: the case of Ghana. Forest Ecology and Management 149, 253–264.

Parviainen, J., Västilä, S., Suominen, S., 2010. Finnish forests and forest management. In: Finland's Forests in Changing Climate. Working Papers of the Finnish Forest Research Institute, No. 159. Finnish Forest Research Institute, Vantaa, Finland.

Perlin, J., 1989. A Forest Journey: The Role of Wood in the Development of Civilization. Harvard University Press, Cambridge, MA, p. 445.

Petry, M., Lei, Z., Zhang, S., 2010. China—Peoples Republic of: Forest Products Annual Report 2010. US Department of Agriculture, Foreign Agricultural Service, Washington, D.C. GAIN Report Number CH100042.

Pirard, R., Irland, L.C., 2007. Missing links between timber scarcity and industrial overcapacity: lessons from the Indonesian pulp and paper expansion. Forest Policy and Economics 9, 1056–1070.

Pool, D.J., Catterson, T.M., Molinos, V.A., Randall, A.C., 2002. Review of USAID's natural forest management programs in Latin America and the Caribbean. In: International Resources Group, Winrock International, and Harvard Institute for International Development. Task Order No. 64.

Portilla, A., Eguren, A., 2007. Update Assessment on Section 118/119 of the FAA Tropical Forestry and Biodiversity Conservation in Peru. US Agency for International Development, Lima, Peru.

Richards, E.G., 2003. British Forestry in the 20th Century, Policy and Achievements. Koninklijke Brill NV, Leiden, The Netherlands, 282 p.

Rodríguez, J.P., 2000. Impact of the Venezuelan economic crisis on wild populations of animals and plants. Biological Conservation 96, 151–159.

Roos, J.A., Brackley, A.M., Sasatani, D., 2011. Trends in Global Shipping and the Impact on Alaska's Forest Products. US Department of Agriculture, Forest Service, Pacific Northwest Research Station, Portland, OR. General Technical Report PNW-839. 30 p.

Rozelle, S., Huang, J., Benziger, V., 2003. Forest exploitation and protection in reform China, Assessing the impacts of policy and economic growth. In: Hyde, W.F., Belcher, B., Xu, J. (Eds.), China's Forests: Global Lessons from Market Reforms, Resources for the Future, Washington, D.C., pp. 109–133.

Rubio, N., 2006. Argentina Solid Wood Products, Argentina's Forestry Sector 2006. US Department of Agriculture, Foreign Agricultural Service, Buenos Aires. GAIN Report No. AR6015.

Sander, K., Peter, C., Gros, C., Huemmer, V., Sago, S., Kihulla, E., Daulinge, E., 2010. Enabling Reforms: A Stakeholder-Based Analysis of the Political Economy of Tanzania's Charcoal Sector and the Poverty and Social Impacts of Proposed Reforms. The World Bank, Washington, D.C. Report Number 55140. 50 pp.

Scott, C.W., 1954. Radiata pine in Chile. Unasylva 8 (4). http://www.fao.org/3/x5373e/x5373e03.htm#radiata%20pine%20in%20Chile (accessed 19.03.20).

Shackleton, C.M., Shackleton, S.E., Buiten, E., Bird, N., 2007. The importance of dry woodlands and forests in rural livelihoods and poverty alleviation in South Africa. Forest Policy and Economics 9, 558–577.

Siiskonen, H., 2007. The conflict between traditional and scientific forest management in 20th century Finland. Forest Ecology and Management 249, 125–133.

Smith, J., Colan, V., Sabogal, C., Snook, L., 2006. Why policy reforms fail to improve logging practices: the role of governance and norms in Peru. Forest Policy and Economics 8, 458–469.

Solberg, B., Moiseyev, A., Kallio, A.M.I., Toppinen, A., 2010. Forest sector market impacts of changed roundwood export tariffs and investment climate in Russia. Forest Policy and Economics 12, 17–23.

Sunseri, T., 2009. Wielding the Ax: State Forestry and Social Conflict in Tanzania, 1820–2000. Ohio University Press, Athens, OH, 293 p.

Teplyakov, V.K., Kuzmichev, Y.P., Baumgartner, D.M., Everett, R.L., 1998. A History of Russian Forestry and its Leaders. Washington State University, Pullman, WA, 77 p.

Tewari, D.D., 2001. Is commercial forestry sustainable in South Africa? The changing institutional and policy needs. Forest Policy and Economics 2, 333–353.

Torniainen, T.J., Saastamoinen, O.J., Petrov, A.P., 2006. Russian forest policy in the turmoil of the changing balance of power. Forest Policy and Economics 9, 403–416.

Tóth, S.F., Ueki, T., McDill, M.E., 2006. Monitoring the forest resources and management of private landowners in Nagano, Japan. Journal of Forest Planning 12, 59–64.

Unidad de Reconstrucción Nacional, 1999. Lineamientos del Sector Forestal. Unidad de Reconstrucción Nacional. Plan Maestro de la Reconstrucción Nacional. Gabinete de la Reconstrucción, Tegucigalpa, Honduras.

United Nations Department of Economic and Social Affairs, 2008. World Statistics Pocketbook 2007, Series V, No. 32. United Nations, Department of Economic and Social Affairs, Statistics Division, New York, 244 p.

U.S. Central Intelligence Agency, 2019. Africa: Congo, Democratic Republic of the. US Central Intelligence Agency, Washington, D.C. www.cia.gov/library/publications/the-world-factbook/geos/cg.html (accessed 20.03.20).

U.S. Department of Agriculture, 2020. Natural Resources Conservation Service. Web Soil Survey. U.S. Department of Agriculture, Natural Resources Conservation Service, Washington, D.C. https://websoilsurvey.sc.egov.usda.gov/App/HomePage.htm (accessed 20.03.20).

Usher, A.D., 2009. Thai Forestry, a Critical History. Silkworm Books, Chiang Mai, Thailand, 238 p.

Van Bottenburg, M., 1952. The forestry situation in Colombia. Unasylva 6 (2). http://www.fao.org/docrep/x5363e/x5363e04.htm (accessed 19.03.20).

Voitleithner, J., 2002. The National Forest Programme in light of Austria's law and political culture. Forest Policy and Economics 4, 313–322.

Warren, D.D., 2011. Harvest, Employment, Exports, and Prices in Pacific Northwest Forests, 1965–2010. US Department of Agriculture, Forest Service. Pacific Northwest Research Station, Portland, OR. General Technical Report PNW-857. 17 p.

Wear, D.N., Greis, J.G., 2011. The Southern Forest Futures Project: Summary Report. US Department of Agriculture, Forest Service, Southern Research Station, Asheville, NC, 79 p.

Weaver, P.L., Lombardo, D.M., Sánchez, J.C.M., 2003. Biodiversity and Tropical Forest Conservation, Protection and Management in Nicaragua: Assessment and Recommendations. US Department of Agriculture, Foreign Agricultural Service, Washington, D.C., 66 p.

Westoby, J., 1989. Introduction to World Forestry. Basil Blackwell Ltd, Oxford, UK, 228 p.

White, A., Martin, A., 2002. Who Owns the World's Forests? Forest Tenure and Public Forests in Transition. Forest Trends, Washington, D.C., 32 p.

Whitesell, C.D., DeBell, D.S., Schubert, T.H., Strand, R.F., Crabb, T.B., 1992. Short-rotation Management of *Eucalyptus*: Guidelines for Plantations in Hawaii. US Department of Agriculture, Forest Service, Pacific Southwest Research Station, Albany, CA. General Technical Report PSW-137. 30 p.

Widmann, R.H., Balser, D., Barnett, C., Butler, B.J., Griffith, D.M., Lister, T.W., Moser, W.K., Perry, C.H., Riemann, R., Woodall, C.W., 2009. Ohio Forests: 2006. US Department of Agriculture, Forest Service, Norther Research Station, Newtown Square, PA. Resource Bulletin NRS-36. 119 p.

Wiggins, S., Marfo, K., Anchirinah, V., 2004. Protecting the forest or the people? Environmental policies and livelihoods in the forest margins of southern Ghana. World Development 32, 1939–1955.

Williams, M., 1982. Clearing the United States forests: Pivotal years 1810–1860. Journal of Historical Geography 8 (1), 12–28.

Willis, K.J., 2011. Evolution and Function of Earth's Biomes: Temperate Forests. Encyclopedia of Life Support System, Earth System History and Natural Variability, vol. III. Eolss Publishers Co. Ltd., Isle of Man, UK.

Wishnie, M.H., Dent, D.H., Mariscal, E., Deago, J., Cedeño, N., Ibarra, D., Condit, R., Ashton, P.M.S., 2007. Initial performance and reforestation potential of 24 tropical tree species planted across a precipitation gradient in the Republic of Panama. Forest Ecology and Management 243, 39–49.

Woodall, C.W., Webb, M.N., Wilson, B.T., Settle, J., Piva, R.J., Perry, C.H., Meneguzzo, D.M., Crocker, S.J., Butler, B.J., Hansen, M., Hatfield, M., Brand, G., Barnett, C., 2011. Indiana's Forests 2008. US Department of Agriculture, Forest Service, Northern Research Station, Newtown Square, PA. Resource Bulletin NRS-45. 56 p.

Forest landowner goals, objectives, and constraints

How a forest is managed is primarily the decision of the owner or group of people entrusted with its care, made within the laws of the country that apply to forest management. These individuals or groups select the types of management activities that will be implemented, and thus determine the availability, quantity, and quality of wood and other natural resources made available to themselves and society. Management decisions are based on the objectives and constraints facing each form of land ownership, and these are a reflection of each landowner's priorities (Widmann et al., 2009). A forest and natural resource management plan, whether developed informally or by formal methods, indicates what a landowner or decision-maker desires to achieve from owning or managing the land (Bettinger et al., 2017). *Goals* are broad vision statements that describe in general what natural resources might produce or become in the future, while *objectives* are the measurable criteria one might use to determine the degree that the goals have been achieved. Objectives are usually the desired outcomes from, or the desired future conditions of, natural resources under the control of the landowner or land manager. *Constraints*, as they relate to forest and natural resource management, are the conditions or forces that limit or restrict the use of management activities. As they relate to natural resource management, constraints can involve the physical environment (soils and typical weather patterns), legal environment, biological environment (tree species, and insect and disease status), economic condition of the landowner or the broader economy, available technology, and social norms and pressures, among others.

This chapter was designed to illustrate the variety of influences on the behavior and motivation of typical landowner groups. Upon completion of this chapter, readers should

- understand the types of goals and objectives characteristic of different forest landowner groups;
- understand the similarities and differences in goals and objectives between public and private landowners, why these might vary, and the key influences upon them;
- understand the types of constraints that landowners and managers face when managing forests and natural resources; and

- understand the relationship between goals, objectives, and constraints.

The values and socioeconomic status of the landowner will influence the development of goals and objectives, but these should be consistent with the productive potential of the resources available (Fig. 3.1) (University of Florida, IFAS Extension, 2014).

3.1 Introduction

Decisions related to the management of forests and natural resources are a reflection of the wishes and desires of the landowner or organization with administrative responsibility over the land or resources. These aspirations guide the selection of actions that ultimately affect the character and condition of the resources. Therefore, the objectives and constraints of a landowner can have a significant impact not only on the outcomes desired by the landowner but also on the economic, environmental, and social conditions of the locale where the land and resources reside, the country of residence, and the broader global environment. Goals and objectives define what people desire in both the short term and the long term. There are many unique combinations of goals and objectives, and many people (and organizations) desire different combinations of these to suit their individual needs. Sometimes goals are relatively simple, such as the desire to lead a happy life, have enough money to go to college, buy a home, or start a family, be promoted, have an interesting career, or go on a

vacation to Spain. Often goals are more complicated and elusive, for example, to determine how much land should be allocated to forest or agricultural uses, to preserve an endangered species, to discover a cure for cancer, to end poverty, or to win the World Cup. Goals and objectives may also change as people or organizations age. Therefore, it is necessary to understand the broad current and future needs of each forest landowner group.

Within the natural environment, goals are generally related to ecological, economic, or social issues (Bettinger et al., 2017). As an example, the primary goal of a timberland investment management organization (TIMO), such as Molpus Woodlands Group in Mississippi, is to obtain the highest financial return for their investors. To obtain this goal, they manage their forestlands in such a way that trees are grown, harvested, and eventually sold to local sawmills or pulp and paper processing facilities. While this may be their primary goal, the Molpus Woodlands Group may have other goals that guide the management of their forests. For example, they may also have the goals of protecting sensitive environments and endangered species such as the red-cockaded woodpecker (*Picoides borealis*). Thus, goals can involve both economic and ecological aspects. Another example would be the HessenForst State Forest Enterprise in Germany, a government forestry agency not unlike the Mississippi Forestry Commission, Pennsylvania Bureau of Forestry, or the Forestry Commission Scotland. The HessenForst State Forest Enterprise manages 351,114 hectares (ha; 867,603 acres [ac]) of state forestlands in Germany, and also provides advice and support in the management of nearby corporate (about 324,320 ha [801,395 ac]) and private (about 218,746 ha [540,521 ac]) forests (HessenForst, 2018). HessenForst is required to manage state forests under a sustainable management system that is both ecologically and economically sound and to conduct research on a variety of ecological issues to build broad interrelationships between communities and organizations across the landscape in which they operate. Thus, its goals include those related to sustainable forest management, research, and outreach.

We begin this chapter by discussing the measurable outcomes (objectives) associated with the diverse goals of forest and natural resource management, which may vary by landowner or land management organization. There are many similarities in objectives among the wide variety of potential landowners and management organizations, as well as some distinct differences. It is important to understand the reasons behind the similarities and differences so that we can work effectively as natural resource professionals in a globally oriented society. We should accept the assumption that legally permissible objectives are valid, even though our personal set of values may suggest otherwise.

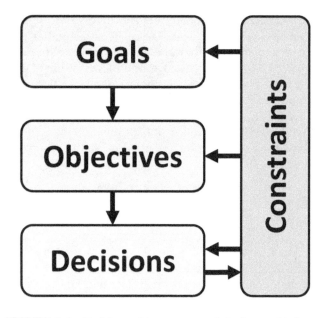

FIGURE 3.1　Decision-making process and the flow and influence of information.

3.2 Objectives of forest landowners

Objectives are the measurable criteria by which one can determine whether or not goals have been achieved. Objectives can vary from those related to commodity production and economic efficiency to those related to ecosystem-oriented processes and social issues. Objectives may be simple or complex and a landowner may be guided by one or by many. Objectives may or may not be related to the sustainability (however defined) of a resource. It is important to understand that landowners and land managers may have different and unique objectives for their land, guided at times by laws and regulations and at other times by their values and personal experiences, and perhaps also by economics, markets, or other external forces.

Objectives for the use of forestlands and associated resources are usually determined by forest landowners, who hold property rights to forests, or by groups of people entrusted with the management of a forest. These rights imply the ability to use, control, and transfer forests or products derived from them. Property rights are often exclusive but seldom absolute (Cubbage et al., 1993). They do not allow unreasonable interference with other private or public property rights. The power to determine property rights, both private and public, rests with governments through land tenure systems. Governments may assign all above-ground and below-surface uses to landowners or they may allow only partial uses, such as selected commercial uses. In some cases, forest landowners may be permitted to sell development rights to retain forestlands in a natural state, and in other cases they may not. The selling of development rights often comes in the form of *conservation easements*. An easement conveys the right to use real property (land) from a landowner to another person or group for a specific purpose, but legal title to the land is retained by the landowner. Conservation easements are legal instruments in which landowners voluntarily surrender their right to change the use of their land to promote conservation (Reeves et al., 2018) usually for a tax advantage or due to inheritance. For example, a conservation easement may prevent the conversion of a forest into an agricultural field or a residential neighborhood. Landowners with conservation easements are often given a financial incentive to undergo this legal change on their ownership documents.

Forests can be owned by entities in the public sector or the private sector, or nongovernmental organizations (NGOs). According to the Food and Agriculture Organization of the United Nations (2016), 76% of the world's forests are actually publicly owned. This category of ownership is broad and represents forests owned by federal, state, and local public administration units, as well as publicly owned institutions and corporations. Although the amount varies from one country to the next (and even from one state or province within a country to the next), only 20% of the world's forests are privately owned. The private ownership group is very diverse and is comprised of individuals, families, communities, firms, businesses, corporations, private educational entities, NGOs, investment funds, churches, and other private entities. In recent years, the level of public ownership of forests has declined slightly, primarily due to deforestation. On the other hand, private ownership of forests has increased slightly, primarily through the establishment of forest plantations.

Public forest ownership is prevalent in all regions of the world, with Africa, Asia, and Europe having the highest proportions in their countries. While public ownership may dominate the land ownership spectrum, ownership patterns can actually vary substantially across regions and among those countries with larger areas of forest (Table 3.1). For example, while all forests in the Russian Federation (the most forested country in the world) are publicly owned, public forest ownership in nearby eastern Europe, and even in western Europe, represents only a small proportion of the whole. Some countries, such as Germany (52% public forests) and Poland (82%) have extensive public forest systems. However, excluding the Russian Federation, public forest ownership accounts for only 45% of European forest area (Food and Agriculture Organization of the United Nations, 2015). North America, South America, and Oceania have a higher proportion of private forest ownership than other regions. Among countries with large forest endowments, the United States is unique because of the dominant presence of private forest ownership.

Property rights continue to change as government policies toward forests and markets evolve. Recent changes in forest ownership (land tenure) include forest privatization efforts in New Zealand or South Africa; forest restitution to former owners in eastern Europe following the collapse of communism; current forest landownership reforms in China; and the rise of institutional investor ownership in the United States. Interestingly, the role of local communities in forest management has been expanding in Africa, Asia, and South America as governments devolve property rights to local people who live in the forests and are thought to have a greater incentive to implement sustainable forest management.

3.2.1 Public ownership

We define public ownership of forests as land that, although access and use may be restricted, is technically the common property of the people that inhabit an area.

TABLE 3.1   Ownership pattern of forests in select world countries.

	Forest area		Ownership (%)	
Country	(ha)	(ac)	Public	Private
Brazil	498,458,000	1,231,689,718	79	21
Canada	347,302,000	858,183,242	92	8
China	200,611,000	495,709,781	57	43
Democratic Republic of the Congo	154,135,000	380,867,585	100	0
Germany	11,409,000	28,191,639	52	48
India	69,790,000	172,451,090	86	14
Indonesia	94,432,000	233,341,472	87	13
Russian Federation	815,135,000	2,014,198,585	100	0
Sweden	28,073,000	69,368,383	25	75
United States	308,720,000	762,847,120	42	58

From Food and Agriculture Organization of the United Nations (2015).

The term *public* in public ownership can be defined as extensively as a nation and as narrowly as a local community. In the forthcoming sections regarding public ownership of forests, we use a hierarchy that begins with perhaps the largest governmental body (a nation or country) and ends with smaller management entities, such as counties, cities, or local communities.

3.2.1.1 Federal or national governments

In many parts of the world, federal or national governments retain ownership of vast areas of forestlands or have obtained control of forests through land purchases, or land tenure and ownership reforms. Typically, federal ownership of land can arise from acquisition programs or colonization efforts, or through warfare. An example of purchased forestland is the United States' acquisition of a large territory from Russia in 1867, which later became the state of Alaska and where now most forests are managed by public agencies. An example of colonization is the Portuguese subjugation of native populations in Brazil. The current delineation of Brazil's national borders was finalized through the Treaty of San Ildelfonso (1777) with the Spanish Empire. With this treaty, Spain gained control of the Banda Oriental, known today as Uruguay, and Portugal gained control of the Amazon basin. Examples of territories (with forestland) acquired through warfare or political strife are the areas commonly known today as Arizona, California, Colorado, Nevada, New Mexico, Oklahoma, Texas, and Utah, which were acquired by the United States after the Mexican-American War (1846–48).

The key point of federal ownership of land and forests is that it has often served as the basis for the development of all other ownership groups as they are found today. Some national governments have sold (or given away) large portions of public land to private entities, such as individuals or businesses, or have transferred ownership to lower governmental structures such as state or provincial governments. This is the case in the United States with acts like the Homestead Act or the Timber and Stone Act and land granted to states upon their entry into the union. In some cases, forestlands have been considered by law to be the property of a central government, yet inadequate institutional arrangements have led to open-access and open-use situations in practice; such has been the case for Vietnam in the second half of the twentieth century (Nguyen et al., 2010). However, federal lands, including those with forests, not distributed to other ownership groups are typically kept as strategic public reserves. Some reasons for maintaining these strategic reserves relate to wood supplies, mineral development, supply of potable water, historical site preservation, recreational opportunities, or the preservation of areas with significant scenic beauty or ecological value (e.g., Yellowstone National Park in Wyoming).

Depending on the type of government structure involved, public forests are managed for a variety of objectives. Governmental structures that are monarchal or dictatorial may commonly define national goals by which public forests and natural resources are managed. They may mandate that forests maximize the production of wood or that certain forests are to be left untouched for watershed protection purposes. For example, the objectives of forest management on federal land in northwestern Russia are to improve forest productivity and efficiency, increase protection efforts against insect and disease problems, and satisfy demands for services from forests in a sustainable manner (Karvinen et al., 2014). Other governmental structures, such as those guided by democratic ideals, may have many different objectives for specific forests, and these objectives may not always be complementary. For instance, managers of certain public forests in the United States may be asked to meet timber production targets, yet may also be asked to protect areas of high aesthetic quality and conserve habitats for threatened species such as the northern spotted owl (*Strix occidentalis caurina*) in the Pacific Northwest of North America or the red-cockaded woodpecker in the southern United States (U.S. Fish and Wildlife Service, 2019). The distinction between federal ownership and local usage of forests may also be unclear in some parts of the world. For example, in Tunisia, while the government technically owns the land, local households have free access to forest resources that allow them to collect firewood, graze livestock, and obtain other nontimber forest products (e.g., cork). Thus, the objectives of federal forest management may also be guided by the needs of local users of the resource (Daly-Hassen et al., 2010).

Many countries that control forests and natural resources at the national level have one or more governmental agencies that manage these resources. As an example, we will delve briefly into the situation in the United States, where several federal agencies control and manage forestlands, including the U.S. Forest Service, the Bureau of Land Management, and the U.S. National Park Service. The U.S. Forest Service is part of the Department of Agriculture and manages approximately 93 million ha (230 million ac) contained within 154 national forests (Fig. 3.2) and 20 national grasslands throughout 43 states and the territory of Puerto Rico (U.S. Forest Service, 2014a). The Forest Reserve Act of

1891 and the Transfer Act of 1905 initiated a process that created the agency. The current mission of the US Forest Service is "to sustain the health, diversity, and the productivity of the Nation's forests and grasslands to meet the needs of present and future generations" (U.S. Forest Service, 2014b). Common management objectives balance resource protection and recreation with resource extraction. As an example of some of the planned, measurable objectives of a U.S. National Forest, the Coconino National Forest (U.S. Forest Service, 2018) in Arizona (Fig. 3.3) has, among others, objectives that concern

- cultural resources
- fire
- forest products
- grasslands
- livestock grazing
- minerals
- recreation
- riparian areas
- roads
- scenic quality
- soils
- springs
- tribal relations
- wetlands
- wildlife and fish

The Bureau of Land Management (BLM) is another federal agency in the United States with land management responsibilities; however, it is contained within the Department of the Interior. The BLM manages approximately 102.3 million ha (253 million ac) of land, of which 22.2 million ha (55 million ac) is forested. Its origins date back to the Land Ordinance of 1785, and the Taylor Grazing Act of 1934 cemented the BLM's modern structure (Bureau of Land Management, 2010). The BLM's mission is "to sustain the health, diversity, and productivity of the public lands for the use and enjoyment of present and future generations" (Bureau of Land Management, 2018). The objectives that the BLM employs to achieve this mission are generally called *multiple use* and *sustained yield management*.

The U.S. National Park Service is a federal agency in the United States charged with managing forestlands and natural resources that are located inside national parks and monuments. The National Park Service is another Department of the Interior agency, and the Organic Act of 1916 established its modern form (U.S. National Park Service, 2016). It manages approximately 34.2 million ha (84.6 million ac) of land within 409 parks and preserves. Its mission is to preserve "unimpaired the natural and cultural resources and values of the National Park System for the enjoyment, education, and inspiration of this and future generations"

FIGURE 3.2 Sign indicating the approximate administrative boundary of the Siuslaw National Forest in Oregon, United States. *Kelly A. Bettinger.*

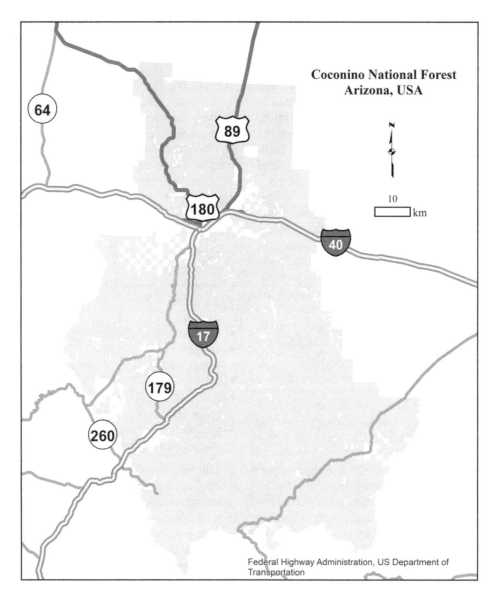

FIGURE 3.3   Extent of the Coconino National Forest in Arizona, United States. *Roads data courtesy of U.S. Department of Transportation, Federal Highway Administration.*

(U.S. National Park Service, 2016). Therefore, the objectives of these public lands lean more toward conservation and social (aesthetics and recreation) concerns than commodity production concerns.

Military installations in some countries may also contain significant reserves of forests and natural resources. In the United States, a few areas reserved for military training exercises encompass very large areas of land; one example is the nearly 113,000 ha (about 279,000 ac) Fort Stewart military installation in southern Georgia (Fig. 3.4). While the main mission of a military base may be to maintain the readiness of troops and prepare for the rapid deployment of forces throughout the world, these lands can also support critical natural resources. Fort Stewart, for example, contains one of the largest areas of longleaf pine (*Pinus palustris*) forestland in the southeastern United States. Most large military installations in the United States also have a forestry or natural resource staff to manage the vegetation and plant species contained in the areas. Besides the overriding military mission, other objectives might relate to habitat development and maintenance, forest health, timber production, watershed protection, and silvicultural or forest development targets.

Public forests in the United States have endured various controversies concerning the management of resources, mainly due to advances in science and changes in societal values. For example, when the first English settlers arrived on the North American continent, they had many views or perspectives on how to use the

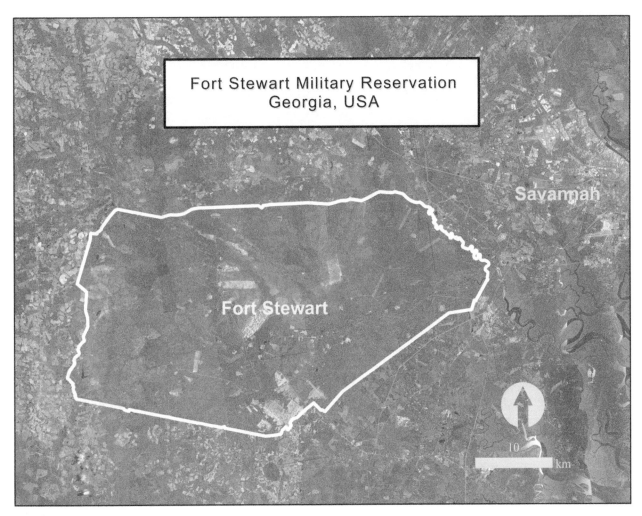

FIGURE 3.4 Fort Stewart military installation, located just west of Savannah, Georgia, United States. *Imagery courtesy of the U.S. Department of Agriculture, National Agriculture Imagery Program.*

vast forest resource in front of them. One such perspective involved converting appropriate pieces of land into agricultural fields. Today, although forests persist, many of the older and larger trees are gone, as are certain species such as the American chestnut (*Castanea dentata*) that composed the original forest. Society now places a value on maintaining biodiversity, protecting endangered plant and wildlife species, and providing recreational opportunities on public lands. Since the public serves as a shareholder of sorts in democratic societies, the objectives and constraints related to the management of forest and natural resources on public lands have become more complex.

3.2.1.2 State or provincial governments

State or provincial governments control and manage lands in a manner similar to federal governments, but generally these forestlands are smaller and are located within smaller geographical boundaries. Exceptions to this can certainly be found, as in the provincial forests in

Canada. As with federal lands, given the public nature of these forests and natural resources, multiple use and forest sustainability are critical concerns for natural resource managers. For instance, in the state of Iowa, broad goals for state-owned lands typically refer to recreation, sustainability, utilization, water quality, and wildlife (Iowa Department of Natural Resources, 2019). Local objectives may be more specific, such as the need to acquire more land adjacent to the existing public forest, to harvest timber using specific silvicultural management systems, to maintain locations for hiking, viewing and backpacking, to manage recreational facilities and structures, to inventory existing forests, and to provide a diversity of forest structures (Byrd, 2010).

Another state, the Commonwealth of Pennsylvania, has a mission to conserve and sustain public forests for the use and enjoyment of present and future generations, and its goals are to conserve, responsibly manage, and efficiently manage state forests (Pennsylvania Department of Conservation and Natural Resources,

2019). As with many states in the United States, Pennsylvania has the goal of protecting not only public forests but also private lands from diseases, fires, insects, and other potential threats (e.g., invasive plant and animal species). One of the most influential issues affecting the management of state forestlands in Pennsylvania was the discovery of Marcellus shale underneath the state's landscape. Marcellus shale is a rock formation that is expected to hold up to 500 trillion cubic feet of natural gas (Institute for Energy Research, 2012). Although accessing this natural gas resource can affect forest and riparian area (for access to water) conditions, considerable efforts are being applied to monitor and minimize impacts to aesthetics and other resources (Pennsylvania Department of Conservation and Natural Resources, 2017). Even though we tend not to discuss mineral resources or other fossil fuel resources in this book, it seems pertinent to briefly mention this example to illustrate the interaction between forests and below-ground natural resources.

In Canada, provincial governments manage public forests with the same type of authority as state governments in the United States. The primary difference is that the Canadian provinces administer a greater extent of public forests than do the states in the United States. For instance, in British Columbia all forestlands must follow the statutes set forth in the British Columbia Forest and Range Practices Act (Queen's Printer, 2019). This Act requires the protection of watersheds and wildlife habitat and sets requirements for planning, road building, logging, reforestation, and grazing. These requirements are typically more restrictive than those commonly found at the state level in the United States. About 89% of the land area in Canada is considered public land to be administered by either the federal Crown or the provincial Crown. A number of provincial parks (Fig. 3.5) are provincial Crown lands and they have similar, though not quite the same, recreational and environmental objectives as national parks in the United

States. However, provincial Crown lands also provide revenue for Canadian provinces through oil and gas leases and timber harvests; therefore, the objectives of these lands can vary considerably.

3.2.1.3 County, parish, or borough government

In the United States, as in other countries, major political divisions (states or provinces) are divided into smaller political districts (counties, parishes, or boroughs) that serve an administrative role for enforcing laws, managing resources, and maintaining local infrastructure. Sometimes county governments own forests, wetlands, and various water bodies, and the size, type, and quality of these will vary considerably. However, not all counties manage such resources. The demographics of a county may have important impacts on the goals and objectives of forest management. For example, rural counties may have a different perspective on forest management, and thus a different set of needs, than counties with forests located near urban areas. An example of an urban county forest is the Forest Preserves of Cook County, Illinois, home to the city of Chicago. These preserves were created after the passage of the Forest Preserve District Act of 1913 (Forest Preserves of Cook County, 2016). Since then, the preserve's area has grown to nearly 28,329 ha (70,000 ac) and consists of a network of forests, lakes, open spaces, prairies, streams, and wetlands; the mission of the preserve is to acquire and hold natural lands to protect and preserve flora, fauna, and scenic beauty (Forest Preserves of Cook County, 2012). The preserve is unlike a park because it does not provide organized recreational facilities. However, the preserve supports bicycling, cross-country hiking, fishing, golfing, picnicking, skiing, and other activities, and is used to maintain the native vegetation and protect many endangered and threatened flora and fauna communities. Despite county ownership of this forest, volunteer groups and the general public assist in developing policies and guiding the management of the resources.

An example of a rural county forest is the Barron County Forest in northwestern Wisconsin. Barron County has a population of 46,000 and lies primarily in the central plain physiographic region, where the topography has a glacial origin. In the early twentieth century, catastrophic wildfires, extensive timber harvesting, and reduced soil fertility contributed to many private landowners losing title to their land because of delinquent property tax payments. The County Board approved the creation of a county forest, and in 1940 they approved the entry of 1312 ha (3242 ac) of land into the management program under the Forest Crop Law (Cisek, 2018). The county forest has since grown to over 6475 ha (16,000 ac). The current goal of the county forest is "to manage, conserve, and protect the

FIGURE 3.5   Pigeon River Provincial Park, Ontario, Canada. *www.goodfreephotos.com.*

natural resources of the county on a sustainable basis for present and future generations" (Cisek, 2018). The objectives within the most recent forest plan are meant to enable and encourage management of these forests for the optimum production of forest products, as well as to attend to rare plant and animal communities, recreation, species diversity, stream flow, watershed protection, and wildlife concerns, while also protecting other public rights (Barron County Government, 2013). Many of these objectives are to be achieved through common forestry and natural resource practices.

3.2.1.4 Municipal or local government

Local or municipal governments are typically thought of as towns or cities. Sometimes referred to as *town forests*, these are usually small, given the relative density of human habitation within their boundaries. Wooded areas within these boundaries may encompass a neighborhood park, hiking trails (Fig. 3.6), or a small water body such as a pond. In addition, these wooded areas may line various walking trails or bicycle paths across the community. Despite this interest, many small communities have little information or experience to guide them in selecting an appropriate course of action for their forests. One of the most significant constraints for small communities, besides a lack of knowledge of forestry practices, is the lack of funding for long-term management of these natural resources (Grado et al., 2006). Objectives of municipal forests frequently involve maintaining high quality water supplies and providing habitat for wildlife, income to the city, and recreational opportunities to city residents. Among many others, two examples of local or municipal forests are the Frederick Municipal Forest (Frederick, Maryland) and the

Newport City Municipal Forest (Newport, Vermont). In fact, in 2015 there were 356 town forests in Vermont, comprising 27,628 ha (68,269 ac), that ranged in size from less than 1 ha to over 1400 ha (MacFaden, 2015).

When considering municipal forests, a distinction should be made between trees located within a defined, enclosed green area and trees situated in easements or rights-of-way. Street trees are trees that line the right-of-way of easements of major or minor roadways; and green areas refer to squares, plazas, parks, and other urban forest vegetation growing on public land. In some cases, both street trees and green areas (enclosed areas) may be managed by municipalities. The management of much of this type of resource is under the purview of an urban forester, and a wide array of laws and ordinances govern their management. Escobedo et al. (2006) describe the management of these types of resources within the local communities that comprise the greater Santiago, Chile, area, and note that the intensity of management of the resource is related to the socioeconomic characteristic of the community. In San Francisco, as in other cities, management of the thousands of individual street trees is challenged by program funding problems and issues related to tree maintenance responsibilities (San Francisco Planning Department, 2014).

3.2.1.5 University forests

While constituting a relatively small percentage of the larger set of land ownership groups, many colleges and universities own and manage forestland (Fig. 3.7). The objectives of these lands generally relate to the research, teaching, and outreach mission of the particular institution, but they may also be related to income generation goals and the desire to provide demonstration areas. For example, the objectives of the Dubuar Forest, which is owned and managed by the State University of New York College of Environmental Science and Forestry, call for the generation of a minimum level of revenue,

FIGURE 3.6 A hiking path through Newlands Forest in Cape Town, South Africa. *S. Molteno through Wikimedia Commons. Image Link: https://commons.wikimedia.org/wiki/File:Woodcutters_path_ through_indigenous_woodland_-_Newlands_Forest_Cape_Town.jpg License Link: Public Domain.*

FIGURE 3.7 A University of Georgia Warnell School of Forestry and Natural Resources, Georgia, United States, visited by Dr. Hayati Zengin, a professor of forestry at Duzce University in Turkey. *Pete Bettinger.*

maintenance of diverse forest structures for research and teaching purposes, maintenance of biodiversity and ecological integrity, and sustainability of forest growth (Savage, 2015).

3.2.1.6 Communal and community forests

Many communal and community forestry strategies shift the control, authority, accountability, responsibility, and costs of public forest management to local communities (Fig. 3.8). A definition of *community forestry* may be elusive, given the cultural and historical contexts within which these systems have evolved (Mustalahti and Nathan, 2009). For example, forms of community forest management may also be referred to as *collective forest management*, *community-based resource management* (Bolivia, Colombia, Mexico, Nepal, the Philippines, Tanzania, and Zambia), *joint forest management* (India), *participatory forest management*, or *social forestry* (Hyde et al., 2003). The Cairnhead Community Forest in Scotland is an example of a partnership between a local community and Forestry Commission Scotland. The objectives of a community forest program vary according to the socioeconomic climate of a particular region. Local empowerment over natural resources and the use of more precise and context-specific management approaches are some of the main advantages of this type of control. However, the opinions of some local groups may be marginalized in the process, which can exacerbate inequalities (Thoms, 2008), and the lack of administrative skills, capital, and technical assistance can pose obstacles to successful programs (Klooster and Masera, 2000).

Community forestry has been practiced in southeastern Asia for centuries and more recently *introduced* community forestry has produced a different form of

FIGURE 3.8 A boundary of the Kumrose Community Forest in Nepal. *U.S. Agency for International Development, Biodiversity, and Forestry.*
Image Link: https://commons.wikimedia.org/wiki/File:USAID_Measuring_Impact_Conservation_Enterprise_Retrospective_(Nepal;_National_Trust_for_Nature_Conservation)_(39404227375).jpg
License Link: Public Domain.

forestry, in which a system of forest management is presented from outside the local community by the government, an international agency, or an NGO. Aside from protecting existing forest and rehabilitating others, poverty alleviation is one of the objectives of these programs (Sunderlin, 2006). China, perhaps, has the most experience of transferring the administration of federal land to local groups (Hyde et al., 2003). Community forest management began in China in the mid-1950s (Rozelle et al., 2003) and now about 60% of China's forests are managed as collective forests. Prior to 1978, collective forests were managed by local communities, and since the onset of economic reforms, individual households now manage a significant portion (about 80%) of the collective forests (Hyde et al., 2003). One condition for allocating land to family plots (*ziliushan*) is that trees need to be planted and grown, and individual households have primary responsibility for management of the land. Another form of collective management involves areas called *responsibility hills* (*zerenshan*), in which households and local communities share the responsibilities and the benefits of management (Liu and Edmunds, 2003). In these cases, the objectives relate to household and community income and watershed rehabilitation. In India, about 17.3 million ha (42.8 million ac) of forestlands are administered by almost 85,000 joint forest management committees (Paul and Chakrabarti, 2011). In the state of Orissa, in eastern India, community forest objectives have ranged from reducing soil erosion and increasing the availability of fuelwood and timber products to minimizing environmental concerns (Sekher, 2001). In Mexico, a number of political agrarian units (*ejidos* and *comunidades agrarias*) hold as common properties forests that range in size from 100 ha (247 ac) to about 100,000 ha (247,100 ac) and, in some cases, the objectives of management are to produce timber products and support the employment of local communities (Klooster and Masera, 2000). In the Lepaterique community forest in Honduras, objectives include generating income from charcoal production, firewood, resin, and timber production, providing opportunities for animal grazing, and allowing the collection of medicinal plants (Nygren, 2005). In some areas of Brazil, the state owns the land and provides concessions to communities, which then develop management plans and allow extraction of resources within individual land plots (*colocações*) managed by households that may or may not cooperate with each other (Goeschl and Igliori, 2004).

In Cameroon, community forestry programs are growing, and each area (no larger than 5000 ha or 12,355 ac) is determined through a negotiation between the government and a community. In these areas, it is agreed that sustainable forest management will be practiced for a period of 25 years. Objectives of community forestry in Cameroon include protecting the forest heritage, the environment, and biodiversity; raising local standards of living; revitalizing the forest sector; and ensuring regeneration of forest resources through plantation. Communal forests in Cameroon are defined as the residual areas of forests not set aside for long-term use and not included in private forest estates (Minang et al., 2007). In the United Kingdom, some of the objectives of England's Community Forest program are to enhance the environment and biodiversity, encourage commercial enterprise initiatives, and provide recreational opportunities through a network of public access routes. These community forests, which encompass about 453,000 ha (a little over 1,118,496 ac), are designed to improve local environmental conditions, enhance local economies, and increase local social capital by raising property values and attracting investments and jobs (Kitchen et al., 2006).

Community-based forestry efforts in North America have been practiced for a long time on a relatively small scale; however, these have expanded since the 1990s in response to the perceived inability of conventional administrative organizations to take into account a wider range of objectives of local citizens (Cheng et al., 2011). Many towns in New England currently own or have owned land (called *common lands*) since colonial times for people to use as a source of firewood and which is now frequently used for recreational purposes. The community forestry concept was more formally reintroduced in North America in the 1990s as a response to a variety of contentious forest management issues. Further, while public land is the primary focus, this form of forest management can involve both public and private land and refers to the guidance of management activities by community residents through participatory processes aimed at the acquisition and maintenance of broader societal benefits. One of the larger examples in North America is the Weaverville Community Forest in northern California, in which about 5260 ha (about 13,000 ac) of public lands are managed cooperatively under a short-term (10-year) agreement with the U.S. Forest Service and the BLM. The Weaverville Community Forest is located within the wildland-urban interface of the town of Weaverville and is managed with the objectives of, among others, controlling invasive species, improving the health of the forest, improving wildlife habitat, increasing recreational opportunities, maintaining or improving water quality, protecting viewsheds and aesthetics, and reducing forest fuel hazards (Trinity County Resource Conservation District, 2010).

In western Canada, the McBride Community Forest in British Columbia is about 60,000 ha (148,260 ac) in size and is managed with timber production as the main economic objective. However, the overall objective of managing the land is to address a large set of potential needs of the community, which include maintaining,

protecting, or enhancing water resources, maintaining recreational opportunities, enhancing and attracting education and research opportunities, maintaining a healthy living environment, and developing opportunities for other nontimber values (Anderson and Yates Forest Consultants, Inc., 2010).

The goals and objectives of community-based forestry may seem extensive and vague at times, but they generally focus on rights and access to forest resources; the distribution of benefits from forests; and the retention, management, and restoration of healthy ecosystems (Christoffersen et al., 2008). Other objectives might focus on increasing land conservation, reducing fragmentation of managed ecosystems, increasing community stability, reducing fire risk, increasing reforestation rates, and acknowledging resources not currently valued in the traditional economic sense.

3.2.2 Private ownership

Private ownership of land implies that formal title to land is recognized by local and national governments. In this section we describe several distinct classes of private landowners that are important managers of forest resources throughout the world.

3.2.2.1 Industrial landowners

It is not uncommon to associate large private forestland holdings with the forest products industry. Forestry companies based in Brazil, Chile, Scandinavia, other parts of Europe, and the United States (Fig. 3.9), currently, or historically, have owned large areas of forests within their home countries and abroad. For example, during the 20th century and the early part of the 21st century, American companies such as Georgia-Pacific, International Paper, Westvaco, and Weyerhaeuser owned vast areas of land in the United States and other countries around the world, such as in Canada, New Zealand, and South America. Sometimes, these types of organizations owned the land and forests outright, yet in numerous situations they owned concession rights to harvest timber on government land, for example, in Africa, Canada, eastern Russia, and southeast Asia. During the early part of the twentieth century, the corporate structure became one that stressed vertical integration, in which a company owned mills and maintained a forestland asset that provided predictable supplies of wood to maintain stable manufacturing flows. Troncoso et al. (2015) showed the operational advantage of approximately 5% increase in net present value from vertically integrated companies in Chile. However, during the first decade of the 21st century, many vertically integrated companies in the United States began to sell their lands and to focus on manufacturing wood-based products, purchasing wood from the open market. This allowed them to take advantage of new business classification with reduced taxes,

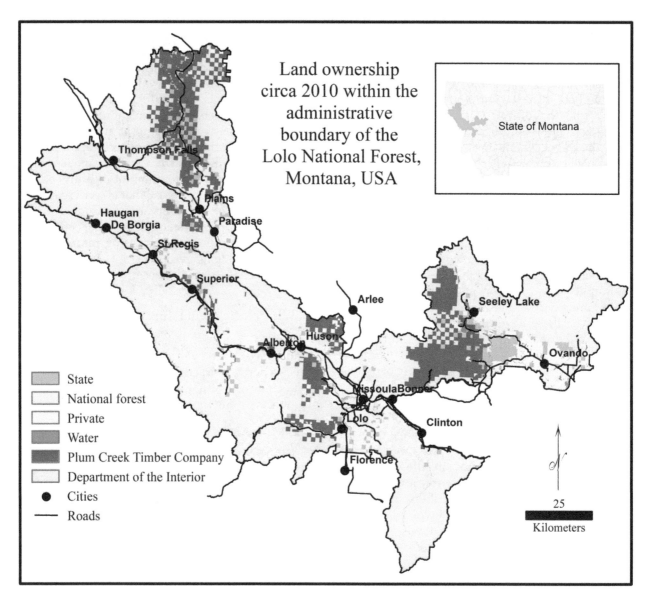

FIGURE 3.9 Forest company lands within the administrative boundary of the Lolo National Forest (Montana, USA), prior to the acquisition of Plum Creek Timber Company by the Weyerhaeuser Company. *U.S. Forest Service, Lolo National Forest (2010).*

real estate investment trusts (REIT). However, Weyer-haeuser did the opposite and sold most of their mills to focus on timberland management. Even though this transition represents the industrial forest owner situation in the United States, one can find similar trends in other countries, such as the situation with Stora Enso, which divested itself of forested lands in Finland and the United States (Sande, 2002). Vertically integrated companies still exist, such as J.D. Irving, Ltd. of eastern Canada This company owns land in the province of New Brunswick, as well as the state of Maine in the United States. Other industrial landowners, such as Alberta−Pacific Forest Industries and Western Forest Products Limited, control (i.e., have leases for) large areas of Canadian governmental (provincial Crown) forestland. In general, the objectives of industrial forest landowners are related to maximizing the efficiency of an organization and might include, among others, maximizing (1) shareholder value (if a company is traded on a stock exchange); (2) the net present value of operations; (3) cash flow; (4) wood production; or (5) an even flow of wood production.

3.2.2.2 Nonindustrial private owners

In many areas of the world, there is a need to balance public values with private property rights and the management objectives of the private landowner. A private forest landowner's attitudes, interests, and values will vary depending on their personal social and economic situation and the larger local, regional, and global economic, environmental, and social conditions. Private forest owners may further be influenced by their age, education, and life experiences (Widmann et al., 2009). In many areas of the world, private ownership of forests is viewed from a more socially oriented perspective, in which the enjoyment of owning land and a concern for future generations may be highly important (Kvarda, 2004). The objectives of private landowners can therefore range from recreational opportunities (fishing, hiking, hunting, relaxation, etc.) to economic or investment opportunities. However, their objectives may be as straightforward as producing fuelwood or income, as in the case of small private landowners in the western Balkan region of Europe (Glück et al., 2010). The economic interests of private landowners may be to obtain a positive cash flow each year or to meet periodic financial needs, such as those related to a child's educational costs (e.g., college). Some landowners may use their land for risk aversion purposes during periods in which they need extra income (e.g., in times of emergencies). If one were to survey private landowners across a broad region, one would probably find that their objectives span the entire range of economic, environmental, and social considerations, and would more than likely be based on their socioeconomic status and inherent value system. Most private landowners have multiple objectives in mind for their land, and some have the objective of actively developing habitat conditions that will promote populations of certain wildlife species and enhance the biodiversity of their land. Income generation, forest health, fuel reduction, nonforest product production (e.g., shiitake mushrooms in Japan and other areas), and charcoal and firewood production are also objectives for many private forest owners. Farmers in western and central Finland have as objectives economic and financial security, employment, nontimber forest product production, conservation, and aesthetics. In addition, a fair proportion of farmers (about 12%) have no specific objective in the management of their forests (Selby et al., 2005). Private forest landowners can also be interested in forests for the contribution to privacy and the continuance of a family legacy that they provide (Butler, 2008). Incentive systems, cost-share programs, and educational opportunities are often available to private landowners to encourage certain forms of forest stewardship, and these may influence the objectives of private forest landowners.

Family forests in the United States (Fig. 3.10) are a good example of a very diverse nonindustrial private forest ownership group. In addition to the immediate family, these owners may include individuals, trusts, estates, family partnerships, and other unincorporated groups of individuals. They account for 92% of private owners and 35% of the entire forest area in the United States (Butler, 2008). Many private landowners do not have formal management plans for their forests. For example, of the 336,000 private landowners in the state

FIGURE 3.10 Small family forest with mixed pine and hardwoods, Kemper County, Mississippi, United States. *Donald L. Grebner.*

of Ohio, only about 8% of these have formal management plans (Widmann et al., 2009). Family forests are mostly located in the southern United States. Family owners hold forests for multiple objectives, including aesthetic values, inheritance, investment, hunting and fishing, nature protection, nontimber forest product collection, recreation, residence, and timber and fuelwood production. While timber production is not rated as the highest priority by most of these landowners, harvesting activities are still quite common. The relatively low priority suggested by landowners for timber production does not imply that landowners will not harvest trees as, when conditions are conducive, many landowners will harvest their trees (Widmann et al., 2009). On average, family owners hold 23.5 ha (58 ac) of forestlands; thus, in the United States there are millions of private forest landowners. The size of a holding appears to correlate with the management objectives of the landowner and the common land-use actions implemented by the landowner, such as conducting commercial timber harvests (Butler, 2008).

The Swedish forest commons are a special case of private forestland and, in theory, this concept is related to collaborative forest management. These areas were created in the late 19th and early 20th centuries by the Swedish government as a way of transferring ownership of forestland to rural residents; however, the government retained regulatory rights to the forest resources, rights that are said to be stricter than those imposed on regular private landowners in Sweden (Holmgren et al., 2010). Initially, the objectives of the forest commons were to provide a sustainable timber supply and sustainable incomes for farmers. However, objectives now also place a high value on environmental goals. The forest commons are interesting in that the lands are technically privately owned, yet the government defines the goals and priorities for their management (Holmgren et al., 2010).

While the stated objectives of private forest landowners may give an indication of their tendency to implement management practices, there may often be a mismatch among the various desires of private forest landowners. For example, in the Republic of Ireland, nearly two-thirds of the forest landowners failed to note timber production as an objective for managing their land, yet many did plan to enact a thinning operation on their land, perhaps to enhance the value or quality of other objectives (Ní Dhubháin et al., 2010).

In the United States, The Nature Conservancy has developed a program called the *Forest Bank* that offers professional forestry services to private landowners (Gilges, 2000). This program is viewed as a conservation tool by The Nature Conservancy, since in return for professional services the *Forest Bank* receives a conservation easement that requires private landowners to maintain their land in a forested condition. While the services are provided by an NGO that has a general goal of protecting habitats, the program guarantees private landowners that some revenue can be generated from forest harvests. Therefore, private landowners with economic or commodity production objectives can enter into this program. However, the main constraint on the management of forests seems to be that only periodic partial harvests or thinnings are allowed.

3.2.2.3 Timberland investment management organizations and real estate investment trusts

Another large private corporate ownership category is the TIMO. The origin of this landowner category dates to the 1970s (Binkley, 2007). Basically, these groups act as brokers for institutional investors and as managers of the timberland investments purchased by institutional investors. The objectives of TIMOs are to meet institutional investors' primary goals, which could be to maximize the return on an investment or to achieve a steady flow of revenue from timber harvesting operations over time. In essence, TIMOs do not actually own the land they manage; they are contracted to manage the land for the investment group that is privately owned. Some examples of TIMOs in the United States include the Hancock Timber Resource Group and the Molpus Woodlands Group. A REIT is a legal organizational business structure where interested businesses can utilize a wide array of investment capital to finance various real estate opportunities. Timber REITs operating in the United States include PotlatchDeltic, Rayonier, and Weyerhaeuser. Many of these were formed from the woodlands divisions of vertically integrated forest products companies.

The rapid rise of TIMOs and REITs since the 1980s provides evidence of a very swift private forestland ownership change in the United States. In the past, primary wood-product manufacturers owned 8% to 10% of United States forests, and they were accordingly classified as forest industry ownership groups (Smith et al., 2009). In the 1980s, they began either selling their forestland holdings to TIMOs or converting their ownership structure to REITs. Most of the industrial forestland in North America was divested from forest products companies between 2000 and 2010, and the largest seller was International Paper Company. It is estimated that by 2006, TIMOs and REITs had acquired 80% of the former traditional forest industry lands (Smith et al., 2009). As of the beginning of 2019, the largest REIT in the United States was Weyerhaeuser (about 4.9 million ha or 12.2 million ac), followed by Rayonier (about 0.9 million ha or 2.2 million ac), and PotlatchDeltic (about 0.8 million ha or 2.4 million ac) (TimberMart-South, 2019). The largest TIMOs in the United States were the Hancock Timber Resource Group (about 1.4 million ha or 3.5 million ac), followed by the Forestland Group (about 1.1 million ha or 2.8 million ac) and the Forest Investment Associates (about 1 million ha or 2.4 million ac). Since 2000, large forestland transactions involved nearly 25.4 million ha (about 63 million ac), representing US$78 billion in value. Most of these transactions (54%) occurred in the southern United States.

The primary reasons behind the vast forestland sell-off were the perceived poor financial performance of large, vertically integrated forest products corporations, and pressure applied by financial markets to *unlock the value* of forestland holdings. At the same time, both TIMOs and REITs received preferential tax treatment in the United States, compared to traditional forest products corporations that own land. Income earned by TIMOs and REITs is subject only to ordinary (personal) income taxes in the United States, while income earned by forest products corporations is subject to both corporate and ordinary income taxes.

Broad ownership changes such as these can potentially have significant implications for forest resources and their management. While TIMOs and REITs are thought to manage their land in a similar manner to forest industry corporations, there are some important differences. Forest products corporations manage their lands to grow timber to supply their mills. They practice traditional timber management that is focused, to a large extent, on tree improvement and intensive management, including the development of plantations and the use of forest protection measures. Some forest products companies also provide technical assistance to nonindustrial private landowners and invest heavily in research activities. TIMOs and REITs treat forestlands more as an investment asset and are focused on (have the objective of) generating the highest financial returns possible, which may result in converting some forestlands to higher valued uses (residential or developed areas). It is likely that these types of private forestlands will be traded more frequently and that fragmentation of forest holdings (with associated management consequences) will continue (Wear et al., 2007).

3.2.3 Nongovernmental organizations

An NGO is a legal organization that is independent of federal, state, or local governments and is typically structured as a not-for-profit entity. Some NGOs can function only within a specific political border, whereas others can participate in activities across international boundaries. With regard to the management of forests and natural resources, NGOs play an important role in the technical assistance they provide to communities in need and the advocacy in which they engage regarding emerging environmental or social issues important to both developed and developing countries. Therefore, many NGOs act either as catalysts (informing and consulting on land management issues) or as cooperators (having a voice and power in the management of land) (Da Veiga Mendonça, 2010). Many NGOs also actively acquire forestland and associated natural resources to achieve organizational goals, such as the preservation of unique landscapes, which is the central topic of this section.

An example of a large NGO is The Nature Conservancy, which originated as a United States charitable environmental organization and was incorporated as a nonprofit organization in 1951. Its first chapter charter was granted for New York in 1954. The main mission, and hence the broad objective of The Nature Conservancy is to conserve ecologically important lands and waters for nature and people (The Nature Conservancy, 2019). Over time, The Nature Conservancy has expended efforts in 72 countries to protect about 48 million ha (119 million ac) of land, 8000 kilometers (km; 5000 miles [mi]) of rivers, and 100 marine conservation projects.

The Nature Conservancy also has other objectives for some of the various forests that they manage. In some areas of the southern United States and Pacific Northwest, for example, timber harvesting is allowed on their land to generate income, enhance habitat quality, or address wood flow agreements with cooperators (Siry et al., 2015). Another type of NGO is a conservation land trust, generally considered to be a private, not-for-profit organization. The specific goals of conservation land trusts can vary across forests, farmlands, prairies, deserts, wildlife habitat, wetlands, coastlines, scenic corridors, archeological sites, and historical locations (Brewer, 2003). An example of this type of NGO is the Natural Lands Trust, based in Media, Pennsylvania. The trust was founded by avid bird watchers in 1953 and its primary goal is to preserve natural assets from development in the Delaware Valley. Their approach involves protecting land through direct acquisition or conservation easements. Their objectives involve managing and restoring long-term ecological health, promoting natural reserves in the region, and supporting local communities with natural resource planning (Natural Lands Trust, 2019a). Natural Lands Trust owns and manages 43 preserves in eastern Pennsylvania and southern New Jersey. Some of these include Harold N. Peek Preserve in Millville, (Cumberland County), New Jersey; the Binky Lee Preserve in Chester Springs (Chester County), Pennsylvania; the Gwynedd Wildlife Preserve in North Wales (Montgomery County), Pennsylvania; the Sadsbury Woods in Sadsbury Township (Chester County), Pennsylvania; and the Wawa Preserve in Media (Delaware County), Pennsylvania (Natural Lands Trust, 2019b).

3.3 Constraints of forest landowners

All owners and managers of forestland, whether public or private, big or small, American, Brazilian, German, or Indian, face *constraints* when they make decisions regarding the management of their forests and natural resources. Basically, constraints limit the achievement of any or all of the goals or objectives stated by the landowners. While many constraints are beneficial to environmental or social goals, in general each constraint added to a forest management plan can reduce the economic value of the plan. From another point of view, economic or commodity production constraints (e.g., minimum achievement levels) can reduce the ecological or social value of a plan. Some constraints on the management of forests are self-imposed. For example, a landowner may want to limit the amount of timber harvesting on their property for aesthetic or recreation-related reasons. Other constraints on the management of forests are imposed by law or regulation. An example of this is Oregon's Forest Practices Act which limits the extent and type of forest management practices that can be employed across a landscape (Oregon Department of Forestry, 2018). The types of constraints forest landowners face vary depending on their geographical, political, and economic situation. Some common types of constraints include or relate to

- aesthetic values,
- availability of resources,
- biodiversity,
- concern for future generations,
- habitat for endangered or threatened plant and wildlife species,
- inherent site limitations,
- invasive species,
- local labor or employment needs,
- market uncertainties,
- mill requirements,
- organizational or regulatory policies,

- recreational opportunities,
- soil erosion or compaction,
- sustainability of resources,
- weather conditions, and
- wood quality.

This is not an exhaustive list, but it provides a general impression of the constraints landowners might face when making natural resource management decisions. In addition, the spatial and temporal scale at which these constraints operate can affect natural resource management decisions. For instance, exceedingly wet weather conditions may make harvesting operations infeasible one season but not the next. Alternatively, the inherent level of productivity of a specific parcel of land for a specific tree species may preclude the development of a forest with a specific tree species composition. One parcel of land in western Washington State might be better suited to grow noble fir (*Abies procera*) than another piece, which then may affect how that land is managed. Finally, if a constraint relates to sustaining or protecting a resource for future generations of society (or families), some types of management activities might be restricted for quite a long period.

Common constraints for private landowners also include equipment, experience, and time limitations (Kvarda, 2004). A common constraint facing all types of landowners relates to the financial budget that is available. Budgets dictate whether a landowner or land manager has access to the financial resources necessary to allow them to perform management actions on the landscape. Without financial resources, a landowner cannot always successfully follow established management plans that, for example, may involve effectively preparing a site for tree planting activities, installing water bars to minimize soil erosion on forest roads, conducting midrotation fertilization treatments, installing duck boxes, creating seed food plots, paying annual land-use taxes, installing a new road gate, hiring a surveyor to locate and mark property boundaries, or paying the salaries of critical staff members. A general lack of interest in forestry can also be viewed as a constraint of private forest landowners. A lack of interest may be related to the age of the landowner and the length of the perceived planning horizon. Together, these may affect a landowner's level of motivation and interest in forestry activities (Selby et al., 2005).

Another common constraint that has been mentioned a few times in this chapter is the concern by private landowners for the welfare of future generations. This concern focuses on leaving forests and natural resources in such a state that landowners feel they have not deprived their children or their children's children (or society in general) of the same opportunities that the landowner enjoyed. A concern for future generations can be expressed in numerous ways. Historically, it involved only harvesting the amount of wood that a forest would grow annually, or hunting only the number of animals (e.g., deer or moose) expected to be born and survive during a typical year. This form of sustainability is moderately complex and includes commodity production (wood), economic (income), ecological (habitat), and social (aesthetic) dimensions. As an aside, famous examples where the sustainability of natural resources was disregarded and subsequently led to extinctions include the passenger pigeon (*Ectopistes migratorius*) and the flightless moa (e.g., *Megalapteryx benhami*). In the case of the passenger pigeon, there were billions of these birds across North America at the time of European colonization (Department of Vertebrate Zoology, National Museum of Natural History, 2001). Extensive hunting often with large guns, punt guns, capable of killing multiple birds in each shot, and habitat loss of the passenger pigeon in the 1800s led to a catastrophic decline in its population and, unfortunately, its extinction in the early 1900s. The flightless moa comprised a group of 11 species of flightless birds found throughout New Zealand before the arrival of humans in the 1200s. The loss of habitat and excessive hunting of the moa by the Māori population led to their extinction by the time European settlers arrived.

Forest landowners interested in managing their property for other natural resources, such as a specific wildlife species, are often constrained by the habitat requirements of the specific species. For instance, if forest landowners in the eastern United States were interested in managing their land for deer and rabbit populations, then they would have to maintain a certain percentage of their property as edge (transition) habitat. If they were interested in managing their land for an interior-dwelling wildlife species, then they would have to maintain a contiguous forest area of a specific size and structure. As suggested, aspects of habitat constraints can relate to whether a forest has the structure necessary for nesting, roosting, and foraging activities. For example, some management systems (e.g., for longleaf pine ecosystems) require repeated prescribed burning that, in turn, promotes herbaceous plant communities that act as suitable forage for certain species of wildlife.

Other constraints common to both small and large forest landowners involve local mill requirements for wood. For small landowners, this could relate to the tree species the mill desires as well as the minimum log dimensions. Sometimes small forest landowners are obligated to offer a *first right of refusal* to a local mill to buy their wood in exchange for assistance in reforesting and managing their property. Larger landowners have many of the same concerns, but they sometimes enter into supply agreements with local forest products industries that may require the landowner to provide a specific level of wood volume every year. This arrangement can be attractive to both a mill and the landowner because it provides a steady flow of inputs to a manufacturing process while generating a steady flow of revenue to the forest landowner. The trade-off is that this type of an arrangement can limit the management options available to a forest landowner during specific periods of time.

Summary

As you may have gathered, the objectives and constraints related to the management of forests and natural resources are strongly influenced by governmental, industrial, and societal goals. It should be no surprise that in many cases, social, economic, and environmental considerations have played important roles in the development of forest management objectives and constraints (Johann, 2007). Managing forests and their associated natural resources is highly dependent on the objectives of the entities that control the resources. As we have described in this chapter, public and private ownership of forestlands and their natural resources place different priorities on specific objectives.

Large private landowners, such as REITs or TIMOs, are primarily focused on profit maximization through timber production and real estate sales. Smaller nonindustrial private landowners may be interested in generating revenue and profits, but they may also have other objectives, such as creating recreational opportunities or maintaining a forest structure that they want to bequeath to their heirs. Public forests managed by federal and state agencies face numerous pressures from many different constituency groups, and this requires a focus on multiple objectives, which can complicate a planning process.

All forest landowners face limitations or constraints on the management of their forests and natural resources. The nature of these constraints can vary from those based on internal organizational policies to those that are required by law. Some of the most important constraints may be related to the need to generate revenue or to adhere to budget allocations that are available for implementing common forestry practices. Other constraints may involve managing a landscape to meet the habitat requirements of certain wildlife species. Whether perceived as well intentioned or as intrusive and overbearing, every constraint, including others not mentioned here, can limit achievement of the objectives of either public or private landowners.

Questions

(1) *National lands.* Write a short essay on the objectives and constraints of a forest owned or managed by a federal natural resource organization that is of interest to you. Describe whether the objectives and constraints are different from what you had expected or from what was discussed in this chapter. If this particular area of national lands is divided into different tracts or management areas, describe whether the objectives and constraints vary for each. Develop a short, professional PowerPoint presentation to illustrate the management situation of the forested area within the national lands and include a map of the area along with relevant pictures of the critical resources.

(2) *Local land trusts.* Create a list of local land trusts in the area where you live. Once you have a list, create a table that identifies differences and similarities between these land trusts. Describe in a short memorandum the land trust that controls the most land and the one that controls the least. In addition, provide a short discussion of conservation practices that are promoted with other forest landowners, and

describe the objectives and constraints associated with this landowner group. Discuss your findings with two or three of your colleagues.

(3) *Private forests.* If there are private forest landowners in your area of the world, locate one or two who are willing to discuss the management of their forests. Visit these people and determine the types of objectives and constraints that seem to guide their management actions. Develop a short report that briefly communicates your thoughts on this issue and provides one or two examples of specific private landowners. Finally, lead a discussion with a small number of other students who have undertaken the same assignment, and compare and contrast the objectives and constraints of the private landowners.

(4) *Industrial forests.* Select an industrial forest landowner who currently owns land in your area of the world. Through an interview with an employee of the organization, or through published information, determine the types of objectives and constraints that seem to guide their management actions. Describe how you think the objectives and constraints may

Continued

Q u e s t i o n s *(cont'd)*

have changed in the last two decades. Develop a short memorandum for your instructor that briefly communicates your thoughts on this issue and that provides one or two distinct examples. Finally, develop a five-minute oral presentation for your class that outlines what you have learned.

(5) *Communal forests.* Locate some information regarding a communal forest in Africa, Asia, or South America.

From your interpretation of the published information, outline the types of objectives and constraints that seem to guide their management actions. In addition, lead a discussion of communal forests with others who have selected a forest located on another continent. From this discussion, develop a short memorandum for your instructor that briefly communicates your combined thoughts on this issue.

References

Anderson & Yates Forest Consultants, Inc., 2010. Community Approaches to Forest Management across Canada: An Analysis of Current Community Forests. Anderson & Yates Forest Consultants, Inc., Corner Brook, Newfoundland.

Barron County Government, 2013. Barron County Forest 15 Year Plan. Barron County Government Center, Barron, W.I. www.barroncountywi.gov/index.asp?Type=B_BASIC&SEC={DA23A2FD-C65C-4B14-9471-0898923297B5}&DE={350204FC-C64D-4C08-803E-53DD99793BF0} (accessed 13.06.19).

Bettinger, P., Boston, K., Siry, J.P., Grebner, D.L., 2017. Forest Management and Planning, second ed. Academic Press, New York.

Binkley, C.S., 2007. The Rise and Fall of the Timber Investment Management Organizations: Ownership Changes in the U.S. Forestlands. 2007 Pinchot Distinguished Lecture. Pinchot Institute for Conservation, Washington, D.C. https://www.pinchot.org/files/Binkley.DistinguishedLecture.2007.pdf (accessed 14.06.19).

Brewer, R., 2003. Conservancy, the Land Trust Movement in America. Dartmouth College Press, Hanover, NH.

Bureau of Land Management, 2010. BLM and its Predecessors. U.S. Department of the Interior, Bureau of Land Management, Washington, D.C. https://web.archive.org/web/20141126221553/http://www.blm.gov/wo/st/en/info/About_BLM/History.print.html (accessed 13.06.19).

Bureau of Land Management, 2018. BLM Mission Statement. US Department of the Interior, Bureau of Land Management, Washington, D.C. https://www.blm.gov/about/our-mission (accessed 13.06.19).

Butler, B.J., 2008. Family Forest Owners of the United States, 2006. US Department of Agriculture, Forest Service, Northern Research Station, Newtown Square, PA. General Technical Report NRS-27.

Byrd, J., 2010. Shimek State Forest Management Plan. Iowa Department of Natural Resources, Farmington, IA.

Cheng, A.S., Danks, C., Allred, S.R., 2011. The role of social and policy learning in changing forest governance: an examination of community-based forestry initiatives in the US. Forest Policy and Economics 13 (2), 89–96.

Christoffersen, N., Harker, D., Lyman, M.W., Wyckoff, B., 2008. The Status of Community-Based Forestry in the United States: A Report to the US Endowment for Forestry and Communities. US Endowment for Forestry and Communities, Greenville, SC.

Cisek, J., 2018. Welcome to the Barron County Forests. Barron County Government Center, Barron, WI. https://www.barroncountywi.gov/index.asp?Type=B_BASIC&SEC=%7B65176268-78C8-4F86-8105-46F22DEE067A%7D (accessed 13.06.19).

Cubbage, F.W., O'Laughlin, J., Bullock, C.S., 1993. Forest Resource Policy. John Wiley & Sons, Inc., New York, NY.

Da Veiga Mendonça, C., 2010. A comparison of the management models of protected areas between China and the southern African region. Forestry Studies in China 12 (3), 151–157.

Daly-Hassen, H., Pettenella, D., Ahmed, T.J., 2010. Economic instruments for the sustainable management of Mediterranean watersheds. Forest Systems 19 (2), 141–155.

Department of Vertebrate Zoology, National Museum of Natural History, 2001. The Passenger Pigeon. Smithsonian Institution, Washington, D.C. https://www.si.edu/spotlight/passenger-pigeon (accessed 13.06.19).

Escobedo, F.J., Nowak, D.J., Wagner, J.E., Luz De la Maza, C., Rodríguez, M., Crane, D.E., Hernández, J., 2006. The socioeconomics and management of Santiago de Chile's public urban forests. Urban Forestry and Urban Greening 4 (3–4), 105–114.

Food and Agriculture Organization of the United Nations, 2015. Global Forest Resources Assessment 2015, Desk Reference. Food and Agriculture Organization of the United Nations, Rome, Italy.

Food and Agriculture Organization of the United Nations, 2016. Global Forest Resources Assessment 2015, How Are the World's Forests Changing? Food and Agriculture Organization of the United Nations, Rome, Italy.

Forest Preserves of Cook County, 2012. Mission and Vision. Forest Preserves of Cook County, River Forest, IL. http://fpdcc.com/about/mission-vision/ (accessed 13.06.19).

Forest Preserves of Cook County, 2016. The Early History of the Forest Preserve District of Cook County, 1869-1922. Forest Preserves of Cook County, River Forest, IL. http://fpdcc.com/about/history/ (accessed 13.06.19).

Gilges, K., 2000. The Nature Conservancy's forest bank: a market-based tool for protecting our working forestland. Corporate Environmental Strategy 7 (4), 371–378.

Glück, P., Avdibegović, M., Čabaravdić, A., Nonić, D., Petrović, N., Posavec, S., Stojanovska, M., 2010. The preconditions for the formation of private forest owners' interest associations in the Western Balkan Region. Forest Policy and Economics 12 (4), 250–263.

Goeschl, T., Igliori, D.C., 2004. Property rights for biodiversity conservation and development: an analysis of extractive reserves in the Brazilian Amazon. In: Horne, P., Tönnes, S., Koshela, T. (Eds.), Proceedings of the Conference on Policy Instruments for Safeguarding Forest Biodiversity—Legal and Economic Viewpoints, the Fifth International BIOECON Conference, Working Papers of the Finnish Forest Research Institute. Finnish Forest Research Institute, Vantaa Research Centre, Helsinki, Finland, pp. 144–156.

Grado, S.C., Grebner, D.L., Measells, M., Husak, A., 2006. Assessing the status, needs, and knowledge levels of Mississippi's governmental entities relative to urban forestry. Journal of Arboriculture and Urban Forestry 32 (1), 24–32.

HessenForst, 2018. Nachhaltigkeitsbericht für 2017. Hessen-Forst, Kassel, Germany. https://www.hessen-forst.de/publikationen/ (accessed 12.06.19).

Holmgren, E., Keskitalo, E.C.H., Lidestav, G., 2010. Swedish forest commons—a matter of governance? Forest Policy and Economics 12 (6), 423—431.

Hyde, W.F., Xu, J., Belcher, B., 2003. Introduction. In: Hyde, W.F., Belcher, B., Xu, J. (Eds.), China's Forests: Global Lessons from Market Reforms, Resources for the Future, Washington, D.C., pp. 1—26.

Institute for Energy Research, 2012. Marcellus Shale Fact Sheet. Institute for Energy Research, Washington, D.C.

Iowa Department of Natural Resources, 2019. Iowa's State Forests. Iowa Department of Natural Resources, Des Moines, IA. https://www.iowadnr.gov/Places-to-Go/State-Forests (accessed 13.06.19).

Johann, E., 2007. Traditional forest management under the influence of science and industry: the story of the alpine cultural landscapes. Forest Ecology and Management 249 (1—2), 54—62.

Karvinen, S., Välkky, E., Gerasimov, Y., Dobrovolsky, A., 2014. Northwest Russian Forest Sector in a Nutshell. Finnish Forest Research Institute, Joensuu Research Unit, Joensuu, Finland.

Kitchen, L., Marsden, T., Milbourne, P., 2006. Community forests and regeneration in post-industrial landscapes. Geoforum 37 (5), 831—843.

Klooster, D., Masera, O., 2000. Community forest management in Mexico: carbon mitigation and biodiversity conservation through rural development. Global Environmental Change 10 (4), 259—272.

Kvarda, M.E., 2004. "Non-agricultural forest owners" in Austria— a new type of forest ownership. Forest Policy and Economics 6 (5), 459—467.

Liu, D., Edmunds, D., 2003. Devolution as a means of expanding local forest management in South China. In: Hyde, W.F., Belcher, B., Xu, J. (Eds.), China's Forests: Global Lessons from Market Reforms, Resources for the Future, Washington, D.C., pp. 27—44.

MacFaden, S., 2015. Vermont Town Forests. University of Vermont, Burlington, Montpelier, VT.

Minang, P.A., Bressers, H.T.A., Skutsch, M.M., McCall, M.K., 2007. National forest policy as a platform for biosphere carbon management: the case of community forestry in Cameroon. Environmental Science and Policy 10 (3), 204—218.

Mustalahti, I., Nathan, I., 2009. Constructing and sustaining participatory forest management: lessons from Tanzania, Mozambique, Laos and Vietnam. Folia Forestalia Polonica, Series A 51 (1), 66—76.

Natural Lands Trust, 2019a. Growing Greener Communities. Natural Lands Trust, Media, PA. https://natlands.org/what-we-do/saving-open-space/growing-greener-communities/ (accessed 14.06.19).

Natural Lands Trust, 2019b. Find a Place. Natural Lands Trust. Media, PA. https://natlands.org/visit/ (accessed 14.06.19).

Nguyen, T.T., Bauer, S., Uibrig, H., 2010. Land privatization and afforestation incentive of rural farms in the Northern Uplands of Vietnam. Forest Policy and Economics 12 (7), 518—526.

Ní Dhubháin, A., Maguire, K., Farrelly, N., 2010. The harvesting behaviour of Irish private forest owners. Forest Policy and Economics 12 (7), 513—517.

Nygren, A., 2005. Community-based forest management within the context of institutional decentralization in Honduras. World Development 33 (4), 639—655.

Oregon Department of Forestry, 2018. Forest Practice Administrative Rules and Forest Practices Act, Chapter 629, Forest Practices Administration. Oregon Department of Forestry, Salem, OR.

Paul, S., Chakrabarti, S., 2011. Socio-economic issues in forest management in India. Forest Policy and Economics 13 (1), 55—60.

Pennsylvania Department of Conservation and Natural Resources, 2017. Shale Gas Monitoring Manual. Pennsylvania Department of Conservation and Natural Resources, Harrisburg, PA.

Pennsylvania Department of Conservation and Natural Resources, 2019. About DCNR. Pennsylvania Department of Conservation and Natural Resources, Harrisburg, PA. https://www.dcnr.pa.gov/about/Pages/default.aspx (accessed 13.06.19).

Queen's Printer, 2019. Forest and Range Practices Act. Queen's Printer, Victoria, British Columbia, Canada. http://www.bclaws.ca/Recon/document/ID/freeside/00_02069_01 (accessed 13.06.19).

Reeves, T., Mei, B., Bettinger, P., Siry, J., 2018. Review of the effects of conservation easements on surrounding property values. Journal of Forestry 116 (6), 555—562.

Rozelle, S., Huang, J., Benziger, V., 2003. Forest exploitation and protection in reform China, Assessing the impacts of policy and economic growth. In: Hyde, W.F., Belcher, B., Xu, J. (Eds.), China's Forests: Global Lessons from Market Reforms, Resources for the Future, Washington, D.C., pp. 109—133.

Sande, J.B., 2002. Restructuring and Globalization of the Forest Industry: A Review of Trends, Strategies and Theories. UNECE/FAO Team of Specialists on Forest Products Marketing & IUFRO Research Group 5.10.00. Forest Products Marketing & Business Development, Louisiana Forest Products Development Center, Baton Rouge, LA. Global Market Enhancement Articles, Others, Number 31.

San Francisco Planning Department, 2014. San Francisco Urban Forest Plan. San Francisco Planning Department, San Francisco, CA.

Savage, J., 2015. Dubuar Memorial forest, New York, United States of America. In: Siry, J.P., Bettinger, P., Merry, K., Grebner, D.L., Boston, K., Cieszewski, C. (Eds.), Forest Plans of North America. Academic Press, New York, pp. 79—86.

Sekher, M., 2001. Organized participatory resource management: insights from community forestry practices in India. Forest Policy and Economics 3 (3—4), 137—154.

Selby, A., Petäjustö, L., Koskela, T., 2005. Forests and Afforestation in a Rural Development Context: A Comparative Study of Three Regions in Finland. Finnish Forest Research Institute, Helsinki Research Unit, Helsinki, Finland. Working Papers of the Finnish Forest Research Institute 92.

Siry, J.P., Bettinger, P., Merry, K., Grebner, D.L., Boston, K., Cieszewski, C. (Eds.), 2015. Forest Plans of North America. Academic Press, New York.

Smith, W.B., Miles, P.D., Perry, C.H., Pugh, S.A., 2009. Forest Resources of the United States, 2007. US Department of Agriculture, Forest Service, Washington Office, Washington, D.C. General Technical Report WO-78.

Sunderlin, W.D., 2006. Poverty alleviation through community forestry in Cambodia, Laos, and Vietnam: an assessment of the potential. Forest Policy and Economics 8 (4), 386—396.

The Nature Conservancy, 2019. Who We Are. The Nature Conservancy, Arlington, VA. https://www.nature.org/en-us/about-us/who-we-are/ (accessed 14.06.19).

Thoms, C.A., 2008. Community control of resources and the challenge of improving local livelihoods: a critical examination of community forestry in Nepal. Geoforum 39 (3), 1452—1465.

TimberMart-South, 2019. TimberMart-South Market News Quarterly, 1st Quarter 2019 24 (1), 54—57 (Frank W. Norris Foundation, Athens, Georgia, USA).

Trinity County Resource Conservation District, 2010. Weaverville Community Forest Strategic Plan 2010-15. Trinity County Resource Conservation District, Weaverville, CA.

Troncoso, J., D'Amours, S., Flisberg, P., Rönnqvist, M., Weintraub, A., 2015. A mixed integer programming model to evaluate integrating strategies in the forest value chain—a case study in the Chilean forest industry. Canadian Journal of Forest Research 45 (7), 937—949.

University of Florida, IFAS Extension, 2014. Your Land Management Plan. IFAS Extension, University of Florida, Gainesville, FL. http://www.sfrc.ufl.edu/Extension/florida_forestry_information/forest_management/plan.html (accessed 12.06.19).

US Fish and Wildlife Service, 2019. ECOS Environmental Conservation Online System: Listed Animals. US Department of the Interior, Fish and Wildlife Service, Washington, D.C. https://ecos.fws.gov/ecp0/reports/ad-hoc-species-report?kingdom=V&king

dom=I&status=E&status=T&status=EmE&status=EmT&stat us=EXPE&status=EXPN&status=SAE&status=SAT&fcrith ab=on&fstatus=on&fspecrule=on&finvpop=on&fgroup=o n&header=Listed+Animals (accessed 12.06.19).

U.S. Forest Service, Lolo National Forest, 2010. Geospatial Data. U.S. Forest Service, Lolo National Forest, Missoula, MT. https://www. fs.usda.gov/detailfull/lolo/landmanagement/gis/?cid=stelprdb5 068292&width=full (accessed 16.06.19).

U.S. Forest Service, 2014a. About the Agency. U.S. Department of Agriculture, Forest Service, Washington, D.C. https://www.fs.fed.us/ about-agency (accessed 12.06.19).

U.S. Forest Service, 2014b. What We Believe. U.S. Department of Agriculture, Forest Service, Washington, D.C. https://www.fs.fed.us/ about-agency/what-we-believe (accessed 12.06.19).

U.S. Forest Service, 2018. Land and Resource Management Plan for the Coconino National Forest, Coconino, Gila, and Yavapai Counties,

Arizona. US Department of Agriculture, Forest Service, Southwestern Region, Albuquerque, AZ. MB-R3-04-31.

U.S. National Park Service, 2016. The National Parks: Index 2012-2016. U.S. Department of the Interior, National Park Service, Office of Communications and the Office of Legislative and Congressional Affairs, Washington, D.C.

Wear, D., Carter, D., Prestemon, P., 2007. The US South's Timber Sector in 2005: A Perspective Analysis of Recent Change. US Department of Agriculture, Forest Service, Southern Research Station, Asheville, NC. General Technical Report SRS-99.

Widmann, R.H., Balser, D., Barnett, C., Butler, B.J., Griffith, D.M., Lister, T.W., Moser, W.K., Perry, C.H., Riemann, R., Woodall, C.W., 2009. Ohio Forests: 2006. US Department of Agriculture, Forest Service, Northern Research Station, Newtown Square, PA. Resource Bulletin NRS-36.

CHAPTER

4

Forest products

Wood is an organic material that, with the assistance of solar energy and the process of photosynthesis, is created from carbon (C), hydrogen (H), and oxygen (O). During the spread of early European colonists in North America, forests provided a wide range of wood products, many of which were essential for survival. Settlers relied on wood to create fences and houses and used wood as fuel for cooking and providing heat during cold weather. Forests also provided the settlers' animal stock with a source of free feed, which may have consisted of acorns, beechnuts, chestnuts, grass, roots, and shoots of various plants (Williams, 1989). More recently, there has been a considerable amount of debate over the use and future of wood and paper in our daily lives, including how we can conserve and recycle these resources, and how we might substitute other products for these resources (e.g., steel studs for wood framing).

Some resources we use in our daily lives are renewable (i.e., we can grow more), and some are nonrenewable (once used they cannot be reproduced); wood products are considered a renewable resource. This implies that, once harvested, wood products can once again be grown on the same plot of land. Our ability to shape and create products from wood, and the cost-efficiency of doing so, facilitates a wide application of wood products in our daily lives. This chapter is designed to provide an overview of the broad array of products that can be derived from wood, trees, or

forests; we simply call these *forest products*. Forest products can be directly associated with the production of lumber or paper, or indirectly associated with the existence of forests. This latter group might be considered *nontimber forest products* (NTFPs), and we allude to a broad set of these in this chapter. Additional chapters in this book are designed to examine certain nontimber values of forests in more detail. These include chapters focusing on wildlife habitat (Chapter 5), ecosystem services (Chapter 6), and recreation (Chapter 7). We briefly address these types of NTFPs in this chapter. Upon completion of this chapter, readers should have acquired an understanding of

- the wide variety of products produced from forests around the world;
- the differences between traditional forest products and NTFPs;
- the historical and modern uses of different products from forests; and
- the relationships between forest inputs, forest management, and forest outputs.

Wood is a material produced by trees that is stored in the stem(s) and branches just inside the cambial layer (Nieuwenhuis, 2010). Trees, of course, produce bark, fruit, leaves, and oxygen that are available outside of the cambial layer and are useful in controlling such things as air temperature, and their roots reduce soil erosion, which are a value to society. Since the late 2000s, wood removed from forests for industrial purposes (i.e., the creation of other products such as paper or lumber) was greatest in North America and Europe. However, around the world, variability in the production of wood resources for human consumptive needs is striking (Table 4.1). For example, North America and Europe combined accounted for about 38% of the total wood removals in 2011 (Food and Agriculture Organization of the United Nations, 2015), much of which was used to make lumber and paper products. The term *roundwood* includes tree stems, with or without bark, that have been delimbed (Nieuwenhuis, 2010). *Fuelwood* refers to wood cut into short lengths or chips that is used to generate heat for personal warmth or cooking. Wood consumed for fuelwood purposes was greatest in Africa and Asia, where it is a

basic necessity as it is involved in food production and may therefore be in high demand. The production of wood for fuelwood purposes on these two continents accounts for nearly 74% of the total forest extractions (Food and Agriculture Organization of the United Nations, 2010). You may have already obtained a sense for this as you read the country profiles in Chapter 2. The types of products that can be derived from trees or forests vary widely and can arise from wood or from roots, leaves, bark, or plants and animals that depend on the existence of the trees. The products range from chemicals and tissue to lumber and the plants and animals that are intimately tied to the existence of forest vegetation. Given uncertainties in the price of petroleum-based products today, options are also being explored for deriving liquid fuels from whole trees (Ganguly et al., 2018). Products such as wood can be an integral part of a management process that produces refined goods (Fig. 4.1). We begin our discussion of forest products with wood-based commodities, then transition to the vast array of nontimber products facilitated by the existence of forests.

TABLE 4.1 Top 10 producers of industrial roundwood and fuelwood.

Country	World share (%)
Industrial roundwood	
United States	27.3
Canada	12.1
Russian Federation	7.6
Brazil	6.6
Sweden	4.3
China	3.6
Germany	3.3
Finland	3.1
India	2.6
Chile	2.0
All other countries	27.5
Fuelwood	
India	18.5
Brazil	8.7
Ethiopia	7.7
Indonesia	6.1
Democratic Republic of the Congo	5.8
Nigeria	5.0
China	4.5
United States	3.6
Russian Federation	3.6
Uganda	3.0
All other countries	33.5

Food and Agriculture Organization of the United Nations (2010).

4.1 Commodities

A *commodity* is a product that has no brand identification but has commercial value or a product transported as part of acts of commerce (U.S. Forest Service, 1989). The development of wood product commodities by humans can require milling, chipping, grinding, tearing, or chemically decomposing trees into smaller pieces and perhaps building a larger product from these pieces. The course we take in describing commodities begins with solid wood products (e.g., lumber and others) and progresses to nonsolid products that fall into the following groups: pulp and paper, composites and engineered wood, chemicals, and residues. When harvested and brought to a mill, woody material (e.g., logs and chips) is measured in a form specific to a mill or a region. These include cubic units (feet [ft] or meters [m]), weights (pounds [lb], kilograms [kg], tons, or metric tons [t]), or other solid measures such as cords (theoretically 3.5 cubic meters [m^3]; 128 cubic feet [ft^3], although a standard cord contains about 2.1–2.6 m^3 [75–90 ft^3] of wood and bark), or board feet (1 inch [in]; 2.5 cm [cm] thick, 12 in [30.5 cm] wide, and 12 in [30.5 cm] tall), or more commonly thousand board feet or *MBF*. *Timber* is typically a category of solid standing or felled wood that will eventually have a use other than for fuel or pulping (Nieuwenhuis, 2010). A *log* can be defined as the stem or limb of a tree that, after felling and delimbing, is suitable for a manufacturing process. Other terms such as *pulpwood*, *sawtimber*, *chip-n-saw* (southern United States), *plywood* or *veneer logs*, *poles*, *bolts* (or *billets* or *blocks*), *sticks* (1.2–2.4 m or 4–8 ft long), and, lately, *woody biomass* have also been used to describe

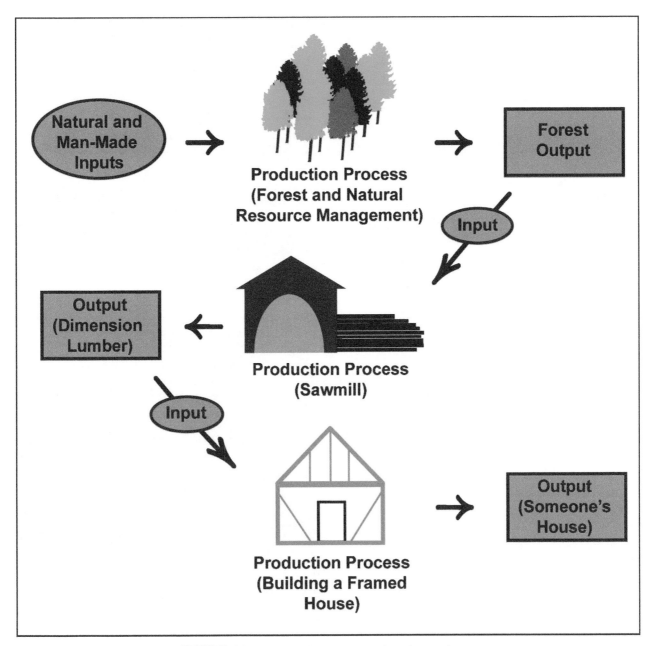

FIGURE 4.1 Forest production process from forest to house.

solid wood products. Sawtimber products can be further differentiated by the grade of wood (peeler, No. 1 sawmill, No. 2 sawmill, and so on), which is based on the diameter of the log, the presence of defects or sweep, and other parameters. Trees or logs that are of a suitable size for a manufacturing process but are unusable due to damage imposed during logging or natural defects, are often called *culls* or *cull wood* (Nieuwenhuis, 2010).

4.1.1 Solid wood products

In comparison to the alternatives (steel and other products), wood is relatively light, depending on the tree species, easy to handle, and, when it has been dried, cost-effective to transport over long distances. One of the major advantages of using wood to build items of value to human society, compared to the alternatives, are that pieces of solid wood can easily be fastened together with glue, nails, staples, or screws. Solid wood is also a poor transmitter of electricity, heat, and sound, and is therefore a decent insulating material. Wood also has the ability to absorb a certain amount of shock and vibration and, unlike some metals, does not rust. Further, in many cases, wood can hold paints and other finishes relatively well owing to its porous nature. All of these valuable qualities of wood have been known for quite

FIGURE 4.2 Stacked lumber used by a U.S. Naval mobile construction battalion in Helmand Province, Afghanistan. *Patrick W. Mullen III, through Wikimedia Commons.*
Image Link: https://commons.wikimedia.org/wiki/File:US_Navy_090527-N-8547M-107_Construction_Electrician_3rd_Class_Olatunde_Lawal,_assigned_to_Naval_Mobile_Construction_Battalion_(NMCB)_5,_directs_Utilitiesman_3rd_Class_Patrick_Krupa_as_he_drives_a_forklift.jpg
License Link: Public Domain

FIGURE 4.3 Edging and molding products for interior home use, Huntsville, Alabama, United States. *Christina E.H. Cooper.*

FIGURE 4.4 A campfire facilitated by fuelwood. *Dirk Beyer, through Wikimedia Commons.*
Image Link: https://commons.wikimedia.org/wiki/File:Campfire_4213.jpg
License Link: https://creativecommons.org/licenses/by-sa/3.0/deed.en

some time (Panshin et al., 1950) and contribute for wood lasting as a building material. While fuelwood is one of the largest uses of solid wood in developing countries, perhaps the single largest use of solid wood products in industrialized countries is in the development of houses and buildings (Youngquist and Fleischer, 1977). Some of the common solid wood products derived from trees include lumber products (Fig. 4.2) of various sizes used for building and construction purposes. In the United States, the 2 × 4 (a little less than 2 in thick and 4 in wide [due to finishing processes] and in various lengths) is perhaps the most common construction product, mainly for the framing of walls. Other dimensions of lumber (e.g., 2 × 6, 2 × 8, and 4 × 4) are also common in North America for roof and floor support necessary for houses and businesses. In summary, a wide variety of board dimensions are milled around the world for numerous purposes. Some of these are standard rectangular shapes, while others are milled along one edge for ornamental or decorative purposes (Fig. 4.3).

Fuelwood (or *firewood*) has been an important solid wood product during the developmental stages of nations and continues to be an important source of heat in rural areas. Technically, fuelwood includes logs and other minor portions of trees. Firewood for heating and cooking purposes (Fig. 4.4) was among the most important forest products in the early development of North America. Among other conifers, pitch pine (*Pinus rigida*) splinters and knots were once commonly used to start fires or to convey fire from one place to another. Fuelwood is now mainly important in rural communities and has been supplanted by coal, electricity, and

petroleum products for these purposes in many urbanized areas. Youngquist and Fleischer (1977) note that the demand for fuelwood was once so high that in 1738 some people in the state of Rhode Island believed that there were no remaining trees suitable for this purpose, as all had been cut. In the early developmental period of North America, firewood was important for household cooking and heating purposes and for various manufacturing processes. For example, for salt extraction from seawater, one cord of fuelwood was needed to produce about 11 bushels of salt. In developed countries today, fuelwood is still needed to support specialized industries. For instance, in a few local pizzerias near Rome, Italy, the branches of European chestnut (*Castanea sativa* Mill.) are used as a special fuelwood in the baking of pizza (Fig. 4.5).

FIGURE 4.5 European chestnut (*Castanea sativa* Mill.) firewood used for baking pizza, just to the south of Rome, Italy. *Donald L. Grebner.*

FIGURE 4.6 Charcoal pile ready for ignition. *Pete Bettinger.*

FIGURE 4.7 Replacement utility poles for tornado-damaged areas in Mena, Arkansas, United States. *Win Henderson, through Wikimedia Commons.*
Image Link: https://commons.wikimedia.org/wiki/File:FEMA_-_40831_-_A_utility_crew_delivers_new_poles_in_Arkansas.jpg
License Link: Public Domain

FIGURE 4.8 Piles are used to support the footbridge to Guards Club Island, in Maidenhead, Berkshire, England. *Andrew Findlay, through Wikimedia Commons.*
Image Link: https://commons.wikimedia.org/wiki/File:Guards_Club_Island_Bridge.jpg
License Link: https://creativecommons.org/licenses/by-sa/3.0/deed.en

Charcoal (Fig. 4.6), which can be considered as either a solid wood product or a composite product, is made by slowly burning wood in a closed environment using a restricted air supply. Charcoal is generally used for outdoor food preparation (grilling or barbecuing) activities in North America but is a common and important resource for other areas of the world. For example, in Tanzania, as much as 80% of the urban domestic fuel usage is based on charcoal made in the rural communities of the East African country (Sunseri, 2009). Historically, charcoal was once widely used in ironmaking (and is still commonly used for this purpose in Brazil today) and glass-making processes, and in the creation of gunpowder and black paint (Youngquist and Fleischer, 1977).

Round timbers are a group of solid wood products that include poles, piles (pilings), and posts. Poles are long, chemically treated logs that are partially sunk into the ground (although the majority of the log is above ground) and support cable, electrical, and telephone wires. Poles are classified according to their length, strength, and the corresponding amount of cracks, sweep, and knots they contain (Dilworth, 1970). Southern pines and Douglas-

fir (*Pseudotsuga menziesii*) are commonly used to make utility and transmission poles (Fig. 4.7) in the United States, although American chestnut (*Castanea dentata*) was once the primary tree species used for this purpose. Some of the wood used in the development of railroad trestles and bridges can also technically be considered round timbers. In assessing the potential of a tree for the development of a pole, specific parameters are employed that assess the presence, density, and quality of checks, shakes, splits, insect damage, knot size, knot angle, and crooks (Panshin et al., 1950). Piles are heavier timbers that provide support for the foundations of buildings, wharves, bridges (Fig. 4.8), and other structures.

FIGURE 4.9 Fence posts lining a property boundary in eastern Oregon, United States. *Kelly A. Bettinger.*

FIGURE 4.10 Virginia rail fence located in the George Washington National Forest in Virginia, United States. *Ed Brown, through Wikimedia Commons.*
Image Link: https://commons.wikimedia.org/wiki/File:Spit_rail_fence_Sherando_Lake.jpg
License Link: Public Domain

Piles are similar to poles but, because they are driven further into the ground, they must be structurally stronger than utility poles to endure the shock and pressure exerted upon them during the driving process, as well as the weight of a large structure once in place (Dilworth, 1970). Posts are principally used in farm and rural areas as fencing material. Fence posts (Fig. 4.9) are usually about 2 m (6.6 ft) in length and are fashioned from tree species that have a relatively slow wood decay rate. In North America, tree species with slow decay rates include black locust (*Robinia pseudoacacia*), eastern red cedar (*Juniperus virginiana*), osage orange (*Maclura pomifera*), redwood (*Sequoia sempervirens*), and western red cedar (*Thuja plicata*). In addition to posts, some types of wooden fences require rails that are about 3 m (9.8 ft) in length. Therefore, depending on the type of fence desired, a considerable amount of wood may be required. For example, the Virginia rail fence (which originated in Scandinavia), a set of interlocking rails placed in a zigzag arrangement on a landscape (Fig. 4.10), may require over 2000 rails per kilometer (km; 3200 per mile [mi]) (Youngquist and Fleischer, 1977).

Other types of solid wood products include railroad ties and mine timbers. Railroad ties include a group of products that are used as crossties, bridge ties, or switch ties. Railroad crossties are the treated timbers that most are probably familiar with; they support the railroad beds (Fig. 4.11). The bending strength and the hardness of the wood are two of the most important properties when considering tree species for railroad tie development. Chemical treatments (historically, creosote-based preservatives) are usually required to reduce the wood decay rate and extend the service life of a tie from about 6 years to nearly 30 years (Webb, 2005). Among the wide variety of tree species that have been used for railroad ties in North America are black locust, gums, elms, hackberry (*Celtis occidentalis*), hickories, maples, oaks (nearly

FIGURE 4.11 Wooden railroad ties that are part of a CSX railroad track in Bay St. Louis, Mississippi, United States. *Stephen C. Grado.*

two dozen species), ponderosa pine (*Pinus ponderosa*), southern pines, western hemlock (*Tsuga heterophylla*), western white pine (*Pinus monticola*), and many others. One kilometer of railroad track is said to require 1600 to 2000 ties (2600–3300 ties per mi) (Panshin et al., 1950); however, many more may be needed in switching stations or terminals (Fig. 4.12). Mine timbers (Fig. 4.13) are the supporting materials used to reduce the potential for caving in underground shafts, tunnels, and other openings. The strength and durability of the wood are two of the most important properties when considering tree species for mine timber development.

Shingles are commonly used to cover the roofs and sides of houses (Fig. 4.14) and other buildings. A shingle is a thin piece of wood with parallel edges that are thicker at one end than the other. Shakes are split boards

FIGURE 4.12 Wooden crossties on Norfolk Southern company railroad tracks adjacent to the Alto interlocking tower in Altoona, Pennsylvania, United States. *Stephen C. Grado.*

FIGURE 4.14 Wood shingles lining the lower 1 to 2 m of a house in Corvallis, Oregon, United States. *Kelly A. Bettinger.*

FIGURE 4.13 Wooden timbers supporting the roof of a mine in Leadfield, California, United States. *Daniel Mayer, through Wikimedia Commons.*
Image Link: https://commons.wikimedia.org/wiki/File:Mine_at_Leadfield.JPG
License Link: https://creativecommons.org/licenses/by-sa/3.0/deed.en

FIGURE 4.15 Red wine oak barrels from the Barbera d'Asti wine region in the Piedmont of northwestern Italy. *Jodi B. Roberts.*

that are usually the same thickness at both ends. Certain straight-grained woods provide a source of solid wood material from which shingles and shakes can be made. Depending on the species used, shingles can have a useful life of 25—35 years. In North America, shingles and shakes have been developed from the heartwood of American chestnut (historically), baldcypress (*Taxodium distichum*), eastern white pine (*Pinus strobus*), northern white cedar (*Thuja occidentalis*), Port-Orford-cedar (*Chamaecyparis lawsoniana*), southern white cedar (*Chamaecyparis thyoides*), redwood, southern pines, western red cedar, and others. Species are selected on the basis of the appearance, durability, stability, strength, and weight of the wood (Panshin et al., 1950). For example, western red cedar is commonly used for shakes and shingles due to its straight grain, dimensional stability,

and durability, the latter of which arises from its heartwood extractives (Stirling, 2010).

The development of barrels, casks, and kegs, collectively known as cooperage, can involve the use of solid wood products (Fig. 4.15). Cooperage includes a wide variety of wood containers that consist of staves (the generally curved wood that forms the sides) and heads (the flat top and bottom) that are securely fastened together with hoops. Similar containers have been created to hold flour, nails, tobacco, and other products. Beer, whiskey, and wine barrels are referred to as *tight cooperage*, while nail kegs and others are referred to as *slack cooperage*. Large tanks or vats are referred to as *heavy cooperage*, and some of these designed to hold water may be found on top of buildings in larger cities or on elevated stands beside rural homesteads. Evidence of cooperage dates back at least 2500 years, and in the colonial period of North America these containers were necessary for the trade of naval stores, oil, rum, and other products (Panshin et al., 1950). In the production of liquor, wine, and other products, barrels and casks are commonly made from a variety of ashes, beeches, gums, red oaks, and white oaks. For example, white oak (*Quercus alba*) is commonly used in the United States and rebollo oak (*Quercus pyrenaica*) is commonly used in Spain to make wine barrels (Adame et al., 2006). In North America, larger tanks and vats are typically constructed with baldcypress, Douglas-fir, redwood, and white oak.

Solid wood has long been used in the development of tools including billhooks, draw knives, dowels, foot axes, hammers, handles, mallets, needles, pitchforks, rulers, sickles, support braces, and wheels (Figs. 4.16 and 4.17). Wooden-wheeled vehicles have included chariots, wagons, bicycles, and the structural frames of cannons (Fig. 4.18). Solid wood products have been used in warfare since early human existence and include weapons such as battering rams, bows and arrows, catapults, javelins, quarterstaffs (a pole weapon), spears, and stone-headed maces. During the medieval era, wood was used to construct trebuchets and catapults, which were used to toss large objects over considerable distances. Other weapons with solid wood components included axes, ballistas, eku (blades on staffs), kanabō (wooden clubs), macana (sword-like weapons of Central and South America), mere (broad-bladed clubs of New Zealand), nunchaku (nunchucks or chain sticks), siege towers (a tower on wheels used to attack forts),

FIGURE 4.17 Chopping and digging implements from Duffy Neubauer's private collection, Starkville, Mississippi, United States. *Donald L. Grebner.*

tomahawks, and various polearms (long shafts with metal weapons on the end, of European use). During the colonial period of the United States, the British Navy reserved large eastern white pine trees as potential masts for their ships. During the 18th and 19th centuries, wood such as white oak was used in the manufacture of muskets, repeating rifles, merchant ships, ironsides (wooden-hulled heavy naval frigates that could resist cannon-fire), frames for artillery pieces, trains, and various other animal-driven vehicles. During the American Civil War at the battle of Vicksburg, wooden mortars were used by the Union Army against besieged confederate soldiers (Hickenlooper, 1888). During the modern era, wood from Sitka spruce (*Picea sitchensis*) was used in the manufacture of propellers of World

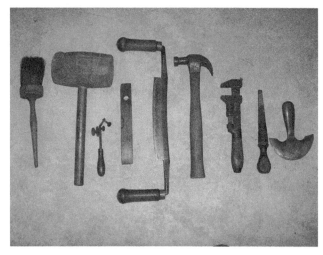

FIGURE 4.16 Wooden tools, or tools with wooden components, from Duffy Neubauer's private collection, Starkville, Mississippi, United States. *Donald L. Grebner.*

FIGURE 4.18 American Civil War cannon from Duffy Neubauer's private collection, Starkville, Mississippi, United States. *Donald L. Grebner.*

War I airplanes and the nose cones of Trident missiles (The New York Times, 1918; Siuru, 1977). Solid wood products were also once used extensively for ship-building purposes. The frames of early American ships were made from white oak and American chestnut, planking was made from southern pine and eastern white pine, and nails were made from black locust or hickory steeped in tar (Youngquist and Fleischer, 1977). Solid wood, along with other composite wood products, was also used in making the patrol torpedo (PT) boats that were used to attack large surface vessels during the Pacific theater of World War II (National WWII Museum, 2007).

Many common indoor furnishings (beds, cabinets, ceiling fans, chairs, dressers, flooring, shelves, and tables) are made from solid wood products (Figs. 4.19–4.25). In exploring some of the wide variety of uses of solid wood products (Table 4.2), you may have noticed that baskets, kitchen implements, tool handles, and even shoes have been and currently are made from solid wood products.

There is an entire class of wood called the tonal woods, which are used to make musical instruments. Often these use a variety of rare woods such as ebony (*Diospyros* spp.) or clarinets and oboes often made of African blackwood (*Dalbergia melanoxylon*). Musical instruments, such as banjos, guitars, and pianos, are made from pieces of solid wood (Figs. 4.26 and 4.27). These can become quite valuable and rare. Unfortunately, some of these woods are often traded illegally. In the United States, the Lacey Act (1900) prohibits the

FIGURE 4.20 Dining room chair made of solid wood. *Donald L. Grebner.*

FIGURE 4.21 Large table made from thick wooden boards. *Donald L. Grebner.*

FIGURE 4.19 Bedroom night table made of solid wood. *Donald L. Grebner.*

FIGURE 4.22 Wooden chest made for storing blankets, clothing, and other personal items. *Donald L. Grebner.*

FIGURE 4.23 Wooden stand designed to hold a television and other electronic devices. *Donald L. Grebner.*

FIGURE 4.25 Ceiling fan, with blades made of solid wood. *Donald L. Grebner.*

importation of wood from illegal sources, and some guitar manufacturers have been fined and had some wood confiscated.

Other items derived from solid wood include the bases for trophies associated with various sports awards, sports equipment and memorabilia (Fig. 4.28), jewelry boxes (Fig. 4.29), and small wooden collectibles such as carved animals or people representing historical cultures (Fig. 4.30). Boxes, clothespins, handles, and washboards are all commonly made from tree species such as birch (*Betula* spp.). Eastern red cedar, commonly found on abandoned farmland in eastern North America, has

FIGURE 4.24 Cabinet designed to store china and glassware. *Donald L. Grebner.*

TABLE 4.2 Examples of solid wood products and wood commonly used in their construction.

Product	Wood
Baseball bats	Ash, maple
Canoes	Cedar, pine
Clogs	Alder, beech, birch, poplar, sycamore, willow
Cricket bats	White willow (*Salix alba*)
Drumsticks	Hickory, maple, oak
Flooring for basketball, racquetball, squash, and volleyball courts	Maple
Furniture	Cherry, maple, oak, pine, poplar, walnut, and others
Guitars	Cedar, ebony, maple, rosewood, spruce
Gunstocks	Birch, maple, mesquite, myrtle, walnut
Ice hockey sticks	Ash, hornbeam, maple, willow, yellow birch
Older golf clubs	Maple, willow
Pencils	Incense-cedar, juniper
Pianos	Basswood, maple, spruce
Pool cues	Ash, maple
Sailing masts	Fir, pine, spruce
Tool handles	Apple, ash, hickory, maple, osage orange
Violins, cellos	Ebony, maple, poplar, spruce, willow
Wine/whiskey barrels	Oak

FIGURE 4.26 Piano made from numerous pieces of solid wood. *Donald L. Grebner.*

been used to make chests, coffins, and furniture. Many of the mechanical devices in the 18th century (balers, gristmills, and spinning and weaving equipment) also included a high content of solid wood.

There are numerous other solid wood products that we have contact with on a daily basis. Solid wood products are extensively used for the development of recreational and hunting equipment, from fishing poles to bird decoys, bows and arrows, gunstocks, and spears. Canoes were once made from the bark of paper birch (*Betula papyrifera*). A wide variety of small toys are made from solid wood, as are baskets for the collection of plants and fruits, and toothpicks and matches (for obvious purposes). In North America, eastern red cedar and incense-cedar (*Libocedrus decurrens*) are often used to make pencils (Fig. 4.31), and in Europe black alder (*Alnus glutinosa*) is commonly used for this purpose.

FIGURE 4.27 Wooden parts of a Gibson five-string banjo. The upper left corner shows the fingerboard made of rosewood (*Dalbergia spp.*), the upper right corner shows the banjo bridge made of European maple (*Acer pseudoplatanus*), the lower left image shows the resonator made from sugar maple (*Acer saccharum*), and the lower right image shows another view of the resonator. *Joshua P. Adams.*

FIGURE 4.28 Trophies with wooden bases and a wooden baseball bat. *Donald L. Grebner.*

FIGURE 4.30 Wooden knickknacks. *Donald L. Grebner.*

FIGURE 4.29 Wooden box for storing jewelry. *Donald L. Grebner.*

FIGURE 4.31 Wooden pencils. *Pete Bettinger.*

4.1.2 Pulp and paper

Paper production is a very old process that began in ancient China nearly 1800 years ago. As time progressed, paper production involved the use of various plant fibers, linens, rags, and other tissues. The production of paper involving wood pulp began around 1840 in Germany and around 1860 in the United States (Panshin et al., 1950). The development of wood-based pulp and paper involves the extraction of wood fiber from wood chips. Tree species typically used in pulp and paper production include many coniferous and deciduous species such as eucalypts (*Eucalyptus* spp.), firs, maples, pines, spruces, and sweetgum. Various chemical, mechanical, and thermal processes or combinations of these can be used in the disaggregation of solid wood and the development of wood pulp. While wood cell walls also contain pectin, proteins, and other trace materials, the objective of pulping is to break down or separate wood fibers into the constituent components of cellulose, lignin, and hemicellulose. These three components account for nearly 80% of the plant biomass on Earth (Bidlack et al., 1992; Persson et al., 2006). Cellulose fibers (which are linear chains of glucose units) comprise about 40% to 50% of wood volume and are not only used in the

Pallets created for the storage and transfer of goods are made from solid wood products, perhaps from maples or oaks. Various eating and drinking utensils (Figs. 4.32 and 4.33) and wooden bowls commonly used for preparing salads and desserts (Fig. 4.34) are made from solid wood. In Japan, wood remnants from the manufacture of other products may be used for making chopsticks (Okazaki, 1964). Chopsticks are also made in Georgia, United States, from the wood of sweetgum (*Liquidambar styraciflua*) trees. Further, small *diameter breast height* (DBH; 5–30-cm or about 2–12-in diameter) hardwood trees are often used as poles for the development of rustic beach huts in tourist areas such as the Yucatan Peninsula of Mexico (Racelis and Barsimantov, 2008). While we have covered many of the typical solid wood products, the complete set of tools and products that humans use containing solid wood is much more extensive.

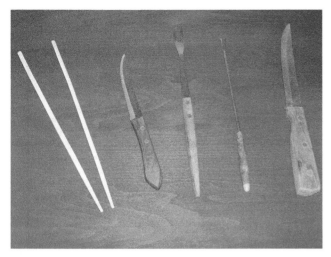

FIGURE 4.32 Eating and cooking utensils with solid wooden handles. *Donald L. Grebner.*

FIGURE 4.34 Wooden bowls and utensils made from kauri (*Agathis* spp.) in New Zealand and *Acacia* spp. in the Philippines. *Donald L. Grebner.*

production of pulp and paper products but also often used in the development of adhesives and films. Cellulose is also often used as a binding agent, emulsifying agent, thickener, or stabilizer in food production (Stern, 2009). Highly purified cellulose has been used in the development of cellophane, lacquers, rayon, sausage casings, and sponges (Panshin et al., 1950). Lignin is an aromatic amorphous molecule present in the cell walls of wood fibers and, while it is said to be of positive benefit in the development of some products such as fiberboard (a composite product), it is regarded to be undesirable for pulp and paper production; thus, processes to delignify wood pulp are common in paper manufacturing (Fromm et al., 2003). Hemicelluloses are polysaccharides primarily located in the secondary cell walls of wood fibers. Although typically degraded during the pulping process, there is interest in using these renewable materials for novel products (Gabrielii et al., 2000).

FIGURE 4.33 Kåsa Camping cups made from birch burl wood in Östersund, Sweden. *Donald L. Grebner.*

The United States, Canada, China, Japan, and several European and Scandinavian countries are the main producers of wood pulp, wood flour, and paper products. *Wood pulp* is used for a variety of end-products that include books, diapers, paper towels, napkins, newspapers, magazines, maps, and toilet tissue, to name just a few (Fig. 4.35). Even automobile and motorcycle air filters and oil filters contain wood pulp. *Wood flour* is a finely ground wood that has the texture of sand or sawdust and is technically a form of wood pulp. Wood flour is often used as a filler in the development of linoleum and plastics and has also been used in the development of explosives.

Paper can be developed from wood pulp, and, when produced, it can be described by a wide variety of weights, colors, and thicknesses. A typical bundle, or *ream*, of printing or copy paper contains 500 sheets. Common types of paper used in our everyday lives include kraft paper, coated paper, construction paper, and newsprint (Fig. 4.36). *Kraft paper* is produced through a chemical process and the end-product has a high tensile strength (i.e., the maximum stress from stretching or pulling that it can endure before failing). Kraft paper is often used in the development of brown paper bags, such as sandwich or grocery bags. *Coated*, or *glossy, paper* has calcium carbonate (or kaolinite clay) applied to it to facilitate high-quality printing processes. Coated paper is often used for the development of calendars, magazines, and books such as this one. *Construction paper* comes in many different colors, is generally coarse and unfinished, and is typically used for artistic projects. Although *newsprint* is a low-cost, lightweight paper, it is strong enough to be used in high-speed printing systems and durable enough to accommodate the application of various colors during the printing process (Noda, 2002).

FIGURE 4.35   Various paper products. The upper left corner shows paper towels, the upper right corner shows toilet paper, and the lower left and lower right images show facial tissues. *Donald L. Grebner.*

Corrugated containers (Fig. 4.37) are made from liner-board and corrugated medium (containerboard), a thick paper manufactured using a process similar to that of kraft paper. The *corrugated medium* (the center part that is fluted) is affixed to two relatively tough and durable outer sheets (*linerboard*) to form a relatively strong material (or *corrugated board*) that can be manipulated into various shapes for the purpose of shipping products around the world. Sheets of containerboard are also used to create packaging for cosmetics, fruit and vegetable products, mechanical apparatus, medicines, pottery, and other miscellaneous goods (Noda, 2002).

Many of the products described here are made with *virgin pulp*, derived directly from a tree or derived as a residue produced from sawmilling. However, recycled pulp can also be mixed with virgin pulp to produce products with similar characteristics. Recycled newspapers, magazines, corrugated containers (boxes), business communications papers, and trimmings from mills and printing offices are considered *wastepaper*. It

is often collected in conjunction with voluntary or mandatory recycling programs, although it can otherwise be delivered to, and purchased by, centralized collection centers. In some areas of the world, wastepaper can be exchanged for other products, such as toilet paper (Noda, 2002).

4.1.3 Composites and engineered wood

Composite wood products, sometimes referred to as *engineered wood products*, are based on wood pieces or wood fibers that have been transformed from the original log, mixed or coated with an adhesive or special glue, then recombined to create the desired product. In contrast with paper and pulp products, the final form of composites or engineered wood products are boards or sheets. In addition, the original solid wood is usually not reduced to individual cellulose or hemicellulose fibers, as may occur in pulp and paper manufacturing processes. The first wood composites, a form of

FIGURE 4.36 Various types of paper. The upper left corner shows printing paper, the upper right corner shows labels and photo quality printing paper, the lower left corner shows notebook paper, and the lower right corner shows construction paper. *Donald L. Grebner.*

FIGURE 4.37 Typical cardboard box with an inset showing a cross-section of the corrugated interior and linerboard exterior. *Pete Bettinger.*

wet-processed fiberboard, were created near the end of the 1800s, and the first hardboard product was created in 1924 (Schniewind, 1989). Composite products developed later include oriented strand board, particleboard, plywood, and steam-pressed scrim lumber. If you examine them closely, you will find that a large variety of household furniture products are created from composites and engineered woods. Composites are made from many of the commercially important tree species that we have previously mentioned. Alternative materials used for composite wood panels include bamboo, rice straw, and rubberwood (*Hevea brasiliensis*) (Jarusombuti et al., 2009). Some composites and engineered woods (plywood and particleboard, in particular) are strong relative to their weight and are therefore very useful in the development of buildings and inexpensive furniture.

Plywood is a type of engineered wood made by adhering several layers of veneer together. The process of veneering dates back nearly 3500 years to the time of Egyptian pharaohs. A *veneer* is a sheet of wood of uniform thickness that has been peeled, sliced, lathed, or sawn from a solid wood product, usually a high-quality log. In the plywood manufacturing process, low moisture content veneer layers are oriented at right angles to one another, a suitable adhesive is applied, heat is sometimes applied, and the layers are pressed and bonded together. The 90-degree orientation of each veneer layer enhances the strength of the plywood. Urea-formaldehyde resins are the most widely used adhesives for bonding together wood products because of their low cost, lack of color, and excellent adhesive and cohesion qualities (Aydin et al., 2006). Three-, five-, and seven-ply plywood sheets are common in home improvement stores throughout North America (Fig. 4.38). During World War II, plywood was used in the construction of some aircraft, such as the British Mosquito bomber, and some boats, such as the PT boat

FIGURE 4.38 Stack of five-ply plywood sheets at a home improvement store in Athens, Georgia, United States. *Pete Bettinger.*

commanded by President John F. Kennedy (John F. Kennedy Presidential Library and Museum, 2020; Naval History and Heritage Command, 2015; Royal Air Force, 2020). In North America, the typical tree species used in plywood manufacturing include birch, Douglas-fir, fir, mahogany, maple, oak, pine, spruce, and a number of other hardwood species. *Block board*, another laminated wood product, is typically composed of birch, pine, poplar, or other hardwoods and is commonly used for decorative projects and furniture manufacturing. Block board contains an inner core of larger solid wood pieces and an outer sheet of veneer and comes in a number of dimensions. It is similar to plywood but is generally thicker in nature.

Particleboard is a term used for a group of panel products that includes a large set of wood-based composites made from chips, sawmill shavings, and sawdust (Fig. 4.39). Particleboard was developed by Max Himmelheber, a German inventor, in 1932. As with other composite products, particleboard is manufactured by decomposing wood into fibers or small chips, mixing these with adhesive, and pressing them together.

Particleboard is a low-cost, highly versatile product that was developed as a replacement to plywood and is commonly used for flooring, cabinets, displays, furniture, kitchen countertops, speakers, and table tennis surfaces, among other products (Nemli and Colakoglu, 2005; Wong and Kozak, 2008). For the most part, particleboard is not appropriate for exterior uses, since some of these products have the tendency to perform poorly when exposed to water, as they can expand and warp. Numerous tree species used in the manufacture of particleboard include alder, beech, birch, pine, and spruce.

Oriented strand board, also referred to as OSB, is a panel product developed from wood particles (wood wafers or strands) (Fig. 4.40). Some have classified these products as a subgroup of particleboard products. The particles employed are organized into thin layers that are intentionally oriented in different directions, which provide these products with a high mechanical strength. Adhesives are then applied and the entire mass is pressed to a predetermined thickness to form a board. OSB is primarily intended for indoor uses, such as construction panels, but has yet to be extensively used for furniture production (Rebollar et al., 2007). In the United States, OSB panels are commonly made from tree species such as aspen (*Populus tremuloides*) and southern yellow pines (Yang et al., 2010).

Hardboard is another type of composite board that was developed in the early 1900s and represents a class of composite wood products that have an especially uniform density throughout a board (Dolezel-Horwath et al., 2005). The wood particles used to make this board are similar to those found in particleboard, but hardboard is much denser and stronger. Hardboard is usually smooth on one side and may have an imprinted fine mesh on the other (Dolezel-Horwath et al., 2005). While

FIGURE 4.39 Close-up of a particleboard sheet. *Pete Bettinger.*

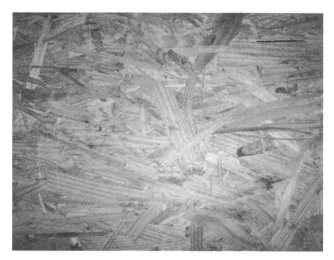

FIGURE 4.40 Overhead view of the face of a sheet of oriented strand board. *Donald L. Grebner.*

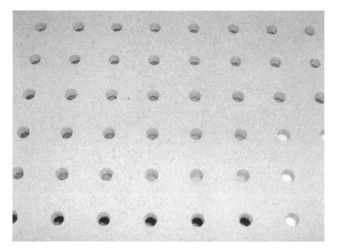

FIGURE 4.41   Example of hardboard pegboard.   *Donald L. Grebner.*

it is typically used for construction and furniture development purposes, hardboard can also be used as a pegboard material, which is often used to organize tools in one's garage (Fig. 4.41). Hardboard composites can be made from hemicellulose fibers, lignin, and a formaldehyde-based adhesive; thus, there is some environmental concern regarding products developed with this type of process, as some products cannot be recycled (Sehaqui et al., 2011). However, one form of hardboard, called *masonite*, is made without the use of additional adhesives and utilizes the inherent strength of wood fiber (after heating and pressing) to provide stability. Masonite was developed in 1924 by William H. Mason (Fig. 4.42) (Schniewind, 1989) and has been commonly used in the production of canoes, doors, desktops, roofing, and walls.

An example of a recently developed engineered wood product is *steam-pressed scrim lumber*. Steam-pressed scrim

FIGURE 4.42   Example of a sheet of masonite.   *Donald L. Grebner.*

FIGURE 4.43   Steam-pressed scrim lumber developed from the TimTek process.   *Donald L. Grebner.*

lumber was developed using the TimTek process, originally developed in Australia but commercialized at the Mississippi State University in Starkville, Mississippi (Mississippi State University, 2010). In the development of steam-pressed scrim lumber (Fig. 4.43), green (not dried), low-value raw material from small-diameter trees is crushed and long fibrous materials (called *scrims*) are derived. After the scrims have been dried, resins or adhesives are applied. The scrims are then stacked into a mat, and heat and pressure are applied to produce engineered beams that can act as a substitute for sawtimber. Tree species used in this process include various eucalypts and southern yellow pines. Dead or dying trees from various parts of the interior northwest of North America are now being evaluated for use in developing this product, which, if successful, may create a market for wood derived from forests where fuel reduction or forest health concerns are great (Linton et al., 2010). *Oriented strand lumber* is a similar type of product in which wood fibers are oriented along the length of the structural timber and bound together with an adhesive (Beck et al., 2009).

Technically, steam-pressed scrim lumber could be considered a form of glue-laminated timber (*glulam*), an engineered product containing two or more layers of wood glued together with the grain of the wood pieces (laminates) running lengthwise along the final product (Kuzman et al., 2010). Like the other products described in this section, glulam is a composite material (Fig. 4.44), but is very rigid and durable and is often employed as the arches, beams, columns, or frames of buildings. Fir, larch, poplar, and spruce are common tree species used in the development of glulam products. In contrast to other products, such as plywood, steam-pressed scrim lumber, and particleboard, laminates used in the construction of glulam products may

FIGURE 4.44 Piece of glulam window framing. *Karpova Yana, through Wikimedia Commons.*
Image Link: https://commons.wikimedia.org/wiki/File:Okon_brus.jpg
License Link: https://creativecommons.org/licenses/by-sa/3.0/deed.en

FIGURE 4.45 Two examples of wood plastic composites, made using hollow and foam technology. *Geolam, through Wikimedia Commons.*
Image Link: https://commons.wikimedia.org/wiki/File:Alveolaire-moussee.jpg
License Link: https://creativecommons.org/licenses/by-sa/3.0/deed.en

be much thicker, perhaps up to 45-millimeters (mm) (1.8-in) thick. Wedged finger joints are commonly used to bind individual laminates together lengthwise. This type of composite wood product can be developed from pieces of weaker wood or even recycled wood (Kuzman et al., 2010).

Another engineered wood product is cross-laminated timber. It is similar to glulam in that multiple layers of wood are glued together, but the direction of grain within each layer alternates by a 90-degree angle. It is a quasi-ridge composite that has an uneven number of layers (Brandner et al., 2016). Cross-laminated timber is a light material that can be designed as prefabricated components when assembling tall mass-timber structures (Brownwell, 2016). Although it can be more expensive than other traditional building materials, it has multiple advantages related to the design of structures. This emerging product has been made possible through changes in building codes in certain locales (American Wood Council, 2019).

Wood plastic (or *polymer*) *composites* (Fig. 4.45) are another form of wood product that have been developed for outdoor uses, including residential decking, seating (chairs and benches), and sidewalks, where a resistance to biodeterioration is welcome. There are a great number of possible combinations of wood and polymers (Schneider and Witt, 2004). For example, wood plastic composites can be made by impregnating veneer with phenol-formaldehyde resin or mixing wood flour and polypropylene together. While extensively used for nonstructural applications, wood plastic composites have recently been assessed for load-bearing applications in the building and construction fields (Tamrakar and Lopez-Anido, 2010).

4.1.4 Chemicals

Chemicals such as acetic acid (for making wood glue or plastic bottles and which gives vinegar its sour taste), acetone (a solvent), and creosote (a wood preservative) have all been distilled from the wood of trees. Wood from coniferous and deciduous trees can also be distilled to create formic acid (a preservative or antibacterial agent), butyric acid (a food and perfume additive, as well as a fishing bait additive due to its pungent odor), propionic acid (a preservative that inhibits the growth of mold), methanol, turpentine, and various other acids and wood-based oils. *Naval stores* include pitch, resin, tar, and turpentine, which were used in shipyards for caulking the seams and protecting the hulls and decks of wooden ships. The use of tar and pitch for these purposes dates back to the times of the ancient Egyptians. In the early 20th century, the southeastern United States produced over half of the world's supply of naval stores (Panshin et al., 1950). In the mid-18th century, a single cord (3.6 m³; 128 ft³) of longleaf pine (*Pinus palustris*) could be used to produce about 189 L (50 gallons [gal]) of tar. *Potash*, once created by boiling down sap-filled hardwood logs into a thick brown salt, is necessary for making soaps and glass (Youngquist and Fleischer, 1977). Potash can also be used as a fertilizer, for bleaching textiles, and for manufacturing gunpowder. *Tung oil*, or China wood oil, is created from the seeds of the tung tree (*Aleurites fordii*), a species native to central and western China. Tung oil is an important component of caulks, paints, and varnishes. Latex is a product from the rubber tree (*Hevea brasilienesis*) that is drained from the trunk of the tree, just below the tree's bark. Rubber trees grow only in the tropics and are common in Asia and South

America. Latex is used in a variety of surgical, public health, and industrial goods. Many of the chemicals described here have been used to create other products (lubricants, soaps, paints, polishes, roofing materials, and varnishes) that we use in our daily lives (Smith, 1940; Williams, 1989), often without knowing how they were created.

4.1.5 Residues

Wood residues include those materials derived from a manufacturing process, those derived from a logging process, and those that are reused or recycled. Wood residues are commonly used for basic fuel purposes in manufacturing facilities (to produce energy through the burning of this waste) but have also been used as sources of raw material bioenergy processes, the development of wood pellets (Fig. 4.46) or smaller wooden articles, and for pulp and paper processes. For example, in the lumber manufacturing process, the bark removed from logs and the edgings, sawdust, shavings, slabs, and trimmings derived from sawing logs are considered residues and are often used as fuel to generate electricity. In addition, the bark from cypress, pine, and redwood trees that may have been removed prior to lumber or pulp processing has been used as landscaping mulch. In veneering processes, the cores and trimmings of logs are considered a wood residue. While many of the wood residues generated from sawmilling and veneering operations in the southern United States are in chip form (Johnson and Steppleton, 2011), *peeler cores* may be used as low-quality lumber, landscaping timbers, or post and rail fencing pieces (Sugiyanto et al., 2010). In the utility pole development process, unused tree butts and tops are considered residues and older, used utility poles may also be considered residues. The reuse and recycling of wooden utility poles has

also been subject to debate, since the wood within the pole needs to be mechanically sound to be of value for specific purposes (Piao and Monlezun, 2010). Planer shavings, sawdust, and wood chips are common residues from carpentry shops (Morais dos Santos Hurtado de Mendoza et al., 2010). Wood flour, which we mentioned earlier, is often considered a residue from furniture manufacturing processes.

Wood residues that are created through logging processes are now being evaluated as a source of biomass for energy production purposes (e.g., Fernandes and Costa, 2010). *Woody biomass* includes the stems, branches, and unused portions of trees and other plants. Often these are left on top of the ground after a forestry operation. One specific definition of woody biomass includes the above-ground dry weight of wood in the stem and limbs of live trees at least 2.5 cm (1 in) in diameter (Conner and Johnson, 2011), which excludes dead vegetation, foliage, seedlings, and other vegetation growing in a forest. To facilitate future management of the land, woody biomass resources are typically piled and perhaps burned, yet they can also be lost through decomposition processes. With pressure mounting for *green energy,* there is a strong interest in collecting and using this material for bioenergy purposes (i.e., the production of electricity or fuels) (Perez-Verdin et al., 2008). Slash bundling is beginning to be evaluated as a way to extract woody biomass from forests for these purposes. In some areas of the world, bundles of slash can be collected and stored in outdoor conditions for several months, while maintaining their structural integrity and losing little of their fuel value (Patterson et al., 2010).

4.2 Nontimber forest products

NTFPs encompass a wide array of resources that have medicinal, nutritional, spiritual, and technical (construction and crafts) value and have been among the most valuable resources for certain human civilizations. For example, during the 19th century, bark, ferns, and fruits were the main products derived from forests on Japan's Kyūshū Island (Okazaki, 1964). Some NTFPs are called *amenities*, or resources or experiences that do not have an established market value (U.S. Forest Service, 1989). Worldwide diversity of marketable or nonvalued NTFPs is contingent on cultural, economic, and environmental factors, as was noted in many of the country profiles in Chapter 2. For example, while more than 80% of the population of Laos relies on forests for food, fuel, shelter, and tools, in some areas of the country the sale of NTFPs (e.g., mushrooms) represents a substantial component of household income, although spiritual values are also highly valued

FIGURE 4.46 Wood pellets produced in southern Mississippi, United States. *Donald L. Grebner.*

FIGURE 4.47 Abetasunk mushrooms (*Phellinus* spp.) found in forests in Armenia. *Zachary A. Parisa.*

(Foppes and Ketphanh, 2005; Hodgdon, 2010). In contrast, berries, honey, mushrooms, and wild fruits are produced in the forests of Belarus, along with other apicultural products such as wax and propolis, a medicinal resin with a number of uses (Krott et al., 2000). In Armenia, the abetasunk mushroom (*Phellinus* spp.) is an important food source (Fig. 4.47). In addition, chestnut and hazel coppice sprouts were once used extensively for crate rods, garden screens, and sheep fencing in the United Kingdom (Richards, 2003).

What should be obvious is that forested areas are diverse enough to facilitate the development of numerous plant-based sources of food for human consumption. For instance, in Africa the fresh flowers of abal (*Calligonum comosum*) are important consumable NTFPs, as are the flowers, leaves, and pods of acacia (*Acacia farnesiana*) in Central America. In North America, the fruit of blueberry (*Vaccinium* spp.), huckleberry (*Gaylussacia* spp.), and *Rubus* spp. (blackberry, dewberry, and raspberry) are important consumable products. Other common consumable forest products are the nuts of the pecan tree (*Carya illinoensis*) and the pinyon (*Pinus edulis*). In Europe, the roots and leaf stalks of the greater burdock (*Arctium lappa*) are important and, in South America and Africa, all parts of the amaranth plant (*Amaranthus* spp.) are edible. Moreover, in Asia, the fruit from the bael tree (*Aegle marmelos*) and breadfruit (*Artocarpus incisa*) are edible NTFPs. These are only a few of the consumable NTFPs.

To further indicate the diverse nature of NTFPs that can be obtained from forests, a number of minor human indulgences are facilitated by woody plants. For example, dyes were once made from butternut (*Juglans cinerea*) and yellow-poplar (*Liriodendron tulipifera*), stains were once developed from tree species such as red oak, and lye has been developed from wood ash as well as

from the leaves of sumac (*Rhus* spp.). All of these products were once necessary for the animal hide tanning process (Youngquist and Fleischer, 1977), a necessary step in the production of leather goods. In parts of Malaysia, bamboo and rattan are favored NTFPs (Svarrer and Olsen, 2005), used to produce small household implements and items such as brooms.

Some of the more popular NTFPs derived from trees are sugars and syrups. Maple sugaring (Fig. 4.48) was once a major activity in the northeastern parts of North America and still continues to a lesser extent, although the northeastern United States and the eastern provinces of Canada remain the largest producers of maple syrup worldwide. The yield of sap from a sugar maple (*Acer saccharum*) tree is highly and positively correlated with the condition of the crown of a tapped tree. As with other NTFPs, sugaring is a seasonal activity, with the sap-collecting period generally ranging from February to April, although this will vary depending on the weather (Panshin et al., 1950).

Forests also produce important nontimber products such as plants containing chemical compounds that are suitable for producing medicines for treating human ailments. Medicinals made from tree bark, leaves, and roots were once widely concocted in North America and are still considered important. For example, aspirin was derived from willow (*Salix* spp.) bark and is one of the most common medicines used in the world. Sassafras (*Sassafras albidum*) leaves were once boiled to make a tonic, and root beers and tonic beer were made from the sap and twigs of birch (*Betula* spp.) and black spruce (*Picea mariana*). Some medicinals that remain important today include taxol, which is derived from the Pacific yew (*Taxus brevifolia*) and used to treat ovarian cancer, and quinine, which is derived from the cinchona tree (*Cinchona* spp.) in South America and used to treat malaria. In addition, vinblastine, derived from Madagascar's endemic rosy periwinkle (*Catharanthus roseus*), is used to treat Hodgkin's disease, and vincristine, also derived from rosy periwinkle, is used to treat leukemia. Further, tubocurarine (derived from the curare [*Chondrodendron tomentosum*] plant of South America) is used as a muscle relaxant and pilocarpine (from jaborandi or *Pilocarpus* spp. in South America) is used to treat glaucoma. Scopolamine (from henbane [*Hyoscyamus niger*] from Eurasia) is used to treat motion sickness and erythromycin (from tropical fungi), which is similar in function to penicillin, is used as an antibiotic (World Resources Institute, 2010). Other drugs such as procaine (commonly known as novocaine) and cortisone are derived from Central and South American plants (Butler, 2012) and are in common use today.

Duke (1997) lists several North American medicinal and consumable NTFPs that are still collected today,

FIGURE 4.48 Maple sugaring in Vermont, United States. The upper left corner shows a hole being drilled to insert the tap, the lower left corner shows the tap being used to drain sap, the upper right corner shows the bucket used for collecting sap, and the lower right corner shows processed maple syrup. *Lynn Sheldon.*

including bloodroot (*Sanguinaria canadensis*), ginseng (*Panax quinquefolius*), goldenseal (*Hydrastis canadensis*), Indian tobacco (*Lobelia inflate*), mayapple (*Podophyllum peltatum*), wild ginger (*Asarum canadense*), witch-hazel (*Hamamelis virginiana*), yarrow (*Achillea millefolium*), and sarsaparilla (*Smilax regelii*), an ingredient in old-style root beer. In Africa, there is the bark of *Erythrophleum* spp., the fruit of the nightshade (*Solanum* spp.), roots of mucherekese (*Bobgunnia madagascariensis*), the seed of jequirity (*Abrus precatorius*), and the stems and leaves of the resurrection bush (*Myrothamnus flabellifolius*) (Cunningham, 1997). In Sri Lanka, there is bael tree fruit, coriander (*Coriandrum sativum*), edible tubers of *Plectranthus* spp., ginger (*Zingiber* spp.), and malabar nut (*Adhatoda vasica*) (de Alwis, 1997). In fact, in Sri Lanka alone there are between 550 and 700 medicinal

plants species traditionally used in local communities. In addition, in China, approximately 1000 plant species are used for medicinal purposes (He and Sheng, 1997). A Food and Agriculture Organization of the United Nations (1994) report listed a number of commonly traded medicinal plants, including the dried leaves of licorice (*Glycyrrhiza glabra*), the roots and rhizomes of snakeroot (*Rauwolfia serpentina*), the roots and rhizomes of *Berberis* spp. and *Mahonia* spp., and the tops and roots of belladonna (*Atropa belladonna*) from India. In the Amazon rainforests of Brazil, important NTFPs are derived from the annatto/lipstick tree (*Bixa orellana*), the cocoa tree (*Theobroma cacao*), clavillia (*Mirabilis jalapa*), suma (*Pfaffia paniculata*), and trumpet tree (*Cecropia peltata*).

Other important NTFPs include corks, greens, mosses, straws, and small trees. Spanish moss (*Tillandsia*

FIGURE 4.49 Spanish moss on oak tree in Beaufort, South Carolina, United States. *J. Taylor Thomas.*

usneoides), an epiphyte that commonly grows in long, hair-like strands on the limbs of trees (Fig. 4.49), such as live oak (*Quercus virginiana*) in the southern United States, has been used as stuffing for dolls, mattresses, pillows, and upholstered furniture. Cork used to seal bottles containing fermented and distilled drinks (Fig. 4.50) can be obtained from the outer bark of the cork oak (*Quercus suber*), a deciduous Mediterranean tree. In fact, France, Italy, Morocco, Portugal, and Spain are perhaps the most important cork-producing countries. Pine straw (fallen needles) is an important landscaping product in

North America. It is raked off the forest floor in areas where long-needled pines, such as longleaf pine or slash pine (*Pinus elliottii*), are grown. Holiday greens are acquired from live forest trees such as the American holly (*Ilex opaca*) or from other plants that grow in forests, including those that live on trees such as the mistletoe (*Phoradendron serotinum*, *Viscum album*, and others). Although Christmas trees (Fig. 4.51) are also an important NTFP, today these trees are commonly derived from Christmas tree farms. Blue spruce (*Picea pungens*), Douglas-fir, Fraser fir (*Abies fraseri*), jack pine (*Pinus banksiana*), and Scotch pine (*Pinus sylvestris*) are all common Christmas trees used in North America.

In the following sections, we focus on a few specific groups of NTFPs. There are a number of other forest-related resources, such as social or spiritual values, that we could delve more deeply. These include *cultural resources* (objects or buildings with historical, prehistoric, or scientific significance (U.S. Forest Service, 1989) that may be found on a landscape. However, we have chosen to provide here only a brief description of a few resources that are either hot topics in today's natural resource management fields or are popular among the general public. These include wildlife habitat and rangeland resources, recreation, water, aesthetics, biodiversity, biofuels, and carbon. Some of these are also described in more detail in other chapters of this book.

4.2.1 Wildlife habitat and rangeland resources

Wildlife habitat is the natural environment of a plant or animal and can be considered an output or product of forest and natural resource management. While we describe this topic in more detail in Chapter 5, wildlife habitat can be created naturally through forest successional processes, such as those we discuss in Chapter 10, or through active manipulation of forest vegetation,

FIGURE 4.50 Cork oak bark harvest, Portugal. *Carsten Niehaus, through Wikimedia Commons.*
Image Link: https://commons.wikimedia.org/wiki/File:Quercus_suber_corc.JPG
License Link: https://creativecommons.org/licenses/by-sa/3.0/deed.en

FIGURE 4.51 Harvesting Christmas trees in the Yale-Myers Forest in northern Connecticut, United States. *David Hobson.*

as we discuss in Chapter 11. Habitats generated in a forested landscape by either process can promote the establishment of a variety of herbaceous and shrub species. This variety of plant communities can yield valuable foraging, nesting, and roosting environments for a wide variety of vertebrate and invertebrate species. This diversity of flora and fauna can improve the species richness and biodiversity of a forest. A suite of these habitats distributed across a landscape can also improve the quality of recreational activities, such as birdwatching or hunting. A diversity of wildlife habitats can be viewed as regulatory and cultural aspects of forest ecosystem services, which we discuss in more detail in Chapter 6. Rangeland resources can also be intermixed with forested areas to provide the forage and resources necessary for livestock or other animals that graze on low-lying vegetation.

4.2.2 Recreation

Although it seems that forests have only become heavily utilized for recreational purposes since the mid-1900s, in some areas of the world (e.g., Japan), forests have been intimately connected with the local or regional human societies for centuries (Okazaki, 1964). In the United Kingdom, as in many other areas of the world, camp sites, car parks, forest roads, forest trails (Fig. 4.52), and picnic spots are considered recreational resources that are important for local communities. The concept of forest bathing, *shinrin-yoku* in Japanese, or taking in the forest as method of relaxation, is a growing trend around the developing world that may benefit human health and welfare (Li, 2010). Natural resource managers are frequently involved in the planning and development of these facilities for recreational opportunities such as fishing, orienteering, and wildlife

viewing within a forested setting (Richards, 2003). Potential recreational opportunities can be delineated through mapping processes, and recreational use can be assessed using metrics such as the *recreation visitor day*, of which one represents 12 visitor hours at a specific location (U.S. Forest Service, 1989). In Chapter 7 we discuss the recreational resources and opportunities associated with forests in more detail.

4.2.3 Water

Clean water is an important output generated from a forested landscape, and we discuss this in Chapter 6 as an ecosystem service facilitated by forests. Many communities worldwide derive their primary drinking water supply from forested watersheds. Growth projections suggest that increases in human population levels will occur around the world, which is likely to increase the importance of water resources over the course of your career. Forested landscapes have two main water resource functions, related to the quantity and quality of water available for our use. First, during heavy rain events, forests can reduce the amount of surface runoff into streams, rivers, and other water bodies, and slow the rate at which the water flows through the watershed. It is not difficult to notice how water bodies collect muddy water after heavy rainstorms, mainly from surface runoff arising from recently developed or deforested areas. However, much of the reduced surface runoff that passes through forests becomes groundwater. The second function of forests is to absorb groundwater and filter it as it moves downward through a soil profile. This process improves the quality of the water that accumulates in stream systems (Fig. 4.53) and other water bodies.

FIGURE 4.52 Forest trail in Saxon, Switzerland, southeast of Dresden, Saxony, Germany. *Florian Geyer.*

FIGURE 4.53 Forest stream in Maine, United States. *Samantha Langley.*

Many communities around the world place a high priority on water quality and quantity and face interesting decisions regarding the management of forest resources with regard to these issues. The Massachusetts Water Resource Authority, for example, manages the drinking water supply for the city of Boston and other nearby municipalities. The water resources they manage include the Quabbin and Wachusett reservoirs in the western and central parts of the state. A policy that began in the 1930s and lasted about 50 years prohibited hunting in these areas. However, during the late 1980s and early 1990s, concern arose by professional staff that the deer population had become an indirect threat to water quality (Massachusetts Department of Conservation and Recreation, 2020). The concern was centered on the notion that the deer population had overgrazed the understory vegetation, which increased the risk of damage to the water quality should the forest canopy become destroyed (by a hurricane, for example). As a result, organized deer hunts were implemented to control the population, although not without significant public controversy over this practice (Shrestha et al., 2010). However, once an appropriately sized deer population was achieved, plant diversity increased. Unfortunately, the moose population also increased, which now poses a new threat (Shrestha et al., 2010).

4.2.4 Aesthetics

Another potential product arising from forests are *aesthetic values*, which are judgments we make regarding how pleasing (e.g., beautiful) their features seem. If a landscape is pleasing to the eye, it is commonly thought of as being aesthetically pleasing. Mature forests are often described as aesthetically pleasing because large trees often seem to stretch to the heavens, and an abundance of wildlife and other plant and animal species can be found in the understory. Aesthetic quality is a value-laden issue, however. For example, if you live in northern California and visit an old-growth redwood forest (Fig. 4.54), you may have the feeling that these forests are very aesthetically pleasing. Some people who live in the southern United States find young pine plantations aesthetically pleasing (Fig. 4.55). However, the westerner visiting a young loblolly pine (*Pinus taeda*) forest in Georgia that is 10 to 15 years old may not have the same feeling as the southerner about the aesthetic quality of the forest. To a great extent, the term *aesthetically pleasing* is specific to each person's point of view. One of the important issues about managing forests for aesthetic values is that people often have different views on what is beautiful and what is not. This has led to

FIGURE 4.54   Redwood forest at Muir Woods National Monument in California, United States. *Kelly A. Bettinger.*

FIGURE 4.55   Loblolly pine (*Pinus taeda*) forest in central Georgia, United States. *Chuck Bargeron, University of Georgia, through Bugwood.org.*

various controversies with regard to the common forestry practices discussed in Chapter 11. For example, during the 1970s, a common forestry practice, clearcutting, began to become controversial because a segment of society expressed the opinion that, aesthetically, it was distasteful. As discussed in Chapter 11, a clearcutting practice facilitates even-aged management of forests containing shade-intolerant tree species and requires the removal of nearly all of the trees in a given management area. Therefore, when people who are unfamiliar with common forest practices (e.g., tourists) view clearcut sites or sites treated with chemicals to control competing brush, they reach the opinion that a forest has been destroyed, without knowing that certain tree species require a significant amount of sunlight to successfully regenerate. Unfortunately, the aesthetic quality of forests can therefore raise concerns over the use of common silvicultural practices in forest and natural resource management.

4.2.5 Biodiversity

Biodiversity commonly refers to the distribution and abundance of plant and animal species within an ecosystem (U.S. Forest Service, 1989). For instance, a forest described as having high biodiversity is one that has a variety of microsites and environments that can (and do) support an abundance of different amphibian, animal, bird, fungal, insect, plant, and microorganism life-forms. In general, higher levels of biodiversity in an ecosystem are often linked to the relative health of the ecosystem. In other words, many suggest that when a forest has a high level of biodiversity, the forested ecosystem is also healthy (see Chapter 6 for more details). However, strategies for maintaining high levels of biodiversity vary greatly. Some natural resource managers prefer to leave forests in a near-wilderness state (Fig. 4.56) and allow natural processes to transform these forests over time. Other natural resource managers prefer to actively manage forests using common forestry or natural resource practices, which to some extent will allow natural processes to transform the forests but which will also influence biodiversity. Both approaches can lead to a level of biodiversity that can maintain a healthy ecosystem. When discussing biodiversity, one must consider the

FIGURE 4.56 Biologically diverse natural forest in Sarawak, Malaysia. *Hugh Bigsby.*

scale and distribution of management activities. On his or her own, a small landowner may not achieve a high level of biodiversity on his or her property, especially if he or she owns less than 40 hectares (ha; about 100 acres [ac]). However, a collection of 100 or more landowners connected across a landscape, each with their own goals and objectives, could facilitate a much greater overall level of biodiversity in that area.

4.2.6 Biofuels

Another NTFP that periodically becomes important is the production of biofuels. In its most basic form, a *biofuel* is any plant species that can combust to produce heat and energy. Therefore, a biofuel could be biological waste, dried grass, dried peat, or a tree. With the invention of automobiles, biofuels were important for propelling these vehicles until liquid petroleum products became available. Today, we find ourselves concerned with the issue of deriving liquid biofuels or bioenergy from forests; however, our concern is not new. During the 1970s, the economy of the United States experienced two oil crises that exposed the nation's dependence on imported oil. Gas stations would commonly post green flags indicating they had gasoline, red flags if they were out of gasoline, or yellow flags if they were nearly out. In 1979, Austria began researching the use of short-rotation plantation forests as a source of renewable energy; this was also stimulated by increases in oil prices. Austria saw the production of energy and raw materials from short-rotation forests as an alternative for the agricultural sector of their economy (Tiefenbacher, 1991). Fuel crises such as these have prompted research into the development of alternative biofuels, such as biodiesel and ethanol, but interest has waxed and waned with the onset and passing of each crisis. During this first decade of the 21st century, we find a renewed interest in developing alternative liquid fuels, such as bio-oil (Fig. 4.57), from woody material. Bio-oils can be made from a variety of feedstocks but, unfortunately, efficient conversion technologies that begin with trees and end with cost-efficient combustible liquids (e.g., the conversion of wood to bioethanol) have been relatively slow to develop, despite advances in similar areas (e.g., the conversion of corn to ethanol).

4.2.7 Carbon

Another NTFP that has recently become important is the storage of carbon within trees (Fig. 4.58). Aside from their water content, trees are composed of approximately 50% carbon; as they grow, they can capture

FIGURE 4.57 Bio-oil (right-most bottle) along with a variety of feedstocks (surrounding the bottles), the feedstocks after being broken down (left-most bottle), and the feedstock after processing (middle bottle). *David Jones.*

FIGURE 4.58 Standing forest carbon in Saxon Switzerland, southeast of Dresden, Saxony, Germany. *Florian Geyer.*

and store (sequester) carbon. The potential to sequester carbon in solid wood products has received quite a lot of attention, since these wood products can be viewed as reservoirs for carbon storage when they remain in solid form (Sarkar and Manoharan, 2009). We classify the process of sequestering carbon as an ecosystem service and discuss this in more detail in Chapter 6. The interest in carbon as a product of forests has been prompted by the growing concern of global climate change, which is linked to increasing levels of carbon dioxide (CO_2) in the atmosphere. In short, large amounts of carbon dioxide are emitted into the atmosphere from the combustion of fossil fuels, such as coal and gasoline. In addition, deforestation contributes to this emission through the loss of sequestration activities, decomposition of woody residues, and fires that may be used to clear areas for agricultural or developmental activities. Trees may be able to capture some of this carbon dioxide through the photosynthetic process.

There are many possible solutions to the problem of global climate change; one of these includes sequestering carbon in living, thriving forests (Noormets et al., 2015). Thus, one worldwide effort to address global climate change involves basically growing more trees. It is important, however, to realize that planting new forests or maintaining existing forests cannot completely address the issue of global climate change. Forests can play an important role in this matter, along with the development of greater fuel-efficient automobiles and of alternative energy sources, such as wind and solar processes, in addition to energy conservation.

Summary

As you may have gathered, forests and their associated natural resources provide a large variety of commercial and noncommercial products for human use and consumption. Of the other sites available for acquiring natural resources (below ground, deserts, lakes, oceans, polar areas, seas, upper atmosphere, and wetlands), forests are perhaps the most important. A common mistake is to view wood as the only product of forest and natural resource management. As we have illustrated, forest and natural resource management can provide a wide variety of products and services, from mushrooms to recreational experiences and sequestered carbon. Forests have been a source of shelter and food for humans since the dawn of time. The products generated from this renewable resource have evolved from food and fuelwood to items such as lumber, particleboard, plywood, and toilet tissue. A number of consumable foods and medicines can be derived from forested landscapes, and the medical research community continues to search for compounds derived from forest vegetation that may be of value in treating or curing diseases. As a society's values change, so do the demands placed on forests and managers are likely to find themselves developing plans to obtain resources such as biodiversity, carbon, timber, water, and perhaps bioenergy. In summary, forest resources are important for many human needs but also for maintaining global ecosystem health and ecological diversity.

Questions

(1) *A quick inventory.* From the chair in which you are sitting and without standing up, look around and develop a quick inventory of the items that you think were derived from forests. Develop a short professional memorandum briefly describing the items, the tree species they might have been derived from, and the location (country or region) where you think they might have been produced.

(2) *Nontimber forest products.* Decide upon the last NTFP that you may have consumed. Using the resources at your disposal, develop a short presentation on the product for your class. In this presentation, briefly describe the product and the forested areas you think it might have come from, along with any other interesting facts regarding its harvest, storage, transport, or processing.

(3) *Nontimber forest products.* Select a medicinal product that is typically grown in a forested area or derived directly from trees. Develop a one-page report regarding its natural distribution and its importance to humans. In a small group that includes three or four other people, discuss the products you have chosen.

(4) *Biofuels and carbon sequestration.* Of these two products or services potentially facilitated by forests, which is more important to you? Develop a short, professional memorandum describing the advantages and disadvantages of producing it. In addition, describe other methods for producing the one you have chosen without using trees, and the trade-offs facing society if it is not produced.

(5) *Aesthetics, recreational opportunities, water, and wildlife habitat.* For each of these four products or services potentially facilitated by forests, detail the last time you actually used them. Develop a short essay that describes when and where each use occurred, and the value you obtained from each experience.

(6) *Commercial timber products.* Visit a manufacturing facility that makes solid wood products, engineered products, or pulp and paper products. Describe using a flow chart how wood is transformed into another product of value to society. Include this flow chart in a short report that details the people, technology, and equipment necessary to create the final product.

References

Adame, P., Cañellas, I., Roig, S., Del Río, M., 2006. Modelling dominant height growth and site index curves for rebollo oak (*Quercus pyrenaica* Willd.). Annals of Forest Science 63, 929–940.

American Wood Council, 2019. 2021 IBC Proposed Code Change Resources. https://www.awc.org/tallmasstimber (accessed 29.07.19).

Aydin, I., Colakuglu, G., Colak, S., Demirkir, C., 2006. Effects of moisture content on formaldehyde emission and mechanical properties of plywood. Building and Environment 41, 1311–1316.

Beck, K., Salenikovich, A., Cloutier, A., Beauregard, R., 2009. Development of a new engineered wood product for structural applications made from trembling aspen and paper birch. Forest Products Journal 59 (7/8), 31–35.

Bidlack, J., Malone, M., Benson, R., 1992. Molecular structure and component integration of secondary cell walls in plants. Proceedings of the Oklahoma Academy of Science 72, 51–56.

Brandner, R., Flatscher, G., Ringhofer, A., Schickhofer, G., Thiel, A., 2016. Cross laminated timber (CTL): overview and development. European Journal of Wood Products 74, 331–351.

Brownwell, B., 2016. T3 Becomes the First Modern Tall Wood Building in the U.S. Architect Magazine. https://www.architectmagazine.com/technology/t3-becomes-the-first-modern-tall-wood-building-in-the-us_o (accessed 29.07.19).

Butler, R., 2012. Medicinal Plants. Mongabay.com, San Francisco, CA. http://rainforests.mongabay.com/1007.htm (accessed 02.04.20).

Conner, R.C., Johnson, T.G., 2011. Estimates of Biomass in Logging Residue and Standing Residual Inventory Following Tree-Harvest Activity on Timberland Acres in the Southern Region. U.S. Department of Agriculture, Forest Service, Southern Research Station, Asheville, NC. Resource Bulletin SRS-169. 25 p.

Cunningham, A.B., 1997. An Africa-wide overview of medicinal plant harvesting, conservation and health care. In: Boedeker, G., Bhat, K.K.S., Burley, J., Vantomme, P. (Eds.), Medicinal Plants for Conservation and Health Care. Food and Agriculture Organization of the United Nations, vol. 11. Non-wood forest products, Rome, Italy, pp. 116–129.

de Alwis, L., 1997. A biocultural medicinal plants conservation project in Sri Lanka. In: Boedeker, G., Bhat, K.K.S., Burley, J., Vantomme, P. (Eds.), Medicinal Plants for Conservation and Health Care. Food and Agriculture Organization of the United Nations, vol. 11. Non-wood forest products, Rome, Italy, pp. 100–108.

Dilworth, J.R., 1970. Log Scaling and Timber Cruising. Oregon State University Book Stores, Inc., Corvallis, OR.

Dolezel-Horwath, E., Hutter, T., Kessler, R., Wimmer, R., 2005. Feedback and feedforward control of wet-processed hardboard production using spectroscopy and chemometric modelling. Analytica Chimica Acta 544, 47–59.

Duke, J.A., 1997. Phytomedicinal forest harvest in the United States. In: Boedeker, G., Bhat, K.K.S., Burley, J., Vantomme, P. (Eds.), Medicinal Plants for Conservation and Health Care. Food and Agriculture Organization of the United Nations, vol. 11. Non-wood forest products, Rome, Italy, pp. 147–158.

Fernandes, U., Costa, M., 2010. Potential of biomass residues for energy production and utilization in a region of Portugal. Biomass and Bioenergy 34, 661–666.

Food and Agriculture Organization of the United Nations, 1994. International Trade in Non-wood Forest Products: An Overview. Food and Agriculture Organization of the United Nations, Rome, Italy. Working paper Misc/93/11. http://www.fao.org/docrep/x5326e/x5326e00.htm#Contents (accessed 02.04.20).

Food and Agriculture Organization of the United Nations, 2010. Global Forest Resources Assessment 2010. Food and Agriculture

Organization of the United Nations, Rome. Italy FAO Forestry Paper 163. pp. 340.

Food and Agriculture Organization of the United Nations, 2015. Global Forest Resources Assessment 2015, Desk Reference. Food and Agriculture Organization of the United Nations, Rome, Italy, pp. 244.

Foppes, J., Ketphanh, S., 2005. Non-timber forest products for poverty reduction and shifting cultivation stabilisation in the uplands of the Lao PDR. In: Bouahom, B., Glendinning, A., Nilsson, S., Victor, M. (Eds.), Poverty Reduction and Shifting Cultivation Stabilization in the Uplands of Lao PDR: Technologies, Approaches and Methods for Improving Upland Livelihoods. National Agriculture and Forestry Research Institute, Vientiane, Laos, pp. 181—193.

Fromm, J., Rockel, B., Lautner, S., Windeisen, E., Wanner, G., 2003. Lignin distribution in wood cell walls determined by TEM and backscattered SEM techniques. Journal of Structural Biology 143, 77—84.

Gabrielii, I., Gatenholm, P., Glasser, W.G., Jain, R.K., Kenne, L., 2000. Separation, characterization and hydrogel-formation of hemicellulose from aspen wood. Carbohydrate Polymers 43, 367—374.

Ganguly, I., Pierobon, F., Bowers, T.C., Huisenga, M., Johnston, G., Eastin, I.I., 2018. 'Woods-to-Wake' Life Cycle Assessment of residual woody biomass based jet-fuel using mild bisulfite pretreatment. Biomass and Bioenergy 108, 207—216.

He, S.-A., Sheng, N., 1997. Utilisation and conservation of medicinal plants in China with special reference to Atractylides lancea. In: Boedeker, G., Bhat, K.K.S., Burley, J., Vantomme, P. (Eds.), Medicinal Plants for Conservation and Health Care. Food and Agriculture Organization of the United Nations, vol. 11. Non-wood forest products, Rome, Italy, pp. 109—115.

Hickenlooper, A., 1888. The Vicksburg mine. In: Battles and Leaders of the Civil War (Thomas Yoseloff, 1956 Edition), vol. III. Century Co., New York, pp. 539—542. https://ehistory.osu.edu/books/battles/vol3 (accessed 02.04.20).

Hodgdon, B.D., 2010. Community forestry in Laos. Journal of Sustainable Forestry 29, 50—78.

Jarusombuti, S., Hiziroglu, S., Bauchongkol, P., Fueangvivat, V., 2009. Properties of sandwich-type panels made from bamboo and rice straw. Forest Products Journal 59 (10), 52—57.

John F. Kennedy Presidential Library and Museum, 2020. John F. Kennedy and PT109. John F. Kennedy Presidential Library and Museum, Boston, MA. https://www.jfklibrary.org/learn/about-jfk/jfk-in-history/john-f-kennedy-and-pt-109 (accessed 02.09.20).

Johnson, T.G., Steppleton, C.D., 2011. Southern Pulpwood Production, 2009. U.S. Department of Agriculture, Forest Service, Southern Research Station, Asheville, NC. Resource Bulletin SRS-168. pp. 38.

Krott, M., Tikkanen, I., Petrov, A., Tunytsya, Y., Zheliba, B., Sasse, V., Rykounina, I., Tunytsya, T., 2000. Policies for Sustainable Forestry in Belarus, Russia, and Ukraine. European Forest Institute Research Report 9, Leiden, The Netherlands, pp. 174. Koninklijke Brill NV.

Kuzman, M.K., Oblak, L., Vratuša, S., 2010. Glued laminated timber in architecture. Drvna Industrija 61, 197—204.

Li, Q., 2010. Effect of forest bathing trips on human immune function. Environmental Health and Preventive Medicine 15 (1), 9—17.

Linton, J.M., Barnes, H.M., Seale, R.D., Jones, P.D., Lowell, E.C., Hummel, S.S., 2010. Suitability of live and fire-killed small-diameter ponderosa and lodgepole pine trees for manufacturing a new structural wood composite. Bioresource Technology 101, 6242—6247.

Massachusetts Department of Conservation and Recreation, 2020. Quabbin Reservation Deer Hunt. Massachusetts Department of Conservation and Recreation, Boston, MA. https://www.mass.gov/service-details/quabbin-reservation-deer-hunt (accessed 02.04.20).

Mississippi State University, 2010. TimTek at Mississippi State University. Forest and Wildlife Research Center, Mississippi State University, Mississippi State, MS. http://www.cfr.msstate.edu/timtek/index.asp (accessed 02.04.20).

Morais dos Santos Hurtado de Mendoza, Z., Evangelista, W.V., de Oliveira Araújo, S., de Souza, C.C., Ribeiro, F.D.L., de Castro Silva, J., 2010. An analysis of the wood residues generated by carpentry shops in Viçosa, State of Minas Gerais. Revista Árvore 34, 755—760.

National WWII Museum, 2007. Boats of Wood: Men of Steel. The National WWI Museum, New Orleans, LA. https://www.nationalww2museum.org/media/press-releases/boats-wood-men-steel (accessed 02.04.20).

Naval History and Heritage Command, 2015. PT109. Naval History and Heritage Command Museum, Washington D.C. https://www.history.navy.mil/research/histories/ship-histories/danfs/p/pt-109.html (accessed 02.09.20).

Nemli, G., Colakoglu, G., 2005. The influence of lamination technique on the properties of particleboard. Building and Environment 40, 83—87.

Nieuwenhuis, M., 2010. Terminology of Forest Management, Terms and Definitions in English, Second revised edition. International Union of Forest Research Organizations, Vienna, Austria. IUFRO World Series Volume 9-en.

Noda, H., 2002. The Japanese pulp and paper industry and its wood use. In: Iwai, Y. (Ed.), Forestry and the Forest Industry in Japan. UBC Press, Vancouver, Canada, pp. 214—229.

Noormets, A., Epron, D., Domec, J.C., McNulty, S.G., Fox, T., Sun, G., King, J.S., 2015. Effects of forest management on productivity and carbon sequestration: a review and hypothesis. Forest Ecology and Management 355, 124—140.

Okazaki, A., 1964. Forestry in Japan. Hill Family Foundation Forestry Series, School of Forestry, Oregon State University, Corvallis, OR, pp. 62.

Panshin, A.J., Harrar, E.S., Baker, W.J., Proctor, P.B., 1950. Forest Products. McGraw-Hill Book Company, Inc., New York, pp. 549.

Patterson, D.W., Hartley, J.I., Pelkki, M.H., Steele, P.H., 2010. Effects of 9 months of weather exposure on slash bundles in the mid-South. Forest Products Journal 60, 221—225.

Perez-Verdin, G., Grebner, D.L., Munn, I.A., Sun, C., Grado, S.C., 2008. Economic impacts of woody biomass utilization for bioenergy in Mississippi. Forest Products Journal 58 (11), 75—83.

Persson, T., Matusiak, M., Zacchi, G., Jönsson, A.-S., 2006. Extraction of hemicelluloses from process water from the production of masonite. Desalination 199, 411—412.

Piao, C., Monlezun, C.J., 2010. Laminated crossarms made from decommissioned chromated copper arsenate-treated utility pole wood. Part I: mechanical and acoustic properties. Forest Products Journal 60, 157—165.

Racelis, A.E., Barsimantov, J.A., 2008. The management of small diameter, lesser-known hardwood species as polewood in forest communities of central Quintana Roo, Mexico. Journal of Sustainable Forestry 27, 122—144.

Rebollar, M., Pérez, R., Vidal, R., 2007. Comparison between oriented strand boards and other wood-based panels for the manufacture of furniture. Materials and Design 28, 882—888.

Richards, E.G., 2003. British Forestry in the 20th Century, Policy and Achievements. Koninklijke Brill NV, Leiden, The Netherlands pp. 282.

Royal Air Force, 2020. de Havilland Mosquito B35. Royal Air Force Museum, London, U.K. https://www.rafmuseum.org.uk/research/collections/de-havilland-mosquito-b35/ (accessed 02.09.20).

Sarkar, A.B., Manoharan, T.R., 2009. Benefits of carbon markets to small and medium enterprises (SMEs) in harvested wood products: a case study from Saharanpur, Uttra Pradesh, India. African Journal of Environmental Science and Technology 3, 219—228.

Schneider, M.H., Witt, A.E., 2004. History of wood polymer composite commercialization. Forest Products Journal 54 (4), 19—24.

Schniewind, A.P., 1989. Concise Encyclopedia of Wood & Wood-Based Materials. The MIT Press, Cambridge, MA, pp. 354.

Sehaqui, H., Allais, M., Zhou, Q., Berglung, L.A., 2011. Wood cellulose biocomposites with fibrous structures at micro- and nanoscale. Composites Science and Technology 71, 382–387.

Shrestha, I., Ha, N., Khan, R., 2010. Deer and Moose Impact on the Quabbin. Amherst College, Amherst, MA. https://www.amherst.edu/people/facstaff/aanderson/sciobservatory/problematicspecies/2010/projects/deer/node/250648 (accessed 02.04.20).

Siuru, W.D., 1977. SLBM—the Navy's contribution to Triad. Air University Review 28 (6), 17–29. http://www.airpower.au.af.mil/airchronicles/aureview/1977/sep-oct/siuru.html (accessed 07.08.11).

Smith, H.F., 1940. Primary Wood-Products Industries in the Lower South. U.S. Department of Agriculture, Forest Service, Southern Forest Experiment Station, New Orleans, LA. Forest Survey Release No. 51.

Stern, T., 2009. Wood for food: wood-based products in the dietary fiber additives market—a branch-analysis approach. Forest Products Journal 59 (1/2), 19–25.

Stirling, R., 2010. Residual extractives in western red cedar shakes and shingles after long-term field testing. Forest Products Journal 60, 353–356.

Sugiyanto, K., Vinden, P., Torgovnikov, G., Przewloka, S., 2010. Microwave surface modification of *Pinus radiata* peeler cores: technical and cost analyses. Forest Products Journal 60, 346–352.

Sunseri, T., 2009. Wielding the Ax: State Forestry and Social Conflict in Tanzania, 1820–2000. Ohio University Press, Athens, OH, pp. 293.

Svarrer, K., Olsen, C.S., 2005. The economic value of non-timber forest products—a case study from Malaysia. Journal of Sustainable Forestry 20, 17–41.

Tamrakar, S., Lopez-Anido, R.A., 2010. Effect of strain rate on flexural properties of wood plastic composite sheet pile. Forest Products Journal 60, 465–472.

The New York Times, 1918. Strong Wood Used in Making Planes. The New York Times, New York. http://query.nytimes.com/gst/abstract.html?res=9F03E4DC113BEE3ABC4851DFB6678383609EDE (accessed 02.05.20).

Tiefenbacher, H., 1991. Short rotation forestry in Austria. Bioresource Technology 35, 33–40.

US Forest Service, 1989. Land and Resource Management Plan. Siskiyou National Forest. US Department of Agriculture, Forest Service, Pacific Northwest Region, Portland, OR.

Webb, D.A., 2005. The Tie Guide: Handbook for Commercial Timbers Used by the Crosstie Industry. Railway Tie Association, Fayetteville, GA, pp. 76.

Williams, M., 1989. Americans & Their Forests. Cambridge University Press, Cambridge, UK, pp. 599.

Wong, D.C., Kozak, R.A., 2008. Particleboard performance requirements of secondary wood products manufacturers in Canada. Forest Products Journal 58 (3), 34–41.

World Resources Institute, 2010. A Short List of Plant-Based Medicinal Drugs. World Resources Institute, Washington, D.C. http://archive.wri.org/newsroom/wrifeatures_text.cfm?ContentID=471 (accessed 07.08.11).

Yang, D.-Q., Wang, X.M., Wan, H., 2010. Optimizing manufacturing conditions for durable composite panels with eastern white cedar and aspen strands. Forest Products Journal 60, 460–464.

Youngquist, W.G., Fleischer, H.O., 1977. Wood in American Life, 1776–2076. Forest Products Research Society, Madison, WI, pp. 192.

Wildlife habitat relationships

Conserving, developing, and maintaining natural areas for a multitude of human uses is an important consideration for many natural resource professionals, and is central in the minds of a large segment of the public (Livingston et al., 2003). Performing these same actions for the benefit of wildlife species is also paramount to many resource managers in a number of areas of the world. The interaction between wildlife species and the natural and managed environment is complex and varies from one species to another, and perhaps even within the region where a wide-ranging species exists. While many ecologists favor the conservation, maintenance, and development of wildlife habitat in wildlands, many believe that there is a need to develop habitat in urban areas as well (Kelcey, 1975). A general understanding of the habitat components of wildlife species can help one become an effective manager when strategies to maintain healthy ecosystems and abundant wildlife populations need to be developed. This chapter, therefore, has a number of objectives, which include enhancing the ability to

- understand the basic requirements of quality wildlife habitat for a given species;
- understand how forest succession affects wildlife habitat over time;
- differentiate between direct and indirect forest and natural resource management impacts on wildlife habitat;

- understand the differences between forest edge and forest interior as they pertain to wildlife habitat; and
- become acquainted with quantitative methods for assessing wildlife habitat quality.

With the discussion provided in this chapter, and the diversity of associated examples, an appreciation for the various challenges and opportunities to the conservation and maintenance of wildlife habitat can be gained. However, the value placed on wildlife and wildlife habitat can vary from one person to the next. Therefore, while some management practices may be guided by laws or regulations that specifically address a wildlife species (and thus are necessary), other management practices may be guided by a landowner or a manager's desire to assist in the development of habitat or populations (and thus are voluntary).

5.1 What is wildlife?

Technically, the term *wildlife* refers to any nondomesticated living organism existing in a natural or managed setting. Wildlife can include plants, animals, or other organisms such as fungi, lichens, bacteria, viruses, fish, insects, and algae. Typically, in forestry and natural resource management, the term *wildlife* pertains to species primarily residing within the mosaic

TABLE 5.1 A sample of large mammals, small mammals, birds, reptiles, and amphibians that use forests.

Common name	Scientific name	Picture
Larger mammals		
Mule deer	*Odocoileus hemionus*	Fig. 5.1
Gray wolf	*Canis lupus*	Fig. 5.2
American black bear	*Ursus americanus*	Fig. 5.3
Elk (wapiti)	*Cervus canadensis*	Fig. 5.4
Moose	*Alces alces*	Fig. 5.5
Forest elephant	*Loxodonta africana cyclotis*	Fig. 5.6
Grizzly bear	*Ursus arctos horribilis*	Fig. 5.7
Wood bison	*Bison bison athabascae*	Fig. 5.8
Pronghorn antelope	*Antilocapra americana*	Fig. 5.9
Smaller mammals		
Raccoon	*Procyon lotor*	Fig. 5.10
Yellow-bellied marmot	*Marmota flaviventris*	Fig. 5.11
American pika	*Ochotona princeps*	Fig. 5.12
Western least chipmunk	*Neotamias minimus*	Fig. 5.13
Eastern cottontail rabbit	*Sylvilagus floridanus*	Fig. 5.14
Birds		
Bald eagle	*Haliaeetus leucocephalus*	Fig. 5.15
Common barn owl	*Tyto alba*	Fig. 5.16
Great blue heron	*Ardea herodias*	Fig. 5.17
Wild turkey	*Meleagris gallopavo*	Fig. 5.18
Willow ptarmigan	*Lagopus lagopus*	Fig. 5.19
Resplendent quetzal	*Pharomachrus mocinno*	Fig. 5.20
Eastern bluebird	*Sialia sialis*	Fig. 5.21
American robin	*Turdus migratorius*	Fig. 5.22
Reptiles and amphibians		
Florida gopher tortoise	*Gopherus polyphemus*	Fig. 5.23
Marbled salamander	*Ambystoma opacum*	Fig. 5.24
Australian green tree frog	*Litoria caerulea*	Fig. 5.25
Cane toad	*Bufo marinus*	Fig. 5.26
Green iguana	*Iguana iguana*	Fig. 5.27
Eastern indigo snake	*Drymarchon corais couperi*	Fig. 5.28
Eastern diamondback rattlesnake	*Crotalus adamanteus*	Fig. 5.29

FIGURE 5.1 A mule deer (*Odocoileus hemionus*) in velvet near Cloudcroft, New Mexico, United States. *Andrew J. Sánchez Meador.*

mammals, birds, insects, and reptiles. Sometimes, freshwater fish are included due to the presence of ponds, lakes, streams, rivers, swamps, marshes, and vernal pools found within terrestrial landscapes. In this chapter, the focus will be directed toward the typical collection of wildlife found in terrestrial environments and those that use forests for one or more aspects of their life requirements. For instance, larger mammals that use forest resources include species such as the eastern white-tailed deer (*Odocoileus virginianus*), the coyote (*Canis latrans*), and a few others noted in Table 5.1. Of the many reasons these species may be found in forests, two obvious ones are that they use forests to forage and to hide from predators. Further, larger mammals may reproduce in forests or use forests as corridors to move between other more favorable habitats. Of course, this list is not exhaustive, but it provides a taste of the diversity of larger mammals that use forests. Smaller-sized mammal species, such as the wolverine (*Gulo gulo*), North American opossum (*Didelphis virginiana*), and eastern gray squirrel (*Sciurus carolinensis*), along with many others, have similar life requirements provided by forests, the most common being snags, standing dead trees, or coarse woody debris (CWD) on the ground that are used for nesting structures.

Birds comprise a large group of wildlife species, and the collection that uses forests includes those as diverse as condors and hummingbirds. Large birds of prey that use forests for nesting, roosting, and foraging activities include eagles, falcons, hawks, vultures, and owls. Water birds often use the forests near water bodies for nesting and roosting activities. These species include ducks, herons, swans, and geese, among a few. Forests also provide CWD as input into water bodies, and this material may be of value to water birds. Ground-dwelling birds include common species such as quail, turkey, pheasant, and ptarmigan, and a large collection of others that feed,

of land and water that can be found across a forested landscape, which includes species of amphibians,

FIGURE 5.2 Gray wolf (*Canis lupus*) with tracking collar in Denali National Park and Preserve, Alaska, United States. *Stephen C. Grado.*

FIGURE 5.4 Elk (*Cervus canadensis*) in Yellowstone National Park, Wyoming, United States. *Stephen C. Grado.*

FIGURE 5.3 Black bear (*Ursus americanus*) in Massachusetts, United States. *Thomas O'Shea.*

FIGURE 5.5 Moose (*Alces alces*) in Denali National Park and Preserve, Alaska, United States. *Stephen C. Grado.*

nest, or hide in forests around the world. Further, a number of neotropical migrants and birds we consider common *backyard* birds use forests to support some aspect of their life requirements. Reptiles and amphibians are another large group of wildlife species and are found in forests on every continent except Antarctica. These include turtles and tortoises, salamanders, toads and frogs, iguanas and lizards, and snakes. Many of these are associated with forested landscapes. For example, the gopher tortoise burrows underground in forested areas of the southern United States, and many species of salamanders and newts prefer the habitat facilitated by CWD (fallen logs and limbs from trees) on a forest

floor. Undoubtedly, if one spends a significant amount of one's life outdoors, some of these species of wildlife will be encountered.

Fish are another common form of wildlife, although at times it is argued that they should not be considered as such due to the production and introduction of some populations through fish hatcheries. In this text, we considered fish as wildlife since many are a natural resource directly or indirectly affected by the management of forests and forested landscapes. Certain species of fish have a significant relationship with forests and freshwater bodies, either because they are freshwater fish species (such as largemouth bass [*Micropterus salmoides*]) or because they are diadromous species such as Atlantic salmon (*Salmo salar*), coho salmon (*Oncorhynchus kisutch*), chinook salmon (*Oncorhynchus tshawytscha*), and rainbow trout (*Oncorhynchus mykiss*) (Fig. 5.30). Diadromous fish species travel between fresh and saltwater bodies during portions of their life cycle.

FIGURE 5.6 Elephant (*Loxodonta africana cyclotis*) in South Africa. *Dawn M. Grebner.*

FIGURE 5.7 Grizzly bear (*Ursus arctos horribilis*) in Yellowstone National Park, Wyoming, United States. *Stephen C. Grado.*

FIGURE 5.8 Bison (*Bison bison athabascae*) in Yellowstone National Park, Wyoming, United States. *Stephen C. Grado.*

FIGURE 5.9 Pronghorn antelope (*Antilocapra americana*) in Yellowstone National Park, Wyoming, United States. *Stephen C. Grado.*

Fish that live mostly in the oceans yet breed in freshwater are considered *anadromous*. Many eels will breed in oceans but spend the majority of their life in freshwater and are especially important food sources for both native peoples and modern societies. The presence of a particular species of fish in a terrestrial water body may be an indicator of habitat quality of the area, and this may be important to forest landowners and managers (Karr, 1981).

Two terms commonly attached to observable wildlife species are the words *threatened* and *endangered*. In one sense, they are legal terms attached to the U.S. Endangered Species Act that describe the need for protection.

Endangered wildlife species are those native species whose prospect for survival is low or those that are in danger and at risk of becoming extinct. This vulnerability may have arisen due to disease, predation, exploitation, pollution, or changes in habitat conditions. The giant panda (*Ailuropoda melanoleuca*) and the California condor (*Gymnogyps californianus*) are two examples of endangered species. Threatened wildlife species are those whose existence is not in immediate jeopardy, yet they may eventually become endangered if the conditions associated with their welfare continue to decline. Other terms commonly associated with wildlife species are *extinct* (a species that has disappeared from its entire range), *extirpated* (a species that has disappeared from a specific area or region), and *special interest* (a species that occurs periodically in an area and may breed there). These five distinctions may be applied universally (worldwide), at a national level, or at smaller jurisdictional levels (states, provinces, etc.). For example, the eastern indigo snake is considered threatened in the

FIGURE 5.12 American pika (*Ochotona princeps*) in Alpine Lakes Wilderness, Washington, United States. *Walter Siegmund, through Wikimedia Commons.*
Image Link: https://commons.wikimedia.org/wiki/File:Ochotona_princeps_26473.JPG
License Link: https://creativecommons.org/licenses/by-sa/3.0/deed.en

FIGURE 5.10 Raccoons (*Procyon lotor*) near Cloudcroft, New Mexico, United States. *Andrew J. Sánchez Meador.*

FIGURE 5.13 Western least chipmunk (*Neotamias minimus*) in Yellowstone National Park, Wyoming, United States. *Stephen C. Grado.*

FIGURE 5.11 Yellow-bellied marmot (*Marmot flaviventris*) in Yellowstone National Park, Wyoming, United States. *Stephen C. Grado.*

5.2 What is wildlife habitat?

states of Georgia and Florida (United States), but not in every state in which it may naturally be found. Other forms and subsets of these may also be communicated among natural resource management professionals. The International Union for Conservation of Nature (IUCN) also maintains a red list of threatened species whose status is assessed against certain criteria, and these species are then placed into one of seven categories: extinct, extinct in the wild, critically endangered, endangered, vulnerable, near threatened, and least concern (International Union for Conservation of Nature, 2019).

Wildlife habitats are areas distributed horizontally and vertically across the landscape that fulfill some or all of the needs of a specific wildlife species for the basic requirements of food, water, reproduction (nesting), and protection against predators and competitors (cover). Habitat provides the space requirements that allow wildlife to occupy, move around, and to generally survive and cope with climatic extremes (Morrison et al., 2006). The concept of wildlife habitat varies according to the needs of each species, and for land managers the concept may be simplified to include a description of those areas that are best suited for a species to

FIGURE 5.14   Eastern cottontail rabbit (*Sylvilagus floridanus*) near the Rideau River, Ottawa, Ontario, Canada.   *D. Gordon E. Robertson, through Wikimedia Commons.*
Image Link: https://commons.wikimedia.org/wiki/File:Eastern_Cottontail_rabbit,_Rideau_River.jpg
License Link: https://creativecommons.org/licenses/by-sa/3.0/deed.en

FIGURE 5.16   Barn owl (*Tyto alba*) in flight.   *Luc Viatour, through Wikimedia Commons.*
Image Link: https://commons.wikimedia.org/wiki/File:Tyto_alba_1_Luc_Viatour.jpg
License Link: https://creativecommons.org/licenses/by-sa/3.0/deed.en

FIGURE 5.15   Bald eagle (*Haliaeetus leucocephalus*) in Yukon, Canada.   *Andreas Trepte, through Wikimedia Commons.*
Image Link: https://commons.wikimedia.org/wiki/File:Haliaeetus_leucocephalus_cayu.jpg
License Link: https://creativecommons.org/licenses/by-sa/2.5/deed.en

FIGURE 5.17   Great blue heron (*Ardea herodias*) in the wetlands of the Ocala National Forest, Florida, United States.   *Ianaré Sévi, through Wikimedia Commons.*
Image Link: https://commons.wikimedia.org/wiki/File:Ardea_herodias_(panting_in_water).jpg
License Link: https://creativecommons.org/licenses/by-sa/3.0/deed.en

successfully nest, roost, forage, and reproduce. Given the wide diversity of wildlife within and across continents of the world, the needs of a particular wildlife species will vary greatly; however, all terrestrial species require food, cover, water, and space (Yarrow and Yarrow, 1999). It is generally accepted that increases in the diversity of vegetation across a landscape will lead to increases in the value of the landscape as habitat for a variety of species (Whitaker and McCuen, 1976). As a result, since the late 20th century, many professionals

have shifted their focus from a narrow view that is guided by the habitat value of one or a few individual species to a broader view that recognizes the multiple values that wildlife habitat can provide for a wider mix of species (Johnson and O'Neil, 2001). One important aspect in understanding whether sufficient wildlife

FIGURE 5.18 Eastern wild turkey (*Meleagris gallopavo*) in Avery County, North Carolina, United States. *Ken Thomas, through Wikimedia Commons.*
Image Link: https://commons.wikimedia.org/wiki/File:Wild_Turkey-27527-1.jpg
License Link: Public Domain

FIGURE 5.20 Resplendent quetzal (*Pharomachrus mocinno*) native to Central America. *Peter Förster, through Wikimedia Commons.*
Image Link: https://commons.wikimedia.org/wiki/File:Pharomachrus_mocinno_-male.jpg
License Link: Public Domain

FIGURE 5.19 Willow ptarmigan (*Lagopus lagopus*) in Yellowstone National Park, Wyoming, United States. *Stephen C. Grado.*

FIGURE 5.21 Eastern bluebird (*Sialia sialis*) in the central United States. *Mongo, through Wikimedia Commons.*
Image Link: https://commons.wikimedia.org/wiki/File:Eastern_bluebird1.jpg
License Link: Public Domain

habitat is available involves understanding the needs and requirements of a species throughout its life cycle. In many cases, healthy and bountiful wildlife populations are dependent on a mosaic of different habitats across a landscape. To complicate matters a bit, habitat requirements may vary by season of year, as some species (particularly birds) may migrate thousands of miles between breeding and wintering grounds.

The availability of food is a basic habitat requirement to which we can all relate. Whether the food is a McDonald's Big Mac, a sushi roll with tuna, gallo pinto (rice and beans), or a pile of acorns, the availability of the food plays a critical role in allowing a species, including humans, to live, grow, reproduce, and survive. The consumption of food allows wildlife species to generate energy, which is critical since they need energy for daily life functions, to reproduce, and to escape predators. Food is also important for predators, since they need energy to hunt prey in the first place. Not having enough food weakens a species' ability to move and to avoid being consumed by predators. A lack of food can also

FIGURE 5.22 American robin (*Turdus migratorius*) in Yellowstone National Park, Wyoming, United States. *Stephen C. Grado.*

FIGURE 5.24 Marbled salamander (*Ambystoma opacum*) in Durham County, North Carolina, United States. *Patrick Coin, through Wikimedia Commons.*
Image Link: https://commons.wikimedia.org/wiki/File:Ambystoma_opacumPCSLXYB.jpg
License Link: https://creativecommons.org/licenses/by-sa/2.5/deed.en

FIGURE 5.23 Gopher tortoise (*Gopherus polyphemus*) in Georgia, United States. *Dirk Stevenson.*

FIGURE 5.25 Australian green tree frog (*Litoria caerulea*). *Liquid-Ghoul, through Wikimedia Commons.*
Image Link: https://commons.wikimedia.org/wiki/File:Australia_green_tree_frog_(Litoria_caerulea)_crop.jpg
License Link: Public Domain

weaken a wildlife species' ability to ward off disease, which can then make it vulnerable to a variety of other threats.

Not all food sources are of the same quality. All wildlife species, as well as humans, have their preferred suite of foods. For example, while some humans may prefer pepperoni pizza, others would rather eat Chinese dumplings or perhaps gallo pinto. In the world of wildlife, food preferences are also observable for each species. For instance, white-tailed deer in the southern United States prefer to consume acorns, which are produced by various species of oaks (*Quercus* spp.). Apples are another favorite food source of the white-tailed deer,

which is a concern of landowners who grow apple trees near wooded areas. White-tailed deer will also consume a wide variety of leaves from bushes and trees, as well as forbs and grasses if a sufficient supply of acorns is not available. However, white-tailed deer tend to avoid eating vegetation with leathery or prickly foliage. Other wildlife species, such as the gray wolf, prefer to consume ungulates such as deer, elk, caribou, or moose, but they will also consume smaller animals, such as hares, badgers, squirrels, and mice, as well as lizards, snakes, and frogs, when larger prey is in short supply.

Even though wildlife species express preferences for food, they will typically consume whatever is available in order to generate the energy needed to reproduce and to survive. Like humans, they would prefer to

FIGURE 5.26 Cane toad (*Bufo marinus*) in Cabo Rojo, Puerto Rico, United States. *Francisco J. Vilella.*

FIGURE 5.27 Green iguana (*Iguana iguana*). *Christian Mehlführer, through Wikimedia Commons.*
Image Link: https://commons.wikimedia.org/wiki/File:MC_GruenerLeguan.jpg
License Link: https://creativecommons.org/licenses/by/2.5/deed.en

FIGURE 5.28 Eastern indigo snake (*Drymarchon corais couperi*) in Georgia, United States. *Dirk Stevenson.*

FIGURE 5.29 Eastern diamondback rattlesnake (*Crotalus adamanteus*). *Tim Vickers, through Wikimedia Commons.*
Image Link: https://commons.wikimedia.org/wiki/File:Crotalus_adamanteus_(5).jpg
License Link: Public Domain

simply consume certain types of food, but generally cannot because a limitless supply of the preferred food is usually unavailable. In many cases, this does not have an adverse impact on wildlife species health (Yarrow and Yarrow, 1999). Unfortunately, if a particular habitat only provides low-quality food sources, then the health and vigor of a particular wildlife species could be adversely affected. Low-quality food supplies could lead to weak individuals and potentially affect or inhibit reproduction processes.

Another key requirement of wildlife habitat is the cover that exists within a landscape. Cover is used by different species of wildlife for many purposes such as nesting, breeding, roosting, rearing young, and escaping predators (Yarrow and Yarrow, 1999). Predators use cover as a venue for creeping up on and stalking potential prey. Cover may also be used as thermal protection during extremely hot or cold periods. Cover requirements of different wildlife species can vary greatly. For instance, white-tailed deer in the southern United States commonly bed down (sleep) in dense coniferous or deciduous forests or in places that contain a dense collection of understory vegetation. In the northeastern United States, it is not uncommon to find deer in coniferous stands of trees in the winter, because these types of

FIGURE 5.30   Rainbow trout (*Oncorhynchus mykiss*) in pond, Kaibab National Forest, north of Grand Canyon, Arizona, United States. *Andrew J. Sánchez Meador.*

FIGURE 5.31   Horny toad (*Phrynosoma platyrhinos*) near Flagstaff, Arizona, United States.   *Andrew J. Sánchez Meador.*

forests exclude more snow than do pure deciduous stands of trees. Deer, however, will venture out into open areas such as clearcuts, fields, or suburban backyards to forage on the various grasses, forbs, shrubs, or herbaceous plants located there. Yarrow and Yarrow (1999) suggest that, when frightened, deer will flee to a wooded area and quickly stop running because they feel more secure within the cover of the surrounding vegetation.

A third requirement of wildlife habitat involves the availability of water. As with humans, all wildlife species require some level of water consumption in order to survive and reproduce. In addition, bodies of water can be the specific places where birds, mammals, or other species, such as the North American river otter (*Lontra canadensis*), can locate the food that they prefer to consume. Perhaps these are the only areas where the sources of food that they consume exist. Water is also critical for regulating body temperature, metabolism, and digestion, and for facilitating the removal of metabolic wastes (Yarrow and Yarrow, 1999). Some wildlife species, such as roadrunners (e.g., *Geococcyx velox* of Mexico and Central America and *Geococcyx californianus* of the southwestern United States) or horny toads (*Phrynosoma platyrhinos*) (Fig. 5.31), have adapted to arid areas with low amounts of annual rainfall, while other species require sufficient water (e.g., red salamanders [*Pseudotriton ruber*] of the United States) and require landscapes that contain depressional wetlands, ponds, and other hydrologic features.

Two example species with their habitat requirements are described next: the gopher tortoise of the southern United States and the northern spotted owl of western North America. The gopher tortoise (Fig. 5.23) is a land turtle whose population is thought to currently be in decline (McCoy et al., 2006). The tortoise is commonly found in dry upland sites within the coastal land areas of the southeastern United States, from Louisiana to Georgia. These land tortoises prefer the habitat provided by longleaf pine sandhills, pine flatwoods, and coastal grasslands (Wilson and Mushinsky, 1997). Suitable habitat must contain well-drained, sandy soils for the development of tortoise burrows, yet they can also survive in pastures and along roadsides (Diemer, 1992). Forested areas that are not frequently burned are considered less suitable as habitat for the gopher tortoise.

The northern spotted owl is both a famous and controversial wildlife species because of the conflict between the species' habitat requirements and the wood production required to maintain employment in local communities along the western coast of North America. The spotted owl prefers the habitat provided by a variety of coniferous forest types, such as those that contain significant portions of ponderosa pine (*Pinus ponderosa*), Douglas-fir (*Pseudotsuga menziesii*), sugar pine (*Pinus lambertiana*), evergreen hardwoods, western hemlock (*Tsuga heterophylla*), grand fir (*Abies grandis*), Sitka spruce (*Picea sitchensis*), and redwood (*Sequoia sempervirens*) (Carey et al., 1992; Diller and Thome, 1999; North et al., 1999; Franklin et al., 2000; Irwin et al., 2004; U.S. Fish and Wildlife Service, 2004; Dugger et al., 2005; Gaines et al., 2010). The spotted owl typically nests in the tops of broken trees, snags (dead trees), and cavities which are commonly present in older, larger, live trees. Although the species can be found in a variety of forest types, it generally prefers a sufficient area of older forests across a landscape for foraging and roosting activities, and in some areas prefers landscapes with gentle slopes (Everett et al., 1997; Gaines et al., 2010). It feeds primarily on arboreal mammals such as flying squirrels (*Glaucomys sabrinus*) or tree voles (*Arborimus longicaudus*); in the southern portion of its range it will consume dusky-footed wood rats (*Neotoma fuscipes*).

5.3 Characterization of habitat types

Habitat characterizations are generally based on scientific principles and expert opinion and can require a significant amount of time and energy to delineate broadly across a large landscape. In their assessment of the states of Oregon and Washington in the United States, Chappell et al. (2001) delineated 32 broad habitat types that ranged from lowland conifer-hardwood forests to shrub-steppe lands, bays, and estuaries. Of these, perhaps only a dozen contained forests. Location-specific habitat types may be more useful in land-use planning, and these would be tailored perhaps to the species and conditions within the land areas that an organization manages. The factors one may use to delineate habitat types include climate, elevation, soils, hydrology, geology, and topography, as was noted in the earlier discussion regarding northern spotted owls. The pattern of vegetation within the landscape is also important, as is the structural condition of the vegetation, the tree species composition, and the successional pathways that are assumed for the vegetative resources. Finally, the assumed natural disturbance regime of the region and the planned or prior anthropogenic activities may influence how habitat is characterized (Chappell et al., 2001). These factors can allow one to develop broad characterizations of habitat types, and if you were to delve into one of the many habitat models designed for a species of wildlife, you may find that habitat can be characterized on a quantitative scale (usually 0 to 1) or qualitative scale (poor to optimal) using measures or assessments of these factors.

5.4 Succession and stand conditions

One misguided assumption about land management is that when suitable wildlife habitat is available, the arrangement of the habitat across a landscape will result in healthy and viable wildlife populations. In general, simply because an area contains what we consider to be high-quality habitat does not ensure that wildlife species of interest will inhabit the area. Further, the concept of *wildlife habitat* does not address uncertainties related to natural disasters or genetics, and at times it may fail to address demographic issues specific to a species or fail to address other environmental issues (e.g., climate, predators) (O'Neil et al., 2001). However, even given the uncertainties and limitations of defining habitat, most

forest-based assessments of habitat quality involve the current structural conditions of the trees. Further, planning requires assessments and projections of forest succession (how forests change over time). Forest management may also be guided by the development and transition of CWD, which include the snags (standing dead trees) and downed logs scattered throughout a forest. Many wildlife species use these resources for various stages of their life cycle, and efforts to model the abundance and the changes in these resources through time are intimately tied to forest succession and current forest conditions (Lester and Beatty, 2002).

The succession of a forest refers to how it may change over time due to human-caused or natural processes, and we discuss this again in Chapter 10 when forest dynamics are presented. The structure and condition of a forest are characteristics that a natural resource manager can manipulate to achieve various management objectives. In forestry, this manipulation is called *silviculture*. The condition of wildlife habitat is also highly dependent on concepts of ecological succession. Ecological succession is the process of orderly change of an area's structure with living organisms becoming established over time and transitioning through different plant, vertebrate, and invertebrate communities until a climax stage is reached. If an area is composed of rocks or sand, or devoid of any living organism, then it will experience what is known as *primary succession*. Primary succession is the process where organisms such as lichens, blue-green algae, lyme-grass (*Leymus arenarius*), marram grass (*Ammophila breviligulata*), and sea couch grass (*Agropyron pungens*) become established and slowly alter the site conditions for the arrival of the next stage of organisms. For instance, lichens will trap debris, which will slowly increase soil depth as well as increase soil moisture content and lead to the accumulation of additional organic matter. Eventually, this process changes the physical environment to favor the establishment of annual herb species. As this process continues, the physical environment becomes more favorable for grass species, shrubs, and eventually large woody plants (such as trees) to colonize. The initial collection of tree species on the site will then transition from shade-intolerant to shade-tolerant tree species. What this implies is that the initial set of tree species on the site prefer to become established in full sunlight conditions (thus are shade intolerant), while later in the development of the forest, other tree species that are able to become established in lower light conditions (thus are shade tolerant) work their way into the canopy of the forest. In the northeastern United States and Canada, this transition might be from eastern white pine (*Pinus strobus*) and birch to sugar maple (*Acer saccharum*) and oaks.

Another form of succession is *secondary succession*; secondary succession could either be a natural event caused by disturbances such as hurricanes, or wildfires, wind, insect outbreaks and can often be a combination of all of these or could occur in areas that have been disturbed by humans. Given the abundance of resources already available on these sites, such as developed soils, organic matter, and local plant seed sources, the disturbed area is relatively quickly colonized by pioneer plant species (grasses, herbs, shrubs, and trees), and then succession proceeds through a series of plant communities until a climax state is reached. Another way to think about primary and secondary succession is that both concepts have most of the same transitional steps, but that primary succession begins earlier in a developmental timeline and secondary succession begins later. The final stage of forest succession is called the *climax* stage. When a forested area reaches this level of succession, it is assumed that the ecosystem will perpetuate itself indefinitely in terms of plant and wildlife species presence and diversity. Unfortunately, achieving the climax stage of succession may be impossible in many instances due to constant change associated with human and natural disturbances.

Odum (1960) noted that in the southern United States the speed of succession of plant communities can vary depending on the productive quality of an area. For example, Odum's research in the southern United States suggested that succession proceeds faster as the percentage of silt and clay in a soil increases and as the depth to a water table decreases. For example, on old, abandoned agricultural fields near the Savannah River in South Carolina, five forbs and grass species were recorded on well-drained soils during the first year after abandonment, yet 12 species were found on the poorly drained soils that contained higher clay and silt content (Odum, 1960). In the second year, six new species were found on the well-drained soils, and 12 new species were found on the poorly drained soils. Further, as an area passes through different stages of succession, the character of habitat conditions for different wildlife species will change (Smith, 1980). For instance, open fields in New York might initially support eastern meadowlarks (*Sturnella magna*), meadow voles (*Microtus pennsylvanicus*), eastern cottontail rabbits, short-tailed shrews (*Blarina brevicauda*), grasshopper sparrows (*Ammodramus savannarum*), and grasshoppers. As these fields transition into mixed herbaceous and shrubby habitat, species such as the ruffed grouse (*Bonasa umbellus*), purple finch (*Carpodacus purpureus*), and American robin begin to replace the voles, meadowlarks, and rabbits. The continued transition into more vertically structured vegetation, such as trees, changes the suite of wildlife species present as well (Smith, 1980). For instance, wildlife species such as the white-tailed deer, American red squirrel (*Tamiasciurus hudsonicus*), red fox, and veery (*Catharus fuscescens*) become more common, while populations of robins, rabbits, song sparrows (*Melospiza melodia*), and purple finches decline.

5.5 Edge versus interior habitats

An important issue in wildlife habitat management concerns the spatial definition of the habitat area and the corresponding edges or interior areas that are present. An edge is an area of forest, for example, that is located next to an open field, a road, a clearcut, or a pasture (Fig. 5.32). Edges are, therefore, the junction of two different landscape features and can be either inherent (a long-term feature based on topography or soils, for example) or induced (short-term) as a result of management or land-use activity (Yahner, 1988). If a specific habitat area includes edges (either naturally or anthropogenically created), then the area will support different types of wildlife species than other areas that have less or no edge. The vegetation found along an edge may consist of numerous plant species that stratify different vertical layers. For instance, one may be able to see a progression of vegetation that begins with grass, then transitions into low shrubs, higher shrubs, and finally trees. Smith (1980) once indicated that in New York edges may be beneficial and support wildlife species such as grasshopper sparrows, meadowlarks, rabbits, and deer. However, edges may also have negative

FIGURE 5.32   A forested edge along Garadhban Forest in Scotland. *Chris Upson, through Wikimedia Commons.*
Image Link: https://commons.wikimedia.org/wiki/File:Edge_of_the_Forest_-_geograph.org.uk_-_79094.jpg
License Link: https://creativecommons.org/licenses/by-sa/2.0/deed.en

consequences for certain wildlife species. For example, bird nest predation may be higher in edges due to the greater density of predators and due to easier access to prey (Hanley et al., 2005). Interior habitat areas are located some distance away from an edge, perhaps in the middle of a large contiguous forest, and contain very little structure that what one would define as edge. Commonly, these are located inside larger expanses of forested areas where the height of the canopy and the density of trees are relatively consistent. Interior forest areas are important for supporting the habitat requirements of wildlife species such as American red squirrels, deer, white-footed mouse (*Peromyscus leucopus*), and red fox.

When managing wildlife habitat across a broad landscape it is important to understand the dynamics between edges and interior areas as two basic wildlife habitat components. Edges are where two types of habitat intersect, such as a forest and meadow; interior forest habitat is an area not influenced by other surrounding habitats. In providing sufficient quality habitat for wildlife species that require interior forests, we need an understanding of the minimum size (hectares or acres) of continuous mature forests for each targeted wildlife species. If an interior forest area is too small, then it will likely be insufficient to support the nesting, foraging, roosting, and breeding activities of a targeted wildlife species that requires interior forest habitat. In many areas of the world the loss of interior forest area has become a major management issue, and across broad areas the situation is commonly referred to as *forest fragmentation*. In parts of the world with a relatively high percentage of private land ownership, this form of land tenure could compound the situation, for as private forest lands are handed down from one generation to the next, they may be divided into even smaller parcels and managed separately and independently. In effect, forest fragmentation may reduce the amount of quality interior forest habitat and increase the amount of edge habitat.

5.6 Riparian zones and wildlife corridors

Riparian zones are strips of vegetation that border water bodies such as rivers, streams, vernal pools, ephemeral creeks, ponds, and lakes. The size and width of the zones can vary tremendously from 3 to 50 meters (m; 9.8–164 feet [ft]) on both sides of the water body. Typically, these zones are left somewhat intact after nearby management actions or disturbances have occurred. For example, nearby actions might include the harvest of an upland forest (Fig. 5.33) or the conversion of higher ground to some other land use, such as an agricultural field, suburban residential area, or parking

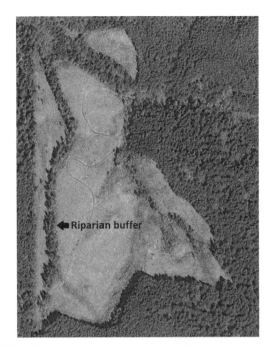

FIGURE 5.33 An aerial view of a riparian buffer along a stream in western Oregon, United States. *U.S. Department of Agriculture, Natural Resources Conservation Service (2020).*

lot. Thus, areas or zones along or near a water body could retain the vegetation that was present prior to the disturbances that occurred in the upland areas. Many forest practice regulations around the world have developed regulations to protect these important habitat types.

The required width and structure of riparian zones are often defined in local, state (provincial), or national laws, and where these do not exist, they are sometimes defined in organizational policies or in guidelines adopted widely and known as *best management practices*. These vegetative strips serve many important functions. First, they can serve as a filter of various types of sedimentation or debris that travels over the landscape during rain or flood events, acting as a tool for reducing erosion within landscapes. Second, the shade from the trees within the riparian zone can mitigate increases in the water temperature of streams and rivers that may otherwise occur if sunlight were allowed to directly strike the surface of the water. This temperature mediation service can then facilitate the maintenance of suitable habitat for various fish species. Third, the overstory vegetation within a riparian zone can provide some of the habitat requirements for various mammal and bird species. Fourth, these zones may act as corridors or pathways for animals to travel from one forested area to another and can serve as forms of cover during travel. Fifth, the leaf litter supports the trophic systems in the streams creating food for a variety of macroinvertebrates that form food chains in mountain streams

(Garcia, 2001). Finally, these zones are generally more biologically diverse than the adjacent areas.

As we suggested, riparian zones can act as wildlife corridors, but the term *wildlife corridor* can also pertain to other land areas that are not directly associated with water bodies. Wildlife corridors are designed with the purpose of mitigating the impacts of habitat fragmentation. The pathway attribute of a corridor is important in preventing certain wildlife species from becoming isolated. Isolation can reduce the local wildlife species and genetic diversity and can result in long-term impacts on species, health, reproduction, and survival (Payne and Bryant, 1998; Draper et al., 2017). Corridors allow wildlife species to move more easily or securely around their home ranges, promote the dispersion of juvenile members of specific animal or insect populations, and facilitate seasonal migration patterns (Payne and Bryant, 1998). However, some wildlife species, such as amphibians, reptiles, and even some small mammals, may require a significant amount of time to utilize a corridor for these purposes.

5.7 Direct and indirect management of habitat

Management of land affects wildlife habitat in a direct manner and affects wildlife species in both direct and indirect manners. All forests support some type of wildlife habitat yet changes in the structure of forests change their character and their ability (positively or negatively) to support the habitat requirements of certain species. Therefore, the alteration of native vegetation may directly determine the number and population of wildlife species found in the area affected (Theobald et al., 1997; Rocha et al., 2018; Shaffer et al., 2019). Developmental actions that do not change the character of a forest, such as the creation of roads or hiking trails, may indirectly affect wildlife. Hiking or camping activities, for example, may cause some wildlife species in the affected area to alter their normal pattern of activities, which may result in higher energetic costs or higher losses of nests due to abandonment or predation (Theobald et al., 1997). In some areas of the world there is a growing interest in testing alternative management actions that would reduce the deleterious effects incurred on wildlife habitat (Hanley et al., 2005). Often the purpose of the test is to determine whether the wildlife habitat present (or projected to be present in the future) will be of sufficient quality for a particular wildlife species. This can be a complex problem depending on the number and diversity of wildlife species that a forest or natural resource manager wishes to cultivate. The objectives of management actions may better suit one species than another, and often managing for a particular species requires not looking specifically at a forest stand

but more broadly across a landscape. Nevertheless, when managing for wildlife species within a specific forest stand, tools are used to modify an existing habitat structure and make it more suitable for specific wildlife species. For instance, eastern white-tailed deer prefer foraging in open areas because it is there where they locate edible plants that require high levels of direct sunlight. The creation of clearcuts or small gaps in mature forest areas can help make available these forage areas.

Silviculture is the art and science of manipulating forest vegetation to enhance forest growth and establish new forest stands and wildlife habitat. Our venture into silviculture is brief since we discuss a number of silvicultural techniques in Chapter 11. Forests are often managed in an even-aged or an uneven-aged system. An even-aged system is one where the dominant trees in the forest are all about the same age. An uneven-aged system is one where there is a greater diversity of tree age classes in the forest. Both of these systems of silvicultural management have advantages and disadvantages for creating or maintaining wildlife habitat. For example, even-aged systems may provide many stages of succession, horizontal diversity, and reductions in management costs while promoting higher-valued timber products (Yarrow and Yarrow, 1999). *Horizontal diversity* is where a large number of habitats exist across a stand or forest. *Vertical diversity* is where a large number of layers exist between the ground and the stand's canopy. While there may be high horizontal diversity across even-ages forests, the disadvantages of even-aged management systems include less vertical diversity within individual forest stands, a general lack of snags and overmature trees, and a low diversity of tree age classes, tree species, and tree sizes. Even-aged systems also tend to contribute to habitat fragmentation since the main processes for reestablishing even-aged forests are clearcutting and other similar actions requiring the removal of the majority of overstory trees. Although cost-effective and beneficial for producing early successional wildlife habitat, clearcutting can be viewed as aesthetically displeasing by some people immediately after a logging operation, and the process can potentially cause soil erosion on steep slopes. This aesthetic concern usually fades once forest vegetation grows to a significant height. Even-aged systems can also be created through natural regeneration processes (after fires, windstorms, insect outbreaks, volcanoes, etc.) and silvicultural systems designed to facilitate regeneration of a forest without the complete removal of overstory trees during the harvesting operation.

With regards to uneven-aged forests, these areas contain multistoried tree canopies and have a wider diversity of tree species, ages, and sizes (Yarrow and Yarrow, 1999). In addition, there are generally more snags and mature trees per unit area in uneven-aged forests. The disadvantage of uneven-aged forests is that management requires more frequent entries of heavy machinery, which can cause damage to residual live trees, reduce their commercial value, and increase their vulnerability to insects and diseases. While there may be high vertical diversity in these forests, some disadvantages of uneven-aged forests are that there may be lower horizontal diversity, and these areas lack an early secondary successional stage which would be beneficial to certain kinds of wildlife. Finally, uneven-aged silviculture is generally more costly to implement than even-aged silviculture.

Other silvicultural techniques that are commonly used in wildlife management include managing tree planting density and conducting selection harvests (Fig. 5.34), thinnings, fertilization treatments, and burning operations. The selection of a planting density is an important decision that forest, and natural resource professionals make when establishing a new stand or regenerating an old one, yet there are trade-offs involved. For example, if a new forest is planted with trees that have a wide spacing (the trees are planted far apart), the development of abundant herbaceous plants can be facilitated, which may be useful as forage for a wide variety of wildlife species. However, since healthy trees grown in open conditions produce a large amount of foliage, the quality of wood grown for timber purposes may be lower because the trees will have many branches and knots. When trees are planted closer together, smaller amounts of foliage are produced, and

one should expect to obtain better wood quality. However, when trees are planted closer together, at some point the herbaceous plant growth will be limited due to lack of direct sunlight.

Thinning operations involve removing a specified number of tree stems from a forest to provide more growing space for the residual (remaining live) trees. This action can produce healthier residual trees as well as open spaces where direct sunlight can reach the forest floor. The openings created in the canopy allow the growth of herbaceous plants, which can be a source of forage for many wildlife species. The openings produced by thinnings promote the establishment of native grasses, forbs, and other nonwoody plants such as raspberries, and therefore may represent good sites for foraging, brooding, and nesting activities. Thinning can also increase the horizontal diversity of a landscape (Yarrow and Yarrow, 1999). The shape and form of openings from a thinning operation must be planned ahead of time to maximize benefit to targeted wildlife species.

Another management activity that improves wildlife habitat is the development of snags (standing dead trees) in an area after a silvicultural or management operation. The level of decay of a snag can vary greatly around the world and is based on a tree species' chemical and physical wood properties and local climatic conditions. Snags provide opportunities for nesting (Fig. 5.35) and roosting of numerous bird species. In addition, they may host insect populations, which are an important food source for many bird species. In North America, bird

FIGURE 5.34   Selective harvest of hazard tree in a ponderosa pine stand near Mayhill, New Mexico, United States. *Andrew J. Sánchez Meador.*

FIGURE 5.35   A bird cavity in a lodgepole pine snag near Flagstaff, Arizona, United States. *Andrew J. Sánchez Meador.*

species that typically use snags include woodpeckers, bluebirds, red-tailed hawks, and chickadees (Yarrow and Yarrow, 1999). In addition to the decay rate, the initial size of snags is another important factor in the development or maintenance of habitat since some wildlife species prefer larger snags over smaller snags.

Burning is a common silvicultural tool used in many areas of the United States and various countries around the world. Prescribed fires (Figs. 5.36 and 5.37) are planned burns that are typically used to reduce fuel loads within a forested landscape or to prepare an area for tree planting activities. Prescribed fires can be beneficial in reducing the risk of uncontrolled wildfires by reducing fuel loads especially in the smaller or finer fuels. With regard to modifying wildlife habitat, fire can help reduce or increase horizontal diversity by discouraging or promoting the growth of forbs, grasses, herbaceous plants, and other trees. Many wildlife species are dependent on periodic forest fires. In the southern United States, animals such as the red-cockaded woodpecker (*Picoides borealis*) and the gopher tortoise

FIGURE 5.36 Initiating a prescribed fire for improving wildlife habitat in Mississippi, United States. *Amy Castle Blaylock.*

FIGURE 5.37 A low-intensity prescribed fire in a western United States ponderosa pine forest. *Andrew J. Sánchez Meador.*

are highly dependent on fire for producing the necessary habitat for nesting and foraging activities. Plant species such as the green pitcher plant (*Sarracenia oreophila*) and wiregrass (*Aristida stricta*) require frequent fires to produce the site conditions necessary for establishment and growth. In the eastern United States, forested habitats that are burned on a three-to five-year cycle may become favorable nesting and foraging habitat for the eastern wild turkey. Further, in the southern United States, annual fires can produce habitat that has an abundant supply of insects and seeds required by the northern bobwhite quail.

5.8 Habitat models

Habitat models are tools that are designed to help assess the quality of an area to act as habitat for a specific wildlife species (Garcia and Armbruster, 1997; Ahmad et al., 2018; Krainyk et al., 2019; Thomasma and Cleveland, 2019). These models can be developed through expert opinion, personal experience, or qualitative and quantitative measurements. The scope of these models ranges from simple qualitative descriptions of relative habitat quality, to complex nonlinear relationships that may require computer programming and data (spatial and nonspatial) management skills. Models are grouped into several categories including habitat evaluation procedures, habitat suitability models, Bayesian and pattern recognition models, species-habitat relationships, statistical models, empirical models, and coarse-filter models (Roloff et al., 2001). Methods for developing a quantitative index of habitat quality were developed during the 1970s (Whitaker and McCuen, 1976), and a number of habitat suitability models have been developed over the last 40 years (Sallustio et al., 2017) including the classic series of models developed by the U.S. Fish and Wildlife Service in the early to mid-1980s. Planners and managers are constantly seeking improved tools or models for understanding landscape-level management activities on wildlife habitat (Santelmann et al., 2006). However, models are simplified representations of the inputs, outputs, and functional relationships involved within a real system, since real-world systems are generally too complex to be completely understood. This statement is not meant to deter the use of models in forestry and natural resource management, since modeling is an important component of the decision-making process (Patton, 2010; Roloff et al., 2001).

One example of a simplified habitat suitability model is the white-tailed deer model for the Piedmont region of the southern United States (Crawford and Marchinton, 1989). The model was based on winter forage requirements for this deer species. Components include estimates of the green herbaceous vegetation available during the winter (weight per unit area), the basal area of oak trees 25 centimeters (cm) (10 inches [in]) in diameter and greater, the number of oak species present in the forest, the productive quality of the forest (as represented by a site index), and the amount and proximity of agricultural land in the nearby area. These components are assessed, and a score between 0 and 1 is assigned to each. The scores for the components are then combined in a nonlinear fashion that produces an overall assessment of the forest ranging from poor deer habitat suitability (a score near 0) to optimal deer habitat suitability (a score closer to 1).

5.9 Threats to wildlife habitat

A threat is anything that could possibly cause harm to an existing condition. In this case, a threat to wildlife habitat is anything that endangers the quality or the existence of wildlife habitat across a landscape. Although there are many possibilities, the most critical threat is the destruction of wildlife habitat through the conversion of land to alternative uses, such as agricultural fields or urban settlements. Another threat is the invasion into existing wildlife habitats of invasive plant and animal species from other parts of the globe. The conversion of forested lands to alternative land uses is not a phenomenon unique to any part of the world. As we discussed in Chapters 1 and 2, the conversion of forested areas to agricultural uses has been an integral part of human history for many thousands of years. For example, some forested areas of Central and South America have been actively converted to cattle ranches since the late 20th century to produce beef for local household consumption as well as for the fast food industry. In the southern United States, developed (structures, driveways, parking lots, sidewalks, etc.) and transportation (roads, railroads, etc.) areas represented about 3.1% of the total land area in 2013, increasing from about 2.9% seven years earlier, from land uses that included forests (Bettinger and Merry, 2019). As world populations continue to grow, more land and new technologies are needed to feed the never-ending supply of hungry human mouths.

Overhunting is another important threat related to the quality of wildlife habitat. Although hunting pressure does not directly change the vegetative state of a forest, it can upset the ecological balance within a biological community. For instance, in North America, gray wolves once ranged across most of the conterminous United States (the lower 48 states). Bounties and extensive hunting nearly eradicated this large carnivore from every state except Minnesota. Although extensive reintroduction programs have been successful in some areas of the western United States, the initial near eradication of the gray wolf has forced us to use hunting as the primary tool for limiting the expanding populations of eastern white-tailed deer. In Yellowstone, the loss of the wolves has led to increase in elk population that resulted in damage to the riparian areas due to elk grazing on the aspens. The reintroduction of wolves has resulted in improved riparian conditions.

As mentioned earlier, another important threat to wildlife habitat is the introduction of invasive plant and animal species into forested communities and landscapes (Benez-Secanho et al., 2018; Zhai et al., 2018). The list of invasive species is quite extensive and concerns (environmental, economic, social, political, etc.) related to these vary widely around the world. As an example of these and their impact on wildlife species, presented below are brief discussions of one invasive amphibian species (cane toad) and two invasive plant species (kudzu and cogongrass).

5.9.1 Cane toad

Cane toads (*Bufo marinus*) are large terrestrial frogs (Fig. 5.26) native to Central and South America, but they are also found naturally in southern Texas. The cane toad has a voracious appetite (it is known for swallowing prey whole) and is a prolific breeder. It was introduced into areas it previously did not occupy for pest control purposes. These included countries bordering the Pacific Ocean (Australia, Fiji, New Guinea, and the Philippines), the Caribbean islands, and Florida. The cane toad will consume almost anything, and its diet includes small rodents, reptiles, amphibians, birds, plants, and many types of invertebrates. Unfortunately, this invasive toad excretes a poison that is toxic to everything else that attempts to subsequently consume it. As an interesting aside, the Embera-Wounan tribe in South America is thought to have once used the toxin from the cane toad for their poison arrows, although this cannot be confirmed.

The cane toad has become an important factor in the reduction of biodiversity in many areas. In Australia, it has adversely affected native populations of some lizard, land snake, and crocodile species (Shanmuganathan et al., 2010). The toxic excretion kills those creatures that attempt to consume it, which has led to a steady decline in frog-eating predators in Australia (Doody et al., 2009). Control measures for the cane toad are evolving. Many people simply trap the toads and destroy them. In Australia, local events are held and prizes are given to whoever eliminates the most cane toads. Unfortunately, this type of control action is only a short-term solution to the problem, since one female cane toad can lay about 30,000 eggs in one deposit. Other methods of control that are suggested include placing nests of meat ants (*Iridomyrmex purpureus*) near cane toad habitat or exposing the population to lung-worm disease. Thus far, no one approach has proven to be the best in controlling this invasive species.

5.9.2 Kudzu

Kudzu (*Pueraria montana* var. *lobata*) is a rapidly growing vine plant species (Fig. 5.38) native to southern and southeastern Asia. Kudzu arrived in the United States in 1876, and many people used it as an ornamental plant (Shurtleff and Aoyagi, 1977; Everest et al., 1999; Mitich, 2000). Over time, kudzu served a variety of roles including fiber production, food supply, medicinal use, and rapid ground cover development. Farmers in the United States were once impressed by kudzu's forage potential, and in the 1930s the United States Soil Conservation Service promoted the use of kudzu in soil conservation efforts (Everest et al., 1999; Britton et al., 2002). This human-induced spread of kudzu across the United States landscape continued until it was eventually listed as a noxious weed in 1970 (Miller and Edwards, 1983; Everest et al., 1999).

Kudzu grows aggressively, can out-compete other plant species for water and nutrients, and can cover an area of land relatively quickly. It is not uncommon to drive through rural parts of the southern United States and see an old homestead, power poles, trees, electrical lines (Fig. 5.39), and much of the surrounding area covered with kudzu vines (Shurtleff and Aoyagi, 1977). Kudzu patches prevent the establishment of other native plants (including all tree species) and can reduce the productive capabilities of an area, resulting in financial losses to a landowner (Beckwith and Dangerfield, 1996; Mitich, 2000; Britton et al., 2002). Studies by Ezell and Nelson (2006) and Grebner et al. (2011) have delved

FIGURE 5.39   Kudzu (*Pueraria montana* var. *lobata*) spreading along telephone pole wires near Lacey's Spring, Alabama, United States. *A. Taylor Hall.*

into the financial trade-offs of alternative chemical approaches to controlling this invasive plant species.

Even though kudzu is generally considered an invasive plant species in North America, it does provide a limited set of wildlife habitat benefits. The leguminous nature of kudzu makes it an excellent source of forage for wildlife. In addition, the dense blanket of kudzu provides cover for various wildlife species. White-tailed deer highly prefer to eat this invasive plant species and will do so until the leaves die in early winter (Jones et al., 2008). Although in the United States the Natural Resources Conservation Service (2010) lists kudzu as having moderate cover potential for large and small mammals, the loss of forgone higher quality wildlife habitat outweighs these benefits, which suggests land managers should eradicate this species across known areas of establishment.

5.9.3 Cogongrass

Cogongrass (*Imperata cylindrica*) is a perennial grass native to Asia that now infests half a billion ha (1.2 billion ac) worldwide (Holm et al., 1977; Dozier et al., 1998). As far as we know, cogongrass arrived unceremoniously in the United States in 1912 as packing material in boxes of imported goods. However, cogongrass is an aggressive plant species and out-competes native vegetation to form single species communities (Fig. 5.40) which are generally not favorable to the development of wildlife habitat (Lippincott, 1997; Dozier et al., 1998; Holzmueller and Jose, 2012). Patches of cogongrass form an extensive rhizome root system. This root system pierces the roots of competing plant species, such as native grasses and tree seedlings (Dozier et al., 1998; Prevost, 2007; Daneshgar et al., 2008). Although it is a sun-loving species, it can be found in areas with low levels of

FIGURE 5.38   Kudzu (*Pueraria montana* var. *lobata*) near Lacey's Spring, Alabama, United States. *A. Taylor Hall.*

FIGURE 5.40 Cogongrass (*Imperata cylindrica*) in Hancock County, Mississippi, United States. *Jon D. Prevost.*

sunlight, such as dense forests, and it can survive temperatures as low as −14°C (6.8°F) (Wilcut et al., 1988). A very hardy species, cogongrass even thrives in a fire-dominated environment.

Cogongrass spreads easily, but bare soil is necessary for seed germination or the translocation of leaf and root fragments (Prevost, 2007; Holzmueller and Jose, 2012). Man-made forest disturbances, such as harvesting activities, site preparation for planting, burning, or mechanical operations, can produce ideal site conditions for the spread of cogongrass (Dickens and Moore, 1974; Holm et al., 1977; Shilling et al., 1997; Prevost, 2007; Holzmueller and Jose, 2012). Small pieces of cogongrass rhizomes that become stuck on the tires of machines, grapples, tractor blades and tracks, and planting machinery are easily transported to new locations where colonization can begin (Willard, 1988; Dozier et al., 1998; Prevost, 2007; Holzmueller and Jose, 2012). Cogongrass can spread into an area freshly tilled using below-ground rhizomes that stretch back to an undisturbed patch (Prevost, 2007).

Currently, efforts are being implemented to control this invasive plant species in a number of areas of the world. For example, in Asia the spread of cogongrass is typically controlled by hand hoeing. In the Mississippi Delta region of the United States, agricultural fields rarely sustain active patches of cogongrass for very long due to constant tilling and disking activities, yet in other vulnerable areas approaches have focused on the use of herbicidal treatments (Prevost, 2007; Grebner et al., 2010; Ramsey et al., 2012). Effective herbicidal control of cogongrass patches can lead to measurable improvements in the recolonization of native plants, and, therefore, improvements in species richness, relative density, evenness, and diversity (Prevost, 2007). In southern Mississippi, recolonizing plants included southern dewberry (*Rubus trivialis*), variable panicgrass

(*Dichanthelium commutatum*), beak-rush (*Rhynchospora* spp.), goldenrod (*Euthamia tenuifolia*), and broomsedge (*Andropogon virginicus*).

Summary

An important set of natural resources that forests provide and facilitate are wildlife species and their associated habitat. In this chapter, we focused on wildlife habitat components and how they can influence the presence of certain species of wildlife. The character and abundance of wildlife habitat managed across a large forested area can also influence the diversity of wildlife species found there. The set of plant, animal, and insect species found around the world is enormous, but their basic requirements involve resources (food, water, cover) that facilitate nesting, roosting, and foraging activities. If we took a picture today of our favorite forest, it would provide us with an idea of the type and amount of habitat present at one point in time. However, we know that forests change over time due to successional processes, natural disasters, and human actions. If our favorite forest were harvested, treated with some intermediate type of silvicultural treatment, or blown down by a hurricane, then succession begins anew, and habitat characteristics will obviously be different. Changes in habitat type over time cause a shift in the number of wildlife species that can be found there. For instance, open fields are preferable to eastern cottontail rabbits, but as open fields change to closed forests, the rabbit population will decline while species such as the American red squirrel will likely increase. As natural resource managers, we need to remember that changes to habitat conditions continually occur. The actions we propose may therefore influence the trajectory of these habitat changes, even if our decision is to do nothing at all.

As the world's human population increases, and as people move around the globe in search of work, political asylum, vacation, or homes, they sometimes introduce plant or animal species that are not native to a specific environment. These nonnative species may have a competitive advantage against native species, and therefore may spread rapidly through an ecosystem. These nonnative introductions may be able to use resources that effectively prevent native species from surviving in localized areas. These nonnative introductions can also be harmful to certain wildlife species because they can directly reduce species numbers or alter habitat conditions, thus reducing the amount of cover, shelter, and food available. The invasive species we mentioned in this chapter are only a few examples that exist around the world. Foresters and natural resource managers should therefore become aware of the threats present in their local communities and forests.

Questions

(1) *Favorite wildlife species.* Choose two wildlife species that interest you and that can also be found in forests. For each species, write a one-to-two-page report that discusses where each typically nests, roosts, and forages, as well as any unique aspects you find about its life cycle. In addition, describe whether the population is threatened or endangered. If this is the case, describe the causal agents for the species' decline. Develop a short, five-minute PowerPoint presentation where you can express your findings to others in your class.

(2) *Edges and interiors.* In this chapter the importance and differences of wildlife species inhabiting forest edges and forest interiors was discussed. Locate and visit a forested edge and a forested interior where you live. Take some time (30 min or so) to quietly watch the edge during daylight hours and during dusk. Record the number and type of mammal and bird species that you see. Develop a short report describing what you found in both locations. Be sure to highlight the time of year (season) and duration of viewing time, as these may matter in the quality and quantity of results you obtain.

(3) *Endangered species.* For a specific area of the world, determine the number of threatened and endangered species that exist there. Why are they endangered? What factors have made them that way? What was the spatial distribution of the species before it became threatened or endangered? What is it now? What are the forest conditions thought to be necessary for nesting, roosting, or foraging activities? Create a table where the rows represent the species and the columns represent the factors you discovered and write a short report describing the findings.

(4) *International Union for Conservation of Nature red list.* Through the Internet, locate the IUCN red list of threatened species. Select an endangered or critically endangered species that uses forests. In a brief, one-page memorandum, describe the causes that you think have led to this distinction. You may to perform more research in order to fully understand the circumstances. In a small group (four to six person) environment, communicate your findings to your colleagues.

(5) *Invasive species.* Visit your state or provincial government's forestry or natural resources organization Internet site and locate information on invasive plant and animal species in your area. How many species can you find for both categories? Did they have a category that you did not expect? For one plant and one animal species, in a one-page report briefly describe it, discuss how it is being spread, and discuss what is currently being done to address the issue.

References

Ahmad, F., Goparaju, L., Qayum, A., 2018. Wildlife habitat suitability and conservation hotspot mapping: remote sensing and GIS based decision support system. AIMS Geosciences 4 (1), 66—87.

Beckwith III, J.R., Dangerfield Jr., C.W., 1996. Forest Resources One Liners. Extension Forest Resources Unit, Georgia Cooperative Extension Service. University of Georgia, Athens, GA. Bulletin FOR 96—040.

Benez-Secanho, F., Grebner, D.L., Ezell, A.W., Grala, R.K., 2018. Financial tradeoffs associated with controlling Chinese privet (*Ligustrum sinense* Lour.) in forestlands in the southern United States. Journal of Forestry 116 (3), 236—244.

Bettinger, P., Merry, K., 2019. Land cover transitions in the United States South: 2007-2013. Applied Geography 105, 102—110.

Britton, K.O., Orr, D., Sun, J., 2002. Kudzu. U.S. Department of Agriculture, Forest Service. In: van Driesche, F., Blossey, B., Hoodle, M., Lyon, S., Reardon, R. (Eds.), Biological Control of Invasive Plants in the Eastern United States. Forest Health Technology Enterprise Team, Morgantown, WV. FHTET-2002—04.

Carey, A.B., Horton, S.P., Biswell, B.L., 1992. Northern spotted owls: influence of prey base and landscape character. Ecological Monographs 62, 223—250.

Chappell, C.B., Crawford, R.C., Barrett, C., Kagan, J., Johnson, D.H., O'Mealy, M., Green, G.A., Ferguson, H.L., Edge, W.D., Greda, E.L.,

O'Neil, T.A., 2001. Wildlife habitats: descriptions, status, trends, and system dynamics. In: Johnson, D.H., O'Neil, T.A. (managing directors) (Eds.), Wildlife-habitat Relationships in Oregon and Washington, Oregon State University Press, Corvallis, OR, pp. 22—114.

Crawford, H.S., Marchinton, R.L., 1989. A habitat suitability index for white-tailed deer in the Piedmont. Southern Journal of Applied Forestry 13, 12—16.

Daneshgar, P., Jose, S., Collins, A., Ramsey, C., 2008. Cogongrass (*Imperata cylindrical*), an alien invasive grass, reduces survival and productivity of an establishing pine forest. Forest Science 54, 579—587.

Dickens, R., Moore, G.M., 1974. Effects of light, temperature, KNO_3, and storage on germination of cogongrass. Agronomy Journal 66, 187—188.

Diemer, J.E., 1992. Home range and movements of the tortoise *Gopherus polyphemus* in northern Florida. Journal of Herpetology 26 (2), 158—165.

Diller, L.V., Thome, D.M., 1999. Population density of northern spotted owls in managed young-growth forests in coastal northern California. Journal of Raptor Research 33, 275—286.

Doody, J.S., Green, B., Rhind, D., Castellano, C.M., Sims, R., Robinson, T., 2009. Population-level declines in Australian predators caused by an invasive species. Animal Conservation 12, 46—53.

Dozier, H., Gaffney, J.F., McDonald, S.K., Johnson, E.R.R.L., Shilling, D.G., 1998. Cogongrass in the United States: history, ecology, impacts, and management. Weed Technology 12, 737—743.

Draper, J.P., Waits, L.P., Adams, J.R., Seals, C.L., Steury, T.D., 2017. Genetic health and population monitoring of two small black bear (*Ursus americanus*) populations in Alabama, with a regional perspective of genetic diversity and exchange. PLoS ONE 12 (11), e0186701.

Dugger, K.M., Wagner, F., Anthony, R.G., Olson, G.S., 2005. The relationship between habitat characteristics and demographic performance of northern spotted owls in southern Oregon. The Condor: Ornithological Applications 107, 863–878.

Everest, J.W., Miller, J.H., Ball, D.M., Patterson, M., 1999. Kudzu in Alabama: History, Uses, and Control. Alabama Cooperative Extension System. Auburn University, Auburn, AL. Circular ANR-65.

Everett, R., Schellhaas, D., Spurbeck, D., Ohlson, P., Keenum, D., Anderson, T., 1997. Structure of northern spotted owl nest stands and their historical conditions on the eastern slope of the Pacific Northwest Cascades, USA. Forest Ecology and Management 94, 1–14.

Ezell, A.W., Nelson, L., 2006. Comparison of treatments of controlling kudzu prior to planting tree seedlings. In: Connor, K.F. (Ed.), Proceedings of the 13th Biennial Southern Silvicultural Research Conference, General Technical Report SRS-92. U.S. Department of Agriculture, Forest Service, Southern Research Station, Asheville, NC, pp. 148–149.

Franklin, A.B., Anderson, D.R., Guitiérrez, R.J., Burnham, K.P., 2000. Climate, habitat quality, and fitness in northern spotted owl populations in northwestern California. Ecological Monographs 70, 539–590.

Gaines, W.L., Harrod, R.J., Dickinson, J., Lyons, A.L., Halupka, K., 2010. Integration of northern spotted owl habitat and fuels treatments in the eastern Cascades, Washington, USA. Forest Ecology and Management 260, 2045–2052.

Garcia, L.A., Armbruster, M., 1997. A decision support system for evaluation of wildlife habitat. Ecological Modelling 102, 287–300.

Garcia, M.A.S., 2001. The role of invertebrates on leaf litter decomposition in streams — a review. International Review of Hydrobiology 86 (4–5), 383–393.

Grebner, D.L., Amacher, G.S., Prevost, J.D., Grado, S.C., Jones, J.C., 2010. Economics of cogongrass control for non-industrial private landowners in Mississippi. In: Gan, J., Grado, S., Munn, I.A. (Eds.), Global Change and Forestry: Economic and Policy Impacts and Responses. Nova Science Publishers, Hauppauge, NY, pp. 87–98.

Grebner, D.L., Ezell, A.W., Prevost, J.D., Gaddis, D.A., 2011. Kudzu control and impact on monetary returns to non-industrial private forest landowners in Mississippi. Journal of Sustainable Forestry 30, 204–223.

Hanley, T.A., Smith, W.P., Gende, S.M., 2005. Maintaining wildlife habitat in southeastern Alaska: Implications of new knowledge for forest management and research. Landscape and Urban Planning 72, 113–133.

Holm, L.G., Plucknett, D.L., Pancho, J.B., Herberger, J.P., 1977. The World's Worst Weeds: Distribution and Biology. University Press of Hawaii, Honolulu, HI, pp. 609.

Holzmueller, E.J., Jose, S., 2012. Response of the invasive grass *Imperata cylindrica* to disturbance in the Southeastern Forests, USA. Forests 3, 853–863.

International Union for Conservation of Nature, 2019. IUCN Red List of Threatened Species, Version 2019.3. International Union for Conservation of Nature and Natural Resources, Cambridge, United Kingdom. http://www.iucnredlist.org (accessed 20.01.20).

Irwin, L.L., Fleming, T.L., Beebe, J., 2004. Are spotted owl populations sustainable in fire-prone forests? Journal of Sustainable Forestry 18, 1–28.

Johnson, D.H., O'Neil, T.A., 2001. Introduction. Wildlife-habitat Relationships in Oregon and Washington. In: Johnson, D.H., O'Neil, T.A. (managing directors) (Eds.), pp. xv–xix Oregon State University Press, Corvallis, OR, pp. 768.

Jones, J.C., Arner, D.H., Byrd, J., Yager, L., Gallagher, S., 2008. Selected non-native plants of rights-of-ways (ROWs) in the southeastern United States and associated impacts. In: Goodrich-Mahoney, J.W., Abrahamson, L.P., Ballard, J.L., Tikalsky, S.M. (Eds.), The Eighth International Symposium of Environmental Concerns in Rights-Of-Way Management. Elsevier, Amsterdam, pp. 547–554.

Karr, J.R., 1981. Assessment of biotic integrity using fish communities. Fisheries 6 (6), 21–27.

Kelcey, J.G., 1975. Opportunities for wildlife habitats on road verges in a new city. Urban Ecology 1, 271–284.

Krainyk, A., Ballard, B.M., Brasher, M.G., Wilson, B.C., Parr, M.W., Edwards, C.K., 2019. Decision support tool: Mottled duck habitat management and conservation in the Western Gulf Coast. Journal of Environmental Management 230, 43–52.

Kudzu, 2009. Mississippi Wildlife. Fisheries, & Parks, Jackson, MS. http://home.mdwfp.com/wildlife/species/deer/articles.aspx?article=192 (accessed 16.07.11).

Lester, A.M., Beatty, I.D., 2002. Northeastern Coarse Woody Debris (NECWD) Model: User's Guide. University of Massachusetts, Amherst, MA, pp.18.

Lippincott, C.L., 1997. Ecological Consequences of *Imperata Cylindrica* (Cogongrass) Invasion in Florida Sandhill. Ph.D. dissertation. Botany Department, University of Florida, Gainesville, FL, pp. 165.

Livingston, M., Shaw, W.W., Harris, L.K., 2003. A model for assessing wildlife habitats in urban landscapes of eastern Pima County, Arizona (USA). Landscape and Urban Planning 64, 131–144.

McCoy, E.D., Mushinsky, H.R., Lindzey, J., 2006. Declines of gopher tortoise on protected lands. Biological Conservation 128, 120–127.

Miller, J.H., Edwards, B., 1983. Kudzu: where did it come from? And how can we stop it? Southern Journal of Applied Forestry 7, 165–169.

Mitich, L.W., 2000. Kudzu [*Pueraria lobata* (Willd.) Ohwi]. Weed Technology 14, 231–234.

Morrison, M.L., Marcot, B.G., Mannan, R.W., 2006. Wildlife-habitat Relationships, Concepts and Applications, third ed. Island Press, Washington, D.C., pp. 520.

Natural Resources Conservation Service, 2010. Plants Profile: *Pueraria montana* (Lour.) Merr. Kudzu. U.S. Department of Agriculture, Natural Resources Conservation Service, Washington, D.C. http://plants.usda.gov/java/profile?symbol=PUMO#wildlife (accessed 20.01.20).

North, M.P., Franklin, J.F., Carey, A.B., Forsman, E.D., Hamer, T., 1999. Forest stand structure of the northern spotted owl's foraging habitat. Forest Science 45, 520–527.

O'Neil, T.A., Bettinger, K.A., Vander Heyden, M., Marcot, B.G., Barrett, C., Mellen, T.K., Vanderhaegen, W.M., Johnson, D.H., Doran, P.J., Wunder, L., Boula, K.M., 2001. Structural conditions and habitat elements of Oregon and Washington. In: Johnson, D.H., O'Neil, T.A. (managing directors) (Eds.), Wildlife-habitat Relationships in Oregon and Washington. Oregon State University Press, Corvallis, OR, pp. 115–139.

Odum, E.P., 1960. Organic production and turnover in old field succession. Ecology 41, 34–49.

Patton, D.R., 2010. Forest Wildlife Ecology and Habitat Management. CRC Press, London, pp. 292.

Payne, N.F., Bryant, F.C., 1998. Wildlife Habitat Management of Forestlands, Rangelands, and Farmlands. Krieger Publishing Company, Malabar, FL, pp. 840.

Prevost, J.D., 2007. Evaluation of Forest Protection Practices on Cogongrass (*Imperata cylindrica* (L.) Beauv.) on the Mississippi Gulf Coast. Master's thesis. Mississippi State University, Starkville, MS, 108 pp.

Ramsey, C.L., Jose, S., Zamora, D., 2012. Cogongrass (*Imperata cylindrica*) control with Imazapyr and Glyphosate combined with and without four adjuvants. Southern Journal of Applied Forestry 36 (4), 204–210.

Rocha, E.C., Brito, D., Silva, P.M., Silva, J., Bernardo, P.V.S., Leandro, J., 2018. Effects of habitat fragmentation on the persistence of medium and large mammal species in the Brazilian Savanna of Goiás State. Biota Neotropica 18 (3), e20170483.

Roloff, G.J., Wilhere, G.F., Quinn, T., Kohlmann, S., 2001. An overview of models and their role in wildlife management. In: Johnson, D.H., O'Neil, T.A. (managing directors) (Eds.), Oregon State University Press, Corvallis, OR, pp. 512–536.

Sallustio, L., De Toni, A., Strollo, A., Di Febbraro, M., Gissi, E., Casell, L., Geneletti, D., Munalo, Vizzarri, M., Marchetti, M., 2017. Assessing habitat quality in relation to the spatial distribution of protected areas in Italy. Journal of Environmental Management 201, 129–137.

Santelmann, M., Freemark, K., Sifneos, J., White, D., 2006. Assessing effects of alternative agricultural practices on wildlife habitat in Iowa, USA. Agriculture, Ecosystems and Environment 113, 243–253.

Shaffer, J.A., Loesch, C.R., Buhl, D.A., 2019. Estimating offsets for avian displacement effects of anthropogenic impacts. Ecological Applications 29 (8), 1–15.

Shanmuganathan, T., Pallister, J., Doody, S., McCallum, H., Robinson, T., Sheppard, A., Hardy, C., Halliday, D., Venables, D., Voysey, R., Strive, T., Hinds, L., Hyatt, A., 2010. Biological control of the cane toad in Australia: a review. Animal Conservation 13 (Suppl. S1), 16–23.

Shilling, D.G., Bewick, T.A., Gaffney, J.F., McDonald, S.K., Chase, C.A., Johnson, E.R.R.L., 1997. Ecology, Physiology, and Management of Cogongrass (Imperata Cylindrica). Final Report. Florida Institute of Phosphate Research, Bartow, FL. Publication No. 03–107–140.

Shurtleff, W., Aoyagi, A., 1977. The Book of Kudzu: A Culinary and Healing Guide. Autumn Press, Brookline, MA, pp. 102.

Smith, R.L., 1980. Ecology and Field Biology, third ed. Harper & Row Publishers, New York, pp. 835.

Theobald, D.M., Miller, J.R., Hobbs, N.T., 1997. Estimating the cumulative effects of development on wildlife habitat. Landscape and Urban Planning 39, 25–36.

Thomasma, S., Cleveland, H., 2019. Wildlife habitat associations in SILVAH and NED. In: Stout, S.L. (Ed.), SILVAH: 50 Years of Science-Management Cooperation. Proceedings of the Allegheny Society of American Foresters Training Session, September 20-22; Clarion, PA, General Technical Report NRS-P-186. U.S. Department of Agriculture, Forest Service, Northern Research Station, Newtown Square, PA, pp. 120–131. https://doi.org/10.2737/NRS-GTR-P-186-Paper11 (accessed 19.01.20).

U.S. Fish and Wildlife Service, 2004. Northern Spotted Owl Five-Year Review: Summary and Evaluation. US Department of the Interior, Fish and Wildlife Service, Portland, OR. http://www.fws.gov/pacific/ecoservices/endangered/recovery/pdf/NSO_5-yr_Summary.pdf (accessed 20.01.20).

U.S. Department of Agriculture, Natural Resources Conservation Service, 2020. Web Soil Survey. U.S. Department of Agriculture, Natural Resources Conservation Service, Washington, D.C. https://websoilsurvey.sc.egov.usda.gov/App/HomePage.htm (accessed 25.02.20).

Whitaker, G.A., McCuen, R.H., 1976. A proposed methodology for assessing the quality of wildlife habitat. Ecological Modelling 2, 251–272.

Wilcut, J.W., Dute, R.R., Truelove, B., Davis, D.E., 1988. Factors limiting the distribution of cogongrass, Imperata cylindrica, and torpedograss, Panicum repens. Weed Science 36, 577–582.

Willard, T.R., 1988. Biology, Ecology and Management of Cogongrass (Imperata Cylindrica (L.) Beauv.). Ph.D. dissertation. Agronomy Department, University of Florida, Gainesville, FL, pp. 113.

Wilson, D.S., Mushinsky, H.R., 1997. Species Profile: Gopher Tortoise (Gopherus polyphemus) on Military Installations in the Southeastern United States. U.S. Army Corps of Engineers, Waterways Experiment Station, Viscksburg, MS. Technical Report SERDP-97-10.

Yahner, R.H., 1988. Changes in wildlife communities near edges. Conservation Biology 2, 333–339.

Yarrow, G.K., Yarrow, D.T., 1999. Managing Wildlife. Sweetwater Press, Birmingham, AL, pp. 588.

Zhai, J., Grebner, D.L., Grala, R.K., Fan, Z., Munn, I.A., 2018. Contribution of ecological and socioeconomic factors to the presence and abundance of invasive tree species in Mississippi, USA. Forests 9 (1), Article 38.

6

Ecosystem services

In previous chapters, we alluded to a number of ecosystem services that forests provide and have therefore designed this chapter to specifically and formally address the various types of ecosystem services that are facilitated by healthy forests. We begin with a definition of ecosystem services and proceed to a discussion of ecosystems in general. After these introductory concepts are presented, we categorize and describe ecosystem services by the function they are perceived to play (cultural, provisioning, regulating, and supporting). Finally, we provide a short discussion of the challenges related to managing forests for a broad suite of ecosystem services. At the completion of this chapter, readers should

- be able to describe the broad array of ecosystem services that are generated by forests;
- understand how to communicate to others the services an ecosystem can potentially provide;
- understand why ecosystem services are important to both human civilizations and the health of the planet;
- understand that there are trade-offs involved in managing forests for different ecosystem services; and
- understand the linkages between different ecosystem services.

The potential services provided by an ecosystem are all-encompassing. To fully value the manner in which they are classified may require us to view systems and processes within forests from a much different perspective than we have used previously to understand them.

6.1 What is an ecosystem?

The underlying concepts and theories that are used to help us understand ecosystem services include many branches of science such as biology, ecology, geology, and meteorology, to name a few. All of these fields are interwoven into a natural system that provides life to countless organisms of all shapes and sizes. However, to understand more thoroughly the ecosystem services that can be derived from forests, an understanding of the concept of a forest ecosystem first needs to be developed. A *forest ecosystem* has been defined by Nieuwenhuis (2010) as "an ecological system composed of interacting biotic and abiotic components of the environment, in which trees are the major component." More simply, many U.S. National Forests have defined an ecosystem as "the complete system formed by the interaction of a group of organisms and their environment" (U.S. Department of Agriculture, Forest Service, 1989). A forest ecosystem can also be viewed as one composed of a set of inputs and outputs: Inputs might include radiant energy (or light) from the sun, carbon dioxide (CO_2), oxygen (O_2), water, and other nutrients. Outputs or outflows might include carbon dioxide, oxygen, water, nutrients, and other by-products of tree growth processes (Smith and Smith, 2000).

The size of an ecosystem can vary and be defined as a log (Fig. 6.1), a pond (Fig. 6.2), a field (Fig. 6.3), a hill (Fig. 6.4), a small parcel of land, a county forest, a national forest, or more broadly, the Earth's biosphere

153

FIGURE 6.1 Decaying log ecosystem, which provides a medium for the establishment of seedlings such as western hemlock (*Tsuga heterophylla*) and is home to lungless salamanders. *T.J. Watt, through Wikimedia Commons.*
Image Link: https://commons.wikimedia.org/wiki/File:Avatar_Grove_Nurse_Log_3.jpg
License Link: https://creativecommons.org/licenses/by-sa/3.0/deed.en

FIGURE 6.3 Field ecosystem near Creag an Righeanan in Scotland. *Lis Burke, through Wikimedia Commons.*
Image Link: https://commons.wikimedia.org/wiki/File:Forest_edge_-_geograph.org.uk_-_915632.jp
License Link: https://creativecommons.org/licenses/by-sa/2.0/deed.en

FIGURE 6.2 Pond ecosystem near Čimelice village in Písek District, Czech Republic. *Petr Brož, through Wikimedia Commons.*
Image Link: https://commons.wikimedia.org/wiki/File:Čimelice-pond.jpg
License Link: https://creativecommons.org/licenses/by-sa/3.0/

FIGURE 6.4 Hill ecosystem, as represented by La Cienega Forest, a mature, unlogged area of Durango pine (*Pinus durangensis*) in the Sierra Madre Occidental in northwestern Mexico. *Pete Fulé.*

(Fig. 6.5) (Helms, 1998). To explore the development of a small forested ecosystem, we begin with some bare ground, which contains nutrients such as calcium, iron, nitrogen, phosphorus, potassium, and other minerals that are vital for plant growth. After primary succession occurs (which we discussed in Chapter 5), various sun-loving plants, such as grasses, forbs, and other herbaceous plants, will colonize the area. The material necessary for this process (plant seeds or vegetative parts) may have traveled to the area with the aid of the wind or water or may have been deposited through animal defecation. Fueled by their initial stores of carbohydrates and by radiant energy from the sun, these plants absorb nutrients and water from the soil,

stimulating biological growth. As these plants grow, they sequester carbon dioxide from the atmosphere and store it as biomass in various parts of their vegetative structure (e.g., trunks, branches, and twigs). Plants also respire and release oxygen back into the atmosphere. Over time, if larger plants such as trees become established on the site, they may envelope the sun-loving, vertically challenged plant species (grasses, forbs, and other herbaceous species) in shade (Fig. 6.6) to create a tree-dominated, or forested, ecosystem.

In temperate areas of the world, as the seasons change from fall to winter, many plants die back to ground level or *brown up*. For instance, many forms of grasses will die back to the root collar in the fall or the winter. Deciduous

FIGURE 6.5   The biosphere called Earth. *National Aeronautics and Space Administration, through Wikimedia Commons.*

FIGURE 6.6   Japanese beech (*Fagus crenata*) forest located in Shirakami Mountains of the Akita prefecture, Japan. *Enoki Yoshio.*

FIGURE 6.7   Deciduous trees, producing brilliant fall colors prior to the loss of their leaves, situated along the scenic Blue Ridge Parkway in the Appalachian Mountains near Asheville, North Carolina. *David W. Wilkinson.*

FIGURE 6.8   Fruiting bodies of the puffball mushroom (*Lycoperdon pyriforme*) growing on a decaying pine log near La Ronge, Saskatchewan, Canada. *Sasata, through Wikimedia Commons.*
Image Link: https://commons.wikimedia.org/wiki/File:Lycoperdon_pyriforme_Sasata_scale.jpg
License Link: https://creativecommons.org/licenses/by/3.0/deed.en

trees (what many typically consider to be *hardwood* trees) form buds and lose their leaves during the fall season (Fig. 6.7). However, most coniferous trees, such as pines and spruces, except for larch (*Larix* sp.) and some ancient species such as dawn redwood (*Metasequoia glyptostroboides*), maintain green foliage throughout the seasons. Some nutrients stored in the dying vegetation of the grasses and deciduous trees are translocated back into the main stem or root mass to be used again the following year when the buds begin to grow (or *leaf out*) (Barnes et al., 1998). The decaying organic matter produced by plants or trees plays an important role within the ecosystem. Most of the decaying plant material that lies on the ground slowly releases nutrients back into the soil. The decay process is facilitated by the activities of numerous types of microorganisms,

bacteria, and fungi (Fig. 6.8). These organisms transform the decaying vegetation into inorganic compounds, which are eventually filtered into the soil and absorbed once again by the roots of growing plants.

Insects and a wide variety of herbivores such as rabbits, deer, moose, elk, and caribou also aid in the transformation of vegetative matter to inorganic matter. They do so by eating vegetative matter, digesting it, and using it for their growth. Eventually, they defecate the material onto the ground and the vegetative material undergoes further transformation. Herbivores that range over large territories, such as deer or moose, may consume vegetation in one ecosystem and defecate the vegetative material in a neighboring ecosystem. This leads to the loss of nutrients by one ecosystem and the gain of nutrients by the other. Interestingly, many plant species have reproductive processes that are facilitated by the birds and animals that eat their fruit. In these cases, seeds may be transported to other locations within or outside of the ecosystem. The digestive acids of these animals scarify the seeds so that when they are defecated, they are more likely to germinate. These acts can assist in the process of succession, as discussed in Chapter 5, since seeds from some plant species, such as those from the pin cherry (*Prunus pensylvanica*), may lie dormant and viable in the soil for decades, awaiting favorable establishment conditions (Bormann and Likens, 1979).

Another input into an ecosystem is precipitation (rain- or snowfall) from the atmosphere. When precipitation occurs, some of the water is intercepted by existing vegetation, whether grass, shrubs, or trees, and may be absorbed by live or dead woody material. Other water will reach the ground and may be directly absorbed by plant roots. Some water will be absorbed by the woody debris and organic layer situated on top of the ground; this water will evaporate, be slowly released into the soil, or be used by various organisms in decomposition processes. As more water enters the ground, the soil will eventually reach saturation point and excess water will flow over the ground to lower elevation points such as ponds, streams, or rivers. If the ground is bare of vegetation or organic debris (i.e., no plant communities exist), rainwater will wash away some of the soil into nearby streams and rivers. This process not only depletes an area of valuable nutrients but also transports soil into water bodies (measured as *sedimentation*), which may adversely impact fish populations and other aquatic life. Sedimentation also acts to reduce water flow at hydroelectric dams and diminishes the quality of human drinking water. Many forested areas are components of the large water supplies of cities such as New York and Boston (Barnes et al., 2009). In essence, vegetation scattered across a landscape can serve as a filter for water as it makes its way from the atmosphere to the places (homes, pools, dams) in which we use it.

At the Hubbard Brook Experimental Forest in New Hampshire, a famous study of forest ecosystems by Bormann and Likens (1979) determined that when trees were removed from within a closed forested ecosystem, summer water flows from the system immediately increased. Without trees to utilize inputs of water to an ecosystem, water moves quicker through (and over) the ground and into stream systems. In addition, higher than normal amounts of nutrients, such as calcium, potassium, magnesium, and nitrates, may also leave the system. As vegetation begins to reestablish itself on a site,

water and nutrient flows out of the system decline. The effects of tree removal disturbances may be used as examples of what can be expected in a forest that has experienced a severe fire or wind (hurricane, cyclone) event. Some ecosystem processes may be disrupted until the vegetative plant communities pass through a reorganization phase and the ecosystem becomes regulated again.

There are many different types of outputs or outcomes from forested ecosystems; some of these have been discussed in Chapter 4 and others will be discussed further in Section 6.3. Perhaps the most important output from forested ecosystems is oxygen, which is needed by most life-forms on Earth. When trees are harvested for the purpose of creating lumber or paper, wood is the output, although, technically, sequestered carbon and certain amounts of nutrients are removed from the site. Wild foods and medicinal plants collected and consumed by humans can also be considered outputs. Although most of this discussion has implied that humans are responsible for removing resources from an ecosystem, nutrients are also imported and removed through the movement of animals between ecosystems.

An important question, when working in any ecosystem, concerns the ecosystem's carrying capacity for plant life. The *carrying capacity* is the maximum number of one or all of a species' life functions that can be supported over long periods. This concept applies to both plants and animals and is highly dependent on the initial soil productivity of the area, as well as local climatic conditions. For instance, a newly established tree seedling does not need a lot of horizontal space to obtain energy from the sun; nor does it need a considerable amount of water and nutrients from the soil. However, as a tree grows it requires more of these resources until it begins to compete with other trees and vegetation (Fig. 6.9). This competition may limit access to resources, which can effectively limit vertical, horizontal,

FIGURE 6.9 Young pine trees facing severe competition for light, nutrients, and water. *Andrew J. Sánchez Meador.*

and below ground growth. Soils with a high productivity potential and situated in wet climates can accommodate more vegetative competition, as plants attempt to locate vertical and horizontal growing space niches in order to utilize the existing resources.

Within an ecosystem, wildlife species also face a carrying capacity. For a given area and a given vegetation type, only a certain amount of food can be produced to meet the various dietary needs of wildlife, only a certain amount of structure will be available to facilitate nesting and foraging activities (Fig. 6.10), and only a certain amount of water will be available. Although these factors are discussed in greater detail in Chapter 5, it is important to link them to the carrying capacity concept. The number of individuals of a species that an ecosystem can support will also limit the number of predators that ecosystem can support. These species interactions are controlled by an ecosystem's carrying capacity and play an important role in sustaining its long-term health. Sometimes, through human intervention or the grace of nature, a particular species population will exceed an ecosystem's carrying capacity; this overpopulation can lead to challenging natural or human responses. For example, overpopulation of one species may stimulate an increase in that species' natural predators (desired or undesired), which may return the population to its normal carrying capacity. Sometimes, however, this adjustment has to be performed through human intervention. For instance, many residential communities in the United States provide the ideal habitat (grass and edge) for deer and, in some cases, there is an overabundance of palatable food. Since humans have decimated the natural predator of deer (i.e., wolves), reductions in deer populations by humans (i.e., hunting) are necessary to address the population problem and prevent damage to existing ecosystems.

FIGURE 6.10 El Tecuán recreational area, located near Durango, Mexico, and containing a forest type composed of Cooper pine (*Pinus cooperi*) and oaks (*Quercus* spp.). *Gustavo Perez-Verdin.*

6.2 What are ecosystem services?

An important topic in academic, research, and environmental communities centers on the potential set of ecosystem services provided by forests and other land areas. *Ecosystem services* refer to the benefits that ecosystems provide for both human societies and Earth itself. This description is admittedly broad, since benefits from forests can be classified as natural capital (i.e., land and existing forest conditions), goods (i.e., derived products or services), socioeconomic outcomes (such as jobs), or environmental or ecosystem services (the focus of this chapter). In addition, these classes of benefits may overlap considerably depending on the focus of discussion. For example, ecosystem services can include marketable goods produced from the forest but can also include nonmarketable items of importance to us (clean air and aesthetics). These can also be characterized as public or private goods. Public goods include clean air, clean water, aesthetic values, and services provided with no associated fee, whereas private goods are typically consumables that have a price (Brown et al., 2008; Binder et al., 2017; Chamberlain et al., 2017; Nowak, 2018). Public goods are often characterized as being *nonrival in consumption*, which implies that they can be enjoyed by many people at the same time, whereas private goods are characterized as being *rival in consumption* (i.e., can be simultaneously enjoyed by one or a few). Further, public goods may have a *nonexclusion possibility*, which suggests that anyone, whether they have paid for the goods or not, can enjoy them. Private goods, on the other hand, prevent those who have not paid for the goods from enjoying them (Ver Eecke, 1999).

Some ecosystems both provide consumable products and facilitate the regulating functions that support life processes, such as carbon sequestration, water purification, decomposition, nutrient cycling, and plant pollination, to name just a few.

In a more specific definition, ecosystem services are described as:

> essentially the environmental benefits enjoyed by households, communities, and economies, including agricultural food and fiber production, fresh water, air quality regulation, climate regulation, pollination, erosion regulation, water purification, soil development, acidity buffering, waste treatment, disease regulation, nutrient cycling, pest regulation, and so forth (Keliang et al., 2010).

From this definition, one can surmise that ecosystem services provide considerable benefits to human civilization that are usually taken for granted even though they may influence the health of our planet. Daily (1997) previously defined ecosystem services as "the conditions and processes through which natural ecosystems, and the species that make them up, sustain and fulfill human life." Certainly, others have defined the scope of ecosystem services, and most have included the same basic components—ecosystem functions and their association with human values. Shortly, we will describe four basic classes of these services (provisioning, regulating, supporting, and cultural) and provide examples to illustrate their importance.

Methods for valuing forest ecosystem services have been evolving since the late 1900s, and often are based on assessments that involve market values, replacement costs, and benefit-transfer methods. In general, however, most ecosystem services are treated as public goods that have no market value (Sell et al., 2006; Chamberlain et al., 2017). The absence of a real market for many ecosystem services implies a certain degree of subjectivity in their valuation, and research has illustrated widely varying valuations for services situated in a single area, when examined at different scales and with different perspectives. Placing a value on an ecosystem service can be a way in which to raise its awareness to the general public; nevertheless, when local communities fail to receive benefits associated with these services, the incentive for managing land sustainably may decline (Wu et al., 2010; Small et al., 2017). However, both the methods and the meaning of valuing biodiversity and ecosystem services are controversial. For example, the notion that landowners can be compensated for any good or service their land provides, or that a good or service can be substituted for (or replaced by) another, may be difficult to accept (Salles, 2011).

Determining which aspects of biodiversity or of ecosystem services can be valued is not easy, and approaches have ranged from empirical (factual) to theoretical. Benefit-cost analyses or willingness to pay analyses are often used, but in many cases the individual components need to be assessed along a consistent scale (e.g., financially). Some suggest a direct extension of financial valuation methods to biodiversity or to ecosystem services, while others suggest addressing the substitutability of natural capital by financially valuing these (Salles, 2011). However, if in the latter case there are no realistic means to use substitution as a way to value biodiversity or ecosystem services, then the process fails and the meaning of any sort of valuation becomes tenuous. Ultimately, if our objectives were to select good, practical choices and actions for a forest landscape, the benefits that ecosystems provide for human societies need to be valued using explicit and contestable methods (Salles, 2011). Recently, initiatives around the world have been developed with the intent of creating a market for ecosystem services (Salzman, 2005; Grima et al., 2016). These have generally been classified as *payments for ecosystem service* programs. These programs are meant to encourage landowners, through direct payments of money, to practice environmentally sound management of land, and thus

the landowner is viewed as a service provider (Van Hecken and Bastiaensen, 2010). Some of these programs are private ventures, while others are supported by public agencies or organizations and nongovernmental organizations (NGOs). Most have focused on well-defined services associated with ecosystems, such as those related to watershed, biodiversity, carbon management, or ecotourism issues (Kosoy et al., 2008). Some programs also have the objective of alleviating poverty in developing areas of the world (Van Hecken and Bastiaensen, 2010).

One example ecosystem service program is Mexico's Program of Payments for Carbon, Biodiversity, and Agroforestry, in which carbon payments are made to households in *ejidos* (sellers of services) with the intent of enhancing organizational skills and adjusting the use of forest management practices. Funds are allocated from Mexico's Congress, and the National Forestry Commission (the buyer of services) manages and administers the program (Corbera et al., 2009). Another example is the Mantadia Project in Madagascar. The services produced in this project are emission reductions (carbon management) and existence values (biodiversity), generated through forest management practices. The government is the seller of the services, since they control the land, and the World Bank's BioCarbon Fund was the initial buyer of the carbon emission reduction services. The government of Madagascar is working with NGOs (e.g., Conservation International) to accomplish the goals of the program, and the revenues are disbursed with the intent of enhancing the livelihood of local communities and encouraging reforestation and forest management activities (Wendland et al., 2010).

6.3 Types of ecosystem services

As alluded to earlier, ecosystem services are products, functions, and processes that have value to human life and to animals, plants, and other organisms. These services can be grouped into four general classes: provisioning, regulating, cultural, and supporting. *Provisioning services* can be viewed as ecosystem outputs that are directly used by humans and other organisms. *Regulating services* relate to the functions that facilitate the maintenance and regulation of ecosystem processes. These include flood and disease regulation and water purification. *Cultural services* are the social benefits that humans value from ecosystems (religious, spiritual, recreational, etc.). *Supporting services* are those deemed essential for the continuation of other ecosystem services, such as primary production and water cycling. Some ecosystem services are also considered *intermediate services* in that they contribute to one of the four main services but do not by themselves constitute the service. Some provisioning services, for example, along with most of the supporting and regulating services, can also be considered as intermediate services (Smart et al., 2010).

6.3.1 Provisioning services

Forests provide a large and diverse array of goods and services for humans to use and consume and, as we previously noted, these are primarily provisioning services (Costanza et al., 2017) and include energy, fiber, food, genetic resources, and water (Sileshi et al., 2007). *Wood* (Fig. 6.11) is the most obvious and common product generated by forested ecosystems. Wood has a variety of uses and, in many cases, represents the basic building material for residential homes and businesses. Wood can be found in the flooring, walls, ceilings, and roofs of many structures. Doors and window frames are also commonly made from wood. Wood is frequently used in the manufacture of furniture for homes and offices, such as bed frames, chairs, chests, dressers, shelves, and tables. Tool handles are typically made from wood; these include axes, chisels, hammers, knives, rakes, shovels, and wheelbarrows. Other items made from wood include bowls, cups, napkin holders, plates, and utensils. Paper products are another set of important wood products that are derived from wood fiber and include books, copy paper, magazines, newspapers, and writing tablets, as well as softer products such as hygiene products, napkins, paper towels, tissue paper, and toilet paper.

Since the moment that fire was discovered to be of value to humans, wood from forested ecosystems has been used for heating and cooking purposes. More recently, technologies have been developed to transform wood into pellet form, which is easier to transport but can only be safely used indoors in specially designed stoves. Efforts have also been initiated (again) to transform wood into liquids that might one day be used to power automobiles, motorcycles, trucks, or other types of transportation systems.

Other provisioning services generated by forested ecosystems are the wild foods and medicines that can be derived from plants and animals found there. Wild foods may include various parts of cattails (*Typha* spp.) (Fig. 6.12), ginger (*Zingiber officinale*) (Fig. 6.13), sassafras (*Sassafras albidum*), blackberries (*Rubus* spp.) (Fig. 6.14), raspberries (e.g., *Rubus strigosus*) (Fig. 6.15), blueberries (e.g., *Vaccinium myrtillus*) (Fig. 6.16), honey (Fig. 6.17), and mushrooms (Fig. 6.18), to name a few. Some of these such as truffles, a mushroom found in oak woodlands and used in gourmet cooking, can be extremely valuable. Although these are noted with reference to their importance for human consumptive purposes, wild foods also play an important role in the diet of many different species of wildlife. Natural medicines have also been derived from forest plants for thousands of years. Extracts from medicinal plants have been used to treat glaucoma, motion sickness, and cancers such as Hodgkin's disease and leukemia, as well as to create antibiotics.

FIGURE 6.11 Stacked sawnwood of Caribbean pine (*Pinus caribaea*) at a sawmill outside Punta Gorda, Belize. *Robert Crook.*

FIGURE 6.12 Cattails (*Typha* spp.) on the edge of a small wetland in Marshall County, Indiana, United States. *Derek Jensen, through Wikimedia Commons.*
Image Link: https://commons.wikimedia.org/wiki/File:Typha-cattails-in-indiana.jpg
License Link: Public Domain

FIGURE 6.13 Ginger (*Zingiber officinale*) roots sold in a local market. *Nino Barbieri, through Wikimedia Commons.*
Image Link: https://commons.wikimedia.org/wiki/File:-_Ginger_-.jpg
License Link: https://creativecommons.org/licenses/by-sa/3.0/deed.en

FIGURE 6.15 Raspberries (*Rubus* spp.) in Jarom ĕř, Czech Republic. *Karel Jakubec, through Wikimedia Commons.*
Image Link: https://commons.wikimedia.org/wiki/File:Malina.jpg
License Link: Public Domain

FIGURE 6.14 Blackberries (*Rubus* spp.) ripening on a bush in Victoria, Australia. *G. King, through Wikimedia Commons.*
Image Link: https://commons.wikimedia.org/wiki/File:Blackberries_on_bush.jpg
License Link: Public Domain

FIGURE 6.16 Blueberries (*Vaccinium myrtillus*) growing under a Scots pine (*Pinus sylvestris*) forest in Satakunta, Finland. *MPorciusCato, through Wikimedia Commons.*
Image Link: https://commons.wikimedia.org/wiki/File:Tuoretta_kangasmetsää_kesällä.jpg
License Link: https://creativecommons.org/licenses/by-sa/3.0/deed.en

Another provisioning service facilitated by forested ecosystems is the water resource that arises from forested areas. Water regulation and supply are suggested to account for nearly 7% of the value of ecosystem services, and many of the world's largest cities partly rely on water derived from forests (Stolton and Dudley, 2007). As an example, the value of water quality in the tropical cloud forests of Mexico was assessed by Martínez et al. (2009) to infer the economic impacts of changes in nutrient and suspended sediment levels resulting from land use changes. We have implied that forested ecosystems play an important role in the local storage and purification of water for domestic human consumption. While the vegetative roots of trees, shrubs, grasses, and other herbaceous plant species actively absorb groundwater, they also actively prevent the surrounding soil from eroding and polluting nearby streams or other water bodies with sedimentation. In addition, shade produced from forested ecosystems can reduce local air temperatures during the summer months and act to mitigate water temperature increases in adjacent stream systems. The resulting lower water temperatures can lead to favorable habitat for certain fish species, such as freshwater trout. In addition, a lack of clean water can lead to both short- and long-

FIGURE 6.17 Honey produced from beehives located among bluebells (*Hyacinthoides non-scripta*) under a coppice overstory in the United Kingdom. *Bob Embleton, through Wikimedia Commons.*
Image Link: https://commons.wikimedia.org/wiki/File:Bee_Hives_and_Bluebells_in_Mayalls_Coppice_-_geograph.org.uk_-_103768.jpg
License Link: https://creativecommons.org/licenses/by-sa/2.0/deed.en

FIGURE 6.18 Shiitake mushrooms (*Lentinula edodes*) growing on a tree trunk. *Keith Weller, through Wikimedia Commons.*
Image Link: https://commons.wikimedia.org/wiki/File:Lentinula_edodes_USDA.jpg
License Link: Public Domain

term human health problems, which can reduce the productivity of a region and place a great deal of pressure on the health services sector of human society (Stolton and Dudley, 2007).

6.3.2 Regulating services

Forested ecosystems provide a number of regulating functions or services that can be viewed as services for

FIGURE 6.19 Metolius River in the Deschutes National Forest, Oregon, USA. *U.S. Forest Service, Pacific Northwest Region.*

human civilization and for the Earth's biosphere or, more generally, as benefits obtained from other processes (Costanza et al., 2017). These services include carbon sequestration, climate regulation, disease regulation, erosion control, flood regulation, and water purification (Fig. 6.19). Carbon storage, flood control, and water filtration are considered the most important ecosystem services in Canadian boreal forests (Schindler and Lee, 2010). An example of an assessment for regulating services (erosion control and climate regulation) facilitated by tropical forests in the Amazon region of Brazil is outlined in the work of Portela and Rademacher (2001). Smart et al. (2010) describe regulating services as intermediate ecosystem services that relate more directly to the continued existence and maintenance of ecosystem processes. Regulating services are also important for human security (or personal safety), resource access, and disaster mitigation. These services are linked to the basic needs of most living creatures for food, shelter, and access to goods. These services can also influence human health indirectly by promoting strength and positive feelings and directly by providing access to clean air and water.

An important regulating service provided by ecosystems is species redundancy. It has been postulated that ecosystems have many different species that perform the same or similar sets of functions. If one species is lost to an ecosystem, then another may take its place. Species redundancy could lead to greater ecosystem resiliency from natural disturbance processes (Naeem, 1998; Goswami et al., 2017). Further, an ecosystem that has high species diversity (i.e., more species) may have stable ecosystem processes and regulating functions if the species that perform similar functions have different levels of environmental sensitivity (Chapin et al., 1997). An ecosystem that has too few species or species with similar functionality may be more vulnerable if the removal of one of these species from the ecosystem has an important impact on the function of the ecosystem (Fonseca and Ganade, 2001; Oliver et al., 2015; Cooke et al., 2019). However, the scientific community does not know enough about species interactions across time and space to be able to predict the impact of species removal; therefore, it may be unwise to rely on this concept (species redundancy) when managing ecosystems (Gitay et al., 1996).

Another important regulating service facilitated by forested ecosystems is plant pollination. Pollination is the process by which whole pollen grains are transported to flowers of the same species for the purpose of sexual reproduction. In some cases, pollen is disseminated by wind, but it can also be disseminated by native pollinators (Fig. 6.20). According to the U.S. Department of Agriculture, Forest Service (2011), 90% of plants require animal pollinators, of which there are approximately 200,000 species. Some examples of native pollinator species are bees, beetles, butterflies, moths, hummingbirds, and bats, many of which use forests for certain aspects of their life cycle. Pollination is not only important for the continued reproduction of plant species within the forested ecosystem but is also important for crops on lands near forests. The species mentioned help pollinate plants such as almonds, apples, bananas, blue agave, blueberries, cocoa, coffee, melons, peaches, potatoes, and pumpkins (U.S. Department of Agriculture, Forest Service, 2011). Native pollinators such as bees can improve the quality and quantity of agricultural crops produced.

Unfortunately, it is not unusual to see news reports decrying the decline of native pollinator populations, such as bees. There are many possible explanations for this vexing problem; these may include the misuse of pesticides during periods of peak flower bloom. Other possibilities include the infestation of native pollinator populations by new parasites and diseases. Increased urbanization around the world has also led to the loss of habitat, resulting in the prevalence of less suitable forage, which diminishes pollinator populations. The loss of habitat can result in beehives being constructed in or around human dwellings, which are often destroyed. Misguided concerns over the possibilities of killer bee colonies have also led to beehive destruction for other species of bees. Native bee species also face competition for resources from invasive honeybees. To complicate matters further, air pollution disrupts the fragrances emitted by a flower to attract native pollinators, such as bees and moths. This disruption reduces the range at which these pollinators can detect flower fragrances, which decreases pollinator efficiency and not only reduces bee health but also adversely affects plants that depend on native pollinators for sexual reproduction. Ultimately, this could lead to a shift in vegetative patterns to native plants that are wind-pollinated or reproduce asexually (McFrederick et al., 2008). There are currently programs being implemented that support the native pollinators by promoting the planting of native, pollen rich plants (Winfree et al., 2011).

Numerous actors or events can affect the regulating services provided by forested ecosystems. For example, one actor involves changes in local land use and land cover, which can have enormous impacts on the regulating functions of an ecosystem. For instance, conversion of forested ecosystems into parking lots or shopping centers can have an adverse impact on the local ecosystem by changing local hydrologic functions. Asphalt covering will largely prevent infiltration of rainwater into the soil below. Wildlife habitat will also be lost, as will nesting, roosting, foraging, and hiding cover for certain species. The potential loss of wildlife species that depend on forest interiors may reduce species richness and biodiversity of the area. Finally, the surface of the parking lot may absorb more heat than a forest normally would, which could elevate local air temperatures and increase the evapotranspiration processes of nearby plants.

The introduction or removal of plant or animal species from an ecosystem can be viewed as a second actor that can reduce the ecosystem's redundancy. As we alluded to earlier, if a high percentage of a particular species is removed from the landscape, then ecosystem redundancy function may be diminished. Although the functions performed by one species may be replaced by another, a continual reduction in species numbers may reach a point at which an ecosystem service has been lost or severely diminished. For instance, if an insect-eating bird is hunted to extinction, then it is possible that another avian species may thrive, since there is less competition for the insects, and the extinct bird's role in the ecosystem may be thus assumed by another. However, if the second avian species is then eliminated, it is possible that no other avian species can fill the role played by both. This could lead to increases in insect infestation because the insects would

FIGURE 6.20   A pollinator.   *Andrew J. Sánchez Meador.*

have fewer natural predators. In addition, if a particular plant species were dependent on these avian species for disbursement of its seeds, then that plant species may experience regeneration success issues.

The introduction of nonnative invasive species into a natural ecosystem can, in some cases, alter natural ecosystem processes. For instance, as mentioned in Chapter 5, if a nonnative invasive plant species such as cogongrass (*Imperata cylindrica*) were to become established in an area of the southern United States, then it would most likely out-compete native plant species for water and nutrients. Cogongrass can also facilitate the spread of fire. If fires occur frequently, these events could change the nature of the original ecosystem. In addition, cogongrass could also reduce the species richness of local plant communities, which would then reduce the foraging opportunities for numerous wildlife species.

A third actor affecting regulating services is the introduction of a new technology, which can have positive or negative impacts on how we manage ecosystems and the services they provide. In the early 1900s, the advent of chainsaws and large machines to haul timber or move soil helped facilitate changes in land use. This argument can also be applied to improvements in medical care and the creation of new medications (sometimes derived from plant material), each of which may help people live longer, which is arguably a good idea. Other technologies, such as global positioning systems, can be used by land management organizations to monitor forest and natural resource management activities more accurately, which may help maintain the sustainable management of these resources.

A fourth potential actor affecting regulating functions of ecosystems is the introduction of external inputs such as fertilizers, pest control measures, irrigated water, and other nonnatural (level or type) inputs associated with intensive land management. Some may argue that introducing genetically modified tree species into forest plantations could have an impact on the local gene pool through reductions in genetic diversity. If this were the case, these areas may become more vulnerable or susceptible to insect and disease infestations.

The harvest and consumption of forests and their associated natural resources is a fifth potential actor affecting the regulating functions of ecosystems. As we have discussed, the removal of tree vegetation can have a significant impact on an ecosystem's carbon sequestration, oxygen production, soil erosion mitigation, water purification potential, and wildlife habitat availability. Coastal forests, in particular, offer protection against erosion and storm surges (arising from tropical cyclones or hurricanes) associated with larger water bodies. Removal of these forests may therefore lead to higher coastal erosion rates.

Climate change is a sixth potential actor that might affect the regulating functions of ecosystems. Although increased concentrations of carbon dioxide in the atmosphere may stimulate plant growth (which is not in itself a bad thing), the increase over time could lead to changes in the distribution of particular plant communities. The possible shifts in plant communities may then have adverse impacts on many wildlife species (Hanson and Hoffman, 2011; Wilkening et al., 2019). Depending on how such shifts occur, some wildlife species may not adjust fast enough, which could lead to population declines and additional losses of the regulating functions of forested ecosystems.

The last set of actors or events that potentially affect regulating functions of ecosystems includes natural and physical events, such as volcanoes, and biological processes, some of which may induce change very quickly. A volcanic eruption, as can be imagined, may very quickly destroy all of a forested ecosystem in the path of the fallout of ash, gases, or associated *lahar* (debris flow). Forested ecosystems near Mount St. Helens in Washington State were destroyed by the volcanic eruption in 1980 (Nash, 2010). However, since the eruption, the denuded landscape has gradually recovered with the spread of plants such as Canada thistle (*Cirsium arvense*), fireweed (*Epilobium angustifolium*), huckleberries (*Vaccinium* spp.), and pearly everlasting (*Anaphalis margaritacea*) (Halpern et al., 1990; Cook and Halpern, 2018). Tree species such as bitter cherry (*Prunus emarginata*), black cottonwood (*Populus trichocarpa*), and red alder (*Alnus rubra*) are also spreading across the blast zone, and Douglas-fir (*Pseudotsuga menziesii*), noble fir (*Abies procera*), and western hemlock (*Tsuga heterophylla*) are beginning to naturally establish themselves (Dale and Adams, 2003). A biological driver such as evolution, on the other hand, would induce relatively slow changes to an ecosystem over a very long period. Another biological driver that can disrupt ecosystem services is the introduction of invasive plant and animal species. As discussed, a moment ago, invasive species can displace native species, thus reducing the quality of wildlife habitat. In addition, some invasive species can alter the natural fire regimes, which can impact many other aspects of the existing ecosystem, such as changing the entire ecosystem over to a fire-dependent type.

6.3.3 Cultural services

Cultural services are a key set of values that forested ecosystems provide or facilitate. Cultural services include nonmaterial benefits to humans; these relate to changes in human welfare (Smart et al., 2010) and are highly dependent on the other ecosystem services mentioned in this chapter. The cultural services that humans derive from forested ecosystems generally include cognitive development, recreation, reflection, spiritual enrichment, and other values gained by viewing aesthetically pleasing vistas. The cultural services (i.e., recreational and tourism opportunities) facilitated by tropical forests of the Tapantí National Park in Costa Rica provide one set of examples (Bernard et al., 2009). The Millennium Ecosystem Assessment (2005) refines the idea of cultural services to more closely address issues such as cultural identity and diversity, and spiritual and religious values (Fig. 6.21), knowledge systems, educational values, aesthetic values, social relations, a sense of space, cultural heritage, and recreation and ecotourism opportunities.

Forest ecosystems all around the world have played an important role in defining cultural identity and diversity across landscapes. Indigenous peoples who live in tropical regions have traditionally manipulated forested ecosystems to suit their cultural needs. Savannah ecosystems in central Asia, for example, have been important to nomadic pastoralists. Human interactions with these different ecosystems have led to the development of human value systems. Generations of people have created and expressed stories of their lives within the environment that surrounds them. These memories and associations over time and space are embedded in the language and history of many cultures.

Unfortunately, changes in ecosystems can greatly diminish the cultural services derived from forested ecosystems. The displacement of forest dwellers to unknown environments can lead to a loss of cultural identity and diversity, as well as to land degradation, which can affect other ecosystem services. In some parts

FIGURE 6.21   God's Forest: Wonju Shillim Village Shrine Forest, South Korea. *Jacek P. Siry.*

of the world, many indigenous languages have not yet been formally cataloged. As these languages, cultural identities, and traditional ways of life are lost, people become estranged from the land, which they may no longer value the range of services it can provide and can lead to overexploitation and ecosystem degradation, and thus the loss of many ecosystem services (Millennium Ecosystem Assessment, 2005).

The spiritual and religious values offered by forest ecosystems have long played an important and sometimes complicated role in human civilizations. Early Celtic and Germanic civilizations placed great importance on forests for religious rituals (Philpot, 1897; Delaney, 1986). Today, many people celebrate Christmas with a tree, which has been a custom in Germany for centuries. In parts of the southern Miombo woodlands of Africa, the marula (*Sclerocarya birrea*) fruit tree is considered to have great cultural value (Sileshi et al., 2007). Forested ecosystems are, therefore, important for providing cultural heritage to local communities. Cultural heritage is obviously linked to cultural identity and is developed over both space and time. Cultural heritage can influence how people learn to manage the forested ecosystems around them. Loss of these heritage areas disconnects people from their past and, in the case of forested ecosystems, leads to important knowledge being lost and less effective management of the surviving resource.

Another cultural service that ecosystems provide relates to the heritage and knowledge of local communities. Local knowledge systems are dependent on human interaction with local forested environments and are developed over several human generations. The high level of indigenous knowledge developed over time not only connects local inhabitants with forests and natural resources but also leads to effective management and an understanding of how a particular ecosystem functions. Although this kind of knowledge is not usually obtained from formal training at universities, knowledge derived from repeated observations of various aspects of the ecosystem can be valuable. Unfortunately, as older generations pass away, much of this knowledge is lost, which may lead to a breakdown in the traditional ways of life or an estrangement from the land. This *knowledge loss* can also slow the process of identifying new chemicals and compounds that may be useful in the development of new medicines.

Aesthetic values, recreational opportunities, and ecotourism are other cultural services that forested ecosystems can provide or facilitate. For example, a walk through the redwood (*Sequoia sempervirens*) grove, such as the one found in the Whakarewarewa Forest outside Rotorua, New Zealand, can be a breathtaking experience. Tall, majestic redwood trees are not only inherently beautiful to many people, but they can also provide a

FIGURE 6.22 Heilige Stiege (Holy Staircase) hiking trail in Saxon Switzerland, Saxony, Germany. *Norbert Kaiser, through Wikimedia Commons. Image Link: https://commons.wikimedia.org/wiki/File:Sächsische_Schweiz_Heilige_Stiege_(01).JPG License Link: https://creativecommons.org/licenses/by-sa/3.0/deed.en*

sense of awe, or peace and well-being. Although aesthetic values can be considered subjective assessments, they can play an important role in people's behavior and in how forested ecosystems are managed. In the western United States, the aesthetic unpleasantness of clearcuts sparked an ongoing controversy that has influenced the management of both public and private forests.

Other cultural services facilitated by forests include recreational opportunities and ecotourism experiences. Many forms of recreation situated in forested ecosystems are linked to the other types of ecosystem services, and these can be intimately tied to a sense of cultural identity and heritage. People can become enamored with nature and develop a connection to the landscape through opportunities to hunt and fish, and through the availability and access to hiking trails (Fig. 6.22). Forest canopy excursions, facilitated by elevated walkways or rope bridges, have also become popular in some areas of the world, and these experiences can influence the management of natural resources. Ecotourism is a form of recreation in which people travel to destinations specifically to visit ecologically unique areas. Forest-based tourism or ecotourism opportunities can be developed to positively benefit the social and environmental aspects of local communities. Forest bathing, the idea of spending time in the forests, is now recognized as a treatment for depression in some countries (Park et al., 2010).

6.3.4 Supporting services

The final key ecosystem service provided or facilitated by forests and related communities are the supporting services. Smart et al. (2010) describe *supporting services* as "intermediate ecosystem services that are necessary for the production of all other ecosystem services." The Millennium Ecosystem Assessment (2005) describes these services as being more clearly indirect than the regulating services. For example, forest termites of tropical savannas facilitate supporting services by withdrawing litter and soil organic matter from normal decomposition pathways, redistributing these to other processes, and creating patches of nutrients such as nitrogen (Jouquet et al., 2011). Supporting services within an ecosystem are essential for the continuation of all other ecosystem services, and include processes such as primary production, photosynthesis, soil formation, nutrient cycling, and water cycling.

Primary production is the energy accumulated in plants (Smith and Smith, 2000), and radiant (electromagnetic) energy is one of the main inputs into a forested ecosystem. This energy is critical for the maintenance of plant health and reproduction. Energy reaching Earth can be used by plants or lost through convective or other processes. All plants utilize radiant energy. Some of this energy is transformed through the process of respiration; the remaining energy, called *net primary production*, is stored as organic material or biomass (Smith and Smith, 2000; Salisbury and Ross, 1985). Plants store biomass internally in roots, stem boles, branches, seeds, flowers, and leaves. The ratio of above ground biomass to below ground biomass can provide an indication of the character of a particular ecosystem (Smith and Smith, 2000). For example, plants situated in dry ecosystems tend to allocate more biomass to root systems in order to search for water so that health and vigor can be maintained. In contrast, some plants found in wet ecosystems have shallower root systems. The allocation of root biomass by a plant species could be extremely important when the plant encounters adverse weather conditions. For example, forest trees with shallow root systems tend to be more susceptible to windthrow events (Bettinger et al., 2010; Merry et al., 2010), such as those that occurred along the Florida coast in 2018 during Hurricane Michael.

The productivity of ecosystems is highly dependent on local and regional rainfall levels and temperature regimes. From a primary production standpoint, tropical forest ecosystems are typically more productive than are forests located in boreal regions. The variation in ecosystem productivity around the world is further illustrated by the quantity of biomass stored in those ecosystems. Litter production in Arctic and alpine regions ranges from about 0 to 200 g per square meter (m^2) per year (0.66 ounce [oz.] per square foot [ft^2] per year) compared to about 200 to 600 g per m^2 per year (0.66 to 1.97 oz. per ft^2 per year) in temperate forest ecosystems, and about 900 to 1500 g per m^2 per year (2.95–4.92 oz. per ft^2 per year) in tropical forest ecosystems (Smith and Smith, 2000, citing earlier work by Bray and Gorham, 1964).

The process that plants employ to grow is photosynthesis. *Photosynthesis* is a process in which radiant energy from the sun is collected, carbon dioxide (CO_2) and water are transformed into chemical energy for plant growth, and oxygen is released into the atmosphere (Kramer and Kozlowski, 1979). A by-product of photosynthesis, oxygen, sustains almost all forms of life on Earth. The photosynthetic process is dependent on the presence of chlorophyll, a light-absorbing pigment. Chlorophyll can be found in leaves, petioles, buds, cotyledons, cortical parenchyma of young twigs, phelloderm of some species, and some roots exposed to light, but it is mostly found in the chloroplasts of leaves (Kramer and Kozlowski, 1979). In plants, green wavelengths of electromagnetic energy have higher reflectance values (i.e., are reflected more frequently) than blue and red wavelengths; thus, blue and red wavelengths have higher relative absorbance rates (Barnes et al., 1998). The greater amount of reflected electromagnetic energy in the green light spectrum results in the greenish color of our vegetative landscapes.

The development and maintenance of chlorophyll are dependent on numerous internal (i.e., to the plant) and environmental factors (Kramer and Kozlowski, 1979). Interference from any of these factors can result in yellowing, or chlorosis, of the leaves or needles. Some of the internal factors include the genetic potential of the plant and the adequacy of the carbohydrate supply. Environmental factors, such as the availability of light, minerals, oxygen, temperature, and water, can limit the formation of chlorophyll. Although radiant energy is critical, sometimes plant species require lower light intensities to produce chlorophyll, as very bright light can destroy chlorophyll. Interestingly, chlorophyll is constantly being formed and destroyed by light (Kramer and Kozlowski, 1979). Another important environmental factor is air temperature. Chlorophyll can be formed at a wide range of air temperatures but, for many species, it tends to decompose more rapidly in colder temperatures. Water is also important because it aids in the movement of nutrients from the soil to the leaves. Unfortunately, too much water can disrupt plant metabolism and cause yellowing of leaves or needles. Insufficient supplies of minerals such as iron, magnesium, nitrogen, and potassium, among others, can reduce or cease the formation of chlorophyll even if sufficient sunlight is available.

Another important function that supports photosynthesis in addition to other ecosystem processes is nutrient cycling. Within an ecosystem context, it is the flow of critical nutrients throughout the system that keeps all critical functions operating. As suggested earlier in this chapter, soils are essentially repositories of nutrients; these nutrients are acquired and used in the process of photosynthesis for growth of new vegetation. As plants grow, nutrients are absorbed by trees and then stored within the wood, bark, and twigs. Nutrients are returned to the soil through ground flora, leaf litter, stemflow, and washing and leaching of the forest canopy (Barnes et al., 1998). Nutrients such as calcium, nitrogen, and potassium are imported into the system from atmospheric processes (e.g., precipitation). Nutrients can also be lost from an area due to animal translocation (defecation), fire, soil erosion, and timber harvesting.

A key supporting service found in all ecosystems is water cycling. The water cycle is critical for transporting nutrients in and out of an ecosystem. Precipitation is the primary form of water input to a forested ecosystem, but periodic floods from nearby river systems (Fig. 6.23) can also bring necessary amounts of water to some forested ecosystems. When water enters the ecosystem in the form of precipitation, it can be intercepted by plants and soil. Water infiltrates the soil and percolates downward until it reaches a saturation point and then flows through a stream system, or perhaps as overland flow, to lower elevations in the landscape (Fig. 6.24). Processes such as evaporation convert exposed water (on the ground, leaves, or other biomass surfaces) into a gaseous form, which is then lost to the atmosphere. In addition, plants lose water through evapotranspiration, or the leakage of water in a gaseous form through leaf stomata into the atmosphere.

FIGURE 6.23 Flooded forest along the River Lagan in Northern Ireland. *Albert Bridge, through Wikimedia Commons.*
Image Link: https://commons.wikimedia.org/wiki/File:The_River_Lagan_in_flood_(3)_-_geograph.org.uk_-_664394.jpg
License Link: https://creativecommons.org/licenses/by-sa/2.0/deed.en

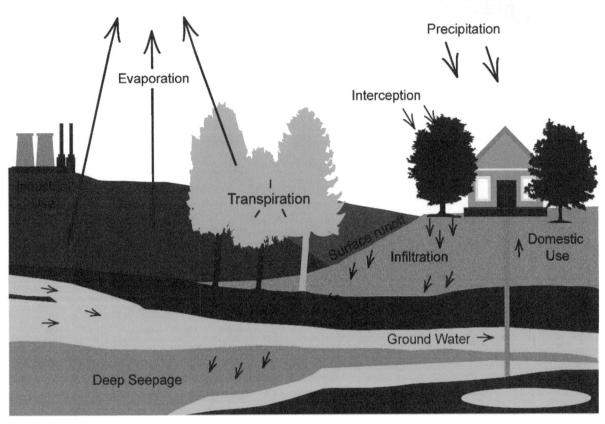

FIGURE 6.24   The water cycle.

The last key supporting service associated with forests that we will discuss in this chapter is the formation of soil. Soil is formed through rock weathering and debris decomposition processes. The rate of rock weathering is relatively slow and is dependent on the type of rock that exists in an area. Some rocks, such as granite, can take extremely long periods of time to break down, whereas limestone can break down in a much shorter period. The weathering process breaks down rocks through exposure to varying temperature levels, which causes rocks to contract and expand. When water seeps into cracks formed in the rocks, it can accelerate the weathering process. Eventually, perhaps over thousands of years, layers of fine, loose material weathered from rocks form *regolith*. Depending on the slope position of the regolith, different types of soils are formed.

Organic material decomposed from surrounding plant and animal biomass can also seep into the upper layers of a soil and influence the type and development of soil present in a landscape. The establishment of plants and subsequent root growth in the regolith further advances the process of soil formation. Minerals are pumped up from lower soil depths by plants and are eventually deposited on the upper soil layer in the form of leaf litter and fallen branches. The layer of leaves and branches provides a suitable habitat for a variety of bacteria, fungi, millipedes, centipedes, mites, springtails, grasshoppers, and other organisms that consume and break them down into a more refined state. This process is known as *decomposition*. Over time, decomposition, along with climatic influences, gradually forms what is commonly known in the profession as *soil horizons*. Soil horizons (or the soil profile) represent an underground stratification of a soil and help define soil types (Fig. 6.25). The topsoil layers contain the highest percentage of organic material, carbon, and biological activity, while the lower soil layers vary in their biological, chemical, and physical properties. The infiltration of water into the soil enables minerals to leach from one layer in the soil horizon to another.

FIGURE 6.25   A Cecil soil profile, typical of the Piedmont uplands of the southern United States. *U.S. Department of Agriculture, Natural Resources Conservation Service (1999).*

6.4 Trade-offs when managing for different forest ecosystem services

In the management of forested ecosystems and their associated natural resources, specific management goals can result in the enhancement of one ecosystem service at the expense of another. This particular issue constantly challenges natural resource managers. Therefore, the actual and potential suite of ecosystem services arising from the management of forest ecosystems needs to be assessed, given that many aspects of human society depend on well-functioning forests (Salles, 2011). As mentioned earlier, early human civilizations often relied heavily on forested ecosystems surrounding their communities for a wide variety of goods. In many cases, these systems are resilient and can revert relatively smoothly to previous forested conditions. However, in some cases, especially environments with low annual precipitation rates, reversion to previous forested conditions is not possible, resulting in a net loss of ecosystem services. A case in point is the arid landscape of many Mediterranean countries that centuries ago experienced deforestation due to frequent wars, ill-fated public policies, and intensive human consumption. Some of these areas have not yet recovered because of extensive soil erosion. Although earlier civilizations in this area benefited from the services provided by these formerly forested ecosystems, current societies are bearing the cost of the loss of ecosystem services. For instance, widespread soil erosion in these ecosystems has led to a decline in soil

productivity, which affects not only tree growth but also the growth of understory vegetation necessary for wildlife species to breed, forage, hide, nest, and roost. In many areas, only rocks and soils with low fertility remain.

A common trade-off to which we have briefly alluded is harvesting goods from forested ecosystems. When human society utilizes one of the provisional services of an ecosystem, it is likely to give up another ecosystem service; hence, the trade-off. For instance, if a landowner wishes to harvest trees because of a need for wood or the revenue generated through a sale, then the landowner is an active participant in receiving the provisioning services from the forested ecosystem. The trade-off of this action is a loss of some regulating and supporting services, such as carbon sequestration and photosynthesis. However, if the trees are all converted to solid wood products it may take 100 years or more for the carbon to reach the atmosphere through natural decay processes. During that time, new trees on the site may be growing and sequestering carbon. Other services lost as a result of the harvest include the formerly (depending on your perspective) aesthetically pleasing environment, flood control, ecotourism potential, water quality, wild food production, and wildlife habitat. An important point to remember is that sometimes the trade-offs in services rendered from ecosystems are temporary, as in many timber harvesting situations.

Another trade-off involves the protection and preservation of forested ecosystems. For example, some governments have reserved large areas of forests for the protection and maintenance of biodiversity, and also perhaps for the availability of remote recreational opportunities. Water quality and availability, carbon sequestration, flood control, and ecotourism potential may be enhanced when this occurs (Millennium Ecosystem Assessment, 2005). However, provisioning services, such as wild food production for human consumption or timber harvests are often diminished or lost. Managing a forested ecosystem in this way may reduce disease potential and enhance biodiversity but may also lead to a greater disease threat if the age and species structure of the dominant overstory trees suggest that the forest is vulnerable.

Examples of trade-offs that are not temporary occur when forested ecosystems are deforested and converted to another land use (e.g., agriculture) or where human structures (e.g., industrial parks, movie theaters, office buildings, residential homes, or shopping malls) are built. Although human communities gain facilities and structures in both horizontal and vertical space that may improve human well-being in the short run, communities sacrifice most other provisioning, regulating, supporting, and cultural services provided by the previous forested ecosystem. Hard surfaces such as roads, parking lots, and sidewalks impede the infiltration of water into the soil below and facilitate rapid rainwater runoff into nearby streams and water bodies. This runoff can transport various harmful chemicals, such as antifreeze, gasoline, and motor oil, into the nearest water system, which may lead to a decline in aquatic vertebrate and invertebrate populations. Other services that are lost include aesthetics, cultural heritage, photosynthesis, pollination, and wildlife habitat. Fortunately, as we will discuss in more detail in Chapter 16, there are many ways to mitigate the adverse impacts of management actions on ecosystem services through activities associated with urban forestry and the development of green spaces, nature trails, parks, and urban gardens. In addition, not all land use changes are equal in their importance or effect. For instance, a shopping mall has much more of an impact on ecosystem services than does a residential neighborhood, where house lots can contain substantial amounts of greenery (i.e., grasses, shrubs, and trees) around each home.

An important issue that all forest and natural resource managers face is whether to manage their forested ecosystems to facilitate the maintenance of multiple ecosystem services on a per-unit area basis (i.e., per acre or per hectare) or to do so piecemeal across a specific landscape. In allocating land to the maintenance of ecosystem services, we may find that there is no easy solution. In some cases, managing a single piece of land for multiple ecosystem services is possible, but extremely difficult to accomplish in practice. A common strategy is to set aside some areas that are most sensitive to structural changes for regulating, supporting, or cultural services. For instance, cultural heritage sites and groves of trees with spiritual value might be reserved (i.e., left alone or managed in a manner that excludes active management), while other areas might be used for developing and maintaining provisioning services.

Although we need to sustain aspects of all four groups of ecosystem services, managing forested ecosystems primarily for one ecosystem service over another, on a site-by-site basis, is not necessarily bad management. In the case of preserving forests (i.e., implementing no actions), some ecosystem services will not be fulfilled, and others may degrade over time. Unless, as a society, we find that we can obtain the unfulfilled ecosystem services elsewhere, we will be faced with trade-offs. In the case of managing forests specifically for extractive purposes, Chapter 4 highlighted a large quantity of products that human society desires from forests. The fact that many forest resources are renewable makes this resource far superior, from a climate change perspective, than resources that are mined (e.g., iron ore) or drilled (e.g., petroleum) and converted to other products (e.g., steel or plastic). The main message from this discussion may be that forest and natural resource managers need to understand the trade-offs when promoting one ecosystem service over another, which may lead to more effective strategies for managing forests and natural resources.

Summary

Forests, as well as other natural ecosystems, are critical landscape elements that are vital for the existence of human civilization. Forests provide a large number of provisioning, regulating, cultural, and supporting services that facilitate human life and the ability to sustain it over time. Forests provide food, water, wood, and numerous other products that human society uses on a daily basis; these are known as provisioning services. Forests also facilitate numerous regulating services, such as carbon sequestration, flood control, plant pollination, and water filtration. These regulating services interact with one another and can promote the development of vibrant and healthy ecosystems, as well as influence human health conditions. Cultural services, such as aesthetics, recreational opportunities, and spiritual values, are also facilitated by forests and can influence human well-being. All of these services are enhanced by the underlying supporting services provided by forests. Supporting services, such as nutrient cycling, photosynthesis, and water cycling, promote plant and animal life by capturing radiant energy from the sun and transporting the necessary nutrients and water through an ecosystem to sustain these processes. Unfortunately, in real life, there are always trade-offs associated with managing land. When humans manage forest resources with a preference for one service over another, the possibility exists that some ecosystem services will be adversely affected. Understanding the possible trade-offs between ecosystem services that may arise through management actions can lead to effective management strategies, which can then lead to the sustainability of forests and natural resources for current and future generations.

Questions

(1) *Species of interest.* Identify a species of interest to you. It can be an animal, bird, fish, fungus, insect, or plant. With this species in mind, determine how it interacts with its environment and how many other species with which it potentially interacts. For each stage of this species' life cycle, determine how might it provide an ecosystem service. Write a two-to three-page essay addressing these questions. Further, develop one or two conceptual diagrams to illustrate the important linkages with other species or services.

(2) *Traditions that involve ecosystem services.* All families have traditions that are passed down through the generations. Using your family as an example, or a friend's family, describe the traditions that utilize provisioning ecosystem services (i.e., products of forested ecosystems and their natural resources). Communicate your findings through a short presentation to your class.

(3) *Hunting.* Hunting has at times been a controversial activity. In a 500-word memorandum, describe the advantages and disadvantages of hunting to a local, rural community. In this summary, provide an example of each of the main species of interest to your local community. If you are a hunter of these species, explain what enjoyment you receive from the experience. If you are not a hunter of these species, try to describe why you might (or might not) support hunting experiences.

(4) *Trade-offs.* Imagine that your local government is about to develop a highway through a forested area where you live. Through a short PowerPoint presentation, discuss the trade-offs involved in developing the proposed highway and what ecosystem services are involved in these trade-offs. Locate photographs and other graphics that will assist in supporting your discussion and include them in your presentation.

References

Barnes, B.V., Zak, D.R., Denton, S.R., Spurr, S.H., 1998. Forest Ecology, fourth ed. John Wiley & Sons, New York, pp. 774.

Barnes, M.C., Todd, A.H., Lilja, R.W., Barton, P.K., 2009. Forests, Water and People: Drinking Water Supply and Forest Lands in the Northeast and Midwest United States. U.S. Department of Agriculture, Forest Service. Northeastern Area State and Private Forestry, Newtown Square, PA. NA-FR-01—08. pp. 71.

Bernard, F., de Groot, R.S., Campos, J.J., 2009. Valuation of tropical forest services and mechanisms to finance their conservation and sustainable use: a case study of Tapantí National Park, Costa Rica. Forest Policy and Economics 11, 174—183.

Bettinger, P., Merry, K.L., Grebner, D.L., 2010. Two views of the impact of strong wind events on the forests of the southern United States. Southeastern Geographer 50 (3), 291—304.

Binder, S., Haight, R.G., Polasky, S., Warziniack, T., Mockrin, M.H., Deal, R.L., Arthaud, G., 2017. Assessment and Valuation of Forest Ecosystems Services: State of the Science Review. U.S. Department of Agriculture, Forest Service. Northern Research Station, Newtown Square, PA. General Technical Report NRS-170.

Bormann, F.H., Likens, G.E., 1979. Pattern and Process in a Forested Ecosystem. Springer-Verlag, New York, pp. 253.

Bray, J.R., Gorham, E., 1964. Litter production in the forests of the world. Advanced Ecological Research 2, 101—157.

Brown, T.C., Kingsley, D., Peterson, G.L., Flores, N.E., Clarke, A., Birjulin, A., 2008. Reliability of individual valuations of public and private goods: choice consistency, response time, and preference refinement. Journal of Public Economics 92, 1595—1606.

Chamberlain, J.L., Frey, G.E., Ingram, C.D., Jacobson, M.G., Downes, C.M.S., 2017. Forest ecosystem services: provisioning of non-timber forest products. In: Sills, E.O., Moore, S.E., Cubbage, F.W., McCarter, K.D., Holmes, T.P., Mercer, D.E. (Eds.), Trees at Work: Economic Accounting for Forest Ecosystem Services in the U.S. South. U.S. Department of Agriculture, Forest Service,

Southern Research Station, Asheville, NC. General Technical Report SRS-226.

Chapin III, F.S., Walker, B.H., Hobbs, R.J., Hooper, D.U., Lawton, J.H., Sala, O.E., Tilman, D., 1997. Biotic control over the functioning of ecosystems. Science 277, 500—504.

Cook, J.E., Halpern, C.B., 2018. Vegetation changes in blown-down and scorched forests 10-26 years after the eruption of Mount St. Helens, Washington, USA. Plant Ecology 219, 957—972.

Cooke, R.S.C., Bates, A.E., Eigenbrod, F., 2019. Global trade-offs of functional redundancy and functional dispersion for birds and mammals. Global Ecology and Biogeography 28, 484—495.

Corbera, E., González Soberanis, C., Brown, K., 2009. Institutional dimensions of payments for ecosystem services: an analysis of Mexico's carbon forestry programme. Ecological Economics 68, 743—761.

Costanza, R., de Groot, R., Braat, L., Kubiszewski, I., Fioramonti, L., Sutton, P., Farber, S., Grasso, M., 2017. Twenty years of ecosystem services: how far have we come and how far do we still need to go? Ecosystem Services 28, 1—16.

Daily, G.C., 1997. Introduction: what are ecosystem services? In: Daily, G.C. (Ed.), Nature's Services: Societal Dependence on Natural Ecosystems. Island Press, Washington, D.C., pp. 1—10

Dale, V.H., Adams, W.M., 2003. Plant reestablishment 15 years after the debris avalanche at Mount St. Helens, Washington. The Science of the Total Environment 313, 101—113.

Delaney, F., 1986. The Celts. Little, Brown and Company, Boston, pp. 240.

Fonseca, C.R., Ganade, G., 2001. Species functional redundancy, random extinctions and the stability of ecosystems. Journal of Ecology 89, 118—125.

Gitay, H., Wilson, J.B., Lee, W.G., 1996. Species redundancy: a redundant concept? Journal of Ecology 84, 121—124.

Goswami, M., Bhattacharyya, P., Mukherjee, I., Tribedi, P., 2017. Functional diversity: an important measure of ecosystem functioning. Advances in Microbiology 7, 82—93.

Grima, N., Singh, S.J., Smetschka, B., Ringhofer, L., 2016. Payment for ecosystem services (PES) in Latin America: analysing the performance of 40 case studies. Ecosystem Services 17, 24–32.

Halpern, C.B., Frenzen, P.M., Means, J.E., Franklin, J.F., 1990. Plant succession in areas of scorched and blown-down forest after 1980 eruption of Mount St. Helens, Washington. Journal of Vegetation Science 1, 181–194.

Hanson, L.J., Hoffman, J.R., 2011. Climate Savvy: Adapting Conservation and Resource Management to a Changing World. Island Press, Washington D.C., pp. 245.

Helms, J.A. (Ed.), 1998. The Dictionary of Forestry. Society of American Foresters, Bethesda, MD, pp. 210.

Jouquet, P., Traoŕe, S., Choosai, C., Hartmann, C., Bignell, D., 2011. Influence of termites on ecosystem functioning. Ecosystem services provided by termites. European Journal of Soil Biology 47, 215–222.

Keliang, Z., Vhugen, D., Hilgendorf, N., 2010. Who Owns Carbon in Rural China? an Analysis of the Legal Regime with Preliminary Policy Recommendations. Rights and Resources Initiative, Washington, D.C., pp. 37.

Kosoy, N., Corbera, E., Brown, K., 2008. Participation in payments for ecosystem services: case studies from the Lacandon rainforest, Mexico. Geoforum 39, 2073–2083.

Kramer, P.J., Kozlowski, T.T., 1979. Physiology of Woody Plants. Academic Press, Inc., New York, pp. 811.

Martínez, M.L., Pérez-Maqueo, O., Vázquez, G., Castillo-Campos, G., García-Franco, J., Mehltreter, K., Equihua, M., Landgrave, R., 2009. Effects of land use change on biodiversity and ecosystem services in tropical montane cloud forests of Mexico. Forest Ecology and Management 258, 1856–1863.

McFrederick, Q.S., Kathilankal, J.C., Fuentes, J.D., 2008. Air pollution modifies floral scent trails. Atmospheric Environment 42, 2336–2348.

Merry, K.L., Bettinger, P., Grebner, D.L., Hepinstall, J., 2010. Perceptions of foresters of wind damage in Mississippi forests. Southern Journal of Applied Forestry 34, 124–130.

Millennium Ecosystem Assessment, 2005. Ecosystems and Human Well-Being: Synthesis. Island Press, Washington, D.C, pp. 155.

Naeem, S., 1998. Species redundancy and ecosystem reliability. Conservation Biology 12, 39–45.

Nash, S., 2010. Making sense of Mount St. Helens. BioScience 60, 571–575.

Nieuwenhuis, M., 2010. Terminology of forest management, terms and definitions in English. In: International Union of Forest Research Organizations, second revised ed., vol. 9. IUFRO World Series, Vienna, Austria.

Nowak, D.J., 2018. Quantifying and valuing the role of trees and forests on environmental quality and human health. In: van den Bosch, M., Bird, W. (Eds.), Nature and Public Health, Oxford Textbook of Nature and Public Health. Oxford University Press, Oxford, UK, pp. 312–326.

Oliver, T.H., Isaac, N.J.B., August, T.A., Woodcock, B.A., Roy, D.B., Bullock, J.M., 2015. Declining resilience of ecosystem functions under biodiversity loss. Nature Communications 6, 10122.

Park, B.J., Tsunetsugu, Y., Kasetani, T., Kagawa, T., Miyazaki, Y., 2010. The physiological effects of Shinrin-yoku (taking in the forest atmosphere or forest bathing): evidence from field experiments in 24 forests across Japan. Environmental Health and Preventive Medicine 15 (1), 18.

Philpot, J.H., 1897. The Sacred Tree. Macmillan and Co., Limited, London, pp. 179.

Portela, R., Rademacher, I., 2001. A dynamic model of patterns of deforestation and their effect on the ability of the Brazilian Amazonia to provide ecosystem services. Ecological Modelling 143, 115–146.

Salisbury, F.B., Ross, C.W., 1985. Plant Physiology, third ed. Wadsworth Publishing Company, Belmont, CA, pp. 540.

Salles, J.-M., 2011. Valuing biodiversity and ecosystem services: why put economic values on nature? Comptes Rendus Biologies 334, 469–482.

Salzman, J., 2005. The promise and perils of payments for ecosystem services. International Journal of Innovation and Sustainable Development 1, 5–20.

Schindler, D.W., Lee, P.G., 2010. Comprehensive conservation planning to protect biodiversity and ecosystem services in Canadian boreal regions under a warming climate and increasing exploitation. Biological Conservation 143, 1571–1586.

Sell, J., Koellner, T., Weber, O., Pedroni, L., Scholz, R.W., 2006. Decision criteria of European and Latin American market actors for tropical forestry projects providing environmental services. Ecological Economics 58, 17–36.

Sileshi, G., Akinnifesi, F.K., Ajayi, O.C., Chakeredza, S., Kaonga, M., Matakala, P.W., 2007. Contributions of agroforestry to ecosystem services in the miombo eco-region of eastern and southern Africa. African Journal of Environmental Science and Technology 1, 68–80.

Small, N., Munday, M., Durance, I., 2017. The challenge of valuing ecosystem services that have no material benefits. Global Environmental Change 44, 57–67.

Smart, S., Maskell, L.C., Dunbar, M.J., Emmett, B.A., Marks, S., Norton, L.R., Rose, P., Henrys, P., Simpson, I.C., 2010. An Integrated Assessment of Countryside Survey Data to Investigate Ecosystem Services in Great Britain. National Environmental Research Council, Center for Ecology & Hydrology, Wallingford, Oxfordshire, UK. CS Technical Report No. 10/07, pp. 230.

Smith, R.L., Smith, T.M., 2000. Ecology and Field Biology, sixth ed. Benjamin Cummings Publishing, San Francisco, CA, pp. 720.

Stolton, S., Dudley, N., 2007. Managing forests for cleaner water for urban populations. Unasylva 58 (4), 39–43.

U.S. Department of Agriculture, Natural Resources Conservation Service, 1999. Cecil soil profile. U.S. Department of Agriculture, Natural Resources Conservation Service, Washington, D.C. https://www.nrcs.usda.gov/wps/portal/nrcs/detail/soils/survey/office/ssr7/profile/?cid=nrcs142p2_047954 (accessed 26.02.20).

U.S. Department of Agriculture, Forest Service, 1989. Land and Resource Management Plan. Siskiyou National Forest. U.S. Department of Agriculture, Forest Service, Pacific Northwest Region, Portland, OR.

U.S. Department of Agriculture, Forest Service, 2011. Brought to You by … Pollinators (Pollinator Factsheet). U.S. Department of Agriculture, Forest Service, Rangeland Management, Botany Program, Washington, D.C. http://www.fs.fed.us/wildflowers/pollinators/documents/factsheet_pollinator.pdf (accessed 13.02.20).

Van Hecken, G., Bastiaensen, J., 2010. Payments for ecosystem services: justified or not? A political view. Environmental Science and Policy 13, 785–792.

Ver Eecke, W., 1999. Public goods: an ideal concept. The Journal of Socio-Economics 28, 139–156.

Wendland, K.J., Honzák, M., Portela, R., Vitale, B., Rubinoff, S., Randrianarisoa, J., 2010. Targeting and implementing payments for ecosystem services: opportunities for bundling biodiversity conservation with carbon and water services in Madagascar. Ecological Economics 69, 2093–2107.

Wilkening, J., Pearson-Prestera, W., Mungi, N.A., Bhattacharyya, S., 2019. Endangered species management and climate change: when habitat conservation becomes a moving target. Wildlife Society Bulletin 43 (1), 11–20.

Winfree, R., Bartomeus, I., Cariveau, D.P., 2011. Native pollinators in anthropogenic habitats. Annual Review of Ecology, Evolution, and Systematics 42, 1–22.

Wu, S., Hou, Y., Yuan, G., 2010. Valuation of forest ecosystem goods and services and forest natural capital of the Beijing municipality, China. Unasylva 61, 28–36.

CHAPTER

7

Forest recreation

Recreational opportunities are facilitated by both public and private forests to meet the physical and socioeconomic needs of human populations around the world. The recreational pursuits of people vary by culture, education, region, wealth, and the resources available. These pursuits can range from hiking a trail in a remote wilderness to viewing endangered woodland animals from the safety of a boat or bus and, in cases such as the latter, recreationists may also be called *tourists*. Others may include hunting as their main form of forest recreation (Recknagel, 1917). One challenge facing future land managers will be to meet the needs of a wide spectrum of recreation enthusiasts in a sustainable manner, as it relates to both the recreational opportunities, the resource being managed, and the potential conflicts over uses. The management of resources with recreation and tourism potential has become a matter of meeting the goals of local communities while addressing the perceptions and attitudes of local populations and nonlocal recreationists and the long-term sustainability of resources (Zacarias et al., 2011). In this chapter, we present several topics that will allow readers to

- develop an understanding of recreational opportunities in general and those facilitated by forests and other natural resources;
- understand the differences between varying forms and types of recreational opportunities;
- understand the evolution of outdoor recreation in human societies; and
- understand some of the factors that influence human participation in forest recreational opportunities.

As current or potential managers of recreational resources, we may seek to provide recreationists with a diverse array of experiences so that they can obtain various levels of satisfaction (educational, emotional, or physical). Experience, laws, policies, and research are used to guide the development of these activities. Forest structural features may also need to be altered in an effort to develop and maintain these opportunities. Ultimately, the recreational experiences provided should result in a benefit to both the individual recreationist and society as a whole (Clark and Stankey, 1979).

7.1 What is recreation?

Although some forms require strenuous effort, recreation is a leisure activity that is unrelated to normal work activity; therefore, it is not directly associated with generating an economic benefit to the person engaged in the activity. However, recreation is dependent on work for its meaning and is often cathartic, therefore refreshing an individual for subsequent work (Smith and Godbey, 1991). Although some people make a living performing recreational activities, in general one should think of recreation as an activity or hobby that is performed before or after normal working hours. It can be used as a means for staying physically fit or as a manner in which to unwind and release stress or challenge one's capacities. What people do for recreation can vary extensively, and what one person thinks is a recreational activity may not be perceived in the same manner by another. For instance, parents might enjoy some quiet time after work to read a newspaper, magazine, or book. However, during the same period, their children may prefer to play video games or chase each other around the house as their form of recreation.

People indulge themselves in numerous forms of recreation, including building bird houses, collecting stamps or coins, cooking, dancing, listening to music, painting, playing a musical instrument, playing cards, playing video games, reading, refurbishing antique furniture, roller skating, shopping, visiting museums, and watching movies. Sporting activities consist of a large group of recreational opportunities, including badminton, baseball, basketball, cricket, football, hockey, polo, racquetball, soccer, squash, tennis, track, and volleyball, to name only a few, some of which are performed outdoors. Other outdoor recreational activities include birdwatching, camping, dog walking, fishing, gardening, golfing, hang gliding, hiking, hunting, ice-skating, mountain biking, paragliding, sightseeing, skiing, snorkeling, star gazing, surfing, swimming, taking photographs, and even drawing sketches of landscapes. Although some recreational activities (e.g., backpacking) have shown recent declines in participation, others (e.g., photography) have shown increases (Table 7.1); overall trends in nature-based recreational activities are changing, and the range of activities is growing (Cordell, 2008) and includes emerging outdoor activities such as geocaching, night running, and open-water swimming (Mackintosh et al., 2018).

The suite of potential recreational activities can also be classified according to whether they are performed communally or pursued alone. Activities may also be planned or undertaken spontaneously. Certain activities can require strenuous physical activity, specific skills (e.g., orienteering), or travel to a specific area, such as to a tobogganing track situated in a specific forest (Fig. 7.1). Recreational activities can be considered either healthy or harmful and may or may not be useful to

TABLE 7.1 Percent change in growth of a set of nature-based recreational activities in the United States, 2000–07.

Activity	Change in number of participants (%)	Change in days enjoyed (%)
Viewing or photographing flowers and trees	25.8	77.8
Viewing or photographing natural scenery	14.1	60.5
Off-road driving	18.6	56.1
Viewing or photographing:		
Wildlife	21.3	46.9
Birds	19.3	37.6
Kayaking	63.1	29.4
Backpacking	−0.6	24.0
Snowboarding	7.3	23.9
Visiting nature centers	5.0	23.2
Visiting wilderness	3.0	12.8

Cordell (2008).

FIGURE 7.1 Summer tobogganing track on Straža Hill in Bled, a municipality in northwestern Slovenia. *Donald L. Grebner.*

society. In addition, recreational activities may be available only during specific seasons of the year or year-round. Examples of communal recreational activities include team sports such as baseball, basketball, football, or hockey, whereas solitary recreational activities may include hiking, playing a musical instrument, running, or skiing. Examples of active recreation include hiking or swimming, whereas passive activities may include reading, playing video games, or watching movies. Physical activities such as running or yoga are commonly thought of as healthy recreational options, whereas attending a fraternity party may not be. Some recreation activities, such as working in a garden, could be helpful to society because the food grown can be sold in a marketplace or consumed by the recreationist and his or her family. In contrast, other recreational activities, such as *four-wheeling* with a truck, can cause ruts in forest soils and stimulate erosion, and thus may not be helpful to society. The key idea about *recreation* is that it is an activity separate from work and other responsibilities and is commonly pursued to provide oneself with physical or mental rejuvenation, or both. Recreation requires a recreationist and a recreational opportunity or experience; it is commonly situated within a broader landscape (small or large) and may have an impact (good or bad) on the forests and natural resources found there (Monz et al., 2010).

7.2 What is forest recreation?

In the previous section we established what, in general, constitutes a recreational experience. However, *forest recreation* includes a broad set of recreational activities that are exclusively performed outside, in a forested setting. For instance, forest recreation generally includes activities within forest settings that many of us tend to take for granted as normal aspects of our lives. These activities may include camping (Fig. 7.2), hiking (Fig. 7.3), fishing (Fig. 7.4), hunting (Fig. 7.5), rock climbing (Fig. 7.6), driving all-terrain vehicles (Fig. 7.7), whitewater rafting (Fig. 7.8), zip lining (Fig. 7.9), horseback riding (Fig. 7.10), cross-country skiing (Fig. 7.11), picnicking (Fig. 7.12), canoeing (Fig. 7.13), mountain biking (Fig. 7.14), mushroom collecting (Fig. 7.15), and many others that are not normal work activities and are not necessarily carried out to meet daily household needs. These opportunities are important, as forest recreation has been shown to relax both the minds and bodies of people (Park et al., 2009).

Outdoor photography has become a popular recreational pursuit; in fact, recent growth in recreational activities in the United States has been greatest for activities that involve photographing or observing nature (Cordell, 2008). Although many forests do not explicitly maintain facilities for photographers, recreational features such as auto tour routes, observation areas, and trails frequently support the desires of photographers to capture images of animals, landscapes, and wild plants. The advantages of recreating in forests include direct contact with nature, the availability of clean air, and the presence of a pleasant environment (Janeczko et al., 2018). Recreational activities can be performed in both public and private forests, depending on the level of access allowed by each management organization. The forest resource is therefore an important input into the supply of recreational opportunities available to society (Cordell et al., 2013).

Some purists view forest recreation as a narrow subset of a broader group of outdoor recreation activities.

FIGURE 7.2 Camping in the Yorkshire Dales in the United Kingdom. *Immanuel Giel through Wikimedia Commons.*
Image Link: https://commons.wikimedia.org/wiki/File:Gordale_Camping_03.jpg
License Link: Public Domain

FIGURE 7.3 A person hiking on the Camp Creek National Recreational Trail in California. *Courtesy San Bernardino National Forest through Wikimedia Commons.*
Image Link: https://commons.wikimedia.org/wiki/File:Camp_Creek_National_Recreation_Trail_(44758795192).jpg
License Link: Public Domain

FIGURE 7.4 A person fly fishing on the Flathead National Forest in Montana, United States. *U.S. Forest Service, Northern Region through Wikimedia Commons.*
Image Link: https://commons.wikimedia.org/wiki/File:Fishing_on_the_Flathead_National_Forest_(34173797351).jpg
License Link: Public Domain

FIGURE 7.5 A hunter surveying the landscape on the Deschutes National Forest in Oregon, United States. *U.S. Forest Service, Pacific Northwest Region through Wikimedia Commons.*
Image Link: https://commons.wikimedia.org/wiki/File:Hunting,_Deschutes_National_Forest_(36169874752).jpg
License Link: Public Domain

FIGURE 7.7 Dirt bike riding on the Ochoco National Forest in Oregon, United States. *U.S. Forest Service, Pacific Northwest Region through Wikimedia Commons.*
Image Link: https://commons.wikimedia.org/wiki/File:ATV_riding_Grassland,_Ochoco_National_Forest_(36456358141).jpg
License Link: Public Domain

FIGURE 7.6 A rock climber on Tonquani Kloof in the Magaliesberg Mountains, South Africa. *Francis Vergunst through Wikimedia Commons.*
Image Link: https://commons.wikimedia.org/wiki/File:Rock_Climber.JPG
License Link: Public Domain

FIGURE 7.8 Whitewater rafting on the Rangitata River, south of Christchurch, South Island, New Zealand. *Donald L. Grebner.*

From this perspective, *outdoor recreation* consists of every nonnormal work activity performed outside (e.g., baseball, cricket, or soccer), which includes all activities situated in deserts, oceans, and water bodies, regardless of the presence of nearby forests. Under this classification, forest recreation is, then, only those activities that are specifically located inside a forest. In this book, we will assume that some outdoor recreational activities

FIGURE 7.9 Zip lining in a tropical forest near Laguna de Arenal and Volcán Arenal in northwestern Costa Rica. *Stacey Herrin*

FIGURE 7.10 Horseback riding in the Ochoco National Forest in Oregon, United States. *U.S. Forest Service, Pacific Northwest Region through Wikimedia Commons.*
Image Link: https://commons.wikimedia.org/wiki/File:Cyrus_Horse_Camp-Ochoco_(23305371804).jpg
License Link: Public Domain

FIGURE 7.11 Cross-country skiing in the Mt. Hood National Forest in Oregon, United States. *U.S. Forest Service, Pacific Northwest Region through Wikimedia Commons.*
Image Link: https://commons.wikimedia.org/wiki/File:Cross_country_skiing_Trillium_Lake,_Mt_Hood_National_Forest-2_(37003596476).jpg
License Link: Public Domain

FIGURE 7.12 Picnic table in the Blue Bay Campground at Suttle Lake on the Deschutes National Forest in Oregon, United States. *U.S. Forest Service, Pacific Northwest Region through Wikimedia Commons.*
Image Link: https://commons.wikimedia.org/wiki/File:BLUE_BAY_CAMPGROUND-DESCHUTES-103_(28653535451).jpg
License Link: Public Domain

FIGURE 7.13 Canoeing on calm waters in Rhode Island, United States. *Lynn Sheldon.*

FIGURE 7.14 Mountain biking in a forest in New Zealand. *U.S. Embassy in New Zealand, through Wikimedia Commons.*
Image Link: https://commons.wikimedia.org/wiki/File:Mountain_Biking,_September_24,_2017_(37435321555).jpg
License Link: Public Domain

FIGURE 7.15 Mushroom harvesting on the Mt. Hood National Forest in Oregon, United States. *U.S. Forest Service, Pacific Northwest Region through Wikimedia Commons.*
Image Link: https://commons.wikimedia.org/wiki/File:Mushroom_harvesting_Mt_Hood_National_Forest_(37019039482).jpg
License Link: Public Domain

that occur on water bodies and in open areas associated with forests will be considered forest recreation activities. We feel that this is a reasonable perspective, since many water bodies and open areas intermixed with public or private forestlands are holistically managed to meet landowner objectives and comply with existing public laws.

7.3 A brief history of forest recreation in the United States

Forested ecosystems provide many provisioning services to humans. Although we can infer from forest patterns and processes the potential recreational opportunities, the complexity of the socioeconomic and political histories and legal systems of each country vary greatly, and these issues can have an impact on the nontimber provisioning services available from forests. In the United States, the forest recreation movement was not widely recognized until the early 1900s, and a significant number of years were required for a legal framework to be developed to assert that these types of activities need to be incorporated into the management plans of public lands (Zinser, 1995). Similarly, although some countries in Europe explicitly incorporated recreation into policies or legislation as far back as the turn of the 20th century, for the most part outdoor recreation concerns have been on the national political agendas of these governments only since the late 20th century (Mann et al., 2010). The recreation movement of the 20th century in Europe and North America arose from social dislocation issues related to the Industrial Revolution. Many recreational programs were developed to increase the quality of life of individuals in an increasingly urbanized working class population (Smith and Godbey, 1991). The amount of leisure time available today may depend on the age, gender, and occupation of an individual, as well as the economic, social, and political conditions of the region or country in which they live (Travis, 1982). Given the broad spectrum of recreational programs around the world, in order to simplify the discussion this section is confined to the emergence and evolution of forest recreational policies and laws in the United States.

The development of a broad recreational program in the United States has followed three distinct waves of political and legal activity over the past 100 years or so (Zinser, 1995). Prior to the first wave came the establishment of two significant national parks. Yellowstone National Park (Figs. 7.16 and 7.17) was established in 1872 by President Ulysses S. Grant (U.S. National Park Service, 2000) and was viewed as a pilot for future systems of conservation. This development signaled the emergence of recreation as a viable provisioning service

FIGURE 7.16   Lower Yellowstone Falls in the Yellowstone National Park, Wyoming, United States. *Stephen C. Grado.*

FIGURE 7.17   Old Faithful geyser in Yellowstone National Park, Wyoming, United States. *Stephen C. Grado.*

from ecosystems. Although established as a national park in 1890, some portions of Yosemite National Park (Fig. 7.18) were set aside in 1864 for preservation and public access by the U.S. Congress. These two parks are very large. Yellowstone National Park covers nearly 900,000 hectares (ha) (over 2.2 million acres [ac]), roughly 151 times the size of Manhattan Island and about the same size as Puerto Rico or Cyprus. Yosemite National Park is over 308,000 ha (over 760,000 ac), about 52 times the size of Manhattan Island. However, the largest (in terms of land area) national park in the United States is the Wrangell-St. Elias National Park in Alaska, which is about six times the size of Yellowstone and larger than either Costa Rica or Bosnia and Herzegovina.

FIGURE 7.18 Half Dome at Yosemite National Park, California, United States. *Stephen C. Grado.*

FIGURE 7.19 Entrance to Devils Tower National Monument, northeastern Wyoming, United States. *Tibor Pechan.*

The first wave of political and legal activity in the United States came in the early 1900s and was associated with President Theodore Roosevelt's advocacy of conservation of American natural resources (Zinser, 1995). President Roosevelt's appointment of Gifford Pinchot as the Chief of the Forest Service in 1905 led to the incorporation of conservation concepts in the management approaches of U.S. National Forests, in contrast to John Muir's preservationist philosophy, which in 1892 led to the creation of the Sierra Club. From the beginning, there were conflicts over the use of forest resources as the philosophical differences between these two people are still evident in some environmental debates over a century later. Aldo Leopold, author of *A Sand County Almanac*, was also influential during this period. A forester by training, through his work experiences he developed an ecological ethic stressing a balance of nature, and later in his career encouraged the development of wilderness areas. Bob Marshall, who was for a brief period head of recreation management for the U.S. Forest Service, was one of Leopold's contemporaries and helped establish the Wilderness Society.

In 1902, the Reclamation Act was enacted in the United States that allowed the construction of reservoirs and other water programs in the western states and territories. The development of these water resources facilitated the expansion of water-based recreational opportunities. In addition, the Antiquities Act of 1906 gave the President of the United States the power to proclaim national monuments on public lands, and this legislation served as a vehicle for the creation of future national parks. The Devils Tower (Fig. 7.19) in Wyoming was the first national monument created through this legislation. Information derived from a broad-scale natural resource inventory prompted President Roosevelt to place about 81 million ha (about 200 million ac) of

FIGURE 7.20 Entrance to Denali National Park, Alaska, United States. *Stephen C. Grado.*

land into federal control, of which about 61 million ha (about 150 million ac) would later serve as the backbone of the U.S. National Forest System. One of the latter acts in this first wave of activities was the passage of the National Park Service Act (1916), which provided a guide for managing the growing number of national parks (Fig. 7.20).

The second wave of political and legal activity influencing the development of a broad recreational program in the United States occurred mainly in the 1930s during the presidency of Franklin D. Roosevelt (Zinser, 1995). During this economically difficult period, a number of environmental policies were enacted and several government agencies were created to focus on land use planning and soil improvement programs. One large program created in 1933 is the Tennessee Valley

Authority (TVA), which is well known in the southern United States. This program manages water projects over an area covering about 129,500 square kilometers (km²; about 50,000 square miles [mi²]) across eight southern states. The water systems managed by TVA have provided numerous recreational opportunities in the southern region. Other recreation-related legislation enacted during this second wave of activity include the Migratory Bird Hunting Stamp Act of 1934, the Historic Sites Act of 1935, and the Wildlife Restoration Act of 1937.

The third wave of activity influencing the development of a broad recreational program in the United States began in the late 1950s and continued throughout the 1960s (Zinser, 1995). This period was marked by several political and legal maneuverings that supported the expansion of recreation in North America. While action on a nationwide highway system had occurred nearly two decades earlier, in 1956 President Eisenhower signed the Federal-Aid Highway Act that prompted the development of the Interstate Highway system in an effort to produce jobs, support national defense, and to open up rural areas to travelers (Weingroff, 1996). In 1958, President Eisenhower also signed the Outdoor Recreation Resources Review Act to address the pent-up demand for forest recreation. This legislation required the determination of the present and future outdoor recreational needs and desires of United States citizens, an inventory of all outdoor recreational resources available to meet American needs by 1976 and 2000, and recommendations of policies and programs that could be used to satisfy present and future public needs. The Multiple-Use and Sustained Yield Act (1960) established a national policy that required national (public) forests to be managed in such a way as to provide a wide variety of natural resources, such as fish, outdoor recreation, range, timber, and wildlife. The passage of this legislation was seen as a turning point in the management philosophy of U.S. National Forests, shifting focus from the sustainability of timber production to the sustainability of multiple products, services, and uses (Fig. 7.21).

In 1963, enactment of the Outdoor Recreation Act (Sharpe et al., 1983) required the U.S. Department of the Interior to prepare and maintain a continuous inventory and evaluation of outdoor recreation needs and resources, and subsequently prepare an outdoor recreation resource classification system for assisting in the management of these resources. This was significant because one of the agencies in this department, the Bureau of Land Management (BLM), is a land management agency that oversees close to one-eighth of the total land area of the United States. Along the way, the legislation required the formulation and maintenance of a comprehensive, nationwide outdoor recreation plan and

FIGURE 7.21 Entrance to Giant Sequoia National Monument, California, United States. *Stephen C. Grado.*

required the Department of the Interior to provide technical assistance and advice to all states, lower levels of government, and the private sector on outdoor recreational issues. The legislation encouraged regional planning, acquisition, and development of outdoor recreational opportunities and the engagement and cooperation of educational institutions.

The Wilderness Act (1964) established a national system of congressionally designated wilderness areas around the United States (Sharpe et al., 1983). Most of the land established as wilderness under this legislation was situated in the western United States and administered by the U.S. Forest Service. The Forest Service's task is to maintain the wilderness attributes of these protected areas and to develop assessments of their level of recreational use. It still allowed some existing extractive uses such as grazing to continue. In the 1970s, the United States Congress changed the definition of what could be considered wilderness; this was controversial, since the change allowed lands that had formerly been impacted by human activity to be designated wilderness areas. The idea behind this definitional change was that some lands with wilderness potential may be intermixed with land in which management activities may have previously occurred and that ecological processes would, over time, erase or obscure the human impact on these land areas. This eventually led to the passage of the Eastern Wilderness Act (1975), which established 16 wilderness areas in the more extensively impacted eastern United States (Douglass, 2000).

In 1964, enactment of the Land and Water Conservation Fund Act (Sharpe et al., 1983) provided funds to federal, state, and local governments to allow these organizations to acquire land and develop outdoor recreational opportunities across the United States in an attempt to meet growing recreational demands. The Act initially authorized funds of U.S.$60 million per year, but later amendments increased the allocation of funds to nearly U.S.$1 billion per year (Douglass, 2000). However, actual appropriations of funds through this legislation have been closer to U.S.$200 million per year. Land and water conservation funds now primarily arise from oil and gas lease revenues from federal lands and require matching funds from state and local governments in the United States (U.S. National Park Service, 2013). The Wild and Scenic Rivers Act (1968) declared that certain rivers and their immediate environments would be preserved and protected into perpetuity for the benefit of the public (Sharpe et al., 1983). A key outcome of this legislation was the creation of the National Wild and Scenic Rivers System, which categorized some rivers as either wild river areas, scenic river areas, or recreational river areas. The National Trails System Act (1968) was enacted to establish trails within established scenic areas or near urban areas (Sharpe et al., 1983). As part of this legislation, the Appalachian Trail and the Pacific Crest Trail were incorporated into this system. The National Trails System classifies trails as national historical trails, national recreational trails, national scenic trails, or connecting and side trails. The National Trails System currently has 30 national scenic and national historic trails. In addition, there are over 1300 national recreational trails located in each of the 50 states, the District of Columbia, and Puerto Rico. These comprise about 81,000 km (50,000 mi) of combined recreational experiences (U.S. National Park Service, 2019a). This distance is over two times the length of the Earth's equator.

After the third wave of political and legal activity, numerous other pieces of national legislation have been enacted that played an important role in the development of forest and natural resource recreation in the United States. Some of these include the Endangered Species Act of 1973, the Forest and Rangeland Renewable Resources Planning Act of 1974, the National Forest Management Act of 1976, the Railroad Revitalization and Regulatory Reform Act of 1976, the Surface Mining and Reclamation Act of 1977, the Endangered American Wilderness Act of 1978, the National Parks and Recreation Act of 1978, and the Alaska National Interest Lands Conservation Act of 1980 (Sharpe et al., 1983; Douglass, 2000).

In other areas of the world, the development and evolution of recreation programs may have differed. In one part of Russia, for example, recreation management began with the need for summer recreation areas for children and evolved through a recognition of the need for health resorts, the need to designate recreation zones, and the development of recreation complexes (Isachenko et al., 2015). These efforts initially focused on satisfying the needs of local residents yet have expanded over time to address the interests of tourists.

7.4 Where are the forest recreational opportunities?

An opportunity to recreate can be found on nearly every ownership category of land throughout the world, even if this simply means you can view from afar the pristine or primeval forests of a preserved area. In the United States, Zinser (1995) categorizes these opportunities by ownership characteristics and suggests that there are five recreational resource sectors, of which three involve public land and two involve private land: public sectors include federal, state, local/regional recreational opportunities, and private sectors include noncommercial recreation and commercial recreation. Regardless of the type of organization managing a particular area of land, events (i.e., management activities or wildfires), or the result of events (i.e., flood damage) can significantly impact the location and availability of recreational use within a forest (Starbuck et al., 2006). Therefore, while the ensuing discussion tends to categorize opportunities by landowner group, spatial and temporal (e.g., economic or environmental) changes to the landscape regardless of landowner group can influence where opportunities may be found.

In countries in which all of the forestland is owned or managed by a central government (e.g., the Republic of Turkey), one can surmise that all of the formalized forest recreational opportunities are located on public lands. In addition, although about 57% of the land in the United States is privately owned, most of the existing formalized recreational opportunities are located on federal or national lands (President's Commission on Americans Outdoors, 1987). The area of rural parks and wilderness in the United States continues to increase (Bigelow and Borchers, 2017). While federal lands compose about 28% of the land of the United States (Vincent et al., 2017), federal lands accommodate nearly 91% of all formal recreational lands available (Table 7.2), and land managed or owned by individual states represents about 8% of the recreational lands. Unsurprisingly, the average size of recreational lands is greatest for those opportunities located on federal lands in the United States (161,444 ha or 398,929 ac), while the average size on municipal lands is much smaller (17.9 ha or 44.3 ac). Of course, simply changing the management philosophy of large expanses of public lands would influence these differences. Further, the location of the lands is important, for in countries in which there is a diverse mixture of public and private land, recreational areas may be intermixed with private or public lands that are managed for objectives other than recreation. In mixed-ownership situations such as these, access to recreational opportunities can become important to society (Barkley et al., 2015).

TABLE 7.2 Distribution of public recreational lands in the United States.

	Federal	State	Regional	Municipal
Number of recreational areas	1774	20,375	19,884	67,685
Percentage of recreational areas	1.6	18.6	18.1	61.7
Total size of land (million ha)	286.4	25.1	2.3	1.2
Total size of land (million ac)	707.7	62.0	5.7	3.0
Percentage of total recreational land	90.9	8.0	0.7	0.4
Average size of recreational land (ha)	161,444	1231	116	18
Average size of recreational land (ac)	398,929	3043	287	44

Adapted from Zinser (1995).

7.4.1 Federal recreational opportunities

Federal lands are lands controlled or owned by the central government of a country. In some countries, there is very little or no private ownership of land, and in others land tenure systems have been designed to encourage or manage private ownership of land. In this section, we refer specifically to recreational opportunities provided on lands that are controlled or owned by the central government of a country. Public access to some of these lands is prohibited, including lands dedicated for military training purposes, lands protected because they pose a danger (e.g., contaminated from prior catastrophes), or lands reserved for use by the head of state or other high-ranking politicians.

In the United States, federal recreational areas can be found in every region, although the largest of these are located in the western states. In some states, federal lands comprise a significant percentage of the total land area. For instance, in Nevada, about 83% of the land is federally owned, and other states such as Alaska (68%), Utah (64%), Idaho (63%), and Oregon (52%) also contain substantial areas of federal land. Some southern states, on the other hand, have less than 10% of their land in federal ownership. Of course, these public lands are not entirely forested. In fact, to a large extent, the public land in some of these states consists of arid or rocky terrain that is sparsely covered in forest vegetation. Most of this public land is managed by the BLM, the U.S. Forest Service, the Fish and Wildlife Service, the National Park Service, the U.S. Army Corps of Engineers, the Bureau of Reclamation, the TVA, or the Bureau of Indian Affairs (BIA).

The U.S. Forest Service is part of the U.S. Department of Agriculture and manages over 78 million ha (193 million ac) of land that is either administered by the National Forest System or National Grassland System (U.S. Forest Service, 2013a). Although most Forest Service lands are located in the western United States, they can be also found in eastern states. For instance, within Mississippi are the Bienville National Forest, the DeSoto National Forest, the Holly Springs National Forest, and the Tombigbee National Forest. Although forest plans are currently guided by a need to sustain ecosystems and social values, Forest Service lands are typically managed for a variety of provisioning services, such as minerals, range, recreation, timber, water, and wildlife. In fact, the Forest Service administers 22 national recreational areas, two national volcanic monuments, four national monuments, and eight national scenic areas (U.S. Forest Service, 2013b). Further, a large portion of the National Wilderness Preservation System is located within the National Forest System.

The U.S. Fish and Wildlife Service, an agency housed in the U.S. Department of the Interior, manages over 36.1 million ha (89.1 million ac) of land within the National Wildlife Refuge System (Vincent et al., 2017). This land is located in 567 separate refuges and 38 wetland management districts (U.S. Fish and Wildlife Service, 2019). On average, approximately 41 million people visit national wildlife refuges each year, mainly for recreational purposes. As with the National Forest System, wildlife refuges are found all across the United States. The first wildlife refuge, Florida's Pelican Island, was founded in 1903 during the presidency of Theodore Roosevelt (U.S. Fish and Wildlife Service, 2015). Although the primary purposes of wildlife refuges are to protect and maintain sustainable populations of various wildlife species, recreational opportunities are also widely available. For instance, in the Iroquois National Wildlife Refuge in western New York, fishing, hiking, hunting, and picnicking activities are all permitted at various times of the year.

The National Park Service is another unit of the U.S. Department of the Interior that manages over 34.4 million ha (85 million ac) of land organized in the national park system (U.S. National Park Service, 2019b). Visits to U.S. national parks have been in the range of 250 to 280 million people per year since the early 1990s (Cordell et al., 2008). The 419 units, often called national parks, of this program include nearly 2600 national historic landmarks (U.S. National Park Service, 2018). In addition to the parks already mentioned, some well-known parks in the United States include Acadia, Crater Lake, Death Valley, Grand Canyon, Grand Teton (Fig. 7.22), Great Smoky Mountains, Isle Royale, Natchez Trace, Joshua Tree, Olympic, and Shenandoah. While parks such as these may accommodate many different types of recreational activities, others, such as the Blue Ridge Parkway in Virginia and North Carolina, specifically target motorized recreational activities (Johnson, 2015).

The BLM is yet another agency in the U.S. Department of the Interior that manages over 100.4 million ha (248

FIGURE 7.22 Mountains of the Grand Teton National Park, Wyoming, United States. *Stephen C. Grado.*

million ac) of land, with a significant quantity of federal recreational lands in the states of Alaska, Arizona, California, Colorado, Idaho, Oregon, Montana, Nevada, New Mexico, Washington, Wyoming, and Utah (U.S. Bureau of Land Management, 2018). Unlike in the National Park Service, the management of these lands is guided by the philosophies of sustainability of multiple uses and yields of various products. These lands are typically not as productive as the lands found in the National Forest System or other federal lands, and include Arctic tundra, forested high mountains, desert landscapes, and grasslands. However, the Oregon and California Railroad (O&C) revested lands in Oregon are some very productive forestlands. Some of these lands form part of the National Wilderness Preservation System (Zinser, 1995). The BLM maintains numerous recreational areas and has a program for continuously enhancing outdoor recreational experiences by improving opportunities for fishing, hunting, and shooting sports. In addition, the BLM is attempting to expand the National Recreation Trails System; it manages 18 National Scenic and Historic Trails that are about 8230 km (approximately 5080 miles) in length (U.S. Bureau of Land Management, 2018).

Another important federal agency in the United States regarding recreational opportunities is the Army Corps of Engineers, part of the U.S. Department of Defense. The Department of Defense manages 4 million ha (11.4 million ac) of land in the United States (Vincent et al., 2017) that contain about 460 hydroelectric dams and numerous navigational and flood control projects (Zinser, 1995). Although the Corps has no official mandate for doing so, it manages for recreational activities on various reservoirs under the provisions of the Flood Control Act of 1944 and the Federal Water Project Recreation Act of 1965. Many of these projects are located near urban areas and therefore experience intensive use (Zinser, 1995).

The Bureau of Reclamation, another agency within the U.S. Department of the Interior, manages 338 reservoirs and 492 dams that support irrigation on about 4 million ha (10 million ac) of farmland, thus affecting the livelihoods of about 31 million people in local communities (U.S. Bureau of Reclamation, 2019). The land and water resources managed by the Bureau of Reclamation are located on 2.47 million ha (6.1 million ac) of land in 17 western states, including the prairie states but not Alaska or Hawaii. Although supplying water to farmlands and local communities is the primary concern of the Bureau of Reclamation, a significant amount of effort is also expended in developing and maintaining recreational opportunities. Currently, the Bureau of Reclamation manages 289 recreational areas and 550 campgrounds, which receive about 90 million visits each year (U.S. Bureau of Reclamation, 2019).

We mentioned earlier that the TVA was created in 1933. This agency currently offers about 93,000 ha (229,000 ac) of land that is available for hunting, camping, and other recreational activities on undeveloped lands that it manages. The TVA administers 80 public recreational areas such as boat launching ramps, campgrounds, and day-use areas (Tennessee Valley Authority, 2019). The water bodies and surrounding lands offer extensive opportunities for fishing, hiking, picnicking, camping, birdwatching, nature photography, water skiing, canoeing, sailing, swimming, and windsurfing.

The last major United States federal agency with a significant recreational agenda is the BIA. This agency is responsible for managing 22.2 million ha (55 million ac) of land in trust for various tribes and provides services to approximately 1.9 million Native Americans (U.S. Bureau of Indian Affairs, 2011a). The BIA has a wildlife and park program that supports tribal level activities such as fisheries, wildlife, outdoor recreation, public use management, and conservation enforcement (U.S. Bureau of Indian Affairs, 2011b). An important difference between other public lands in the United States and those managed by the BIA is that the latter can be considered private land, because this land is held in trust and considered the sovereign land of the various tribes. Only since the last decades of the 20th century have Indian communities opened these lands to broader recreational and tourism activities, mainly to enhance the economic development of local communities and pueblos (Zinser, 1995). Examples include the White Mountain Apache Tribe that offer guided hunts to the public for a variety of species.

7.4.2 State or provincial recreational opportunities

Many states and provinces also administer public lands, and these provide a wealth of recreational opportunities. Often on these lands, recreational activities are encouraged if they are not in conflict with other uses of the land and do not disrupt environmental conditions (Mueller et al., 2015). Visits to state parks in the United States grew rapidly in the mid-20th century and have been relatively steady at about 710 to 740 million people per year since 2000 (Cordell et al., 2008). In general, state or provincial recreational lands (whether forests, parks, or water resources) are similar to federal recreational lands (e.g., national parks) in that they are created to preserve areas of natural beauty and historic interest; however, they are administered at a lower governmental level. Examples include the Humboldt Redwoods State Park in northern California (Fig. 7.23) and the Algonquin Park in Ontario, Canada. State lands in the United States are publicly held areas that play an important role in providing recreational opportunities, even though the total area involved is considerably smaller than the total area of federal lands. Some state university forests also encourage recreational activity if it is not in conflict with the primary uses of the land (Savage, 2015). U.S. states manage approximately 25 million ha (61.8 million ac) of forest, recreational areas, and wildlife areas across 5300 state parks, 721 state forests, 7400 wildlife areas, and numerous historical, cultural, and water-based sites (Zinser, 1995).

The State of Mississippi provides an interesting example of public land held and managed at the governmental level. Mississippi has 25 state parks and 4 state golf courses (Mississippi Department of Wildlife, Fisheries, and Parks, 2019). Many of these parks have various forms of camping opportunities, such as tent camping and developed camping areas, as well as boat launches, picnic areas, playgrounds, pools, tennis courts, and water skiing. Another important state agency involved in providing recreational opportunities in Mississippi is the Mississippi Forestry Commission. This agency provides technical assistance to private landowners but has management authority over 194,249 ha (480,000 ac) of nonfederal public forestlands (Mississippi Forestry Commission, 2018). Public lands managed by the agency include Kurtz and Camden State Forest and school trust lands, also known as 16th Section Lands. These recreational areas are found in 67 of the 82 state counties and are managed for a variety of objectives. The production of timber is important to the state, as it generates revenue to support local schools, but concern is also given to aesthetics, maintaining wildlife habitat, and recreational opportunities.

The state of New York manages 215 state parks that are open to the public for recreational purposes. The two largest and best-known parks are the Adirondack and Catskill Parks. The Catskill Park is 283,280 ha (approximately 700,000 ac) in size and is managed by the Department of Environmental Conservation. It supports numerous recreational opportunities such as camping, fishing, hiking (Fig. 7.24), hunting, and skiing. Interestingly, the Catskill Park (Fig. 7.25) is actually a mixture of public and private lands, and approximately 50,000 people reside there year-round; the Catskill Forest Preserve, for example, is the state land (116,350 ha; 287,500 ac) within the park (New York Department of Environmental Conservation, 2018). The Adirondack Park covers approximately 2.4 million ha (about 6 million ac), which is roughly the size of the nearby state of Vermont and includes 10,000 lakes and 48,600 km (30,000 mi) of streams and rivers (Adirondack Park Agency, 2018). Similar to the Catskill Park, the Adirondack Park is a mixture of public and private lands, and within its boundaries are 103 towns and villages that support 132,000 residents, farms, small businesses, working forests, and open spaces. The park provides numerous recreational opportunities that range from developed interpretive areas to wilderness areas (Fig. 7.26).

FIGURE 7.23 Entrance to Humboldt Redwoods State Park in California, United States. *Stephen C. Grado.*

FIGURE 7.24 A sign indicating the presence of a trail to the hamlet of Grand Gorge, New York, at the headwaters of the Delaware River. *Courtesy of Rock NJ through Wikimedia Commons.*
Image Link: https://commons.wikimedia.org/wiki/File:CST_Sign.JPG
License Link: Public Domain

FIGURE 7.25   Overlook Mountain vista in Catskill Park, New York, United States. *Courtesy of Maximusnukeage, through Wikimedia Commons. Image Link: https://commons.wikimedia.org/wiki/File:Catskills_from_Overlook_Mountain.jpg License Link: Public Domain*

FIGURE 7.26   Backpackers on Algonquin Mountain, Adirondack Park, New York, United States. *Courtesy of Anne LaBastille through Wikimedia Commons. Image Link: https://commons.wikimedia.org/wiki/File:BACKPACKERS_DESCENDING_THE_SUMMIT_OF_ALGONQUIN_MOUNTAIN_IN_THE_HIGH_PEAKS_REGION_OF_THE_ADIRONDACK_FOREST_PRESERVE..._-_NARA_-_554504.jpg License Link: Public Domain*

7.4.3 Urban and municipal recreational opportunities

Another group of areas in which one may find recreational opportunities are urban forests or municipal lands. These areas typically consist of open spaces, parks, ponds, scenic parkways, trails, and walkways, which may provide distinctly different recreational opportunities (e.g., dog walking or jogging) than those available, for example, in a remote forested backcountry. Urban forests may be managed by a mixture of public agencies and private property owners, as an urban forest can be composed of trees located on commercial and

residential property in addition to parks and along roads (Swae, 2015). Although small in total area compared to recreational areas in federal or state lands in the United States, the number of visits to urban and municipal forests and parks for recreational purposes is far greater owing to their proximity to population centers (Zinser, 1995). Urban forests typically incur more intense recreational use than forests situated in more remote areas, and the resulting pressure placed upon these resources varies depending on the character of the surrounding urban area, which in turn may influence the level and type of recreational opportunity available (Arnberger, 2006). Depending on the size of the community, there may be only one or two recreational areas within the administrative boundary of a small town. For instance, one small town, Clinton in Massachusetts, has a central park downtown that is commonly used for picnicking, skateboarding, or walking. Clinton also has a series of recreational fields used by various local sports leagues. In larger urban areas, more extensive systems of parks, forests, trails, and recreational areas are commonly found. Counties, which are subdivisions of states or provinces, can also play important roles in providing recreational opportunities for people to pursue. For example, the Sawyer County Forest in northern Wisconsin is a relatively large county forest (over 46,000 ha) that maintains systems of trails for walking, hiking, and biking activities, as well as for motorized activities that involve the use of all-terrain vehicles and snowmobiles.

Examples of urban and municipal recreational areas in a forested setting include the Atsuta Jingu sacred deciduous forest in Nagoya, Japan, the Khimki Forest in Moscow, Russia, and the Tijuca Forest in Rio de Janeiro, Brazil. Cities may own large tracts of land that they consider *community forests* and manage them in ways that provide local residents with various recreational opportunities and that further attract nonresident tourist activity (Andre, 2015). Some of the oldest city parks in the United States (Plaza de la Constitución in St. Augustine, Florida, Boston Common in Boston, Massachusetts, and New Haven Green in New Haven, Connecticut) were established over 370 years ago (The Trust for Public Land, 2010). Two examples of large municipal recreational areas in the United States are Central Park in Manhattan and Forest Park in Portland, Oregon. The most visited municipal park in the United States, Central Park (Fig. 7.27), was purchased in 1858 (The Trust for Public Land, 2010) and is used extensively for biking, birdwatching, boating, ice-skating, jogging, picnicking, rock climbing, swimming, and walking activities. In addition, the park hosts horse carriage rides, outdoor theatrical plays and concerts, playgrounds, and a small zoo (Greensward Group, LLC, 2019). Central Park has also been extensively used as a backdrop for scenes included in television shows (e.g., *Law & Order*) and movies such that permit private ownership of land. Invariably, concerns arise over the ability of public lands to meet the growing demand for certain forms of recreation (e.g., hunting). Private forestlands can play an important role in alleviating these concerns. In countries that have private land tenure systems, the vast majority of forest recreational opportunities located on public lands are typically relatively far from urban centers, although the majority of private forestlands are located much closer to urban centers, which enhances the importance of this resource. Hunting is often one of the most important recreational activities on privately owned lands (Munn et al., 2015). Some other common forms of forest recreational opportunities available on private lands include berry and mushroom collecting, birdwatching, boating, camping, canoeing, cross-country skiing, fishing, hiking, horseback riding, rock climbing, and all-terrain vehicle riding. In some cases, overcrowding of public forestlands increases the attractiveness of private forest recreational opportunities.

In some countries, such as Scotland and Finland, the *everyman's right* or a set of *outdoor access codes* allow certain public recreation activities on private resources as long as the recreational activities respect the privacy, safety, and livelihood of people living or working there (Scottish Natural Heritage, 2017; Lankia et al., 2015). In other cases where public rights to access and engage in recreation on private lands is restricted, private recreational opportunities can be categorized as either noncommercial or commercial, with the presence of a profit motive of the private landowner being the key difference between the two groups (Zinser, 1995). Boy Scouts of America camps are an example of a nonprofit forest recreational opportunity. Typically, these camps are developed to address the needs of adolescent boys and girls who are Cub Scouts (younger) or Scouts (older) and who have the desire to learn how to camp, fish, and interact with nature. Parents of children can accompany them to these recreational areas and, in many instances, camping events on these sites may not only represent a Cub Scout's first overnight experience in the forest but may also represent the first opportunity of this type for their parents. These camps are also available as recreational opportunities for nonaffiliated groups in order to help generate revenue to maintain camp facilities. One example is Camp Kern, situated at Huntington Lake in the Sierra Nevada Mountains of southern California. This camp is quasi-private, since it is situated on federal land and operates under a permit issued by the U.S. Forest Service. However, it operates independently of the federal government and provides opportunities for hiking, horseback riding, sailing, and swimming (Southern Sierra Council, Boy Scouts of America, 2019). Another example of a Boy Scout camp is Camp Seminole, just north of Starkville, Mississippi, which is owned by the Boy Scouts of America Pushmataha Area Council. Camp Seminole is a 117 ha (288 ac) facility that hosts many of the local scouting events in the

FIGURE 7.27 Aerial view of Central Park in New York. *Carol M. Highsmith Archive collection at the U.S. Library of Congress, through Wikimedia.*
Image Link: https://commons.wikimedia.org/wiki/File:Aerial_view_of_Central_Park,_New_York,_New_York_LCCN2011631310.tif
License Link: Public Domain

FIGURE 7.28 Scenic trail in Forest Park, Portland, Oregon, United States. *Nickpdx through Wikimedia Commons.*
Image Link: https://commons.wikimedia.org/wiki/File:Path_in_Forest_Park_(Portland).jpg
License Link: Public Domain

as *Breakfast at Tiffany's, Bride Wars, Ghostbusters, Home Alone II, Kramer vs. Kramer, Spider-Man III, The Day the Earth Stood Still,* and *When Harry Met Sally.* Forest Park was officially designated in 1948 and is situated on the northwestern outskirts of Portland, Oregon. The park covers nearly 2100 ha (5157 ac) (The Trust for Public Land, 2010) and represents a key component of the network of natural areas, parks, and trails (Fig. 7.28) that can be found in the Portland metropolitan area.

7.4.4 Private recreational opportunities

Private lands are another key source of recreational opportunities in countries with land tenure systems

Starkville area, as well as those sponsored by other community groups. The type of educational and recreational opportunities facilitated by this privately owned camp includes navigational skills, rappelling, and team-building exercises (Boy Scouts of America, 2011).

Other noncommercial recreational opportunities facilitated on lands owned by private individuals can be second homes or holiday getaways. If a private individual owns more than about 10 ha (25 ac) surrounding their primary place of residence, these lands can also be considered to facilitate forest recreational opportunities. However, these properties are commonly used only by the individuals who own them, and thus access for the general public is usually limited. Often, nonindustrial private landowners in North America are hesitant to allow unfamiliar people access to their land for fishing, hunting, or other activities because of liability concerns stemming from damage or accidents. Private landowners may also be concerned about littering, property damage, or theft, as well as the economic or legal implications of these acts. Lawsuits aimed at private forest landowners related to injuries or accidents incurred by others while visiting their land can deplete the private landowner's personal wealth. Whatever the perceived risk, there seem to be relatively few instances of people injuring themselves or others on private property during recreational events (Sun et al., 2007).

Two nonprofit organizations in the United States that offer forest recreational opportunities include the National Audubon Society and The Nature Conservancy. The National Audubon Society is a nonprofit organization whose mission is, in part, to protect bird populations and their habitats for the benefit of humanity (National Audubon Society, 2019). The society has 500 chapters within the United States and is active in providing birdwatching opportunities in conjunction with other conservation efforts. The society also supports numerous educational centers and sanctuaries that attract millions of visitors each year in their quest to discover and understand the natural world. As an example, the National Audubon Society operates the Starr Ranch sanctuary, a 1618 ha (4000 ac) preserve in Orange County, California. The sanctuary was designed to protect native vegetative types such as chaparral, coastal sage scrub, grasslands, oak woodlands, and riparian woodlands (National Audubon Society, 2018). The sanctuary also supports a variety of native wildlife such as fence lizards (*Sceloporus occidentalis*), canyon tree frogs (*Hyla arenicolor*), red-shouldered hawks (*Buteo lineatus*), and mountain lions (*Puma concolor*). Although technically a sanctuary for plant and animal species, the facility offers a variety of educational programs for young people and their families. The Nature Conservancy is a nonprofit organization with an international reach. The Nature Conservancy's mission is "to conserve the land and waters on which all life depends" (The Nature Conservancy, 2019). The conservancy helps advance conservation throughout the United States and 71 other countries. Many of these lands are available to the public for certain recreational uses. One such property is the Bridgestone Nature Reserve at Chestnut Mountain, located on the Cumberland Plateau in Tennessee. This property of over 2300 ha (5700 ac) helps protect a unique ecological area for plant and animal life, acts as a carbon sequestration project, and facilitates some forms of low-impact public recreation (The Nature Conservancy, 2018).

Commercial forest recreational opportunities vary in the size and the types of activities they offer. A small commercial operation may consist of a forest landowner who offers a hunting lease to a local *hunting club*. A hunting club is typically a group of people who pay an annual membership fee, which can then be used to acquire a lease on a private landowner's property for the privilege of hunting ducks, deer, turkey, or other game over a given period. Although small landowners may engage in leasing their property to hunting clubs as a method of generating income, large landowners, such as forest products companies, timberland investment groups, and real estate investment trusts, may also lease their lands for these purposes. For example, the Weyerhaeuser Company, the largest private landowner in the United States, with land in states from Maine to Washington and across the southern region, has a well-developed recreation program that complements their business objectives.

Other types of commercial forest recreational activities can be found through organizations that own land and offer specialized opportunities. One such organization is Tara Wildlife, located about 49 km (about 30 mi) by road northwest of Vicksburg, Mississippi, in what is known as the Mississippi Delta. Tara Wildlife maintains a large conference center and owns about 3600 ha (9000 ac) of bottomland hardwood forests and oxbow lakes that facilitate various types of consumptive activities, such as deer and turkey hunting, and nonconsumptive activities, such as birdwatching, canoeing, fishing, nature photography, nature walks, regulation skeet shooting, river tours, and wildlife tours (Tara Wildlife, 2019). In addition, conference facilities can be used to host business meetings and Tara Wildlife also offers lodging, meals, and facilities to host naturalist camps supported by the National Audubon Society and other organizations.

7.5 Recreation opportunity spectrum

One of the most widely used systems of land classification for recreational purposes is the recreation

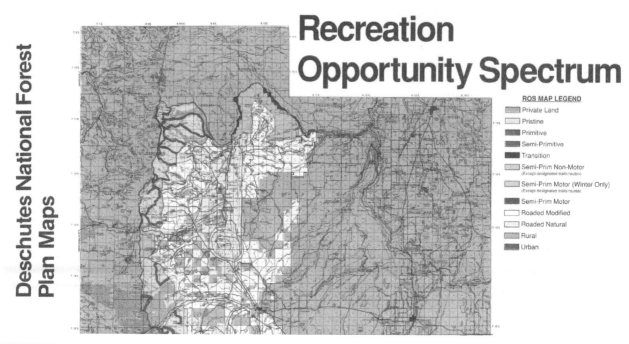

Deschutes National Forest Plan Maps

Recreation Opportunity Spectrum

ROS MAP LEGEND
- Private Land
- Pristine
- Primitive
- Semi-Primitive
- Transition
- Semi-Prim Non-Motor
  (Except designated trails/routes)
- Semi-Prim Motor (Winter Only)
  (Except designated trails/routes)
- Semi-Prim Motor
- Roaded Modified
- Roaded Natural
- Rural
- Urban

FIGURE 7.29   A portion of a recreation opportunity spectrum (ROS) for a national forest in central Oregon. *U.S. Department of Agriculture, Forest Service (1982).*

opportunity spectrum (ROS). The ROS is based on the remoteness of an area (e.g., distance to roads), its natural condition (e.g., type of forest), and the social experiences that might be accommodated there (Clark and Stankey, 1979). It was originally designed for the extensive areas of public land in the western United States. And, it is a tool often used by public agencies for planning, inventorying, and managing recreation resources (Gundersen et al., 2015). Various physical, social, and managerial settings of a landscape are assessed, often within a geographic information system, to determine the ROS class for land areas (Fig. 7.29). At one end of the spectrum are mainly natural features that facilitate infrequent social interaction and solitude. At the other end of the spectrum are mainly managed landscapes that facilitate frequent social interaction. The class names typically include primitive, semiprimitive, roaded natural, rural, urban, and others. A modification extends the ROS to water resources (Haas et al., 2011).

7.6 Factors that affect participation in forest recreation

Despite the economic slump that many sectors of the economy experienced during the economic recession of the late 1970s and early 1980s, the aggregate demand for outdoor recreation increased in the United States (Douglass, 2000). Similar trends in leisure time available to people occurred in Europe as well (Travis, 1982). The demand for recreation has been increasing

almost at the rate of the growth of the human population. Given the worldwide economic downturn in the early 21st century, it will be interesting to see whether people have responded similarly to recreational opportunities. The condition of recreational areas can affect participation in recreational activities. Poorly maintained areas, inadequate facilities, and pollution have been noted as indicators of structural constraints (Ghimire et al., 2014). In addition to a general lack of opportunities, numerous other factors affect participation in forest recreational opportunities; four of these are people, money, time, and communications (Douglass, 2000).

7.6.1 People

The first of four key factors affecting participation in forest recreational opportunities is influenced by the characteristics of the living area, and the average age, ethnicity and culture, and educational level of the nearby human population. Generally, as the nearby population increases, so does the aggregate demand for recreational opportunities. Historically, urban centers were intensely inhabited communities constrained by poor transportation systems. However, with the advent of better transportation systems (e.g., trains and highways), people began to move into surrounding areas and beyond. These new inhabitants may have had different recreational demands than the original inhabitants of the community. In many rural communities, recreational activities such as hunting and fishing are

FIGURE 7.30 Resting place commemorating mountain climbers at a hilltop in the Primorska region of Slovenia. *Donald L. Grebner.*

FIGURE 7.31 Hiking trail along the Cliffs of Moher in County Claire, Ireland. *Zennure Ucar.*

common. Immigrants from metropolitan areas to rural areas often become concerned over the potential harm caused by firearms, which may lead to restrictions on hunting activities. These new suburbanites generally prefer other recreational opportunities, such as biking, camping, hiking, skiing, and swimming and conflicts can occur. Another way in which the living area affects participation in recreation involves topographic or physiographic differences between regions. Regions that contain mountains and significant changes in topography will attract individuals who enjoy recreational activities within these ecosystems (Fig. 7.30). This type of rationale is also manifested in people who prefer to live near water bodies such as lakes and oceans.

Age is also an important factor that influences participation levels in forest recreation activities (Williams and Shaw, 2009). In general, as people become older, they slow down and become less physically active or are less able to perform strenuous activities. Therefore, the types of recreational opportunities a person desires may change as that person ages. For example, participation in strenuous forest recreational activities such as mountaineering, rock climbing, backpacking, camping, and hiking declines as people age (Douglass, 2000), and participation in nonstrenuous activities such as sightseeing may increase. Interestingly, in the United States, many of the existing recreational facilities are tailored toward young adults, although the largest generational group of recreationists consists of older adults who also have the greatest share of disposable income to devote to recreational activities.

Ethnicity and race may play an important role with regard to participation in forest recreational activities (Tierney et al., 2001). For example, Caucasian Americans seem to prefer to travel farther away from their homes for longer periods than do others (Douglass, 2000).

These issues are also important in other parts of the world, such as Europe (Gentin, 2011). Further, across most minority groups in the United States, including older people and people from urban environments, safety and accessibility (Fig. 7.31) seem to be important constraints to participation in recreational activities (Ghimire et al., 2014). Educational levels of recreationists can explain some, but not all, of the participation patterns in forest recreation (Scholes and Bann, 2018). Educational levels can also influence how an individual perceives the recreational opportunities available from forested ecosystems. However, social support may mitigate some of these issues, as it may promote engagement in physical leisure time activity (Dellaserra et al., 2018).

7.6.2 Money

The second key factor that affects participation in forest recreation is the money or funding available to a person to cover recreational expenses. On average in the United States, of the expenditures incurred by a person on an annual basis, as much as 9% may go to recreation-related purchases, memberships, goods, and services (U.S. Census Bureau, 2015). In general, the more money an individual has access to, the greater the suite of choices that exists among the various forest recreational opportunities. As we have suggested, as people age they tend to have more discretionary income, which might be used for recreational activities. Thus, a person's age and discretionary income may be related to the number and types of recreational activities they choose (Williams and Shaw, 2009). This factor is also related to educational levels of recreationists, since on average college graduates earn more income per year than do individuals with only a high school degree (Swanson, 1955; Day and Newburger, 2002). However, regardless of

FIGURE 7.32 A view from a hiking trail within Yosemite National Park, California. *Evelyn Martin.*

earning potential, inadequate transportation options have been suggested to be a barrier to participation in recreational activities for minority groups in the United States (Ghimire et al., 2014). Further, the overall cost of some recreational activities, such as boating and skiing, can be very expensive and may require significant funds for equipment, lodging, lift tickets, and meals. These relative costs may constrain participation in recreational opportunities. The required travel distance can also dramatically influence the cost of a recreational experience. For example, if a person living in New York City were interested in hiking and sightseeing in the Yosemite National Park (Fig. 7.32), the travel costs involved would be much greater than those of a person living in nearby San Francisco. However, the advent of low cost airlines may mitigate this issue.

7.6.3 Time

Activities undertaken by humans in the course of their everyday life require the spatial and temporal coordination of home life, school life (perhaps), recreational pursuits, and work life (Gren, 2009); therefore, the notion and value of *time* is critical to many. People are born, they grow to an adult age, they mate to create young that continue the species, they work to produce income to support themselves, and, as they do all of these things, they age and eventually pass away. A person's life may be short relative to the lifespan of some tree species and is merely a fleeting moment relative to the lifespan of large landmasses. Entire human civilizations may come and go before a mountain is eroded by wind and water. Since humans only exist for a finite period, many view time as being extremely valuable. When people are not performing the tasks necessary to stay alive, they may find the time to unwind and relax.

This is commonly known as *leisure time*. Along with mobility, leisure time affects a person's willingness to engage in forest recreation (Douglass, 2000). The amount of leisure time available to a person or a family can also dictate the types of activities in which they become engaged. During a typical Monday to Friday work week, the average person has little time to engage in forest recreation activities because most of their time may be spent sleeping, preparing for work, taking children to school, going to work, working, retrieving children from school, returning home, performing chores, consuming dinner, and so on. On weekends, when a person or family may have no work responsibilities, they often have more leisure time to apply to forest recreational opportunities. Another time-related aspect of forest recreation is the amount of yearly vacation or holiday time a person accrues, which can vary by the type of job a person holds, the organization that employs them, and even the country in which they reside. For instance, 20 days of vacation and holiday time is common in the United States, but the amount of time away from work is nearly double this in some European countries (Huberman and Minns, 2007). Some studies even suggest that the rate of decline of a person's memory (Richards et al., 2003) and perhaps even their average life expectancy (Fransson and Hartig, 2010) can be correlated to the amount of leisure time a person is permitted during their life.

A person's ability to travel is important, as those who have a limited amount of leisure time may not be willing to travel long distances to engage in certain kinds of forest recreation (Douglass, 2000). In some cases, the time required to travel may be greater than the benefit of the recreational activity at the targeted location (Williams and Shaw, 2009). In these cases, the cost of time can be greater for individuals who do not have, for instance, direct access to a motorized vehicle. The development of national railroad and highway systems in the United States has connected many remote corners of the country, thus facilitating mobility and increasing recreational opportunities for many urban communities (Douglass, 2000). In addition, the expansion of the commercial aviation industry over the past 100 years has allowed people with sufficient leisure time and financial means to travel relatively quickly to locations abroad, such as the tropical forests of Brazil, the Himalayan highlands of Nepal, the savanna forests of Kenya, or the boreal forests of Canada.

7.6.4 Communications

Communication involves a set of methods and tactics for presenting forest recreational opportunities to the public to encourage their participation (Douglass,

2000). However, the effectiveness of signs posted at recreational areas is a challenge to potential visitors in the United States (Ghimire et al., 2014). Communicating with ethnic groups using their native language, for example, can be an effective approach for promoting forest recreational opportunities (Winter et al., 2004). Before the advent of the Internet, television, newspapers, bulletins, magazines, journals, books, mailing lists, bulletin boards, fliers, and a wide range of paper-based products were used to advertise forest recreational opportunities. Other routes included old standbys such as personal experiences passed down by generations, word of mouth recommendations from friends and acquaintances, radio programs, or telephone conversations. The advent of the Internet now allows many people to access material promoting forest recreational opportunities in a much less costly and more expedient manner. For example, prior to the creation of the Internet, learning of whitewater rafting opportunities available in New Zealand or Costa Rica would have been relatively costly, from both a time and money perspective, for a person living in Omaha, Nebraska. In addition, the Internet allows interested individuals to simultaneously view text and video descriptions of potentially exciting new recreational opportunities. The development of new technologies, such as smartphones, also allows people the opportunity to not only communicate orally but now through video that allow the viewing of information from the Internet and to perhaps navigate through the landscape using built-in global positioning system technology.

7.7 Recreational planning and development

Recreational habits, prior recreational experiences, and emotional attachments are all influential in the use of a landscape for recreational purposes (Hailu et al., 2005). Recreational demand for various activities arises from a population source and is measured as the amount of a certain activity that would occur if the necessary facilities were available in sufficient supply (Saunders et al., 1981). The *recreational carrying capacity* concept has been proposed as a method by which planners can help manage the negative aspects of recreational opportunities and to estimate usage limits (Wang et al., 2017). A *carrying capacity* can be described as the optimal level of activity, beyond which the underlying resources will be irreversibly damaged. In terms of forest recreation, the carrying capacity may be described as the maximum number of people who can visit a destination and engage in the same activity at the same time without leading to a decline in the quality of the environment. Carrying capacity can also relate to the maximum capacity (of recreationists) that

recreational facilities are designed to support, given the level of maintenance provided. One could apply this concept to developed hiking or biking trails, for example. Alternatively, carrying capacity can relate to congestion within a recreational area, that is, the number of recreationists beyond which no further overall enjoyment is gained (Schwarz et al., 1976). Again, the level of use along a trail may at some point exceed what some have determined to be the carrying capacity, given that too many group encounters can lower the enjoyment (however defined) of others. Along these lines is the managerial carrying capacity, which relates to the number of recreationists per day that the managerial staff can adequately and safely manage within a given recreational area.

While there are many benefits to individual recreationists from a recreational opportunity, there may be negative externalities of recreational visits on the area supporting the activity. With extensive use, site hardening is often used to reduce the impacts, and these activities include improvements to route patterns, additions of rock or other material to surface a heavily trafficked area, or other means to enhance or maintain the visual quality of a recreation area. In less visited areas, site hardening may be limited to the development of features such as boardwalk trails, boat moorings, or public recreation cabins. Negative externalities relate to the degradation of a resource, and, therefore, terms such as recreation carrying capacity have been advanced to assess the number of visitors a particular recreational area can accommodate. However, it is difficult, if not impossible, to develop a single, clear benchmark to describe the carrying capacity of an area (Fleishman and Feitelson, 2009).

In addition to using the carrying capacity of a recreational area to determine when and where activities might be allowed, other forms of planning processes have been used. For example, Klisky (2000) illustrated how one could use geographic information systems (GIS) databases to map recreation terrain suitability. In analyses such as these, the suitability of a number of spatial variables for recreation could be determined, weighting factors for each variable can be developed, and an overlay process can be used to develop a map portraying the suitability and spatial extent of recreational opportunities across a landscape. Site planning, or site design, is another form of planning. Here careful attention is paid to developing a recreational design that works well within the forest setting and the user expectations. Site planning includes the engineering, architectural, and landscaping design aspects that are necessary for a single project that is under the control of one land management organization. The planning process concerns design issues such as the relationship and continuity between components of the recreational site (natural

Recreation opportunity spectrum class

Encounters with other people

Primitive
Semi-Primitive, non-motorized
Semi-Primitive, motorized
Roaded, natural
Roaded, modified
Rural
Urban

Few

Many

FIGURE 7.33   Relative amount of human interaction within each recreation opportunity spectrum (ROS) class.

and developed), and the placement of recreational features (e.g., buildings, trails) with respect to land limitations and aesthetic quality (Schwarz et al., 1976). This is the most detailed level of planning, sometimes referred to as operational planning, and often includes features as small as a single building or trail. Certainly, within lands that are considered *working forests*, the potential conflicts between potential recreational activities and other uses of the land can require a significant planning effort, particularly when considering forest management activities that may occur during the busiest part of a recreation season (Davidson, 2015).

A more widely known geographical model for assessing recreational potential of a landscape is the ROS, which was developed in the late 1970s (Clark and Stankey, 1979). In the process of mapping the ROS for a landscape, biological, managerial, physical, and social factors are employed, spatial analyses are performed, and each area of a landscape is assigned a recreational value that ranges from primitive experience potential to motorized experience potential (Fig. 7.33). This manner of describing the potential of a landscape for providing recreational opportunities takes into account the services provided by nature, conditions provided by management of the land (e.g., road density and forest character), and potential social interactions with other recreationists. *Limits of acceptable change* of the landscape are often based on monitoring resource use and are related to ROS classes. The type of management conditions, resource conditions, and social conditions that are acceptable for each class are determined, and then variables that reflect those conditions are identified and measured. For example, the physical carrying

capacity of recreationists in a recreation area, perhaps described as the number of group encounters, can be used to help define the ROS class that the area should represent. Visitors of a primitive ROS class of land should expect fewer group encounters than should visitors of a semiprimitive ROS class of land. The social carrying capacity is similar and reflects the maximum number of encounters that can occur before the recreational experience becomes impaired. These types of measures and analyses allow recreation managers to assess the effectiveness of management actions on recreation quality and help to determine whether a limit of acceptable change has occurred and whether management intervention is necessary. The ROS analysis process has been used to inform the development of forest plans, particularly where the need to provide recreational opportunities is among the main issues facing the management of public lands such as the Chattahoochee—Oconee National Forest in the eastern United States (Bettinger et al., 2015b).

Finally, recreational user groups have been formed in a number of areas to represent a variety of recreational interests. These groups can be influential in the planning and development of recreational opportunities, and some are very formal in the development of their mission. For example, the Mauna Kea Recreational Users Group in Hawaii has goals that include maintaining a right of public access to the volcano, ensuring that recreationists are involved in the planning of the area, and educating the public on their responsibilities when using the land (Mauna Kea Recreational Users Group, 2019). The relationship between a recreation user group and the landowner or land management organization is important. For example, a user group may be independent of the landowner or land management organization. In this situation, it may simply provide advice and feedback on resource use (perhaps through surveys) or may also lobby landowners and land managers for (or against) proposed actions that are important to the group as a whole. Alternatively, a user group may be composed of people associated with the landowner or land management organization and be directly involved in the planning of the recreational opportunities within a resource area. At times, a recreation user group may be composed of people with a wide variety of interests, for example horseback riding and mountain biking. These interests may conflict and may need to be reconciled for the user group to successfully influence recreational development across a landscape. Surveys of recreationists can also inform land managers of their preferences for landscape features. For example, a survey of motorcyclists who travel through rural, managed, southern United States landscapes revealed a preference for water features in the forefront, along the roads (Merry et al., 2020).

Some of the trade-offs that landowners and land managers must consider, and perhaps reconcile, include problems of aging facilities, overcrowding of recreational areas, overuse of trails, species extirpation, and water pollution. Further, forest health can be an important factor in engaging people, as mature, healthy forest areas may be preferred over forests affected by insect and disease issues, and those containing substantial amounts of dead and dying trees (Arnberger et al., 2018). These complex issues might be addressed through collaboration with user groups. The connection between the natural system and organizational strategies may need to be explored to provide management direction and assess trade-offs. Along these lines, the ROS model can also be employed to assess locations in which visitors can obtain diverse experiences while sensitive areas are also protected. However, planning and implementation processes may require additional financial or personnel resources to manage the tension between the land resource and the visitor. For example, after years of consideration, the U.S. National Park Service decided to shoulder the financial burden of maintaining a portion of the Smoky Mountain National Park on a sustainable basis through a set of ecological restoration activities and mechanical interventions designed to restore the natural plant and animal communities (Young, 2006). Economic analysis is often employed in the development of alternatives for managing lands with recreational importance, and may involve the projection or estimation of discounted net present value, consumer surplus, employment, and willingness to pay for recreational experiences (Bettinger et al., 2015a), and the use of choice experiments

and travel cost methods to understand the recreational value of forests (Grilli et al., 2014). Issues of financing recreational opportunities are faced by many public organizations, and at times the funds arise not only from recreation fees, but also timber sales and the sales of minor forest products (Eng, 2015). For certain areas, the development of a conservation plan is also necessary when high-intensity active recreation, such as summer camp operations, are served on a property (Kallesser, 2015). These types of plans help the landowner to consider actions that minimize impacts from recreational activities.

Summary

Forest-based recreational opportunities can consist of camping, fishing, hiking a secluded trail (Fig. 7.34), hunting, or a multitude of other activities unrelated to one's normal work activities. Forest recreation is an important aspect of many people's lives, as emotional and physical well-being can be enhanced through these experiences. Therefore, recreational opportunities are an important consideration for land managers and landowners who allow public access to the forests they own or manage. In addition, the management of recreational opportunities can involve developing options that allow people to derive educational experiences. If permitted, the types of recreational activities located in a previously undeveloped region may evolve from those reflecting a primitive environment to those typical of a highly developed, modern setting for recreation. Clark and Stankey (1979) suggest that several questions should be pondered by managers of these resources:

- What opportunities for recreation is the land area best suited to provide?

FIGURE 7.34 Forest trail leading to the Blue Pools in Mount Aspiring National Park, South Island, New Zealand. *Donald L. Grebner.*

- What range or mix of opportunities might be developed in conjunction with other developmental activities in the nearby area?
- What effects have past management decisions had on the potential future recreational opportunities for the area?
- How will future changes in the nature of the recreational opportunities provided affect the kinds of experiences the land provides to people?

Although providing recreational opportunities for the general public to experience and enjoy may seem easy at first glance, there are a number of deeper issues to consider. These issues range from the impact of recreation on the ecosystem and natural resources contained therein to the effect of the recreational opportunity on the recreationist. These issues may also involve delving into other economic, legal, and social considerations associated with forest recreation.

Questions

(1) *Connecting recreation to forests.* From the variety of opportunities that are available, select a recreational activity you feel can be supported in a forested environment. Develop a short (10 slides or less) professional PowerPoint presentation that describes how forest conditions can influence or detract from the quality of the experience to the recreationist. Finally, deliver the presentation to your class.

(2) *Your recent recreational experiences.* Thinking back over the last year, list all of the recreational activities in which you participated, whether they were formal trips to faraway places or simple, short diversions within your local area. Develop a short 250-word report that (1) describes whether forests or forest conditions influenced the choice of location for these activities; (2) describes the value of the activities to you personally; and (3) describes how much these cost (time and actual monetary value).

(3) *A single recent recreational experience.* Try to identify the most recent forest recreational experience you had. Develop a short memorandum that describes the value of this experience to you, the costs you may have incurred, and the type of landowner that provided the experience. In a group setting, discuss your experience with four or five of your classmates.

(4) *Recreation opportunity spectrum.* Perform some investigative work regarding the ROS, using library or Internet resources. Select one of the ROS classes and describe how it is formally defined and delineated on a landscape. Select a recreation area and, either by hand on a map or with the assistance of a GIS, develop a representation of what you think is the boundary of the selected ROS class and estimate its area. Present this to your class in a formal two-to-three-minute presentation using only the map you created.

(5) *Planning recreational trips.* Locate the national park closest to your home and determine how far away it is and how much money would be required to travel to it. Now, pick a national park that you would most like to visit, and determine how far away it is and how much money would be required to travel to it. In a short one-page summary, describe the trade-offs of visiting each park (including travel time and cost).

(6) *Constraints to taking recreational trips.* Choose a forested recreational area that you would one day like to visit. In a short one-page summary, describe the constraints that might be preventing you from visiting this place in the near future.

References

Adirondack Park Agency, 2018. New York State Adirondack Park Agency 2018 Annual Report. State of New York Adirondack Park Agency, Ray Brook, NY. https://apa.ny.gov/Documents/Reports/2017ApaAnnualReport.pdf (accessed 07.07.19).

Andre, M., 2015. Arcata Community Forest, California, United States of America. In: Siry, J.P., Bettinger, P., Merry, K., Grebner, D.L., Boston, K., Cieszewski, C. (Eds.), Forest Plans of North America. Academic Press, New York, pp. 53–59.

Arnberger, A., 2006. Recreation use of urban forests: an inter-area comparison. Urban Forestry and Urban Greening 4 (3–4), 135–144.

Arnberger, A., Ebenberger, M., Schneider, I.E., Cottrell, S., Schlueter, A.C., von Ruschkowski, E., Venette, R.C., Snyder, S.A., Gobster, P.H., 2018. Visitor preferences for visual changes in bark beetle-impacted forest recreation settings in the United States and Germany. Environmental Management 61 (2), 209–223.

Barkley, J., Bodine, J., Hoffman, C., Koslowski, J., Stevens, L., Schwantes, J., Severt, J., 2015. Bayfield County Forest, Wisconsin, United States of America. In: Siry, J.P., Bettinger, P., Merry, K., Grebner, D.L., Boston, K., Cieszewski, C. (Eds.), Forest Plans of North America. Academic Press, New York, pp. 265–275.

Bettinger, P., Merry, K., Demirci, M., Klepacka, A.M., 2015a. Tongass National Forest, Alaska, United States of America. In: Siry, J.P., Bettinger, P., Merry, K., Grebner, D.L., Boston, K., Cieszewski, C. (Eds.), Forest Plans of North America. Academic Press, New York, pp. 413–421.

Bettinger, P., Merry, K., Mavity, E., Rightmyer, D., Stevens, R., 2015b. Chattahoochee-Oconee National Forest, Georgia, United States. In: Siry, J.P., Bettinger, P., Merry, K., Grebner, D.L., Boston, K.,

Cieszewski, C. (Eds.), Forest Plans of North America. Academic Press, New York, pp. 277—284.

Bigelow, D.P., Borchers, A., 2017. Major Uses of Land in the United States. U.S. Department of Agriculture, Economic Research Service, Washington, D.C. Economic Information Bulletin Number 178.

Boy Scouts of America, 2011. Pushmataha Area Council, Camp Seminole, Starkville, MS. http://www.campseminole.org/ (accessed 07.07.19).

Clark, R.N., Stankey, G.H., 1979. The Recreation Opportunity Spectrum: A Framework for Planning, Management, and Research. U.S. Department of Agriculture, Pacific Northwest Forest and Range Experiment Station, Portland, OR. General Technical Report PNW-98. 32 p.

Cordell, H.K., 2008. The latest on trends in nature-based outdoor recreation. Forest History Today, 4—10. Spring 2008.

Cordell, H.K., Betz, C.J., Green, G.T., 2008. Nature-based outdoor recreation trends and wilderness. International Journal of Wilderness 14 (2), 7—13.

Cordell, H.K., Betz, C.J., Zarnoch, S.J., 2013. Recreation and Protected Land Resources in the United States, A Technical Document Supporting the Forest Service 2010 RPA Assessment. U.S. Department of Agriculture, Forest Service, Southern Research Station, Asheville, NC.

Davidson, B., 2015. French-Severn Forest, Ontario, Canada. In: Siry, J.P., Bettinger, P., Merry, K., Grebner, D.L., Boston, K., Cieszewski, C. (Eds.), Forest Plans of North America. Academic Press, New York, pp. 367—376.

Day, J.C., Newburger, E.C., 2002. The Big Payoff: Educational Attainment and Synthetic Estimates of Work-Life Earnings. U.S. Department of Commerce, Economics and Statistics Administration, U.S. Census Bureau, Washington, D.C., pp. 23—210. Current Population Reports.

Dellaserra, C.L., Crespo, N.C., Todd, M., Huberty, J., Vega-Lopez, S., 2018. Perceived environmental barriers and behavioral factors as possible mediators between acculturation and leisure-time physical activity among Mexican American adults. Journal of Physical Activity and Health 15 (9), 683—691.

Douglass, R.W., 2000. Forest Recreation, fifth ed. Waveland Press, Inc., Prospect Heights, IL.

Eng, H., 2015. Jackson Demonstration State Forest, California, United States of America. In: Siry, J.P., Bettinger, P., Merry, K., Grebner, D.L., Boston, K., Cieszewski, C. (Eds.), Forest Plans of North America. Academic Press, New York, pp. 177—187.

Fleishman, L., Feitelson, E., 2009. An application of the recreation level of service approach to forests in Israel. Landscape and Urban Planning 89 (3—4), 86—97.

Fransson, U., Hartig, T., 2010. Leisure home ownership and early death: a longitudinal study in Sweden. Health and Place 16 (1), 71—78.

Gentin, S., 2011. Outdoor recreation and ethnicity in Europe—a review. Urban Forestry and Urban Greening 10 (3), 153—161.

Ghimire, R., Green, G.T., Poudyal, N.C., Cordell, H.K., 2014. An analysis of perceived constraints to outdoor recreation. Journal of Park and Recreation Administration 32 (4), 52—67.

Greensward Group, LLC, 2019. Welcome to CentralPark.Com. Greensward Group, LLC, New York. www.centralpark.com/ (accessed 07.07.19).

Gren, M., 2009. Time geography. International Encyclopedia of Human Geography. 279—284.

Grilli, G., Paletto, A., de Meo, I., 2014. Economic evaluation of forest recreation in an Alpine valley. Baltic Forestry 20 (1), 167—175.

Gundersen, V., Tangeland, T., Kaltenborn, B.P., 2015. Planning for recreation along the opportunity spectrum: the case of Oslo, Norway. Urban Forestry and Urban Greening 14, 210—217.

Haas, G., Aukerman, V.G., Jackson, J., 2011. Water and Land Recreation Opportunity Spectrum Handbook, second ed. U.S. Department of the Interior, Bureau of Reclamation, Program and Administration, Denver Federal Center, Denver, CO.

Hailu, G., Boxall, P.C., McFarlane, B.L., 2005. The influence of place attachment on recreation demand. Journal of Economic Psychology 26 (4), 581—598.

Huberman, M., Minns, C., 2007. The times they are not changin': days and hours of work in Old and New Worlds, 1870—2000. Explorations in Economic History 44 (4), 538—567.

Isachenko, T.E., Sevast'yanov, D.V., Guk, E.N., 2015. Emergence and evolution of recreation nature management in the Noril'sk Region. Geography and Natural Resources 36 (2), 179—186.

Janeczko, E., Tomusiak, R., Woźnicka, M., Janeczko, K., 2018. Preferencje społeczne dotyczące biegania jako formy aktywnego spedzania czasu wolnego w lasach (Social preferences regarding running as a form of active leisure time in the forests). Sylwan 162 (4), 305—313.

Johnson, G.W., 2015. Blue Ridge Parkway, Virginia and North Carolina, United States of America. In: Siry, J.P., Bettinger, P., Merry, K., Grebner, D.L., Boston, K., Cieszewski, C. (Eds.), Forest Plans of North America. Academic Press, New York, pp. 129—138.

Kallesser, S.W., 2015. Camp No-Be-Bo-Sco, New Jersey, United States of America. In: Siry, J.P., Bettinger, P., Merry, K., Grebner, D.L., Boston, K., Cieszewski, C. (Eds.), Forest Plans of North America. Academic Press, New York, pp. 1—9.

Klisky, A.D., 2000. Recreation terrain suitability mapping: a spatially explicit methodology for determining recreation potential for resource use assessment. Landscape and Urban Planning 52 (1), 33—43.

Lankia, T., Kopperoinen, L., Pouta, E., Neuvonen, M., 2015. Valuing recreation ecosystem service flow in Finland. Journal of Outdoor Recreation and Tourism 10, 14—28.

Mackintosh, C., Griggs, G., Tate, R., 2018. Understanding the growth in outdoor recreation participation: an opportunity for sport development in the United Kingdom. Managing Sport and Leisure 23 (4—6), 315—335.

Mann, C., Pouta, E., Gentin, S., Jensen, F.S., 2010. Outdoor recreation in forest policy and legislation: a European comparison. Urban Forestry and Urban Greening 9 (4), 303—312.

Mauna Kea Recreational Users Group, 2019. Mauna Kea Recreational Users Group, about. Mauna Kea Recreational Users Group, Hilo, HI. http://maunakearug.com/ (accessed 07.07.19).

Merry, K., Bettinger, P., Siry, J., Bowker, J.M., 2020. Preferences of motorcyclists to views of managed, rural southern United States landscapes. Journal of Outdoor Recreation and Tourism 29, Article 100259.

Mississippi Department of Wildlife, Fisheries, and Parks, 2019. Parks and Destinations. Mississippi Department of Wildlife, Fisheries, and Parks, Jackson, MS. http://mdwfp.com/parks-destinations/ (accessed 07.07.19).

Mississippi Forestry Commission, 2018. Mississippi Forestry Commission, Protection-Management-Information, 2018 Annual Report. Mississippi Forestry Commission, Jackson, MS. www.mfc.ms.gov/sites/default/files/MFC%202018%20Annual%20Report%20Web%20Compressed.pdf (accessed 08.02.20).

Monz, C.A., Marion, J.L., Goonan, K.A., Manning, R.E., Wimpey, J., Carr, C., 2010. Assessment and monitoring of recreation impacts and resource conditions on mountain summits: examples from the northern forest, U.S.A. Mountain Research and Development 30 (4), 332—343.

Mueller, S., Marshall, G., Held, K., Solin, J., 2015. Rib Lake School Forest, Wisconsin, United States of America. In: Siry, J.P., Bettinger, P., Merry, K., Grebner, D.L., Boston, K., Cieszewski, C. (Eds.), Forest Plans of North America. Academic Press, New York, pp. 41—46.

Munn, I., Wright, W.C., Hunter, W., Bentley, G., 2015. Willow Break LLC, Mississippi, United States of America. In: Siry, J.P., Bettinger, P., Merry, K., Grebner, D.L., Boston, K., Cieszewski, C.

(Eds.), Forest Plans of North America. Academic Press, New York, pp. 119–128.

National Audubon Society, 2018. Starr Ranch Sanctuary. National Audubon Society, Starr Ranch Sanctuary, Trabuco Canyon, CA. http://www.starrranch.org/ (accessed 07.07.19).

National Audubon Society, 2019. History of Audubon and Science-Based Bird Conservation. National Audubon Society, New York. https://www.audubon.org/about/history-audubon-and-waterbird-conservation (accessed 07.07.19).

New York Department of Environmental Conservation, 2018. Catskill Forest Preserve. New York Department of Environmental Conservation, Albany, New York. http://www.dec.ny.gov/lands/5265.html (accessed 07.07.19).

Park, B.-J., Tsunetsugu, Y., Kasetani, T., Morikawa, T., Kagawa, T., Miyazaki, Y., 2009. Physiological effects of forest recreation in a young conifer forest in Hinokage Town, Japan. Silva Fennica 43 (2), 291–301.

President's Commission on Americans Outdoors, 1987. Americans Outdoors: The Legacy, the Challenge. Island Press, Washington, D.C.

Recknagel, A.B., 1917. The Theory and Practice of Working Plans (Forest Organization). John Wiley & Sons, Inc., New York.

Richards, M., Hardy, R., Wadsworth, M.E.J., 2003. Does active leisure protect cognition? Evidence from a national birth cohort. Social Science and Medicine 56 (4), 785–792.

Saunders, P.R., Senter, H.F., Jarvis, J.P., 1981. Forecasting recreation demand in the upper Savannah River basin. Annals of Tourism Research 8 (2), 236–256.

Savage, J., 2015. Dubuar Memorial Forest, New York, United States of America. In: Siry, J.P., Bettinger, P., Merry, K., Grebner, D.L., Boston, K., Cieszewski, C. (Eds.), Forest Plans of North America. Academic Press, New York, pp. 79–86.

Scholes, S., Bann, D., 2018. Education-related disparities in reported physical activity during leisure-time, active transportation, and work among U.S. adults: repeated cross-sectional analysis from the National Health and Nutrition Examination Surveys, 2007 to 2016. BMC Public Health 18. Article 926.

Schwarz, C.F., Thor, E.C., Elsner, G.H., 1976. Wildland Planning Glossary. U.S. Department of Agriculture, Forest Service, Pacific Southwest Forest and Range Experiment Station, Berkeley, CA. General Technical Report PSW-13. 259 p.

Scottish Natural Heritage, 2017. Scottish Outdoor Access Code - Rights and Responsibilities. Scottish Natural Heritage, Inverness, Scotland. www.outdooraccess-scotland.scot/act-and-access-code/scottish-outdoor-access-code-rights-and-responsibilities (accessed 14.07.19).

Sharpe, G.W., Odegaard, C.H., Sharpe, W.F., 1983. Park Management. John Wiley & Sons, Inc., New York.

Smith, S.L.J., Godbey, G.C., 1991. Leisure, recreation and tourism. Annals of Tourism Research 18 (1), 85–100.

Southern Sierra Council, Boy Scouts of America, 2019. SSCBSA Home. Southern Sierra Council, Boy Scouts of America, Bakersfield, CA. http://www.sscbsa.org/ (accessed 07.07.19).

Starbuck, C.M., Berrens, R.P., McKee, M., 2006. Simulating changes in forest recreation demand and associated economic impacts due to fire and fuels management activities. Forest Policy and Economics 8 (1), 52–66.

Sun, C., Pokharel, S., Jones, W.D., Grado, S.C., Grebner, D.L., 2007. Extent of recreational incidents and determinants of liability insurance coverage for hunters and anglers in Mississippi. Southern Journal of Applied Forestry 31 (3), 151–158.

Swae, J., 2015. City of San Francisco, California, United States of America. In: Siry, J.P., Bettinger, P., Merry, K., Grebner, D.L., Boston, K., Cieszewski, C. (Eds.), Forest Plans of North America. Academic Press, New York, pp. 285–292.

Swanson, E.O., 1955. Is college education worth while? Journal of Counseling Psychology 2 (3), 176–181.

Tara Wildlife, 2019. Home. Tara Wildlife, Vicksburg, MS. http://www.tarawildlife.com (accessed 07.07.19).

Tennessee Valley Authority, 2019. Recreation. Tennessee Valley Authority, Knoxville, TN. www.tva.com/Environment/Recreation (accessed 07.07.19).

The Nature Conservancy, 2018. Chestnut Mountain: A Gift for All of Tennessee. The Nature Conservancy, Arlington, VA. www.nature.org/en-us/about-us/where-we-work/united-states/tennessee/stories-in-tennessee/chestnut-mountain-a-gift-for-all-of-tennessee/ (accessed 07.07.19).

The Nature Conservancy, 2019. About Us: Who We Are. The Nature Conservancy, Arlington, VA. www.nature.org/en-us/about-us/who-we-are/ (accessed 07.07.19).

The Trust for Public Land, 2010. The 150 Largest City Parks. The Trust for Public Land, Center for City Park Excellence, Washington, D.C. http://cloud.tpl.org/pubs/ccpe-largest-oldest-most-visited-parks-4-2011-update.pdf (accessed 07.07.19).

Tierney, P.T., Dahl, R., Chavez, D., 2001. Cultural diversity in use of undeveloped natural areas by Los Angeles county residents. Tourism Management 22 (3), 271–277.

Travis, A.S., 1982. Leisure, recreation and tourism in western Europe. Tourism Management 3 (1), 3–15.

U.S. Bureau of Indian Affairs, 2011a. About Us. U.S. Department of the Interior, Bureau of Indian Affairs, Washington, D.C. www.bia.gov/about-us (accessed 07.07.19).

U.S. Bureau of Indian Affairs, 2011b. Branch of Fish, Wildlife, and Recreation. U.S. Department of the Interior, Bureau of Indian Affairs, Washington, D.C. www.bia.gov/bia/ots/division-natural-resources/branch-fish-wildlife-recreation (accessed 07.07.19).

U.S. Bureau of Land Management, 2018. Public Land Statistics 2017, vol. 202. U.S. Department of the Interior, Bureau of Land Management, Washington, D.C.

U.S. Bureau of Reclamation, 2019. Bureau of Reclamation Facts & Information. U.S. Department of the Interior, Bureau of Reclamation, Washington, D.C. https://www.usbr.gov/main/about/fact.html (accessed 07.07.19).

U.S. Census Bureau, 2015. Statistical Abstract of the United States: 2012. U.S. Department of Commerce, Economic and Statistics Administration, Census Bureau, Washington, D.C.

U.S. Department of Agriculture, Forest Service, 1982. Deschutes National Forest, Recreation Opportunity Spectrum. U.S. Department of Agriculture, Forest Service, Pacific Northwest Region, Portland, OR.

U.S. Fish and Wildlife Service, 2015. Pelican Island. U.S. Department of the Interior, Fish and Wildlife Service, Washington, D.C. www.fws.gov/refuge/pelican_island/about/history.html (accessed 07.07.19).

U.S. Fish and Wildlife Service, 2019. Public Lands and Waters. U.S. Department of the Interior, Fish and Wildlife Service, Washington, D.C. www.fws.gov/refuges/about/public-lands-waters/index.html (accessed 07.07.19).

U.S. Forest Service, 2013a. By the Numbers. U.S. Department of Agriculture, Washington, D.C. https://www.fs.fed.us/about-agency/newsroom/by-the-numbers (accessed 17.06.19).

U.S. Forest Service, 2013b. Other Congressionally Designated Areas. U.S. Department of Agriculture, Forest Service, Washington, D.C. http://www.fs.fed.us/recreation/programs/cda/special-areas.shtml (accessed 17.06.19).

U.S. National Park Service, 2000. Yellowstone National Park: Its Exploration and Establishment—Part III: The Park Movement. U.S. Department of Interior, National Park Service, Washington, D.C. http://www.cr.nps.gov/history/online_books/haines1/iee3c.htm (accessed 07.07.19).

U.S. National Park Service, 2013. Land and Water Conservation Fund, State and Local Assistance Program 2012 Annual Report. U.S. Department of the Interior, National Park Service, Washington, D.C.

U.S. National Park Service, 2018. National Historic Landmarks Program. U.S. Department of the Interior, Washington, D.C. www.nps.gov/orgs/1582/index.htm (accessed 07.07.19).

U.S. National Park Service, 2019a. Reference Manual 45, National Trails System, Chapters 1 through 10 & Appendices. U.S. Department of the Interior, Washington, D.C.

U.S. National Park Service, 2019b. National Park System. U.S. Department of the Interior, Washington, D.C. www.nps.gov/aboutus/national-park-system.htm (accessed 07.07.19).

Vincent, C.H., Hanson, L.A., Argueta, C.N., 2017. Federal Land Ownership: Overview and Data. Congressional Research Service, Washington, D.C. CRS Report R42346.

Wang, E., Wang, Y., Yu, Y., 2017. Assessing recreation carrying capacity of the environment attributes based on visitors' willingness to pay. Asia Pacific Journal of Tourism Research 22 (9), 965–976.

Weingroff, R.F., 1996. Federal-aid highway act of 1956: creating the interstate system. Public Roads 60 (1). www.fhwa.dot.gov/publications/publicroads/96summer/p96su10.cfm (accessed 14.07.19).

Williams, A.M., Shaw, G., 2009. Future play: tourism, recreation and land use. Land Use Policy 26 (1), 326–335.

Winter, P.L., Jeong, W.C., Godbey, G.C., 2004. Outdoor recreation among Asian Americans: a case study of San Francisco Bay Area residents. Journal of Park and Recreation Administration 22 (3), 114–136.

Young, T., 2006. False, cheap and degraded: when history, economy and environment collided at Cades Cove, Great Smoky Mountains National Park. Journal of Historical Geography 32 (1), 169–189.

Zacarias, D.A., Williams, A.T., Newton, A., 2011. Recreation carrying capacity estimations to support beach management at Praia de Faro, Portugal. Applied Geography 31 (3), 1075–1081.

Zinser, C.I., 1995. Outdoor Recreation: United States National Parks, Forest, and Public Lands. John Wiley & Sons, Inc., New York, p. 898.

CHAPTER

8

Forest measurements and forestry related data

Information used to assist in the management of forests can consist of observations, statistical summaries, or qualitative opinions that are generated from a variety of data. This data is then analyzed and in doing so creates information often needed by forest managers to make both routine and extraordinary decisions regarding the management of forests and natural resources. Information could also be used to avoid making inefficient and ineffective decisions regarding the allocation of resources and may therefore help prevent the occurrence of undesired impacts from a particular activity that has taken place. A forester or natural resource manager can spend a significant amount of time collecting forest measurements (data). In an effort to use time wisely, we must first understand the type of data to collect, how this data is typically collected, and for what purposes the data will be used to generate information. The objective of this chapter is therefore to provide an overview of many of the basic forest measurements that are collected by professional natural resource managers. Upon completion of this chapter, readers should be able to

- recognize and understand the measurement techniques, instruments, and tools commonly used to describe the condition of trees and forests;
- understand the technologies that can be used to describe the recreational, vegetation, water, wildlife, and other resources found in an area at different scales (tree, stand, forest, and landscape); and
- effectively communicate the types of remote sensing products that are available for obtaining data related to forests and other landscapes.

Topics presented in this chapter represent both basic tree measurements and a small sample of the wide variety of other measurements that can be collected from forest resources. Many upper-level or advanced natural resource management courses provide an in-depth analysis of one or more of these topics. For example, a small portion of this chapter concerns geographic information systems (GIS), yet most college-level forestry and natural resource programs allocate an entire term (semester or quarter) delving into capabilities of these types of

Introduction to Forestry and Natural Resources, Second Edition
https://doi.org/10.1016/B978-0-12-819002-9.00008-0

mapping systems. With the discussions provided in this chapter, readers should acquire a sense of the commonly collected forest-related measurements that natural resource professionals may encounter in their careers.

8.1 Measuring trees and the forest

Of all the resources found in wildland and urban forests, the tree is most often used to help describe the ability of an area to produce or extract provisional and cultural values of interest to a society. Besides the obvious connection between tree characteristics and wood production values such as lumber or carbon production, tree characteristics have also been used to quantitatively describe scenic beauty (e.g., Brown and Daniel, 1986; Gong et al., 2015) and wildlife habitat suitability (e.g., Crawford and Marchinton, 1989; McIntyre et al., 2019). Ideally, if we were interested in the total value of a forest (ecological, economic, or otherwise) with respect to a land management objective, we would intricately measure the shape and size of every tree in the forest and apply the appropriate valuation methods. However, time constraints usually prevent such an effort from being undertaken. Therefore, sampling schemes

have been devised based on the measurement of certain parts of trees in a certain portion of a collection of trees in a forest. These sampling techniques, if performed appropriately, can be used to provide an estimate of a mean or average value for a population, or a predesignated area to be sampled, with some level of confidence. Some of the common sampling schemes will be presented shortly, but first the typical tree and forest measurements collected by foresters and natural resource professionals are presented.

A living tree consists of a root mass, one or more main stems, branches, and either leaves or needles. The latter two (i.e., branches and leaves [or needles]) are collectively known as the *tree crown* (Fig. 8.1). In developing estimates of forest density, foresters and natural resource managers are interested in the number of trees of a certain size or species within a given area. At times, this type of measurement consists of tallying (counting) only the mature or merchantable trees within a given land area; at other times, it consists of tallying all trees, even seedlings only a few centimeters tall, within a given land area. The count or tally is then adjusted to a per-unit-area value (e.g., trees per hectare [ha] or trees per acre [ac]). Forest density, as with wood volume, can also be a function of the diameter of a tree.

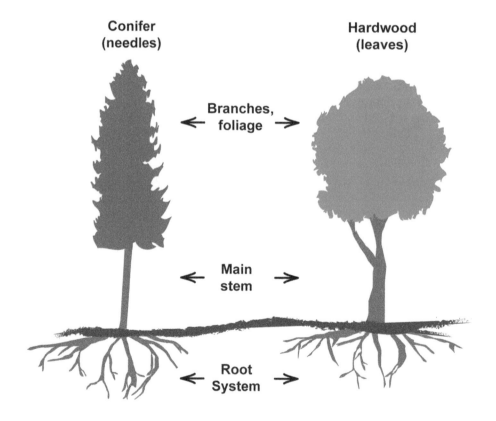

FIGURE 8.1   Basic structure of a tree.

In North America, tree diameters are usually measured at 1.37 meters (m; 4.5 feet [ft]) above the ground on the uphill side of a tree. This is called the *diameter at breast height* (DBH). A number of specific rules must be followed for leaning trees or trees that have more than one main stem (Avery and Burkhart, 2002). To measure the diameter of a tree, calipers or DBH tapes are commonly used and, in some cases, a specialized measuring stick called a Biltmore stick or another solid object may be employed. Calipers and tapes can provide fairly precise measures of the size of the tree bole, even in situations where the bole is not perfectly round.

A forester or natural resource manager may use measurements to estimate the *basal area* per unit area, another measure of forest density. To determine the basal area, the diameter (DBH) of each tree is used to determine the surface area of the tree bole at a point 1.37 m above ground (Fig. 8.2), which can be visualized as if the tree were cut horizontally at this point. Another way to view this measure is to visualize making the tree into a table at this height and determining the area of the top of the table.

The surface areas of all tree boles within a given land area are then added together to arrive at an average per-unit-area value. The basal area is commonly used in forestry and natural resource professions to provide a target density for silvicultural treatments and is often used as an input for habitat models (e.g., Crawford and Marchinton, 1989; McIntyre et al., 2019). The most widely communicated basal area units are square meters (m^2) per ha or square feet (ft^2) per ac. If quite a bit of variation in tree bole diameters (differences among the values from each element in the sample) exists in a forest, it would be necessary to sample a sufficient number of trees to arrive at an estimate of basal area with which a high level of confidence could be applied. In measuring the basal area for a single tree, we assume that the cross-section of the tree bole is circular. This is not necessarily so in nature but is necessary here to simplify the mathematical relationships. If the DBH measurement is collected in units of centimeters (cm) and an estimate of basal area in units of m^2 is desired, then the basal area equation can be reduced to a simple form that does not require the use of π or the equation for the area of a circle:

$$\text{Basal area (m}^2) = 0.00007854 \times \text{DBH}^2$$

If the DBH measurement is collected in units of inches (in) and an estimate of basal area in units of ft^2 is desired, then the basal area equation can be reduced to another simple form that does not require π:

$$\text{Basal area (ft}^2) = 0.005454 \times \text{DBH}^2$$

At times, upper-stem diameter measurements of trees may be needed. These measurements may involve estimates of the diameter of a tree bole as low as 5 m (16 ft) or as high as 50 m (164 ft) or more above ground, or a series of upper-stem diameters for an individual tree will be needed to accurately determine a tree's volume. Upper-stem diameters may be important for understanding the length of a tree bole that meets certain merchantability (or milling) standards or more closely resembles potential lumber yield. For example, some mills may impose a requirement on the size of the small end diameter for logs they will purchase. In the United States, the common small end log diameter ranges

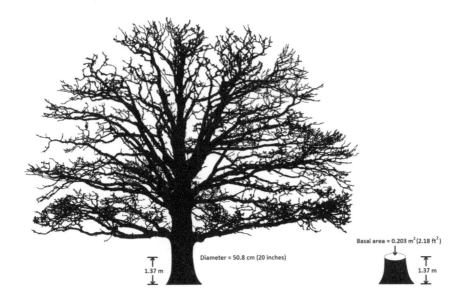

FIGURE 8.2 Conceptual model of a basal area measurement. *Basic image of tree courtesy of Karen Arnold, through PublicDomainPictures.net.*

between 10 and 15 cm (4—6 in). To obtain the length of a tree bole up to a minimum diameter, a tree-climbing ladder may be used in conjunction with a tree caliper or a DBH tape; however, ladders are difficult to transport through woods and there is a limit to the height at which they may be used. There are also a number of remote reading instruments that can be used to determine the upper-stem diameter of a tree. These instruments include the Wheeler pentaprism, various brands of dendrometers, and electronic measuring instruments that use sound waves or lasers.

Bark thickness is another tree characteristic that may be important to measure along with a tree bole's diameter as it supports the improved estimates of solid wood in the tree, excluding the bark. A measure of DBH includes both the solid wood portion of a tree bole and the bark of the tree. From a commodity production perspective, the solid wood portion of a tree bole may be one of the two most important components of a tree's size (the other being the tree height). To determine the thickness of bark on a tree, the bark can be cut away from the bole until the cambial layer is reached, and then the thickness can be measured directly with a scale. Alternatively, a bark gauge could be used to measure bark thickness. This is an instrument that is punched into a tree bole at the same height above ground as the DBH measurement. The tip of the gauge is sharp and cuts through the bark when it is punched into a tree; it stops abruptly once the solid wood portion of the tree bole is encountered. A scale on the gauge is then read to determine the depth or thickness of the bark. In determining a diameter inside bark (dib), twice the thickness of the bark is subtracted from the DBH measurement in order to account for the bark on both ends of the tree diameter.

The circumference of a tree bole is a measurement that is valuable for both natural resource managers and the general public. The *National Register of Big Trees* in the United States, a registry of the largest tree of each species, uses measures of the circumference, total height, and crown spread (or crown diameter) of a tree for computing a score. The circumference of a tree bole is measured at the same point above ground as the tree's DBH. For example, the largest green ash (*Fraxinus pennsylvanica*), which is located in Hocking, Ohio, has a circumference of 584 cm (230 in) (American Forests, 2020). The relationship between a tree's DBH and its circumference is:

$$\text{Circumference} = \pi \times \text{DBH}$$

Therefore,

$$\pi \times \text{DBH} = 584 \text{ cm}$$

and

$$\text{DBH} = 584 \text{ cm}/\pi$$

Thus, the diameter of this champion green ash tree is

$$\text{DBH} = 186 \text{ cm (or about 73.2 in)}$$

Tree volumes and weights are generally a function of both the DBH and the total or merchantable height of a tree; therefore, measuring the appropriate height of a tree is an important task. Height measurements can be related to a tree's crown structure and shape (Fig. 8.3). One of the most common tree height measurements is the total tree height, which represents a distance from a place where the tree bole meets the ground, thus not including the root system, to the tip of the tree. This measurement is moderately difficult to collect for trees with rounded crowns or those that are very tall. With practice, a person can estimate total tree heights relatively accurately using tools such as a clinometer or an Abney level (Fig. 8.4). Height poles can also be used to measure tree heights, but generally require two people to take the measurement, and are only practical (if not preferred) for measuring shorter trees. A merchantable tree height represents the length of the tree bole from an imagined stump to a place where the diameter of a tree bole becomes so small that the remainder of the bole is not useful for making a product. Most sawmills define a small end diameter for logs or tree stems, and this can be used as a proxy for the upper-stem diameter that defines the limit of the merchantable tree height. The *crown length* represents the length of the live tree crown from the tip of the tree to the lowest live branch, although in some cases this is subjectively defined and may require professional judgment. The *crown ratio* is the crown length divided by the total tree height. Crown ratio measurements are often used in tree growth models to represent the potential for a tree to grow, or as proxies for a competition index (Soares and Tomé, 2001).

Tree age is a measure of interest to many land managers and landowners, as it is often used to help understand the productivity of a stand by assessing volume produced over time. Age can be expressed in at least four different ways:

(1) The age of a tree since the moment of seed germination.
(2) The time since a tree emerged from the ground or sprouted from a stump.
(3) If it was planted, the period since the time of planting of a tree.
(4) The breast height age of a tree, which is the number of years after a tree was as tall as the point where the DBH was measured (1.37 m above ground).

The last method is used in cases where tree ages are measured at the same point above ground as where the DBH is measured. A number of methods can be used to determine a tree's age, from information gathered through historical evidence or management reports to cursory evidence provided by the tree itself or from

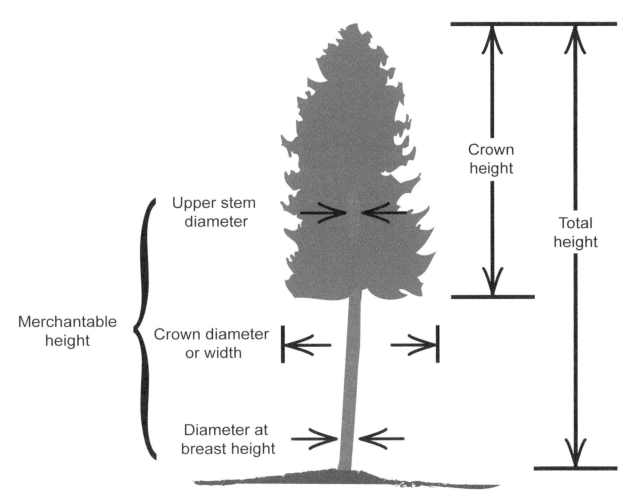

FIGURE 8.3   Measurements that can be developed for a typical tree.

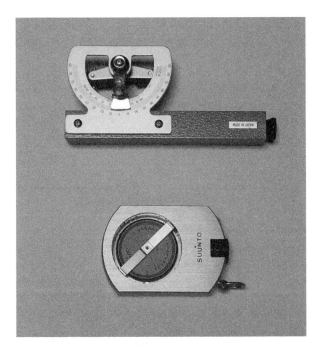

FIGURE 8.4   Tools used to measure tree heights and ground slopes: an Abney level (top) and a Suunto clinometer (bottom).

invasive procedures that enable tree ring information to be obtained. Management records or historical evidence (e.g., dated aerial photos) can help determine the length of time that has passed since an area of land was cleared. In many silvicultural management systems, tree planting activities occur a year or two following a final harvest, unless natural regeneration is used to establish the ensuing forest, in which case regeneration may be assumed to begin immediately. Outward or cursory evidence to determine a tree's age is only possible for tree species that provide distinct physiological traits related to each year of their life. For example, eastern white pine (*Pinus strobus*) produces a whorl of limbs near the tip of the tree each year. By simply counting the whorls of limbs along a tree's bole, one can closely determine the age of an eastern white pine tree. However, this count becomes problematic as the tree gets taller, since whorls 20 to 30 m (65.6–98.4 ft) above ground can be difficult to see.

Many trees in temperate climates provide a record of their growth history or age in the annual rings they produce. Each year, earlywood (formed in the spring and early summer) and latewood (formed in late summer and fall) cells are created in a tree. The width, or spacing, of the earlywood or latewood indicates the growth rate of the tree. These annual rings emanate outward from the pith, or center, of a tree. To view these rings, one would either need to cut or sever a tree bole or extract a core from a tree. Cutting the tree bole is a permanent event that leads to the death of the tree and should only be used to measure a tree's age if the purpose of cutting is to harvest the tree, and not simply to determine its age. However, viewing a tree's annual rings is quite easy when this process is employed (Fig. 8.5). Alternatively, tree coring is an invasive procedure that uses an increment borer to extract a pencil-sized sample of tree rings from a living tree (Fig. 8.6). In this procedure, the hollow borer is screwed into a tree at a height along the bole where the DBH is normally measured until it is determined that the pith has been reached. An extractor is then inserted into the borer, and a core of wood that contains the tree rings is extracted. The rings can be counted to determine how many years have passed since the tree was 1.37 m in height. While this is an invasive procedure, one study of conifers in northern California suggests that tree mortality rates are not influenced by tree coring procedures such as these (van Mantgem and Stephenson, 2004).

Tree volumes are frequently measured for purposes related to the production of wood products. Volumes can be expressed in cubic units (m^3 per ha, or ft^3 per ac), weight (tons), or other production-based units, such as the board foot (1 in [2.54 cm] thick × 12 in [30.5 cm] wide × 12 in [30.5 cm] tall) or the cord (theoretically, 4 ft [1.22 m] wide × 4 ft [1.22 m] tall × 8 ft [2.44 m] long). Tree volumes are estimated in standing, live trees and are generally based on the relationship between the diameter (or DBH) and a measurement of the height. Obviously, if there are two trees with the same DBH,

but one is shorter than the other, then the shorter tree will contain less wood volume. The height measurement might involve the total tree height or a height lower on a tree bole beyond which merchantable products would not be produced. This is called the *merchantable height* and might reflect, for example, the location on a tree bole where the diameter has tapered to a minimum diameter such as 15 cm (about 6 in). As with other measures (e.g., basal area), the volume of trees on each plot or point is summarized to arrive at a per-unit-area measure. We then express the volume of a forest as an average value per unit area, for example 115 m^3 per ha (1644 ft^3 per ac). The volume of dead standing trees or of coarse woody debris (fallen tree boles) is often a necessary ingredient of some wildlife habitat models and is often used to describe fuel loads on the forest floor (with respect to wildfire potential). In these cases, the length and diameter (perhaps at the small and large ends of snags or logs) are used to estimate wood volume.

A summary of structural characteristics of a sample of a mixed conifer forest in the western United States is presented in Table 8.1. From this summary, we can see that the average diameter of the trees is nearly 10 in (25 cm) and that the distribution of diameters is somewhat normal. If one were to graph the diameter distribution as a histogram, it would be somewhat *bell-shaped* (Fig. 8.7). This type of diameter distribution is illustrative of an even-aged forest, where most of the trees are of the same age class or age cohort. The snags per unit area are presented to illustrate how the sample can facilitate an estimate of habitat quality. A number of habitat suitability models use estimates of snags per unit area to characterize habitat quality. Basal area per unit area is, as we noted, a measure of the density of the forest. This measure should be compared to the typical range of basal area values for the region of the world within which they were measured. In this case, the basal area value (166 ft^2 per ac) is in the mid to upper range of basal area values for mixed conifer forests in the western United States. A land manager associated with this forest might therefore assess whether the stand of trees is overstocked, and whether a treatment (harvest or otherwise) is necessary to reduce competition among trees. The volume of live trees per unit area is useful for understanding the current value of the forest. If volume estimates from previous years were also available, then

TABLE 8.1 An example summary of a sample from a mixed conifer forest in eastern Oregon, United States.

DBH class (in)	Average trees per acre	Average height (ft)	Average snags per acre
2	–	–	0.1
3	–	–	0.5
4	–	–	1.9
5	–	–	2.0
6	4.4	60.3	0.8
7	23.5	63.8	0.7
8	44.5	68.4	0.3
9	42.0	71.1	0.3
10	74.0	74.0	0.1
11	65.5	75.4	–
12	29.2	78.7	–
13	11.1	84.5	0.1
14	5.5	87.6	0.1
15	1.8	89.6	–
Total	301.5		6.9

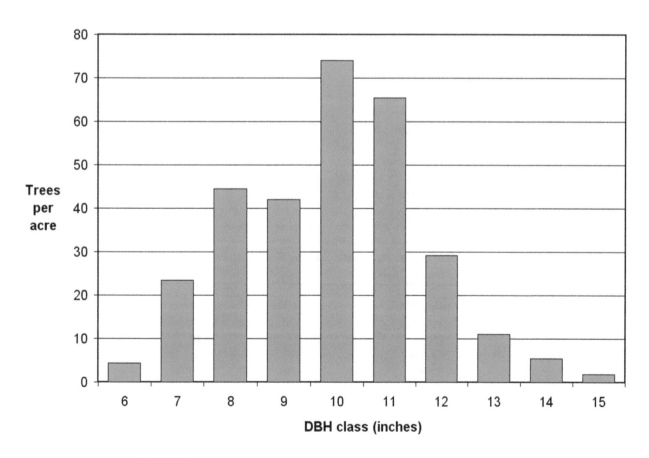

FIGURE 8.7 Diameter distribution of a mixed conifer forest in eastern Oregon.

the value growth rate of the forest could be determined. At times, the value growth rate of the forest is used to inform harvest decisions. In essence, if economic objectives guide the management of the forest and if the forest grows too slowly in value (or does not grow in value at all), this may prompt a decision to harvest.

The quality of a forest can be described both qualitatively and quantitatively, from both a wood-growing perspective (total annual wood volume produced) and an ecosystem perspective (total annual plant and animal volume produced). Measures of quality are generally attached to a site or particular location and are commonly called *site quality* or *site index* measures. From a qualitative perspective, sites can be described as poor, good, or excellent, or in some other manner that facilitates a description of relative differences. For example, the Bienville Parish, Louisiana, soil survey describes site quality for important trees in terms of their potential productivity as: very high, high, moderately high, moderate, and low (Cooley et al., 2002). From a quantitative perspective, a *site index* is often used as a surrogate measure of site quality. The site index is a value (i.e., a number) that describes the average height of the dominant and codominant trees (with respect to canopy position) at a specific base age. For example, if the metric system was being employed, the base age used was 25 years, and the site index value was 33 (i.e., $SI_{25} = 33$), we would expect that the dominant and codominant trees in the area measured would be 33 m tall when they become (or when they were) 25 years old. Dominant and codominant trees are selected for sampling because they are assumed to have been free to grow throughout their lives, and thus their height development has been relatively unimpeded by other trees. When developing site index relationships, the age and total height of sample trees need to be collected. While not an exact measure of productivity, we would expect areas with higher site index values to be relatively more productive than areas with lower site index values, from a wood production point of view.

Site quality can also be inferred from the presence, absence, or abundance of understory plants. For example, as the abundance of deciduous shrubs, forbs, and vascular plants that produce neither flowers nor seeds (i.e., pteridophytes) increases in forests of Alberta, Canada, lodgepole pine (*Pinus contorta*) site index values increase. Conversely, as the abundance of lichens and nonvascular plants (e.g., mosses) increases, lodgepole pine site index values decrease (Szwaluk and Strong, 2003). Thus, a measurement of the plant types found in a forest can infer the overall productivity of the trees.

Many of the measures that have been mentioned are related to individual trees, although some have been developed to describe large groupings of trees or stands of trees. A site index is an example of a measure best presented for a stand rather than a single tree. Sampling processes are necessary to develop a reasonable estimate of the mean and variation of measurements within a stand or forest. Unless the resource (i.e., trees) are very highly valued and the area involved is somewhat small (40 ha [98.8 ac] or less), a complete measurement of every tree is usually not worth the time or effort exerted. Therefore, measurements of a portion of the tree population are used to infer characteristics of the whole population. In essence, a sample of tree measurements is collected rather than a complete enumeration of the characteristics of all trees. Two general types of sampling systems are *systematic sampling* and *random sampling* and both are used in forestry. However, systematic sampling procedures are performed more frequently than are random sampling procedures. In a systematic sample, a person navigates through a forest in a regular pattern along straight lines (or along or within regular intervals) and periodically stops and measures resources such as trees. Random sampling involves developing a scheme where the sample units (plots, prism points, or trees, etc.) are randomly selected from the tree population. To implement random sampling for forest management purposes, it may be necessary to navigate through a forest in an irregular pattern in order to visit each randomly suggested sample location.

Besides individual trees, two other common sampling units are plots and points. Circular sampling plots are most commonly used in forest inventory processes. When using these, all of the trees within a plot are measured, and a summary of the tree characteristics are provided for the plot. These measures are then expanded to a per unit area basis (per ha or per ac), regardless of the plot size. For example, 0.04 ha (0.10 ac) circular plots are commonly used in the southern United States in efforts to develop an inventory of forest resources of mature forests. If 17 trees were measured within a given 0.10-ac plot, then an estimate of trees per unit area around that plot would be 420 per ha (170 per ac). Even smaller plots (0.004 ha, or 0.01 ac) are used to measure the success of regeneration efforts. Trees in this case may be very short and small, and perhaps less than 1 year old. If seven trees were measured in one of these 0.01-ac plots, an estimate of trees per unit area around the plot would be 1730 per ha (700 per ac). Square plots are often used for research purposes, in which the same trees are measured repeatedly over a number of years. Permanently establishing the corners and the center of square plots with stakes or metal rods assists in precisely identifying the trees to be measured (using a sighting from one corner to the next). When using plots as sample units, sampling is based on the frequency of occurrence of trees across the landscape, and an expansion factor is used to determine whether per unit area estimates are the same for each tree and each plot of a given size.

Points are sample units associated with prism sampling. Prism sampling is also called point sampling, variable radius plot sampling, plotless sampling, or Bitterlich sampling (after the person who developed the technique, Walter Bitterlich), depending on the region of the world in which it is employed. With this sampling method, there is no formal measurement of plot area, just a point on the landscape around which some of the trees will be measured based on their bole size (diameter). A device such as a glass prism is used to view a projected angle from the sampling point to the bole of each tree at the location where the DBH is measured. If the angle projected by the prism, when it reaches a tree bole, is greater than the diameter of the tree bole, the tree is not sampled. When using a glass prism, the portion of the tree bole viewed from inside the prism will be shifted outward such that it no longer overlaps the portion of the tree bole viewed outside of the prism. If the portion inside the prism overlaps the portion seen outside, then the tree is sampled (Fig. 8.8). Each prism will have a basal area factor (BAF) associated with it, which indicates how much basal area, per unit area, each sampled tree represents, regardless of the individual tree's size. For example a BAF 10 prism is commonly used in the southern United States. When using this prism, each sampled tree represents 10 ft^2 of basal area per ac (2.3 m^2 per ha) no matter the size of each tree's bole. If eight trees were sampled at a given sample point using a BAF 10 prism, then the estimate of basal area for the point would be 80 ft^2 per ac (18.4 m^2 per ha). In the western United States, larger BAF prisms are used in forests that are denser and have greater basal areas per unit area. For example, a BAF 40 prism might be used in an old-growth Douglas-fir (*Pseudotsuga menziesii*) forest. If seven trees were sampled at a given sample point using a BAF 40 prism, then the estimate of basal area for that point would be 280 ft^2 per ac (64.3 m^2 per ha). The goal is to sample between 3 and 11 trees per point; thus, this form of sampling is relatively fast compared to plot sampling, and some measures (e.g., basal area) are readily determined using these methods.

Measuring trees and forests requires one to work in the setting (the woods) in which the resources reside (Fig. 8.9). As a result, it is essential that natural resource professionals become comfortable navigating alone in rough terrain and varying weather conditions. However, Paine and Kiser (2012) describe methods by which one can make a number of measurements remotely, from vertical aerial photographs. Therefore, map interpretation, photo interpretation, and distance and area measurement skills are also important tools for foresters and natural resource professionals to acquire. Field computers, perhaps in conjunction with other technologies, are now frequently used to store information measured by forest and natural resource managers (Fig. 8.10). By entering collected data directly into a field computer, immediate summaries of forest conditions can be obtained, and fewer errors in data collection are obtained (due to data filters employed).

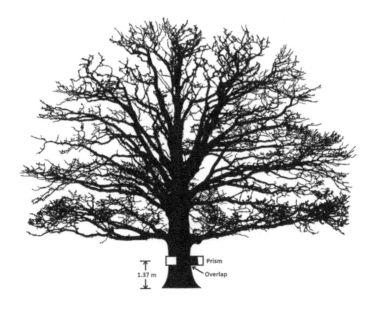

FIGURE 8.8 Conceptual model of a tree that should be sampled during a prism sample process. *Basic image of tree courtesy of Karen Arnold, through PublicDomainPictures.net.*

FIGURE 8.9 Use of an angle gauge to measure the basal area of trees. *Photo courtesy of Mckees, through Wikimedia Commons.*
Image Link: https://commons.wikimedia.org/wiki/File:Angle_gauge_use.JPG
License Link: Public Domain

FIGURE 8.10 A laser rangefinder, GPS receiver, and field computer used for forest and natural resource data collection efforts. *Claudiusmm, through Wikimedia Commons.*
Image Link: https://commons.wikimedia.org/wiki/File:Field-Map_birdie.jpg
License Link: Public Domain

8.2 Measuring other resources

In developing a description or characterization of natural resources that are dependent on or that coexist with forests, a forester or natural resource professional may need to collect or acquire diverse sets of information. For example, a forester may need to inspect the soil upon which a stand of trees resides or a hydrologist may need to assess the quality of water derived from a forest. A land manager may need to estimate the potential smoke-related impacts of burning a forest given projected weather conditions, or a park manager may need

to estimate the amount of use a recreational area may incur given the resources provided. There are a vast number of potential decisions to make in forestry and natural resource management and a wide variety of informational needs, each of which may require collecting data or measuring resources in the forest.

8.2.1 Soil resources

The study of soils is called *pedology* and involves collecting measurements of soil resources. Forest soils facilitate the regulation of water flow, filter and immobilize pollutants, store and cycle nutrients, and provide structural support for plants. There are 12 general orders of soils, some of which are commonly associated with forests (Table 8.2). At an individual site, or within a stand or forest, soil measurements might include the depth of each soil layer (horizon), along with qualitative or quantitative descriptions of the soil color, texture, consistency, structure, and chemical composition. Salinity and acidity (or *pH*) are two other measures that reflect the character of a soil. Soil salinity reflects the salt content within a soil, and high levels can affect plant growth and may lead to plant mortality or significant changes in plant communities to those that can tolerate high salt conditions. For example, storm surges associated with hurricanes (tropical cyclones) may deposit large amounts of salt-laden seawater several kilometers inland. In the southern United States, this type of event can increase soil salinity level and stress and kill baldcypress (*Taxodium distichum*), blackgum (*Nyssa sylvatica*), and pines (Gardner et al., 1992; Conner, 1995). Soil acidity influences the weathering rate of soil parent material and soil formation but, perhaps more importantly, affects in various ways the availability of the main nutrients necessary for plant growth. Quantitatively, soil acidity, or pH, is measured as the negative logarithm of the hydrogen ion (H^+) concentration in water.

$$pH = -\log(H^+)$$

On this logarithmic scale, pH values range from 1 to 14. At 25°C, a water-based solution derived from a soil that has a pH value from 1 to 6 is generally considered acidic (i.e., there are higher concentrations of hydrogen ions in the soil solution), with a pH value of 1 being the most acidic within this range. A water-based solution derived from a soil that has a pH value ranging from 8 to 14 is generally considered alkaline or basic (i.e., there are lower concentrations of hydrogen ions in the soil solution), with a pH value of 14 being the most basic within this range. The pH value can exceed 14 or fall below 0 in cases involving very strong bases and acids, respectively. A pH value of 7 is considered neutral, and it is at this pH that most nutrients in soils are at their highest level of availability (Barnes et al., 1998). Meters and probes can be used to measure soil salinity and acidity, and litmus papers can also be used to measure the pH of soil or a water body.

TABLE 8.2 Twelve soil orders.

Order	Description
Alfisols	Soils found in semiarid to moist condition environments; formed primarily in hardwood forests or where vegetative cover is mixed; highly fertile.
Andisols	Soils commonly found in cool areas with moderate to high levels of precipitation; young, weathered soils formed from volcanic ash; highly fertile.
Aridisols	Soils formed in arid or semiarid climates; dry soils with limited soil development; common in deserts; low levels of organic matter.
Entisols	Soils with little or no soil horizon development; found in areas with recently deposited parent material or where deposition or erosion rates are faster than soil development rates; areas include dunes, flood plains, and steep slopes.
Gelisols	Soils in very cold climates with permafrost near the surface; organic matter accumulates near the surface; common in higher latitudes or higher elevations; not highly fertile.
Histosols	Soils with a high amount of organic matter and saturated nearly year-round; formed through decomposition of plant materials; poorly drained soils found in bogs and moors, sometimes referred to as peats or mucks.
Inceptisols	Soils found in semiarid to humid environments; no accumulation of clay or organic matter; only a moderate degree of soil development.
Mollisols	Soils that have a dark surface horizon and high levels of organic matter; formed under grasses in climates with moderate to high levels of seasonal moisture deficits; found in valleys and savannas.
Oxisols	Soils found in tropical or subtropical regions on lands that have been stable for long periods; highly weathered, indistinct horizons, low fertility; red or yellow in color due to high iron and aluminum concentrations.
Spodosols	Soils formed through weathering processes, where organic matter and aluminum are leached downward to the B horizon, sometimes referred to as a hardpan, which is below a grayish eluvial layer; common in coniferous forests of humid regions; acidic.
Ultisols	Soils formed from intense weathering processes that result in a clay subsoil; red in color due to high iron concentrations; no calcium carbonate; acidic.
Vertisols	Soils with a high percentage of expansive clay materials that change in volume with changes in soil moisture content; they form deep cracks in dry seasons; dark soils; highly fertile.

U.S. Department of Agriculture, Natural Resources Conservation Service (1999).

Another important measure related to soil resources concerns the amount of water contained in a soil, or the soil moisture level. Soil moisture can be described as the available water above the permanent wilting point for a particular plant, or can be described as the actual soil moisture content of a soil. The *field capacity* of a soil is the upper limit of available soil moisture for that soil. Any further addition of moisture to a soil once it arrives at its field capacity will result in water percolation (i.e., downward movement) or runoff (i.e., nearly horizontal movement). The *soil moisture deficit* is the amount of water required to move a soil from its current soil moisture content level to field capacity. The water-holding force of a soil is measured as the *soil moisture tension* and is described in terms of centibars of pressure. Greater levels of soil moisture tension apply greater levels of stress to the plants growing in those soil conditions. Soil-related measurements can be obtained using sophisticated equipment such as compaction testers, moisture meters, conductivity meters, salinity testers, nuclear densiometers, probes or basic pH testing equipment, profile sampling equipment, or augers (Fig. 8.11). Seasoned soil surveyors rely heavily on their judgment (touch and feel) and extensive experience, in addition to precise measurements.

Soil erosion (Fig. 8.12) can be measured quantitatively or qualitatively, and either at the local level or for a larger watershed. Quantitatively, soil can be measured by collecting samples of deposited material and making volumetric estimates of the amount that has moved away from an area of interest. Benchmarks, pipes, and pins could also be placed in the ground of an eroding landscape to record the amount of soil lost over a certain period. Qualitatively, features exposed through the loss of soil can be used to estimate the amount of soil lost. Debris dams, exposed roots, and exposed layers of soil can be used for estimating the amount of soil resource lost.

8.2.2 Aquatic resource quality

Stream temperature and stream sediment are two common measures of aquatic habitat quality. They are often the two measures of aquatic systems that forest practices influence the most. Stream temperature levels are mainly a function of the amount of solar radiation reaching the water surface (Fig. 8.13). Stream temperatures are generally measured in degrees Celsius or Fahrenheit and can be ascertained using temperature meters. Stream sediment levels are a function of natural background input, runoff caused by management activities and road traffic,

FIGURE 8.11 Extracting soil sample using a soil auger in a forest in Virginia, United States. *Hammer (2018).*

FIGURE 8.12 Extensive soil erosion. *Rivas (2006).*

FIGURE 8.13 Partially shaded stream. *Reeves et al. (2016).*

and sediment already within the system. Probes can be used to measure the depth of fine sediment in streambeds, and suspended sediment in stream water can be measured with turbidity meters. Other types of meters can be used to measure the heavy metal, organic material, and salt content (salinity) of water, as well as the dissolved oxygen (O_2) level of water. Probes, meters, or litmus paper can be used to estimate the pH of a water sample. These measurements can be obtained at individual locations within a stream system (i.e., within a stream reach), or estimated for larger watershed through sampling at various strategic locations.

Another measure of aquatic resources is the electrical conductivity of water, which represents the ability of a dissolved material in an aqueous solution to conduct an electrical current. Electrical conductivity is measured in Siemens per unit area or milliSiemens per centimeter (mS/cm). In general, the higher the electrical conductivity of a material, the greater the amount of that material that can be dissolved in water. Electrical conductivity can be measured with a meter or a probe specifically designed to pass a voltage between electrodes. The amount of electrical current flowing through water when the meter or probe is immersed can be used to determine the electrical conductivity of the water, given the materials dissolved in it.

8.2.3 Wildlife resources and habitat quality

The habitat requirements of target wildlife species are necessary for the development of land use plans, and for mitigation and monitoring programs. Further, wildlife habitat measurements are required to provide the necessary inputs for wildlife habitat suitability models. Chapter 5 described the relationships between wildlife and their habitats. In many cases, these relationships are associated with the structural condition of forests. Since the early 1980s, a number of relationships have been developed to link forest conditions to the suitability or capability of habitat to support populations of various wildlife species. The quality of habitat for inland, freshwater fish, for example, can be a function of the amount of shade provided to the stream system, the character of streamside vegetation, dissolved oxygen content, pool density, stream depth (or thalweg), stream temperature, substrate type, water acidity, and water velocity in the spawning areas. Many of these functions directly relate to the condition of the adjoining forests. For example, pools may be created by fallen trees and shade may be provided by taller, living trees that can be a good distance from the edge of a stream. Some functions require landscape-level measurements such as buffer sizes, the distribution of forage and cover areas, and interior habitat sizes, which can be accommodated by GIS analyses.

If a forested area is being assessed for habitat quality for amphibians, factors of importance may include the canopy closure, density of medium or large trees per unit area, distance to water, herbaceous ground cover, or the percentage of the canopy that consists of deciduous trees. For some bird species, the percentage of canopy cover for a tree species group (deciduous or coniferous) or a specific tree species may be important, as well as the understory condition and the density of mast-producing trees per unit area (perhaps from trees of a given size). Further, the average diameter of trees of a given species or species group or the average diameter of overstory trees may be important for describing the quality of habitat for a given bird species. In addition, some habitat might be described using variables such as density of snags (standing dead trees) per unit area, the level of decay of the wood in the snags, successional stage of the forest, tree canopy diversity, and perhaps of a given size, soil drainage class, or a combination of all of these variables (Fig. 8.14). Soil conditions, for example, might relate to food (earthworm) availability for birds such as the American woodcock (*Scolopax minor*) and other species (Cade, 1985; Jirinec et al., 2016). The quality of habitat for many mammal species may also be described using many of the measures described above, along with distances to suitable forage and cover areas. Suitable forage areas may be described according to both their herbaceous or understory vegetative content and their size (a minimum number of contiguous hectares or acres), while

FIGURE 8.14 A baldcypress forest in the Bayou Cocodrie National Wildlife Refuge in Louisiana, United States, which is an ideal habitat for herons, egrets, and wood ducks (*Aix sponsa*). *U.S. Fish and Wildlife Service, through Wikimedia Commons.*
Image Link: https://commons.wikimedia.org/wiki/File:Bayou_Cocodrie.jpg
License Link: Public Domain

suitable cover areas may be described by the size (diameter and height) and density of trees.

Unfortunately, some forest characteristics that are associated with wildlife habitat quality are generally not accounted for during a typical forest inventory. These can include detailed nontimber field measurements such as understory compositions, down woody debris levels, sediment deposition, and plant species diversity. While canopy cover is often used as a measure of habitat quality, measurements of canopy cover for wildlife management purposes are often performed for various strata of vegetation. Techniques described earlier for trees are also appropriate for determining the relative cover of larger vegetation; however, the cover of shrubs and herbs might be of specific interest. For these, methods have been developed to determine the vegetation present along a line transected at various sampling intervals. Cover can also be provided by stems, flowers, other plant parts, and perhaps litter, and this type of plant cover measurement is often used in range management to assess erosion potential, and in ecological studies to assess primary production levels (Hayes et al., 1981).

8.2.4 Air quality and atmospheric conditions

Air quality is of particular concern when land managers are considering the use of prescribed fires as silvicultural treatments or when managing wildfires (Fig. 8.15). Atmospheric conditions will dictate how smoke can be dispersed; therefore, land managers and meteorologists must closely monitor these conditions. The three air quality concerns for both prescribed fires and wildfires are the

FIGURE 8.15 Smoke from a forest fire near Los Alamos, New Mexico, United States in 2000. *Andrea Booher and the U.S. Federal Emergency Management Agency, through Wikimedia Commons.*
Image Link: https://commons.wikimedia.org/wiki/File:FEMA_-_3492_-_Photograph_by_Andrea_Booher_taken_on_05-04-2000_in_New_Mexico.jpg
License Link: Public Domain

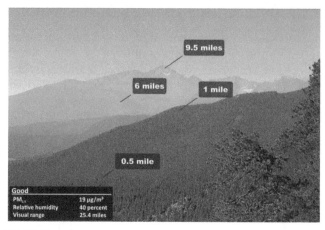

FIGURE 8.16 Simulated visual perspective of a part of the Rocky Mountain National Park, Colorado, United States, given an assumed level of particulate matter. *Hyde et al. (2016).*

particulate matter introduced into the air, the resulting visibility, and the amount of ozone (O_3) pollution. Soot and ash from fires that rise into the air and disperse across the landscape may annoy neighboring landowners and also cause a number of health-related issues, such as respiratory problems and eye and throat irritations. In addition, fires contribute suspended particulate matter into the air in the form of hydrocarbons and organic particles from uncombusted fuels, which can range in size from 0.1 to 10 micrometers (μm; one-millionth of a meter). Particulates that are 2.5 μm or smaller are of the highest concern for human health issues and are measured in micrograms per cubic meter ($\mu g/m^3$). In the United States, the U.S. Environmental Protection Agency has developed standards for annual and daily average levels for particulates in the 2.5—10 μm range, although monthly averages are sometimes reported for urbanized areas (Tian et al., 2008). Individual states in the United States have also developed standards for particulates in the atmosphere that are consistent with the national standards, developed in response to the Clean Air Act of 1970 (42 USC. § 7401—7431). Particulate matter monitors are sensitive receptors used to measure deposition on a sample collection filter in order to provide mass concentration values over a given period (e.g., minutes, hours, or days). This data can be used to not only assess current visual impairment associated with a given level of observed particulate matter but also to simulate potential particulate matter conditions and their effects on a viewshed (Fig. 8.16).

Visibility is an important issue for public transportation systems (e.g., motor vehicles and airplanes), as well as for the overall aesthetic quality of our lives. As

a public resource concern, visibility can be impaired by haze caused by light scattering and absorption by fine particles suspended in the air. Fine particles are very efficient per unit mass at scattering light and, as humidity increases, water adsorption by these particles increases their light scattering efficiency. Sulfates, nitrates, organic compounds, soot, and dust from soils are mainly responsible for the impairment of visibility (U.S. Environmental Protection Agency, 1998). Visibility is generally expressed as the distance one can see across a landscape and may be compared to the estimated visibility in the absence of human-caused air pollution. As an example, naturally occurring visibility ranges from 105 to 190 kilometers (km; 65—118 miles [mi]) in the eastern United States to 190 to 270 km (118—167 mi) in the western United States (U.S. Environmental Protection Agency, 1998). Smoke and haze from fires can reduce these visibility levels considerably.

Other air and atmospheric measurements that are important when planning prescribed fires and for wildfire management include dew point, humidity, ozone levels, temperature, and wind speed. Ozone is a product of the combustion of forest fuels and, while it may be dispersed quickly in the air, it may contribute to the suite of greenhouse gases in the atmosphere. Many of these measurements can be obtained in forests through the use of sensors contained in inexpensive handheld devices. For example, the sling psychrometer (Fig. 8.17), a device containing two thermometers that is spun in the air for a few minutes, is used to determine relative humidity in the field. However, electronic devices are becoming more common for obtaining field measurements of atmospheric conditions. Various types of meters can be used to measure air temperature, humidity, and wind speed (Fig. 8.18).

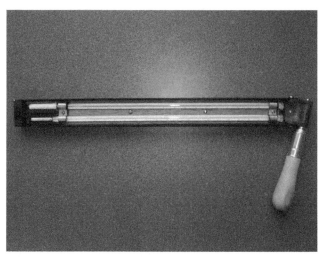

FIGURE 8.17 Sling psychrometer for measuring relative humidity. *Cambridge Bay Weather, through Wikimedia Commons.* *Image Link: hhttps://commons.wikimedia.org/wiki/File:Sling_psychrometer.JPG* *License Link: Public Domain*

FIGURE 8.18 Air velocity and air temperature meter. *Donald L. Grebner.*

8.2.5 Recreation-related resources

Recreation-related resources are often measured using human activity participation rates or assessments of costs and benefits incurred per recreational visitor day. These are measures of recreation demand; the former involve sampling participation in activities, while the latter involve the analysis of travel costs and perhaps the willingness of recreationists to pay for these opportunities. Surveys of recreationists (Fig. 8.19) can be conducted at the site of the activity or through printed or online survey questionnaires. Measurements of recreation supply, like timber supply, have focused on resource inventories at specific locations, but now involve spatial analyses of opportunities that can be provided across the landscape, using models such as the Recreation Opportunity Spectrum (ROS; Garber-Yonts, 2005).

FIGURE 8.19 U.S. National Park Service employees conducting a visitor use survey. *U.S. Department of the Interior, National Park Service.*

In managing recreational areas, it may be important to determine whether sites in which the areas are located are overused or underused by people, or whether the use of the areas is well-balanced. If concerns arise, it may be necessary to evaluate and understand the possible causes of human behavior or factors affecting this situation. Measuring the carrying capacity of a recreational area is one aspect of this assessment. As mentioned in Chapter 7, *carrying capacity*, as it relates to recreation resources, is the maximum number of people that can be accommodated by a facility or location where a social benefit or enjoyment is provided. This implies that once the number of people exceeds the carrying capacity, the overall enjoyment of the recreational opportunity begins to decline. From a biological perspective, if the number of humans participating in a recreational opportunity exceeds the carrying capacity, then damage may occur to the land and vegetation and overall user satisfaction is likely to decline (Schwarz et al., 1976). Factors that affect carrying capacity include climate, the level of development of the recreational opportunity, proximity to other resources, and topography. Some measures that may be used to assess when the carrying capacity has been exceeded include conflicts among recreationists, complaints by recreationists, loss of or damage to vegetation, soil compaction or erosion, and changes in the quality of other resources in the area.

Natural resource managers may also measure the usage of recreational areas and then calculate the economic impacts and benefits of the use to determine the efficiency and effectiveness of the recreational opportunities provided. Recreational area efficiency and effectiveness are important factors when land managers are considering how to allocate resources to the development of specific projects. Efficiency measures can be determined by assessing how well resources (funds, time, people, etc.) have been used for meeting the needs of recreational users. A benefit-cost analysis may be used to compare the outputs generated with the inputs required to offer the recreational opportunity (Lawrence and Titre, 1984). Ideally, we want to provide an opportunity using the lowest input cost per unit of output or recreational experience, as measured by the cost of providing one recreational day of use, for example. This should not imply that recreational opportunities will be provided at the lowest cost, but that the resources available (i.e., people, time, and funds) are used wisely to develop the best suite of experiences for potential recreationists. Recreational effectiveness relates to the overall goals of the recreational program of an organization and how well the opportunities provided meet these goals. An assessment of visitor satisfaction might be one method employed to assess the effectiveness of a recreational program.

The aesthetic quality of a recreational area can be assessed with the assistance of forest resource information. Further, the attractiveness of an area may be related to vegetation diversity and the presence of older trees (Jaakko Pöyry Consulting, Inc., 1993). For example, Brown and Daniel (1986) developed scenic beauty relationships that are a function of forest measurements. The relationships basically indicate that scenic beauty increases as both the amount of vegetation and the size of trees increase. An example of such a relationship is:

Scenic beauty estimate = −32.47 + 4.70 × (number of large ponderosa pine sawtimber trees per acre) + 0.37 × (pounds of herbage per acre)

Like any other model (e.g., a wildlife habitat model), some variation may not be explained by the variables employed in a human's (or animal's) response to the environment. However, these types of models can still be useful for estimating the effects of management activities on the aesthetic quality of the a relationship.

8.2.6 Rangeland resources

Rangelands include grasslands, woodlands, and other areas that are often intimately associated with forests (Fig. 8.20). Measurements that may be necessary in a rangeland area include sampling the vegetative cover of a particular plant species or the density per unit area, plant species diversity, total plant biomass, and perhaps a number of soil properties. These measures can also pertain to invasive species of concern and can be used to assess changes in rangeland resources over time. At specific locations, and in addition to the inventory and classification of rangeland conditions, the carrying capacity of rangeland vegetation may also be measured or estimated. By assessing the carrying capacity of a rangeland area, we inherently measure and estimate the productivity of rangeland plant communities. Rangeland measurements may also involve assessing the foraging and browsing activities of various animal species, which may also

FIGURE 8.20 Elk (*Cervus canadensis*) on rangeland in Yellowstone National Park, Wyoming, United States. *DHeyward, through Wikimedia Commons.*
Image Link: https://commons.wikimedia.org/wiki/File:Several_elk_in_Yellowstone_National_Park.jpg
License Link: Public Domain

involve conducting a census of wildlife populations. At the broader, landscape scale, rangeland measurements can involve assessing the level of recreational use or of erosion occurring within a watershed.

8.2.7 Forest fuels

A fuel is any material that, by one process or another, can be converted into energy. Fuels of interest to forest and natural resource managers are those that reside in the forest (Fig. 8.21) and that can be converted to energy through combustion (fire). Unwanted fires are typically

FIGURE 8.21 Forest fuels within a Douglas-fir forest in the western United States. *Ottmar et al. (1990).*

referred to as *wildfires*, while planned fires are typically called *prescribed burns*. In either case, knowledge of the level of fuels available is of importance to land managers. Forest fuels include both live and dead vegetation that can be burned, and two basic measurements of these within a particular area are the *total fuel level* and the *available fuel level*. Total fuel levels comprise the vegetation that would burn under the most severe environmental conditions, while available fuel levels consist of vegetation that may actually combust under a given condition. When assessing forest fuels, a resource manager will collect measurements of the moisture content of the fuel, the abundance (in weight per unit area) and size of the fuel, the compaction level of the fuel, and the amount of fuel that consists of dead vegetation. These measurements may be also collected for different classes (e.g., sizes) of forest fuels. The moisture content of fuel lying in a forest is determined by weighing a sample of collected fuel, drying it, and then comparing the wet weight to the dry weight. The abundance and size of a fuel are measured by collecting small samples or comparing fuel loads to reference areas, and these measures may need to be collected for various strata, such as ground fuels, mosses and lichens, larger down woody debris, herbs and shrubs, and the forest. A variety of instruments and sampling methods can be employed to obtain these measures. For example, canopy fuels may be estimated using hemispherical photography of crowns or using samples provided by Light Detecting and Ranging (LiDAR) measurements (Skowronski et al., 2007). Line transect methods may also be used to select a sample of debris from the forest floor to provide both a volume per unit area and number of pieces per area. At a landscape scale, radar technology has been used to assess fuel loads across large areas, such as portions of Yellowstone National Park and northwestern Oregon (Saatchi et al., 2007; Hermosilla et al., 2014).

8.2.8 Biodiversity

Biodiversity is a very broad concept, which is difficult to capture and describe using a single measure or index. When biodiversity is discussed, it may refer to the diversity of habitat conditions, plant or animal species, genetic material, or some other set of resources. There are times when we need to understand the relative level of biodiversity for small areas (a small forest) or for larger landscapes. Contemporary measurements of biodiversity typically involve counts of unique individuals and assessments of the dissimilarity among individuals. Along these lines, two of the most common surrogate measures of biodiversity are *species richness* and *relative abundance*. Species richness of a forest is

typically defined as the number of different plant or animal species found within a forested community or landscape. It may be estimated by sampling the forest or landscape and counting the number of unique plants, for example. Fixed area plots can be used to collect data for measuring species richness at a local level; however, it can become a challenge to identify each unique plant species within a plot. Remote sensing techniques have been used to estimate biodiversity at larger landscape scales (Turner et al., 2003) using spectral reflectance levels of landscape features. To understand how common species are dispersed within a forested environment, the relative abundance of species can be described using metrics such as the Shannon—Weiner Information Index (H'):

$$H' = -\sum_{i=1}^{n} p_i \ln(p_i)$$

Here, p_i represents the proportion of species i present in the area of interest with respect to all of the species present. If we were measuring the diversity of forest types across a broad landscape, p_i would represent the proportion of the landscape occupied by forest type i. The Shannon—Weiner Information Index equals zero when there is only one type of plant or animal species present within the area of interest and increases as the number of different species present increases. Other metrics, such as the Simpson Diversity or the Shannon Diversity Indices can be used to describe aspects of both richness and evenness within a forested environment. There are now a large number of patch-level and landscape-level indices that can be employed (e.g., McGarigal and Marks, 1995) to describe the heterogeneity of an area with respect to plant or animal life. These measures can be used to infer a level of biodiversity.

8.2.9 Nontimber forest products

Nontimber forest products (NTFPs) are resources of value to human society that arise from forests, yet are not based explicitly on the wood that is produced. Many of the products are of high value to Native Americans such as material used for medicinal purposes or for basketry. Due to the wide variety of NTFPs, the types of measurement techniques used to assess their availability also vary considerably. Further, the spatial scale used can vary from local, small areas to large regions. Local area inventories of NTFPs usually focus on management of the resources and the markets to which they are oriented. Inventories can be designed to determine the quantity of products available or the quantity of the products used, the latter of which involves market studies regarding the flow of products through a human system (Kleinn et al., 1996). Regional knowledge of the resource, as well as plant databases, can be used to inform assessments of quantity and availability. However, any inventory system must be flexible in order to respond quickly to trends in demand. General forest inventories may be used to develop information on the availability of some NTFPs at a local scale, as these can provide information of forest structure and tree diversity. For example, tree species, tree ages, soil conditions, and recent management history can all be important factors in assessing the suitability of a forest to produce wild mushrooms. Remote sensing techniques can also be applied to assess potential availability across broad landscapes. For example, Russell-Smith et al. (2006) describe methods for assessing wild populations of medicinal plants in Sri Lanka using Landsat satellite imagery as one of the data sources.

8.3 Geographic information systems

A GIS is a set of tools and services that allows one to collect, organize, manage, analyze, and display georeferenced information (Wing and Bettinger, 2008). Computerized mapping and spatial analysis programs trace their roots to the Harvard University Laboratory for Computer Graphics and Spatial Analysis, where one of the first GIS programs (SYMAP) was developed in the mid-1960s. GIS is closely associated and tightly integrated with various aspects of computer-aided drafting, computer-based cartography, remote sensing, and database management. Since the early 1980s, a number of GIS programs have been used for natural resource management purposes, including those developed by esri, Intergraph Corporation, and Pitney Bowes. ArcGIS, an esri product, is perhaps the most widely used GIS program in natural resource management, although a number of organizations in North America may also use Erdas Imagine, GRASS, Idrisi, MapInfo, or other commercial or proprietary systems. SuperMap is a computer program commonly used in China.

GIS is widely used in natural resource management, and recent graduates of forestry and natural resource programs are expected to be fairly proficient with this tool (Merry et al., 2016). Since de Steiguer and Giles (1981) first described the potential uses of GIS in forestry and natural resource management, the overall concept has not changed to a great extent. Foresters and natural resource managers are often asked to locate and map past or future activities such as wildfires, silvicultural practices, stream systems, timber sales, urban forests, areas for wildlife or aquatic habitats, and other landscape features. In doing so, the structural conditions of the vegetation that reside in these areas may need to be analyzed and presented, so that alternative plans of action can be assessed. While some management-related maps may still be hand-drawn, computerized mapping systems facilitate the rapid analysis of many natural resource management issues and allow the creation of professional maps to guide the management of natural resources.

GIS allows a forester or natural resource manager to use two distinctly different data structures: vector (points, lines, and polygons) and raster (grid cells) (Fig. 8.22). The choice of data structure employed is generally based on the information available, the purpose of the analysis, and the organizational standards within which one must work. Early in their development, GIS programs had to be operated on specialized workstations; however, they can now be installed and used on laptops and personal computers. In fact, some handheld devices now enable us to use GIS outdoors in order to display, collect, and manipulate geographic data in real time. Some common data collection methods associated with GIS include traditional field data collection processes (i.e., measuring land and trees), scanned images, and digitized features (*heads-up* using personal computers or more traditionally with a digitizing table). Data collected with satellite-based positioning systems (i.e., global positioning systems [GPS]) or with any of a number of remote sensing devices (e.g., Landsat, Ikonos, IRS, or SPOT [*Système Pour l'Observation de la Terre*] satellites) are also commonly used within a GIS. A wide variety of GIS data are available over the Internet from both private and public sources. Some types of data that cannot easily be created by a natural resource manager but may be essential for natural resource management purposes, such as digital elevation models and digital orthophotographs, can be acquired from other sources or created in GIS. Some examples of processes that can be performed with GIS are:

- *Creating a thematic map.* For example, a forester or natural resource manager may need to develop a timber sale map, wildlife habitat map, trail system map (Fig. 8.23), wildfire map, or a general land ownership map. Several variations or versions of the map can be made quickly using a computerized mapping system and then printed.
- *Querying a database for features of interest.* For example, a wildlife biologist may be interested in understanding the location of forests that have structural conditions highly desirable for an endangered species such as the red-cockaded woodpecker (*Picoides borealis*). In this case, forests with a specific density, size, and species of trees would be of interest. Assuming that appropriate GIS data are available, suitable habitat areas could be located by querying the attribute data describing the forest.
- *Defining areas in proximity to significant features.* For example, a number of land management issues

FIGURE 8.22 Raster (digital aerial photograph) and vector (stands, streams, and survey points) data for a forest in northeastern Georgia, United States.

require defining and locating riparian areas or areas near streams. Given a digitized stream system database, the arcs or lines that define the streams can be buffered by fixed or variable distances to provide an approximation of the extent and amount of land within which management activities should be limited or excluded.

- *Overlaying two distinctly different spatial databases to address a management issue.* One example involves understanding the potential issues that may arise

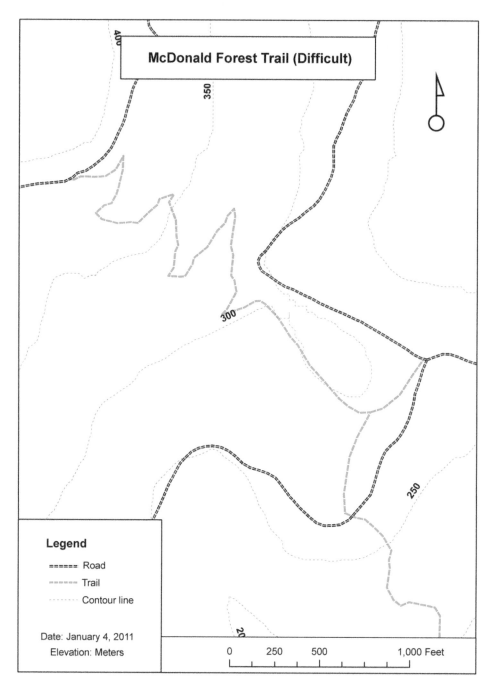

FIGURE 8.23   Map of a trail system on the McDonald-Dunn Forest in western Oregon, United States. *Map courtesy of Pete Bettinger.*

when scheduling management activities within a heavily recreated area. In this case, one might first define viewsheds from roads or buffer zones around hiking trails and then intersect these with a database of areas that require a form of silvicultural treatment. Areas of concern would be identified in the resulting overlay process and these would need to be addressed prior to implementing the treatments.

8.4 Global satellite—based positioning and navigation systems

Many people now use satellite-based positioning systems for navigational and data collection purposes. Satellite-based positioning systems are commonly available in motor vehicles, boats, and airplanes, and a wide variety of specialized handheld devices are available for purchase on the open market. While we typically focus on the receiver that is used to determine a position on Earth, a complete satellite-based positioning system actually consists of control stations, base stations, satellites, receivers, and software. In natural resource management, foresters use satellite-based positioning systems to delineate management areas and collect data regarding roads, streams, and other resources. Wildlife biologists use satellite-based positioning systems to document the locations of wildlife species, their nesting points, and their critical habitat characteristics. Recreational managers use satellite-based positioning systems to map, among other things, trail systems. The satellite-based positioning system developed by the United States is officially called the NAVSTAR GPS program but is more commonly referred to as the GPS. To put the chronological development of this technology into perspective, the GPS system only became fully functional in the mid-1990s. GPS satellites are located approximately 20,800 km (12,900 mi) above the surface of Earth and orbit Earth in one of six orbital planes.

Other countries or organizations have also developed or are currently developing satellite-based positioning systems. The Global Orbiting Navigation Satellite System (GLONASS) was originally designed by the former USSR and is now maintained by the Russian Federation. Signals from this system are now being used in conjunction with NAVSTAR GPS signals by GPS receivers worldwide. The European Union developed the European Navigation System (GALILEO), which became fully operational in 2016. China currently operates the BeiDou Navigation Satellite System, also called COMPASS, that will be fully operational in 2020. Finally, in 2018 Japan completed the development of the Quasi-Zenith Satellite System (QZSS), also called Michibiki, which complements the NAVSTAR GPS system in Asia, and address shortfalls in signal visibility in urban canyons and steep terrain.

Each satellite continuously emits a coded signal using energy in the microwave portion of the electromagnetic spectrum; signals contain information that allows a receiver to determine the time difference between signal emission and signal reception. Time is measured very precisely by each satellite using atomic clocks. *Coordinated Universal Time*, which is a close approximation to Greenwich Mean Time, is used as a standardized time reference. With knowledge of satellite positions in the sky, a receiver can trilaterate its current position if four signals can be obtained from different satellites.

In general, there are three classifications of GPS receivers: recreation or consumer grade, mapping grade, and survey grade. Consumer-grade receivers are the least costly (US$100—US$1000) and provide the least accurate positional information. These receivers are popular among outdoor enthusiasts for navigational

and informational purposes, and a wide variety are available for purchase on the open market. Cell phones and tablets that provide GPS services would also fall into this category (Merry and Bettinger, 2019). Mapping-grade receivers are moderate in price (US$1000—US$10,000), and in some cases can provide 1 to 2 m (3.3—6.6 ft) accuracy in forested conditions (Ransom et al., 2010). These receivers are commonly used for the development of base maps and for research applications. Survey-grade receivers, as suggested by their name, are mainly used for land and water surveying purposes and can meet the higher requirements for professional surveying work. These receivers are more expensive (>US$10,000) and in most cases can provide centimeter-level accuracy. The choice of which receiver to use will be a function of the applications to which they will be employed, along with other budgetary considerations.

Satellite-based positioning system technology is constantly changing, and an awareness of the accuracy of positions suggested by the technology should be central to their use. Some applications of satellite-based positioning systems (e.g., roughly where are we?) need only relative accuracy (5—20 m; 16.4—65.6 ft), while other applications (property boundary location or research studies) may require higher levels of precision. Recent studies of the accuracy of satellite-based positioning systems have been performed in a number of areas of North America and Japan. For example, Ransom et al. (2010) found that commonly used mapping-grade GPS receivers operating in southern Piedmont forests of the United States (both hardwood and pine stands) may report positions, on average, within 2 to 3 m (6.6—9.8 ft) of true ground positions without using differential correction. For comparison purposes, Bettinger and Fei (2010) tested a consumer-grade GPS receiver on the same area over the course of a year and found reported positions to be, on average, within 5 to 12 m (16.4—39.4 ft) of true ground positions. Accuracy will vary depending on whether data are collected in open areas or in closed canopy forests, accuracy may also vary when using this technology in different forest types, and accuracy may vary using this technology during different seasons of the year.

8.5 Aerial photographs and digital orthophotographs

Aerial photographs (Fig. 8.24) are captured by cameras installed in (or attached to) small airplanes, helicopters, unmanned aerial vehicles (UAVs), balloons, kites, and other systems. These photos are often printed onto photo paper for use in both the office and outdoors in the natural environment. Digital orthophotographs

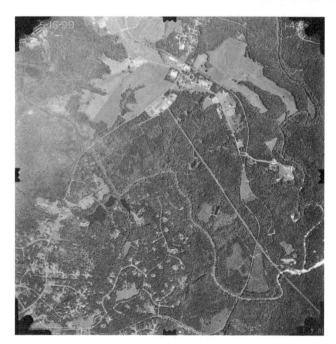

FIGURE 8.24 True color aerial photograph of the University of Georgia Whitehall Forest, Georgia, United States. *University of Georgia.*

FIGURE 8.25 A 2015 panchromatic digital orthophotograph of a 3600 ha (about 8900 ac) area of central Iowa, United States. *U.S. Department of Agriculture, Farm Service Agency, Aerial Photography Field Office, image 4109109 nw.*

(Fig. 8.25) are either single aerial photographs or photo composites that have been converted to digital form (if initially captured on film) and georeferenced using ground control points for use in GIS or GPS receivers. Digital orthophotographs are different from aerial photographs in that a large proportion of the tilt and topographic displacement of features inherent in a single aerial photograph are removed. A digital orthophotograph composite is essentially a collection of individual aerial photographs that have been *stitched* together and georeferenced. These are commonly used in GIS and Internet-based mapping applications such as Google Earth. Upon close examination of these images, the edges of individual aerial photographs are apparent.

The effective use of aerial photographs and digital orthophotographs for natural resource management purposes requires interpretation of resources, which involves examining images, identifying features, and judging their significance toward your objectives. A number of basic principles for the effective interpretation of aerial photographs or digital orthophotographs have been suggested, such as assessing absolute and relative sizes of features in the images, recognizing shapes of features, locating shadows, recognizing differences in the tone, color, and texture of features, locating patterns on the landscape (e.g., man-made vs. natural), and using a reasoning process to relate features to their surroundings (Paine and Kiser, 2012). Each of these principles can be employed to infer something about the characteristics of a resource at the time the image was captured.

In forestry and natural resource management, the most common aerial photographs or digital orthophotographs are captured and stored as true color, black and white (panchromatic), or color infrared images. Thermal images have also been used in wildlife, stream, and wildfire surveys to detect animals or hot spots, depending on the objective of the survey. Photographic film and digital sensors can both be used to record reflected or emitted energy. Visible light wavelengths are in the range of 0.4 to 0.7 μm. To put visible wavelengths of light into perspective, along the electromagnetic spectrum visible light is located between the shorter high-energy wavelengths associated with X-rays, γ rays, and cosmic rays, which are all harmful to humans, and the longer low-energy wavelengths associated with microwaves (used in satellite-based positioning system technology), radar, television, and radio technology, which are generally not harmful to humans (Fig. 8.26).

When using a printed version of an aerial photograph, the scale of the image is important for making area or distance measurements of the resources presented. Scale is commonly described using a representative fraction (i.e., 1:24,000), in which one unit of any measure on the photo represents X of those *same units* on the ground (in this case, 1 cm on a photo represents 24,000 cm on the ground). Scale can also be presented using an equivalence relationship, such as $1'' = 10$ chains (where one chain is 20.1 m [66 ft]). With

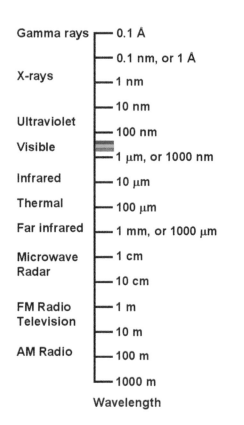

Gamma rays	— 0.1 Å
	— 0.1 nm, or 1 Å
X-rays	— 1 nm
	— 10 nm
Ultraviolet	— 100 nm
Visible	— 1 μm, or 1000 nm
Infrared	— 10 μm
Thermal	— 100 μm
Far infrared	— 1 mm, or 1000 μm
Microwave Radar	— 1 cm
	— 10 cm
FM Radio Television	— 1 m
	— 10 m
AM Radio	— 100 m
	— 1000 m

Wavelength

FIGURE 8.26 The electromagnetic spectrum. *Pete Bettinger.*

knowledge of the scale of an aerial photograph, a measurement from the photograph can be converted to a ground measurement.

8.6 Other remotely sensed imagery

Aerial photographs and digital orthophotographs comprise a subset of remotely sensed imagery; however, the focus of this section is on digital remotely sensed data collected from other sensors and used for natural resource management purposes. Inherently, the products or images obtained from many remote sensing devices that can be used for mapping and classification purposes are stored digitally in a raster data structure. The basic unit of these images, as with digital orthophotographs, is the *picture element* or *pixel*. A number of sensors and devices can be used to create raster databases, including those contained in satellites, digital cameras, scanners, and computer screen-capturing algorithms. Computer programs can also be developed to create raster databases.

Remote sensors, such as cameras, can be housed in aircraft, although many are now present in satellites positioned over Earth. In fact, there are a number of satellite imagery products that may be of value for forestry and natural resource management purposes. Each

---

EXAMPLE

The original color infrared aerial photograph of Rock Eagle Lake in Georgia (Fig. 8.27) has a scale of about 1: 12,365. However, when printed in this book, the scale is much different. Using a dot grid or a planimeter (a mechanical device used for measuring areas) on the original photograph, we find that the area of the main portion of the lake on the photograph is 3.33 in^2 in size. Since the original photograph has a scale of about 1:12,365, this suggests that 1 in on the photo represents 12,365 in on the ground. Therefore, 1 in^2 on the photo represents 152,893,225 in^2 on the landscape (12,365 in × 12,365 in) or 24.375 ac. If the area of the lake on the photograph is 3.33 in^2, then the area of the lake in real life is about 81 ac (3.33 in^2 × 24.375 ac per photograph in^2) or nearly 33 ha.

Geometric relationships have also been developed to allow heights of features to be estimated from vertical aerial photographs (those with less than 3° tilt). In these cases, precise measurements of the displacement of a feature on a photograph, the measurements of the distance of the feature from the photograph's *principal point* (or geometric center), and the flying height of the airplane are needed. The scale of a photograph is, therefore,

important for converting measurements made on the photo itself to characteristics of the resource. In addition, smaller-scale photos (e.g., 1:80,000) may prevent one from locating important features of natural resources that may be easily discernible on larger-scale photos (e.g., 1:10,000). Pelletier and Griffin (1988) describe a study in which they evaluated the effect of scale on identification of conservation practices and noted that only about 29% of conservation practices could actually be located on 1: 80,000 scale photos, whereas about 90% could be located on 1:10,000 scale photos.

Aerial photographs and digital orthophotographs can be obtained from a number of sources, either free or requiring payment, and many can now be obtained via the Internet from platforms such as those provided by the National Map (U.S. Geological Survey, 2020a). In terms of governmental programs, the National Agriculture Imagery Program and the National Aerial Photography Program provide complete coverage of the United States once every few years. Current and historical aerial photographs or county mosaics (digital orthophotographs) can be acquired through these programs. Further, Natural

Continued

EXAMPLE   *(cont'd)*

Resources Canada maintains a National Air Photo Library, where one can purchase aerial photographs or digital orthophotographs of Canadian provinces. In addition, a number of private contractors throughout the world provide services that range from the acquisition of a few photos during a specialized reconnaissance survey to the acquisition and processing of a large number of photos for the development of georeferenced digital orthophotographs. High altitude digital aerial photography is a specialized technology that uses cameras to collect visible and near-infrared reflected energy over a broad area, given the wide angle of coverage possible. Aircraft such as the U-2 reconnaissance plane were employed to fly at very high altitudes (20,000 m or more) for collecting these images, a much higher altitude than is used during a typical aerial photo mission. The United States government operated a high-altitude aerial photography program in the late twentieth century, called the National High Altitude Program. The state of Alaska had a similar program during the same period, which acquired nearly 70,000 images.

Aerial photographs and digital orthophotographs are widely used for forest management purposes such as assessing wildfires, designing timber sales, inventorying forest conditions, mapping vegetation, and monitoring insect and disease problems. With either of these tools, some knowledge of biological sciences and human development patterns, and some photo interpretation experience, one should be able to delineate land features, and determine whether portions of a landscape are relatively flat or steep and whether significant relief is present. Stream systems and their drainage patterns should also be evident, if one can interpret the vegetation (Fig. 8.28) and geological patterns associated with these features. Soil mapping professionals also use drainage patterns, along with the tone and texture of landscape features inherent in aerial photographs, to make inferences regarding soil conditions. Further, rangeland vegetation patterns, wildlife habitat conditions, and even archeological features (Fig. 8.29) can be ascertained using these remote sensing tools.

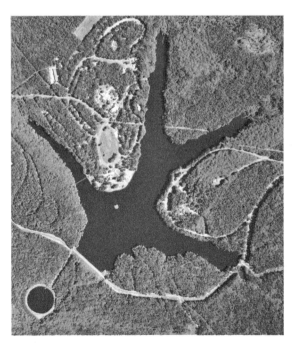

FIGURE 8.27   Portion of a color infrared aerial photograph of Rock Eagle Lake in Georgia, United States. *U.S. Department of Agriculture, May 10, 1982, USDA 12 61803V roll 482, image 40.*

satellite system was designed to collect data representative of the *spectral reflectance* of features, and to store data in grid cells that have spatial resolutions of various sizes, depending on the data capture protocol employed. Spectral reflectance is a measure of the wavelength of the electromagnetic energy collected, and spatial resolution refers to the size of the raster grid cells that represent spectral data. Raster data are commonly stored as regular, two-dimensional tessellations, of which grid cells or pixels of a common size and shape cover an entire area. Each grid cell therefore provides information on the spectral reflectance or energy emittance of the part of the landscape that it spatially represents.

Perhaps the most widely used satellite-derived data come from the United States' Landsat program, which has been operating since the 1970s and is currently managed by the U.S. National Aeronautics and Space Administration (NASA) and the U.S. Geological Survey (USGS). Landsat satellites orbit Earth at an elevation of 705 km (437 mi) above ground in a sun-synchronous fashion; therefore the mean sun time for each individual position in a satellite's orbit is about the same with each pass of a satellite. One orbit of the Earth by a Landsat satellite requires about 99 min and, since Earth is constantly revolving, a satellite will pass over the exact same place

FIGURE 8.29 Fort Loudoun, burned in 1760 and reconstructed as a historical area in 1933, eastern Tennessee, United States. *U.S. Department of Agriculture, National Agriculture Imagery Program, image 2812719.*

FIGURE 8.28 True color image of differing vegetation patterns in a western United States forest. The lighter tones in the middle of the image represent an area in which deciduous forests are present and signify the presence of a stream. *U.S. Department of Agriculture, National Agriculture Imagery Program, image 2226808.*

TABLE 8.3 Characteristics associated with the Landsat 8 satellite sensors.

Band	Spectral resolution (µm)	Spatial resolution (m)	Description
1	0.43—0.45	30	Coastal/aerosol
2	0.45—0.51	30	Blue
3	0.53—0.59	30	Green
4	0.64—0.67	30	Red
5	0.85—0.88	30	Near infrared
6	1.57—1.65	30	Short-wave infrared
7	2.11—2.29	30	Short-wave infrared
8	0.50—0.68	15	Panchromatic
9	1.36—1.38	30	Cirrus
10	10.6—11.19	100	Thermal
11	11.5—12.51	100	Thermal

U.S. Geological Survey (2020b).

on the landscape about once every 16 days. The current series of Landsat satellites capture reflected or emitted energy in 10 bands of electromagnetic energy (Table 8.3). Landsat satellites also provide a panchromatic (black and white) image, which combines the visible reflectance data (blue, green, and red bands). Landsat data are provided to the public in a series of *scenes*, approximately 185 km (about 115 mi) wide and 180 km (about 112 mi) long. In terms of accessibility, Landsat data can be acquired through the Internet from the USGS Earth Resources Observation and Science Center Global Visualization Viewer (GloVis) system.

Landsat data have a wide variety of uses, from the identification of vegetation and topography (Fig. 8.30) to the identification of spatial patterns of disturbances (Masek et al., 2008), forest structures (Hall et al., 2006),

FIGURE 8.30 Composite Landsat 7 ETM+ image for path 112, row 38, illustrating the Sakura-jima volcano in Japan. *U.S. Geological Survey (2011).*

FIGURE 8.31 MODIS image of Cyclone Favio as it approached the Mozambique coast in 2007. *U.S. National Aeronautics and Space Administration (2007).*

and biomass (Tangki and Chappell, 2008; Gasparri et al., 2010). Changes over time in forest cover and fragmentation have also been estimated using Landsat data (e.g., Huang et al., 2009; Coops et al., 2010), as have the distribution of wildlife species and their habitat (Laurent et al., 2005; McDermid et al., 2009). Further, Landsat data have been used to study aquatic vegetation (Torbick et al., 2008), assist in the identification of recreation opportunity classes (Joyce and Sutton, 2009), and assess wildfire severity (Wimberly and Reilly, 2007).

Other satellite-based remote sensing programs include the ASTER, Ikonos, and QuickBird, and SPOT programs. The ASTER program, a joint effort between the United States and Japan, provides data in a wider range of spectral bands of energy than the Landsat or SPOT programs. Compared to the Landsat program, in which the spatial resolution is generally about 30 m (98.4 ft) for most spectral bands, the Ikonos (1 m [3.3 ft]) and QuickBird programs (2.4 m [7.9 ft]) provide higher spatial resolution data. Various radar programs, some developed by NASA, can provide data with a spatial resolution comparable to SPOT or Landsat programs, but with sensitivity to longer wavelengths of energy that may be useful for topographic or meteorological purposes. The Moderate Resolution Imaging Spectroradiometer (MODIS) program also produces raster databases of value for meteorological purposes or broad-scale assessments of large areas (Fig. 8.31). Data derived from the MODIS program have a

wider range of spectral bands of energy than the Landsat or SPOT programs, and the spatial resolution is larger (250–10,000 m [820.2–3280.8 ft] grid cells). The Advanced Very High Altitude Resolution Radiometer (AVHRR) program has been providing nearly continuous global data since 1979 and is used widely in meteorological applications. The spatial resolution of AVHRR data is similar (1.1 km [0.7 mi]) to the largest pixel sizes derived from the MODIS program.

UAVs, often called drones, are offering a new platform for the collection of data. Grenzdörffer et al. (2008) describe the advantages of using these vehicles such as increased affordability and flexibility. These platforms, described in more detail shortly, deploy a variety of sensors from natural light to produce natural looking photography. Thermal sensors can be used to locate hot spots along fire lines to support fire suppression efforts. LiDAR sensors can now be mounted on the platform to collect data that has a variety of uses (described in the next section). Many of the processes used in human-operated aircraft can be applied to UAVs; however, the smaller aircraft are often less stable in high wind environment and may result in higher camera tilt from vertical. However, as the lift of these machines has increased, gimbals are being added to improve the vertical orientation of the sensors. Additionally, improvements in the miniaturization and accuracy of GPS devices are allowing for greater development of mapped products with similar accuracy found in the commercially available products from human-operated vehicles. Continual improvements in battery life will lead to longer flight times, which can further enhance the use of UAVs.

8.7 Laser and LiDAR technology

In natural resource management, laser technology can be used to measure ground slope, vertical and horizontal distances, and, if compasses are incorporated into their design, direction. Lasers use emitted and reflected light energy to determine distances and angles. Their value in measuring the diameters and heights of trees for forestry and natural resource management purposes is increasing (Wing et al., 2004; Weaver et al., 2015), and they can be employed to effectively map the spatial location of trees within sample plots. Unfortunately, as the density of vegetation increases, the efficiency and suitability of using laser technology in forested conditions may be affected (Kalliovirta et al., 2005). Some typical types of lasers are called rangefinders or laser dendrometers. Calipers have also been developed to allow the diameter of trees to be measured without having to physically touch each tree; one simply aims the calipers at a tree of interest and adjusts the caliper tongs so that the laser beams barely touch each side of the tree. The resulting diameter of the tree can then be read from the caliper scale, as one would normally do using a non-laser caliper. Remotely measuring tree diameters with a technique such as this is beneficial for purposes that require measurements at tree bole heights well above the place where DBH is measured (1.37 m or 4.5 ft).

LiDAR is a remote sensing technique similar to radar, which uses laser pulses to determine characteristics of landscape or vegetative features. The development of a LiDAR image is based on the time required for pulses to return to the system that emitted them. With LiDAR technology, a device is needed to emit wavelengths of electromagnetic energy in the ultraviolet to near-infrared range of energy (0.25–10 μm), and a photodetector is needed to collect the reflected energy. LiDAR devices can be mounted in airplanes, UAVs or positioned on the ground. Some common applications of LiDAR in natural resource management include the development of a ground surface represented as a digital elevation model, the development of a canopy surface, a description of tree structure within a forest such as snags (Zimble et al., 2003; Wing et al., 2015), and the estimation of biomass (Nelson et al., 2007). Sub-canopy vegetation height and ground cover from shrubs can also be estimated using this technology (Dubayah and Drake, 2000; Wing et al., 2012). A number of archeological applications are also facilitated using LiDAR

technology, since LiDAR data can be used to locate subtle changes in ground surface conditions. Thus, LiDAR data have been used to locate cultural features that are obscured by dense vegetation (Gallagher and Josephs, 2008). LiDAR has also been used to detect the smoke plumes of forest fires (Fernandes et al., 2006) and to help describe wildlife habitat characteristics (Graf et al., 2009). However, there are challenges associated with using this technology; for example, LiDAR data are relatively expensive to capture and to use, and the accuracy of some LiDAR data (e.g., tree heights) may vary with forest conditions (Gatziolis et al., 2010). However, laser and LiDAR technology, in conjunction with GIS and satellite-based positioning systems, may provide opportunities for precise and efficient forest and natural resource data collection efforts.

8.8 Unmanned aerial vehicles: drones

In the last 10 years, there has been a significant increase in the deployment of UAVs, often called drones, in forestry. Initially, there was a variety of formats that mimicked modern aircraft, with some designed to look like B-2 bat-wing bombers, and others designed to look like model Cessna airplanes or helicopters. Today, it appears that the multiple horizontal rotor vehicle is currently the most adopted style. These vehicles are designed to collect data over local areas and have been adapted to carry a variety of sensors including those that capture visible light, infrared energy, longer wavelength thermal energy, and hyperspectral energy (many different ranges, or bands). Further, UAVs can be equipped with LiDAR sensors that are used in a variety of the remote sensing forestry applications.

The best use of these applications is for small area applications due the limited flight times that generally range from 10 to 30 min before the batteries must be changed. As a comparison, the maximum flight times for conventional aircraft (Cessnas) that are used for larger format data collection is several hours. Thus, UAVs are best suited for small format, small area applications. These systems are flexible in that data can be collected by most people, although many jurisdictions (including federal governments) are requiring operators to obtain a form of pilot's license to fly these vehicles, in an effort to protect the public safety. Data collected by a UAV is stored on-board the vehicle, then easily transferred to a computer. The time between data acquisition and data inspection is almost instantaneous with flying the vehicle; there is no waiting time for film to be developed and processed, as with aerial photography of the past.

The applications of UAVs in society are vast, and we have only begun to exploit the uses of the UAVs in forestry. They are commonly used as a method to count seedlings following planting, to ensure that a harvested area has been fully stocked with healthy and growing seedlings (Feduck et al., 2018). Others have used thermal cameras on UAVs to locate hot spots of fires, to direct fire suppression efforts (Christensen, 2015). Those who are managing sophisticated supply chains can use drones to track harvest activities within a logging unit. Thus, the status of a harvest can be updated relatively quickly to inform managers about the flow of wood to a mill or a to a woodyard (Scholz et al., 2018). Another forestry application is to locate and estimate the size of biomass piles that are potentially available for bioenergy purposes (Bedell et al., 2017). There are also a variety of forest inventory applications, with individual tree recognition a goal in complex sampling schemes. Thus, a variety of tree characteristics that can influence log value might now be estimated with UAVs. Some of these include estimating the live versus dead branch proportion, branch angles, and the average distance between branch whorls, also called the internode length (Torresan et al., 2017). There are also applications being developed to monitor postharvesting impacts. For example, Talbot et al. (2018) used UAVs to monitor postlogging damage in Norway.

UAVs are becoming a more common tool for the modern forester. The technology can provide an immediate view of a forest resource that can be used to improve the effectiveness of the forest management. For better or worse, becoming a licensed UAV pilot may become a necessary job skill in the future.

Summary

Making informed forest and natural resource management decisions requires information about the resources being managed. The collection of information is therefore a key task for many foresters and natural resource managers. In the forestry and natural resource management professions, collecting measurements of tree diameters, heights, and ages is a common activity. These measurements can be used to determine estimates of density (trees per unit area and basal area) and volume, which in turn may be used to assess forest character and guide management decisions. Measures of other resources (air, aquatic, soil, etc.) can be used along with forest measurements to help determine appropriate courses of action for forests or natural resources. In collecting measurements of forests or other natural resources, a sample is usually collected rather than measuring everything in a population. While the mean value of resource conditions is informative, reports of forest and natural resource conditions often require descriptions of the variability inherent in the measurements collected. Therefore, while data collection processes are important, a natural resource professional also needs to understand how to determine and describe the variability associated with the sample. As science and computer technology evolves, we can take advantage of new developments to support the description and analysis of forest resources. GIS, GPS, laser technologies, satellite image products, and LiDAR are now frequently being used to support forestry and natural resource management or are seriously being assessed for use in natural resource management. Young foresters and natural resource professionals should be able to understand how and when to use these technologies, and to understand their strengths and weaknesses for assisting in the collection and analysis of natural resource measurements.

Questions

(1) *Habitat model.* Use your nearby library or conduct an Internet search to locate a habitat model that relates to a wildlife species that uses forested areas for some portion of its lifespan. In a short report, describe the forest measurements that would need to be collected in order to assess the quality of an area with respect to habitat for the species you selected. Include pictures to show examples of the resources that need to be measured.

(2) *Forest sample.* Imagine developing a sampling scheme in which you navigate through a forest to specific locations, stop, and collect information about the conditions in the nearby area. Describe the type of information you might collect at each sampling location. Depending on your field of interest, describe how you would use the information that you collect. Develop a short PowerPoint presentation that illustrates the tools and measures related to the sample.

(3) *New field measurement technology.* Locate a catalog of new technology offered for sale by a forestry equipment vendor. Select one form of new technology, such as a GPS receiver, a laser distance-measuring device, or a water quality probe. In a one-page report,

QUESTIONS *(cont'd)*

describe the capabilities of the technology and how they might be useful in collecting forestry or natural resource data. Finally, describe the current (or older) technology that might be replaced by adopting new technology, and the costs and benefits of making a change in data collection procedures.

(4) *Remote sensing technology.* Select an aerial photo or satellite-based remote sensing program used frequently in your area of the world, and describe the current status of the program, the products provided by the program, and how one could obtain these products. In a small group of four to five people from your class, compare and contrast the programs selected after each describing (in 2–3 min) the program you selected.

(5) *Forest measurement summary.* Given the data provided below, from a sample of diameters of trees from six 0.04-ha (0.10-ac) circular plots, calculate the estimates of basal area and volume per unit area.

DBH	Tally
6	13
8	19
10	41
12	49
14	32
16	10
17	3

Use the following relationships to determine the height and the volume of the trees:

$$\text{Tree height (ft)} = \exp[4.8 - 3.47 \, (1/\text{DBH})]$$

$$\text{Volume (ft}^3) = 0.34864 + 0.00232 \, (\text{DBH}^2 \times \text{Tree height})$$

Remember: these are data collected from *six* plots.

References

American Forests, 2020. Green Ash (*Fraxinus pennsylvanica*). American Forests, Washington, D.C. https://www.americanforests.org/big-trees/green-ash-fraxinus-pennsylvanica/ (accessed 16.01.20).

Avery, T.E., Burkhart, H.E., 2002. Forest Measurements, fifth ed. McGraw-Hill, New York.

Barnes, B.V., Zak, D.R., Denton, S.R., Spurr, S.H., 1998. Forest Ecology, fourth ed. John Wiley & Sons, Inc., New York.

Bedell, E., Leslie, M., Fankhauser, K., Burnett, J., Wing, M.G., Thomas, E.A., 2017. Unmanned aerial vehicle-based structure from motion biomass inventory estimates. Journal of Applied Remote Sensing 11 (2), 026026.

Bettinger, P., Fei, S., 2010. One year's experience with a recreation-grade GPS receiver. Mathematical and Computational Forestry and Natural-Resource Sciences 2, 153–160.

Brown, T.C., Daniel, T.C., 1986. Predicting scenic beauty of timber stands. Forest Science 32, 471–487.

Cade, B.S., 1985. Habitat Suitability Index Models: American Woodcock (Wintering). U.S. Department of the Interior, Fish and Wildlife Service, Washington, D.C. Biological Report 82(10.105). 23 p.

Christensen, B.R., 2015. Use of UAV or remotely piloted aircraft and forward-looking infrared in forest, rural and wildland fire management: evaluation using simple economic analysis. New Zealand Journal of Forestry Science 45 (1), 16.

Conner, W.H., 1995. Woody plant regeneration in three South Carolina *Taxodium/Nyssa* stands following Hurricane Hugo. Ecological Engineering 4, 277–287.

Cooley, M.C., Burke, P.M., Henry Jr., C., Scott, E.D., Godfrey, C.L., Stephens, W.W., Wilson, J., Rosenburg, G., Fontenot, S., Edwards, J., 2002. Soil Survey of Bienville Parish, Louisiana. U.S. Department of Agriculture, Forest Service, Natural Resources Conservation Service, Washington, D.C.

Coops, N.C., Gillanders, S.N., Wulder, M.A., Gergel, S.E., Nelson, T., Goodwin, N.R., 2010. Assessing changes in forest fragmentation following infestation using time series Landsat imagery. Forest Ecology and Management 259, 2355–2365.

Crawford, H.S., Marchinton, R.L., 1989. A habitat suitability index for white-tailed deer in the Piedmont. Southern Journal of Applied Forestry 13, 12–16.

de Steiguer, J.E., Giles, R.H., 1981. Introduction to computerized land-information systems. Journal of Forestry 79, 734–737.

Dubayah, R.O., Drake, J.B., 2000. LiDAR remote sensing for forestry. Journal of Forestry 98 (6), 44–46.

Feduck, C., McDermid, G.J., Castilla, G., 2018. Detection of coniferous seedlings in UAV imagery. Forests 9 (7), 432.

Fernandes, A.M., Utkin, A.B., Lavrov, A.V., Vilar, R.M., 2006. Optimization of location and number of LiDAR apparatuses for early forest fire detection in hilly terrain. Fire Safety Journal 41, 144–154.

Gallagher, J.M., Josephs, R.L., 2008. Using LiDAR to detect cultural resources in a forested environment: an example from Isle Royale National Park, Michigan, USA. Archaeological Prospection 15, 187–206.

Garber-Yonts, B.E., 2005. Conceptualizing and Measuring Demand for Recreation on National Forests: A Review and Synthesis. US Department of Agriculture, Forest Service. Pacific Northwest Research Station, Portland, OR. General Technical Report PNW-645.

Gardner, L.R., Michener, W.K., Williams, T.M., Blood, E.R., Kjerve, B., Smock, L.A., Lipscomb, D.J., Gresham, C., 1992. Disturbance effects of hurricane Hugo on a pristine coastal landscape: North inlet, South Carolina. Netherlands Journal of Sea Research 30, 249–263.

Gasparri, N.I., Parmuchi, M.G., Bono, J., Karszenbaum, H., Montenegro, C.L., 2010. Assessing multi-temporal Landsat 7 ETM+ images for estimating above-ground biomass in subtropical dry forests of Argentina. Journal of Arid Environments 74, 1262–1270.

Gatziolis, D., Fried, J.S., Monleon, V.S., 2010. Challenges to estimating tree height via LiDAR in closed-canopy forest: a parable from western Oregon. Forest Science 56, 139–155.

Gong, L., Zhang, Z., Xu, C., 2015. Developing a quality assessment index system for scenic forest management: a case study from Xishan Mountain, suburban Beijing. Forests 6, 225–243.

Grenzdörffer, G.J., Engel, A., Teichert, B., 2008. The photogrammetric potential of low-cost UAVs in forestry and agriculture. The International Archives of the Photogrammetry, Remote Sensing and Spatial Information Sciences 31 (B3), 1207–1214.

Graf, R.F., Mathys, L., Bollmann, K., 2009. Habitat assessment for forest dwelling species using LiDAR remote sensing: Capercaillie in the Alps. Forest Ecology and Management 257, 160–167.

Hall, R.J., Skakun, R.S., Arsenault, E.J., Case, B.S., 2006. Modeling forest stand structure attributes using Landsat ETM+ data: application to mapping of aboveground biomass and stand volume. Forest Ecology and Management 225, 378–390.

Hammer, G., 2018. New USDA Pathways Employee Soil & Wetland Training. U.S. Department of Agriculture, Natural Resources Conservation Service, Office of the Chief, Soil Science and Resource Assessment, Soil Science Division, Beltsville, MD. Weekly update, July 13, 2018.

Hayes, R.L., Summers, C., Seitz, W., 1981. Estimating Wildlife Habitat Variables. U.S. Department of the Interior, Fish and Wildlife Service, Western Energy and Land Use Team, Washington, D.C. FWS/OBS-81-47.

Hermosilla, T., Ruiz, L.A., Kazakova, A.N., Coops, N.C., Moskal, L.M., 2014. Estimation of forest structure and canopy fuel parameters from small-footprint full-waveform LiDAR data. International Journal of Wildland Fire 23 (2), 224–233.

Huang, C., Kim, S., Song, K., Townshend, J.R.G., Davis, P., Altstatt, A., Rodas, O., Yanosky, A., Clay, R., Tucker, C.J., Musinsky, J., 2009. Assessment of Paraguay's forest cover change using Landsat observations. Global and Planetary Change 67, 1–12.

Hyde, J.C., Blades, J., Hall, T., Ottmar, R.D., Smith, A., 2016. Smoke Management Photographic Guide: A Visual Aid for Communicating Impacts. U.S. Department of Agriculture, Forest Service. Pacific Northwest Research Station, Portland, OR. General Technical Report PNW-GTR-925.

Jaakko Pöyry Consulting, Inc, 1993. Recreation and Aesthetics: A Technical Paper for a Generic Environmental Impact Statement on Timber Harvesting and Forest Management in Minnesota. Jaakko Pöyry Consulting, Inc., Raleigh, NC, 217 p.

Jirinec, V., Isdell, R.E., Leu, M., 2016. Prey availability and habitat structure explain breeding space use of a migratory songbird. The Condor: Ornithological Applications 118 (2), 309–328.

Joyce, K., Sutton, S., 2009. A method for automatic generation of the Recreation Opportunity Spectrum in New Zealand. Applied Geography 29, 409–418.

Kalliovirta, J., Laasasenaho, J., Kangas, A., 2005. Evaluation of the laser-relascope. Forest Ecology and Management 204, 181–194.

Kleinn, C., Kindt, R., Islam, S.S., Lund, G., Munyanziza, E., Sah, S.P., Schreckenberg, K., Taylor, F., Temu, A.B., Weinland, G., 1996. Assessment and Monitoring of Non-timber Forest Products. Domestication and Commercialization of Non-timber Forest Products in Agroforestry Systems. Food and Agriculture Organization of the United Nations, Rome, Italy. Non-Wood Forest Products 9.

Laurent, E.J., Shi, H., Gatziolis, D., LeBouton, J.P., Walters, M.B., Liu, J., 2005. Using the spatial and spectral precision of satellite imagery to predict wildlife occurrence patterns. Remote Sensing of Environment 97, 249–262.

Lawrence, L., Titre, J., 1984. Measuring recreation area O&M efficiency and effectiveness. U.S. Army Corps of Engineers, Waterways Experiment station, Vicksburg. MS. Recnotes Volume R-84–5, pp. 1–3.

Masek, J.G., Huang, C., Wolfe, R., Cohen, W., Hall, F., Kutler, J., Nelson, P., 2008. North American forest disturbance mapped from a decadal Landsat record. Remote Sensing of Environment 112, 2914–2926.

McDermid, G.J., Hall, R.J., Sanchez-Azofeifa, G.A., Franklin, S.E., Stenhouse, G.B., Kobliuk, T., LeDrew, E.F., 2009. Remote sensing and forest inventory for wildlife habitat assessment. Forest Ecology and Management 257, 2262–2269.

McGarigal, K., Marks, B.J., 1995. FRAGSTATS: Spatial Pattern Analysis Program for Quantifying Landscape Structure. U.S. Department of Agriculture, Forest Service. Pacific Northwest Research Station, Portland, OR. General Technical Report PNW-351.

McIntyre, R.K., Conner, L.M., Jack, S.B., Schlimm, E.M., Smith, L.L., 2019. Wildlife habitat condition in open pine woodlands: field data to refine management targets. Forest Ecology and Management 437, 282–294.

Merry, K., Bettinger, P., 2019. Smartphone GPS accuracy study in an urban environment. PLoS ONE 14 (7), e0219890.

Merry, K., Bettinger, P., Grebner, D.L., Boston, K., Siry, J., 2016. Assessment of geographic information system (GIS) skills employed by graduates from three forestry programs in the United States. Forests 7 (12). Article ID 304.

Nelson, R.F., Hyde, P., Johnson, P., Emessiene, B., Imhoff, M.L., Campbell, R., Edwards, W., 2007. Investigating RaDAR-LiDAR synergy on a North Carolina pine forest. Remote Sensing of Environment 110, 98–108.

Ottmar, R.D., Hardy, C.C., Vihnanek, R.E., 1990. Stereo Photo Series for Quantifying Forest Residues in the Douglas-fir-hemlock Type of the Willamette National Forest. U.S. Department of Agriculture, Forest Service. Pacific Northwest Research Station, Portland, OR. General Technical Report PNW-GTR-258.

Paine, D.P., Kiser, J.D., 2012. Aerial Photography and Image Interpretation, third ed. John Wiley & Sons, New York.

Pelletier, R.E., Griffin II, R.H., 1988. An evaluation of photographic scale in aerial photography for identification of conservation practices. Journal of Soil and Water Conservation 43, 333–337.

Ransom, M.D., Rhynold, J., Bettinger, P., 2010. Performance of mapping-grade GPS receivers in forested conditions. Rurals: Review of Undergraduate Research in Agricultural and Life Sciences 5 (1). Article 2.

Reeves, G.H., Pickard, B.R., Johnson, K.N., 2016. An Initial Evaluation of Potential Options for Managing Riparian Reserves of the Aquatic Conservation Strategy of the Northwest Forest Plan. U.S. Department of Agriculture, Forest Service. Pacific Northwest Research Station, Portland, OR. General Technical Report PNW-937.

Rivas, T., 2006. Erosion Control Treatment Selection Guide. U.S. Department of Agriculture, Forest Service, National Technology & Development Program, San Dimas, CA, pp. 0677–1203 (SDTDC).

Russell-Smith, J., Karunaratne, N.S., Mahindapala, R., 2006. Rapid inventory of wild medicinal plant populations in Sri Lanka. Biological Conservation 132, 22–32.

Saatchi, S., Halligan, K., Despain, D.G., Crabtree, R.L., 2007. Estimation of forest fuel load from radar remote sensing. IEEE Transactions on Geoscience and Remote Sensing 45, 1726–1740.

Scholz, J., De Meyer, A., Marques, A.S., Pinho, T.M., Boaventura-Cunha, J., Van Orshoven, J., Rosset, C., Künzi, J., Kaarle, J., Nummila, K., 2018. Digital technologies for forest supply chain optimization: existing solutions and future trends. Environmental Management 62 (6), 1108–1133.

Schwarz, C.F., Thor, E.C., Elsner, G.H., 1976. Wildland Planning Glossary. U.S. Department of Agriculture, Pacific Southwest Forest and Range Experiment Station, Berkeley, CA. General Technical Report PSW-13. 252 p.

Skowronski, N., Clark, K., Nelson, R., Hom, J., Patterson, M., 2007. Remotely sensed measurements of forest structure and fuel loads in the Pinelands of New Jersey. Remote Sensing of Environment 108, 123–129.

Soares, P., Tomé, M., 2001. A tree crown ratio prediction equation for eucalypt plantations. Annals of Forest Science 58, 193–202.

Szwaluk, K.S., Strong, W.L., 2003. Near-surface soil characteristics and understory plants as predictors of *Pinus contorta* site index in southwestern Alberta, Canada. Forest Ecology and Management 176, 13–24.

Talbot, B., Rahlf, J., Astrup, R., 2018. An operational UAV-based approach for stand-level assessment of soil disturbance after forest harvesting. Scandinavian Journal of Forest Research 33 (4), 387–396.

Tangki, H., Chappell, N.A., 2008. Biomass variation across selectively logged forest within a 225 km^2 region of Borneo and its prediction by Landsat TM. Forest Ecology and Management 256, 1960–1970.

Tian, D., Wang, Y., Bergin, M., Hu, Y., Liu, Y., Russell, A.G., 2008. Air quality impacts from prescribed fires under different management practices. Environmental Science and Technology 42, 2767–2772.

Torbick, N., Hu, F., Zhang, J., Qi, J., Zhang, H., Becker, B., 2008. Mapping chlorophyll-*a* concentrations in West Lake, China, using Landsat 7 ETM+. Journal of Great Lakes Research 34, 559–565.

Torresan, C., Berton, A., Carotenuto, F., Di Gennaro, S.F., Gioli, B., Matese, A., Miglietta, F., Vagnoli, C., Zaldei, A., Wallace, L., 2017. Forestry applications of UAVs in Europe: a review. International Journal of Remote Sensing 38 (8–10), 2427–2447.

Turner, W., Spector, S., Gardiner, N., Fladeland, M., Sterling, E., Steininger, M., 2003. Remote sensing for biodiversity science and conservation. Trends in Ecology and Evolution 18, 306–314.

U.S. Department of Agriculture, 1999. Natural Resources Conservation Service. Soil Taxonomy: A Basic System of Soil Classification for Making and Interpreting Soil Surveys. U.S. Department of Agriculture, Natural Resources Conservation Service, Washington, D.C. Agriculture Handbook 436.

U.S. Environmental Protection Agency, 1998. Interim Air Quality Policy on Wildland and Prescribed Fires. US Environmental Protection Agency, Washington, D.C., 38 p.

U.S. Geological Survey, 2011. Landsat Missions. U.S. Department of the Interior, Geological Survey, Washington, D.C. https://commons. wikimedia.org/wiki/File:Sakurajima_Landsat_image.jpg (accessed 25.01.20).

U.S. Geological Survey, 2020a. The National Map. U.S. Department of the Interior, Geological Survey, Washington, D.C. https://www. usgs.gov/core-science-systems/national-geospatial-program/nation al-map (accessed 24.01.20).

U.S. Geological Survey, 2020b. Landsat-Earth Observation Satellites. U.S. Department of the Interior, Geological Survey, Earth Observation and Science (EROS) Center, Sioux Falls, SD.

U.S. National Aeronautics and Space Administration, 2007. Earth Observation. Goddard Space Flight Center, Greenbelt, MD. https://earthobservatory.nasa.gov/images/7428/tropical-cyclone-favio (accessed 25.01.20).

van Mantgem, P.J., Stephenson, N.L., 2004. Does coring contribute to tree mortality? Canadian Journal of Forest Research 34, 2394–2398.

Weaver, S.A., Ucar, Z., Bettinger, P., Merry, K., Faw, K., Cieszewski, C.J., 2015. Assessing the accuracy of tree diameter measurements collected at a distance. Croatian Journal of Forest Engineering 36 (1), 73–83.

Wimberly, M.C., Reilly, M.J., 2007. Assessment of fire severity and species diversity in the southern Appalachians using Landsat TM and ETM+ imagery. Remote Sensing of Environment 108, 189–197.

Wing, M.G., Bettinger, P., 2008. Geographic Information Systems: Applications in Natural Resource Management. Oxford University Press, Canada, Don Mills, ON.

Wing, B.M., Ritchie, M.W., Boston, K., Cohen, W.B., Gitelman, A., Olsen, M.J., 2012. Prediction of understory vegetation cover with airborne lidar in an interior ponderosa pine forest. Remote Sensing of Environment 124, 730–741.

Wing, B.M., Ritchie, M.W., Boston, K., Cohen, W.B., Olsen, M.J., 2015. Individual snag detection using neighborhood attribute filtered airborne lidar data. Remote Sensing of Environment 163, 165–179.

Wing, M.G., Solmie, D., Kellogg, L., 2004. Comparing digital range finders for forestry applications. Journal of Forestry 102, 16–20.

Zimble, D.A., Evans, D.L., Carlson, G.C., Parker, R.C., Grado, S.C., Gerard, P.D., 2003. Characterizing vertical forest structure using small-footprint airborne LiDAR. Remote Sensing of Environment 87, 171–182.

Tree anatomy and physiology

The development of a forest is intimately tied to the growth or demise of individual trees. The mechanisms that affect the future status of individual trees can help us understand how forest structure might change over time (Ward and Stephens, 1993). Models of forest dynamics may also help us to understand these transitions and to visualize potential future conditions in the light of anthropogenic (human-caused) or natural disturbances. The main focus of this chapter is tree anatomy and physiology, and biological processes operating at the scale of a cell up to the size of a tree. One goal of this book is to help readers learn, comprehend, and articulate the essential principles of forestry and natural resource management. Therefore, as outcomes of reading this chapter, readers should be able to understand

- the structure of trees and describe their various functions;
- the needs of trees for survival and the processes by which materials (water, minerals, etc.) move through trees;
- the photosynthetic process and how trees use carbon dioxide (CO_2), water (H_2O), and minerals to create carbon building blocks and oxygen (O_2);
- how and why trees respire, how they respond to signals, why they have different tolerances to shade, and what nutrients are necessary for unrestricted growth;

- the function of tree root systems; and
- the various regeneration processes trees employ.

Ultimately, through studying the topics presented in this chapter, readers should gain a basic understanding of the composition of a typical tree, how it grows, and how it maintains itself to contribute to the wide suite of ecosystem services that forests provide.

9.1 Tree anatomy

If one could ascribe a goal to a forest tree or plant, it would be to grow large enough and live long enough to produce viable seed that perpetuates its species. In addition, trees provide a number of provisional and regulatory ecosystem services that are beneficial to humans and other plant and animal life. In order to facilitate these services, a tree must compete with other vegetation for resources. Thus, it attempts to secure resources by growing upward and outward, filling the space with foliage, and growing downward and outward, filling the space with a root system (Lanner, 2002). To survive, a tree must absorb energy (sunlight), carbon dioxide, water, and nutrients, all the while competing for these resources with neighboring trees and plants, and respiring oxygen and other organic compounds (Šimpraga et al., 2019) back into the atmosphere. In addition, a tree needs defense mechanisms to be able

FIGURE 9.1 Elk damage on conifers in Lahemaa National Park, Estonia. *Athanasius Soter, through Wikimedia Commons.*
Image Link: https://commons.wikimedia.org/wiki/File:Põdrakahjustus.JPG
License Link: Public Domain

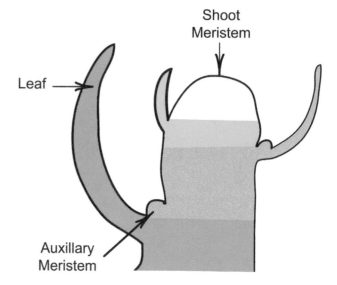

FIGURE 9.2 Conceptual model of the apical or shoot meristems at the tip of a tree leader.

to survive droughts, resist insects and diseases, survive severe wind events and heavy accumulations of water and ice, and recover from animal damage (Fig. 9.1). In other words, trees must be capable of adapting and taking advantage of resources when they become available in order to be capable of self-healing and to develop mechanical structures that resist and distribute stresses (Bejan et al., 2008). All of this is further complicated by the fact that, once established and without assistance from humans, a tree cannot move naturally, on its own, to a more suitable location. Of course, the requirements of trees are the same for many other types of plants: nutrients in the soil, water, and access to light.

The primary growth of trees is vertical, through division and enlargement of cells in the vertical shoots or leaders of a tree. More precisely, vertical growth occurs in the apical or shoot meristems of tree leaders, located in the apex of dormant buds (Fig. 9.2). Apical meristems are very small sections of undifferentiated tissue (i.e., young tissue for which the structural role cannot be determined) found in the growing tips of buds or roots of plants. At some point, this tissue differentiates into xylem and phloem (secondary meristem or procambium), pith (ground meristem), or epidermis (protoderm). Leaf primordia also form in the apical meristem, forming lateral outgrowths that eventually develop into leaves or needles. In temperate forests, dormant buds are protected by bud scales, and in tropical forests the apical meristem in trees is generally protected by a looser rosette of embryonic leaves. In the spring, when the bud axis expands, the primordia enlarge and growth directly below the bud axis forces the apex upward in a surge of growth. Once growth surge has ceased, the apical meristem forms a new

bud for the next cycle of growth, and the supporting tissue is lignified, thus making it woody (Lanner, 2002).

Secondary growth in trees is horizontal in nature and, put simply, as a tree grows in height, it must also grow in girth in order to have the strength to maintain itself upright between its peers (which are all competing for light by growing their canopies as high as possible above ground) and against the forces of wind. Trees grow in diameter through the enlargement of cells in the vascular cambium (a lateral meristem). The cambium is an undifferentiated multicell layer located between the inner bark and the outermost portion of an annual ring. The cambium comprises two basic types of cells: elongated fusiform initials, which produce xylem and phloem; and ray initials, which give rise to rays. In this *cambial zone*, cells produced toward the inside of the tree become differentiated into mother xylem (wood) cells (Fig. 9.3). Mitosis (separation of chromosomes into sets of nuclei) and cytokinesis (division of the nuclei) are the processes involved in the mitotic phase of cell division from mother cells to daughter cells. These cells (i.e., the initials) may divide several times before differentiating into distinct, mature xylem cells (Lachaud et al., 1999). Mature xylem cells, in addition to providing strength, conduct water and minerals (or sap) that may also contain organic compounds, amino acids, and proteins. Interestingly, the ascending pathway for sap within a tree has been understood for over 200 years (Pickard, 1981). The long-term sustainability of some forestry and agroforestry systems may be threatened in the future by regional climatic changes in which rainfall is predicted to decline up to 30% from current rates during certain seasons of the year. Therefore, it is important to understand the role of fluid

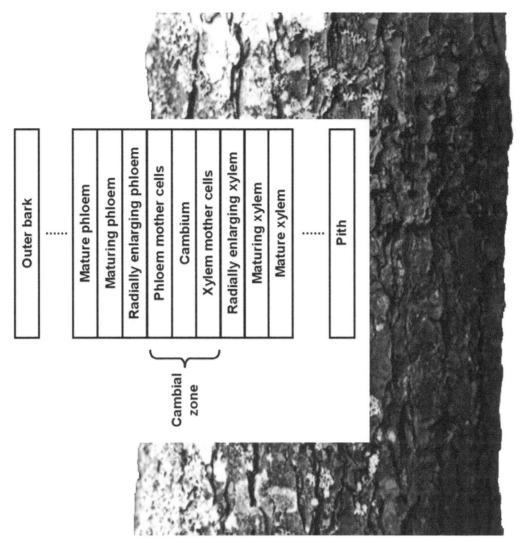

FIGURE 9.3 Conceptual model of the cambium and the production of secondary conducting tissues. *Adapted from Lachaud et al. (1999); image of tree courtesy of Bill Cook, Michigan State University, through Bugwood.org.*

transport in trees and how it may vary according to the season (David et al., 2004).

Xylem cells produced early in a growing season are large in diameter, have thin cell walls, and are thus relatively soft and weak. These are typically called *earlywood*, in comparison to cells produced later in the season (with a smaller diameter and thicker cell wall and are tougher and stronger) that are called *latewood*. If the earlywood is comprised of cells with diameters that are distinctly larger than the cell diameters of the latewood, the resulting type of wood is considered *ring porous*. The contrast between earlywood and latewood in many temperate tree species results in an annual ring from which one might be able to date the age of a tree (Fig. 9.4). If there is little or no differentiation in the size of the earlywood and the latewood cell diameters, the wood is considered *diffuse porous*. This type of growth means that the contrast between

FIGURE 9.4 Growth rings associated with a southern United States pine tree. *Tor Schultz.*

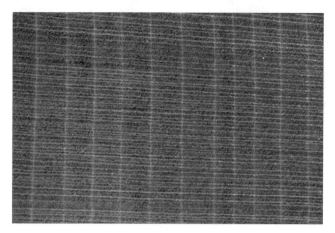

FIGURE 9.5 Transverse view of yellow-poplar (*Liriodendron tulipifera*). The *bright line* of terminal parenchyma at the annual ring boundary is characteristic of this diffuse porous hardwood. 10× magnification. *Tor Schultz.*

earlywood and latewood in some temperate hardwoods may be less distinct (Fig. 9.5). Since the growth of trees is nearly continuous (or episodic) in tropical forests, annual rings may be absent (Lanner, 2002). If present, the thickness of each ring is representative of the resources available during the growth period; therefore, one can deduce periods of drought and periods of intense competition for light and nutrients from the annual rings. The central portion of a tree bole is inactive and has the sole function of supporting the canopy. Although not all trees have it, when present, the *heartwood* of a tree is a central core of dead xylem in which the cell walls are filled with defensive or waste materials, tannins, dyes, resins, oils, and other organic compounds.

The *pith* of a tree is a residual core of cellular material that is generated by a growing shoot, and xylem is placed over this tissue to form the bole of a tree. As we mentioned, the initial xylem cells produced by the cambium are considered mother cells, as are phloem cells, which are produced just to the outside of the cambium and form the inner bark. As with mother xylem cells, the mother phloem cells may divide several times before differentiating into distinct, mature phloem cells (Lachaud et al., 1999). Mature phloem is a group of sieve or tube cells that serve to distribute photosynthate (i.e., products of photosynthesis, such as a sugar) and other substances produced by the leaves or needles to the rest of the tree. In contrast to xylem cells, which must be dead and emptied of their contents in order to function, phloem cells must be alive to perform their function. However, they live for only 1 year and are crushed between the expansion of xylem and the existing wall of bark in the second year (Lanner, 2002). As a result, bark is essentially an assorted layer of crushed phloem cells and other cortical remnants or protective

cells generated by the cambium (Coder, 1999). It is pushed outward by the formation of new bark and new wood (xylem), and the outer layers stretch, crack, and eventually slough off (Forest Products Laboratory, 1987).

Although we have briefly described cell structure, before we delve too far into tree growth and growth dynamics, we need to step back and describe the basic building blocks of wood. Below the bark, a tree is composed mainly of hollow, elongate, spindle-shaped cells that are parallel to one another within the bole, limbs, and branches. Customarily, wood cells are called libriform fibers, tracheids, or vessels, and they vary in length both within a tree and between tree species, with the age of a tree, the season in which they were formed, and perhaps also due to other environmental factors. Tracheids are the primary water-conducting cells in conifers, cycads, and ginkgo (*Ginkgo biloba*) trees and other gymnosperms (naked seed plants). Tracheids have thick cell walls and provide structural support to trees. Vessels are the water-conducting elements of angiosperms (flowering plants). Both are located in the xylem area of a tree, and their lignified walls are perforated to allow the flow of water and minerals from one to the next. The presence of these tracheids and vessels differentiates vascular plants (those with xylem and phloem) from nonvascular plants. Libriform fibers, which are thin cells with simple pits, are the major component of wood in deciduous (hardwood) trees. The general range of fiber lengths for softwood (coniferous) trees is 3 to 8 mm, while the range of fiber lengths for hardwood (deciduous) trees is generally 1 to 2 mm, but again there is considerable variation. For example, in one analysis the fiber length of poplar hybrids (a hardwood) grown in Washington State ranged from 0.4 to 1.0 mm and seemed to increase with the age of the trees (DeBell et al., 2002).

The main chemical components of fibers and tracheids are cellulose, hemicellulose, and lignin, which we described in Chapter 4. Cellulose and hemicellulose are polymers of glucose: cellulose is a linear chain of glucose units, and hemicellulose is a polysaccharide. During the growth of a tree, cellulose fibers are arranged into ordered strands of helically bound (twisted together) fibrils, which are then organized into larger structural elements that form the walls of wood fibers. Cellulose molecules are surrounded by lignin and hemicellulose. One of the main roles of hemicellulose is to facilitate fiber-to-fiber bonding. Lignin is a natural phenolic resin (an aromatic amorphous molecule, or three-dimensional phenyl-propane polymer) found within cell walls but concentrated toward the outside and between cells. The main role of lignin is to hold the fibers and tracheids together. In terms of structure, a tree cell consists of a thin primary wall (Fig. 9.6) and

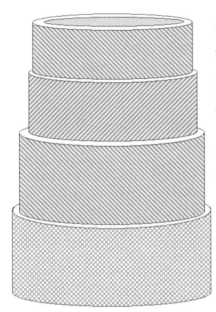

Inner secondary wall layer (S3)

Middle secondary wall layer (S2)

Outer secondary wall layer (S1)

Primary wall

FIGURE 9.6 Conceptual model of a woody cell wall showing the primary wall and the three secondary wall layers. *Pete Bettinger.*

a thicker secondary cell wall consisting of three layers with cellulose fibrils helically wound at different angles in the various layers. The walls of adjacent cells are bonded together by the middle lamella (Fig. 9.7), which consists of lignin, calcium, and pectin substances (Schwarze, 2007).

Xylem rays radiate outward from the pith of a tree and are used for resource storage, defense, and radial transport of materials between the xylem and phloem. Rays typically account for 10% to 25% of tree cells (Lachaud et al., 1999). They can be heterocellular, as in firs, oaks, and pines, or homocellular, as in alders, maples, and sycamores. A heterocellular ray system is one in which the wood is characterized as having more than one type of ray cell (e.g., parenchyma, sheath cells, or tracheids). A homocellular ray system contains only one type of ray cell (Zuckerman and Davidson, 2020). Parenchymal cells are those whose purpose (e.g., storage, assimilation, or wound healing) relates to the function of a tree rather than its structure. Parenchyma is the food storage area of a tree and is more prominent in angiosperms than in gymnosperms. In conifers, these are represented by resin ducts. In some trees, a thin band of parenchymal cells is produced at the end or beginning of a growth season and is filled with calcium oxalate crystals (interestingly, a constituent of human kidney stones and the scale that forms on beer kegs) (Gourlay et al., 1996). Plasmodesmata (plasma membrane-lined channels in connected cell walls) facilitate direct intercellular transport of nutrients, signals necessary for growth and development, proteins, and viruses (Cilia and Jackson, 2004).

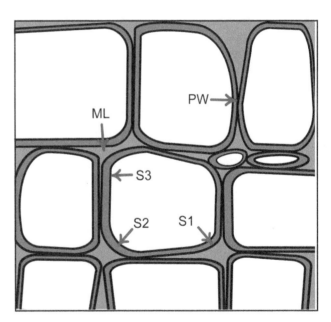

FIGURE 9.7 Conceptual model of a transverse section of early-wood tracheids. *ML*, middle lamella; *PW*, primary wall; *S1*, outer secondary wall layer; *S2*, middle secondary wall layer; *S3*, inner secondary wall layer.

Apical meristems comprise two groups of cells used to develop new tree structures. The apical *shoot meristem* is responsible for the production of above-ground structures, and branches, leaves, pollen cones, and seed cones (and other organs) originate from nearby *axillary meristems* as small primordia. The shoot meristem consists of three layers (Fig. 9.8), and layering is maintained by cell divisions that are perpendicular to the surface in

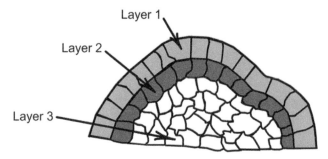

FIGURE 9.8 Conceptual model of a shoot apical meristem.

the outer two layers (Simon, 2001). The apical *root meristem* is responsible for the production of below-ground structures; thus, this is where the primary axis of the root system is produced, with lateral roots originating a short distance behind. The root meristem is characterized as having a root cap (or root tip), an elongation zone, and a differentiated zone. The presence of mycorrhizal fungi, which have a symbiotic relationship with the root system of a plant, can influence water and nutrient uptake by trees (Peterson et al., 1984). For most temperate and some tropical trees, mycorrhizal fungi are found in the rhizosphere of the root system, a narrow region of soil surrounding the mycorrhizal roots that contains bacteria and sloughed plant cells and that is influenced by root development. A tree obtains a more efficient method of obtaining soil, water, and nutrients from its symbiotic relationship with mycorrhizal fungi. Plant growth promotes the development of diverse bacteria communities, which can either facilitate or inhibit the development of mycorrhiza. Plants allocate a good portion (10%—20%) of their photosynthetic assimilate (i.e., consumed nutrients) to mycorrhizal partners, which represents a major route of carbon transfer between the plant and the soil (Bending et al., 2002).

In the bole of a tree, *lateral meristems*, comprising vascular cambium and cork cambium, are the two places where cell division occurs. The cork cambium is a one-cell thick layer situated among the phloem that produces the cork (bark) cells that are used to protect the tree bole, branches, and limbs and to reduce water and sugar loss. The vascular cambium is a thin layer of cells that generate both xylem and phloem for the transport of water and sugars in a tree: xylem is produced to the inside of the tree bole and phloem is produced to the outside of the tree bole. As we discussed earlier, the woody part of a tree is composed of xylem and the bark or cork part of a tree is composed of phloem and other tissue material. The tree stem and branching system of a tree have the principal function of supporting and positioning the canopy in such a way as to maximize the production of photosynthate. However, these

portions of a tree also act as a conducting pathway that allows sugars produced from photosynthesis to be transported downward to the root system and water upward to the canopy (Pickard, 1981).

Leaves or needles of a tree contain chlorophyll and comprise the organ of the tree in which photosynthesis takes place. Pickard (1981) asserts about leaves that:

> their basic job is to hold a disperse photosynthetic pigment system somewhat perpendicular to an incident photon beam so that the pigment can transduce the energy of the incoming photons into a form in which it can be used to synthesize sugar from water and atmospheric carbon dioxide.

Stomata are cell structures in the epidermis of tree leaves and needles that are involved in the exchange of carbon dioxide and water between plants and the atmosphere. Thus, stomatal conductance (i.e., the ability to convey or transmit) is an important issue in achieving a balance with environmental conditions. Stomata are protected by specialized sensory organs (guard cells) that respond to environmental conditions and adjust their turgor (i.e., tension created by fluid content) to help control conductance (Mansfield, 1998). Stomatal guard cells respond to carbon dioxide, hormones, light, ozone (O_3) levels, and vapor pressure deficits to adjust the aperture (opening size) of the stomata (Chavez et al., 2011). Stomatal conductance varies among and within tree species based on age, chlorophyll concentration, leaf (or needle) temperature, nitrogen content, the amount of energy being received, soil moisture, and the vapor pressure deficit between the leaf (or needle) and the atmosphere (Matsumoto et al., 2005).

9.2 Transpiration and sap flow in a tree

Trees need water, nutrients, and sunlight in order to survive. These resources are critical for the metabolic activity of wood cells, primordia, and parenchyma during primary or secondary growth of a tree. Water is necessary for most of the metabolic processes within a tree. Water must migrate through the soil to the root hairs of a tree, through the cortex of the roots, across an endodermis (the Casparian strip) or the periderm of the roots, and into the stele, which contains vascular tissue. As we noted, the primary components of vascular tissue in trees are xylem and phloem, and the tracheids and vessels of the xylem assist in the transport of water and minerals from the ground. Tree transpiration is a key factor in both water and carbon exchange (Gartner et al., 2009). Transpiring plants lose water through the cell walls of leaves or needles, and this loss is replaced with water transported through the xylem. The cohesion-tension theory was developed to describe water movement up a tree stem and it suggests that water movement is determined by water loss through transpiration, which creates tension in the xylem tissue within a tree. However, this theory has been challenged because the tension involved does not seem to be adequate for moving water upward against gravity (Pickard, 1981; Meinzer et al., 2001). This theory has been debated and discussed for over a century, and it is now suggested that other processes may also be involved. Not all trees are alike in their rate of transpiration. Transpiration rates seem to adjust to climatic conditions, and substantial differences in stomatal regulation of water use at the leaf (needle) level can be found among different tree species (Čermák et al., 1995). In addition, a study of maritime pine (*Pinus pinaster*) in southwestern France found that transpiration rates declined with the age of trees, due to reduced levels of stomatal conductance in taller trees and a reduction in leaf area in older trees (Delzon and Loustau, 2005).

Two types of processes, osmotic and physical, seem to regulate transpiration, which is often measured by the flow of sap in trees. The osmotic process is based on anatomical barriers to sucrose movement and the ability of a tree to adjust the pressure needed to prevent or facilitate the flow of water. The resulting turgor pressure exerted outward on cell walls makes plants rigid, thus allowing them to stand upright. The physical process is based on changes in air temperature and the resulting effects on tree liquids. For example, when maples (*Acer* spp.) are dormant and air temperatures fluctuate above and below freezing point, the alternating pressure placed on the trunk and branches causes either freezing or thawing of sap. When warming occurs, sap pressure increases and is forced out of small wounds.

Humans have capitalized on these natural processes and devised ways to collect and use the sap flowing through trees, for example collecting maple sap, birch sap, and latexes from tropical plants. Maple sap is characterized as a dilute solution of water (95%–99%), sugar (1%–5%), and trace amounts of other substances. Sucrose is the primary sugar found in maple sap, potassium and calcium comprise the majority of the inorganic elements, and alanine, asparagine, glycine, and others represent free amino acids. A few organic acids are found in maple sap, as are a wide range of phenolic compounds, most of which appear to be derived from lignin (Perkins and van den Berg, 2009). In contrast, the main sugar components in birch sap are glucose and fructose (Kallio and Ahtonen, 1987). Latexes of tropical plants, such as the Indian mango (*Mangifera indica*), papaya (*Carica papaya*), and rubber tree (*Hevea brasiliensis*), are saps that contain proteins, resins, sugars, tannins, and other minerals and compounds. In many tropical trees, sap flows freely when fruit are cut and harvested (Saby John et al., 2003).

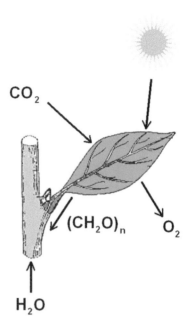

FIGURE 9.9 The photosynthetic process. *Image of the sun, stem, and leaf courtesy of Stepa, through Wikimedia Commons.*
Image Link: https://commons.wikimedia.org/wiki/File:Fotosynteza2.png
License Link: Public Domain

9.3 Photosynthesis

Photosynthesis is a process that occurs in leaves or needles of trees through which electromagnetic energy is captured and used to convert carbon dioxide into complex organic compounds, including sugars. In the presence of enzyme systems associated with chlorophyll in plants, electromagnetic energy (usually sunlight), CO_2, and H_2O combine to form glucose ($[CH_2O]_n$) and oxygen (O_2; Fig. 9.9). During this reaction, light energy (photons) absorbed by chlorophyll produces the energy to split water (H_2O) into hydrogen (H) and oxygen (O_2) (Barnes et al., 1998). As a result, oxygen is released by the plant, and carbon dioxide is converted to sugars through a process called carbon fixation. Most plant species require only simple inorganic nutrients to accomplish this process and therefore are considered autotrophic.

The process of photosynthesis provides the carbon substrate (or building blocks) for the development of roots, vascular tissue, and other parts of a tree. Chlorophyll and other proteins in chloroplasts absorb more blue (0.4–0.5 µm wavelength) and red light (0.6–0.7 µm wavelength) than they do green light (0.5–0.6 µm wavelength). As a result, more green light energy is reflected by the crowns of trees. This explains why tree crowns appear greenish even though the reflectance of green light energy may be relatively low (about 10%). Obviously, if electromagnetic energy were

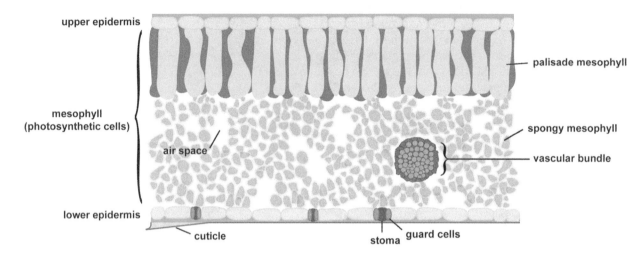

upper epidermis

palisade mesophyll

mesophyll
(photosynthetic cells)

spongy mesophyll

air space

vascular bundle

lower epidermis

cuticle

stoma

guard cells

FIGURE 9.10   Cross-sectional conceptual model of a deciduous tree leaf.

limited, the rate of photosynthesis would decline dramatically. The rate of photosynthesis in a plant can also be reduced due to stresses such as injured leaves (or needles), water deficits, temperature extremes (hot or cold), and nutrient deficiencies (Méthy et al., 1994). Long-term exposure to elevated rates of ozone can also reduce the rate of photosynthesis and thus reduce the growth rate of forests (Matyssek et al., 2007).

Leaves of temperate deciduous trees can change color in the latter part of the growing season. In the fall, the level of chlorophyll produced in leaves slows and eventually ceases. Chlorophyll is then destroyed, and the carotenoids present are allowed to express their color. Carotenoids are organic pigments within chloroplasts, and they facilitate the energy transfer process during photosynthesis. Carotenoids absorb blue light and therefore produce shades of yellow, orange, and brown. While chlorophyll is being destroyed, anthocyanins are synthesized to further alter the color of leaves and plant tissue. Anthocyanins are pigments that act as a sunscreen and an antioxidant, and help produce shades of red, blue, and purple in plant tissue. The difference in the timing of leaf color change and the amount of color expressed in leaves of tree species is both genetically controlled and related to temperature and moisture conditions of the environment (U.S. Forest Service, 2011).

Leaves and needles of trees have a variety of shapes, arrangements, and surface textures. These organs are enclosed in a thin layer of cuticle cells, below which is a set of epidermis cells surrounding palisade and spongy mesophyll cells (Figs. 9.10 and 9.11). In addition to providing protection to the enclosed mesophyll cells, the cuticle also acts to absorb ultraviolet electromagnetic energy and as a defense mechanism (Kinnunen et al., 2001). Chloroplasts are located in the mesophyll cells located between the protective upper and lower epidermis, and most photosynthesis occurs in the palisade mesophyll cells. The spongy mesophyll cells absorb air and perform light-independent reactions of photosynthesis. Vascular tissue networks throughout the leaves contain both xylem and phloem cells to facilitate the transport of water and nutrients. We mentioned earlier that stomata are scattered across the surface of a leaf or needle, linking the outside world with the parenchyma, and that their function is regulated by guard cells. Stomata can be very densely packed on a leaf or needle surface. In silver fir (*Abies alba*) seedlings, stomatal density is reported to be in the range of 6000 to 7500 per square centimeter (cm^2; about 38,700–48,400 per square inch [in^2]) (Robakowski et al., 2004); in silver birch (*Betula pendula*) it is reported to be around 13,000 per cm^2 (83,900 per in^2) (Pääkkönen et al., 1997); and in Scots pine (*Pinus sylvestris*) the density is around 9700 per cm^2 (about 62,600 per in^2) (Turunen and Huttunen, 1996).

In the fall, a layer of new cells forms at the base of leaf stems of temperate, deciduous tree species, initiating the process of abscission (i.e., the shedding of leaves). As this layer of cells grows, it clogs and closes the leaf veins and blocks fluid transport into and out of leaves. Sometime after the connective tissues are sealed, leaves will separate and fall (abscise) from a tree. Trees such as water oak (*Quercus nigra*) can retain their dead leaves for several months after they have effectively died. Although often called evergreens, most coniferous trees have needles that will not remain indefinitely attached to a tree; in fact, older needles will change color and eventually abscise and fall from a tree. Because most conifers retain green needles of staggered ages that may be one, two, or three (or greater) years old, the loss of older needles sometimes goes unnoticed. The lifespan of a coniferous needle is 2 to 5 years, depending on the species and climatic conditions (e.g., droughts may

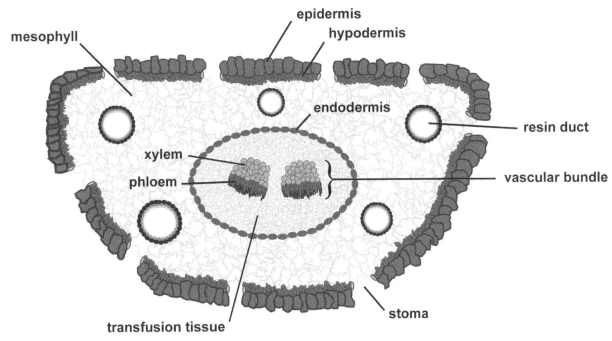

FIGURE 9.11 Cross-sectional conceptual model of coniferous tree needle.

precipitate the loss of conifer needles). For example, Austrian pine (*Pinus nigra*), eastern white pine (*Pinus strobus*), and Scots pine all retain needles for about 3 years, while red pine (*Pinus resinosa*) retains needles for approximately 4 years (Heimann and Pellitteri, 2006). Needle retention is often used a proxy for the tree vigor.

9.4 Tree respiration

Trees both produce and consume oxygen. In conjunction with photosynthesis, carbon fixation occurs in trees only when sunlight is available and, with respect to deciduous trees, only when leaves or needles are available. However, respiratory activity occurs continuously in the living tissues of trees and over the course of a year respiratory activity from the woody part of a tree can exceed the respiratory activity of foliage (leaves, needles) (Edwards and Hanson, 1996). For example, glucose created through photosynthesis produces water and carbon dioxide during cellular respiration, creating energy that is necessary for certain cellular growth activity. In order to measure total respiration for a whole tree or stand of trees, both leaf and woody organ (root, trunk, and branch) respiration rates need to be considered. There are two basic respiration components: growth-related respiration (for construction of new tissue) and maintenance-related respiration (used to maintain ion gradients and repair membranes) (Levy and Jarvis, 1998). These are a function of tree growth rate (growth-related respiration) or tree bole biomass (maintenance-related respiration); however, estimates are complicated by the activities of mature and growing woody tissue (Ryan et al., 1994).

Although variation exists vegetatively and geographically, respiration rate per unit volume in a tree seems to decrease with increasing tree diameter and maintenance-related respiration seems to increase with increases in leaf area index, but to decrease with tree age and geographic latitude (Anekonda et al., 2000; Kim et al., 2007). Respiration also is affected by the availability of photosynthetic substrates and the season in boreal forests and is perhaps weakly correlated with air temperature (Griffiths et al., 2004). Maintenance-related respiration rates also positively correlate with tree stem temperature and tree volume (Lavigne et al., 1996). Growth-related respiration should increase with increasing availability of nitrogen, mainly because growth rates increase as the supply of nitrogen increases (Maier, 2001). Interestingly, increases in ozone concentrations in the atmosphere may either lead to increases in respiration rates (due to increases in metabolic activity in order to repair injuries) or decreases in respiration rates (due to decreases in growth due to injury) (Kellomäki and Wang, 1998).

9.5 Tree growth

Trees essentially grow upward and outward through primary and secondary growth processes. Therefore, every growth period can be described as a coordinated longitudinal and radial expansion of woody tissue (Coder, 1999). Growth periods can be initiated by transitions out of dormancy, changes in temperature, and precipitation events. With each period of growth, a new sheath of woody material is produced as a layer covering the previous collection of woody material. Growth occurs in the buds (shoot tips), root tips, and the cambium portion of a tree (Coder, 1999). New wood cells (xylem) are formed on the inside of the cambium, and new bark cells (phloem) are formed on the outside, and no growth in diameter or length of cells continues after they are initially formed. In other words, an increase in the size of a tree is due to the addition of new wood cells, not to enlargement or elongation of old cells (Forest Products Laboratory, 1987).

The *sapwood* of a tree is the portion of the xylem between the cambium and heartwood that contains the majority of live-cell volume in a tree stem. The sapwood contributes to water and sap transport processes, and food is also stored in this region. In general, the more vigorously growing tree species will have wider regions of sapwood (Forest Products Laboratory, 1987). Cells in the *heartwood* area of a tree, which ranges from the edge of the sapwood to the pith, are inactive and do not function in water transport or food storage and are limited to support of the tree. Extractives are stored in the heartwood, which results in darkened wood in some tree species. The heartwood of some tree species (e.g., western red cedar [*Thuja plicata*]) is also resistant to decay, but this is not a universal condition of tree heartwood. However, the sapwood of all tree species is susceptible to decay. Some species, such as white oak (*Quercus alba*), have heartwood that is plugged with ingrowths called tyloses. The resulting tightly plugged cell structure prevents the passage of liquids through the cell pores. From a forest products point of view, this condition makes the wood favorable for the production of tight cooperage (e.g., barrels, kegs, and vats) (Forest Products Laboratory, 1987).

The cell division, expansion, and differentiation processes of a tree influence its stature (overall size) and form (architecture). The meristematic control of seasonal growth of bud set and bud flush, along with cambial growth initiation and cessation, also influence tree growth processes. Other traits of a tree that influence its growth capacity include its water use efficiency, the carbon storage and allocation processes employed (above and below ground), and its pathogen and insect stress resistance capabilities. Of course, growth rates are important; the growth rate of the shoot meristem influences height growth and the growth rate of the vascular cambium influences diameter growth (or girth). The condition of the root system is also an important influence, as the rate of movement of nutrients and water from the neighboring soil resource into the tree will ultimately affect tree growth (Grattapaglia et al., 2009).

9.6 Tree roots

Tree root systems are the primary link between the soil and water resources of the land and the water and nutrient requirements of the tree bole, limbs, and foliage. Root systems generally form symbiotic relationships with mycorrhiza-forming fungi or nitrogen-fixing bacteria, or actinomycetes. These relationships increase root system efficiency (Lanner, 2002). The outer layer of a root is an epidermis (Fig. 9.12), which may contain root hairs (tubular outgrowths) that extend the surface area of the root and therefore increase the amount of contact with soil and water resources. The epidermal surface near the tip of a root is where water enters the tree (Pickard, 1981). Below the epidermal layer lie a few layers of cells collectively known as the cortex. This is where starches (foods) are stored, and it is bound on the inside of the root by the endodermis. The endodermis is a thin layer of tightly packed cells that act as an impermeable barrier or water repellant layer (the Casparian strip) through which water can move only one way, toward the center of the root system. Therefore, a tree root system is a conduit for water to be transported from the soil into the bole of the plant (Bejan et al., 2008).

The portion of the soil profile in which plant roots are found is called the *rooting zone* (Ford-Robertson, 1971), which on average is just 1 to 2 meters (m; 3.3—6.6 feet [ft]) below the soil surface. However, in some tree species and in some conditions, the rooting depth may vary widely and extend 20 to 30 m (65.6—98.4 ft) or more below the surface. For example, the maximum depth of ponderosa pine (*Pinus ponderosa*) root systems has been estimated to be between 1 and 24 m (Stone and Kalisz, 1991). The root system can thus be considered a sink or storage location for carbon acquired through photosynthesis (Tasser and Tappeiner, 2005). The amount of root biomass of a tree is positively correlated with the availability of water and nitrogen in the soil; however, a negative correlation between root biomass and soil water was recently observed in South African savannas, which are tropical summer rainfall ecosystems (February and Higgins, 2010). Further, a limited rooting zone can reduce sap flow within a tree when soil moisture levels decrease due to, for example, droughts (Gartner et al., 2009).

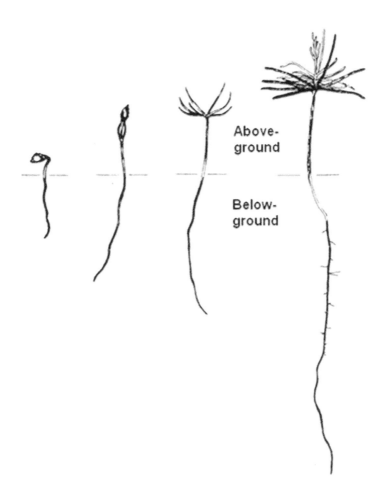

Above-
ground

Below-
ground

FIGURE 9.12 Douglas-fir seedling development, 2, 5, 8, and 22 days after emergence. *Adapted from Owsten and Stein (1974).*

The formation and establishment of tree root systems depends largely on the soil in which the trees have germinated or in which they have been planted (Puhe, 2003). The root cap is responsible for determining the direction of growth, and it is responsive to both stimuli and irritation. Although a number of factors affect the growth direction of tree roots, they generally grow in a direction consistent with the direction that they emerge from a parent root. However, tree root growth is affected by the local environment. Roots can become deflected by barriers, stunted by obstacles, affected by anoxia, or impeded by various soil layers. Factors that impede deep rooting of trees may therefore increase the vulnerability of trees to windthrow (Puhe, 2003). The taproot is the main descending root of a tree, although not all trees have one. A taproot can assist in the mechanical stability of a tree and in development of a straight tree bole, although trees without taproots can also develop straight tree boles (Krause and Plourde, 2008). A taproot can facilitate the exploitation of resources that are deep underground (Doi et al., 2008). Lateral roots that perhaps extend outward from a taproot actually stabilize trees and act as anchors; on

average, they extend about 10 to 15 m (32.8–49.2 ft) from a tree bole but may reach as far as 30 to 60 m (98.4–196.9 ft) and are important actors in water and nutrient acquisition processes (Stone and Kalisz, 1991).

The root density of a tree is a function of the underlying soil characteristics, environmental variables, and tree competition, and is affected by compaction created by heavy machinery. For example, radiata pine (*Pinus radiata*) root systems were shown to be affected by the competition imposed from nearby Australian oak (*Eucalyptus obliqua*) root systems in an Australian study (Bi et al., 1992). A root system is an extensive array of woody material underneath the soil. As an example, the density of roots in jack pine (*Pinus banksiana*) stands in Alberta was estimated to be about 47,000 root pieces per square meter (m^2; about 4400 per square feet [ft^2]) of soil, and nearly 52,000 per m^2 (about 4800 per ft^2) for an aspen (*Populus tremuloides*) forest and about 28,000 per m^2 (about 2600 per ft^2) for a white spruce (*Picea glauca*) forest (Strong and La Roi, 1985). The linear density of fine roots has been shown to range from about 8 centimeters (cm) of root per cubic centimeter (cm^3) of

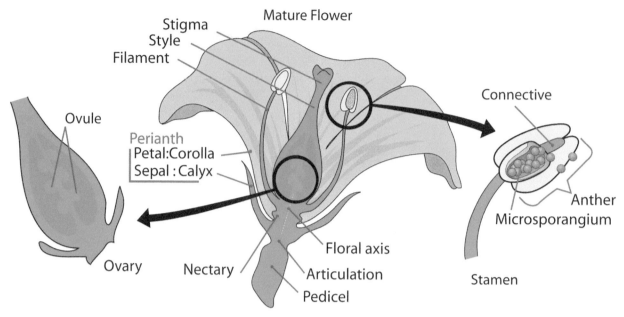

FIGURE 9.13 Single carpel female reproductive organ of an angiosperm. *Mariana Ruiz, through Wikimedia Commons. Image Link: https://commons.wikimedia.org/wiki/File:Mature_flower_diagram-es.svg License Link: Public Domain*

soil (about 4.3 ft per cubic inch [in³]) in pin oak (*Quercus palustris*) plantations in the midwestern United States (Watson and Kelsey, 2006), but only about 1 cm of root per cm³ of soil (about 0.5 ft per in³) in radiata pine forests in Australia (Bi et al., 1992).

Interestingly, although we typically imagine root systems as being located within soils and under the ground surface, roots systems have also been found in the canopies of tropical and temperate trees (Nadkarni, 1981; Sanford, 1987). These types of root systems can occur through the formation of adventitious roots arising from damage or injury to above-ground tree organs or as an adaptation to limited soil resources. The contribution of these to the total fine root system of a tree and to its function is probably negligible, but much more needs to be understood about their role in forest dynamics. Coincidentally, a rediscovery of canopy roots in windthrown, old-growth European beech (*Fagus sylvatica*) located in central Germany was recently reported (Hertel, 2011). The root systems were noticed in humus pockets that had accumulated in the forks of major branches of tree bole in the central part of a tree crown. The chemical and morphological traits of these above-ground root systems are different from those of below-ground root systems, perhaps due to differences in water availability (which is greater above ground than below ground).

9.7 Tree regeneration

Tree regeneration is a process whereby plants maintain and perhaps expand their population. Regeneration can occur through the creation of seeds or the liberation of vegetative shoots. The life of a seed is best expressed as a complex series of events that begins with the initiation of a flower, proceeds through stages of seed development, ripening, and dispersal, and ends with seed germination. Flower initiation, rate, and timing vary with each tree species. For example, some trees flower in the early spring (e.g., cherry) while others flower in late summer (e.g., pine). Most trees native to temperate forests flower only once per year, while some native tropical forest trees flower several times per year (Krugman et al., 1974).

Seed-producing plants (spermatophyta) are divided into two botanical groups: angiosperms and gymnosperms. A number of characteristics are used to describe each group, but the main distinction is that angiosperms (maple, oak, willow, etc.) have seeds enclosed in a carpel, the female reproductive organ of the flower (Fig. 9.13), while gymnosperms (e.g., fir, pine, and spruce) have seeds that are not enclosed in a carpel (Fig. 9.14), and are generally spirally arranged along a central axis to form a cone (Krugman et al., 1974). Gymnosperms are therefore said to have *naked* seeds. Of course, there are some exceptions to these general rules. For example, one coniferous gymnosperm (yew; *Taxus brevifolia*) produces seeds that are not located within cones but that are enclosed within a berry-like, cup-shaped disk called an aril (Bolsinger and Jaramillo, 1990).

Although flowers are characteristic reproductive units of trees, biologists cannot agree on exactly what they consist of (Melzer et al., 2010). Flowers form when a vegetative meristem of a tree changes its pattern

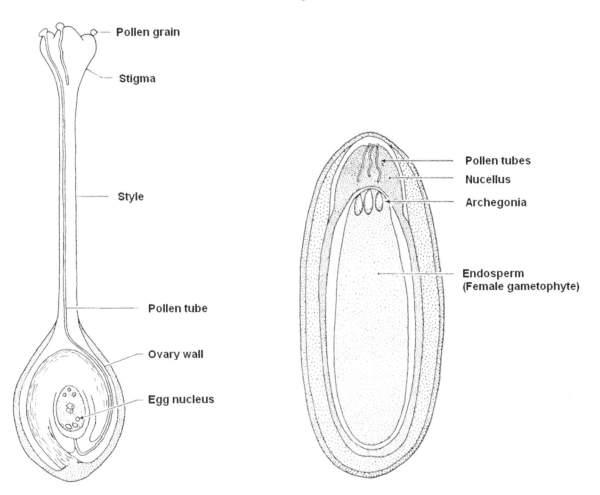

FIGURE 9.14   Longitudinal sections through the pistil of an angiosperm (left) and the ovule of a gymnosperm (right). *Adapted from Krugman et al. (1974).*

of cell development and begins to form reproductive organs (Krugman et al., 1974). Tree flowers vary in their arrangement, color, components, odor, size, and shape. A typical angiosperm flower may have a stalk (peduncle), a calyx (composed of sepals), a corolla (composed of petals), stamens (with anthers and filaments), and one or more pistils (containing a stigma, style, and ovary). A bisexual, or perfect, flower has both a stamen and a pistil, while a unisexual, or imperfect, flower has only one of these. The calyx and corolla are designed to protect the delicate stamen and pistil structures and their color or odor may attract the necessary pollinators to ensure transfer of pollen to the ovary. Embryo sac formation is facilitated by meiosis in the megasporangia. In angiosperms, pollen is formed through meiosis in the microsporangia, or anthers, of male flowers. A typical gymnosperm flower consists of a small cone (or strobilus) that lacks the calyx, corolla, stamen, and pistil structures of angiosperm flowers. In gymnosperms, pollen is generally formed through meiosis in pollen sacs found beneath each staminate

cone scale (Krugman et al., 1974). Some species of trees, such as quaking aspen (*Populus tremuloides*), may only produce unisexual male or female reproductive organs on different individual trees; thus, the trees may be considered either male or female. This is called a *dioecious* condition; however, for a given tree species, a certain small percentage of trees may actually produce flowers that contain both male and female organs. If the male and female reproductive organs can be found on the same plant, they are called *monoecious* plants. Interestingly, most gymnosperms are monoecious.

A gametophyte is the phase in the life cycle of a tree in which sperm and eggs are produced. In general, male and female gametophytes in both angiosperms and gymnosperms are only a few cells in size, and each cell contains a single set of chromosomes. Gametophytes produce male and female gametes through cell division (mitosis). The combination of male and female gametes (at fertilization) produces a multicellular zygote containing two sets of chromosomes. The gamete-bearing phase of reproduction in gymnosperms is relatively short.

Pollen grains produce sperm cells and then pollen is transported to unfertilized seeds of the same species. Once inside the ovule, fertilization (or ovule development) occurs, and the resulting zygote develops into an embryo and eventually a mature seed. Flowers of angiosperms produce both microspores, which become pollen grains, and megaspores, which become the egg cells contained in the ovules. Pollen grains are the male cells situated on the stamen of the flower and are produced by the anther. Tree pollen is dispersed mainly by wind and insects, since the source and the desired destination of pollen are usually in close proximity. Ideally, pollen finds its way to the carpel, or pistil, of a flower, which contains egg cells. Gymnosperm reproductive cones are either male or female organs and, as we noted, the ovules are not enclosed in a carpel (Melzer et al., 2010). Gymnosperms typically develop a temporary herbaceous male cone that produces and releases pollen and a more permanent woody female cone containing the ovules. For species of pines (e.g., *Pinus* spp.), the latter are called *pine cones* (Fig. 9.15). Tree pollen is again dispersed mainly by the wind and insects, although, since the source and the desired destination of pollen are not as close in proximity, as in angiosperm flowers, birds and other mammals may also facilitate transport of the pollen. In addition to pines, various species of fir, redwood (*Sequoia sempervirens*), spruce, and other coniferous trees with hardened cones are representative of this group.

Seed germination includes events leading to the emergence of the embryo and its subsequent development to the point where it becomes independent of its food reserves. If a seed does not germinate shortly after dispersal from the tree, it may remain in a quiescent state (i.e., a persistent state of viability). Food reserves in seeds are mainly lipids (fats) and carbohydrates, and the concentrations of each vary by tree species. The events leading to the emergence of an embryo include the absorption of water and subsequent swelling of the seed coat; enzymatic activity that signals the use of stored food reserves; respiration and assimilation of nutrients; and cell enlargement. The latter events represent the plant growth that we typically observe, including the growth of the root system and the plumule (lower hypocotyl stem, upper epicotyl stem, and leaves). After the roots begin to establish themselves (Fig. 9.16), either a hypocotyl emerges upward with cotyledons attached (typical of epigeal germination) or the cotyledons remain below the ground surface (typical of hypogeal germination) (Krugman et al., 1974).

Seed production strategies can be sexual or asexual. Sexual processes, in general, involve male sperm and female egg cells that unite to form a zygote (Barnes et al., 1998). Typically, when viable seeds have been developed, they are dispersed from the tree, germinate in the soil or litter layer on the forest floor, and become established as young tree seedlings. While asexual reproduction processes are important in the development of fast-growing forest plantations, sexual reproduction of tree species seems to be essential for the natural survival, as low gene flow between plant populations can cause inbreeding or poor adaptation

FIGURE 9.15 Loblolly pine (*Pinus taeda*) cones, Starkville, Mississippi, United States. *H. Alexis Londo.*

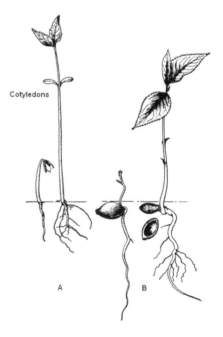

FIGURE 9.16 Epigeal germination (A) of pin cherry (*Prunus pensylvanica*) and hypogeal germination (B) of Allegheny plum (*Prunus alleghaniensis*). *Adapted from Krugman et al. (1974).*

FIGURE 9.17 Seeds (acorns) of willow oak (*Quercus phellos*). *U.S. Department of Agriculture, through Wikimedia Commons.*
Image Link: https://commons.wikimedia.org/wiki/File:Quercus_phellos_acorns.jpg
License Link: Public Domain

FIGURE 9.18 Beech seedlings that have become established through seed germination on an old spruce trunk. *Don Manfredo, through Wikimedia Commons.*
Image Link: https://commons.wikimedia.org/wiki/File:KadVerjuengung.JPG
License Link: https://creativecommons.org/licenses/by-sa/3.0/deed.en

to a changing environment (Rasmussen and Kollmann, 2004). In a natural environment, core populations of tree species generally exhibit higher levels of genetic diversity than do peripheral populations, although there are many exceptions to this rule (Pautasso, 2009). In sexual reproduction of trees, the distance of seed dispersal from a parent tree can be extensive or limited (i.e., locally or widely dispersed), based on seed weight and buoyancy in the air (Barnes et al., 1998). The timing of reproduction by trees is controlled by environmental factors such as air temperature and wind speed, as both pollination and dispersal of seed are affected by weather conditions. Some seeds, for example, are designed to float on wind currents across the landscape. Other seeds, such as those of eastern red cedar (*Juniperus virginiana*), may be digested by birds and then transported considerable distances from the parent tree before being deposited on the ground. Some seeds are relatively heavy (Fig. 9.17) or encased in heavy fruit and, therefore, once detached from the parent tree, may be dispersed only very short distances. The mechanism used for seed dispersal is one of the traits of a tree species that enables it to compete successfully in its ecological niche (Barnes et al., 1998).

Seed can become available through an *active* seed bank (i.e., viable for about a year) or a *dormant* or *persistent* seed bank (i.e., viable for more than a year). Seed from some tree species can remain viable in the soil for many years after dispersal; however, the role of a persistent seed bank varies, and germination of seed may only occur in the soil (for persistent seed) or in the litter layer (for new seed) of a forest floor (Deiller et al., 2003), or on hosts such as logs or stumps (Fig. 9.18). Tree species that regenerate through seed in areas where disturbances such as fire or clearcuts have occurred are typically called *pioneer species*. Tree species that become established through seed under an existing forest canopy and work their way into the canopy when opportunities become available are typically called *gap-phase species*. Seedlings of other tree species may become established through seed and persist in the understory of a forest canopy for many years before opportunities to grow into the canopy become available (Barnes et al., 1998).

Trees must emerge from a juvenile life phase before they are capable of producing seed. They must also obtain a certain size and structure (determined by their species) before they are able to reproduce, and the amount of time required to reach this condition therefore varies by species. Some trees, such as jack pine and lodgepole pine (*Pinus contorta*), may be able to reproduce before they are 10 years old, while other pines may require 20 to 30 years to mature. The middle ages of the typical lifespan of a tree species are generally the most productive from a seed generation perspective. Seed production will vary on an annual basis and environmental conditions may delay flowering (Krugman et al., 1974). Sexual reproduction processes can be affected by both biotic (e.g., insect pollination) and abiotic (e.g., climate and soil conditions) factors (Rasmussen and Kollmann, 2004). For example, the volume of annual crops of oak (*Quercus* spp.) seeds (or acorns) is usually not constant from one year to the next. Further, Ayari et al. (2011) suggest that the seed

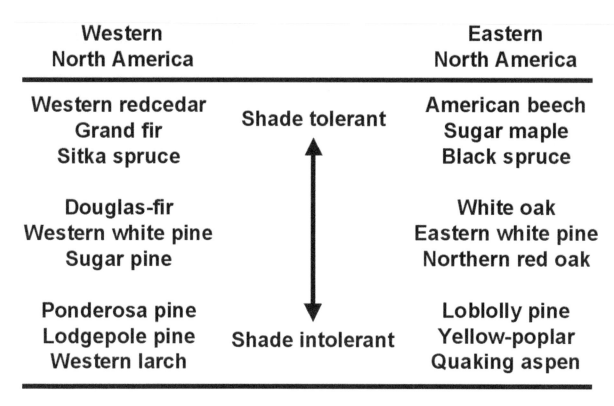

FIGURE 9.19 Simple continuum of a few shade-tolerant trees of North America.

production of tree species such as Aleppo pine (*Pinus halepensis*), which grows naturally in the Mediterranean basin, can be affected by geographical factors such as latitude, longitude, and elevation. Low air temperatures may be necessary to condition a tree to respond normally to flower initiation processes and adequate levels of photosynthesis may be necessary for flower initiation and development. In addition, the quality of the underlying soil is often positively correlated with seed production (Krugman et al., 1974). Therefore, flowering may be restricted in certain forest conditions and in areas with low soil quality.

Vegetative reproduction processes are typically considered to be asexual reproduction processes. These include cloning processes and the development of vegetative shoots from existing plants. A cloning process is usually conducted in a laboratory or tree nursery. Here, genetically identical stems are derived from a plant and each becomes a ramet (an independent member) of the original plant (Barnes et al., 1998). Ramets are allowed to grow a root system and are then planted into the ground during a forest regeneration process. For some commercially important coniferous species, cloning and rooted cutting regeneration processes have been devised (Bettinger et al., 2009). In the forest, vegetative shoots can also form in the roots and boles of many deciduous trees after damage or after harvesting

activities. Regeneration may occur as a result of root shoots, or meristems, and this process may facilitate rapid development of clonal colonies after the removal of an existing forest through disturbance (e.g., wildfire) or harvest activity (Stone and Kalisz, 1991). One form of reproduction through stump sprouts (coppice) is often used to regenerate deciduous forests. Dispersal of genetic material is severely limited in this case; however, the regenerated seedlings may be less vulnerable to stresses than those created from seed due to the soil, water, and nutrient resources that become available through the established root and stem systems of a larger plant (Deiller et al., 2003). All angiosperms have the capability to reproduce vegetatively once they become established in a forest, but most coniferous trees (gymnosperms) are unable to naturally reproduce in this manner. Redwood is an exception; it is a coniferous tree species with the ability to sprout from stumps during any season of the year through the use of dormant or adventitious buds (Olson et al., 1990).

9.8 Tree tolerance to shade

The future status of a live tree is influenced by many natural and anthropogenic factors. In natural settings, two of the main factors of influence are the current canopy position of the tree and its ability to tolerate levels of light suboptimal for tree growth (Ward and Stephens, 1993). Trees require electromagnetic energy (from light) in order to grow and survive, yet many species have adapted to conditions where the availability of this resource is low; thus, to some extent they are tolerant of low levels of light. The term *shade tolerance*, therefore, refers to the ability of a tree to continue to become established, survive, and thrive in shaded conditions, where the availability of direct electromagnetic energy is low. Shade tolerance can also refer to the ability of a species of tree to continuously and successfully compete with other trees for resources and to regenerate under a contiguous canopy. Alternatively, shade tolerance may represent the ability of a tree species to persist in a low-resource environment (Ward and Stephens, 1993). Tree species that can compete well in fully shaded conditions are *shade tolerant*, while those that require full sunlight and limited competition are *shade intolerant*. Between these two extremes are trees that are *intermediate* in their ability to develop and compete for resources in shaded conditions (Martin and Gower, 1996).

While we tend to describe the ability of trees to tolerate shade on a simple continuum (Fig. 9.19), in reality a number of factors are important for determining whether a tree will survive and grow under partial sunlight conditions. In nature, low-light conditions suggest that shade is being created by other tree or plant species, but the concept of shade tolerance in trees could be extended to conditions where shade is created by man-made structures. Shade-tolerant tree species do not require full sunlight for their survival. However, some of these tree species will grow better in partially shaded conditions than when completely exposed to the sun. Other factors, such as the age of a tree, the quality of the site where the tree is growing, and even the region it is growing in may influence its tolerance to shaded conditions. For example, an eastern white pine tree may be more tolerant to shade when it is younger, while a black spruce (*Picea mariana*) tree may acquire shade tolerance as it ages (Martin and Gower, 1996).

Shade-tolerant trees include sugar maple (*Acer saccharum*), basswood (*Tilia americana*), black spruce, eastern hemlock (*Tsuga canadensis*), European beech, American beech (*Fagus grandifolia*), redwood, grand fir (*Abies grandis*) (Fig. 9.20), Sitka spruce (*Picea sitchensis*), and other commercially important tree species with crowns in the upper reaches of a canopy, along with big-leaf maple (*Acer macrophyllum*), dogwood (*Cornus florida*), holly (*Ilex opaca*), yew, and other tree species with crowns typically found below the main tree canopy. Shade-intolerant trees include commercially important species such as hickories (*Carya* spp.), a number of pines, e.g., loblolly pine (*Pinus taeda*), quaking aspen, black cherry (*Prunus serotina*), black walnut (*Juglans nigra*), paper birch (*Betula*

FIGURE 9.20 Grand fir regeneration in shaded conditions, Idaho, United States. *Chris Schnepf, University of Idaho, Bugwood.org.*

papyrifera), and western larch (*Larix occidentalis*), among others, along with sassafras (*Sassafras albidum*) and willows (*Salix* spp.). The intermediate shade tolerance class includes important commercial tree species such as the ashes (*Fraxinus* spp.), many oaks and elms (*Ulmus* spp.), Douglas-fir (*Pseudotsuga menziesii*), eastern white pine, western white pine (*Pinus monticola*), and sugar pine (*Pinus lambertiana*), among others.

Two key determinants in forest stand structure and forest dynamics are tolerance of trees to shade and competition for limited light resources (Zavala et al., 2007). If we were to imagine the natural establishment (primary succession) of a forest after a harvest, fire, or wind-driven disturbance, colonization by shade-intolerant tree species would lead to the diameter distribution of trees eventually becoming a normal (bell-shaped) or flattened form of a bimodal distribution representative of an even-aged forest (Fig. 9.21). In this case, most of the trees would be about the same relative size (diameter and height) and there would be very few smaller trees because these would not be able to survive in limited light conditions. If a forest were colonized mainly by shade-tolerant tree species, then the diameter distribution of trees would eventually have a downward-sloping, reverse J-shape representative of uneven-aged forest conditions. Here, the establishment and growth of smaller, younger trees would continue through time as conditions allow. In forests with trees of mixed tolerance to shade, secondary successional processes would favor the decline of shade-intolerant trees or trees of intermediate tolerance

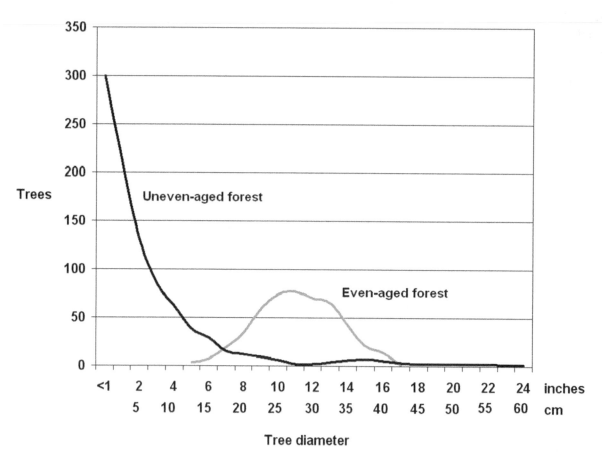

FIGURE 9.21   Diameter distributions representative of even-aged and uneven-aged forests.

due to mortality and a decrease in recruitment (lower regeneration success). In addition, when shade-tolerant trees share a canopy with other tree groups, the sensitivity to shading and the resulting stress and reduced growth rates exhibited by trees with intermediate tolerance and shade-intolerant trees generally results in a decline in their numbers per unit area (Yoshida and Kamitani, 1999). As a result, natural succession processes are strongly related to the shade tolerance levels of trees occupying a site. Shade-intolerant tree species may occupy a site first after a disturbance creates an opportunity for tree establishment, but they are likely to be followed and replaced by intermediate and shade-tolerant tree species as time progresses (Martin and Gower, 1996).

Some silvicultural systems take advantage of variations in shade tolerance. For example, in Germany, the long-rotation management of two shade-intolerant tree species, pedunculate oak (*Quercus robur*) and sessile oak (*Quercus petraea*), may require the incorporation of a shade-tolerant species to suppress the development of epicormic branches. The most frequently recommended admixture species for this purpose is European beech, although European hornbeam (*Carpinus betulus*)

has also been used on high-quality sites (von Lüpke, 1998). In the management of central United States hardwood forests, some oak species are more shade intolerant than others. The presence of oak seedlings in a forest understory is important for the natural regeneration of these forests, since this condition favors the successful reestablishment of an area with this group of tree species (Larson and Johnson, 1998).

9.9 Tree nutrition

Plant development and biomass production depend on the availability of nutrients, such as inorganic nitrogen, in the soil. When considering the growth of trees, nitrogen is usually a limiting factor; thus, nitrogen assimilation and metabolism by trees is important. For wood supply concerns, the manipulation of nitrogen metabolism in trees may impact their wood production potential (Gallardo et al., 2003). The ability to use nutrients contained in soils is likely to vary by the species, age, the current nutritional status, and the physiological development of the tree. For example, Table 9.1 illustrates the distribution of nutrients such as nitrogen,

TABLE 9.1 Distribution of nutrients (kg/hectare) by tree component for a 16-year-old loblolly pine (*Pinus taeda*) plantation.

Nutrient	Needles	Live branches	Dead branches	Tree bole	Bark	Roots
Nitrogen	82	34	26	79	36	64
Phosphorus	10	5	2	11	4	17
Potassium	48	24	4	65	24	61
Calcium	17	28	30	74	38	52
Magnesium	8	6	3	23	7	22

Wells et al. (1975).

phosphorous, potassium, calcium, and magnesium by tree component for a 16-year-old loblolly pine plantation (Wells et al., 1975). Of the five nutrients listed, nitrogen is the most commonly found nutrient throughout a pine tree, particularly in the needles, tree bole, and root system. Calcium is the next most abundant nutrient, with the largest concentrations of this nutrient generally being found in the tree bole, bark, and root system. Substantial amounts of potassium are also prevalent throughout a pine tree, and the largest potassium concentrations are located in the tree bole, root system, and needles. Phosphorous and magnesium are also mainly found in the tree bole and root system of a pine tree. Deficiencies in one or more of these main elements can adversely affect plant growth (Kramer and Kozlowski, 1979; Taiz et al., 2015). For example, nitrogen is essential for the development of amino acids and lack of nitrogen can lead to a failure to produce chlorophyll and cytochromes, which can result in leaf chlorosis (chlorotic leaves are yellow or pale in color). A deficiency in magnesium can also lead to leaf chlorosis. A deficiency of phosphorous can lead to the stunting of height growth in a tree and a potassium deficiency can cause problems in water vapor transfer and the carbon dioxide absorption necessary for photosynthesis (Kramer and Kozlowski, 1979; Taiz et al., 2015). Further, a deficiency of calcium in a tree can affect cell wall and root tip development.

Deficiencies in other essential micronutrients, including boron, copper, iron, manganese, molybdenum, and zinc, can also adversely affect plant growth (Kramer and Kozlowski, 1979; Taiz et al., 2015). For example, boron deficiency can affect the development of apical buds (affecting height growth), root tips, and mycorrhiza, which can also affect the cycling of other nutrients such as carbon (Lehto et al., 2010; Taiz et al., 2015). Manganese and zinc deficiencies can lead to leaf malformation, copper deficiency can affect the function of plant enzymes, and molybdenum deficiency can affect nitrogen fixation by leguminous trees. Aside from other factors that may influence their growth and survival, trees and plants maintaining an appropriate balance of these nutrients are capable of acceptable growth rates and successful reproduction.

9.10 Tree responses to signals

Broadly speaking, signals within plants, such as changes in hormone levels or water pressure, help initiate the formation of stems, leaves (and needles), flowers, and fruit, and have an important influence on tree growth. For example, communication of water deficit information between roots and shoots is thought to occur through hydraulic signals (Chavez et al., 2011). Further, changes in hormone concentrations seem to correlate with changes in metabolic activity in the growth of plant tissue around seeds (Krugman et al., 1974). For both angiosperms and gymnosperms, chemical changes that occur during seed germination are essentially the same. During germination, stored food material in the cotyledons is mobilized and transferred to the embryo, and oxygen, water, and minerals are acquired. The products of metabolic activity provide the energy and carbon fragments required for the development of new plant material. In this process, starches, sugars, and fats are converted to soluble sugars; some proteins are broken down to form amino acids; and cellular growth is facilitated with the assistance of ribonucleic acid (RNA) and deoxyribonucleic acid (DNA). RNA and DNA are long chains or sequences of nucleotides that encode genetic information about a plant or animal. Some RNA and proteins actively sense and communicate cellular-level responses, particularly those involved in the construction of other proteins.

Although root processes, such as carbon storage and soil water exploration capabilities, may be important in determining drought resistance in trees (Klein, 2020), root-to-shoot communication can play a vital part in a tree's defense against drought (Mansfield, 1998). Partial stomatal closure and reduced rates of expansion of new leaves have been observed when water availability in the soil is reduced, long before the leaves become stressed due to reduced water availability in the plant. It is suggested that the sensing of dry soil around the root system and the communication of information to the shoot are facilitated in part by the plant hormone (or phytohormone), abscisic acid. Thus, stomatal responses to drought and salinity are thought to be controlled through a signal provided by changes in levels of abscisic acid (Chavez et al., 2011). In addition, auxins seem to be one of the main classes of plant hormones associated with the regulation of cambial activity (Lachaud et al., 1999). During growth periods, cambial cells must be able to receive hormonal signals, nutrients, and various metabolites. These activities may be facilitated by changes in the pH gradient within and between cells. Auxins, cytokinins, and gibberellins are all classes of hormones that are also influential during the reproduction phases of a plant. The timing of plant flowering, for example, may be manipulated by the external application of gibberellins (Krugman et al., 1974). While these are only a few examples, one should recognize that signals within plants (e.g., changes in hormone levels) can influence the onset of many metabolic processes and the development of stems, leaves (and needles), flowers, and fruit.

Summary

Trees situated in a forested environment may be competing with other trees and plants for nutrients, light, and water. Trees are complex organisms that absorb energy, carbon dioxide, water, and nutrients in their quest to survive. They transpire, respire, and flower, and may eventually die if they are not successful competitors. The role of a tree in the ecological framework of a given forest varies depending on the health of the tree. The tree growth is generally directed upward and outward, and growth mechanisms of tree cells are important to understand, since photosynthesis not only allows trees to survive and thrive, but the process also provides oxygen for us to breathe. Tree rooting behavior is also important to understand, since it facilitates the uptake of nutrients from the soil, and helps trees withstand severe wind events. Reproductive processes vary between angiosperms and gymnosperms, and simply understanding the timing of pollen release and flower production may help one to estimate the extent of potential cone or fruit crops. Tree shade tolerance is an important issue in the management of forests, since advanced reproduction of a set of tree species may be necessary; further, interactions among tree species and their responses to available resources along environmental gradients may influence management decisions. Although many of the topics discussed in this chapter seem to have delved deeply into the technical processes of tree anatomy and physiology, we have barely scratched the surface of the science of forest biology. The processes by which trees live and grow are essential biological concepts for forest and natural resource managers. This knowledge forms an essential toolkit for a professional and we encourage you to explore further topics that have piqued your interest as silviculture uses this understanding to manipulate stand growth to achieve the landowner's objectives.

Questions

(1) *Tree root development.* During a recent family gathering, a distant family member described how a large oak tree in their yard fell during a recent windstorm, pulling with it much of the root system. They noticed that the tree had no taproot and wondered why. In a short paragraph, describe how you would contribute to the conversation with information regarding the depth and radius of root systems, as well as the presence of taproots and other types of rooting systems.

(2) *Plant signals.* In a short paragraph of five sentences or less, describe some of the mechanisms within plants that act as signals to change plant development or functional behavior.

(3) *Seed development.* Choose a tree species that is of interest to you. Develop a short, five-slide PowerPoint presentation that describes the salient aspects of plant flowering and seed development of this species. For readers in North America, the Silvics of North America (http://www.na.fs.fed.us/spfo/pubs/silvics_manual/table_of_contents.htm) might be of value.

(4) *Material transport.* In a short paragraph, describe the different functions of xylem and phloem, and how they might contribute to material transport within a tree. Through a search of the Internet, locate examples that refer to specific tree species and locate information not provided in this chapter. Share your findings with a small group of others and discuss the similarities and differences of your collective findings.

(5) *Photosynthesis.* The photosynthetic process is intimately related to a hot topic of natural resource management, global climate change. In a one-page memorandum to your instructor, describe the main inputs and outputs of the photosynthetic process, and how these relate to the climate change issue.

References

Anekonda, T.S., Criddle, R.S., Bacca, M.J., Hansen, L.D., 2000. Influence of age on dark respiration in eucalypts. Thermochimica Acta 343, 11–17.

Ayari, A., Moya, D., Rejeb, M.N., Mansoura, A.B., Albouchi, A., De Las Heras, J., Fezzani, T., Henchi, B., 2011. Geographical variation on cone and seed production of natural *Pinus halepensis* Mill. forests in Tunisia. Journal of Arid Environments 75, 403–410.

Barnes, B.V., Zak, D.R., Denton, S.R., Spurr, S.H., 1998. Forest Ecology, fourth ed. John Wiley & Sons, Inc., New York, 774 p.

Bejan, A., Lorente, S., Lee, J., 2008. Unifying constructal theory of tree roots, canopies and forests. Journal of Theoretical Biology 254, 529–540.

Bending, G.D., Poole, E.J., Whipps, J.M., Read, D.J., 2002. Characterisation of bacteria from *Pinus sylvestris - Suillus luteus* mycorrhizas and their effects on root-fungus interactions and plant growth. FEMS Microbiology Ecology 39, 219–227.

Bettinger, P., Clutter, M., Siry, J., Kane, M., Pait, J., 2009. Broad implications of southern United States pine clonal forestry on planning and management of forests. The International Forestry Review 11 (3), 331–345.

Bi, H., Turvey, N.D., Heinrich, P., 1992. Rooting density and tree size of *Pinus radiata* (D. Don) in response to competition from *Eucalyptus obliqua* (L'Herit). Forest Ecology and Management 49, 31–42.

Bolsinger, C.L., Jaramillo, A.E., 1990. Pacific yew. In: Conifers, Burns, R.M., Honkala, B.H. (technical coordinators), Silvics of North America, vol. 1. U.S. Department of Agriculture, Forest Service, Washington, D.C. Agriculture Handbook 654.

Čermák, J., Cienciala, E., Kučera, J., Lindroth, A., Bednářová, E., 1995. Individual variation of sap-flow rate in large pine and spruce trees and stand transpiration: a pilot study at the central NOPEX site. Journal of Hydrology 168, 17–27.

Chavez, M.M., Costa, J.M., Madeira Saibo, N.J., 2011. Recent advances in photosynthesis under drought and salinity. Advances in Botanical Research 57, 49–104.

Cilia, M.L., Jackson, D., 2004. Plasmodesmata form and function. Current Opinion in Cell Biology 16, 500–506.

Coder, K.D., 1999. Secondary Growth Anatomy & Tree Rings. University of Georgia, School of Forest Resources. Extension Publication, Athens, GA. FOR99–19.

David, T.S., Ferreira, M.I., Cohen, S., Pereira, J.S., David, J.S., 2004. Constraints on transpiration from an evergreen oak tree in southern Portugal. Agricultural and Forest Meteorology 122, 193–205.

DeBell, D.S., Singleton, R., Harrington, C.A., Gartner, B.L., 2002. Wood density and fiber length in young *Populus* stems: relation of clone, age, growth rate, and pruning. Wood and Fiber Science 34, 529–539.

Deiller, A.-F., Walter, J.-M.N., Trémolières, M., 2003. Regeneration strategies in a temperate hardwood floodplain forest of the Upper Rhine: sexual versus vegetative reproduction of woody species. Forest Ecology and Management 180, 215–225.

Delzon, S., Loustau, D., 2005. Age-related decline in stand water use: sap flow and transpiration in a pine forest chronosequence. Agricultural and Forest Meteorology 129, 105–119.

Doi, Y., Mori, A.S., Takeda, H., 2008. Conifer establishment and root architectural responses to forest floor heterogeneity in an old-growth subalpine forest in central Japan. Forest Ecology and Management 255, 1472–1478.

Edwards, N.T., Hanson, P.J., 1996. Stem respiration in a closed-canopy upland oak forest. Tree Physiology 16, 433–439.

February, E.C., Higgins, S.I., 2010. The distribution of tree and grass roots in savannas in relation to soil nitrogen and water. South African Journal of Botany 76, 517–523.

Ford-Robertson, F.C. (Ed.), 1971. Terminology of Forest Science Technology Practice and Products. Society of American Foresters, Washington, D.C., 370 p.

Forest Products Laboratory, 1987. Wood Handbook: Wood as an Engineering Material. U.S. Department of Agriculture, Forest Service, Washington, D.C. Agricultural Handbook 72. 466 p.

Gallardo, F., Fu, J., Jing, Z.P., Kirby, E.G., Cánovas, F.M., 2003. Genetic modification of amino acid metabolism in woody plants. Plant Physiology and Biochemistry 41, 587–594.

Gartner, K., Nadezhdina, N., Englisch, M., Čermak, J., Leitgeb, E., 2009. Sap flow of birch and Norway spruce during the European heat and drought in summer 2003. Forest Ecology and Management 258, 590–599.

Gourlay, I.D., Smith, J.P., Barnes, R.D., 1996. Wood production in a natural stand of *Acacia karroo* in Zimbabwe. Forest Ecology and Management 88, 289–295.

Grattapaglia, D., Plomion, C., Kirst, M., Sederoff, R.R., 2009. Genomics of growth traits in forest trees. Current Opinion in Plant Biology 12, 148–156.

Griffiths, T.J., Black, T.A., Gaumont-Gray, D., Drewitt, G.B., Nesic, Z., Barr, A.G., Morgenstern, K., Kljun, N., 2004. Seasonal variation and partitioning of ecosystem respiration in a southern boreal aspen forest. Agriculture and Forest Meteorology 125, 207–223.

Heimann, M.F., Pellitteri, P.J., 2006. Evergreen condition: Seasonal needle drop. University of Wisconsin, Cooperative Extension, Madison, WI. Publication A2614.

Hertel, D., 2011. Tree roots in canopy soils of old European beech trees—an ecological reassessment of a forgotten phenomenon. Pedobiologia 54, 119–125.

Kallio, H., Ahtonen, S., 1987. Seasonal variations of the sugars in birch sap. Food Chemistry 25, 293–304.

Kellomäki, S., Wang, K.-Y., 1998. Growth, respiration and nitrogen content in needles of Scots pine exposed to elevated ozone and carbon dioxide in the field. Environmental Pollution 101, 263–274.

Kim, M.H., Nakane, K., Lee, J.K., Bang, H.S., Na, Y.E., 2007. Stem/branch maintenance respiration of Japanese red pine stand. Forest Ecology and Management 243, 283–290.

Kinnunen, H., Huttunen, S., Laakso, K., 2001. UV-absorbing compounds and waxes of Scots pine needles during a third growing season of supplemental UV-B. Environmental Pollution 112, 215–220.

Klein, T., 2020. A race to the unknown: contemporary research on tree and forest drought resistance, an Israeli perspective. Journal of Arid Environments 172. Article 104045.

Kramer, P.J., Kozlowski, T.T., 1979. Physiology of Woody Plants. Academic Press, Inc., New York, 811 p.

Krause, C., Plourde, P.-Y., 2008. Stem deformation in young plantations of black spruce (*Picea mariana* [Mill.] B.S.P.) and jack pine (*Pinus banksiana* Lamb.) in the boreal forest of Quebec, Canada. Forest Ecology and Management 255, 2213–2224.

Krugman, S.L., Stein, W.I., Schmitt, D.M., 1974. Seed biology. In: Schopmeyer, C.S. (technical coordinator), Seeds of Woody Plants of the United States. U.S. Department of Agriculture, Forest Service, Washington, D.C., pp. 5–40. Agriculture Handbook 450.

Lachaud, S., Catesson, A.-M., Bonnemain, J.-L., 1999. Structure and functions of the vascular cambium. Comptes Rendus de l'Académie des Sciences - Series III - Sciences de la Vie 322, 633–650.

Lanner, R.M., 2002. Why do trees live so long? Ageing Research Reviews 1, 653–671.

Larson, D.R., Johnson, P.S., 1998. Linking the ecology of natural oak regeneration to silviculture. Forest Ecology and Management 106, 1–7.

Lavigne, M.B., Franklin, S.E., Hunt Jr., E.R., 1996. Estimating stem maintenance respiration rates of dissimilar balsam fir stands. Tree Physiology 16, 687–695.

Lehto, T., Ruuhola, T., Dell, B., 2010. Boron in forest trees and forest ecosystems. Forest Ecology and Management 260, 2053–2069.

Levy, P.E., Jarvis, P.G., 1998. Stem CO$_2$ fluxes in two Sahelian shrub species (*Guiera senegalensis* and *Combretum micranthum*). Functional Ecology 12, 107–116.

Maier, C.A., 2001. Stem growth and respiration in loblolly pine plantations differing in soil resource availability. Tree Physiology 21, 1183–1193.

Mansfield, T.A., 1998. Stomata and plant water relations: does air pollution create problems? Environmental Pollution 101, 1–11.

Martin, J., Gower, T., 1996. Tolerance of Tree Species. College of Agriculture and Life Sciences, Department of Forest Ecology and Management, University of Wisconsin, Madison, WI. Forestry Facts No. 79.

Matsumoto, K., Ohta, T., Tanaka, T., 2005. Dependence of stomatal conductance on leaf chlorophyll concentration and meteorological variables. Agricultural and Forest Meteorology 132, 44–57.

Matyssek, R., Bytnerowicz, A., P.-Karlsson, E., Paoletti, E., Sanz, M., Schaub, M., Wieser, G., 2007. Promoting the O$_3$ flux concept for European forest trees. Environmental Pollution 146, 587–607.

Meinzer, F.C., Clearwater, M.J., Goldstein, G., 2001. Water transport in trees: current perspectives, new insights and some controversies. Environmental and Experimental Botany 45, 239–262.

Melzer, R., Y.-Wang, Q., Theien, G., 2010. The naked and the dead: the ABCs of gymnosperm reproduction and the origin of the angiosperm flower. Seminars in Cell and Developmental Biology 21, 118–128.

Méthy, M., Olioso, O., Trabaud, L., 1994. Chlorophyll fluorescence as a tool for management of plant resources. Remote Sensing of Environment 47, 2–9.

Nadkarni, N.M., 1981. Canopy roots: convergent evolution in rainforest nutrient cycles. Science 214, 1023–1024.

Olson Jr., D.F., Roy, D.F., Walters, G.A., 1990. Redwood. In: Conifers, Burns, R.M., Honkala, B.H. (technical coordinators), Silvics of North America, vol. 1. U.S. Department of Agriculture, Forest Service, Washington, D.C. Agriculture Handbook 654.

Owsten, P.W., Stein, W.I., 1974. *Pseudotsuga* Carr. Douglas-fir. In: Schopmeyer, C.S. (technical coordinator). Seeds of Woody Plants of the United States. U.S. Department of Agriculture, Forest Service, Washington, D.C., pp. 674–683. Agricultural Handbook 450.

Pääkkönen, E., Holopainen, T., Kärenlampi, L., 1997. Differences in growth, leaf senescence and injury, and stomatal density in birch (*Betula pendula* Roth.) in relation to ambient levels of ozone in Finland. Environmental Pollution 96, 117–127.

Pautasso, M., 2009. Geographical genetics and the conservation of forest trees. Perspectives in Plant Ecology. Evolution and Systematics 11, 157–189.

Perkins, T.D., van den Berg, A.K., 2009. Maple syrup—production, composition, chemistry, and sensory characteristics. Advances in Food and Nutrition Research 56, 101–143.

Peterson, R.L., Piché, Y., Plenchette, C.C., 1984. Mycorrhizae and their potential use in the agricultural and forestry industries. Biotechnology Advances 2, 101–120.

Pickard, W.F., 1981. The ascent of sap in plants. Progress in Biophysics and Molecular Biology 37, 181–229.

Puhe, J., 2003. Growth and development of the root system of Norway spruce (*Picea abies*) in forest stands—a review. Forest Ecology and Management 175, 253–273.

Rasmussen, K.K., Kollmann, J., 2004. Poor sexual reproduction on the distribution limit of the rare tree *Sorbus torminalis*. Acta Oecologica 25, 211–218.

Robakowski, P., Wyka, T., Samardakiewicz, S., Kierzkowski, D., 2004. Growth, photosynthesis, and needle structure of silver fir (*Abies alba* Mill.) seedlings under different canopies. Forest Ecology and Management 201, 211–227.

Ryan, M.G., Linder, S., Vose, J.M., Hubbard, R.M., 1994. Dark respiration in pines. In: Gholz, H.L., Linder, S., McMurtrie, R.E. (Eds.), Environmental Constraints on the Structure and Productivity of Pine Forest Ecosystems, Ecological Bulletins 43, pp. 50–63.

Saby John, K., Bhat, S.G., Prasada Rao, U.J.S., 2003. Biochemical characterization of sap (latex) of a few Indian mango varieties. Phytochemistry 62, 13–19.

Sanford Jr., R.L., 1987. Apogeotropic roots in an Amazon rain forest. Science 235, 1062–1064.

Schwarze, F.W.M.R., 2007. Wood decay under the microscope. Fungal Biology Reviews 21, 133–170.

Simon, R., 2001. Function of plant shoot meristems. Cell and Developmental Biology 12, 357–362.

Šimpraga, M., Ghimire, R.P., Van Der Straeten, D., Blande, J.D., Kasurinen, A., Sorvari, J., Holopainen, T., Andriaenssens, S., Holopainen, J.K., Kivimäenpää, M., 2019. Unravelling the functions of biogenic volatiles in boreal and temperate forest ecosystems. European Journal of Forest Research 138, 763–787.

Stone, E.L., Kalisz, P.J., 1991. On the maximum extent of tree roots. Forest Ecology and Management 46, 59–102.

Strong, W.L., La Roi, G.H., 1985. Root density—soil relationships in selected boreal forests of central Alberta, Canada. Forest Ecology and Management 12, 233–251.

Taiz, L., Zeiger, E., Moller, I.A., Murphy, A., 2015. Plant Physiology and Development, sixth ed. Sinauer Associates, Inc., , Sunderland, MA, 761 p.

Tasser, E., Tappeiner, U., 2005. New model to predict rooting in diverse plant community compositions. Ecological Modelling 185, 195–211.

Turunen, M., Huttunen, S., 1996. Scots pine needle surfaces on radial transects across the north boreal area of Finnish Lapland and the Kola Peninsula of Russia. Environmental Pollution 93, 175–194.

U.S. Forest Service, 2011. Why Leaves Change Color. U.S. Forest Service, Northeastern Area State and Private Forestry, Newtown Square, PA. http://www.fs.usda.gov/naspf/sites/default/files/naspf/pdf/why_leaves_change_color_508_20170808.pdf (accessed 03.03.20).

von Lüpke, B., 1998. Silvicultural methods of oak regeneration with special respect to shade tolerant mixed species. Forest Ecology and Management 106, 19–26.

Ward, J.S., Stephens, G.R., 1993. Influence of crown class and shade tolerance on individual tree development during deciduous forest succession in Connecticut, USA. Forest Ecology and Management 60, 207–236.

Watson, G.W., Kelsey, P., 2006. The impact of soil compaction on soil aeration and fine root density of *Quercus palustris*. Urban Forestry and Urban Greening 4, 69–74.

Wells, C.G., Jorgensen, J.R., Burnette, C.E., 1975. Biomass and Mineral Elements in a Thinned Loblolly Pine Plantation at Age 16. U.S. Department of Agriculture, Forest Service. Southeastern Forest Experiment Station, Asheville, NC. Research Paper SE-126. 10 pp.

Yoshida, T., Kamitani, T., 1999. Growth of a shade-intolerant tree species, *Phellodendron amurense*, as a component of a mixed-species coppice forest of central Japan. Forest Ecology and Management 113, 57–65.

Zavala, M.A., Angulo, O., de la Parra, R.B., López-Marcos, J.C., 2007. An analytical model of stand dynamics as a function of tree growth, mortality and recruitment: the shade tolerance-stand structure hypothesis revisited. Journal of Theoretical Biology 244, 440–450.

Zuckerman, L.D., Davidson, M.W., 2020. Microscopy of Tree Thin Sections, Glossary of Terms. Florida State University, Tallahassee, FL. http://micro.magnet.fsu.edu/trees/glossary.html (accessed 03.03.20).

10

Forest dynamics

The manner in which forests grow and change, and how the trees within them react to different environmental pressures, helps us understand their health, vigor, and potential. The *dynamics* of forests concerns how the collection of trees changes in response to events that are either unplanned (e.g., fires or windstorms) or planned (e.g., harvest activities). As forest and natural resource managers, we have the ability to manipulate forest composition and structure through silvicultural activities. As mentioned in the previous chapter, models of forest dynamics may also help us to understand these transitions and anticipate future conditions in light of anthropogenic (human-caused) or natural disturbances. Therefore, upon completing this chapter, readers should be able to understand

- the definition of a forest community;
- how trees interact and compete for resources;
- the concepts of gradients and niches among forests;
- the various successional stages of forests;
- how forest succession evolves after natural disturbances; and
- gap processes and shade tolerance relationships of trees and forest ecosystems.

Usually, after implementing a management activity, we have an expectation or vision of what the future forest will hold in terms of tree species and stocking levels. However, we also need to understand how forests will then change over time in response to unplanned events that can significantly affect forest structure or composition. For example, the widespread loss of a dominant

tree species from a forest can have significant impacts on a forest community by affecting the composition, size distribution, and age structure of woody vegetation (Brown and Allen-Diaz, 2009). Historical examples from North America include chestnut blight and Dutch elm disease, and a contemporary example is the spread of sudden oak death (caused by the pathogen *Phytophthora ramorum*) along the coastal forests of California.

Although our main focus in this chapter is the dynamics of forestry systems, the dynamics of agroforestry systems may also be of interest, as these are becoming increasingly important. For example, Woodall and Ward (2002) describe tree belts being planted within agricultural systems in southwestern Australia to act as windbreaks. However, the trees in the windbreaks effectively reduce the moisture content in the soils nearby, which can in turn reduce agricultural crop yields. These interactions illustrate the competitive processes within an agroforestry system, and how crops or trees naturally react (or are managed) can affect the services provided by the system. Some extensions to the concept of forest dynamics can be made for agroforestry systems; however, we caution that this may not be completely appropriate. We therefore encourage readers to explore their interests in the area of agroforestry systems independently.

10.1 Forest communities

A *community*, as it relates to forestry and natural resource management, is an assembly of plant organisms

FIGURE 10.1   Forest of Farges fir (right, *Abies fargesii*) and dragon spruce (left, *Picea asperata*) in the Jiuzhaigou Valley, Sichuan, China. *Ken Marshall, through Wikimedia Commons.*
Image Link: https://commons.wikimedia.org/wiki/File:Jiuzhaigou6.jpg
License Link: https://creativecommons.org/licenses/by/2.0/deed.en

FIGURE 10.2   Dense canopy composed of different vertical layers of beech and spruce trees, in the Nature Reserve Malý Blaník, Benešov District, Czech Republic. *Juan de Vojníkov, through Wikimedia Commons.*
Image Link: https://commons.wikimedia.org/wiki/File:Přírodní_rezervace_Malý_Blaník,_koruny_stromů.JPG
License Link: https://creativecommons.org/licenses/by-sa/3.0/deed.en

living together and perhaps utilizing the available resources such that they exclude other plant organisms from entering the area (Ford-Robertson, 1971). Both forest physiognomy (outward appearance) and vertical structure (tree, shrub, herb, and moss) have been used to describe forest communities. For example, Qian et al. (2008) used these measures to describe forests in central China (Fig. 10.1) by their physical characteristics, tree density per structural class, and tree density per niche type. It should seem obvious that both natural events and management activities can affect community dynamics through changes in nutrient, water, and light availability, and through disturbances to the soil resource. Given the wide range of forest conditions possible within a group of plants that might be considered a *community*, a firm and precise definition is usually elusive. In practice, one is more likely to use a rather general definition of a forest community. Paraphrasing Kelly et al. (2000), an example community description is provided below for a scrub oak forest along the eastern coast of North Carolina.

A scrub oak forest is characterized by widely scattered longleaf pine (*Pinus palustris*) and sparsely vegetated ground cover that is generally dominated by wiregrass (*Aristida stricta*). The ground cover is also composed of perennial forbs such as narrowleaf silkgrass (*Pityopsis graminifolia*), perennial grasses such as bluestem (*Andropogon* spp.), and shrubs such as dwarf huckleberry (*Gaylussacia dumosa*).

10.2 Interaction, competition, and strategy among tree species

Interaction, competition, and survival strategies are important for determining whether a tree, once established, will continue to grow given competition for resources with other trees and plants. The size and condition of a tree's canopy and root system are likely to determine how well the tree can compete with other plants for the resources necessary to survive. Physiologically, a tree allocates carbon to four different compartments: shoots, stem, roots, and reproductive features (i.e., flowers and fruits). To survive, a tree, as with humans, must consume at least as much nutrients and water as it needs to expand through physiological and growth processes. During periods of low water supply, individual trees have the ability to adjust their transpiration to accommodate their individual water supply requirements to maximize carbon assimilation and avoid mortality (Zavala and Bravo de la Parra, 2005). At the tree-level, above-ground competition (in the canopy) suggests that larger trees have a disproportionate effect on smaller trees through reducing the availability of light (i.e., by shading the smaller trees). This is called a *size-asymmetric* or *nonequivalent competitive interaction*. We previously discussed the different layers of a forest canopy (dominant, codominant, intermediate, and suppressed trees) and how some tree species have adapted to these vertical layers within a canopy (Fig. 10.2). However, if an individual of a shade-intolerant tree species were located in the lower classes, it would be subject to intense competition for light resources. The competition for below-ground resources (water and nutrients) is typically proportional to the size of the root system and is called a *size-symmetric* or *equivalent competitive* interaction (von Oheimb et al., 2011). As with tree canopies, there may be some stratification of root systems at various depths below ground, which implies that some species may have a competitive advantage for the acquisition of resources. For example, in mixed Norway spruce (*Picea abies*) and European beech (*Fagus sylvatica*) forests, Schmid (2002) found limited spruce root systems in the uppermost soil layers due to competition from beech trees.

Competition among trees for resources can weaken some individuals and thus increase their vulnerability to other life-threatening agents, such as insects. In addition, competition among trees for resources is a well-known direct cause of tree mortality and can be a significant factor affecting growth dynamics at any stage during the lifespan of a tree, even in older forests (Das et al., 2011). The period of competition among trees can be determined by examining a tree's growth rings. Assuming that resources were constant in their availability, cases where tree rings are very tightly compressed would suggest competition among the trees is occurring and that for some trees, a resource (light, water, nutrient) was in short supply for several years. Trees in the upper portion of a canopy (i.e., the dominants and codominants) are better positioned to compete for light energy resources through direct access. Canopy position is therefore important for tree species that are considered shade intolerant. Shade-intolerant trees that have crowns located either in the intermediate or suppressed (overtopped) canopy positions face difficulties in obtaining the necessary light resource to compete effectively with other nearby trees and thus are more vulnerable to life-threatening agents.

10.3 Gradients and niches

Environmental gradients relate to variations in site characteristics and may be described by changes in elevation, changes in site index, or variation in other factors such as soil acidity, soil composition, soil fertility, soil moisture, exposure, edges and interior areas, and even land use history. Some tree species are found in greater abundance on sites with a particular aspect (e.g., southern or southeastern); with a higher or lower slope position (e.g., ridge vs. valley); that are more productive or have greater soil moisture availability; or within a specific elevational range. White oak (*Quercus alba*) in the eastern United States, for example, generally prefers south-facing slopes, high slope positions, and xeric (i.e., dry) soils (Collins and Carson, 2004). Trees may respond differently to the resources available in different gradients at different life stages. For example, oak trees at an intermediate life stage (e.g., saplings) may be more sensitive to environmental gradients than oak trees at a seedling or mature life stage (Collins and Carson, 2004). In some areas of the world, such as eastern Finland (Hokkanen, 2006), the ecological gradients within which trees and plants are distributed are still relatively unknown. However, in other areas of the world, changes in vegetation character and composition along forest gradients have been widely studied. As an example, Jose et al. (1996) described vegetation changes in high altitude tropical forests of India and noted that in this region soil conditions heavily influence secondary succession of forests.

From a forestry and natural resource perspective, a *niche* is a place occupied by an individual, live tree, and represents the sum of the resources required for the tree to survive in the environment; thus, a niche indirectly represents the response of the tree to the resources available. Where, when, and how a tree species becomes genetically adapted to live and exist on a site with other individuals forms part of the description of a *niche* (Barnes et al., 1998). The space occupied, the role played, and the functions facilitated by a tree species during the various plant successional stages of a forest also help define its niche. The complete range of resource conditions in which a tree species can persist, and thus where competition from other tree species is minimized, is called the *fundamental niche*, while the actual range of conditions, given competition from other trees, is called the *realized niche*. The realized niche represents the natural distribution of tree species, given the physical and biological constraints imposed on its various life-stage processes (Booth et al., 1988). For example, the seed dispersal processes of certain tree species may limit their distribution, as was suggested in the discussion of mixed boreal forests of British Columbia by Albani et al. (2005).

Niches may be wide or narrow and may be influential to different degrees during the various life stages of a tree. For example, in northeastern Germany, the germination and establishment of sessile oak (*Quercus petraea*) span a wide niche of environmental variables (water, light, and nutrients) compared to that of European beech (Mirschel et al., 2011). Niches can be extensive or as small as an uprooted tree, a stump (Fig. 10.3), or a downed log, which seems to be the preferred niche for naturally regenerating Norway spruce in parts of Europe (Kuuluvainen and Kalmari, 2003). Further, while some tree species may be able to germinate seed in a wide variety of niches, the resulting saplings may only become successfully established in a smaller subset of niches, as is the case for Italian maple (*Acer opalus*) in the Iberian Peninsula. Here, microhabitats containing high levels of nonpalatable shrubs and a deep litter layer, as well as habitats without bare ground or dense tree cover, are best for initiating stands containing this tree species (Quero et al., 2008). Niches may contract during the lifespan of a tree species, suggesting that a change in the tolerance of the tree species to environmental variables is occurring.

At times, niches of a single species can vary by their scope or extent of their resource utilization horizontally and vertically across a landscape. A tree species will

FIGURE 10.3 Nurse log (stump) supporting Norway spruce reproduction, in the Harz Mountains of Germany. *Gerhard Elsner, through Wikimedia Commons.*
Image Link: https://commons.wikimedia.org/wiki/File:Kadaververjuengung. JPG
License Link: https://creativecommons.org/licenses/by-sa/3.0/deed.en

FIGURE 10.4 Landslide in a forested area of western Oregon, United States. *U.S. Department of Agriculture, Natural Resources Conservation Service (2020).*

reach its maximum density per unit area when its resource requirements are optimally met (Leuschner et al., 2009). Niches of more than one species can be described according to similarities in their resource utilization (Peng et al., 2009), and coexistence of tree species can occur where their niches overlap (Silvertown, 2004). Soil nutrient availability may affect light-dependent growth (associated with shade tolerance), which may therefore affect the ability of a tree species to take advantage of a niche (Bigelow and Canham, 2007). Niche characteristics gradually differ when moving outward from tree population centers to marginal areas. In the case of European beech and sessile oak in central Europe, these tree species seem to prefer less acidic soils along the margins of their natural distribution (Leuschner et al., 2009). The ability of plant and animal species to modulate or modify the availability of resources for themselves and other species has been described as a *niche construction process* (Vandermeer, 2008). Perhaps the most widely understood example of an animal or plant modifying its environment is the beaver (*Castor canadensis*), which, through dam-building, can affect the quality of water resources, the condition of the riparian zone, nutrient cycling processes, and other resources and functions that are important to other plant and animal organisms (Laland and Boogert, 2010).

10.4 Forest succession

Succession, as it relates to forestry and natural resource management, is the orderly supplanting of one community of plants by another, or the change in the character or composition of an area of ecological community over time. Succession is the sequential change in the number, composition, and structure of species within a given area of land. A sequence of successional stages is typically called a *sere* and each stage within a sere is called a *seral stage* (Ford-Robertson, 1971). Succession is *autogenic* when the developing vegetation itself is the main cause of the change (i.e., trees grow in size) and *allogenic* when outside factors (e.g., harvests or natural disasters) influence, or are the direct cause, of the change. Succession is influenced by natural and anthropogenic events, the quality of a site, the state of soil development, seed availability, vegetative competition, and weather conditions. Some of these factors are random (stochastic) in nature, while other factors are relatively predictable. Therefore, the transition from one state in a set of potential successional pathways to another is partly a matter of chance. When succession is initiated in an area in which the formation of communities occurs on a barren, unoccupied (by vegetation) substrate where no soil is present, it is referred to as *primary succession*. These areas might include landscape surfaces that had all soil removed by landslides (Fig. 10.4) or volcanoes and can include areas of bare rock or sand (e.g., surface mines) with virtually no soil formation. Primary succession processes are generally independent of historical or continuous human management of the landscape (Wiegleb and Felinks, 2001). Succession that is initiated by events that do not result in the formation of barren unoccupied substrate is

FIGURE 10.5 Secondary succession of trees in fields no longer cultivated for agricultural purposes near Bolimowski Park Krajobrazowy in central Poland. *Tomasz Kuran, through Wikimedia Commons. Image Link: https://commons.wikimedia.org/wiki/File:Secondary_succesion_cm02.jpg License Link: https://creativecommons.org/licenses/by-sa/3.0/deed.en*

referred to as *secondary succession*. This process is characteristic of areas where soil resources are present and usually begins in areas where there once was vegetation, but the vegetation has been removed or killed by natural or anthropogenic events. Wildfires, tropical cyclones (hurricanes), and other natural disturbances, along with final harvests and other human actions, can initiate secondary succession processes. Forests grown on land once used to produce an agricultural crop (Fig. 10.5) or for pasture purposes can be considered to begin with secondary succession processes. For example, in the mid to late twentieth century, a large area of marginal agricultural land in the southern United States reverted either naturally or artificially returned to forests through secondary succession processes, some areas through the assistance of government programs. Similarly, in the late twentieth century, a decline in traditional agricultural practices in western Europe resulted in secondary succession occurring on infrequently used pastures, land situated on steep slopes, and land containing poor soil resources (Susyan et al., 2011).

Disturbances and local climate can influence the type of species that return to a site (Reyna et al., 2020). In the Upper Mississippi River floodplain of the United States (in Illinois, Iowa, Minnesota, Missouri, and Wisconsin), primary or secondary succession on areas considered to be subject to periodic flood events begins with forests composed of cottonwood (*Populus* spp.), willow (*Salix* spp.), and other pioneer species. In general, these forests transition into communities represented by various ash (*Fraxinus* spp.), elm (*Ulmus* spp.), hickory (*Carya* spp.), maple (*Acer* spp.), and oak species (Yin et al., 2009). For both primary and secondary forest succession, pioneer tree species are influential in the early development of an area. The transition from one species group to another is not necessarily standard. For example, on volcanic deposits of Hawaii, a rainforest becomes established from primary successional processes dominated by a pioneer tree species, the 'ōhi'a lehua (*Metrosideros polymorpha*). In subsequent successional stages, this tree creates conditions favorable for its own regeneration and survival, even though it is a shade-intolerant species, and these processes lead to a forest consisting mainly of the same species (Mueller-Dombois, 2000). Ultimately, growth functions, survival strategies, and emergent properties will vary by tree species in response to the events that led to the initiation of primary or secondary succession (Yin et al., 2009). Further, the rate of invasion of pioneer plant species may increase if a native dispersal agent is present (Mueller-Dombois, 2000).

After a disturbance, plant species can begin to occupy a site by developing stems that arise from seed, root sprouts, and stump sprouts. Trees that exhibit the fastest growth, and thus appear as the dominant tree species, are often called the *pioneer tree species*. Further, advanced regeneration or individual trees growing in the understory of the previous forest can take advantage of the light, water, and soil resources made available through a disturbance to utilize the newly available growing space (Oliver, 1980). If developed from seed, most seedlings in a forest initiate during a relatively short period following a disturbance called the *stand initiation stage*, the length of which varies according to the disturbance intensity, disturbance size, tree species growth rates, regeneration method, density of predators or vegetative competitors, and weather conditions. This stage is representative of the beginning of an even-aged forest, although over time it may transition to an uneven-aged or all-aged forest. The stand initiation stage is important, as recruitment of new seedlings and saplings into a forest often follows disturbances and rarely occurs on a continuous basis. This stage could also be prompted through the loss of individual trees in an uneven-aged forest. However, small gaps such as those created through the loss or removal of a few trees may only serve to allow adjacent living trees to enlarge their crowns and fill the available growing space rather than allowing new tree regeneration to fill the gap (Oliver, 1980).

A combination of natural and human-caused events, such as windstorms, insect and disease outbreaks, and salvage logging operations, can occur during the early developmental stages of a forest and prolong the tree recruitment period by one or more decades (Svoboda et al., 2010). For example, in southeastern Arkansas, naturally regenerated pine forests are generally in the stand initiation stage of development for about 12 years and achieve complete crown closure by the 17th year (Cain and Shelton, 2001). However, variations in climatic conditions can affect the length of time required to complete this stage. The substrate available for the development of new trees can also influence the progress of this stage. For example, in central Europe the abundance of Norway spruce seedlings and saplings positively correlates with presence of down woody debris, thus underscoring the importance of fallen logs and decayed stumps as nurse logs in some areas of the world (Svoboda et al., 2010). Differences in the timing and nature of structural development, and hence the stages evident in succession, have also been attributed to localized differences in soil conditions (Harper et al., 2005). Other influences on stand initiation include seed viability, seed germination success, survival during the early stages of development owing to carbohydrate reserves and mineral availability, and other environmental conditions (Kozlowski, 2002). Seed dispersal processes can also influence the early development of a forest. Tree seeds can be naturally dispersed by wind, water, and animals. For example, in the temperate biome of the southern United States, wind commonly disperses the relatively light pine and maple seed from mature trees to potential seedling establishment areas. Conversely, heavier oak seeds (acorns) may not travel far from a host tree unless the tree is situated on a slope, where gravity may play a role in moving the seed to a seedling establishment area. Further, birds are commonly involved in the transport of seed from trees such as eastern redcedar (*Juniperus virginiana*), which explains why these trees are often found along fence lines. In tropical biomes, birds, insects (e.g., ants), and mammals may be the most important seed dispersal agents (Howe and Smallwood, 1982).

New tree stems occupy a growing space until one or more resources become limiting, which basically prevents the development or initiation of others. This stage is called the *stem exclusion stage*, sometimes characterized by intense competition among the growing seedlings and saplings (Kozlowski, 2002). During the stem exclusion phase, shade-intolerant herbaceous plants can disappear from the forest floor (Cain and Shelton, 2001). Subsequent vertical stratification of trees allows some species or individuals to overtop others, thus expressing their dominance and effectively suppressing the growth of other individuals below (Oliver, 1980). Dense, subalpine forests of the Swiss Alps are often characterized by self-thinning processes that effectively result from the death of trees that are suppressed (Krumm et al., 2011). Tree species that appear to be dominant early in the development of a forest (the pioneers) may eventually be overtopped by other species that are more shade tolerant, which may relegate the pioneers to the understory or lead to their death owing to intense competition for limited resources. Growth and survival of seedlings and saplings can also be

affected by air and soil pollution, drought, flooding, insect and disease outbreaks, and soil compaction (Kozlowski, 2002).

Several years after stand initiation, as the overstory of a forest matures, shrubs and advanced regeneration reinvade the understory. This is often called the *understory reinitiation* phase of forest development; forests at this stage are likely to have significant structural diversity and high levels of above-ground biomass. Some researchers have characterized this as simply a transitional period between the stem exclusion stage and the development of older forest characteristics (Kozlowski, 2002). At some point, the overstory trees begin to die and the vegetation in the understory takes advantage of the gaps created. The long time required to reach the older forest stage, and the probability that a disturbance will affect the forest along the way, makes the achievement of older forest conditions an uncommon event (Oliver, 1980). The latter developmental stages of a forest are often characterized by large amounts of live tree volume and down woody debris (fallen tree boles, limbs, etc.) that are important for biological diversity, soil development, seedling development, and other protective functions (Krumm et al., 2011). For example, a Norway spruce forest in central Europe transitioning from the stem exclusion stage to the understory reinitiation phase is likely to have a relatively high live tree count per unit area and a corresponding relatively high tree basal area (Svoboda et al., 2010). Unfortunately, in fire-prone landscapes, large amounts of down woody debris in a forest also represent a high risk of fire if the debris dries sufficiently to enable combustion.

The culminating (or last) stage in the development of a forest is called the *climax*; however, periodic disturbances make the achievement of an ecological equilibrium (a *climax forest*) nearly impossible to obtain. This is considered an idealized state by many forest management organizations, and management units have sometimes been described by the *potential vegetation* they could support for the appropriate climax forest condition of the area. Further, several different forest communities could be present on a site for an indefinite length of time, rather than a single set of climax vegetation that dominates the others and maintains a steady state (Oliver, 1980). As an example of climax forest conditions near Hong Kong in southern China, the primary climax forest is described as a subtropical evergreen broad-leaved rainforest, composed of trees that have buttressed trunks that are capable of producing flowers (Wang et al., 2006). For northern hardwood forests in Michigan, the overstory and understory vegetation composition has been used to describe climax forest types (Coffman and Willis, 1977). Unfortunately, areas representative of the climax forest state may be difficult to locate in today's world (e.g., the seven-year search by

FIGURE 10.6 Old-growth grand fir (*Abies grandis*) forest on the Umatilla National Forest, Oregon, United States. *Dave Powell, through Bugwood.org.*

Heinselman (1973) in the Boundary Waters Canoe Area of northern Minnesota). If forests are allowed to progress naturally and without disturbances, the path to a climax forest condition can require hundreds of years.

Other terms for areas containing advanced stages of forest development are *late-successional forests* and *old-growth forests*. An old-growth forest (Fig. 10.6) is similar to a climax forest and is maintained in an oscillating steady state through thinning, gap formation, and regeneration (Kozlowski, 2002). Although we lack a general definition for old-growth forests, these are essentially older forests that have obtained certain characteristics relating to, perhaps, the vertical and horizontal structural condition of the live trees, composition of dead trees and fallen woody material, presence of seedlings and saplings under older trees (advanced regeneration), and level of spatial heterogeneity in the distribution of these resources (Bauhus et al., 2009). Sometimes, these conditions are simply represented by the relative age and density of a portion of the trees in a forest and their age distribution. For example, in the northern hardwood forests of North America, old-growth forests might be considered as uneven-aged (i.e., containing two or more general ages of tree cohorts) and containing a cohort of trees over 150 years old (Keeton, 2006). In British Columbia, old-growth forests might simply be defined as those over 140 years old (Nigh, 1998). The condition of older trees and the understory resource have also been used to distinguish old-growth forests from their less-developed counterparts. For example, the presence of unusual bark features (large clumps or strips of hanging pieces) and higher densities of ferns have been suggested to be representative of old-growth forest conditions in the Central Highlands of Victoria, Australia (Lindenmayer et al., 2000). An old-growth Mongolian oak (*Quercus mongolica*) forest in Japan was described as a discontinuous distribution

of trees with five general age classes, each separated by about 100 years (Sano, 1997). Late-successional forests are those that are typically, at a minimum, as old as an economically or biologically mature forest (from a harvesting perspective), but not as old or as structurally diverse as an old-growth forest. For example, in the northern hardwood and spruce/fir temperate and boreal forests of eastern North America, these might include forests that are at least 100 years old but not considered to be old-growth forests (Whitman and Hagan, 2007). Achieving older forest conditions from a young, managed forest may involve allowing the forest to grow undisturbed for many decades; however, some researchers have proposed accelerating the development of old-growth or late-successional forests using group selection harvests (small patch cuts). Chen et al. (2003) suggest that if the necessary tree species are introduced and the necessary biological processes (e.g., natural competition) are reestablished, then forest managers could accelerate the development of older forest conditions. This, of course, would be of interest to land managers or landowners who want to restore ecological processes and ecological services more quickly than nature can achieve alone, which is what Chen et al. (2003) were suggesting for areas of northeast China.

The dynamic development of forests is influenced by many different natural and human-caused events, and management, or the abandonment of management, can alter the trajectory of forest development (Mei et al., 2020). For example, in northern Minnesota, forests initiated after human-caused disturbances (presumably harvesting activities) had stand initiation processes that were initially dominated by aspen (*Populus tremuloides*). However, these forests also had an extended recruitment period (10 years) that facilitated the introduction of shade-tolerant tree species such as balsam fir (*Abies balsamea*), red maple (*Acer rubrum*), and white spruce (*Picea glauca*). During the stem exclusion stage (age 10–40 years), tree competition along with insect outbreaks caused significant tree mortality. This created gaps in the forest canopy and allowed a new cohort of aspen, balsam fir, and red maple trees to colonize the area. Further insect-related outbreaks have continued to open the forest canopy, mainly through losses of individual aspen and balsam fir trees, thus forming a complex mixed-species, multicohort forest (Reinikainen et al., 2012).

10.5 Stand dynamics following major disturbance

While the character of many forests may be considered to be primarily shaped by anthropogenic or small gap dynamics, the role of natural disturbances in the development of forests across the landscape may also

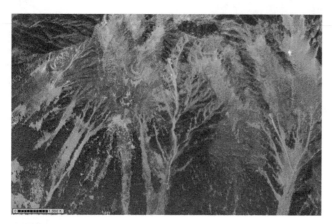

FIGURE 10.7   Avalanche-prone area of subalpine forests in north-eastern Switzerland. *U.S. Department of Agriculture, Natural Resources Conservation Service (2020).*

influence the function and structure of forest ecosystems (Panayotov et al., 2011). Discussed in more detail in Chapter 14, natural disturbances such as fires and windstorms markedly affect the structure and composition of a forest which can dramatically alter the successional stage and successional trajectory of forests. For example, in the Alps, frequent disturbances created by avalanches (Fig. 10.7) generally result in forest communities comprised of smaller, shorter trees of shade-intolerant species, which have low tree densities and greater structural diversity than nearby unaffected forests (Bebi et al., 2009).

Disturbance severity can determine the structure and composition of a post-event forest (Oliver, 1980). As a land manager or owner, it is important to understand the important natural disturbance regimes of the region within which a forest is situated, and how disturbances may affect the character of the forest. For example, in the Alps and Balkan Mountains of Europe, Norway spruce is one of the most important tree species for timber production purposes, yet these forests are vulnerable to windthrow from severe windstorms and attack from the European bark beetle (*Ips typographus*) (Panayotov et al., 2011). These disturbances can have a significant effect on both the ecology and economics of a forest. From an ecological point of view, a rather swift transition from a mature forest to the early stages of secondary succession could occur in the worst-case scenario. In other less dramatic cases, the snag (standing dead tree) and down wood volume of a forest could increase substantially as a result of these types of disturbances, which may benefit many wildlife species but may also increase fire danger levels. In another example, Nelson et al. (2008) described the effects of a tornado that passed through southern Illinois in 2003. Subsequent events (natural and anthropogenic) represented a severe and swift departure from a low-intensity disturbance rate in a bottomland hardwood

forest. However, in the years that followed, the affected forests became dominated by a fully stocked mosaic of diverse bottomland hardwood tree species. As one might expect, the release of advance reproduction of shade-tolerant tree species was high, and the reproduction of shade-intolerant tree species was a function of the degree of canopy removal through these events.

In many cases the general course of land use history of tropical forests is similar to that of higher latitudes in North America and Europe. It is also broadly accepted that most tropical forests have experienced some form of disturbance, and that disturbance and recovery occur at various spatial and temporal scales. In general, tropical forest recovery following disturbances that affect the canopy is relatively fast, particularly if the disturbed forests are near others that can provide seed for regeneration purposes. When damage to the soil is evident, the recovery of forests from disturbances is generally slower. Regardless, the role of residual vegetation is critical in transitioning from the early stages of secondary succession back to a mature forest canopy (Chazdon, 2003). In some tropical areas, the recruitment of seedlings into a disturbed area may also be a function of the timing of the disturbance with respect to climatic or environmental conditions. For example, dry-season droughts, high temperatures, and low humidity cause water stress in seedlings, resulting in seedling mortality. In addition, the seasonal conditions that favor the development and transport of fire are also detrimental to the development of seedlings during primary or secondary successional stages. However, the relative importance of regional or local climatic processes on forest dynamics in the tropics is not well understood (Zimmer and Baker, 2009) and provides a wealth of opportunities for graduate study.

10.6 Gap dynamics

Holes in the forest canopy (e.g., gaps) are often created through the death of individual trees that have dominant or codominant crown positions (Fig. 10.8). The process in which one or more trees in a forest die, leaving a hole in the canopy and making available both light and nutrient resources for seedlings and saplings, is called *gap dynamics* (Brokaw and Busing, 2000). Strong winds, climate extremes (temperature and precipitation), and other events (e.g., lightning) can result in individual tree death. The forest gaps created can be as small as 25 to 40 m^2 (about 1/100 to 1/160 of an acre [ac]) (Kathke and Bruelheide, 2010). Natural gap sizes vary; average reported gap sizes are about 200 m^2 (2152.8 ft^2) in the beech (*Fagus orientalis*) forests of northern Iran (Sefidi et al., 2011) and 100 m^2 (1076.4 ft^2) in natural spruce forests of Germany (Kathke and Bruelheide,

FIGURE 10.8 Canopy gap in a coniferous forest of Warren Wood, near Norwich, England. *Evelyn Simak, through Wikimedia Commons. Image Link: https://commons.wikimedia.org/wiki/File:Fallen_trees_-_geograph. org.uk_-_639276.jpg License Link: https://creativecommons.org/licenses/by-sa/2.0/deed.en*

2010). As individual events, these are considered relatively minor disturbances compared to the damage inflicted on forests by severe wildfires, tropical cyclones (hurricanes or typhoons), or volcanoes.

Multicohort forest tree systems based on large canopy gaps (>200 m²) might be used as an alternative approach for approximating long-term developmental dynamics of a forested area (Reinikainen et al., 2012). The subsequent closing of a forest gap either progresses simply by chance or is influenced by the specialization of the remaining trees, which take advantage of the light and nutrient resources that have been made available. The lifespan of a gap is debatable and may be as long as 50 years in some tropical forests, but the footprint or spatial pattern may persist for well over 150 years in a natural forest (Pagnutti et al., 2011). In areas of the world such as Germany or Japan, the gap formation rate and the gap closure rate are about 0.3% to 1.0% per year (Kathke and Bruelheide, 2010).

Gaps created as a result of small-scale disturbances allow seedling recruitment and growth in forests where large-scale disturbances are infrequent. Wind and fire patterns, stand age, stand composition, hydrology, and perhaps topography contribute to gap regime heterogeneity (abundance, area, and frequency) (Almquist et al., 2002). Although areas along edges of gaps may perhaps contain more available seed for tree regeneration, these areas may also have less available light (depending on the aspect) and higher rates of litterfall from mature trees. These factors and the intraspecific competition from adults could affect the survival and growth of seedlings along the edges of gaps (Brockway and Outcalt, 1998). The dynamic processes within the tree initiation and stem exclusion phases of a gap might also promote the coexistence of competing tree species and therefore contribute to forest diversity (Brokaw and Busing, 2000). However, responses of trees to discrete events, such as the creation of a gap, differ according to the surrounding forest composition, structure, and age class distribution. For example, when an individual tree or two with a significant position in the canopy dies, the adjacent trees may be able to quickly utilize the vacant growing space by lateral crown expansion, and therefore close the gap (Richards and Hart, 2011). This may discourage the process of seedling recruitment and tree regeneration through seeds of shade-intolerant tree species.

Summary

Trees located collectively in a forest are likely to be competing with each other for nutrients, light, and water. As a group, they form a community and interact in a manner partly controlled by the environment, the substrate on which they are located, and their position in the canopy. How trees have adapted to limited light resources helps determine the probable ability and success of remaining within or below their dominant and codominant neighbors. Tree species can express preferences for niches and may likely be found growing within distinct environmental gradients that are perhaps defined by water and nutrient availability and elevational ranges. However, ultimately the ability of a forest to persist through time is a function of both site conditions and long-term environmental and management history (Iglesias and Whitlock, 2020).

A forest may begin with primary or secondary succession processes. Once established, they generally transition through stand initiation, stem exclusion, and understory reinitiation stages. Theoretically, if a forest is allowed to grow very old, it can reach a climax stage, although for practical purposes management of older forests aims to achieve late-successional or old-growth forest conditions. Natural disturbances can force forest succession to return to primary or secondary establishment processes. Disturbances such as tropical cyclones or fires can be broadly influential on forest condition and development. Smaller disturbances that involve the loss of one or two trees result in forest gaps. An understanding of how trees interact as forest communities and how gaps are formed and filled comprises an essential set of tools needed by forest and natural resource managers.

Questions

(1) *Shade tolerance.* Neighbors recently purchased a small tract of recently harvested forestland in central Alabama. They are somewhat frustrated by their repeated attempts to grow pine trees on the land, in spite of the advanced regeneration of sweetgum and other hardwoods that are now 2 to 3 m (6.5—9.8 ft) high. They have no forestry background nor have they sought professional advice. In a short paragraph, describe how you would respond, using the shade tolerance of trees as the main topic.

(2) *Observations of forest dynamics.* Take a walk or travel to a nearby forest (young, old, or otherwise). Venture into the forest, away from the influences of any sort of edge. Once there, try to describe the dynamic forest processes that you feel may be occurring. Try also to determine which stage of succession the forest may be encountering. In a short, professional presentation to your class, and with a few images captured during your visit, describe these processes.

(3) *Climax and old-growth forests.* Some people suggest that the notion of a climax forest is outdated. In a short paragraph, describe your thoughts on the matter. Also develop a second paragraph that compares the notion of a climax forest with the often-used term old-growth. How do these differ, if at all? In a small group of five or six other students, share your opinion of these two issues.

(4) *Shade tolerance.* Select a group of tree species that commonly grow in your area. For example, in the southern United States one might select pines, which would include loblolly pine (*Pinus taeda*), longleaf pine, shortleaf pine (*Pinus echinata*), slash pine (*Pinus elliottii*), pond pine (*Pinus serotina*), sand pine (*Pinus clausa*), and perhaps others. Using a resource that describes the physiological traits of trees (e.g., in North America, perhaps *Silvics of North America*), describe in a short memorandum the shade tolerance of the trees you selected and how this might differ among the group.

References

Albani, M., Andison, D.W., Kimmins, J.P., 2005. Boreal mixedwood species composition in relationship to topography and white spruce seed dispersal constraint. Forest Ecology and Management 209, 167—180.

Almquist, B.E., Jack, S.B., Messina, M.G., 2002. Variation of the treefall gap regime in a bottomland hardwood forest: relationships with microtopography. Forest Ecology and Management 157, 155—163.

Barnes, B.V., Zak, D.R., Denton, S.R., Spurr, S.H., 1998. Forest Ecology, fourth ed. John Wiley & Sons, Inc., New York, p. 774.

Bauhus, J., Puettmann, K., Messier, C., 2009. Silviculture for old-growth attributes. Forest Ecology and Management 258, 525—537.

Bebi, P., Kulalowski, D., Rixen, C., 2009. Snow avalanche disturbances in forest ecosystems—state of research and implications for management. Forest Ecology and Management 257, 1883—1892.

Bigelow, A.W., Canham, C.D., 2007. Nutrient limitation of juvenile trees in a northern hardwood forest: calcium and nitrate are preeminent. Forest Ecology and Management 243, 310—319.

Booth, T.H., Nix, H.A., Hutchinson, M.F., Jovanovic, T., 1988. Niche analysis and tree species introduction. Forest Ecology and Management 23, 47—59.

Brockway, D.G., Outcalt, K.W., 1998. Gap-phase regeneration in longleaf pine wiregrass ecosystems. Forest Ecology and Management 106, 125—139.

Brokaw, N., Busing, R.T., 2000. Niche versus chance and tree diversity in forest gaps. Trends in Ecology and Evolution 15, 183—188.

Brown, L.B., Allen-Diaz, B., 2009. Forest stand dynamics and sudden oak death: mortality in mixed-evergreen forests dominated by coast live oak. Forest Ecology and Management 257, 1271—1280.

Cain, M.D., Shelton, M.G., 2001. Secondary forest succession following reproduction cutting in the Upper Coastal Plain of southeastern Arkansas. USA. Forest Ecology and Management 146, 223—238.

Chazdon, R.L., 2003. Tropical forest recovery: legacies of human impact and natural disturbances. Perspectives in Plant Ecology. Evolution and Systematics 6, 51—71.

Chen, X., Li, B.-L., Lin, Z.-S., 2003. The acceleration of succession for the restoration of the mixed-broadleaved Korean pine forests in Northeast China. Forest Ecology and Management 177, 503—514.

Coffman, M.S., Willis, G.L., 1977. The use of indicator species to classify climax sugar maple and eastern hemlock forests in Upper Michigan. Forest Ecology and Management 1, 149—168.

Collins, R.J., Carson, W.P., 2004. The effects of environment and life stage on *Quercus* abundance in the eastern deciduous forest, USA: are sapling densities most responsive to environmental gradients? Forest Ecology and Management 201, 241—258.

Das, A., Battles, J., Stephenson, N.L., van Mantgem, P.J., 2011. The contribution of competition to tree mortality in old-growth coniferous forests. Forest Ecology and Management 261, 1203—1213.

Ford-Robertson, F.C., 1971. Terminology of Forest Science Technology Practice and Products. Society of American Foresters, Washington, D.C., p. 370

Harper, K.A., Bergeron, Y., Drapeau, P., Gauthier, S., De Grandpré, L., 2005. Structural development following fire in black spruce boreal forest. Forest Ecology and Management 206, 293—306.

Heinselman, M.L., 1973. Fire in the virgin forests of the boundary waters Canoe area, Minnesota. Quaternary Research 3, 329—382.

Hokkanen, P.J., 2006. Environmental patterns and gradients in the vascular plants and bryophytes of eastern Fennoscandian herb-rich forests. Forest Ecology and Management 229, 73—87.

Howe, H.F., Smallwood, J., 1982. Ecology of seed dispersal. Annual Review of Ecology and Systematics 13, 201—228.

Iglesias, V., Whitlock, C., 2020. If the trees burn, is the forest lost? Past dynamics in temperate forests help inform management strategies. Physiological Transactions of the Royal Society B 375, Article 20190115.

Jose, S., Gillespie, A.R., George, S.J., Kumar, B.M., 1996. Vegetation responses along edge-to-interior gradients in a high altitude tropical forest in peninsular India. Forest Ecology and Management 87, 51—62.

Kathke, S., Bruelheide, H., 2010. Gap dynamics in a near-natural spruce forest at Mt. Brocken, Germany. Forest Ecology and Management 259, 624−632.

Keeton, W.S., 2006. Managing for late-successional/old-growth characteristics in northern hardwood-conifer forests. Forest Ecology and Management 235, 129−142.

Kelly, L.A., Wentworth, T.R., Brownie, C., 2000. Short-term effects of pine straw raking on plant species richness and composition of longleaf pine communities. Forest Ecology and Management 127, 233−247.

Kozlowski, T.T., 2002. Physiological ecology of natural regeneration of harvested and disturbed forest stands: implications for forest management. Forest Ecology and Management 158, 195−221.

Krumm, F., Kulakowski, D., Spiecker, H., Duc, P., Bebi, P., 2011. Stand development of Norway spruce dominated subalpine forests of the Swiss Alps. Forest Ecology and Management 262, 620−628.

Kuuluvainen, T., Kalmari, R., 2003. Regeneration microsites of Picea abies seedlings in a windthrow area of a boreal old-growth forest in southern Finland. Annales Botanici Fennici 40, 401−413.

Laland, K.N., Boogert, N.J., 2010. Niche construction, co-evolution and biodiversity. Ecological Economics 69, 731−736.

Leuschner, C., Köckemann, B., Buschmann, H., 2009. Abundance, niche breadth, and niche occupation of Central European tree species in the centre and at the margin of their distribution range. Forest Ecology and Management 258, 1248−1259.

Lindenmayer, D.B., Cunningham, R.B., Donnelly, C.F., Franklin, J.F., 2000. Structural features of old-growth Australian montane ash forests. Forest Ecology and Management 134, 189−204.

Mei, G., Pesaresi, S., Corti, G., Cocco, S., Colpi, C., Taffetani, F., 2020. Changes in vascular plant species composition, top-soil and seed-bank along coppice rotation in an Ostrya carpinifolia forest. Plant Biosystems 154 (2), 259−268.

Mirschel, F., Zerbe, S., Jansen, F., 2011. Driving factors for natural tree rejuvenation in anthropogenic pine (Pinus sylvestris L.) forests of NE Germany. Forest Ecology and Management 261, 683−694.

Mueller-Dombois, D., 2000. Rain forest establishment and succession in the Hawaiian Islands. Landscape and Urban Planning 51, 147−157.

Nelson, J.L., Groninger, J.W., Battaglia, L.L., Ruffner, C.M., 2008. Bottomland hardwood forest recovery following tornado disturbance and salvage logging. Forest Ecology and Management 256, 388−395.

Nigh, G.D., 1998. Site Index Adjustments for Old-Growth Stands Based on Veteran Trees. Ministry of Forests Research Branch, Victoria, B.C. Working Paper 36/1998.

Oliver, C.D., 1980. Forest development in North America following major disturbances. Forest Ecology and Management 3, 153−168.

Pagnutti, C., Anand, M., Azzouz, M., 2011. Estimating gap lifetime and memory from a simple model of forest canopy dynamics. Journal of Theoretical Biology 274, 154−160.

Panayotov, M., Kulakowski, D., Laranjero Dos Santos, L., Bebi, P., 2011. Wind disturbances shape old Norway spruce-dominated forests in Bulgaria. Forest Ecology and Management 262, 470−481.

Peng, Y., Chen, G., Tian, G., Yang, X., 2009. Niches of plant populations in mangrove reserve of Qi'ao Island, Pearl River estuary. Acta Ecologica Sinica 29, 357−361.

Qian, Y., Zongqiang, X., Gaoming, X., Zhigang, C., Jingyuan, Y., 2008. Community characteristics and population structure of dominant species of Abies fargesii forests in Shennongjia National Nature Reserve. Acta Ecologica Sinica 28, 1931−1941.

Quero, J.L., Gómez-Aparicio, L., Zamora, R., Maestre, F.T., 2008. Shifts in the regeneration niche of an endangered tree (Acer opalus ssp.

granatense) during ontogeny: using an ecological concept for application. Basic and Applied Ecology 9, 635−644.

Reinikainen, M., D'Amato, A.W., Fraver, S., 2012. Repeated insect outbreaks promote multi-cohort aspen mixedwood forests in northern Minnesota. USA. Forest Ecology and Management 266, 148−159.

Reyna, T.A., Martínez-Vilalta, J., Retana, J., 2020. Regeneration patterns in Mexican pine-oak forests. Forest Ecosystems 6, Article 50.

Richards, J.D., Hart, J.L., 2011. Canopy gap dynamics and development patterns in secondary Quercus stands on the Cumberland Plateau, Alabama. USA. Forest Ecology and Management 262, 2229−2239.

Sano, J., 1997. Age and size distribution in a long-term forest dynamics. Forest Ecology and Management 92, 39−44.

Schmid, I., 2002. The influence of soil type and interspecific competition on the fine root system of Norway spruce and European beech. Basic and Applied Ecology 3, 339−346.

Sefidi, K., Marvie Mohadjer, M.R., Mosandl, R., Copenheaver, C.A., 2011. Canopy gaps and regeneration in old-growth oriental beech (Fagus orientalis Lipsky) stands, northern Iran. Forest Ecology and Management 262, 1094−1099.

Silvertown, J., 2004. Plant coexistence and the niche. Trends in Ecology and Evolution 19, 605−611.

Susyan, E.A., Wirth, S., Ananyeva, N.D., Stolnikova, E.V., 2011. Forest succession on abandoned arable soils in European Russia—impacts on microbial biomass, fungal-bacteria ratio, and basal CO_2 respiration activity. European Journal of Soil Biology 47, 169−174.

Svoboda, M., Fraver, S., Janda, P., Bače, R., Zenáhlíková, J., 2010. Natural development and regeneration of a Central European montane spruce forest. Forest Ecology and Management 260, 707−714.

U.S. Department of Agriculture, 2020. Natural Resources Conservation Service. Web Soil Survey. U.S. Department of Agriculture, Natural Resources Conservation Service, Washington, D.C. https://websoilsurvey.sc.egov.usda.gov/App/HomePage.htm (accessed 22.03.20).

Vandermeer, J., 2008. The niche construction paradigm in ecological time. Ecological Modelling 214, 385−390.

von Oheimb, G., Lang, A.C., Bruelheide, H., Forrester, D.I., Wäsche, I., Yu, M., Härdtle, W., 2011. Individual-tree radial growth in a subtropical broad-leaved forest: the role of local neighbourhood competition. Forest Ecology and Management 261, 499−507.

Wang, D.P., Yi, S.Y., Chen, F.P., Xing, F.W., Peng, S.L., 2006. Diversity and relationship with succession of naturally regenerated southern subtropical forests in Shenzhen, China and its comparison with the zonal climax of Hong Kong. Forest Ecology and Management 222, 384−390.

Whitman, A.A., Hagan, J.M., 2007. An index to identify late-successional forest in temperate and boreal zones. Forest Ecology and Management 246, 144−154.

Wiegleb, G., Felinks, B., 2001. Primary succession in post-mining landscapes of Lower Lusatia—chance or necessity. Ecological Engineering 17, 199−217.

Woodall, G.S., Ward, B.H., 2002. Soil water relations, crop production and root pruning of a belt of trees. Agricultural Water Management 53, 153−169.

Yin, Y., Wu, Y., Bartell, S.M., Cosgriff, R., 2009. Patterns of forest succession and impacts of flood in the Upper Mississippi River floodplain ecosystem. Ecological Complexity 6, 463−472.

Zavala, M.A., Bravo de la Parra, R., 2005. A mechanistic model of tree competition and facilitation for Mediterranean forests: scaling from leaf physiology to stand dynamics. Ecological Modelling 188, 76−92.

Zimmer, H., Baker, P., 2009. Climate and historical stand dynamics in the tropical pine forests of northern Thailand. Forest Ecology and Management 257, 190−198.

11

Common forestry practices

This chapter presents an overview of the suite of common forestry practices that are used throughout the world to establish and maintain trees, extract commodities from trees, promote the development of productive and healthy forests and manipulate forest habitat. Active management of the world's forests increased significantly in the 20th century for many socioeconomic and political reasons. For example, in Japan there was a need to increase the national income and provide a stable supply of wood for local and regional markets during a period of high economic growth. As a result, 40% of the current forests in Japan were planted, 90% of which were established after World War II (Fujiwara, 2002). Management practices may return revenue to a landowner, and therefore from an economic perspective may be self-supporting. Some management practices may also be seen as a cost to a landowner and must be considered alongside previous and future revenues to determine the appropriateness of the practice. Every practice used has a biological, economic, and social impact on the land and the landowner. Valuing and understanding these impacts will allow one to become an effective land or resource manager. The overall objective of this chapter is to present a broad, general set of forestry practices and to provide examples of how this may be used around the world to maintain the resources of the forest. Upon completion of this chapter, readers should be able to

- understand the terminology commonly used to describe forestry practices within the broad set of natural resource management professions;
- understand the general purposes and uses of common forestry practices;

Introduction to Forestry and Natural Resources, Second Edition
https://doi.org/10.1016/B978-0-12-819002-9.00011-0

- acquire a perspective on the use of forestry practices within the development of a forest from seedling to sawlog or from recently burned landscape to rejuvenated habitat; and
- become familiar with practices where trees are managed in conjunction with agricultural crops and livestock.

We begin this chapter with an overview of site preparation and reforestation practices and proceed through intermediate and final harvest practices of forest management. In the development of forests in which all of the trees begin life at about the same year (even-aged forests), site preparation and reforestation activities may be necessary to ensure that the new forest contains the stocking and species desired by the landowner or land manager. In this discussion we also address natural regeneration through coppicing and other methods, and afforestation, or the development of forests on land not recently forested. Early tending of forests, and thinning, pruning, and fertilization activities are also described, as are methods employed to remove older trees and begin a new forest once again. There are a variety of methods for developing and maintaining uneven-aged forests, and these activities are also described in this chapter. A discussion of prescribed burning and fuel reduction treatments is presented, and we conclude the chapter with a general treatment of agroforestry practices and biomass harvesting activities.

11.1 Site preparation

For forestry purposes, a *site* is an area of ground on which a forest is, or will soon be, residing. Given the physical and biological factors associated with a site, a determination of the type of forest it can support may be made (Nieuwenhuis, 2010). For example, a site with less moisture may support pine while a wetter site may favor a true fir. *Site preparation*, as the name suggests, involves making the site suitable for the establishment of a new forest and, in most cases, an even-aged forest. In many cases of forest establishment, the success of reestablishing a new forest depends on the level of competition encountered by tree seedlings and, in the case of seeding, whether seeds are in contact with mineral soil. Methods for removing ground vegetation and debris prior to the establishment of a new forest fall into the category of site preparation activities, which vary by vegetation type, ecosystem, and socioeconomic conditions (Pancel, 1993). In preparing a site, unwanted logging debris (slash, which may now be considered a biomass resource), roots, stumps, rocks, and other features may need to be removed or pushed aside to facilitate planting activities. Vegetation that exists on a

site prior to the establishment of a new forest may ultimately compete with new trees for nutrients, water, and sunlight and is often removed or reduced. A consideration of whether this vegetation will adversely affect the growth and survival of the preferred trees and whether it is necessary to meet the biological requirements of other objectives will help one decide how or if certain site preparation methods are necessary. The cost of the site preparation activity and the feasibility of conducting it should also be considered when contemplating the preparation of a site for the establishment of a new forest. For example, average costs of site preparation activities in the coastal plain of the southern United States can accumulate quickly when additional treatments are necessary (Table 11.1).

Site preparation methods can be implemented manually, mechanically, or aerially. Some of the more common practices used in North America include burning, chopping, raking, plowing, and bedding. Burning as part of site preparation was used extensively in the southern United States toward the end of the 20th century but has since declined in use as a method for clearing debris from a logged area. Burning piles of debris is still a relatively common practice in the western United States. Site preparation burning is perhaps the most effective method for clearing debris in forested areas of the tropics (Pancel, 1993). Chopping or crushing practices (Fig. 11.1) are commonly used to break down larger residual material to facilitate burning and planting activities. As in the southern United States, burning, raking, and bedding are also common site preparation activities on poorly drained sites in South Africa. Raking (Fig. 11.2) involves using a bulldozer or skidder with a blade containing teeth attached to the front of the vehicle; the teeth can penetrate and mix the soil and disrupt root systems from the previous forest. Windrows (Fig. 11.3) are often developed by raking together

TABLE 11.1 Average costs for site preparation activities in the coastal plain southern United States.

Site preparation component	Cost (US$ per ac)	Cost (US$ per ha)
Chopping	85.76	211.91
Bedding	211.65	522.99
Aerial chemical site preparation application	43.27	106.92
Prescribed fire after chemical application	38.65	95.50
Aerial chemical herbaceous weed control application	13.39	33.09

For chopping and bedding, Barlow and Dubois (2011); for other practices, Barlow and Levendis (2015).

FIGURE 11.1 Tracked machine pulling a chopper, which is breaking down material for site preparation purposes. *Jack Chappell.*

FIGURE 11.2 Raking logging debris in western Oregon, United States. *Scott Roberts, Mississippi State University, Bugwood.org.*

FIGURE 11.3 Windrowed slash on the DeSoto National Forest in Mississippi, United States. *Scott Roberts, Mississippi State University, Bugwood.org.*

FIGURE 11.4 Bedding or creating a raised planting area. *John D. Hodges, Mississippi State University, Bugwood.org.*

felled debris and are often burned when conditions are suitable (Pancel, 1993). Bedding involves the creation of long, linear planting mounds (or beds; Fig. 11.4). Like raised beds in your gardens, these beds reduce seedling mortality due to waterlogged soils. Subsoiling or ripping is sometimes used on soils where potential tree root growth can be restricted due to soil compaction or because the soil has a natural hardpan. As in other areas of the world, drawbacks of these intensive site preparation methods include soil compaction and the removal of topsoil (Dupuy and Mille, 1993). Aerial applications of herbicides can be used to kill undesired woody and herbaceous vegetation. Through the late 1980s in the southern United States, prescribed fires, often called broadcast burns, were frequently used for site preparation purposes, regardless of whether a chemical treatment was applied. However, site preparation prescribed fires in the southern United States are now primarily used in conjunction with chemical treatments and are timed to occur after the treated vegetation has died (Barlow and Dubois, 2011). The dead vegetation can facilitate improved burning; thus, the combined

chemical and fire treatment is more effective than either treatment alone.

The application of herbicides can result in public opposition to the practice based on concern over potential effects to human health and ecosystem processes. Concern over the use of herbicides in forestry seems to be greater than concern arising from the use of herbicides in agriculture, perhaps because herbicide use in forestry is not directly related to the production of food and other site preparation methods seem to be available (Ghassemi et al., 1982), regardless of cost or efficacy. The aerial drift of herbicides during their application is another concern that is often encountered and can be reduced through use of best practices. Examples include buffer zones (unsprayed strips between the application area and neighboring areas) which are frequently used in forestry and agricultural practices as a measure of protection (Carlsen et al., 2006). Other social concerns relate to the residual fate and toxicity of herbicide concentrations in soils and waterways, the latter of which may affect marine and wetland organisms (Lewis et al., 2009).

As might be gathered through this discussion, site preparation methods vary around the world. Site preparation is also called *cutting-area cleaning* in the Russian Federation, and activities are chosen that will create the conditions necessary for forest regeneration and reduce fire and disease risks (Krott et al., 2000). In South Africa, site preparation activities for the establishment of pine plantations have included both manual and mechanical techniques such as pitting, auguring, ripping, and plowing the ground to develop areas for planting trees. A pitting technique involves digging a hole about 45 cm (about 18 in) wide and 20 cm (about 8 in) deep with a hoe. A soil auger may be used to develop a planting pit about the same size. Ripping to a depth of about 60 cm (about 24 in) along parallel lines spaced at about 2.7 m (about 9 ft) apart is accomplished using a bulldozer. Ripping and plowing treatments are generally preceded by removal of logging slash and stumps (Zwolinski and Donald, 1995). In the development of eucalyptus fuelwood or pulpwood plantations in Africa (Togo and Congo), sites are typically cleared, stumps are removed, and the remaining debris is placed into windrows, which may be subsequently burned.

11.2 Forest regeneration

Forest regeneration can proceed naturally or artificially with the assistance of seeds, seedlings, or coppice (stump sprouts). In some areas of the world, the term *forestation* is used to represent an artificial method of reforesting or restocking an area, which includes both reforestation and afforestation activities. *Reforestation* involves establishing a new forest on an area of land that has recently supported a mature forest, whereas *afforestation* involves establishing a new forest on an area of land that has not recently supported a forest. Pancel (1993) suggests that the length of time that defines *recent residency* of forests is about 50 years. From a general point of view, reforestation activities help maintain the forest area, while afforestation activities help increase the forest area. Reforestation activities are often used in the development of even-aged forests, as most of the newly established trees will be about the same age. However, reforestation processes may also be important in the development of uneven-aged (or mixed-aged) forests. Natural regeneration is an important component of uneven-aged forests that ensures the continued

replenishment of trees as older trees die or are harvested. Further, small group selection harvests may be employed to increase the heterogeneity of forest conditions in a management area. Within the small patches that are created using these harvests, regeneration (natural or artificial) of trees is expected. In addition to facilitating the development of future forest products, reforestation activities can be viewed as activities performed to prevent or control soil erosion or to increase the amount of carbon sequestered from the atmosphere.

11.2.1 Natural regeneration

Natural regeneration, which Germans call *naturverjüngung*, involves the establishment of a new forest from self-sown seed, coppice shoots, or root suckers (Nieuwenhuis, 2010). As we will discuss later in this chapter, shelterwood or seed tree harvests can be used to promote the development of self-sown seed from which a new forest can emerge. Natural regeneration from the seed of mature live trees that remain within or on the outskirts of cutting areas is a practice frequently used when the economic or logistical issues of more intensive regeneration practices prevent the direct planting of seedlings or the combination of the surrounding trees, climate, and environment can produce an adequately stocked forest. For example, the taiga forests of the Russian Federation are mainly naturally regenerated due to high labor and equipment costs, and natural regeneration is prevalent in the Ukraine in areas where there is a sufficient quantity of seedlings of the desired tree species (Krott et al., 2000).

Coppice shoots are new growth (stems) arising from dormant or adventitious buds near the base or stump of a tree, where the previous tree was cut (Fig. 11.5).

FIGURE 11.5 Sweet chestnut (*Castanea sativa*) coppice in the New Forest Hampshire, southern England. *Clive Perrin, Wikimedia Commons. Image Link: https://commons.wikimedia.org/wiki/File:Coppice_stool_in_Hasley_Inclosure_New_Forest_Hampshire_-_geograph.org.uk_-_686199.jpg License Link: https://creativecommons.org/licenses/by-sa/2.0/deed.en*

Root suckers arise from shallow or exposed root systems that have been wounded, perhaps within a short distance of a tree's stump (Wagner et al., 2010). Coppicing is often assumed to be part of the reforestation process in naturally regenerated hardwood or mixed pine-hardwood forests of the southern United States, mainly for regenerating the deciduous (hardwood) tree species component. However, the redwoods in California can reproduce through sprouting and have accelerated growth (Berrill and Boston, 2019). Coppicing has also been used as a regeneration method in the steep, deciduous forests of central Japan for several hundred years (Okazaki, 1964) and is a common practice for the regeneration of Mediterranean oaks (*Quercus* spp.) in Spain (Adame et al., 2006). Coppice systems are also used for the development of short-rotation eucalyptus or cottonwood (*Populus* spp.) fiber plantations. Sprouting is another method for naturally establishing a new forest. In this practice, trees are felled, but some live branches are left intact with the stump to become the seedlings and poles of the regenerated forest. White cedars (*Chamaecyparis* spp.) in Japan have been regenerated in this manner, with densities ranging from 4000 to 5000 stems per ha (about 1600–2000 per acre [ac]) (Okazaki, 1964).

11.2.2 Artificial regeneration

Artificial regeneration involves using seed, seedlings, or rooted cuttings to establish a new forest. Seeding is usually performed aerially, with planes or helicopters, across a recently harvested area of land. The rate of seeding varies depending on the assumed success rate of seed germination and therefore the survival of the expected seedlings. For example, in Minnesota, United States, artificial regeneration using direct seeding is suggested at a rate of 61,776 seeds per ha (25,000 seeds per ac; Minnesota Department of Natural Resources, 2020), and rates much higher than these have been used for other species, such as Lebanon cedar (*Cedrus libani*) on karst (weathered rock or limestone) topography in Turkey (Boydak, 2003). Directly seeding an area for forest establishment can increase the chances of successful reestablishment of recently harvested areas and can be applied immediately after a harvesting operation. However, the density of a reestablished, directly seeded forest may be too high and require additional thinning or consumption of seed by rodents may be a problem, advanced plant competition may influence survival rates, and availability of appropriate seeds may be limited, seeds gathered from similar sites. Some of these problems may require early tending practices in the forest to adjust tree density or tree species composition. Further, to ensure the best probability of success, seeds should be collected from forested areas near the planting

site or from an area with a compatible seed zone (Rose and Haase, 2006).

A seedling is a very young tree, perhaps 1 or 2 years old, that has been grown from seed in a tree nursery and transplanted to the area where the new forest will emerge. When developed in a tree nursery (Fig. 11.6), seedlings may be fertilized, top-trimmed, and undercut prior to being lifted from the nursery bed and packaged in bundles (bags or boxes) for transport to the forest. Some seedlings are now grown in containers designed specifically for a single seedling (Fig. 11.7). When lifted, these containerized seedlings are transported to the forest in the containers in which they have been growing, and the container is removed prior to planting the seedling. A seedling can also be developed from a rooted cutting of an older plant. A rooted cutting is a developed shoot cut from a hedge or donor tree that is planted in soil to allow the development of a root system. Rooted cuttings can be developed from several tree species, but in general the donor should be a younger tree, as cuttings derived from older trees may not have the capacity to develop a root system. Almost the entire pine-planting program

FIGURE 11.6 Pine tree nursery in the southern United States. *U.S. Forest Service, Southern Region.*

FIGURE 11.7 Longleaf pine (*Pinus palustris*) containerized seedlings. *U.S. Forest Service, Southern Region.*

in subtropical Queensland (Australia) consists of trees developed through rooted cutting methods (Trueman, 2006). Reforestation using rooted cuttings from live trees has been practiced in Japan for over 200 years (Okazaki, 1964), particularly for the management of sugi or Japanese cedar (*Cryptomeria japonica*) (Fujiwara, 2002). Depending on the type of seedlings or rooted cuttings being planted, they should be handled with care when they are transported from a tree nursery to a planting site. Special handling and storage precautions need to be taken with bare-root seedlings to ensure that they do not dry out prior to insertion into the ground. In some areas of the world, regenerated seedlings may need to be protected from mice, rabbits, deer or elk, and other animals that browse on low-lying vegetation.

Tree-breeding programs have been developed throughout the world in an effort to increase the productivity of commercial tree species and to decrease the susceptibility of these species to diseases. For example, pathogens such as fusiform rust (*Cronartium quercuum f. sp. fusiforme*) can affect the growth and stem form of southern United States pine plantations in such a way as to result in losses as large as perhaps 10% of the annual harvest levels (Geron and Hafley, 1988). However, the incidence of fusiform rust infection can be reduced considerably by using species varieties that exhibit resistance to the pathogen (Isik et al., 2005). Four general methods of producing seedlings for tree planting efforts are

- collecting seed from seed production areas or untested orchards and sowing them in tree nurseries;
- collecting seed from well-tested, open-pollinated families of trees from seed orchards and sowing them in tree nurseries. Here, the genetic identity of only one parent is known;
- collecting seed from trees with well-tested parents, propagated from seed and grown in tree nurseries or through vegetative multiplication. Here, the genetic composition of both parents is known through controlled cross-selection; and
- cloning trees from a desired parent using advanced vegetative propagation methods (McKeand et al., 2003).

The development of clones is a recent change in the development of tree seedlings and may help reduce the number of reproductive cycles necessary to obtain desirable genetic traits (Kriebel, 1983). In the southern United States, pine clones are being developed by a few organizations and planted as trials of new technology (Bridgwater et al., 2005). Currently, however, the supply of cloned seedlings is limited and is focused on pines in the southern United States and radiata pine (*Pinus radiata*) in Australasia (McKeand et al., 2003) and eucalyptus in South America and Asia. When

employed, pine plantations in the southern United States developed from clones might result in improved wood quality and productivity (wood volume at the time of harvest) gains of up to 50% over plantations developed from untested or open-pollinated seed orchards (Fox et al., 2007). Controlled cross-selections of southern pines can also result in as much as 30% gains in productivity (Martin et al., 2005).

The selectivity employed in the development of regeneration stock through seed orchards results in genetically improved seedlings rather than genetically engineered seedlings. One genetically engineered tree species currently planted for commercial forestry use is the *Bacillus thuringiensis* (Bt)-resistant European black poplar (*Populus nigra*) that is used in China (Hu et al., 2001; Kellison, 2007). Others include a frost-tolerant eucalyptus that is planted in the southern United States. There are several criticisms against the use of genetically improved seedlings in forestry activities, from potential effects on biodiversity to insect and disease resistance (and hence plantation failure). However, tree breeding is now seen by many as an essential aspect of commercial forestry programs that increases the economic attractiveness of the investment and helps to reduce pressure on the remaining natural forests (Sedjo, 2001). For example, results of a tree breeding program in Australia, where hybrid eucalypts (*Eucalyptus* spp.) were developed for their growth potential and ability to withstand high salinity levels in former agricultural landscapes, suggest that gains in the volume growth of these over pure eucalypt tree species can be at least 40% (Dale and Dieters, 2007). Further, tree breeding programs have been developed to address challenges in renewable energy and bioenergy initiatives that are important in areas of the world that rely heavily on offshore fossil fuel supplies. In Greece, for example, tree-breeding programs have been pursued to investigate the potential of eucalypts, poplars, willows, and other tree species to meet renewable energy supply needs through biomass production (Aravanopoulos, 2010).

Planting density is an important silvicultural assumption that varies according to the tree species, planting site, and organizational standard. Planting density is generally a compromise between the silvicultural requirements of each tree species planted and other economic and operational considerations. In many areas, planting density is designed with the goal of creating a young forest in which the tree canopy closes rapidly in order to limit the development of other vegetation. In addition, planting density may be designed with the goals of ensuring an adequate crown shape for individual trees and ensuring that natural pruning of lower branches will occur in a timely manner (Dupuy and Mille, 1993). In some silvicultural systems, high-density plantings are used to produce high-quality logs over long periods. One silvicultural system in Japan proposed planting 10,000 trees per ha (about 4000 trees per ac) of sugi and hinoki (*Chamaecyparis obtusa*). In such cases, straight, knotless boards may be desired for construction timber, and the production of boards with a specific number of annual rings per inch of wood may require some control of tree growth. Others consider planting density to be sufficient for stocking at final harvest to avoid high cost of thinning treatments.

Tree seedlings are either planted by hand (Fig. 11.8) or with planting machines (Fig. 11.9). Planting machines can be used effectively on flat terrain with limited logging debris. In other areas, planting crews attempt to plant seedlings by hand in a systematic manner that is labor-intensive but effective where debris or terrain conditions prevent the use of a machine. Ideally, plantations are established in a regular pattern (Fig. 11.10) that facilitates efficient thinning and tending operations. However, the planting rows of many plantations may conform to the shape of a parcel or be a function of the terrain or other operational decisions made at the time of planting (Fig. 11.11). In the development of seed orchards and plantations that may require intensive care, the pattern of planting may be very precise. For example, in some eucalyptus plantations in Africa, a precise grid pattern of planting is developed using a compass and distance measurements. Here, planting sites are staked, holes are excavated, and eucalyptus seedlings then inserted (Dupuy and Mille, 1993).

11.2.3 Afforestation

Afforestation is an internationally accepted term for the practice of planting trees on land that has not *recently* been used to grow a crop of trees (Richards, 2003). Over the last 50 years, afforestation of abandoned or marginally productive agricultural land has been a common practice in many areas of the world, including the southern United States. Interestingly, afforestation in arid or semiarid landscapes can be traced to China as far back as about 300 BC (Wang et al., 2010). As was noted in Chapter 2, afforestation in some areas of the world has ebbed and flowed according to political and socioeconomic conditions. For example, large-scale afforestation programs were prominent in United Kingdom forestry during the 20th century to reduce reliance on

FIGURE 11.8   Hand planting trees on the Umatilla National Forest in eastern Oregon, United States. *Dave Powell, United States Department of Agriculture Forest Service, Bugwood.org.*

FIGURE 11.9   Machine planting sessile oak (*Quercus petraea*) seedlings in Hungary. *Norbert Frank, University of West Hungary, Bugwood.org.*

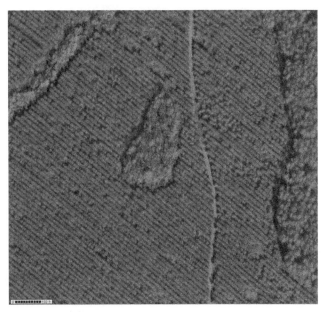

FIGURE 11.10   Aerial view of a regularly arranged, with respect to rows, pine forest plantation in southern Georgia, United States. *U.S. Department of Agriculture, Natural Resources Conservation Service (2020).*

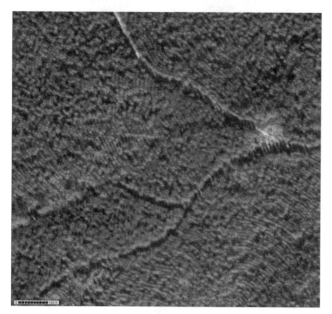

FIGURE 11.11 Aerial view of an irregularly arranged, with respect to rows, pine forest plantation in southeastern Virginia, United States. *U.S. Department of Agriculture, Natural Resources Conservation Service (2020).*

FIGURE 11.12 Shelterbelt plantings in an agricultural landscape in Cottle County, Texas, United States. *U.S. Department of Agriculture, Natural Resources Conservation Service (2020).*

imported wood. Afforestation programs have been used to expand the forest resources in Brazil, Chile, New Zealand, Senegal, and other countries. Currently, afforestation rates are rather high in China as a result of the Upper/Middle Yangtze River Valley afforestation program and the *Great Green Wall* initiative in Inner Mongolia (Rozelle et al., 2003; Ratilla, 2019). Afforestation rates are also high in regions of the Ukraine, particularly the Carpathian Mountains (Krott et al., 2000). There is hope that large-scale afforestation can reduce the effects of climate change with the ability of the new forest to sequester carbon (Doelman et al., 2020).

An afforested plantation composed of trees planted to shelter farmland and agricultural crops from the effects of wind, potentially reduce soil erosion, or serve as a source of fuelwood or income in an arid or semiarid agricultural environment is commonly called a *shelterbelt*. In this context, shelterbelts are commonly established as an element of an agroforestry system. Shelterbelts are often designed to facilitate irrigation and mechanization of agricultural practices and arranged in a manner to provide the associated agricultural crops protection from prevailing winds (Fig. 11.12). Shelterbelts can serve to decrease wind velocities in agricultural fields, reduce the evaporation rate of water, increase soil moisture content, and reduce extreme summer and winter temperatures. Used mainly in arid and semiarid landscapes, these tree and shrub plantings also provide protection to homes and livestock from harsh environmental conditions. Shelterbelts are also used for aesthetic and noise abatement purposes along

roads or highways. Shelterbelts can provide nesting, roosting, and foraging habitat for certain wildlife species, as well as travel corridors for others; thus, they are often used to promote biodiversity. Shelterbelts have been developed widely throughout Africa, Australia, China, Europe, India, Russia, and other areas of the world. Shelterbelt plantings originated in Scotland in the mid-15th century and in Russia in the 18th century. In North America, shelterbelts were employed in the mid- to late-19th century as prairie settlers valued trees for consumptive, aesthetic, and protective purposes (Droze, 1977). National programs such as the Great Plains Shelterbelt Project (United States), the Prairie Farm Rehabilitation Administration (Canada), the Great Plan for Transformation of Nature (Russia), the Three North Shelterbelt Forest Program (China), and the Green Wall of China have been used to promote the development of shelterbelt projects. While many shelterbelt practices have met the needs and objectives of local communities, the benefits and costs of large-scale shelterbelt projects for controlling desertification and dust storms are debatable (Wang et al., 2010).

In developing a shelterbelt, some site preparation and weed control (e.g., mulching) measures may be necessary and, since these are typically established in arid or semiarid environments, drought-resistant tree or shrub species are preferred. The spacing of trees in a shelterbelt is typically tighter than given in rules designed for a forest plantation and can range from one or two rows of trees to 20 or more rows. The orientation and location of shelterbelts are important considerations. For example, a north-south orientation may prevent permanent shading of agricultural crops and allow sunlight to reach the crops at different times of the day (Johnson and Brandle, 2009). Thinning, pruning, or coppice management may be needed over time to maintain the designed function of shelterbelts (Stange et al., 1998). Osage orange (*Maclura pomifera*) hedges were among the first to be developed as shelterbelts in the United States, since their rapid growth and sharp thorns deter domestic animals from agricultural fields. Black locust (*Robinia pseudoacacia*), black walnut (*Juglans nigra*), boxelder (*Acer negundo*), cottonwood (*Populus deltoides*), European larch (*Larix decidua*), green ash (*Fraxinus pennsylvanica*), willow (*Salix* spp.), and other species have also been used in the United States for shelterbelt plantings. Perhaps some of the most vulnerable of plantings, shelterbelts may be damaged by grazing animals, extreme drought, or lack of care or may be removed to accommodate new agricultural methods (Droze, 1977).

FIGURE 11.13 Weeding treatment in a young loblolly pine (*Pinus taeda*) plantation in the southern United States, where undesirable hardwood species are being removed with a brush saw. *James H. Miller, United States Department of Agriculture Forest Service, Bugwood.org.*

11.3 Early tending

During the early developmental stages of an even-aged forest, some practices may be employed to manipulate the vegetative conditions and influence the character of the forest. These practices are designed to affect the stocking of plants, and thus competition among plants, with the intent of enhancing the success of the reforestation effort. In fact, the survival and growth of a new even-aged forest often may depend on follow-up activities performed shortly after its establishment (Pancel, 1993). However, even though they are called *early tending* practices, they are not necessarily limited to even-aged forests; they may also be needed to control the understory vegetative condition (smaller trees, plants, shrubs, etc.) of uneven-aged forests. In some parts of the world, early tending practices are often considered to be those implemented during the prethicket or thicket stages to improve the quality of young trees and to encourage even development of the trees (Nieuwenhuis, 2010). These and other similar treatments are generally considered *release* activities. A release activity can involve freeing desired young tree species from overstory competition (e.g., brush, shrubs, and vines), preventing the development of undesirable plant species, or removing brush, grasses, weeds, or other plant competition from the area around small, developing trees (Schwarz et al., 1976).

The emergence of competing vegetation toward desirable crop trees may be seasonal or continuous and often a struggle for light, water, and nutrient resources (Pancel, 1993). *Weeding* is one type of early tending activity applied to regenerated forests (Fig. 11.13) to eliminate or suppress undesirable vegetation growing alongside the desired tree seedlings. It is similar to a cleaning treatment (Section 11.17) and is often used interchangeably with this term (Tryon and Hartranft, 1977). For our purposes, weeding frees trees from undesirable competition and is usually performed early in the life of a forest. We describe cleaning treatments as those that generally result in an open or cleared understory, with a purpose that does not necessarily relate to freeing trees from undesirable competition, and that can be performed at any forest age. During the first decade or so of a newly established forest, weeding undesirable vegetation may be performed using physical or chemical means to favor the growth and development of certain tree species. Weeding can be performed using herbicides, hand tools (brush knives or axes), or power tools that mow or cut undesired vegetation (Pancel, 1993). In some areas, livestock, especially goats, are used to reduce the vegetation and have reduced fire risks (Hart, 2001). In Japan, the weeding of undergrowth and vines may be performed numerous times (10–20) within a single even-aged rotation of Japanese cedar (Okazaki, 1964). Weeding can be performed up to four or five years after reestablishment of a new forest, and, within a single year, may be applied three or four times depending on the quality of the site and the rate of growth of undesirable species during the different seasons of a year.

In the early stages of the life of a tree, animals such as cows, deer, elk, and rabbits may browse on the seedlings of certain tree species, thus reducing the leaf area and growth potential of a plant or causing physical damage to the plants. The browsing of plants can result in the development of multiple plant stems, an increased susceptibility of trees to frost damage, weakened branches,

FIGURE 11.14 Douglas-fir (*Pseudotsuga menziesii*) seedling enclosed within a mesh tube to protect it from browsing. *Scott Roberts, Mississippi State University, Bugwood.org.*

and open pathways for disease or insect infestation (Kopp, 2007). Protective devices, physical barriers, and chemicals can all be employed on or around tree seedlings (Fig. 11.14) to decrease mortality rates and increase tree growth rates. Significant and selective levels of animal browsing activity can lead to tree regeneration failure (in the case of an even-aged forest) and changes in forest composition (in the case of uneven-aged forests). Mechanical damage to young trees can also occur through the bedding and trampling activities of larger animals. The trade-offs and strategies employed in areas where regeneration success is tenuous because of animal browsing activities can be extensive and costly. Analyses of options will need to consider the objectives of the landowner, the density of the animals causing the damage, and the expected damage to the regenerating trees (Kopp, 2007).

11.4 Precommercial thinning

In many areas of the world, thinnings of forests are made in dense, even-aged stands at an age (generally 3–20 years old depending on the forest region considered) where trees are not large enough to have sufficient commercial value to fund the practice. These *precommercial thinnings* are designed to remove trees of the desirable tree species using practices similar to those used in weeding operations and are meant to facilitate accelerated diameter growth, maintain desirable tree stocking levels,

regulate tree species composition, and improve the form and quality of the remaining trees (Nieuwenhuis, 2010; Ford-Robertson, 1971; U.S. Department of Agriculture, Forest Service, 1989). In some areas of the world, these are also called early *cleanings* or *liberation cutting* activities and are followed by conventional forest management regimes (e.g., Rytter, 2006).

Precommercial thinning may be necessary in over-stocked, naturally regenerated, even-aged deciduous and coniferous forests or in even-aged forest plantations with an abundance of natural regeneration intermixed with the planted trees. If not employed, the growth of these dense young forests may eventually stagnate (Grano, 1969). This can impact forest health, making these trees more susceptible to insects, fungus, or fire. If used, precommercial thinning should occur at an age where costs can be minimized and the best growth response can be obtained from the residual trees (Mann and Lohrey, 1974). The rate that a tree can grow is based on how much energy it needs for respiration, metabolism, and the transport of water and nutrients within the tree, and, since these resources are generally finite, controlling their availability to the crop trees may be important. As a result of increased growth rates in residual live trees, the practice of precommercial thinning can potentially increase the profitability of forest management activities (Smith and Beckham, 2007). The logical trees selected for precommercial thinning are those that are damaged (e.g., by fire), deformed, diseased, or unable to compete with the surrounding trees. At times, however, entire rows of trees are thinned to enhance the efficiency of the practice. Tree density, species composition, height, crown condition, and stand uniformity are also used to determine the intensity of precommercial thinning operations. Mann and Lohrey (1974) once suggested that this treatment should be applied to all young pine stands in the southern United States with a stocking level of over about 12,000 trees per ha (around 5000 per ac). Guidelines for landowners considering this practice have been developed for numerous areas of the world, such as the direction provided by the New Brunswick Department of Energy and Resource Development (2019) for coniferous and deciduous tree species typically found in northeastern North America (Table 11.2).

In a precommercial thinning, trees are generally cut and left on the ground where they have fallen, thus enriching the soil and providing the surrounding trees with additional light, water, and nutrients to enhance their growth. However, there has been interest of late in using the material felled through a precommercial thinning as a feedstock for biomass energy production. The recent attention to biomass utilization has prompted a number of field studies (e.g., Bolding and Lanford, 2005) designed to determine whether precommercial

TABLE 11.2 A precommercial thinning guideline for coniferous and deciduous forests of northeastern North America.

Pretreatment forest selection criteria	Posttreatment forest conditions
Forests with greater than 5000 crop trees per ha (about 2000 trees per ac)	Coniferous forests: 2000−3500 crop trees per ha (about 800−1400 trees per ac)
Forests in which >60% of the trees present are desired crop trees	Deciduous forests: 2000−3500 crop trees per ha (about 800−1400 trees per ac)
Forests in which the average height of coniferous trees is 2 −7 m (6.6−23 feet) or where the average height of deciduous trees is 4−9 m (13.1−29.5 ft)	White pine forests: 3000−4000 crop trees per ha (about 1200−1600 trees per ac) Mixed species forests: 2000−3500 crop trees per ha (about 800−1400 trees per ac)
Forests with acceptable tree species: ash, aspen, cedar, fir, hemlock, jack pine, larch, oak, red maple, red pine, spruce, sugar maple, yellow birch, white pine, and white birch	All forests: >60% of the trees remaining are desired tree species. All forests: >85% of the trees must meet quality standards

New Brunswick Department of Energy and Resource Development (2019).

FIGURE 11.15 Precommercial thinning of pine trees in Virginia, United States, with a brush saw. *Rich Reuse, Virginia Department of Forestry.*

FIGURE 11.16 A precommercial thinning (release) treatment being aerially applied to a 10-year-old slash pine (*Pinus elliottii*) plantation in Georgia, United States. *Ron Halstead, Halstead Forestry and Realty, Inc., Bugwood.org.*

thinnings can be cost-effective through generating woody material useful for generating energy and reducing fuel loads. Rytter (2006) describes a management regime for aspen in southern Sweden, where precommercial treatment within the first few years removes a significant amount of woody biomass for bioenergy production, leaving the remaining forest for the production of pulpwood or sawlogs. Site conditions, forest health conditions, biodiversity objectives, and subsequent commercial forest management objectives are often used to determine the residual live tree density from precommercial thinnings.

Precommercial thinnings can be accomplished using manual methods such as cutting or girdling with chainsaws or brush saws (Fig. 11.15) or other mechanical methods that employ large machines designed to chop, cut, or mulch trees. In many cases, precommercial thinnings involve the use of chemicals applied either manually or aerially, perhaps with the use of helicopters (Fig. 11.16). The openings and debris created by precommercial thinnings may temporarily improve the habitat for deer, rabbits, and some bird species. Concerns over changes in vascular plant diversity have been noted but have also been found to be insignificant in some areas of the world (Neyland and LaSala, 2005). When the woody material cut through a precommercial thinning is left in a forest, the resultant fuel load could become a fire hazard until it decomposes; thus, piling and burning of thinned trees is also common in some

areas of the world, but care is needed not to damage the remaining trees.

11.5 Pruning

Pruning is another practice that may be applied early in the life of a forest to improve the quality of the wood in the main stem of a tree (Fig. 11.17). One goal of pruning is to limit the number and size of knots in the bole of a tree so that high-quality and higher-value boards can be milled. However, in urban forests, trees may also be pruned for safety reasons (Dujesiefken and Stobbe, 2002). During the process of pruning, the lower branches (both live and dead) of a tree are removed, using

FIGURE 11.17 First pruning of a young pine plantation near Mount Gambier, South Australia, Australia. *Donald L. Grebner*

pruning ladders and hand or power saws. The lower branches of a tree may die naturally and fall once the canopy above is well formed; however, the time required to complete this natural process depends on the tree species and environment. Certain tree species, such as pines, may have an extensive system of lower branches and may therefore need to be pruned to produce knot-free wood (Pancel, 1993). Except for urban forests, the pruning of deciduous forests is not as widespread as the pruning of coniferous forests and is mainly performed as a corrective measure related to tree form or as a safety measure. Pruning has been suggested for white pine to reduce the lower branches that might be infected with white pine blister rusts from its dual host *Ribes* spp. (Costanza et al., 2019).

The practice of pruning can be employed in either even-aged or uneven-aged forests to promote the health and quality of the pruned trees. Pruning has been employed in commercial teak (*Tectona grandis*) and klinki (*Araucaria hunsteinii*) plantations in the tropics (Pancel, 1993); in coniferous plantations in Japan (Fujiwara, 2002); in Douglas-fir plantations of the Pacific Northwest of the United States; and in many other areas of the world. In Japan, pruning practices are sometimes employed a dozen or more times within a single even-aged rotation of Japanese cedar (Okazaki, 1964). Often, the first lift (first entry) of a pruning operation is at a height of only around 2.4 meters (m; 8 feet [ft]) above ground, which allows access to the trees and lowers the risk of fire (Pancel, 1993). At other times, the first

lift may be as high as 5 m (17 ft) above ground, to the minimum length of an average log. In the former case, the pruning operation can be applied to trees that are 5 to 7 m (16.4–23 ft) in height, while in the latter case trees would normally be 10 to 15 m (32.8–49.2 ft) tall before pruning is conducted. Additional lifts are usually performed at later stages in the development of a commercial forest, when the trees have grown taller. In these cases, a pruning ladder is needed to reach the lower branches of the trees. The process of pruning only the tree branches that are within one's reach is often called *brashing*, *brushing up*, or *low pruning*: pruning tree branches that are above one's reach is called *high pruning* (Ford-Robertson, 1971). When a branch is pruned, the cut should be clean resulting in a stub that is flush with branch collar (rather than flush with the tree bole), thus resulting in wounds that are smaller, that heal faster (Dujesiefken and Stobbe, 2002), and that are free from splinters or protruding pieces that could prolong the occlusion of the wound.

11.6 Commercial thinning

Essentially *commercial thinning* forestry practices are timber sales employed in immature and nearly mature stands of even-aged trees. Commercial thinnings may also be employed in uneven-aged forests to correct deficiencies in tree size classes; however, we mainly refer to these types of harvests in uneven-aged forests as partial cutting practices (see Section 11.12). In a commercial thinning, individual trees are selectively removed not only to promote the quality and growth of the trees that remain (U.S. Department of Agriculture, Forest Service, 1989) but also to salvage trees that may die before the next thinning or before the final harvest occurs (Nieuwenhuis, 2010). In contrast to precommercial thinning, which is usually viewed as a cost (i.e., no revenue is received) to a landowner or a land management organization, commercial thinning is generally expected to either produce a positive revenue or to break even (revenues equal costs) when the operation is complete. The age at which a forest should be thinned is approximately when tree crowns begin to compete for light or soon after the forest crown *closes* (i.e., when canopy closure approaches 90%–100%). In some forest plantations or otherwise even-aged forests, delaying a thinning practice could result in reduced growth rates or growth stagnation (Pancel, 1993). A *thinning entry* is represented by the age at which a thinning occurs, and a *thinning cycle* is the number of years between subsequent thinnings, if more than one is scheduled during the rotation. The *thinning intensity* is a combination of the frequency in which the practice is employed and the amount of removals at one time. Finally, a *thinning regime* represents the type, degree, and frequency of thinning practices that are employed in a forest.

Because there are many types of commercial thinning practices, foresters or natural resource managers should become acquainted with the practices that are common to the region in which they want to work. Generally, a *selective thinning* removes the inferior trees in all tree size classes and promotes the continued development of the superior trees. However, *selection thinning* can also involve the removal of superior trees to promote the development of smaller trees into merchantable product size classes by creating growing space for the smaller, often younger trees. A *thinning from above* could also be performed, where trees from the dominant, codominant, and suppressed tree classes are removed (Nieuwenhuis, 2010). This practice is sometimes used to promote the long-term improvement and development of a forest and is called *high thinning*. In a *thinning from below* (*low thinning*), trees from the subdominant and suppressed tree classes are removed. *Thinning proportional to the diameter distribution* requires that an equal number of trees are removed from all diameter size classes. Other types of thinnings, particularly for areas that have been planted in rows, include third- or fifth-row thinnings with selection of trees in between the rows. In *row thinning with selection*, every *X* number of rows are completely removed (Fig. 11.18) with a selective thinning in the rows that remain. This is called *systematic thinning* in some areas of the world.

FIGURE 11.18 Aerial view of the result of a fourth-row thinning with selection practice applied to a pine forest in Gulf County, Florida, United States. *U.S. Department of Agriculture, Natural Resources Conservation Service (2020).*

Thinning type can be expressed as a ratio of the average volume of the trees to be thinned (v) to the average volume of trees in the forest prior to thinning (V), or v/V. If this ratio is less than 1.0, a low thinning probably is suggested. In other words, the trees removed are lower in volume (smaller in size) than the average tree prior to thinning. Similarly, if the ratio is greater than 1.0, a high thinning probably is suggested (Nieuwenhuis, 2010). In some areas of the world, the desired spacing of residual trees is used as a guide for determining which tree to remove and which to leave. When a thinning prescription is developed, the tree species, classes, and forms to remove are defined, as well as the expected residual density of trees expressed in basal area. For example, a common, medium-intensity thinning prescription for loblolly pine (*Pinus taeda*) stands in the southern United States may be to remove enough trees so that 17.2 square meters (m^2) per ha (about 75 square feet [ft^2] per ac) of live tree basal area remain throughout the forest. The trees removed would likely comprise suppressed, intermediate, and, as a last resort, dominant and codominant crown classes, and include trees with poor stem form.

11.7 Fertilization

Various fertilization practices are used throughout the world to increase the productivity of forests by eliminating a shortage of a critical nutrient. In general, fertilization practices are often applied in short-rotation biomass forests (e.g., eucalypt), and sometimes in even-aged commercial forests (e.g., pine plantations throughout the world). Fast-growing forests have nutrient requirements that may extend beyond those available from the soil. Therefore, depending on need, various levels of nitrogen, phosphorous, potassium, and other nutrients are applied to forests. The amount of nutrients needed for vegetation development is dependent upon several factors (University of Florida, Institute of Food and Agricultural Sciences Extension, 2008), including

- the level or concentration of nutrients in the soil,
- the inputs and losses from other organisms,
- the inputs and losses related to atmospheric processes,
- the status of organic matter,
- the soil moisture content,
- the soil texture and structure,
- the soil weathering rates,
- the root system development of forest plants, and
- the mycorrhizal associations.

Fertilization practices can occur at time of tree planting, shortly after tree planting has occurred, or later in a forest rotation. In many cases, the decision to fertilize a forest is based on nutrient deficiencies that have been found in the foliage of trees or in a soil sample. At other times, forests are fertilized after a thinning has occurred in order to instigate a growth spurt in the residual trees and thus increase the economic efficiency of a forestry operation. Application rates of fertilizers vary. For example, for young southern pine forests in the United States the application rate of both phosphorous and nitrogen may be around 50 kilograms (kg) per ha (about 45 pounds [lb] per ac). Nitrogen fertilization has been tested on radiata pine in New Zealand at a rate of 50 kg per ha (about 45 lb per ac) (Smith et al., 2000). A rate of 200 kg per ha (178 lb per ac) of nitrogen is recommended for coastal forests of British Columbia (British Columbia Ministry of Forests and Range, 1995). For interior forests of British Columbia, the application rate for nitrogen is slightly less. Application rates for potassium may be double the rates of phosphorous and nitrogen. Phosphorous fertilization has also been tested in many timber production areas of the world including *Acacia kao* tropical forests in Hawaii (Meason et al., 2009), slash pine (*Pinus elliottii*) forests in South Africa (Wienand and Stock, 1995), and others. The addition of minor amounts of boron, copper, magnesium, sulfur, and zinc has been suggested to ensure that deficiencies in these elements will not limit the effect of additional nitrogen. When a nitrogen fertilization application is necessary, urea granules (or prills), which are high in nitrogen content, are commonly applied. When both nitrogen and phosphorous applications are necessary, diammonium phosphate, which has a 1:1 ratio of nitrogen to phosphorous, is commonly applied. Potassium chloride (potash), potassium sulfate, and potassium nitrate are the main fertilizers used when potassium levels are low in forests.

Fertilizers are generally applied every five to eight years either using ground-based tractor systems (on flatter ground, clearer sites, and younger forests) or helicopters (Fig. 11.19). Fixed-wing aircraft and small handheld devices have also been used; however, the use of tractors or helicopters is usually more cost efficient when large areas need to be fertilized. Fertilizers are best applied when tree roots are actively growing, air temperatures are low, and soil is wet. These conditions reduce the chance of losing the applied nutrients through volatilization (British Columbia Ministry of Forests and Range, 1995). A buffer zone around streams and water bodies should remain untreated to prevent fertilizers from entering water systems, preventing a violation of water quality laws.

With some forest practices, there are legitimate concerns regarding nutrient depletion, particularly after logging operations, where the entire tree, including the stump, is removed. It is a concern with biomass

FIGURE 11.19 Application of fertilizer by helicopter on the Wallowa–Whitman National Forest in eastern Oregon, United States. *Dave Powell, United States Department of Agriculture Forest Service, Bugwood.org.*

harvesting operations where the material is removed including the limbs and tops as well as the foliage from the site. Some site preparation methods and most harvesting activities increase the levels of available nitrogen to forests; however, the excess nutrient levels decline over time. Conversely, the combustion of fossil fuels, along with intensive agriculture and livestock operations, have altered the global nitrogen cycle, resulting in increased concentrations of nitrogenous compounds in the atmosphere and high levels of nitrogen deposition in some areas, such as southeast China (Wei et al., 2008). In an extensive list of studies regarding forest fertilization, scientists have examined the effects of fertilization practices on bacterial and fungal communities (Demoling et al., 2008), soil-decomposing animals, such as worms (Haimi et al., 2000), understory vegetation (Fraterrigo et al., 2009), and water quality (Binkley et al., 1999). Given the varied ecological, economic, and social concerns, the trade-offs involved in using fertilizers to enhance tree growth or correct nutrient deficiencies should be carefully considered.

11.8 Final harvest

A final harvest, in terms of this summary of forestry practices, is a *clearcut* practice, which involves the removal of virtually all of the trees in an area (Fig. 11.20) in one continuous harvesting operation (Schwarz et al., 1976). A final harvest can be implemented in both even-aged forests and uneven-aged forests, but generally acknowledges the initiation of a new even-aged forest. This type of practice can also include the removal of only the merchantable trees with the expectation that most of the remaining trees will be removed, killed, or otherwise knocked down through

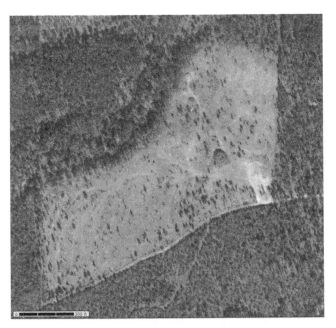

FIGURE 11.20 Aerial view of a small clearcut in Flathead County, Montana, United States. *U.S. Department of Agriculture, Natural Resources Conservation Service (2020).*

subsequent site preparation activities. However, in some final harvest practices, the viable undergrowth, seed trees, and saplings may be retained to help reestablish a new forest (Karvinen et al., 2006), often called advanced regeneration. A clearcut practice may also be called *area-wide felling* (Nieuwenhuis, 2010), *clean felling*, *complete cutting*, or *complete felling* (Ford-Robertson, 1971). From a silvicultural perspective, a clearcutting operation facilitates the development or reestablishment of an even-aged forest. To successfully establish a new forest of some shade-intolerant tree species (such as pines [*Pinus* spp.]), either naturally or artificially, both bare ground and a significant amount of sunlight may be necessary. A final harvest practice should not be confused with a *final felling* or *final cutting* practice, the last in a series of regeneration harvests associated with a seed tree or shelterwood silvicultural systems, which we discuss shortly. In those systems, one or more partial harvests are used to establish the regenerated forest, whereas these actions are absent when using the final harvest practice.

Clearcutting is used extensively in the Russian Federation for final harvests, although the size and arrangement of areas is based on the width of the cuts, economic factors, and local weather risks. In some areas, uncut strips are left between harvests to promote natural regeneration from the seed of mature trees (Krott et al., 2000), although regeneration in the uncut strips could also arise from artificial means. This practice is called *alternate clear-strip cutting* or *alternate clear-strip felling* and when regeneration in the previously harvested

strips is sufficiently established, the uncut strips are harvested (Nieuwenhuis, 2010) and often used with natural regeneration. In these cases, the maximum width of the clear-felled strips can range between 100 m (328 ft) and 500 m (16,411 ft) (Karvinen et al., 2006).

The *felling age*, *exploitation age*, or *rotation age* of a forest is the average age of the trees at the time of the final harvest. In the southern United States, the felling age can range from 20 to 35 years in pure pine plantations and from 30 to 50 years in naturally regenerated mixed pine-hardwood forests. In the Pacific Northwest of the United States, the felling age can range from 35 to 55 years on industrial or private land and from 80 or more years on public land. California regulates the minimum final harvest age to between 50 and 90 years depending on forest growth.

Final harvest ages for eucalyptus plantations in Brazil are generally less than 20 years and can be as short as 5 to 6 years. Standard final harvest ages are 50 to 60 years for some forests of Japan; longer rotations have been proposed because of a perception by some private landowners that active forest management is no longer economically feasible (Fujiwara, 2002). In Germany, rotation ages can reach up to 100 years in some locations. The average rotation length of coppiced stands of Mediterranean oaks in Spain has ranged between 20 and 30 years, although due to decreases in firewood and charcoal use since the middle of the last century, the practice of coppicing is becoming less prevalent across the Spanish landscape (Adame et al., 2006). In reality, felling age is a function of landowner objectives, economics, growth stage of the forest, and perhaps other social or environmental considerations. The size and pattern of final harvests can have a significant effect on the perception of people regarding this practice and in general human perception of scenic quality declines when the practice is employed (Palmer, 2008). However, some tree species require the early successional conditions presented by forest openings to successfully establish a new forest and some wildlife species require the early successional conditions presented by forest openings for one or more nesting, roosting, or foraging activities (DeGraaf and Yamasaki, 2003).

11.9 Group selection harvests

A forest regeneration system that opens the canopy through small groups of fellings (small harvests) that create gaps where regeneration can occur is often called a *group selection harvest* practice. Group selection harvests (Fig. 11.21) are small, cleared areas meant to promote natural regeneration from nearby mature live trees and reduce the aesthetic issues related to clearcutting. Here, the openings are much smaller than those normally recognized for mapping and management purposes (Ford-Robertson, 1971) as the entire patch is influenced by the edge of the adjacent forest patches. Patches as small as 0.2 ha (0.5 ac) and as large as 2 to 4 ha (5—10 ac) have been associated with this practice. Operationally, numerous small patches are removed from larger stands of trees to make the practice economically viable. This type of forest practice is generally applied to uneven-aged forests but can also be used in even-aged forests to prompt the development of uneven-aged or multi-aged cohorts in a forest. The aesthetic value of group selection harvests has been considered in several areas around the world (Okazaki, 1964), and artificial regeneration (planting) is often used since the amount of available shade affects the type of tree that can be naturally regenerated through seeding. With group selection harvests, a *return interval* or *cutting cycle* is usually prescribed, defining the number of years before the next set of patches is removed. Further, the intensity of the practice may be defined by the volume or amount of basal area removed from the entire stand or forest. For example, some coniferous forests in northern Italy are managed with group selection harvests that attempt to achieve timber production, ecological (e.g., soil protection), recreational, and aesthetic objectives. In this system, the typical cutting cycle ranges from 10 to 20 years and the intensity of harvest is about 20% (or one-fifth) of the standing tree volume (Grassi et al., 2003), which suggests that five entries are needed to completely remove the older forest.

In central European beech, fir, and spruce forests, a form of irregular shelterwood that is similar to group selection harvesting, the *femelschlag* system has been used. Here, a forest canopy gap is created by harvesting a small group of trees. Once regeneration of the preferred tree species has reached an acceptable level, the canopy gap is widened through the removal of more mature trees. This process continues until the original forest has been completely removed, leaving a forest containing ecologically different blocks or age classes of trees (Windhorst, 1976).

11.10 Seed tree harvests

A *seed tree* is a tree left standing (and alive) on purpose during a final harvest as a source of seed for natural regeneration of a new, even-aged forest (Schwarz et al., 1976). Therefore, a *seed tree harvest* practice is the removal of all trees (or all merchantable trees in a forest) with the exception of a collection of seed trees (Fig. 11.22). Seed tree harvests are a form of final harvest practice that promotes regeneration by leaving mature, live trees either scattered throughout a site or in clumps (about 15 trees per ha, or 6 per ac) on the site. Trees that remain following a seed tree harvest also act as the gene source for the newly established forest. Therefore, seed trees should be healthy, dominant, or codominant trees that can provide a suitable amount of high viability seed. Additional forms of site preparation may be needed in these areas to properly prepare the ground for natural regeneration. For example, seeds needing bare mineral soil to become established may also require a prescribed fire to scarify the ground or herbicide treatment before seed fall to facilitate their germination and survival. There are two stages generally used to implement the practice. The first stage removes all standing live trees

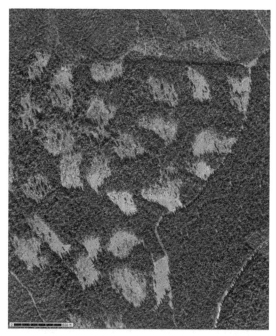

FIGURE 11.21 Small, 0.8—1.2 ha (2—3 ac) group selection patches in Lincoln County, Montana, United States. *U.S. Department of Agriculture, Natural Resources Conservation Service (2020).*

FIGURE 11.22 Recently logged stand with Scots pine seed trees (*Pinus sylvestris*) near Aegviidu, Estonia. *Hannu, Wikimedia Commons. Image Link: https://commons.wikimedia.org/wiki/File:Pine_stand_near_ Aegviidu,_Jan_2010.jpg License Link: Public Domain*

other than the seed trees. A few years later, the second stage involves entering the regenerated stand to remove the mature trees that served as seed trees for the regenerated stand (Schwarz et al., 1976). The second entry takes place when the regenerated, desirable trees are of sufficient stocking and height. However, the second stage is not always implemented since some landowners prefer to leave the larger, older seed trees on the site for aesthetic reasons.

If the risk of windthrow is high, a seed tree harvest practice should be used with caution when soil conditions suggest the presence of a shallow root-restriction layer. The type of seed produced should also be considered, since this will help determine the number of seed trees to leave per unit area. Seed that can be transported easily by winds may require fewer seed trees per unit area. Other considerations when using a seed tree harvest practice involve the arrangement of seed trees on the landscape (e.g., clumped, scattered, or in strips). Guidelines for the number of mature, live seed trees to leave on a site following a final harvest have been developed for numerous areas of the world. In Alabama (southern United States), for example, guidelines have been developed based on tree species composition and average seed tree diameter (Table 11.3).

11.11 Shelterwood harvests

One method to reestablish a new even-aged forest after a final harvest is to use a *shelterwood* harvest practice, in which the new, even-aged forest is naturally regenerated and established using seeds of trees from the previous forest. It is completed using two or three entries into the stand. This practice is similar to the seed tree harvest practice except that the mature overstory trees that remain during the reestablishment phase provide protection to the seedlings as they become established (Nieuwenhuis, 2010). In practice, two or more entries

FIGURE 11.23 Shelterwood practice in a coniferous forest in Idaho, United States. *Chris Schnepf, University of Idaho, Bugwood.org.*

are made into a mature stand of trees during the transition from older forest to regenerated forest. The first entry into the mature forest removes most of the overstory but leaves a number of residual trees scattered throughout the logged area (Fig. 11.23). Over the ensuing years, these residual trees will ideally provide seed from which new seedlings will arise. When the newly established younger forest is sufficiently developed, a second entry onto the site is conducted to remove the mature trees left over from the prior mature forest. Although all of the new seedlings may not have germinated in the same year, the reestablished forest is considered an even-aged forest, since the ages of all the trees are likely to be within a few years of each other. Prior to the first entry of a two-entry shelterwood practice, the understory vegetation may need to be controlled, perhaps with a prescribed fire or precommerical thinning, to reduce the density of undesirable trees or shrubs. Trees designated to remain after the first entry are likely be high quality and disease-free, and

TABLE 11.3 A sample set of guidelines for seed tree density per unit area in Alabama, by average tree diameter and main tree species.

Average seed tree diameter (cm/in)	Number of seed trees needed per unit area									
	Shortleaf pine (*Pinus echinata*)		Loblolly pine (*Pinus taeda*)		Slash pine (*Pinus elliottii*)		Longleaf pine (*Pinus palustris*)		Virginia pine (*Pinus virginiana*)	
	TPH	TPA	TPH	TPA	TPH	TPA	TPH	TPA	TPH	TPA
25.4/10	49.4	20	29.7	12	29.7	12	135.9	55	12.4	5
30.5/12	34.6	14	22.2	9	22.2	9	93.9	38	9.9	4
35.6/14	29.7	12	14.8	6	14.8	6	69.2	28	9.9	4
40.6/16	29.7	12	9.9	4	9.9	4	51.9	21	9.9	4

Alabama Forestry Commission (2004). TPH = trees per hectare; TPA = trees per acre.

therefore care should be taken to avoid damaging them. After the first entry, a prescribed fire might be used to establish a seedbed (bare ground) and reduce fuel levels. As with a seed tree harvest practice, if the regenerated seedlings become too dense after this practice, a precommercial thinning may be needed to reduce the stocking levels. There may also be a three-cut shelterwood system, in which the first entry removes much of the mid- and understory trees, the second entry removes most of the overstory trees, and the third entry removes the remaining parent trees.

In comparison to the seed tree harvest practice, the shelterwood practice generally leaves a higher density (basal area or number of trees per unit area) of standing live trees after the first entry, perhaps to enhance the economic value of subsequent entries or dispersal of the seeds throughout the forests. The first-entry residual basal area of Douglas-fir and western larch (*Larix occidentalis*) in southeastern British Columbia, as suggested by Hawe (1996), is around 13 to 25 m^2 per ha (57−109 ft^2 per ac). In the southern United States, the recommended residual tree density is 49−124 well-spaced trees per ha (20−50 trees per ac). The benefits of using a shelterwood practice are that reestablished trees will be somewhat protected by the residual trees for the first few years of their lives (Macisaac and Krygier, 2009) and that residual trees can grow more freely for 5 to 15 years owing to the availability of additional light, water, and nutrients. This practice favors tree species that normally become established under partial shading conditions and is sometimes used to rehabilitate degraded forests (Ontario Ministry of Natural Resources, 2020) or limit damage to existing forests from insect or disease problems such as the gypsy moth (*Lymantria dispar*) (Minnesota Department of Natural Resources, United States Department of Agriculture Forest Service, and University of Minnesota, 2003).

Variations to the shelterwood practice have been employed around the world. For example, an *irregular* or *progressive shelterwood system* (the *femelschlag* system) employs the cutting of trees along the edges of regeneration areas to advance the spread of the regeneration process. In a fashion similar to group selection harvests, a gap in a forest canopy may be opened through harvesting activities and, as the regeneration within the gap develops (usually naturally from seed of the surrounding trees), the gap is widened or enlarged through subsequent entries (Nieuwenhuis, 2010). While used pervasively across North America, shelterwood harvests have also been attempted in the tropical forests of Africa (Lowe, 1978), coniferous forests of Scandinavia (Holgén and Hånell, 2000), hardwood forests of Japan (Nagaike et al., 1999), eucalyptus forests of Australia (Dignan et al., 1998), and in other areas of the world.

FIGURE 11.24 Partial cutting operation in an uneven-aged conifer forest in southern Colorado, United States. *Dave Powell, United States Department of Agriculture Forest Service, Bugwood.org.*

11.12 Uneven-aged partial selection harvests

In uneven-aged forests, where the intent is to retain a canopy of trees of various ages over time, the annual or periodic removal of trees may occur individually or in groups (Fig. 11.24) through a *partial harvest* or a *selection harvest*. Selection harvest practices can be performed in both even-aged and uneven-aged forests, where the selection of trees is based on goals such as maintenance of structure or viability of the forest (Karvinen et al., 2006). In uneven-aged forests, this practice may be used to achieve a balance in tree diameter classes (a proxy for age) that would lead to a sustained yield of products from a forest (Nieuwenhuis, 2010) and help to establish or maintain the irregular character of an uneven-aged forest. Some suggest that group selection harvests should be categorized as a partial selection harvest; however, we make the distinction here that this practice involves removing individual trees or trees in patches smaller than the typical group selection patch. In addition, this practice may promote the regeneration of shade-tolerant tree species, whereas the group selection harvest practice generally favors the regeneration of shade-intolerant tree species. A partial selection harvest practice usually involves removing a scattered arrangement of the oldest and largest trees, which enhances the ability of smaller trees to grow into canopy gaps, while also enhancing the economic value of the operation. Selection harvests are conducted in entries that are separated in time by 5 to 20 years and continue indefinitely (Schwarz et al., 1976). The continuous establishment of naturally regenerated trees maintains and encourages the uneven-aged character of the forest.

The continuous application of single-tree selection harvest activities has been used to promote or maintain the development of coniferous forests in central Europe. In an example of the management of Calabrian pine (*Pinus nigra*) in southern Italy, the cutting cycle is around 20 years long and with each entry the amount of tree volume removed is slightly less than the amount of growth of the forest over the time period (Ciancio et al., 2006). Uneven-aged silvicultural practices have been suggested as a means for promoting and maintaining late-successional and old-growth characteristics of shade-tolerant sugar maple (*Acer saccharum*) and American beech (*Fagus grandifolia*) forests in northern North America (Keeton, 2006). The amount of tree volume removed during the single-tree selection treatment is based on a desired residual basal area of live trees (18.4 m^2 per ha, or about 80 ft^2 per ac).

FIGURE 11.25 Salvage operation in a southern United States pine forest affected by the southern pine beetle (*Dendroctonus frontalis*). *Terry S. Price, Georgia Forestry Commission, Bugwood.org.*

11.13 Partial selective harvests

In *selective harvest* practices, trees are removed or extracted according to their age, quality, size, or value with little regard for the remaining or residual forest character. These forest practices should not be confused with the selection or partial harvests that have been described previously. Selective harvests are practices commonly referred to as *creaming*, *culling*, or *high grading* and are not generally recommended, since silvicultural goals and the sustainability of forest yields are generally ignored (Ford-Robertson, 1971). These practices could be applied to any type of forest, whether even-aged or uneven-aged.

11.14 Salvage or sanitation harvests

A *salvage harvest* practice involves the removal of dead, dying, or deteriorating trees from a forest before the value of the wood products deteriorates below the extraction costs (Ford-Robertson, 1971; Nieuwenhuis, 2010). However, it may be used to remove trees with no economic value to promote a healthy regenerated forest. Technically, these types of practices could be applied to even-aged or uneven-aged forests. The trees removed may be overmature and therefore subject to damage imposed by insects and diseases or by wind, fire, or other natural events (Fig. 11.25). In this operation, the timing of the harvest is not consistent with the timing of a partial or a final harvest operation. Since 2000, a considerable debate has arisen over the postdisturbance salvage harvest of dead and dying trees in areas subjected to wildfire in North America. The practice must therefore be considered with care, given the goals and objectives of the landowner and the broader economic,

ecological, and social dimensions associated with salvaging forest products from forests damaged by natural events. Although the intent of a salvage harvest practice is admirable, Klingenberg et al. (2010) illustrate how a management response (salvage harvest and subsequent reforestation activity) to an insect outbreak can increase the vulnerability of the new forest to additional insect outbreaks.

A *sanitation harvest* practice involves the removal of trees with the intent of improving the overall health of the forest. In doing so, the goal is to stop, reduce, or prevent the spread of damage imposed by insects or diseases (Nieuwenhuis, 2010). For example, Bull et al. (2004) describe a sanitation harvest conducted in eastern Oregon in which, regardless of size, all trees with a witches' broom, that is, a manifestation of dwarf mistletoe (*Arceuthobium* spp.), greater than 25 cm (10 inches [in]) in diameter were removed in an effort to minimize future forest damage.

11.15 Prescribed burning

Prescribed burning is the intentional, controlled application of fire to a forest to accomplish the objectives of a landowner or land manager (U.S. Department of Agriculture, Forest Service, 1989). Prescribed burning includes fires designed for site preparation purposes and for forest understory maintenance (Fig. 11.26). In this section, prescribed burning refers to a surface fire that burns under the canopy of an established stand of live trees (Carter and Foster, 2004). These types of fires may be used in either even-aged or uneven-aged forests. Stand conditions, weather conditions, fuel moisture conditions, and soil conditions must be considered to determine the likelihood that a fire would be (1) confined to

FIGURE 11.26 Prescribed fire in action in a young southern United States pine forest. *U.S. Forest Service, Southern Region.*

FIGURE 11.27 Oak/maple/cherry forest in Pennsylvania, United States, containing about 32.6 tonnes per ha (14.5 tons per ac) of woody fuel. *Brose (2009).*

the area of interest once lit; and (2) conducted with the intensity of heat and rate of spread that would meet its objectives (Ford-Robertson, 1971). Prescribed burning could be used to assist in the development of a forest with a preferred species overstory, a midstory free of undesirable plant vegetation, and an understory composed of desirable herbaceous and woody plants (Haywood, 2009). In other words, prescribed fires can be used to promote or maintain the development of wildlife habitat characteristics.

Prescribed burning activities are generally moderate- to low-intensity endeavors (as it relates to upward release of energy and subsequent damage to tree crowns) that are repeated 1 to 5 years in a forest following the first burn. However, the objectives of a landowner or land manager would suggest whether prescribed burning activities are even necessary. At times, prescribed fires are used to reduce fuel loads and thus act as a wildfire prevention tool. Other times periodic fires are needed to maintain forest types, such as the longleaf pine (*Pinus palustris*) forest type of the southern United States (Bruce, 1951). Prescribed fires can also benefit the growth of forests by controlling diseases such as the brown spot needle blight (McCulley, 1948; Bruce, 1951). Prescribed fires should be designed to meet the specified silvicultural objectives while minimizing negative impacts on off-site social values. The smoke produced by fires may not only be a nuisance for nearby communities and cause health concerns in people, but may also increase the risk of accidents on roads and harm poultry farms. Within the area being burned, prescribed fires can result in a short-term increase in mineral nitrogen in the soil surface and an increase in phosphorous, the level of which is a function of the duration and intensity of the fire (Galang et al., 2010). However, over time, prescribed burning can prevent or reduce accumulations of nutrient capital that would otherwise occur naturally (Carter and Foster, 2004).

11.16 Forest fuel reduction treatments

The vulnerability of forests and local communities to wildfires is a major problem in many areas of the world due to the increase and expansion of the human population, long periods of fuel accumulation from wildfire suppression, changes in the frequency of wood-burning activities, and climatic variability. In some areas of the world, such as the southern United States, most wildfires are initiated by humans; thus, wildfire occurrence is exacerbated by expanding human populations and the spatial juxtaposition of developed areas and forests. In other areas of the world, lightning is a major cause of wildfires. In either case, the character and quantity of the fuel load will in part determine the severity of a wildfire. A *forest fuel load* consists of the live and dead fuels across a landscape (Fig. 11.27). The three most important considerations in fuel reduction practices are the current and projected fuel loads, the surface area to volume ratios of fuels, and the fuel depth (Burgan, 1987). Fuels are often discussed in terms of vertical layers, such as ground fuels, surface fuels, and elevated ladder and crown fuels (Pyne et al., 1996). *Ground fuels* are located either below the soil surface or within the organic layer and include the duff, organic soils, roots, stumps, and buried logs of a forest. *Surface fuels* include forest litter, branches, logs, and other dead vegetation lying on the surface of the ground. These can be further classified by their lag-time class that is related to their size. For example, a 1-hour time-lag dead class includes needles and leaf litter, grasses, and small twigs that require about 1 hour to dry before they are considered combustible. Fuels in the 10-hour and 100-hour time-lag dead classes consist mainly of dead branches and woody stems (Marshall et al., 2008). *Ladder fuels* are the live and dead branches that provide a path for fires to reach the crowns of trees.

Several types of forest practices are used to control the character and quantity of forest fuels. As noted earlier, prescribed burning is a practice used for managing forest fuels because of its relatively low cost per unit area and because the practice reduces fuel levels rather than just rearranging them on the landscape. The amount of forest fuel consumed by a prescribed fire depends on the fuel moisture, the structure of the fuel bed, and the firing technique employed. The level of fuel reduction accomplished with a prescribed fire can also be attributed to climatic variables such as air temperature, relative humidity, and wind speed. A low-intensity prescribed fire is likely to consume most of the 1-hour time-lag dead fuels (fine fuels); whereas multiple prescribed fires may be needed to consume larger fuels. Repeated prescribed fires can also be used to reduce live fuel loads, but fires need to occur frequently enough to either exhaust the root reserves of the undesired vegetation or affect short-lived vegetation before it can produce seed (Marshall et al., 2008).

Mechanical fuel reduction practices can be used to rearrange biomass across a landscape or remove it entirely. In the former case, mulching (mastication) is one practice that can be used to convert live fuels into smaller pieces and therefore to rearrange these fuels horizontally and vertically within a landscape, effectively moving them closer to the ground surface. A crush and chop practice may also be used for reducing fuel loads, in which the weight of a chopper (pulled along by a tracked machine) repositions fuels closer to the ground. Machines have also been developed to bundle small stems and branches of vegetation for removal from a forest. These types of practices may be cost-prohibitive for private landowners; however, biomass energy markets are developing that may allow these practices to continue without government subsidy. Chipping operations can also be used to reduce fuel loads and perhaps to provide a resource for nearby wood-using industries. However, the value of chips is a function of the species of tree, chip size, shape, percent contamination (with dirt, bark, etc.), and moisture content (Sessions et al., 2013). Many mechanical fuel reduction practices are actually precursors for a subsequent prescribed fire program. For live or dead fuels situated in a forest canopy, pruning and thinning operations can be used to disrupt the ladder and crown fuel structure of a forest. Of the mechanical treatments, the cost per unit area, the productivity, the potential soil damage issues, and the access considerations are the major concerns. Mechanical fuel reduction practices can also increase sediment production in streams if a significant amount of soil disturbance occurs (Marshall et al., 2008).

Herbicide treatments are an alternative practice for reducing fuels within forests. However, in order to remove undesired vegetation situated below a forested canopy without harming the overstory trees, this practice may require a ground-based application (rather than an aerial application) that targets specific types of vegetation. For example, if the live, midstory vegetation of a forest has become so dense that a prescribed fire would have little effect on the fuel load, an herbicide application might be used as a preliminary practice to kill or suppress unwanted live fuels. An herbicide practice might also be employed to kill resprouting vegetation that arises after a prescribed fire or mechanical fuel reduction practice. The effectiveness of herbicide treatments for reducing fuel loads will depend on the type of vegetation currently occupying a site, terrain considerations for equipment, and soil conditions (Marshall et al., 2008). Unfortunately, herbicide practices only reduce the amount of *live* fuels, thus they may gradually increase dead fuel loads over time (Gagnon and Jack, 2004).

In many areas of the world, livestock have been used to suppress forest fuel loads. For example, with this partly in mind, an extensive area of the pine and mixed pine-hardwood forests of the southern United States was grazed throughout the mid-20th century (Campbell, 1948). Cattle or sheep can be used to suppress fine fuels (grasses), and the incidental trampling of animals across the landscape can break small, nonedible fuels and reposition them closer to the ground (Zimmerman and Neuenschwander, 1983). Goats have also been employed to reduce fine fuels and to break down 1-hour and 10-hour time-lag dead fuels through trampling (Tsiouvaras et al., 1989). Livestock are likely to seek vegetation of high nutritional value and avoid areas of dense vegetation; therefore, their impact on fuel levels may be uneven. The effective use of this practice for forest fuel management is contingent on having a sufficient number of livestock that will consume less-palatable vegetation, perhaps of low nutritional value. As with herbicide practices, fuel reduction treatments that rely on animal consumption target live rather than dead fuels; therefore, the practice may be of limited value in areas with high levels of dead fuels. Livestock can also disrupt ladder fuels to a height of about 1 to 2 m (3.3–6.6 ft) above the ground (Marshall et al., 2008).

Unintended side effects of fuel reduction practices include changes to forest structure components that are important habitat considerations for various species of wildlife. For example, the loss of down, coarse woody debris (CWD) could affect habitat conditions favored by ground-dwelling animals such as salamanders or shrews. Unfortunately, CWD is often the target of mulching and prescribed fire practices. The loss of snags (standing dead trees) can also affect the habitat conditions favored by birds, such as woodpeckers. A number of nontimber forest products (NTFPs) may also be affected by fuel reduction practices, including edible mushrooms, floral greens, game species (e.g., quail, turkey, or deer), medicinal plants, pine straw resources, and specialty wood products (e.g., burls, twigs, or branches), which may all be impacted by fire, herbicide applications, livestock grazing practices, or mechanical operations (Marshall et al., 2008). However, some fuel reduction practices can actually increase the quantity and quality of woody and herbaceous food for deer; unfortunately, these practices can also reduce the vegetation required for hiding purposes (Maas et al., 2003).

11.17 Understory cleaning

The clearing of forest litter and understory firewood is considered an important, and often controversial, practice in even-aged and uneven-aged forests of some areas of the world. Like weeding practices, understory cleaning practices remove biomaterial from the forest floor, perhaps completely or only within a certain distance of desired trees. In an intensive cleaning, all woody stems within a forest are cut other than desired crop trees (Della-Bianca, 1969). Cleaning practices can also be used to facilitate other forest management activities, such as the collection and harvest of nontimber forest resources (Fig. 11.28). These practices may cause soil compaction, which can affect soil respiration, soil temperature, soil moisture, and other soil features (Wenjie et al., 2008), in addition to removing potential nutrients from a site. However, cleaning practices can affect the abundance and character of desired tree species; thus, cleaning has been suggested to be important for the growth of trees such as yellow birch (*Betula alleghaniensis*) in northern North America (Erdmann, 1990).

Cleaning treatments are widely applied in coniferous and deciduous forest plantations in Japan to reduce competition for regenerated trees (Hirata et al., 2011). In pine and oak forests of northeastern Spain, cleaning practices have, interestingly, been shown to lead to increases in large diameter trees, shrub species diversity, and tree species diversity (Torras and Saura, 2008).

11.18 Biomass harvesting

One of the most heavily studied operational forestry practices of late, primarily driven by high fossil fuel prices of the early 21st century (Baker et al., 2010), has involved the removal of biomass from forests for bioenergy purposes. These types of practices are similar to precommercial thinning and cleaning practices. However, in contrast to precommercial thinning, biomass harvesting generally implies that a commercial goal exists (e.g., earning money from the sale of forest biomass). In contrast to cleaning practices, the amount of vegetation from a site may not be completely removed and its removal may not focus specifically areas around desired crop trees. Biomass harvesting may involve not only the removal of live vegetation from the understory of a forest but also the removal of logging residues (tops, limbs) that are created during a normal harvesting operation (Fig. 11.29) (Baker et al., 2010). This material has traditionally been left on site. This practice could be employed within even-aged or uneven-aged forests, depending on the goal of the landowner. Several studies are under way around the world to assess the usefulness of traditional forestry equipment and new technology in collecting this resource (e.g., Yoshioka et al., 2006).

FIGURE 11.28   Young longleaf pine (*Pinus palustris*) stand that was cleaned prior to the raking of pine straw (pine needles). *Hans Rohr, through Wikimedia Commons.*
Image Link: https://commons.wikimedia.org/wiki/File:Young_Straw_Stand.JPG
License Link: https://creativecommons.org/licenses/by-sa/3.0/deed.en

FIGURE 11.29   Processing logging slash into wood chips, Uinta National Forest, Utah, United States. *Doug Page, U.S. Department of the Interior, Bureau of Land Management, Bugwood.org.*

11.19 Agroforestry

Although it may seem that agroforestry is a recent phenomenon, it has been a common land use for thousands of years (MacDicken and Vergara, 1990). Only since the middle of the 20th century have forest and natural resource professionals attempted to formalize a description of these systems. When we think of agroforestry, we often think of common forestry practices performed in conjunction with agricultural or livestock production activities. These practices come in many forms and vary greatly depending on the part of the world where they occur. Before we continue our discussion of common agroforestry practices, it is important to explore a few definitions of agroforestry. Bene et al. (1977) defines agroforestry as "[a] sustainable management system for land that increases overall production, combines agricultural crops, tree crops, and forest plants and/or animals simultaneously or sequentially, and applies management practices that are compatible with the cultural patterns of the local population." A later definition by the Society of American Foresters defines agroforestry as "[a] land use system that involves deliberate retention, introduction, or mixture of trees or other woody perennials in crop and animal production systems to take advantage of economic or ecological interactions among the components" (Helms, 1998). This definition is a modification of the one put forth by Nair (1984) and MacDicken and Vergara (1990). As these two definitions indicate, agroforestry is essentially a system of land management using trees and either agricultural crops or animals, and that attempts to produce more of a given product (or set of products) on a given unit of land over a distinct period.

There are many agroforestry systems around the world. For instance, in Costa Rica, it is not uncommon to see trees planted in coffee plantations. Tree species such as bucare ceibo (*Erythrina poeppigiana*), ingá-branco (*Inga laurina*), and roble de sabana (*Tabebuia rosea*) are often found above coffee plantations (Haggar et al., 2011). During the 1930s, the Prairie States Forestry Project was initiated in the United States and led to the establishment of 200 million trees and shrubs within 29,934 kilometers (km; 18,600 miles [mi]) of windbreaks (shelterbelts) adjacent to agricultural fields in the Great Plains region of the country (Baer, 1989). In Kenya, dairy farmers plant fodder shrubs such as red calliandra (*Calliandra calothyrsus*) around homesteads, intercropped with napier grass (*Pennisetum purpureum*), for erosion control and as a source of fodder for cows to boost milk production (Franzel, 2004). In Micronesia, home gardens (or multistoried tree gardens) may contain breadfruit (*Artocarpus altilis*) and coconut (*Cocos nucifera*) at the highest point in the vertical structure of vegetation; the next layer down may contain betel nut (*Areca catechu*), yams (*Dioscorea* spp.), and ylang-ylang (*Cananga odorata*). Below this horizontal strata, another layer of vegetation might contain bananas and plantains (*Musa* spp.), hibiscus (*Hibiscus tiliaceus*), Indian mulberry (*Morinda citrifolia*), soursop (*Annona muricata*), and yam vines (*Dioscorea* spp.). At the ground level, there may be sakau (*Piper methysticum*), swamp taro (*Cyrtosperma chamissonis*), sweet taro (*Colocasia esculenta*), and wild taro (*Alocasia macrorrhiza*) (Raynor and Fownes, 1991a,b; Drew et al., 2004).

Although many different agroforestry systems exist on Earth, they can be classified into two key systems: trees with crops and trees with pastures and livestock (MacDicken and Vergara, 1990). According to MacDicken and Vergara (1990), the first system, trees with crops, can be divided into categories that include trees on cropland, tree gardens, taungya (Menzies, 1988), improved fallows, plantation crop combinations, alley cropping, windbreaks, and shelterbelts. The trees on cropland category involves widely spaced trees that are naturally regenerated among agricultural crops. Examples may include the gum arabic tree (*Acacia nilotica*) intermixed with rice, and paulownia (*Paulownia* spp.) growing among rice, soybeans, and wheat (MacDicken and Vergara, 1990). Tree gardens are vertically structured systems that include not only trees and shade-tolerant crops but also livestock. Taungya is a system whereby land is cleared and trees are planted along with crops. Improved fallows include fallowed land that is enhanced with either planted or naturally regenerated trees before the next cultivation. Alley cropping occurs where crops are planted between rows of trees. Windbreaks and shelterbelts, which were described earlier in this chapter, are rows of trees planted next to agricultural fields.

The trees with pastures and livestock system is simpler in that it hosts three practices: trees on rangelands or pastures; plantation crops with pastures; and live fences. Live fences are formed when trees are planted in a row next to a field and, typically but not always, delineate a property boundary. Plantation crops with pastures are areas in which livestock grazing is of secondary importance. The main goal of trees grown on rangelands or pastures is to provide shade for livestock. Given the plethora of agroforestry practices, we will only illustrate three in this chapter: silvopastoral systems, alley cropping, and windbreaks.

11.19.1 Silvopastoral systems

Silvopastoral systems are those that combine tree growing with the production of livestock. These systems typically include pasture systems containing trees that are widely spaced or planted in clusters throughout

FIGURE 11.30 Sheep sheltering beneath a hawthorn (*Crataegus monogyna*) tree in the United Kingdom. *Pauline Eccles, through Wikimedia Commons.*
Image Link: https://commons.wikimedia.org/wiki/File:Sheep_shelter_beneath_a_hawthorn_tree_-_geograph.org.uk_-_483978.jpg
License Link: https://creativecommons.org/licenses/by-sa/2.0/deed.en

FIGURE 11.31 Alley cropping of coffee (*Coffea* spp.) trees and black turtle beans (*Phaseolus vulgaris*) in Costa Rica. *Donald L. Grebner*

the pasture. This system can be found widely around the world. Although the system may be used primarily to provide shelter for animals (Fig. 11.30), it can also be used as fodder to enhance livestock feed. Some are very effective at providing more favorable conditions for livestock with increased survival and body weight of the animals when using an effective silvopastoral system (Pezo et al., 2018). Husak and Grado (2002) evaluated silvopastoral systems with other single land-use systems such as cattle, rice, and soybean production, as well as pine plantation management, and modeled a silvopastoral system focused on planting loblolly pine using data developed by Clason (1998). They found the silvopastoral system to be profitable when it also provides opportunities for hunting leases, which can boost the financial return to landowners.

11.19.2 Alley cropping

Alley cropping is part of a classification of practices that deals with trees and crops growing on the same unit area of land. Alley cropping is a specific practice in which trees or shrubs and agricultural crops are grown in alternate rows (Fig. 11.31). The trees are commonly pruned to limit the shading of the agricultural crop. Alley cropping can also contribute to nutrient cycling and erosion control. In these systems, one may often see a row of trees followed by multiple rows of agricultural plants such as corn; however, numerous other examples can be found in the literature (e.g., Barker, 1990; MacDicken, 1990; Van Den Beldt, 1990). For instance, Kang and Gutteridge (1994) evaluated alley cropping systems involving pruned hedges of

leucaena (*Leucaena leucocephala*) and crops of maize. Citing earlier work, they summarized that maize yield could experience additional gains of 1.6 ton per ha (0.6 ton per ac) compared to yields from control plots of maize not fertilized with leucaena mulch. However, despite the gains and various other potential benefits of alley cropping, planned increases in crop yields are not always met (Kang and Gutteridge, 1994).

11.19.3 Windbreaks

Although windbreaks and shelterbelts are discussed at length in Section 11.2.3, they can also be considered a form of agroforestry and need to be mentioned again in that context. As we know, windbreaks are groups of trees that are planted in rows to not only protect the production of agricultural crops but to also serve as shelters for livestock (Fig. 11.32). The number of rows of trees will vary due to the area allocated for this particular purpose or effect desired. Typically, windbreaks form a vegetative ladder that helps wind move up and above ground level. The idea is to protect the crops planted behind the break. However, it is common for windbreaks to provide some wind permeability in order to avoid the negative consequences of potential of wind turbulence immediately behind the vegetative barrier. When using this practice, soil loss due to wind erosion may be reduced dramatically and yields of dryland crops can increase by about 25% (Baer, 1989). In one case, the yield of maize growing behind windbreaks increased sufficiently to compensate for the loss of land planted with trees, as windbreaks

FIGURE 11.32 Field windbreaks protect the soil against wind erosion in North Dakota, United States. *Erwin Cole, United States Department of Agriculture Natural Resources Conservation Service, through Wikimedia Commons.*
Image Link: https://commons.wikimedia.org/wiki/File:FieldWindbreaks.JPG
License Link: Public Domain

(Grala and Colletti, 2003). In some areas of the midwestern United States, windbreaks and shelterbelts can also provide landowners with the opportunity to earn additional revenue through hunting leases (Grala et al., 2008).

11.20 Clonal forestry

Improvements in tree growth can be important to society and to landowners, as faster-growing trees may help reduce CO_2 from the atmosphere, while helping a landowner earn a higher rate of return from an investment in forestry. Many of our tree improvement efforts have been aimed at the development of seedling to be planted in plantations. Nowadays, tree seedlings can be developed from the seed of trees in open-pollinated seed orchards, where only one parent is known, from the seed of trees grown in nurseries through vegetative multiplication, where both parents are known, or from clones created through vegetative propagation (McKeand et al., 2003). In the southern United States, these forms of genetic improvement of trees have progressed significantly over the last half of the 20th century (Merkle and Dean, 2000). The development of tree clones does not involve sexual reproduction (as is the case with pollination), and allows an organization to purposefully propagate highly desired types of trees, either through the use of somatic embryogenesis, a process for creating embryos from cells of plants other than gametes, or through the development of rooted cuttings (Bettinger et al., 2009). Compared to 30% gains in productivity from vegetative multiplication, pine

plantations developed from high quality clones can result in 50% productivity gains in wood volume (Fox et al., 2007). Greater tree growth from improved seedlings might imply that today's harvest levels could be obtained from less land, and although certain segments of society might find these methods controversial (e.g., not natural), these forms of tree improvement do not involve genetic modification of trees.

11.21 Forest protection

Many practices deal with protecting forests from external threats such as invasive plant and animal species, as we mentioned in Chapters 5 and 14. Forest protection attempts to prevent a forested landscape from becoming degraded due to these external sources. This protection can come in many forms such as the enactment and enforcement of policies that require inspection and potential denial of entry into a port for certain cargo ships that may be harboring egg clusters from an invasive insect that could adversely harm the surrounding forested landscape (Stanaway et al., 2001). More localized practices may be the establishment and maintenance of fire breaks across a forested property to prevent the spread of wildfires. Prescribed fires may be another preventative activity that can reduce fuel loads across a forest as well as eliminate ladder fuels that could link ground fires to the crown layer of a forest. Annual herbicide spraying of invasive plants may limit the spread of invasive plants into and across forested areas (Miller et al., 2015). Other practices such as scheduled thinning operations can reduce not only ladder fuels within a stand, but promote better tree growth, which is a mechanism for reducing a forest stand's vulnerability to potential insect and disease threats (U.S. Department of Agriculture, Forest Service, 2003; Spyksma et al., 2015). Forest protection activities should be described within a forest management plan and their survey of conditions that make the property vulnerable should be part of the inventory system. For instance, the Revelstoke Community Forest in British Columbia, Canada, has several protection strategies noted in their forest plan, such as aggressive brush control operations to enhance stand health and vigor, annual flights over the landscape to monitor pests, and establishment of fire breaks using broadcast burning (Spyksma et al., 2015). Another example is the Yale-Myers Forest plan in northern Connecticut. This forest plan suggests the use of a selection harvest system that utilizes small patches to sustain eastern hemlock (*Tsuga canadensis*) on moist sites because they appear to be either more resistant to the hemlock woolly adelgid (*Adelges isugae*) or the insect is less prevalent in these areas (Ashton et al., 2015).

FIGURE 11.33 Boundary line marking in January in Maine, United States. *Donald L. Grebner*

Summary

In the normal course of business operations, forestry practices are conducted to address economic, environmental, and social objectives of the landowner or land manager. They can be employed to improve habitat conditions or to respond to natural disasters (e.g., fires, hurricanes, or insects and diseases). This chapter has briefly described many of the common forestry practices used around the world. Some practices that are tied directly to information (or data) management or land ownership, such as forest inventories, property surveys, or boundary line marking (Fig. 11.33), were not covered in this chapter, but can technically be considered as forestry practices. One should review Chapter 18 (Forestry and Natural Resource Management Careers) to gain insight into the common work practices of the wide variety of professionals associated with forestry and natural resource management. We hope it was clear in our discussion of forestry practices that the economic, ecological, and social aspects of each practice influence the choice of practice to employ within a forest. Further, while most practices can be applied in both even-aged and uneven-aged forests, some practices are more appropriate to use in one type than the other. The aesthetics, costs, equipment, materials, personnel, paperwork, noise, and time associated with a practice should all be assessed when contemplating the management of a forest. Forest and natural resource managers should also be aware that whether on purpose or unintentionally, a shift in plant species composition and size is likely to occur with each forest practice that is implemented. In fact, some practices are designed specifically to achieve shifts in the character of the collection of plant species located on a site. The selected practice should have a purpose, whether to enhance the growth of a forested area, develop and maintain habitat structure requirements for certain wildlife species, or achieve some other desired forest condition.

Questions

(1) *Land management.* Imagine that a relative of yours from another region of your country has recently purchased some bare (cutover) ground and is interested in growing a forest there for both profit and recreational enjoyment. They understand that you have some knowledge of forestry and natural resource management and are therefore interested in your opinion of how the forest should be managed. Select one or more tree species that could logically grow on this site and provide, in a short memorandum, a review of the types of forest practices your relative should consider implementing over the next 20 to 30 years and the potential revenues and costs associated with these.

(2) *Regeneration of forests.* Assume that an area of land in your part of the world has been destroyed by a wildfire. In a short report, discuss the various means by which forest managers could reestablish a new forest and focus on the advantages and disadvantages of the approaches.

(3) *Seed tree versus shelterwood practices.* A local landowner and friend of yours is interested in naturally regenerating his land after a forthcoming final harvest of an even-aged stand of pine trees. In a one-page memorandum, compare and contrast seed tree and shelterwood as forest practices that might be used to reestablish even-aged forests.

(4) *Selection versus selective harvest practices.* Selection and selective harvest practices are often used interchangeably in conversation; however, they have different meanings. In a single paragraph, compare the similarities and differences between selection harvest and selective harvest practices. Discuss your conclusions with a group of four or five others.

(5) *Fuel management practices.* As a resident of a community, imagine that you are actively involved in local politics or land use planning. Other members of the community have become concerned about the threat of wildfire to the health and well-being of the people and forests in this community. Develop a short

QUESTIONS *(cont'd)*

PowerPoint presentation that describes the trade-offs of using various fuel management treatments in the nearby forests to address the forest fuel issue.

(6) *Prescribed fire.* As a current or future land manager, you may contemplate the use of prescribed fire to maintain or enhance the natural appearance of a forested area. You may also become concerned about feedback you are likely to receive from neighboring landowners. To illustrate that you have considered this practice thoroughly, prepare a one-page

memorandum that describes the objectives of using prescribed fire and some of the trade-offs involved.

(7) *Group selection harvest practices.* A neighboring landowner is interested in using group selection harvest practices on his land because he prefers small openings over larger final harvest areas. Develop a short summary of 5 to 10 key points that describe the advantages and disadvantages of employing group selection harvest practices in your area of the world.

References

Adame, P., Cañellas, I., Roig, S., Del Río, M., 2006. Modelling dominant height growth and site index curves for rebollo oak (*Quercus pyrenaica* Willd.). Annals of Forest Science 63, 929–940.

Alabama Forestry Commission, 2004. Harvesting for Natural Pine Regeneration. Alabama Forestry Commission. Publication HFNPR030204, Montgomery, AL. http://www.forestry.alabama.gov/Pages/Informational/Forms/Timber/Harvesting_for_Natural_Pine_Regeneration.pdf (accessed 12.02.20).

Aravanopoulos, F.A., 2010. Breeding of fast growing forest tree species for biomass production in Greece. Biomass and Bioenergy 34, 1531–1537.

Ashton, M.S., Duguid, M.C., Barrett, A.L., Covey, K., 2015. Yale School Forest, New England, United States of America. In: Siry, J.P., Bettinger, P., Merry, K., Grebner, D.L., Boston, K., Cieszewski, C. (Eds.), Forest Plans of North America. Academic Press, London, pp. 255–340.

Baer, N., 1989. Shelterbelts and windbreaks in the Great Plains. Journal of Forestry 87 (4), 32–36.

Baker, S.A., Westbrook Jr., M.D., Greene, W.D., 2010. Evaluation of integrated harvesting systems in pine stands of the southern United States. Biomass and Bioenergy 34, 720–727.

Barker, T.C., 1990. Agroforestry in the tropical highlands. In: MacDicken, J.G., Vergara, N.T. (Eds.), Agroforestry Classification & Management. John Wiley & Sons, Inc. New York, p. 382.

Barlow, R.J., Dubois, M.R., 2011. Cost & cost trends for forestry practices in the South. Forest Landowner 70 (6), 14–24.

Barlow, R., Levendis, W., 2015. 2104 cost and cost trends for forestry practices in the South. Forest Landowner 73 (5), 22–31.

Bene, J.G., Beall, H.W., Cote, A., 1977. Trees, food and people: land management in the tropics. IDRC-084e, International Development Research Centre. Canada, Publication IDRC-084e, Ottawa, Ontario, 54 p.

Berrill, J.P., Boston, K., 2019. Conifer retention and hardwood management affect harvest volume and carbon storage in Douglas-fir/tanoak. Mathematical and Computational Forestry and Natural-Resource Sciences 11 (2), 286.

Bettinger, P., Clutter, M., Siry, J., Kane, M., Pait, J., 2009. Broad implications of southern United States pine clonal forestry on planning and management of forests. The International Forestry Review 11, 331–345.

Binkley, D., Burnham, H., Allen, H.L., 1999. Water quality impacts of forest fertilization with nitrogen and phosphorous. Forest Ecology and Management 121, 191–213.

Bolding, M.C., Lanford, B.L., 2005. Wildfire fuel harvesting and resultant biomass utilization using a cut-to-length/small chipper system. Forest Products Journal 55 (12), 181–189.

Boydak, M., 2003. Regeneration of Lebanon cedar (*Cedrus libani* A. Rich.) on karstic lands in Turkey. Forest Ecology and Management 178, 231–243.

Bridgwater, F., Kubisiak, T., Byram, T., McKeand, S., 2005. Risk assessment with current deployment strategies for fusiform rust-resistant loblolly and slash pines. Southern Journal of Applied Forestry 29, 80–87.

British Columbia Ministry of Forests and Range, 1995. Forest Fertilization Guidebook. British Columbia Ministry of Forests and Range, Victoria, B.C. https://www2.gov.bc.ca/gov/content/industry/forestry/managing-our-forest-resources/silviculture/silvicultural-systems/silviculture-guidebooks/forest-fertilization-guidebook (accessed 12.02.20).

Brose, P.H., 2009. Photo Guide for Estimating Fuel Loading and Fire Behavior in Oak Forests of the Mid-Atlantic Region. U.S. Department of Agriculture, Forest Service, Northern Research Station, Newtown Square, PA. General Technical Report NRS-45. 104 p.

Bruce, D., 1951. Fire, site, and longleaf height growth. Journal of Forestry 49, 25–28.

Bull, E.L., Heater, T.W., Youngblood, A., 2004. Arboreal squirrel response to silvicultural treatments for dwarf mistletoe control in northeastern Oregon. Western Journal of Applied Forestry 19, 133–141.

Burgan, R.E., 1987. Concepts and Interpreted Examples in Advanced Fuel Modeling. U.S. Department of Agriculture, Forest Service, Intermountain Forest and Range Experiment Station, Ogden, UT. General Technical Report INT-238. 40 p.

Campbell, R.S., 1948. Forest grazing work in the Deep South. In: Proceedings of the Society of American Foresters. Society of American Foresters, Bethesda, MD, pp. 216–222.

Carlsen, S.C.K., Spliid, N.H., Svensmark, B., 2006. Drift of 10 herbicides after tractor spray application. 2. Primary drift (droplet drift). Chemosphere 64, 778–786.

Carter, M.C., Foster, C.D., 2004. Prescribed burning and productivity in southern pine forests: a review. Forest Ecology and Management 191, 93–109.

Ciancio, O., Iovino, F., Menguzzato, G., Nicolaci, A., Nocentini, S., 2006. Structure and growth of a small group selection forest of calabrian pine in southern Italy: a hypothesis for continuous cover forestry based on traditional silviculture. Forest Ecology and Management 224, 229–234.

Clason, T.R., 1998. Silvopastoral practices sustain timber and forage production in commercial loblolly pine plantations of northwest Louisiana, United States. Agroforestry Systems 44, 293–303.

Costanza, K.K., Crandall, M.S., Rice, R.W., Livingston, W.H., Munck, I.A., Lombard, K., 2019. Economic implications of a native tree disease, Caliciopsis canker, on the white pine (*Pinus strobus*) lumber industry in the northeastern United States. Canadian Journal of Forest Research 49 (5), 521–530.

Dale, G., Dieters, M., 2007. Economic returns from environmental problems: breeding salt- and drought-tolerant eucalypts for salinity abatement and commercial forestry. Ecological Engineering 31, 175–182.

DeGraaf, R.M., Yamasaki, M., 2003. Options for managing early-successional forest and shrubland bird habitats in the northeastern United States. Forest Ecology and Management 185, 179–191.

Della-Bianca, L., 1969. Intensive Cleaning Increases Sapling Growth and Browse Production in the Southern Appalachians. U.S. Department of Agriculture, Forest Service, Southern Forest Experiment Station, Asheville, NC. Research Note SE-110.

Demoling, F., Nilsson, L.O., Bååth, E., 2008. Bacterial and fungal response to nitrogen fertilization on three coniferous soils. Soil Biology and Biochemistry 40, 370–379.

Dignan, P., King, M., Saveneh, A., Walters, M., 1998. The regeneration of *Eucalyptus regnans* (F. Muell.) under retained overwood: seedling growth and density. Forest Ecology and Management 102, 1–7.

Doelman, J.C., Stehfest, E., van Vuuren, D.P., Tabeau, A., Hof, A.F., Braakhekke, M.C., Gernaat, D.E., van den Berg, M., van Zeist, W.J., Daioglou, V., van Meijl, H., Lucas, P.L., 2020. Afforestation for climate change mitigation: potentials, risks and trade-offs. Global Change Biology 26, 1576–1591.

Drew, W.M., Alavalapati, J.R.R., Nair, P.K.R., 2004. Determining agroforestry profitability using the policy analysis matrix. In: Alavalapatti, J.R.R., Mercer, D.E. (Eds.), Valuing Agroforestry Systems. Kluwer Academic Publishers, Dordrecht, the Netherlands, pp. 59–78.

Droze, W.H., 1977. Trees, Prairies, and People: A History of Tree Planting in the Plains States. Texas Woman's University, Denton, TX.

Dujesiefken, D., Stobbe, H., 2002. The Hamburg tree pruning system—a framework for pruning of individual trees. Urban Forestry and Urban Greening 1, 75–82.

Dupuy, B., Mille, G., 1993. Timber Plantations in the Humid Tropics of Africa. Food and Agriculture Organization of the United Nations, Rome, Italy. FAO Forestry Paper 98.

Erdmann, G.G., 1990. Betula alleghaniensis (Britton), yellow birch. In: Burns, R.M., Honkala, B.H. (technical coordinators). Silvics of North America: 2. Hardwoods. U.S. Department of Agriculture, Forest Service, Washington, D.C. Agriculture Handbook 654.

Ford-Robertson, F.C., 1971. Terminology of Forest Science Technology Practice and Products. Society of American Foresters, Washington, D.C., 370 p.

Fox, T.R., Jokela, E.J., Allen, H.L., 2007. The development of pine plantation silviculture in the southern United States. Journal of Forestry 105, 337–347.

Franzel, S., 2004. Financial analysis of agroforestry practices: fodder shrubs in Kenya, woodlots in Tanzania, and improved fallows in Zambia. In: Alavalapatti, J.R.R., Mercer, D.E. (Eds.), Valuing Agroforestry Systems. Kluwer Academic Publishers, Dordrecht, The Netherlands, pp. 9–37.

Fraterrigo, J.M., Pearson, S.M., Turner, M.G., 2009. The response of understory herbaceous plants to nitrogen fertilization in forests of different land-use history. Forest Ecology and Management 257, 2182–2188.

Fujiwara, M., 2002. Silviculture in Japan. In: Iwai, Y. (Ed.), Forestry and the Forest Industry in Japan. UBC Press, Vancouver, Canada, pp. 10–23.

Gagnon, J.L., Jack, S.B., 2004. A comparison of the ecological effects of herbicide and prescribed fire in a mature longleaf pine forest: response of juvenile and overstory pine. In: Connor, K.F. (Ed.), Proceedings of the 12th Biennial Southern Silvicultural Research Conference, General Technical Report SRS-71. U.S. Department of Agriculture, Forest Service, Southern Research Station, Asheville, NC, pp. 304–308.

Galang, M.A., Markewitz, D., Morris, L.A., 2010. Soil phosphorus transformations under forest burning and laboratory heat treatments. Geoderma 155, 401–408.

Geron, C.D., Hafley, W.L., 1988. Impact of fusiform rust on product yields of loblolly pine plantations. Southern Journal of Applied Forestry 12, 226–231.

Ghassemi, M., Quinlivan, S., Dellarco, M., 1982. Environmental effects of new herbicides for vegetation control in forestry. Environment International 7, 389–401.

Grala, R.K., Colletti, J.P., 2003. Estimates of additional maize (*Zea mays*) yields required to offset costs of tree-windbreaks in midwestern USA. Agroforestry Systems 59, 11–20.

Grala, R.K., Colletti, J.P., Mize, C.W., 2008. Willingness of Iowa agricultural landowners to allow fee hunting associated with in-field shelterbelts. Agroforestry Systems 76, 207–218.

Grano, C.X., 1969. Pre-commercial thinning of loblolly pine. Journal of Forestry 67, 825–827.

Grassi, G., Minotta, G., Giannini, R., Bagnaresi, U., 2003. The structural dynamics of managed uneven-aged conifer stands in the Italian eastern Alps. Forest Ecology and Management 185, 225–237.

Haggar, J., Barrios, M., Bolaños, M., Merlo, M., Moraga, P., Munguia, R., Ponce, A., Romero, S., Soto, G., Staver, C., Virginio, E. de M.F., 2011. Coffee agroecosystem performance under full sun, shade, conventional and organic management regimes in Central America. Agroforestry Systems 82, 285–301.

Haimi, J., Fritze, H., Moilanen, P., 2000. Responses of soil decomposer animals to wood-ash fertilisation and burning in a coniferous forest stand. Forest Ecology and Management 129, 53–61.

Hart, S.P., 2001. Recent perspectives in using goats for vegetation management in the USA. Journal of Dairy Science 84 (E. Suppl.), E170–E176.

Hawe, A., 1996. Shelterwood Harvesting in Root Disease Infected Stands. EP 1186 Preliminary Results—Ice Road Site. British Columbia Forest Service, Forest Sciences Section, Nelson, B.C. Extension Note 030.

Haywood, J.D., 2009. Eight years of seasonal burning and herbicidal brush control influence sapling longleaf pine growth, understory vegetation, and the outcome of an ensuing wildfire. Forest Ecology and Management 258, 295–305.

Helms, J.A., 1998. The Dictionary of Forestry. Society of American Foresters, Bethesda, MD, 210 p.

Hirata, A., Sakai, T., Takaashi, K., Sato, T., Tanouchi, H., Sugita, H., Tanaka, H., 2011. Effects of management, environment and landscape conditions on establishment of hardwood seedlings and saplings in central Japanese coniferous plantations. Forest Ecology and Management 262, 1280–1288.

Holgén, P., Hånell, B., 2000. Performance of planted and naturally regenerated seedlings in *Picea abies*-dominated shelterwood stands and clearcuts in Sweden. Forest Ecology and Management 127, 129–138.

Hu, J.J., Tian, Y.C., Han, Y.F., Li, L., Zhang, B.E., 2001. Field evaluation of insect-resistant transgenic *Populus nigra* trees. Euphytica 121, 123–127.

Husak, A.L., Grado, S.C., 2002. Monetary benefits in a southern silvo-pastoral system. Southern Journal of Applied Forestry 26 (3), 159–164.

Isik, F., Goldfarb, B., LeBude, A., Li, B., McKeand, S., 2005. Predicted genetic gains from testing efficiency from two loblolly pine clonal trials. Canadian Journal of Forest Research 35, 1754–1766.

Johnson, H., Brandle, J., 2009. Shelterbelt Design. Department of Primary Industries, Victoria, Australia. http://agriculture.vic.gov.au/agriculture/farm-management/soil-and-water/erosion/shelterbelt-desig (accessed 12.02.20).

Kang, B.T., Gutteridge, R.C., 1994. Forage tree legumes in alley cropping systems. In: Gutteridge, R.C., Shelton, H.M. (Eds.), Forage Tree Legumes in Tropical Agriculture. Tropical Grassland Society of Australia, Inc.

Karvinen, S., Välkky, E., Torniainen, T., Gerasimov, Y., 2006. Northwest Russian Forestry in a Nutshell. Finnish Forest Research Institute. Working Papers of the Finnish Forest Research Institute, Helsinki, Finland.

Keeton, W.S., 2006. Managing for late-successional/old-growth characteristics in northern hardwood-conifer forests. Forest Ecology and Management 235, 129–142.

Kellison, R., 2007. Forest biotechnology: its place in the world. In: Byram, T.D., Rust, M.L. (Eds.), Proceedings of the 29th Southern Forest Tree Improvement Conference, Southern Forest Tree Improvement Committee. Sponsored Publication No. 51, pp. 7–14.

Klingenberg, M.D., Lindgren, B.S., Gillingham, M.P., Aukema, B.H., 2010. Management response to one insect pest may increase vulnerability to another. Journal of Applied Ecology 47, 566–574.

Kopp, G., 2007. Reducing Deer Browse Damage. U.S. Department of Agriculture, Natural Resources Conservation Service, St. Paul, MN. Forestry Technical Note 44.

Kriebel, H.B., 1983. Breeding eastern white pine: a world-wide perspective. Forest Ecology and Management 6, 263–279.

Krott, M., Tikkanen, I., Petrov, A., Tunytsya, Y., Zheliba, B., Sasse, V., Rykounina, I., Tunytsya, T., 2000. Policies for Sustainable Forestry in Belarus, Russia, and Ukraine. European Forest Institute Koninklijke Brill NV, Leiden, The Netherlands. Research Report 9. 174 p.

Lewis, S.E., Brodie, J.E., Bainbridge, Z.T., Rohde, K.W., Davis, A.M., Masters, B.L., Maughan, M., Devlin, M.J., Mueller, J.F., Schaffelke, B., 2009. Herbicides: a new threat to the Great Barrier Reef. Environmental Pollution 157, 2470–2484.

Lowe, R.G., 1978. Experience with the tropical shelterwood system of regeneration in natural forest in Nigeria. Forest Ecology and Management 1, 193–212.

Maas, D.S., Musson, R.L., Hayden, T.J., 2003. Effects of Prescribed Burning on Game Species in the Southeastern United States. U.S. Army Corps of Engineers, Engineer Research and Development Center, Construction Engineering and Research Laboratory, Champaign, IL. ERDC/CERL TR-03–13. 72 p.

MacDicken, J.G., 1990. Agroforestry management in the humid tropics. In: MacDicken, J.G., Vergara, N.T. (Eds.), Agroforestry Classification & Management. John Wiley & Sons, Inc. New York, p. 382.

MacDicken, J.G., Vergara, N.T., 1990. Introduction to agroforestry. In: MacDicken, J.G., Vergara, N.T. (Eds.), Agroforestry Classification & Management. John Wiley & Sons, Inc. New York, 382 p.

Macisaac, D.A., Krygier, R., 2009. Development and long-term evaluation of harvesting patterns to reduce windthrow risk of understory spruce in aspen-white spruce mixed wood stands in Alberta, Canada, Forestry 82, 323–342.

Mann Jr., W.F., Lohrey, R.E., 1974. Pre-commercial thinning of southern pines. Journal of Forestry 72 (9), 557–560.

Marshall, D.J., Wimberly, M., Bettinger, P., Stanturf, J., 2008. Synthesis of Knowledge of Hazardous Fuels Management in Loblolly Pine Forests. U.S. Department of Agriculture, Forest Service, Southern Research Station, Asheville, NC. General Technical Report SRS-110. 43 p.

Martin, T.A., Dougherty, P.M., Topa, M.A., McKeand, S.E., 2005. Strategies and case studies for incorporating ecophysiology into southern pine tree improvement programs. Southern Journal of Applied Forestry 29, 70–79.

McCulley, R.D., 1948. Effect of Uncontrolled Fires on Longleaf Pine Seedlings. U.S. Department of Agriculture, Forest Service. Southeastern Forest Experiment Station, Asheville, NC. Research News No.1–2.

McKeand, S.E., Mullin, T., Byram, T., White, T., 2003. Deployment of genetically improved loblolly and slash pines in the South. Journal of Forestry 101 (3), 32–37.

Meason, D.F., Idol, T.W., Friday, J.B., Scowcroft, P.G., 2009. Effects of fertilisation on phosphorous pools in the volcanic soil of a managed tropical forest. Forest Ecology and Management 258, 2199–2206.

Menzies, N., 1988. Three hundred years of taungya: a sustainable system of forestry in south China. Human Ecology 16 (4), 361–376.

Merkle, S.A., Dean, J.F.D., 2000. Forest tree biotechnology. Current Opinion in Biotechnology 11, 298–302.

Miller, J.H., Manning, S.T., Enloe, S.F., 2015. A Management Guide for Invasive Plants in Southern Forests. U.S. Department of Agriculture, Forest Service, Southern Research Station, Asheville, NC. General Technical Report SRS-131. https://www.srs.fs.fed.us/pubs/gtr/gtr_srs131.pdf (accessed 25.04.20).

Minnesota Department of Natural Resources, 2020. Silviculture Program, Field Practices. Minnesota Department of Natural Resources, St. Paul, MN. https://www.dnr.state.mn.us/forestry/ecs_silv/fieldpractices/index.html (accessed 11.02.20).

Minnesota Department of Natural Resources, United States Department of Agriculture Forest Service, University of Minnesota, 2003. Gypsy Moth Silvicultural Considerations for Minnesota. Tatum Guide. Minnesota Department of Natural Resources, St. Paul, MN. http://files.dnr.state.mn.us/input/mgmtplans/gypsymoth/gm_tatumguide.pdf (accessed 15.02.20).

Nagaike, T., Kamitani, T., Nakashizuka, T., 1999. The effect of shelterwood logging on the diversity of plant species in a beech (Fagus crenata) forest in Japan. Forest Ecology and Management 118, 161–171.

Nair, P.K.R., 1984. Soil Productivity Aspects of Agroforestry. International Council for Research in Agroforestry, Nairobi, Kenya, 91 p.

New Brunswick Department of Energy and Resource Development, 2019. New Brunswick Private Woodlot Silviculture Program, 2019-2020. New Brunswick Department of Energy and Resource Development, Fredericton, NB, 36 p.

Neyland, M.G., LaSala, A.V., 2005. Response of understory floristics to pre-commercial thinning and fertilisation in even-aged eucalypt regeneration. Tasforests 16, 71–82.

Nieuwenhuis, M., 2010. Terminology of Forest Management, Terms and Definitions in English, 2nd revised edition. International Union of Forest Research Organizations, Vienna, Austria. IUFRO World Series Volume 9-en.

Okazaki, A., 1964. Forestry in Japan. School of Forestry, Oregon State University. Hill Family Foundation Forestry Series, Corvallis, OR, 62 p.

Ontario Ministry of Natural Resources, 2020. Forest Management Guide to Silviculture in the Great Lakes-St. Lawrence and Boreal Forests of Ontario. Queen's Printer for Ontario, Toronto. https://www.ontario.ca/page/forest-management-guide-silviculture-great-lakes-st-lawrence-and-boreal-forests-ontario (accessed 15.02.20).

Palmer, J.F., 2008. The perceived scenic effects of clearcutting in the White Mountains of New Hampshire, USA. Journal of Environmental Management 89, 167–183.

Pancel, L. (Ed.), 1993. Tropical Forestry Handbook. Springer-Verlag, Berlin, pp. 645–725.

Pezo, D., Ríos, N., Muhammad, I., Gómez, M., 2018. Silvopastoral Systems for Intensifying Cattle Production and Enhancing Forest Cover: The Case of Costa Rica. International Bank for Reconstruction and

Development/The World Bank, Washington, D.C. https://www.profor.info/sites/profor.info/files/Silvopastoral%20systems_Case%20Study_LEAVES_2018.pdf (accessed 15.02.20).

Pyne, S.J., Andrews, P.L., Laven, R.D., 1996. Introduction to Wildland Fire. John Wiley & Sons, New York, 769 p.

Ratilla, D.C., 2019. A tale of two walls: a comparison of the Green Wall projects in Inner Mongolia and Sahelo-Saharan regions. Budhi 23 (1), 55–85.

Raynor, W.C., Fownes, J.H., 1991a. Indigenous agroforestry of Pohnpei: 1. Plant species and cultivars. Agroforestry Systems 16, 139–157.

Raynor, W.C., Fownes, J.H., 1991b. Indigenous agroforestry of Pohnpei: 2. Spatial and successional vegetation patterns. Agroforestry Systems 16, 159–165.

Richards, E.G., 2003. British Forestry in the 20th Century, Policy and Achievements. Koninklijke Brill NV, Leiden, The Netherlands, 282 p.

Rose, R., Haase, D.L., 2006. Guide to Reforestation in Oregon. College of Forestry, Oregon State University, Corvallis. Research Contribution 48.

Rozelle, S., Huang, J., Benziger, V., 2003. Forest exploitation and protection in reform China: assessing the impacts of policy and economic growth. In: Hyde, W.F., Belcher, B., Xu, J. (Eds.), China's Forests: Global Lessons from Market Reforms. Resources for the Future, Washington, D.C, pp. 109–133.

Rytter, L., 2006. A management regime for hybrid aspen stands combining conventional forestry techniques with early biomass harvests to exploit their rapid early growth. Forest Ecology and Management 236, 422–426.

Schwarz, C.F., Thor, E.C., Elsner, G.H., 1976. Wildland Planning Glossary. U.S. Department of Agriculture. Forest Service, Pacific Southwest Forest and Range Experiment Station, Berkeley, CA. General Technical Report PSW-13. 252 p.

Sedjo, R., 2001. Biotechnology's Potential Contribution to Global Wood Supply and Forest Conservation. Resources for the Future, Washington, D.C. Discussion Paper 01–51, Paper 37.

Sessions, J., Tuers, K., Boston, K., Zamora, R., Anderson, R., 2013. Pricing forest biomass for power generation. Western Journal of Applied Forestry 28 (2), 51–56.

Smith, N., Beckham, C., 2007. Financial Analysis of Precommercial Thinning, Low-Density Planting, and Cost Share Programs. South Carolina Forestry Commission, Columbia, SC, 5 p.

Smith, C.T., Lowe, A.T., Skinner, M.F., Beets, P.N., Schoenholtz, S.H., Fang, S., 2000. Response of radiata pine forests to residue management and fertilisation across a fertility gradient in New Zealand. Forest Ecology and Management 138, 203–223.

Spyksma, R., Brown, C., Williams, D., Bollefer, K., 2015. Revelstoke Community Forest - Tree Farm License (TFL) 56, British Columbia, Canada. In: Siry, J.P., Bettinger, P., Merry, K., Grebner, D.L., Boston, K., Cieszewski, C. (Eds.), Forest Plans of North America. Academic Press, London, pp. 333–340.

Stanaway, M.A., Zalucki, M.P., Gillespie, P.S., Rodriguez, C.M., Maynard, G.V., 2001. Pest risk assessment of insects in sea cargo containers. Australian Journal of Entomology 40, 180–192.

Stange, C., Wilson, J., Brandle, J., Kuhns, M., 1998. Windbreak Renovation. University of Nebraska Cooperative Extension, Lincoln, NE. EC 98–1777-X.

Torras, O., Saura, S., 2008. Effects of silvicultural treatments on forest biodiversity indicators in the Mediterranean. Forest Ecology and Management 255, 3322–3330.

Trueman, S.J., 2006. Clonal propagation and storage of subtropical pines in Queensland, Australia. Southern African Forestry Journal 208, 49–52.

Tryon, T.C., Hartranft, T.W., 1977. Field Experience Silvicultural Cleaning Project in Young Spruce and Fir Stands in Central Nova Scotia.

Proceedings of the Symposium on Intensive Culture of Northern Forest Types: Translating Forestry Knowledge into Forestry Action. U.S. Department of Agriculture, Forest Service, Northeastern Forest Experiment Station, Upper Darby, PA, pp. 151–158. General Technical Report NE-29.

Tsiouvaras, C.N., Harlik, N.A., Bartolome, J.M., 1989. Effects of goats on understory vegetation and fire hazard reduction in a coastal forest in California. Forest Science 35, 1125–1131.

University of Florida, Institute of Food and Agricultural Sciences Extension, 2008. Fertilization. Institute of Food and Agricultural Sciences Extension. University of Florida, Gainesville, FL. http://www.sfrc.ufl.edu/Extension/florida_forestry_information/forest_management/fertilization.html#stand (accessed 15.02.20).

U.S. Department of Agriculture, Forest Service, 1989. Land and Resource Management Plan. Siskiyou National Forest. U.S. Department of Agriculture, Forest Service, Pacific Northwest Region, Portland, OR.

U.S. Department of Agriculture, Forest Service, 2003. A Strategic Assessment of Forest Biomass and Fuel Reduction Treatments in Western States. U.S. Department of Agriculture, Forest Service, Research and Development, Washington, D.C. https://www.fs.fed.us/research/pdf/Western_final.pdf (accessed 25.04.20).

U.S. Department of Agriculture. Natural Resources Conservation Service, 2020. Web Soil Survey. U.S. Department of Agriculture, Natural Resources Conservation Service, Washington, D.C. https://websoilsurvey.sc.egov.usda.gov/App/HomePage.htm (accessed 12.02.20).

Van Den Beldt, R.J., 1990. Agroforestry in the semiarid tropics. In: MacDicken, K.G., Vergara, N.T. (Eds.), Agroforestry Classification and Management. John Wiley & Sons, Inc., New York, 382 p.

Wagner, S., Collet, C., Madsen, P., Nakashizuka, T., Nyland, R.D., Sagheb-Talebi, K., 2010. Beech regeneration research: from ecological to silvicultural aspects. Forest Ecology and Management 259, 2172–2182.

Wang, X.M., Zhang, C.X., Hasi, E., Dong, Z.B., 2010. Has the Three Norths Forest shelterbelt program solved the desertification and dust storm problems in arid and semiarid China? Journal of Arid Environments 74, 13–22.

Wei, Z., Jiangming, M., Yunting, F., Xiankai, L., Hui, W., 2008. Effects of nitrogen deposition on the greenhouse fluxes from forest soils. Acta Ecologica Sinica 28, 2309–2319.

Wenjie, W., Wei, L., Wei, S., Yuangang, Z., Song, C., 2008. Influences of forest floor cleaning on the soil respiration and soil physical property of a larch plantation in Northeast China. Acta Ecologica Sinica 28, 4750–4756.

Wienand, K.T., Stock, W.D., 1995. Long-term phosphorous fertilization effects on the litter dynamics of an age sequence of Pinus elliottii plantations in the southern Cape of South Africa. Forest Ecology and Management 75, 135–146.

Windhorst, H.-W., 1976. Usage and economic forms of the forests of the Earth. Geoforum 7, 31–39.

Yoshioka, T., Aruga, K., Nitami, T., Sakai, H., Kobayashi, H., 2006. A case study on the costs and the fuel consumption of harvesting, transporting, and chipping chains for logging residues in Japan. Biomass and Bioenergy 30, 342–348.

Zimmerman, G.T., Neuenschwander, L.F., 1983. Fuel-load reductions resulting from prescribed burning in grazed and ungrazed Douglas-fir stands. Journal of Range Management 36, 346–350.

Zwolinski, J.B., Donald, D.G.M., 1995. Differences in vegetation cover resulting from various methods of site preparation for pine plantations in South Africa. Annals of Forest Science 52, 365–374.

CHAPTER

12

Forest harvesting systems

Whether alive or dead, trees provide many services to the human population while also contributing to the health and welfare of other species. For example, houses and shelters are built with lumber derived from trees, and houses are invariably built around living trees. There is a new movement to build commercial and multifloor building using engineered-wood products such as structural panels and glue-laminated beams. As we noted in Chapter 4, trees provide crops of value to human society and the health and welfare of a wide range of other animal species. In meeting human needs, it may be necessary to remove trees from a location in which it is growing and transport them to a processing facility to create the products that we value. Forest harvesting systems are employed to meet these societal needs and are continuously evolving to meet the economic, environmental, and social needs of society. The basic forest harvesting process, however, is relatively straightforward, as Simmons (1979) suggested: "Logging … consists of going into the forest, cutting down trees, cutting them up into pieces suitable for local markets, then transporting them to places of ultimate use."

Early logging activities in North America were conducted in areas situated close to mills and where topographic conditions were favorable for the transport of wood. The logging activities were largely conducted by settlers who harvested a few trees with hand tools and animals in the fall and winter months when agricultural activities were low. The equipment used was rudimentary: an ax, a saw, a harness, and water

transport, which was common (Bryant, 1923). Since the beginning of the 20th century, forest harvesting systems have significantly evolved in some developed areas of the world, although others remain rather rudimentary in some developing areas of the world. The social, economic, and environmental conditions of a region can directly affect the adoption of innovations and ideas that have been developed elsewhere. For example, the transformation of harvesting systems in the southern United States since the mid-20th century was initially slow and was limited by cheap labor and traditionally accepted production methods, leaving few resources or incentives for innovation (MacDonald and Clow, 2010). However, this has now changed in the southern United States and many regions are hoping to adopt to self-driving or autonomous equipment due to possible labor shortages in the future.

Foresters and natural resource managers should be aware of the phases and processes of forest harvesting systems, particularly those used in the region in which they work. Therefore, this chapter aims to provide a brief overview of forest harvesting systems, since it is important to understand how forest products are moved from the woods to a processing facility. More in-depth analyses of forest harvesting systems and logging, and the economic, environmental, and social aspects of these can be found in books dedicated to the subject (e.g., Conway, 1982; Stenzel et al., 1985; Sessions et al., 2007a, b; and Uusitalo, 2010). Upon completion of this chapter, readers should be able to understand

Introduction to Forestry and Natural Resources, Second Edition
https://doi.org/10.1016/B978-0-12-819002-9.00012-2

- the various methods used to fell a tree and the systems employed to process a tree prior to relocating it from the area where it was grown;
- the processes that are used to move a tree from where it was grown to a central processing area (e.g., a landing, deck, or roadside); and
- the processes that are employed on logs or trees at a landing, deck, or roadside before they are transported to a wood processing facility.

There are many controversial issues surrounding logging and harvesting trees from forested landscapes. Many of these involve conflicts between the need to maintain and protect social and environmental values and the need to meet demands placed on forests for wood products. For example, social unrest is one of the major factors affecting logging operations in Africa (Ruiz Pérez et al., 2005). From our perspective, human society will continue to require and use a wide variety of wood products for the foreseeable future, and an understanding of how these products are moved from the place where they were grown (the forest) to the place where they are transformed into products (the mill) is therefore important. Additionally, wood is regaining its popularity as a building material due to its lower carbon footprint than steel or concrete materials. This chapter presents a brief, generalized picture of the primary activities associated with forest harvesting activities.

12.1 Tree felling and processing in the woods

Whether performed with an ax (Fig. 12.1), a chainsaw, or a tree-processing machine, felling, limbing, and bucking are the primary activities used to convert a tree into products that humans require daily. Felling, of course, is the process of cutting down a standing tree, which leaves the severed bole lying on the ground. The stump remains, and for coppice, species such as many hardwoods and redwood (*Sequoia sempervirens*) can regenerate from the stump. Limbing is the process of removing both the small and the large limbs from a merchantable tree bole. Limb removal can be performed manually with an ax or saw. Felling machines have knives that can delimb a tree bole when the stem is pulled through the machine. For small-diameter limbs, chain flails, chains spun using a high-speed axel, are used to knock the limbs from the tree.

Bucking is the process of segmenting the limbed tree bole into logs of various lengths. Variations in tree felling and processing systems are based on the size of a product or the accepted norms of a particular region of the world. For example, trees with very large diameters are usually too wide and heavy for a mechanized felling

Axe - A forged and tempered steel, single- or double-bit cutting instrument attached to a wooden handle.

Peavey - A tool used to pry or roll logs, consisting of a 0.8 to 2.4 m handle (2.5 to 8 feet) with a metal point, and a hook (dog).

Cant hook - A tool designed like a peavey, yet with no point on the end. It is fitted with a metal thimble to grip logs or cants.

FIGURE 12.1　Three common hand tools that have been, and perhaps continue to be, used in forest harvesting operations. *Simmons (1951).*

FIGURE 12.2　Feller-buncher working in a coniferous forest in Idaho, United States. *Chris Schnepf, University of Idaho, through Bugwood.org.*

machine to process, and are thus felled, limbed, and bucked into sections by hand using a chainsaw. Smaller diameter trees can be felled by mechanized harvesters (Fig. 12.2). This material can often be limbed and bucked by a single machine that delimbs, measures the length and diameter, and determines the best set of logs to make from the bole. In other cases, trees may simply be felled whole-stem, and bucking may occur at the landing or processing facility.

Felling processes are designed to sever a tree from either its stump or root system at minimal cost while preserving the maximum future value and with the least amount of damage to the tree and other live trees remaining in the area, while using the safest methods possible. The individual conducting the felling operation is often called the *faller*. In felling trees, a faller assesses the physical and ambient environment, as well as the condition of the tree to be felled, to determine the likely safety risks and the potential damage that might occur to the tree (Bentley et al., 2005). This determination involves understanding how the limbs are concentrated on the tree and general lean of the tree. Consideration is also given to the direction in which the trees or logs will be removed from the woods and the space necessary for effective limbing and bucking processes if they are to be employed (Stenzel et al., 1985). In some areas of the world, prefelling steps are necessary to minimize tree damage during the falling stage. For example, in tropical Africa, lianas (long, woody vines that are rooted in the soil and extend into the tree canopy) can influence a tree's falling direction and when entangled with neighboring trees can cause them to fall at the same time. Thus, they may need to be removed before felling (Parren and Bongers, 2001). The time of the year when felling occurs can also be important. For example, if regeneration is based on coppice (stump sprouts), felling activities conducted during the winter or early spring may produce thriftier sprouts while minimizing bark damage on the coppiced stump (Bryant, 1923).

Manual felling can be accomplished with axes, crosscut saws, or chainsaws although mechanization of felling activities has made these operations more efficient and safe. The chainsaw continues to be an indispensable tool in forest harvesting operations for harvesting trees larger in size than the machines can process, which is still common in the redwood region in California. At times the decision to use a chainsaw for felling is based on the limited finances of logging operators, and at other times it is based on the topography (some areas are too rugged or steep to use a mechanized felling machine; Simmons, 1979). Manual felling using chainsaws is still common in forests containing large trees or trees of very high value. In the redwood region of California, hydraulics jacks and wedges are used to direct felled trees into prepared beds to minimize damage to the tree bole. LeDoux (2010) estimated that 70% of felling, bucking, and limbing processes in the northeastern United States still requires the use of a chainsaw. The use of manual felling also dominates activities in eucalyptus plantations around the world (except for Brazil and Chile) due to steep terrain and limited investment capabilities

that prevent mechanization of the harvesting process (Spinelli et al., 2009). When felling trees by hand, a two-step process involving the use of both an undercut (Fig. 12.3) and a backcut is typically employed. The tree falls using a hinge created between the undercut and the backcut (Conway, 1982).

Felling machines are tracked or wheeled vehicles that harvest trees. Two common types of felling machines are feller-bunchers and harvesters. Feller-bunchers are machines that have felling heads consisting of either shears or disc saws (Fig. 12.4). The operator of a feller-buncher grabs the tree with the felling head, then cuts the tree. The operator may elect to carry the tree until enough are gathered or bunched and then lower them to a horizontal position (Bell, 2002). Harvesters use a chainsaw attached to the felling head (Fig. 12.5) that facilitates both the felling and the bucking process. Tree boles are pulled through the harvester felling head which has knife blades attached to complete the

FIGURE 12.3   Beginning the backcut on a mature hardwood tree in the northeastern United States. *U.S. Department of Agriculture, Forest Service, Northeastern Area Archive, through Bugwood.org.*

FIGURE 12.4   Disc saw feller-buncher. *Randy Westbrooks, U.S. Geological Survey, through Bugwood.org.*

FIGURE 12.5 Harvester mounted on the boom of a tracked machine. *Anne Burgess, through Wikimedia Commons.*
Image Link: https://commons.wikimedia.org/wiki/File:Tree_Harvester_-_geograph.org.uk_-_371849.jpg
License Link: https://creativecommons.org/licenses/by-sa/2.0/deed.en

FIGURE 12.6 Prototype walking harvester for rough or steep terrain.
U.S. Department of Agriculture, Technology and Development Program.

delimbing process. While these machines have been designed to fell single trees that are fairly large (up to 75 centimeters [cm] or 30 inches [in] in diameter near the stump), smaller versions of these machines that can accumulate small tree stems have been developed for biomass and pulpwood harvesting programs (Spinelli et al., 2007). Walking machines, with either two legs (and two wheels) or four legs (Fig. 12.6), have also been developed and are being tested for felling operations situated in difficult or sensitive terrain, but are currently unavailable.

Limbing can be a tedious, labor-intensive activity (Fig. 12.7) when a mechanized process is unavailable. Limbing involves chopping, sawing, or otherwise removing limbs from the bole of a tree. The goal is to remove the limb as close to the bole without damaging the bole. Felling machines such as single- or double-grip harvesters use delimbing knives to cut limbs off a tree bole. A single-grip harvester has a felling head that both delimbs and bucks logs as they are pulled through the head; the exterior knives sever the limb from the bole. A double-grip harvester has a felling head and a separate processing unit that delimbs and bucks logs (Kellogg et al., 1993). The delimbing process can be performed in the woods or after a tree stem has been transported to a landing area (see Section 12.3). For larger logs that are limbed at the location where a tree has been felled, it may be necessary to limb the tree bole a second time at the landing to cut off limbs and branch stubs that were not removed in the woods (Stenzel et al., 1985) because they were in the ground and it was impossible to rotate the log. As we noted earlier, specialized machines that use chains hanging from an axel rotated at high speeds, chain flails, can remove the limbs of small trees with suitable quality for pulpwood logs (Mooney et al., 2000).

FIGURE 12.7 Limbing a tree in northwest Oregon, United States.
Rvannatta, through Wikimedia Commons.
Image Link: https://commons.wikimedia.org/wiki/File:Chainsaw1.jpg
License Link: https://creativecommons.org/licenses/by-sa/3.0/deed.en

Bucking processes can also be performed either at the location where a tree is felled, at a landing or deck. While bucking may occur at a landing just before loading of logs onto trucks, owing to tradition, limitations of transport machines and the need to keep unmerchantable material away from the landing, bucking processes are often performed immediately after a tree has been felled (Simmons, 1979). When bucking is performed at the location where the tree was felled, it is usually done to prepare the log(s) for primary transportation to a landing, where they can then be sorted into merchantable groups, those with similar size and wood characteristics. As the assortment of logs that can be produced increases, more complicated bucking assessments are necessary, which can lead to higher felling costs (Carlsson and Rönnqvist, 2005). Bucking processes are also used to remove portions of logs that contain excessive defects and breakages that eliminate economic uses of that portion of the tree (Fig. 12.8). In addition, changing or different market demands may require logs to be bucked to specific lengths for each customer. The number of different bucking patterns that can be applied to a single tree often creates a difficult decision-making environment for the faller (Martell et al., 1998). Mechanized machines have computers that can optimize the bucking of the tree into logs that will yield the greatest return.

In the past, both felling and bucking were performed by hand with crosscut saws, but nowadays both of these processes, if performed manually, involve the use of a chainsaw. Single- or double-grip harvesters can measure the lengths of logs after felling (and during delimbing); however, these machines need continual calibration to maintain their length and diameter accuracy (Boston and Murphy, 2003).

Recent advances in biomass harvesting of small trees, either from very dense forests or during what could be considered a first thinning practice, use a modified form of bucking. In *bundle harvesting* (Fig. 12.9), small trees are felled. A bunch of these are lifted onto the table of the bundling unit (situated at the location of felling), and the bunch is cut to a specific length (e.g., 2.6 meters [m] or 8.5 feet [ft]) with a chainsaw device. Once bucked in this manner, the bundle is compressed, tied together with sisal cord, and dropped onto the ground for primary transportation to a landing (Kärhä et al., 2011). This practice improves the efficiency of handling small logs. Both delimbing and bucking processes produce logging residues that are now considered usable forest products (biomass) in some areas of the world, rather than logging waste (Yoshioka et al., 2006).

FIGURE 12.8 Bucking a tree in northwest Oregon, United States. *Rvannatta, through Wikimedia Commons.*
Image Link: https://commons.wikimedia.org/wiki/File:Bucker2.jpg
License Link: https://creativecommons.org/licenses/by-sa/3.0/deed.en

FIGURE 12.9 Energy wood harvester (bundler) that produces a bundle of biomass from the residual slash in a harvest area. *U.S. Department of Agriculture, Technology and Development Program.*

12.2 Primary transportation

After felling and perhaps some processing, logs or trees are transported from the location where the trees were grown to a central area (a *landing* or log *deck*) where they then may be further processed or loaded onto trucks or trailers for transport to a processing facility. The movement of trees from the location at which they grew to the landing or log deck requires a form of primary transportation (Table 12.1), which can include

TABLE 12.1 Primary transportation modes and methods for wood extraction purposes.

Mode	Method
Carrying	Humans or draft animals
Dragging	Tracked or wheeled machines (skidders) Clambunk skidders Draft animals
Forwarding	Forwarders Sleds
Yarding	Cable-based systems: Shovel loader jammer, high lead, live skyline, north bend, running skyline Truck-mounted yarding systems
Floating	Flumes individual logs in rivers or streams Log rafts in rivers or seas
Flying	Helicopters Balloons

carrying, dragging, forwarding, skidding, yarding, floating, or flying the logs or trees. A rudimentary form of primary transportation still used in many parts of the developing world involves the simple carrying of wood products (Fig. 12.10). Small pieces of mahogany that are used for guitars are still transported by hand. In the very early phases of the development of North America, logs were pushed with peaveys, a specialized pole with a hook, carried with cant hooks, or slid down crude flumes (Fig. 12.11) to a river, where they would be floated (a form of secondary transportation)

to a mill. Animals such as horses, mules, or oxen were used to drag logs down. In Southeast Asia, elephants can be used (Bist et al., 2002). In many cases, water flow, water depth, and ice prevented successful application of this form of transportation during certain periods of the year. Railroad logging was also once used in various parts of the world, involving a system whereby logs were dragged (skidded) or carried from the location of felling to a railroad car using steam-powered winches and cables (Fig. 12.12). In the United States, this form of primary transportation was used from about 1880 to the 1930s. In theory, the landing or deck was not stationary but was a moving system of rail cars. If one examines the landscape closely, you may be able to see railroad logging branch lines that fanned out into remote areas, such as the swamps of Florida (Fig. 12.13) or the pine forests in eastern California.

Skidding is a frequently used form of primary transportation that requires the availability of enough power to drag logs from the woods to a landing area. Skidding can be accomplished by machines or draft animals that include horses, oxen, mules, or elephants. The route that a log or tree takes from the field to the landing or

FIGURE 12.10   Collecting firewood in India. *Gilhampshire, through Wikimedia Commons.*
Image Link: https://commons.wikimedia.org/wiki/File:Collecting_firewood._-_Flickr_-_gailhampshire.jpg
License Link: https://creativecommons.org/licenses/by/2.0/deed.en

FIGURE 12.12   Log being placed in a railroad car in Batottan, British North Borneo, 1926. *Lieutenant (j.g.) Leonard Johnson, United States Coast and Geodetic Survey.*

FIGURE 12.11   Log flume in central Oregon, United States. *Ivo Shandor, through Wikimedia Commons.*
Image Link: https://commons.wikimedia.org/wiki/File:BV-flume-from-sawmill-at-Pa.jpg
License Link: Public Domain

FIGURE 12.13   Branch lines from a former railroad logging operation in Florida, United States. *U.S. Department of Agriculture, Natural Resources Conservation Service (2020).*

roadside (Figs. 12.14 and 12.15) is often called the *skid road* but is also referred to as the *extraction road*, *extraction line*, *extraction rack*, or *skid trail* (Nieuwenhuis, 2010). Modern skidders use grapple hooks mounted on the rear of a tracked or wheeled machine to grab logs (or log bunches) and drag them to a landing (Fig. 12.16). Previously, they used wire ropes, called chokers. Some larger machines, clambunk skidders, have a large grapple turned upward to allow a larger volume of logs or trees to be carried. Cable systems, primarily in mountains where the slopes are too steep for

equipment to operate safely, use a specialized type of crane called a *yarder* to skid logs to a landing.

Ground-based machines are in contact with the forest floor with their tires, bogies, or tracks, and soil disturbance and soil compaction are critical issues, particularly if a machine uses the same trail multiple times. Machines that can process trees (e.g., fell, delimb, and buck) may also be able to deposit the severed limbs onto the skid trails creating a mat that results in lower levels of soil disturbance and compaction compared to traditional ground-based systems (LeDoux, 2010). These soil impacts may be higher when the soils are wet and more susceptible to compaction. Purposefully scheduling when these operations occur may reduce these impacts.

Mules, horses, oxen, and elephants are still used in localized areas around the world to skid logs to landings. Often touted as a *green* alternative to the movement of logs from a forest to a landing, horse logging (or mule or oxen logging) continues to be used in some parts of the world, including North America (Fig. 12.17). However, any trainable, heavy working animal could be employed to skid logs from a forest. In several Asian countries, such as India and Myanmar, elephants are used to extract logs (Bist et al., 2002; Mar 2002). Horses can skid about 1.5 tons of wood nearly a half of a kilometer (km; about 1640 ft); however, the average skidding distances should be around 150 m (about 500 ft). Also, the conditions under which draft animals can move logs may be limited. Slope conditions need to be relatively gentle (3% or less for extended

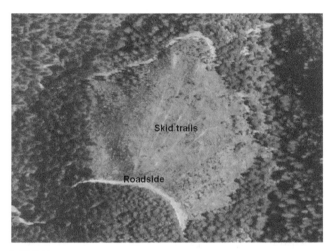

FIGURE 12.14 Skid trails to a roadside created by a cable logging system in southwest Oregon, United States. *U.S. Department of Agriculture, Natural Resources Conservation Service (2020).*

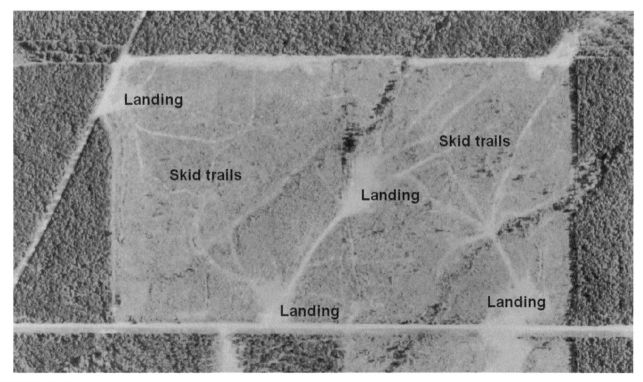

FIGURE 12.15 Skid trails to landings created by a ground-based logging system in northern California, United States. *U.S. Department of Agriculture, Natural Resources Conservation Service (2020).*

FIGURE 12.16 Grapple skidder at a landing in the southern United States. *Mississippi State University.*

FIGURE 12.17 Log skidding using a mule. *James H. Miller, United States Department of Agriculture Forest Service, through Bugwood.org.*

FIGURE 12.18 Forwarder transporting hardwood logs, operating in Sweden. *Christian Dahlqvist, through Wikimedia Commons. Image Link: https://commons.wikimedia.org/wiki/File:Malwa_460_forwarder_ d.jpg License Link: https://creativecommons.org/licenses/by-sa/2.0/deed.en*

distances), ground conditions need to be favorable, downhill (no rocky ground or swampy conditions), and the logs need to be either short or small in diameter (Stenzel et al., 1985). While draft animals have been suggested as an alternative to the use of machines for skidding logs, and may perhaps reduce soil compaction and damage to residual live trees, lower levels of productivity limit their application to specialized instances. In some cases (as with elephants), the practice may be controversial as it can be considered as animal cruelty.

Primary transportation systems that move logs by carrying or towing them were once known as *prehauling* in the southern United States but are now known as *forwarding* in many areas of the world (Stenzel et al., 1985). Prehauling involved filling large U-shaped metal pallets with pulpwood logs, which were then towed or carried to a central landing where they would then be placed onto trucks for transport to a mill. Forwarding involves completely suspending logs over the ground; thus, it differs from skidding, which drags logs on the ground. A modern forwarder (Fig. 12.18) is a rubber-tired, articulated machine used to transport shortwood or cut-to-length logs to a landing, deck, or roadside and is equipped with a grapple loader for loading and unloading activities (Kellogg et al., 1993). Many forwarder tires have a flexible track, a bogie, wrapped around the outside to improve traction. These types of machines are usually used in conjunction with single- or double-grip harvesters in a *cut-to-length logging*

operation. Farm tractors can also be used to forward logs to a deck and are often used in small family farming operations in conjunction with grapples or sleds.

Cable logging is another form of primary transportation that involves a collection of methods whereby logs or trees are transported from the location where they were grown to a landing area (or roadside) primarily using steel cables or wires, with the load partially or fully lifted off the ground (Nieuwenhuis, 2010). It is often called yarder logging in North America or hauling in Oceana. A cable logging system consists of a cable yarder, a loader, and perhaps other equipment used to process logs once they arrive at the landing. A yarder is a machine comprising a tower, a set of winch drums, a power source such as a diesel engine and perhaps a carriage that, once assembled, can pull logs using the winches from the woods to the landing or roadside. Towers range in height from 5 to 40 m (16.4–131.2 ft); some are fixed, and others are telescopic, and generally need guylines to stabilize them during yarding operations (Studier and Binkley, 1974). A carriage is a device used in conjunction with skyline systems to assist in primary transportation (Studier, 1993). The carriage essentially rides on the skyline and is pulled back from (and perhaps out to) the felled logs using the yarder's winch drums. A grapple or a set of chokers (wire ropes or chains) are extended from the carriage to collect logs and move them to a landing either partially or fully suspended. These types of primary transportation systems are mainly used in mountainous or rugged terrain where ground slopes exceed 35% and in areas where logs need to be transported over swamps, mud, rocks, and broken topography (Stenzel et al., 1985). In general, cable logging requires larger crews, longer set-up times, and heavier equipment than ground-based logging systems. However, cable logging can also be employed in areas where the road density is low as these systems are able to span over 600 m (2000 ft) in some cases, or where ground disturbance needs to be minimized (Simmons, 1979). Many different arrangements of cable logging systems have been developed over the last century (e.g., Washington State Legislature, 2019). These include uphill, sidehill, and downhill yarding systems, and standing, running, and slack skyline systems are the most commonly used in North America. Some of the specific names given to these systems include *jammer*, *high lead* (Fig. 12.19), *live skyline*, *north bend*, and *running skyline* systems. Smaller, truck-mounted yarding systems have been developed for thinning and may be useful for fuel reduction purposes (Cluster, 1978) and logging lower valued hardwoods in the eastern United States (Sloan, 1982).

When logging in rugged terrain, tropical swamps, or areas with limited road access, trees and logs can also be flown from the location in which they were felled to a landing or deck. These types of primary transportation

systems use either balloons or helicopters and are classified as *aerial logging* methods. In these systems, a grapple or a set of chokers is suspended by a line from the aerial platform (balloon or helicopter) and carried to a landing. Balloon systems (Fig. 12.20) are similar to cable yarding systems in that they require a system of tethers and wire ropes to send the balloon out to the log locations and to pull the balloon back to the landing once the logs or trees are attached to the chokers or grapple. The balloon supplements the lift generated by the geometry of the terrain. Although once used in the Pacific Northwest of the United States, it seems that the only interest in balloon logging is now in eastern Russia through the Far Eastern Aerostatic Center. Helicopter logging (Fig. 12.21) is used in various parts of the world to extract highly valued logs from remote areas. Since the vehicle is neither tied to the ground nor tethered, some suggest this is one of the most flexible primary transportation systems (Stenzel et al., 1985). Disadvantages of helicopter logging are the amount of fossil fuel consumed during operations and the cost required to operate the helicopter itself. Neither balloon logging nor helicopter logging are as competitive (from an economic perspective) as other types of primary transportation systems but may be useful for specific circumstances. One such example was removal of logs

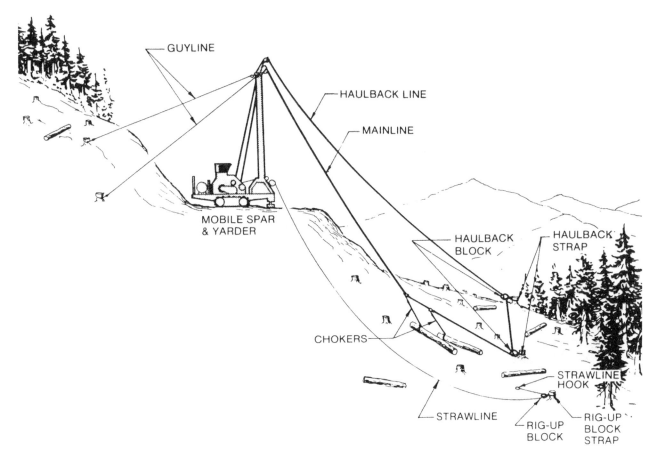

FIGURE 12.19   Basic high lead cable logging design. *Simmons (1979).*

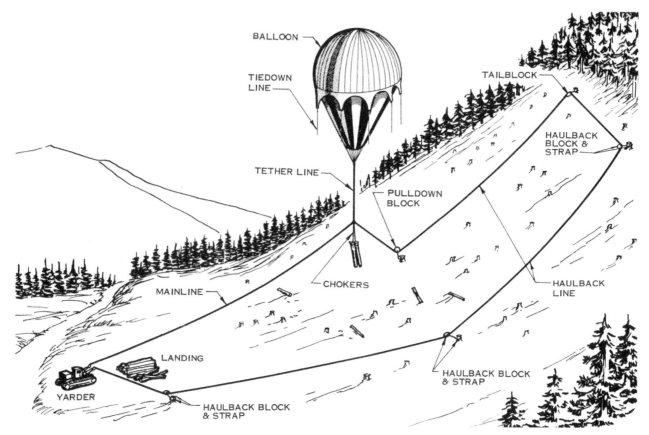

FIGURE 12.20   Basic balloon logging design. *Studier and Binkley (1974).*

FIGURE 12.21   Helicopter logging near Wellington, New Zealand. *Phillip Capper, through Wikimedia Commons.*
Image Link: https://commons.wikimedia.org/wiki/File:Logging_the_Town_ Belt,_Wellington_18_April_2005_02.jpg
License Link: https://creativecommons.org/licenses/by/2.0/deed.en

from Mount St. Helens after the volcanic eruption where a large volume of logs needed to be removed immediately to prevent major losses in timber value. Advantages of using a helicopter for primary transportation are the reduced need for building roads, and the reduced impact on soils and other resources (Stenzel et al., 1985).

12.3 Tree processing at a landing

A *landing* or *deck* is a place where logs or trees are collected, stacked, and prepared for further transportation. A *deck* may also refer to the physical pile of logs; however, we use it here to refer to the area in which both piles of logs or trees and loaders, trucks, and trailers (and other equipment) may be located. A landing must be large enough to provide sufficient room for equipment to process and load wood for the next stage of the harvesting activity (Studier and Binkley, 1974). The number of log sorts (different classes of logs) can dictate the size of the landing. Landings should be designed in such a way as to limit their number and placement within a harvest area, as the soil compaction associated with heavy machine activity can limit the potential growth of future forests in these areas. However,

FIGURE 12.22 Mechanical delimber used at a landing area in Utah, United States. *Doug Page, USDI Bureau of Land Management, through Bugwood.org.*

they need to be big enough to allow for safe work environment as workers and large machines will be present and result in a hazardous work area. Future use of a landing, whether as a forest, a recreation area (parking for a trailhead), or as a wildlife opening, will affect the size and placement of these across the landscape (Minnesota Forest Resources Council, 1999). Landings are generally connected to travel routes (woods roads), and the topographic character of an area can limit where they are placed.

Once trees or logs have reached a landing or deck, additional processing may be employed before traveling to a processing facility. For example, while some bucking and delimbing processes may be performed at the location where a tree was felled, further bucking and delimbing of a log may be necessary before shipping to a mill. These activities can generally be accomplished with a single person and a chainsaw; however, delimbing can also be performed by delimbing machines stationed at landings. For example, slide-boom or stroke-boom delimbers (Fig. 12.22), which use cutting knives to cut limbs off tree boles, can be used to delimb logs or trees at a landing. These machines can pick up individual logs and slice off the limbs and may also be used to merchandize the stem into logs by removing to top and bucking the stem. Chain flail delimbers (used to physically beat limbs off logs or trees (Mooney et al., 2000), comb delimbers (used to rake limbs off), and gate delimbers (used to break limbs off) can all be used on multiple trees stem to remove excess limbs (Stenzel et al., 1985). Other devices can also be employed to debark tree boles before their transportation to a wood-using facility.

Wood chips are often used for fuel and fiber production purposes. The chipping of tree boles and branches with machines (Fig. 12.23) is another form of additional processing that might occur at a landing or deck before transport of the wood (now in chip form) to a processing facility. Trees brought to a landing with a primary

transportation system can be chipped or ground up using a heavy-duty chipping or grinding machine and loaded into trailers by being blown or allow gravity to deposit the comminuted material into trailers. Although stationary when in use, comminution machines are movable and thus can be positioned at a landing or deck to accommodate work processes best. Many comminution machines can also debark wood before chipping; however, wood chips created at a landing may contain a higher level of contaminants such as bark, dirt, or other material, and thus are not as clean as chips created from solid wood at a pulp mill, which can limit their uses.

Another important process that may be employed at a landing or deck is sorting logs or tree stems into different classes (Fig. 12.24). Product separation may be based on the quality, size, species, or value of the logs or trees brought to the landing. This operation is

FIGURE 12.23 Chipping process at a landing located in southeastern Washington, United States. *Dave Powell, United States Department of Agriculture Forest Service, through Bugwood.org.*

FIGURE 12.24 Logs sorted by size and quality prior to transport to a processing facility. *Chris Reynolds, through Wikimedia Commons. Image Link: https://commons.wikimedia.org/wiki/File:Cut_Logs_sorted_by_size_-_geograph.org.uk_-_1205393.jpg License Link: https://creativecommons.org/licenses/by-sa/2.0/deed.en*

generally performed by someone (the loader operator) operating the same machine that will load trucks prior to the secondary transportation process. Utilizing many log sorts can ensure that the correct products are sent to the appropriate facility, although the practice of sorting generally increases truck loading time, and additional sorts can increase the size of a landing (Cass et al., 2009).

12.4 Secondary transportation

In general, secondary transportation involves moving wood from a landing, deck, or roadside to a processing facility, a wood buying point, or a wood delivery point (wood yard). Water movement of wood to a processing facility was the most widespread method of secondary transportation before the advent of railroads. In fact, using the current of a river to transport logs from a central location to a mill was once very widely used in North America, New Zealand, and Europe. Mills would employ log drivers (Fig. 12.25) to navigate freely drifting

logs to a mill. Log drivers would use a pike pole (a 3–4 m or 10–12 ft pole with a pick at one end) to control floating logs (Simmons, 1979). These types of activities took place when water levels were high or swift enough, and when frozen river surfaces were no longer an issue. Towboats, raft boats, steamers, and tugboats have all been used to pull rafts of logs down rivers, across lakes, or through coastal waterways of seas or oceans (Fig. 12.26). With the advent of railroads for both primary and secondary transportation purposes, logging activities were able to move away from waterways. However, secondary transportation today can still involve moving wood through waterways in both developing (Fig. 12.27) and developed (Fig. 12.28)

FIGURE 12.27 Log rafts of teak on the shore of the Irrawaddy River, Myanmar, 2009. *Anne-Carole Fooks, through Wikimedia Commons. Image Link: https://commons.wikimedia.org/wiki/File:Mandalay_32_TeakIndustry_g.jpg License Link: Public Domain*

FIGURE 12.25 Log drivers from Finnskoga, working at the Lusten separation point, Forshaga, Värmland, Sweden, 1918. *Obli, through Wikimedia Commons. Image Link: https://commons.wikimedia.org/wiki/File:Loggers_klaralven.jpg License Link: Public Domain.*

FIGURE 12.26 Benson sea-going log raft, c. 1908. *Doug Coldwell, through Wikimedia Commons. Image Link: https://commons.wikimedia.org/wiki/File:Securing_a_Benson_raft.jpg License Link: Public Domain*

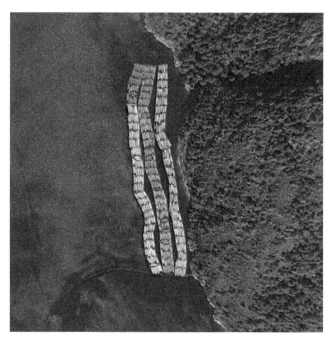

FIGURE 12.28 Log rafts in Vancouver, British Columbia, Canada. *U.S. Department of Agriculture, Natural Resources Conservation Service (2020).*

countries. This mode of secondary transport requires that the logs can float and that there is enough water current to move the logs to a processing facility successfully. Today, secondary transportation typically involves the movement of logs or trees from a landing to a mill along forest roads (haul roads), small rural roads, and highways and railroads. However, large volumes of logs are placed on barges and hauled down the Columbia River in the Pacific Northwest or the Mississippi River.

12.5 Forest roads

Forest roads (Fig. 12.29) are often called *woods roads*; they can have a natural surface (grass or dirt) or can be designed and developed with layers of rock or aggregate to provide stability to the trucks and machines that use them. Forest roads are typically designed by forest engineers or civil engineers, and require the analysis of road grade, surface, alignment, width, maximum speeds, and development costs (Byrne, 1960). In many cases, forest roads are considered to be more than temporary trails that were developed for a single purpose. Skid roads, for example, are temporary trails used to drag or forward wood to a landing. A typical one-lane forest road is about 3 m (about 10 feet) wide, while a two-lane forest road is twice this width. These roads provide vehicular access to forest resources that would otherwise be unavailable and are often used for recreational access. Skid roads, on the other hand, should be designed for temporary uses (the primary transportation of wood products by forest machines or draft animals) and designed in such a way as to minimize

ground disturbance, soil compaction, and damage to other resources.

Forest roads can impact many aspects of the environment (Boston, 2016). The placement and arrangement of roads within a forest are based on timber production, recreational access, topography, environmental impact, economics, and other considerations. Landings are created along forest roads as places to prepare and sort wood for secondary transport. Road standards are based on many factors and are expressed in terms of maximum road gradient, the minimum curve radius, type of road surface, and road width. Forest roads should, therefore, be designed with a commonly used logging truck and trailer configurations (Fig. 12.30) and harvesting equipment in mind. Surfacing with sand, rock, or other material may be necessary to reduce dust and sediment runoff, to protect the subgrade from rutting or potholes, and to provide improved traction for the vehicles. In addition, both the drainage of water from roads and the interaction of roads with water draining from the broader landscape are important. Thus, one goal with roads might be to hydrologically disconnect them from nearby streams to reduce the sedimentation from the roads (Jones et al., 2000).

Further, features such as brush or trees that might obstruct a view of traffic at an intersection or around a sharp curve should be removed (Simmons, 1951). In many cases, roads need to have a crowned shape, fitted with culverts, or otherwise designed in such a way as to facilitate effective transport of water over and under their structure. All tractive (ground-based) logging systems are more efficient in primary transportation processes when logs are moved downhill. Therefore, a transportation system might place roads at the bottom

FIGURE 12.29 Forest road in the Kunratický forest, Prague, Czech Republic. *Packa, through Wikimedia Commons.*
Image Link: https://commons.wikimedia.org/wiki/File:Prague_Kunraticky_Forest_Road.jpg
License Link: https://creativecommons.org/licenses/by-sa/2.5/deed.en

FIGURE 12.30 Logs being loaded into a trailer for secondary transportation, Kershope Forest, England. *Walter Baxter, through Wikimedia Commons.*
Image Link: https://commons.wikimedia.org/wiki/File:Loading_logs_for_transportation_-_geograph.org.uk_-_569472.jpg
License Link: https://creativecommons.org/licenses/by-sa/2.0/deed.en

of valleys or at lower elevations across a landscape. However, many cable logging systems are more efficient when yarding material uphill. Thus, a transportation system with roads placed at higher elevations and on ridge crests may be beneficial in these cases (Stenzel et al., 1985).

Depending on the road surface, sediment can be produced by vehicular traffic (Reid and Dunne, 1984). In order to reduce such sediment, the surface of a road could be changed or a road could be closed for certain periods of time (or removed completely). Further, it may be necessary to lower log-truck tire pressures through central tire inflation devices to reduce the amount of sediment produced (Foltz, 1994). The use of a high-quality aggregate rock surface can lessen the sediment produced from the road surface. The production of sediment from forest roads has received considerable interest in the United States. Recent legislation and court rulings have centered on whether permits should be required for forest roads under the Clean Water Act (33 USC. § 1251). The Act defines two categories of pollut-ants: *point sources* (from a discrete conveyance) that require a permit and *nonpoint sources* (i.e., arising from diffuse sources such as stormwater runoff) that do not require a permit. It also regulates overall water quality. Those stream segments that do not meet the desired water quality have total maximum daily load limits, the total amount of pollution allowed on a stream segment (from both point sources and nonpoint sources) until water quality objectives are achieved (Boston, 2012).

12.6 Safety

Forestry, and logging in particular, has some the high-est fatality rates of any business in many parts of the world; thus, safety is a major concern of forest workers and users. Both federal and state organizations have created specialized rules to promote safety in the forestry environment. For example, the U.S. Department of La-bor's Occupational Safety and Health Administration has developed a standard (1910.266) that establishes safety practices for logging practices. Training and super-vision on safe practices and the use of safety equipment all contribute to a safer work environment (Pearson, 1984). Further, standards and protocols have been designed to promote the safe development of resources that will be used by forest workers. In at least two studies (Thelin, 2002; Neely and Wilhelmson, 2006) half or more of the accidents among forest workers were due to the inability of workers to follow established safety rules or safety gear recommendations.

Safety is not limited to just the forestry workforce, as forest roads can provide access to parts of a landscape for recreational purposes. With respect to the development of forest roads, important considerations

in developing and maintaining a safe road system include the road grade, roadside clearance, guardrails, lighting, and surface maintenance.

12.7 Harvesting other products

When one considers nontimber forest products (NTFPs), we often think of mushrooms, greens, or even animals hunted for consumption. However, other products include exploiting the unique forage opportu-nities offered by forests, or the production of honey or gathering of materials for floral arrangements. Further, some may gather specialty plants that are associated with medicinal purposes, such as ginseng or wild ginger roots. The value of these nontraditional forest products can be significant. Truffles, a fungus often found in a variety of hardwood forests around the world, can fetch $1000 per pound for a premium product. Ginseng can fetch similar prices. Iberian hams, from pigs that forage on acorns, can yield a leg worth $4000. Initially, they were limited to the Iberian Peninsula, but they are now being raised in the oak woodlands of California, with many restaurants vying for this valuable product. Thus, there can be significant wealth opportunities for the harvesting of these crops. The ability to harvest NTFPs can be confusing, particularly on public lands where permits or licenses are often required. On private lands, the permission of the landowner is often a mini-mum requirement, but licenses or permits may still be required to sell these goods.

For harvesting wild resources, the process is very similar to the harvesting of trees. The resources must first be located, captured, and then transported to mar-kets. Often, the most difficult part is locating the resource. For locating truffles, for example, pigs or dogs with their keen ability to smell are used to find the fungus. In locating mushrooms, skill must be used to properly identify each as many mushroom species are toxic. Thus, an experienced mushroom harvester will walk a route through the woods where they believe these mushrooms exist, often near deciduous stands, and collect the fungal bodies. Many collectors record detailed notes on where they find their valuable quarry. Typical tools used are a small shovel, a soil trowel, or shears to dig and cut these products. Often, small brushes are used to remove soil and other material from the delicate fungal bodies. Once mushrooms are located and properly identified, they are severed and transported in a backpack or bag (Fig. 12.31) to the roadside. If the harvester has a relationship with local markets and restaurateurs, they may sell the mushrooms directly to them; otherwise, they may sell the mushrooms to an aggregator. This person buys mushrooms from several harvesters and is able to offer markets and restaurants (or other high-demand

FIGURE 12.31 Moshe Basson, returning from picking mushrooms in Jerusalem Forest, Israel. *Lior Basson through Wikimedia Commons. Image Link: https://commons.wikimedia.org/wiki/File:Moshe_Basson_Picking_Mushrooms.jpg License Link: https://creativecommons.org/licenses/by/3.0/deed.en*

consumers) the convenience of a larger supply at any one time. This is similar to wood dealers in the southern United States who accumulate harvests from smaller loggers for aggregate sales to the larger sawmills. Additionally, for mushrooms at least, the larger aggregators will sort, clean, and package the material to wholesalers or other buyers. Some may even and ship them to markets. Thus, the processes involved for some NTFPs can resemble those used for the harvest, transport, and marketing of logs.

Unfortunately, there are times when NTFP harvesting processes fail to adequately compensate the landowner for the harvested resource, and the ability to monitor the harvest and the harvesting activities can be difficult (Molina et al., 1993). Thus, if a private landowner allows foragers onto their land to search for the products, or to raise bees or to graze animals, how are the landowners compensated for the services? Some of these issues are easy to resolve, as rent can be charged for each grazing animal or for a hunting lease. But the harvest of small goods, such as mushrooms or ginseng, can be quietly conducted and the landowner may not know their value or the fact that the activity is actually taking place on their property. Thus, there is a strong need to inform landowners about the potential value of NTFPs. Further, landowners interested in NTFP options need to control access to these resources, and to protect their property from exploitation and damage incurred during NTFP collection, such as when all-terrain vehicles are used.

One disturbing aspect of NTFP harvest activities is that there is often an increase in crime associated with the harvest of these crops. The money earned can be significant, and the harvesters tend to work secretly to protect sites vulnerable to theft (Amaranthus and Pilz, 1996). Thus, there may be a need to cooperate with local authorities in patrolling areas where NTFPs are being collected, to protect both the resource and the citizens engaged in these activities.

Summary

Harvesting forest products involves selecting the trees to be removed, severing them from their stump, and moving the whole tree or the processed logs to an area where they will be combined with others for transport to a mill. Understanding the route from the forest to the mill and the organizations that are involved in the process is the chain-of-custody of forest products that are often tracked by certification systems and forest legality systems. Although in many ways we have modernized harvesting systems by developing machines that are capable of measuring diameters and lengths and accurately bucking logs to desired specifications, many harvesting systems today still involve people or animals for the felling, bucking, and transport processes. At times, forest harvesting is a difficult and dangerous activity, but it is driven by human demand for wood products. Harvesting systems need to be safe (from both a human and an environmental perspective), efficient (from an economic perspective), and acceptable (from a social or cultural perspective) in addressing our needs for wood products. While many different combinations of equipment are used today, mechanized harvesting systems can generally be classified as whole tree, log length, and cut to length. The differences among these have been described by Kellogg et al. (1993):

(1) Whole tree
  (a) Trees are felled
  (b) Trees are skidded or yarded to a landing or deck
  (c) Trees are processed into logs, chipped, or transported as whole trees to processing facilities
(2) Tree length
  (a) Trees are felled, delimbed, and topped
  (b) Tree-length stems are skidded or yarded to a landing or deck
  (c) Trees are processed into logs or transported as whole trees to processing facilities
(3) Cut to length
  (a) Trees are felled, delimbed, and bucked
  (b) Logs are skidded or forwarded to a landing or deck
  (c) Logs are transported to processing facilities

The selection of systems to use in different regions of the world depends on tradition, equipment (and parts) availability, and the availability of trained personnel, as well as cost factors (purchase price, insurance rates, and operating costs). The environmental impact of logging systems, particularly on soil and water resources, is also an important system selection consideration. Concerns such as soil compaction and water quality may limit the systems selected for use in various silvicultural operations or alter the timing when such systems can be used. Further, the aesthetic appeal of harvest areas (i.e., slash and views of the area from travel routes) may influence the type of system that is chosen. In addition, in some areas of the world, certain types of harvesting systems may be limited to certain seasons. For example, some northern regions require heavy, ground-based logging operations to be completed before the spring thaw. Therefore, distinct operating seasons may be determined for certain soil conditions.

Finally, when used in partial harvesting situations, the potential damage to the residual trees (Bettinger and Kellogg, 1993) may be an important consideration in choosing the harvesting system or the pieces of the system to employ, as some systems may be more easily maneuverable among residual live trees than others. When advanced regeneration is important for the creation of a new forest (through even-aged forest development processes) or maintenance of an uneven-aged forest, the impact of harvest systems on seedlings and saplings can be of great concern. All of these concerns can make the task of harvest system selection challenging for the forester or natural resource manager.

Questions

(1) *Primary transportation of wood products.* When extracting wood products from a forested area, two general forms of transportation are employed: primary and secondary. Of the primary transportation systems, which would you consider for forests situated on very steep ground? Which would you consider for forests situated on relatively flat ground? Develop a short paragraph describing why you arrived at these conclusions.

(2) *Secondary transportation of wood products.* For the area where your family lives, describe in a short memorandum the types of secondary transportation processes used to move trees or logs from the forest to a mill. Provide pictures or conceptual models that illustrate the main types of processes that are used.

(3) *Tree felling.* For your area of the world, do some investigative work on local logging processes and describe in a brief paragraph the types of tree felling processes commonly employed.

(4) *Forest biomass.* Perform an Internet search on one type of forest biomass harvesting system used or studied somewhere in the world. In a short memorandum, describe the collection, processing, and transportation processes used to move the biomass from the forest to a biomass-using facility. In a small group of five or six people, discuss the systems, their differences, and their effectiveness in harvesting, collecting, and transporting forest biomass.

(5) *Nontimber forest products.* Select an extractable nontimber forest product of interest to you. Describe in a short PowerPoint presentation the work processes that might be employed in moving the product from the location in which it is found to a final destination where it will be consumed or converted to other products.

References

Amaranthus, M., Pilz, D., 1996. Productivity and sustainable harvest of wild mushrooms. In: Pilz, D., Molina, R. (Eds.), Managing Forest Ecosystems to Conserve Fungus Diversity and Sustain Wild Mushroom Harvests. U.S. Department of Agriculture, Forest Service, Pacific Northwest Research Station, Portland, OR. General Technical Report PNW-371, pp. 42–61.

Bell, J.L., 2002. Changes in logging injury rates associated with use of feller-bunchers in West Virginia. Journal of Safety Research 33, 463–471.

Bentley, T.A., Parker, R.J., Ashby, L., 2005. Understanding felling safety in the New Zealand forest industry. Applied Ergonomics 36, 165–175.

Bettinger, P., Kellogg, L.D., 1993. Residual stand damage from cut-to-length thinning of second-growth timber in the Cascade Range of western Oregon. Forest Products Journal 43 (11/12), 59–64.

Bist, S.S., Cheeran, J.V., Choudhury, S., Barua, P., Misra, M.K., 2002. The domesticated Asian elephant in India. In: Baker, I., Kashio, M. (Eds.), Giants on Our Hands: Proceedings of the International Workshop on Domesticated Asian Elephant. Food and Agriculture Organization of the United Nations, Regional Office for Asia and the Pacific, Bangkok, Thailand.

Boston, K., 2012. Impact of the ninth circuit court ruling (Northwest Environmental Defense Center v. Brown) regarding forest roads and the Clean Water Act. Journal of Forestry 110 (6), 344–346.

Boston, K., 2016. The potential effects of forest roads on the environment and mitigating their impacts. Current Forestry Reports 2 (4), 215–222.

Boston, K., Murphy, G., 2003. Value recovery from two mechanized bucking operations in the southeastern United States. Southern Journal of Applied Forestry 27 (4), 259–263.

Bryant, R.C., 1923. Logging, the Principles and General Methods of Operation in the United States. John Wiley & Sons, Inc., New York, 556 p.

Byrne, J.J., 1960. Logging Road Handbook: The Effect of Road Design on Hauling Costs. U.S. Department of Agriculture, Forest Service, Washington, D.C. Agricultural Handbook No. 183.

Carlsson, D., Rönnqvist, M., 2005. Supply chain management in forestry—case studies at Södra Cell AB. European Journal of Operational Research 163, 589–616.

Cass, R.D., Baker, S.A., Greene, W.D., 2009. Costs and productivity impacts of product sorting on conventional ground-based timber harvesting operations. Forest Products Journal 59 (11/12), 108–114.

Cluster, A., 1978. Cummins Yarder Design Verification Test. U.S. Department of Agriculture, Forest Service, Equipment Development Center, Missoula, MT. ED&T 8035 Cable Systems for Forest Residues 25 p.

Conway, S., 1982. Logging Practices: Principles of Timber Harvesting Systems. Miller Freeman Publications, Inc., San Francisco, CA, p. 431.

Foltz, R.B., 1994. Reducing Tire Pressure Reduces Sediment. U.S. Department of Agriculture, Forest Service, San Dimas Technology and Development Center, San Dimas, CA. Publication 9477 1306-SDTDC.

Minnesota Forest Resources Council, 1999. Sustaining Minnesota Forest Resources: Voluntary Site-Level Forest Management Guidelines for Landowners, Loggers and Resource Managers. Minnesota Forest Resources Council, St. Paul, MN.

Jones, J.A., Swanson, F.J., Wemple, B.C., Snyder, K.U., 2000. Effects of roads on hydrology, geomorphology, and disturbance patches in stream networks. Conservation Biology 14 (1), 76–85.

Kärhä, K., Jylhä, P., Laitila, J., 2011. Integrated procurement of pulpwood and energy wood from early thinnings using whole-tree bundling. Biomass and Bioenergy 35, 3389–3396.

Kellogg, L.D., Bettinger, P., Studier, D., 1993. Terminology of Ground-Based Mechanized Logging in the Pacific Northwest. Oregon State University, Forest Research Laboratory, Corvallis, OR. Research Contribution 1. 12 p.

LeDoux, C.B., 2010. Mechanized Systems for Harvesting Eastern Hardwoods. U.S. Department of Agriculture, Forest Service, Northern Research Station, Newtown Square, PA. General Technical Report NRS-69. 13 p.

MacDonald, P., Clow, M., 2010. "Things was different in the South:" the industrialization of pulpwood harvesting systems in the Southeastern United States 1945–1995. Technology in Society 32, 145–160.

Mar, K.U., 2002. The studbook of timber elephants of Myanmar with special reference to survivorship analysis. In: Baker, I., Kashio, M. (Eds.), Giants on Our Hands: Proceedings of the International Workshop on Domesticated Asian Elephant. Food and Agriculture Organization of the United Nations, Regional Office for Asia and the Pacific, Bangkok, Thailand.

Martell, D.L., Gunn, E.A., Weintraub, A., 1998. Forest management challenges for operational researchers. European Journal of Operational Research 104, 1–17.

Molina, R., O'Dell, T., Luoma, D., Amaranthus, M., Castellano, M., Russell, K., 1993. Biology, Ecology, and Social Aspects of Wild Edible Mushrooms in the Forests of the Pacific Northwest: A Practice to Managing Commercial Harvest. U.S. Department of Agriculture, Forest Service, Pacific Northwest Research Station, Portland, OR. General Technical Report PNW-309.

Mooney, S.T., Boston, K.D., Greene, W.D., 2000. Production and costs of the chambers delimbinator in first thinning of pine plantations. Forest Products Journal 50, 81–84.

Neely, G., Wilhelmson, E., 2006. Self-reported incidents, accidents, and use of protective gear among small-scale forestry workers in Sweden. Safety Science 44, 723–732.

Nieuwenhuis, M., 2010. Terminology of Forest Management, Terms and Definitions in English, 2nd revised edition. International Union of Forest Research Organizations, Vienna, Austria. IUFRO World Series Volume 9-en.

Parren, M., Bongers, F., 2001. Does climber cutting reduce felling damage in southern Cameroon? Forest Ecology and Management 141, 175–188.

Pearson, M.W., 1984. Safety. In: Wenger, K.F. (Ed.), Forestry Handbook, second ed. John Wiley & Sons, Inc., New York.

Reid, L.M., Dunne, T., 1984. Sediment production from forest road surfaces. Water Resources Research 20, 1753–1761.

Ruiz Pérez, M., de Blas, D.E., Nasi, R., Sayer, J.A., Sassen, M., Angoué, C., Gami, N., Ndoye, O., Ngono, G., Nguinguiri, J.-C., Nzala, D., Toirambe, B., Yalibanda, Y., 2005. Logging in the Congo Basin: a multi-country characterization of timber companies. Forest Ecology and Management 214, 221–236.

Sessions, J., Boston, K., Murphy, G., Wing, M.G., Kellogg, L., Pilkerton, S., Zweede, J.C., Heinrich, R., 2007a. Harvesting Operations in the Tropics. Springer, Heidelberg, 2007.

Sessions, J., Boston, K., Wing, M., Akay, A., Theissen, P., Heinrich, R., 2007b. Forest Road Operations in the Tropics. Springer, Berlin.

Simmons, F.C., 1951. Northeastern Loggers' Handbook. U.S. Department of Agriculture, Forest Service, Washington, D.C. Agricultural Handbook No. 6.

Simmons, F.C., 1979. Handbook for Eastern Timber Harvesting. U.S. Department of Agriculture, Forest Service, Northeastern Area, State & Private Forestry, Broomall, PA, 180 p.

Sloan, H., 1982. Cable logging systems to match our timber, terrain, and markets. In: Proceedings Appalachian Cable Logging Symposium. Virginia Polytechnic Institute and State University, Blacksburg, VA, pp. 53–62.

Spinelli, R., Cuchet, E., Roux, P., 2007. A new feller-buncher for harvesting energy wood: results from a European test programme. Biomass and Bioenergy 31, 205–210.

Spinelli, R., Ward, S.M., Owende, P.M., 2009. A harvest and transport cost model for Eucalyptus spp. fast-growing short rotation plantations. Biomass and Bioenergy 33, 1265–1270.

Stenzel, G., Walbridge Jr., T.A., Pearce, J.K., 1985. Logging and Pulpwood Production, second ed. John Wiley & Sons, Inc., New York, 358 p.

Studier, D., 1993. Carriages for Skylines. Oregon State University, Forest Research Laboratory, Corvallis, OR. Research Contribution 3. 14 p.

Studier, D.D., Binkley, V.W., 1974. Cable Logging Systems. U.S. Department of Agriculture, Forest Service, Division of Timber Management, Portland, OR, 211 p.

Thelin, A., 2002. Fatal accidents in Swedish farming and forestry, 1988–1997. Safety Science 40, 501–517.

U.S. Department of Agriculture, 2020. Natural Resources Conservation Service. Web Soil Survey. U.S. Department of Agriculture, Natural Resources Conservation Service, Washington, D.C. https://websoilsurvey.sc.egov.usda.gov/App/HomePage.htm (accessed 12.02.20).

Uusitalo, J., 2010. Introduction to Forest Operations and Technology. JVP Forest Systems Oy, Tampere, Finland, 287 p.

Washington State Legislature, 2019. Appendix 4—Various Types of Cable Logging Systems. Washington State Legislature, Olympia, WA. WAC 296–54–99013. https://apps.leg.wa.gov/wac/default.aspx?cite=296-54-99013&pdf=true? (accessed 13.02.20).

Yoshioka, T., Aruga, K., Nitami, T., Sakai, H., Kobayashi, H., 2006. A case study on the costs and the fuel consumption of harvesting, transporting, and chipping chains for logging residues in Japan. Biomass and Bioenergy 30, 342–348.

13

Forest and natural resource economics

Economics is a social science that involves studying the choices that individuals, businesses, institutions, and societies make when prioritizing the use of scarce resources to meet their needs. Scarcity implies real trade-offs, in which the costs and benefits of various uses of limited resources must be compared. Scarcity is very evident in forest and natural resources management. Since we need forest products and services in our everyday lives, the relevant economic question is not whether or not we use them but rather how much of these limited resources we choose to use. In economics, incremental decision-making processes, such as considering the use of one more roll of paper towels or the need for spending one more day of recreation in a national park, are called *marginal decisions*. As it turns out, most analytical tools of economics are readily applicable to forest and natural resource management. Some additional considerations, however, need to be addressed to account for the special characteristics of forest resources. These include, among others, long production periods, multiple potential outputs, ownership characteristics, renewability, and uncertainty. The objective of this chapter is to provide an overview of the application of economic analyses to forest and natural resources, and to facilitate a better understanding of the economic factors that influence forest and natural resource management decisions. This chapter will expose readers to the economic concepts and methods needed to solve forest and natural resource management

problems and to analyze current forestry and natural resources management issues. Upon completion of this chapter, readers should have an understanding of

- economics as a method of contemplation, in which incremental costs and benefits are compared to maximize net benefits;
- the economic and financial tools needed to assess, analyze, and solve real-world forest and natural resource management problems; and
- the decision-making skills necessary to effectively address private and public goals and evaluate the impacts of policies and markets on natural resources.

Economic methods are often applied to the analysis of forestry and natural resource management activities, and they are important in helping people sort through the myriad of available choices and facilitate the selection of those that are most desirable. Economic methods are not only used to assess the benefits and costs of individual projects but are also used to assess the impact of management activities on the health and well-being of local communities, regional development, and national economies. For example, Crowley et al. (2001) estimated that for one particular working forest in Ireland, nearly one job was created for every 150 hectares (ha; about 370 acres [ac]) of forest. Therefore, while ecological, hydrologic, recreational, and other values of forests are important, so too are the economic values.

13.1 Introduction

Are you a rational human being? Rationality is an important assumption underlying the economic thinking espoused in this chapter. Rationality implies that a person acts in his or her own self-interest or the interest of the public at large, and perhaps on occasion even acts in a selfish manner. It also means that if given the choice, a person would prefer more of the things that he or she, or the larger set of people in their community, values over fewer of those things. Most importantly, it implies that a person responds to incentives, compares the benefits (carrots) and costs (sticks) of potential actions, and then chooses the course that is most beneficial. Consequently, a change in either the benefits or costs of potential decisions can change a person's behavior. Therefore, at least in principle, solving a perplexing environmental problem could involve changing the related costs and benefits to alter the behavior of those contributing to the problem. This is how most natural resource policies are designed to work.

In general, resource scarcity is central to economics and is particularly so to forest and natural resource economics. The very reason the field of economics exists is because of resource scarcity, a situation which gives rise to all economic questions. In other words, without scarcity there would be no economics. Further, humans generally desire more of a resource than the amount available. Consider, for example, a forest that is capable of producing wood products through harvests or that if left undisturbed could support certain recreational experiences. On one hand, there is a need to continue using paper, lumber, and other wood products; on the other hand, there is a need for recreational areas. However, the wood products need implies that trees must be felled and converted to end-use products. Further, the demand for wood products often requires reliance on intensively managed and uniform tree plantations (Fig. 13.1), which may leave the area forested, yet will change the character of the area. At the same time, many members of the public enjoy vast expanses of undisturbed, old-growth forests (Fig. 13.2) that are reserved for biodiversity, recreation, water resources, and wildlife. Therein lies the dilemma—managing a forest to provide competing needs to a human society. The acknowledgment of the vast carbon that these forests hold is another service provided by these forests. Interestingly, although these two needs are used as examples here, they have influenced the management of forest and forest policy in North America. During the 20th century, political and economic forces in some areas of Europe have also influenced the development of forest policies addressing these issues (Nijnik and van Kooten, 2000). While these types of desires for forests are often incompatible, choices have to be made regarding which to pursue in certain areas of the landscape. Sometimes,

FIGURE 13.1   Forest plantation near Gleann Glas Dhoire, Scotland. *Ian Shiell through Wikimedia Commons.*
Image Link: https://commons.wikimedia.org/wiki/File:Forest_plantation_on_north_side_of_Gleann_Ghlas_Dhoire_-_geograph.org.uk_-_946716.jpg
License Link: https://creativecommons.org/licenses/by-sa/2.0/deed.en

such choices are much starker. In numerous developing countries, many communities must choose between leaving forests intact or using forests to provide the basic needs of families. One would think, since forests are renewable resources and can be managed sustainably, that some of these conflicts could be managed, or possibly avoided altogether.

Scarcity implies that we have to make choices. Even procrastinating to avoid making difficult decisions involves making a choice—that is, deciding to do nothing. Every nontrivial choice involves trade-offs. In general, a person must give up something to receive something in return. Each and every time a person makes a choice,

FIGURE 13.2   Old beech forest at the Urwald Sababurg nature reserve, in the district of Kassel, Hessen, Germany. *Szent István through Wikimedia Commons.*
Image Link: https://commons.wikimedia.org/wiki/File:SababUrw1180500.jpg
License Link: https://creativecommons.org/licenses/by-sa/3.0/deed.en

they face trade-offs. When attending college, these trade-offs may be as simple as, Shall I spend my money on food or fun? Shall I study during the weekend for an exam? or Shall I enjoy the free time with my friends? Later in life, a person may be faced with more difficult trade-offs such as, Shall I invest my funds in a tree-growing endeavor? or Shall I place my funds in a bank savings account? Considering choices in terms of the trade-offs involved emphasizes the need to give up something to receive something else. *As many people say, there is no such thing as a free lunch.* In some sense, what a person must give up represents the cost of what the person hopes to receive. Economists describe this as the *opportunity cost* of choices, and it represents the next best alternative that must be forgone to obtain something of value. Consider, for example, the opportunity cost of attending a college or a university. Although there are exceptions, when doing so, a person generally cannot work full-time (40 h per week) or earn a significant amount of income (a measurable trade-off). As a result, a person probably does not have enough free time to spend on his or her favorite leisure activities (another trade-off, yet more challenging to measure). Further, the money earned from a part-time job may need to be spent on tuition and books (a measurable trade-off) rather than on other items. Of course, the hope is that after graduation, a rewarding career will begin, which will generate sufficient income to alleviate the financial burden of having attended college, thus providing sufficient compensation for the sacrifice incurred.

In this chapter, economic concepts will be used to illustrate how and why people interact with forests and natural resources. An attempt will be made to provide economic reasons why forests in some regions are managed sustainably, while those in other regions have significant levels of deforestation. Through examples of successes and failures, the impact of economic signals on forest landowner behavior will be noted.

For example, changes in wood prices may affect the timing of forest harvests. This casual statement is descriptive of human behavior but also can be tested based on actual, observed behavior. Such assessments represent a branch of economics called *positive economics*, which help describe the way the economy or the financial side of our world works.

There is a chance that some readers of this book will choose forestry as a career and, if so, opinions will be formed regarding how forestry should be practiced. For example, one person may suggest that landowners should allow their trees to grow to very old, biologically efficient ages in order to sustain wood resources for future generations. Another person may suggest that landowners should harvest trees at younger, economically efficient ages for the same reason. Such normative statements involve value judgments and ethical precepts and can only be resolved by discussions and political debate (Samuelson and Nordhaus, 1995). This branch of economics is called *normative economics*.

13.2 Why forest and natural resource economics?

Thus far in this chapter, only a few general economic concepts, such as marginal analysis and opportunity costs, have been mentioned. These and other general economic tools and models readily apply to forestry and natural resource management problems. However, forest and natural resource economics is a subdiscipline of economics, and several distinct characteristics set it apart from the economic analyses of other resources. First and foremost, forests provide multiple products and fulfill multiple functions. They supply food, forage, medicine, timber, water, and wildlife habitat and store much of the Earth's biodiversity. Fig. 13.3 provides an example of the availability over time of multiple forest products. Assuming that the origin of the graph

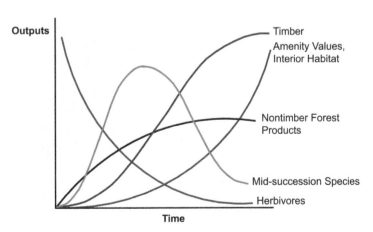

FIGURE 13.3 Multiple products associated with growth and development of an even-aged forest.

represents the initiation (reforestation or afforestation) of a forest, the availability of products or services are associated with the changes in forest growth. Note how the output of particular products changes over time. This is one reason why *time* is such an important consideration in forest and natural resource management. When a forest is young, there is at first little timber but plenty of herbivore habitat, but as forests age there is more timber and closed-canopy forest habitat. Old-growth forests will have higher timber volumes and amenity values; after all, most of us enjoy viewing large monumental trees. However, if we were to harvest the trees for the production of forest products, these amenity values would disappear.

Another interesting characteristic of forests is that the potential outputs fall into various resource and access categories (Fig. 13.4). Forests, wildlife, and nontimber forest products (NTFPs) are generally considered renewable. The fact that resources are renewable, however, does not mean that they cannot be destroyed. However, some may argue that because of the time required to develop old-growth ecosystems, these are not renewable. Access is an important characteristic of outputs, which can be either restricted (private) or unrestricted (public). The ability to exclude others from using one's forest resources is of critical importance for providing resource owners with incentives to conserve and invest in management. As some have found, when forest landowners benefit from sustainable forest management (SFM), they are more likely to use sustainable forest and natural resource management practices (Pearce and Turner, 1991).

Forest production is generally characterized by long (perhaps a decade) to very long (perhaps a century) production periods. This, too, is a distinct characteristic that sets forests apart from other resources. While intensively managed eucalyptus (*Eucalyptus* spp.) plantations in

Brazil (Fig. 13.5) can reach maturity for the production of paper products in seven years, in many other situations, trees must be allowed to grow for several decades before they can be harvested. For example, as a general rule, pine forests in central Europe are not eligible for a final harvest treatment (clearcut) until they are at least 80 years old. The time required to achieve old-growth forest conditions may also be important for developing plans of action for maximizing habitat development. The time required to fully restore a tropical rainforest (if at all possible) may also be important and further illustrates why time is such an important forest and natural resource management consideration.

A further distinct characteristic of forest management that emphasizes commodity production is that the inputs (trees) and the outputs (trees) are the same. As long as trees continue to grow, they are considered the *factory*. The moment trees are harvested, they become the *product*. For this reason, the question of when to cut trees is central to forest and natural resource economics. These characteristics make forest production different from other industries.

As noted in Chapter 6, many forest outputs and functions can be considered ecosystem services or public goods. Some of the ecosystem services can potentially be valued, such as conservation and protection of water resources (Dong et al., 2011) or the ability of forests to store carbon (Corbera et al., 2009). For forestry and natural resources purposes, a public good is a forest product or function that is *nonrival in consumption* and is *nonexcludable from enjoyment*. The first attribute of a public good (nonrival in consumption) suggests that the consumption of a good or a service by one person

Natural characteristics

	Nonrenewable	Renewable
Can Restrict (private)	Minerals, Old Growth	Forestry, Agriculture
Unrestricted (public)	Oil Pools, Aquifers, Tropical Forests	Wildlife, Fisheries, Nontimber Forest Products

Access (ownership)

FIGURE 13.4   Categorization of natural resources by access and natural character.

FIGURE 13.5   Eucalyptus forest located in Brazil. *Denis Rizzoli through Wikimedia Commons.*
Image Link: https://commons.wikimedia.org/wiki/File:Eucaliptal_Aracruz.JPG
License Link: https://creativecommons.org/licenses/by-sa/3.0/deed.en

does not reduce the amount of the good or the service that is available to another person. Consider, for example, a beautiful vista. The fact that a person may enjoy it does not make it less available for others to also enjoy it, unless the first person is creating a nuisance or blocking the view of others. The second attribute of a public good (nonexcludable in enjoyment) implies that people may not be excluded from enjoying certain goods and services, such as carbon storage or aesthetic values, even when they have not contributed to the development or management of those goods or services. When people can freely enjoy these goods and services, a free-rider problem can occur and eventually a market failure can occur. A market fails when the allocation of goods and services in a free market is not deemed efficient, and thus there are situations where one person (or a group of people) may improve their position in life without having paid for the beneficial aspects or improvements. Public goods are closely related to *externalities* that represent impacts of one person's actions (costs or benefits) on others; impacts that are not reflected in prices. Interestingly, externalities can be either negative (pollution) or positive (fire prevention management).

Finally, when a person walks through a forest, especially if beginning at a low point along a creek and traveling upwards toward higher ground, they may find that the general landscape and the specific soil and vegetation resources will change (Fig. 13.6). Even over a short distance, a high level of forest heterogeneity, characterized by varying plant growth rates and numerous plant species, might be found. This heterogeneity makes forest and natural resource management and economics challenging. Further, a person can manage a forest for numerous objectives, some of which are complementary, while others are mutually exclusive. In a forest composed of hundreds of individual stands, the number of possible management alternatives can exceed

comprehension. Forest and natural resource economics can help managers and landowners sort through the alternatives by evaluating the costs and benefits of various management alternatives.

Our approach to forest and natural resource economics relies on understanding individual decisions, in which choices and preferences determine outcomes. While this approach can help understand the observed human behavior, it may suffer by failing to acknowledge group or cultural effects. Values are derived from human wants and desires and the ability of human systems to meet those desires (supply and demand). This forces the study and application of economics to be anthropocentric. While intrinsic, existence, or other values can be important, they are difficult to measure and are not easily incorporated into economic analyses.

13.3 Markets, efficiency, and government

Given the many difficult choices that must be made in forest and natural resource management, the least we can do is to ensure that the resources available to us are used wisely and efficiently. Being efficient, in the most basic sense, implies that we are not wasting resources. While this goal seems straightforward and easy to achieve, in real life it is difficult to accomplish. In a broader sense, efficiency implies that all of the opportunities available to improve the livelihoods of everyone have been fully exploited. In many cases, markets work well to ensure this. When resources become increasingly scarce, their prices in their respective markets rise. According to the *Law of Demand*, when the price of a good rises, less of the product is consumed because of the increased expense and limits on the funds available. In other words, our consumption decisions are based on what a person must give up (often money) to obtain a desired good or service. At the same time, the *Law of Supply* states that when the price of a good rises, its supply rises as well. If prices are high, producers attempt to supply more of the particular desired goods and services in an effort to increase their profits, if that is possible. For example, high timber prices may cause forest landowners to harvest more timber and perhaps invest in intensive management, which can increase the availability of wood products now (Fig. 13.7) and in the future. Markets, some of which are composed of millions of consumers and producers, respond to price signals. Consequently, economic efficiency tends to be associated with positive economics. Rising prices signal consumers to demand less and producers to supply more. In this give-and-take process, markets can achieve efficient resource allocation and reach equilibrium.

Sometimes, however, markets fail to deliver efficient outcomes, as is often the case with natural resources. These

FIGURE 13.6 Lochsa River in the Clearwater National Forest, Idaho, United States. *Idaho Travel Council through Wikimedia Commons. Image Link: https://commons.wikimedia.org/wiki/File:Lochsa-clearwater.jpg License Link: https://creativecommons.org/licenses/by/2.5/deed.en*

FIGURE 13.7   Wood ready to be delivered to a market from Condie Wood, near Edinburgh, Scotland. *Jim Bain through Wikimedia Commons. Image Link: https://commons.wikimedia.org/wiki/File:Harvest_time_-_geograph. org.uk_-_158747.jpg License Link: https://creativecommons.org/licenses/by-sa/2.0/deed.en*

FIGURE 13.8   Chopped and stacked fuelwood in Mali. *M. Poudyal through Wikimedia Commons. Image Link: https://commons.wikimedia.org/wiki/File:Mali_firewood.jpg License Link: https://creativecommons.org/licenses/by-sa/2.0/deed.en*

failures can occur because the link between benefits and costs is, for some reason, broken. Sometimes, for example, a person chooses not to incur certain costs because they are uncertain of the ability to enjoy the benefits of what must be sacrificed (funds, or time). For instance, assume that a person is enrolled in an academic course and, oddly enough, the final grade they are to receive is not related to how long they studied to master the course's material. How likely would this person be to allocate large amounts of study time in a situation such as this?

Markets also may fail in the presence of external benefits and costs or public goods. These conditions often arise from natural resource characteristics and property rights arrangements that are not well defined or exercised. Consider, for example, certain wildlife resources. By their very nature, it is difficult to assign property rights to animals such as deer, since they are quite mobile. It is hard to assign enforceable property rights to deer because to do so would be very expensive. Imagine that a person was given ownership of free-ranging deer. How would they track their deer and at the same time ensure that no one else claims them? These classes of resources are often considered *open-access resources*. In many cases, resources such as wildlife may be owned by no one and, therefore, a government may oversee the management of their populations. Another example of resources with open-access characteristics are the forests of some developing countries. In forests such as these, local human populations can access the resources and extract various products in order to meet basic needs (Fig. 13.8). In some cases, this type of resource use can be unsustainable if it is left unregulated. Since a local populace does not own the forest, and since they have an incentive (survival) to harvest and collect wood and other forest products, they may not have an incentive to conserve the forest resources. Unfortunately, because everyone else in the local area has

access to the same resources, the products might be extracted without limits. As a result, local communities that are most reliant on obtaining forest resources in this manner have no assurance that conservation efforts will ever yield any benefits.

When markets fail to achieve efficient outcomes or to meet important social objectives, government intervention may be necessary to improve the way the natural resources are being used. While efficiency is an important goal in itself, concern should be placed on the distribution of resources among members of our society or, more simply, the equity of resource use. Fairness and social justice imply that every member of a society, no matter how rich or poor they may be, should have access to certain resources such as clean air, clean water, healthy food, and access to forests and any recreational opportunities that are offered. While markets are generally well equipped for achieving efficiency, they may not be well equipped for achieving equity, which remains the domain of government and provides the government with the rationale for its involvement in the management of forest and natural resources. Achieving equity clearly requires value judgments and tends to be associated with normative economics. Given the number of heated public disputes on the management of forest and natural resources, value judgments can be of great importance in forest and natural resource debates.

A government may also be interested in promoting a community's stability in order to prevent financial boom-and-bust cycles caused by unsustainable natural resource extraction. For example, in the 20th century some communities in the western United States were heavily reliant on the harvest of federal timber resources for employment opportunities, tax revenue, and other federal payments. When a timber resource was reduced or when harvesting slowed or ceased for political or environmental reasons, some communities suffered severely because they had no other resources to rely upon. For example, with the harvest limits that are now imposed in U.S. National Forests for the protection of the northern spotted owl (*Strix occidentalis caurina*), many local communities have been struggling desperately to survive. To provide stability to a community, some portion of the forest resources available may need to be managed according to sustained wood yield principles to ensure that a certain level of timber harvest will continue to be available.

Governments have available a range of policy tools for regulating resource use. The first group of tools includes taxes, subsidies, or assistance designed to induce a desired behavior. Taxes are often used as a source of revenue for governments of all sizes and can influence management decisions. In order to reduce resource extraction, taxes may be imposed to increase extraction costs and reduce profit incentives. Taxes on timber harvests have been applied at local, state, and federal levels in North America. Governments may also develop programs that subsidize tree-planting expenses, assist in forest and natural resource management planning efforts, or facilitate wildlife habitat improvement projects. These types of subsidies and assistance are meant to stimulate behavior toward efforts that are deemed important from a natural resource management perspective. A second set of tools includes direct controls or regulations on acceptable and unacceptable uses of resources. Of course, regulations need to be enforced to have an impact, and enforcement may be extremely costly at times, perhaps rendering them ineffective. While government actions may be justified based on market failures, unfortunately, governmental efforts can also fail. The fact that deforestation continues to be such a great challenge in so many countries suggests that poorly designed governmental policies are in effect in those nations. However, a lack of adequate resources (funds, trained personnel, and time) may be the reason for continued resource management problems. In these cases, an examination of the causes of market or government failures and the processes available to eliminate them may be of value.

13.4 The time value of money

Most economists would say that people display a time preference for the benefits and costs associated with their actions. In many cases, when a person must decide on the time to consume desirable goods and services, their actions suggest that they prefer to do this sooner rather than later. If consumption were to be postponed, then a person would generally expect to be compensated in some manner for having to wait. Consider, for example, the act of depositing money in a savings account. Suppose that a person was either fortunate or thrifty and had $1000 to spare. If the person decided to put the money in a savings account rather than spend it right away, he or she would expect the bank to pay interest on the deposit and holding of the funds, say at a rate of 3% per year. The 3% interest rate (0.03 in decimal notation) is the compensation paid by the bank for holding the funds while the person waits 1 year to spend them. The interest amounts to $30 ($0.03 \times \$1000 = \$30$), as the bank pays 3% of the principal sum for the opportunity to use the money for 1 year. An *interest rate* is thus the price of money that is deposited in a savings account for a specific period. The only difference is that it is expressed in percentage terms (3%) rather than in dollars ($30). If 3% interest is acceptable, then an economist would also state that the fortunate or thrifty person is indifferent with regard to the choices of spending $1000 today or $1030 1 year later.

One major component of the interest rate is its time preference. Other major components are risk and inflation, which receive a more complete treatment in other forest economics and management texts (e.g., Klemperer, 1996; Rideout and Hesseln, 2001; Tietenberg, 2006; Bettinger et al., 2017). Risk reflects the potential for losing an investment. In forestry, one must be concerned with biological risks (e.g., diseases, fires, and hurricanes), market risks (e.g., decreasing wood prices), and political risks (e.g., management regulations that increase costs or lower property values). If a person places his or her money in an investment that is not guaranteed to earn a certain rate of return, and therefore may lose value, they need to be compensated for the possibility of an unpleasant outcome. Therefore, in general, as perceived risk increases, so does the interest rate assumed or desired. Inflation, in turn, refers to a gradual rise in prices. As time passes, a unit of currency (e.g., a dollar or a euro) will generally lose its purchasing power, allowing the purchase of less and less of a product or service over time. If funds are placed into a savings account, a person would like to be compensated for inflation. After all, what is the point of placing funds into a savings account if the money will lose value and reduce a person's purchasing power? The higher the time preference, assumption of inflation, and perceived risk, the higher the interest rate a person would need to receive from an investment in order to compensate for postponing the consumption of other products or services with the same funds.

The major implication of time preference is that the value of money (e.g., a dollar) today is not the same as its value in the future. Therefore, in order to meaningfully evaluate benefits and costs that could occur at different points in the future, we need to use the interest rate desired by a person to adjust the values of benefits and costs over time. The term *future value* represents the actual value of benefits and costs received or incurred at some time in the future. The term *present value* also represents the value of future benefits and costs, but these are expressed in terms of their worth today. The interest rate therefore represents an investor's opportunity cost of time, the minimum return necessary to forgo consumption products and services now, or the expected return one can obtain through investments in alternatives.

When we calculated the value of the hypothetical $1000 savings deposit 1 year from now at a 3% interest, we calculated its future value 1 year from today. However, when a person places money into a savings account and leaves it there for more than 1 year, they will typically earn what is called *compound interest*, which means that periodic interest will accrue based on the initial investment (the principal) and all previous interest payments that are earned along the way. The addition of interest to the principal is termed *compounding* and works as shown below with the funds deposited into a savings account and held there for 3 years:

Now (Initial investment): *$1,000.*

After 1 year: $1,000(1 + 0.03) = $1,000(1 + 0.03)^1 = *$1,030,*

or $1,000 + (0.03 × $1,000) = $1,030;

After 2 years: $1,000(1 + 0.03) (1 + 0.03) = $1,000 (1 + 0.03)^2 = *$1,061,*

or $1,030 + (0.03 × $1,030) = $1,061;

After 3 years: $1,000(1 + 0.03) (1 + 0.03) (1 + 0.03) = $1,000(1 + 0.03)^3 = *$1,093,*

or $1,061 + (0.03 × $1,061) = $1,093.

In general, this relationship can be stated as:

$$\text{Future value} = \text{Present value} \times (1 + r)^t$$

where r = the interest rate assumed and t = the number of time periods (years).

Simple interest is different from compound interest in that previously earned interest is not added to the principal prior to determining the current interest earned. With simple interest, after 3 years a person would only accumulate $90 in interest (3 × $30). While in our example the difference between compound and simple interest is only $3, the longer that time passes the larger the difference between the two (compound and simple interest calculations) becomes. Fig. 13.9 illustrates compounding (calculating the future value of) a $1000 investment for as many as 20 years using three distinct interest rates. Assuming an interest rate of 3%, the $1000 investment 20 years from now would grow to $1806; assuming an interest rate of 10%, the $1000 investment would grow to $6727 over the same period.

In assessing forestry and natural resource management projects, a project's costs and benefits are usually expressed in present value terms (their value today). In expressing the benefits and costs in terms of value today, an operation called *discounting* is used, and due to the long investment periods assumed in the management of forests, this operation plays an important role in forest and natural resource economics (Hepburn and Koundouri, 2007). The interest rate used for discounting benefits and costs is called the *discount rate*. For example, if a person were to receive $1030 a year from now and the discount rate (i.e., the interest rate used in discounting applications) was 3%, then the present value of that sum of money would be $1000 (= $1030/1.03). In general, the present value can be calculated by rearranging the future value formula:

$$\text{Present value} = \left(\frac{\text{Future value}}{(1 + r)^t} \right)$$

Fig. 13.10 illustrates the results of calculating the present value of $1000 received 20 years from now using various discount rates. For example, using a 10% discount rate, the present value of the payment amounts to

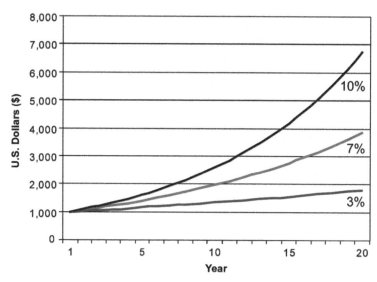

FIGURE 13.9 Compounding $1000 for 20 years at various interest rates.

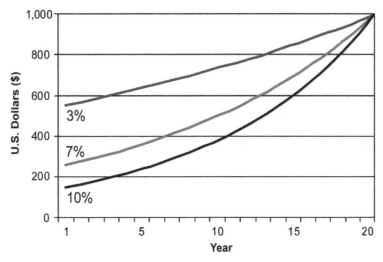

FIGURE 13.10 Discounting $1000 for 20 years at various interest rates.

only $149. This also suggests that if a person were to place $149 today in an investment that will earn 10% per year for 20 years, they would have $1000 at the end of 20 years. These examples demonstrate (Figs. 13.9 and 13.10) that the choice of interest rate is very important, as the use of different rates can result in remarkably different valuations. In calculating future values, the higher the interest rate assumed, the higher will be the future value. When calculating present values, the higher discount rate assumed, the lower the present value will be. The length of a project and the discount rate exert a powerful influence on present and future values. For example, investments in forestry that may yield large benefits (revenues) many years in the future may, in present value terms, seem to have a very low value.

As suggested, selection of the discount or interest rate to be used in evaluating an investment is an important process. For some investors, the choice of an interest rate may be guided by returns on alternative investments of similar duration, risk, and scale (Bullard et al., 2002). Individuals with discretionary money often have several investment opportunities. As a new project is evaluated, the hope is that returns will be generated that are at least as high as returns from other projects in which the funds might have been invested. For other investors, the choice of interest rate may be guided by the cost of borrowing money (e.g., the cost of obtaining a bank loan). In large organizations, such as the U.S. Forest Service or a large forest products company, these issues are likely to have been assessed previously by others, and an analyst will simply be informed of the interest rate to use in the investment analysis. United States government agencies receive specific guidelines regarding the interest rates to use when evaluating their projects. For regulatory or public investment analyses, the Office of Management and Budget (2015) requires

a real discount rate of 7%, noting that the rate reflects real, pretax rates of return on private investments. A real rate of return, or *real interest rate*, is the rate of return expected after removing inflation. Subtracting inflation from a nominal interest rate defines the real rate of return or *real interest rate*. For certain types of investments, rates are related to the government's cost of borrowing money and based on long-term Treasury bond yields. For example, nominal interest rates on U.S. Treasury bonds in 2015 ranged from 2.0% to 3.5%, and real interest rates on U.S. Treasury bonds in 2015 ranged from 0.3% to 1.5%, depending on the bond maturity date (Office of Management and Budget, 2015). In contrast, prior to the economic downturn of the early 21st century, nonindustrial private landowners in Mississippi had minimum acceptable rates of return that ranged from 8% to 13% in nominal terms (including inflation) before taxes, depending on the project type and length and household income (Bullard et al., 2002).

If a person had a good feel for the present and future values of an investment, they could solve the below formula for the value of r to calculate their *rate of return*:

$$r = \left(\frac{\text{Future value}}{\text{Present value}} \right)^{1/t}$$

When calculated, r becomes the rate of return of the investment or the internal rate of return (IRR). This rate provides information regarding how quickly the investment will grow in value if applied to a particular project. In the savings account example presented earlier, the IRR equals 3%. While the formulas presented here can help determine the future and present values of a single cost or benefit (payment), extensions are readily available for multiple-period cash flows that may be incurred annually or periodically (e.g., Klemperer, 1996; Bettinger et al., 2017).

13.5 Forest and natural resource investment evaluation criteria

Once the present value of the costs and benefits associated with forestry or natural resource management projects has been determined, a measuring stick is needed to help determine whether or not a project should be implemented. Economists, planners, land managers, and even private individuals commonly try to maximize some forms of economic return or minimize some forms of cost. Two types of commonly used investment evaluation criteria are net present value (NPV) and IRR. The NPV, also sometimes referred to as present net worth, of an investment project is the difference between the present value of the benefits and the present value of the costs. The goal of this criterion is to indicate whether or not a project will increase someone's economic wealth. In other words, the NPV represents the net gain (or loss) in economic value adjusted for the timing of revenues or costs. At a minimum, one would invest in projects in which the present value of benefits exceeds the present value of costs. To satisfy this condition, the NPV of the project must be positive. One way to calculate NPV is:

$$NPV = \sum_{t=1}^{T} \left(\frac{B_t - C_t}{(1+r)^t} \right)$$

where t = a time period (e.g., a year), T = the total number of time periods (e.g., years), B_t = benefits (revenues) received in time period t, C_t = costs incurred in time period t, and r = the assumed discount rate.

One decision rule would be to accept any project containing a positive NPV, as long as funds are available. In this case, an investment would earn a rate of return at least as great as the assumed discount rate (r). If two investments are mutually exclusive and funds are limited,

then another rule would be to choose the investment that seems to earn the higher NPV. A third rule for use when funds are limited might be to select a combination of investments that provides the highest total NPV. The advantage of using NPV as a decision criterion is that it provides an unambiguous answer: a positive NPV indicates that the project will increase a person's or an organization's wealth, while a negative NPV indicates that the project will decrease a person's or an organization's wealth. However, comparing potential investments for an investment opportunity that has varying operational lengths, such as a short-rotation eucalypt plantation and a long-rotation longleaf pine (*Pinus palustris*) plantation (Fig. 13.11), is difficult.

The other commonly used criterion for investments is the IRR, which we introduced in Section 13.4. The IRR is formally defined as the interest rate required for the NPV to equal zero. This also implies that the present value of the benefits is equal to the present value of the costs. The IRR represents the interest earned when funds are invested in a particular project or investment. One way to derive the IRR is by solving the following equation:

$$NPV = 0 = \sum_{t=1}^{T} \left(\frac{B_t}{(1+IRR)^t} \right) - \sum_{t=1}^{T} \left(\frac{C_t}{(1+IRR)^t} \right)$$

An investment with a relatively high IRR is considered superior to an investment with a relatively low IRR. One advantage of the IRR is that only costs and benefits are needed to calculate it; another is that discount rate assumptions are not necessary. The IRR also indicates the likelihood of a positive NPV. If the IRR is greater than the discount rate typically used in practice to calculate an NPV, then the associated NPV must be positive.

The decision rules with regard to using an IRR to assess alternatives are straightforward. One should

FIGURE 13.11 Eucalyptus (*Eucalyptus* spp.; left) and longleaf pine (*Pinus palustris*; right) forests in the southern United States. *Edward L. Barnard, Florida Department of Agriculture and Consumer Services, United States Department of Agriculture Forest Service Archive, through Bugwood.org.*

invest in a project with an IRR that is greater than the cost of capital (i.e., the cost of borrowing money), as long as the funds are available. Further, if two investments are mutually exclusive, the investment with the higher IRR should be chosen when all else is equal such as time to receive the returns or amount of capital invested. When funds are limited, one should also choose the investments with the higher IRR values. However, one disadvantage in ranking projects by their IRR is that the investments selected may not have the highest NPVs. Other problems with using the IRR to select investments are that it favors rapid payoffs (i.e., short projects) and smaller-scale projects with larger returns, as opposed to larger projects with moderate returns.

We have emphasized throughout this chapter the notion that management activities incur *costs* and can obtain *benefits*. Many times, these are expressed in financial terms (e.g., dollars, euros, yen, yuan, etc.). Like the NPV, the benefit-cost ratio can be used to assess the suitability of proposed projects of investments. Here, the present value of the sum of the benefits is divided by the present value of the sum of the costs. If the ratio is above 1.0, then the present value of the benefits exceeds the present value of the costs in situations using the assumed discount rate. Further, in this case, the NPV is greater than zero. If the ratio is less than 1.0, the present value of the benefits does not exceed the present value of the costs, and the investment would be discouraged on economic grounds. However, if public goods are to be considered as either a benefit or a cost, a social benefit-cost ratio may be necessary. Public goods or physical externalities can at times be expressed in monetary terms, as in the case described by Wei et al. (2010) of an agricultural system in China, in which the costs of recharging groundwater and nitrous oxide mitigation were valued alongside typical yields and costs of agricultural practices. A social benefit-cost ratio can be estimated with an analysis of total direct and total indirect contributions or costs arising from the proposed actions. A straightforward benefit-cost analysis uses economically quantifiable benefits and costs (Duerr, 1993). However, if aspects of a proposed action or investment are not expressed in monetary terms, these intangible benefits can be quantified using rating systems such as the process described by the U.S. Department of State (2012), which suggests quantifying intangible benefits using a 5-point rating system and perhaps applying a weighting system to emphasize certain criteria according to their perceived importance.

For decisions involving a short period or a single rotation of a typical even-aged forest, the NPV appears to be the primary choice for assessing a project's investment potential and is usually considered a better criterion for evaluating forest projects (Lutz, 2011). Individuals who make difficult budgetary decisions would ideally

select projects that maximize net worth. However, the IRR is heavily emphasized in business programs, even though it can lead to erroneous conclusions.

13.6 Economics of a forest rotation

Three interesting stand-level problems have captured the attention of forest planners and analysts over the last three centuries: how to determine the length of an even-aged forest rotation; when to schedule intermediate harvests (thinnings); and how to maintain forest densities that will ensure the development of a healthy forest (Bettinger et al., 2017). A forest rotation represents the number of years between the establishment of a stand of trees and a final harvest and is associated with even-aged forest management (Fig. 13.12). A forest rotation age therefore represents the age at which a stand is scheduled for a final harvest or clearcut. There are several ways in which a rotation age can be defined, such as through its biological or reproductive potential, but one way involves economic analysis (Amacher et al., 2009; Bettinger et al., 2017). To maximize the NPV of an even-aged forest, the time of the final harvest (the forest rotation age) becomes a decision variable. The goal is to find the point in time when the NPV of managing the forest is maximized. The rotation age can also be determined in terms of marginal benefits and marginal costs. In this case, one would determine the optimum time to cut a stand of trees by assessing their value growth rate. In essence, one would decide between cutting a stand of trees now or allowing the stand to continue to grow for another year or more, which represents incremental

FIGURE 13.12 Young Scots pine (*Pinus sylvestris*) forest near Creag Mhigeachaidh in the Cairngorm Mountains of Scotland. *Richard Webb, through Wikimedia Commons.*
Image Link: https://commons.wikimedia.org/wiki/File:Inshriach_Forest_-_geograph. org.uk_-_319239.jpg
License Link: https://creativecommons.org/licenses/by-sa/2.0/deed.en

thinking. If the decision is made to wait one year before harvesting a forest, the trees should grow and increase in volume and value, which would represent the benefit of waiting. The opportunity cost of waiting would be the cost of maintaining the stand inventory (tree volume) and land for an additional year. Given available markets, a landowner could choose to cut all the trees in a stand, sell them along with the land, and then deposit the proceeds into a bank account to earn interest. Alternatively, a landowner could harvest the trees and establish a new forest. In this case, the opportunity cost of waiting would be expressed as the lost growth of a new forest. By comparing marginal costs and benefits, one could determine the optimal rotation, or the harvest age, to maximize the net benefits associated with growing trees.

In evaluating forest rotations, foresters and natural resource managers traditionally assume that the land will be occupied for forestry use in perpetuity. Under this assumption, harvested stands would be immediately replaced with new ones indefinitely; often, this is required by law. An example of such a forest production cycle is presented in Fig. 13.13. To illustrate the various approaches used by foresters and natural resource managers to determine optimal rotation, we will use an example of loblolly pine (*Pinus taeda*) stand-level data from the southern United States (Table 13.1). The pine stand grows on a site characterized by site index (SI_{25}) 70, which indicates that at age 25 the dominant and codominant trees are approximately 70 feet (ft; about 21 meters [m]) tall. This is one way that foresters and natural resource managers describe site productivity (i.e., the height trees reach by a specific age). In order to keep the example simple, assume that the stand is managed for only one wood product, pulpwood, which is used by the pulp and paper industry to produce various grades of paper and paperboard. For this simple analysis, the pine pulpwood stumpage (i.e., standing trees) price is assumed to be constant throughout time at $11 per ton. To further simplify the problem, assume that there is only one cost, $247 per ha ($100 per ac), to regenerate the stand following harvest. To determine the amount of revenue that might be obtained through

a final harvest at any given point in time, we would simply multiply wood volume by pine pulpwood stumpage price. Therefore, at age 40, the potential pine pulpwood harvest would be valued at $4900 per ha ($1983 per ac; i.e., 180.3 tons per ac × $11 per ton = $1983 per ac). Finally, assume that the interest rate (discount rate) is 4.5%, which is used to calculate the present values of costs and benefits. Shortly, we will come back to the economic assumptions and apply them to the task of determining a rotation age. However, we will first assess the biological rotation age, which is based simply on the average growth rate of a forest.

In the early days of modern forest management science, dating back to early 19th century Germany, foresters determined rotation length based on the biological information of a forest stand's growth. At that time, transportation networks were not developed and trade in wood products was limited. As a result, villages and towns (Fig. 13.14) relied on nearby forests for the provision of essential wood and nonwood products. Trees from forests were (and still are in many cases) necessary for community survival, as they provided a source of building materials, as well as fodder, food, fuel, and medicine. The forested habitat provided an opportunity for hunting. Since wood was such an important product, foresters wanted to maximize wood production (i.e., obtain the highest wood volume growth). Their goal was to produce maximum wood volume per unit area into perpetuity, as measured by average annual growth. Average growth is represented by the *mean annual increment* (MAI) and is determined by dividing the standing live volume by the number of years needed to obtain that volume. For the pine stand described earlier, MAI at age 30 equals 12.1 tons per ha (4.90 tons per ac) per year (147 tons/30 = 4.90 tons). To maximize wood production, we might compare the average growth and marginal (period-to-period) growth of a forest stand. Marginal growth is often called *current annual increment* (CAI) and represents current (from one period to another) stand volume growth. In our example, CAI equals 13.8 tons per ha (5.60 tons per ac) per year at age 23 (114.1 − 108.5 tons = 5.60 tons). In practice, forests are rarely measured every year. Instead, a standing inventory is determined periodically, every 5 or 10 years or perhaps at even longer intervals depending on the growth rate, through the collection of sample tree measurements. With this, the *periodic annual increment* can be measured, which is similar to CAI except that time periods are usually longer than one year. Current annual growth can also be approximated using various forest growth models after a sample has been collected.

CAI and MAI values for our pine stand are calculated in Table 13.1 and presented in Fig. 13.15. Note that as long as the current growth rate is greater than the

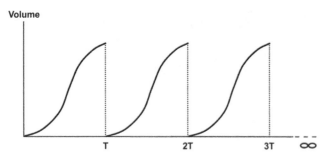

FIGURE 13.13 Conceptual representation of the forest growth among multiple cycles of even-aged forest management.

TABLE 13.1 Southern United States pine stand growth and values.

Age	Volume (tons/ac)	Stumpage value (US$/ac)	MAI (tons/ac/yr)	CAI (tons/ac/yr)	NPV (US$/ac)	BLV (US$/ac)
10	23.3	257	2.33	–	65	183
11	30.9	339	2.81	7.53	109	285
12	38.6	425	3.22	7.78	151	367
13	46.5	511	3.58	7.84	189	433
14	54.3	597	3.88	7.78	222	483
15	61.9	681	4.13	7.63	252	521
16	69.3	763	4.33	7.43	277	548
17	76.5	842	4.50	7.19	298	566
18	83.4	918	4.64	6.93	316	577
19	90.1	991	4.74	6.66	329	581
20	96.5	1061	4.82	6.39	340	581
21	102.6	1129	4.89	6.12	348	577
22	108.5	1193	4.93	5.86	353	569
23	114.4	1255	4.96	5.60	356	559
24	119.4	1314	4.98	5.35	357	547
25	124.5	1370	4.98	5.11	356	533
26	129.4	1424	4.98	4.89	353	518
27	134.1	1475	4.97	4.67	349	503
28	138.6	1524	4.95	4.47	344	486
29	142.9	1571	4.93	4.28	338	469
30	147.0	1617	4.90	4.10	332	452
31	150.9	1660	4.87	3.93	324	435
32	154.7	1701	4.83	3.77	316	418
33	158.3	1741	4.80	3.63	307	401
34	161.8	1780	4.76	3.49	298	385
35	165.1	1817	4.72	3.36	289	368
36	168.4	1852	4.68	3.24	280	352
37	171.5	1887	4.64	3.13	270	336
38	174.5	1920	4.59	3.02	260	321
39	177.4	1952	4.55	2.92	251	306
40	180.3	1983	4.51	2.83	241	291

average growth rate, the average growth rate increases each year. Conversely, as long as the current growth rate is lower than the average growth rate, the average growth rate decreases each year. Annual growth is maximized when it equals average growth, which in our example corresponds to 26 years. In this example, the average annual growth is maximized at 12.3 tons per ha (4.98 tons per ac) per year. This biologically determined rotation age corresponds to the time point at which the highest possible volume of wood grown per year is reached and is called the point of maximum sustained yield (MSY) or the culmination of MAI.

Note that this rotation criterion relies exclusively on biological information and, hence, it leads to what many assume to be the biologically optimal rotation age. Economic information regarding prices, costs,

FIGURE 13.14   Town in the middle of the Black Forest, on Lake Titisee, Baden-Württemberg, Germany. *Ignaz Wiradi, through Wikimedia Commons. Image Link: https://commons.wikimedia.org/wiki/File:4_of_10_-_Lake_Titisee,_ Black_Forest_-_GERMANY.jpg License Link: https://creativecommons.org/licenses/by-sa/3.0/deed.en*

and interest (discount) rates is entirely absent from the analysis. One must assume constant prices and costs and very low interest rates to justify using the MSY for determining an even-aged forest rotation length. Despite these concerns, the MSY is still widely used, primarily in government-owned forests throughout the world, including U.S. National Forests. The Multiple-Use and Sustained Yield Act (MUSY) mandates the U.S. Forest Service to maintain in perpetuity a high-yield of several renewable forest products, including timber, which in practice amounts to the MSY (Cubbage et al., 1993).

The MSY may have been more applicable to forestry and natural resource management in the past; its usefulness today is limited because interest rates are not equal to zero (since we have a time preference, we demand positive interest rates) and prices and costs change with the changing market conditions. Further, in today's global economy, with highly developed trade and transportation links, a substantial share of the wood produced is traded internationally. For example, wood products consumed in Europe may have been produced from forests grown in Brazil, the southern United States, or the far reaches of Russia (Fig. 13.16). As a result, if wood can be grown more efficiently in Brazil than in Europe, it may be rational for European markets to import wood products from Brazil, while devoting their forests to other, higher-valued uses.

To develop an economically optimal rotation age for even-aged forests, we need to incorporate information on prices, costs, and interest rates into the analysis. German forester Martin Faustmann (1995) has been widely credited with formulating and solving this forest optimization problem mathematically in 1849. His solution was reaffirmed by the renowned American economist Paul Samuelson (1976).

Faustmann's starting point for the economic analysis was bare forestland, which was regenerated and managed in even-aged stands in perpetuity, with final harvests immediately followed by forest regeneration activities. This approach assumes that land value is equal to the stream of discounted net benefits (or the NPV) of an infinite series of rotations. The solution to the Faustmann problem is the rotation length (harvest age) that maximizes the bare land value (BLV).

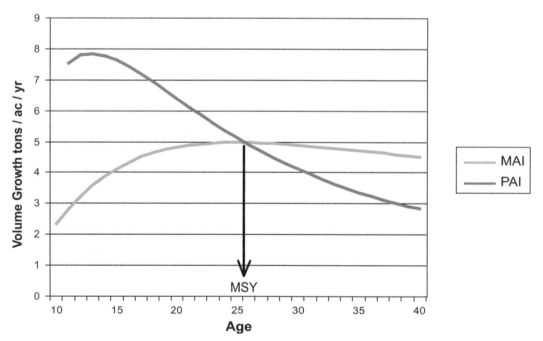

FIGURE 13.15   Timber average volume growth (MAI), marginal growth (periodic annual increment; PAI), and maximum sustained yield (MSY).

FIGURE 13.16 Birch (*Betula* spp.) forest near Novosibirsk, Russia.
Brian Jeffrey Beggerly, through Wikimedia Commons.
Image Link: https://commons.wikimedia.org/wiki/File:Birches_near_Novosibirsk_in_Autumn.jpg
License Link: https://creativecommons.org/licenses/by/2.0/deed.en

To better understand Faustmann's approach, let us first consider the NPV of a single rotation and then expand it to an infinite series of rotations. We start with bare forestland, which is regenerated at the beginning of the rotation and harvested at the end of the rotation. Note that in this single rotation problem there is no opportunity cost associated with delaying future rotations. We can calculate the NPV of a single forest rotation with no intermediate treatments (e.g., thinnings, fertilizations, etc.) as:

$$\text{NPV} = \left(\frac{p_t Q_t}{(1+r)^t}\right) - RC$$

where p_t = the price per unit volume, per unit area, at time t, Q_t = the quantity of merchantable wood volume per unit area at time t, and RC = the regeneration cost per unit area.

To account for economic variables, we consider regeneration costs at the beginning of the rotation ($100 in our

example) and the harvest value ($p_t Q_t$) at the end of the rotation (year t). Since the regeneration activities occur at the beginning of the rotation, this cost is not discounted. Therefore, we only need to discount the value of the planned final harvest. The solution to this problem is the rotation length [age or time (t)] that maximizes the NPV. In our example (Table 13.1; Fig. 13.17), the NPV for the pine stand is maximized at 24 years and equals $882 per ha ($357 per ac).

Stated in marginal terms below, the benefit of waiting to harvest the forest (the marginal benefit) can be expressed as the percentage change of stand value and the cost of waiting (marginal cost) in terms of opportunity cost of capital or the interest rate.

$$\left(\frac{p\Delta Q}{pQ}\right) = r$$

or

$$p\Delta Q = r(pQ)$$

As long as value growth rate of the forest exceeds the interest rate assumed (the discount rate), the stand should be allowed to continue growing because the NPV will increase. The stand should be harvested when the value growth rate equals the interest rate. In our pine stand example, the point at which the value growth rate of the stand falls below 4.5% is age 24. This solution to the single rotation problem is also called the Fisher solution, after the famous American economist Irving Fisher (Samuelson, 1976). It is also an optimal decision rule for other nonrenewable resources, for example, wine aging. The fact that trees are renewable resources complicates the rotation age decision because this analysis ignores the opportunity costs associated with future harvests. However, this problem is avoided by extending this solution to a perpetual (or infinite) series of rotations. Since Faustmann's starting

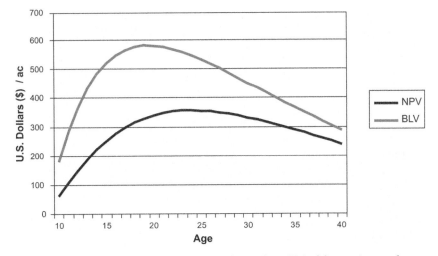

FIGURE 13.17 Net present value (NPV) and bare land value (BLV) of a southern United States pine stand.

point was bare land and his task was to assess the value of that land if used for forestry purposes in perpetuity, the NPV of the infinite series of rotations is called the BLV. At times, it is also called the soil expectation value (SEV) or the land expectation value (LEV). The Faustmann approach has also been used to evaluate optimal management regimes.

$$BLV = \left(\frac{\left(\left(\frac{p_t Q_t}{(1+r)^t} \right) - RC \right)(1+r)^t}{(1+r)^t - 1} \right)$$

The optimal solution will occur at time t when BLV is the highest, i.e., when the marginal benefit equals the marginal cost.

$$p\Delta Q = r(pQ) + r(BLV)$$

Note that the first two terms of marginal conditions are the same as those for the single rotation (Fisher solution). The Faustmann solution, in addition to the opportunity cost of trees, also includes the opportunity cost of land, which is a major innovation. As a result, it includes all relevant costs and benefits and is relatively easy to adapt to more complex scenarios. This approach provides an indication of the value of the land in forest production. In our example, BLV is maximized at age 19, reaching $1436 per ha ($581 per ac) (Table 13.1; Fig. 13.17). This BLV indicates the value of the land if it were devoted to forestry in perpetuity. In other words, if a person were to buy this bare land and manage it as described in our example, then the maximum price they should pay for the land would be $1436 per ha ($581 per ac) if their assumed rate of return for investments was 4.5%.

The Faustmann solution will result in shorter rotations (19 years in our pine stand example) than the Fisher solution (24 years) and the MSY solution (26 years). Note that by including the land value in the marginal conditions above, there is a higher opportunity cost placed on the decision to delay harvest (i.e., a person will demand a higher growth rate in order to allow a stand to continue to grow). Because the cost of waiting is higher, the rotation is shorter. An increase in the interest rate will also tend to shorten the rotation length because the forest owner becomes more impatient and the opportunity cost of waiting is higher. An increase in regeneration or harvesting costs will have an impact analogous to lower timber prices; as a result, the rotation becomes longer as harvesting trees becomes less profitable.

While we have presented here a very simple model, the Faustmann method can be extended to include multiple forest products, nontimber products, and a broad range of environmental benefits. For example, Hartman (1976) extended the Faustmann model to include environmental benefits, finding that their inclusion increased the optimal rotation and had the potential to delay future harvests indefinitely (i.e., potentially never harvesting the trees).

Two examples of incorporating economics into managing forests come from Weyerhaeuser, Inc. and Rayonier, Inc. Both organizations are considered Real Estate Investment Trusts (REITS) in the United States, and both utilize economic methods for determining forest rotations and optimal plans for their forests. Weyerhaeuser owns and manages 12 million ac (4.86 million ha) of commercial forest land around the world (TimberMart-South, 2020). They organize their forests as profit centers to generate revenue over both the short and long run. Although many companies do not reveal much about their resources and internal planning processes, Bigelow et al. (2015) describes their holdings in eastern North Carolina, United States, and their planning environment for managing timberlands. Their primary objective is to maximize financial returns, but they also want to maintain environmental stability. They utilize cutting edge technologies to measure their resources as well as planning tools to evaluate timber harvest flows over time. Management activities are constrained in various ways to also achieve nonmarket products and outcomes. In North Carolina, Weyerhaeuser works cooperatively with the North Carolina Wildlife Resource Commission to sustainably manage habitat for the American black bear (*Ursus americanus*), and the company also adheres to the standards of the Sustainable Forestry Initiative (SFI) certification program (Bigelow et al., 2015). Rayonier is also considered a REIT, and among other land holdings, manages 2.19 million ac (886,261 ha) of timberland across the southern United States (TimberMart-South, 2020). Their primary objective is to maximize NPV across their timberlands. They attempt to achieve this at the stand level, using optimal rotation ages for different products, and at the forest level. According to McTague and Oppenheimer (2015), Rayonier is focused on sustaining a steady flow of harvest revenue income on both a quarterly and an annual basis. In achieving nonmarket objectives, they are also guided by the standards set forth by the SFI certification program.

13.7 Nonmarket forest products

As we have noted in several chapters of this book, forests provide numerous goods and services for human and animal populations. Some of these goods and services are marketed (e.g., timber) but many are not. Consider, for example, aesthetics, biodiversity, recreation, water resources, and wildlife habitat. These goods and services are often unpriced. In other words, they are often not valued in our traditional markets. However, specific examples of NTFPs marketed regionally or globally include wild berry jams from boreal forests of Canada (Boxall et al., 2003), Brazil nuts from western Amazonian forests of Bolivia (Guariguata et al., 2008), and rattan cane from lowland evergreen forests of Indonesia (Widayati and Carslisle, 2012). Some of the obstacles to the valuation of nonmarketed forest outputs are externalities in production and consumption, irreversibility, open-access characteristics (nonexcludability), public goods characteristics (nonrivalry in consumption), and uncertainty. These characteristics can prevent markets from working well in the forestry context and can cause market failures.

A forest or natural resource manager should be concerned with the lack of market values for a range of forest goods and services because they need to make decisions, even to do nothing at all, and these decisions have economic, ecological, and social impacts, many of which cannot be valued in the traditional financial sense. In forestry, the production and consumption of some goods and services will influence the production and consumption of others. To make informed decisions, alternatives need to be evaluated and compared before optimal decisions or strategies can be identified. Consider, for example, a timber harvest, which may increase water yields (through reduction in plant use of water and increased runoff) but may reduce recreational values and transform wildlife habitat. Therefore, to evaluate various management alternatives, we need to have a set of tools to more fully compare their outcomes. In other words, a proverbial yardstick (metric) is needed to more completely measure and compare the results of various management actions. Stated differently, a basic standard is needed that can be used to measure and compare a broad range of forest outputs. As it turns out, money is such a convenient standard. However, if forest outputs are not marketed, then it is difficult to value a resource financially when no money is transferred. Therefore, we need methods to attribute monetary values to nonmarket forest outputs.

Monetary values are usually necessary in evaluating forest outputs. We can devise other methods that are not based on financial values. However, in all likelihood they could be even more complex than assigning monetary values to forest outputs. Our goal here is not to simply suggest placing a price tag on natural resources. For marginal analysis and optimal decision-making, forest and natural resource managers need to understand and evaluate the trade-offs between various forest goods and services and many other resources of value to people (Turner et al., 2003). Money is simply a convenient metric for this task. Rather than trying to evaluate trade-offs in physical units (or any other units for that matter), such

as tons of wood or hectares harvested, cubic feet of water, number of recreational visits, and habitat suitability indices, we may use a common and easy to use, for comparative purposes, metric such as money.

More importantly, economic analyses are now an inherent part of most decision-making processes, even in the management of publicly owned natural resources (Loomis, 1993). While final management decisions may not be based solely on economic criteria, these criteria will be examined and considered. Therefore, determination of the weight (value) assigned to unpriced forest outputs in a typical economic analysis can become important. In many cases, unpriced forest outputs will not be accounted for at all in an economic analysis. In these cases, operational decisions related to the management of the land may neglect the nonvalued outputs, or else addressing these values will be left to the discretion of the land manager. In other words, a market may not provide incentives to producers to supply nonvalued resources in sufficient quantities. It is for precisely these reasons that economists have been developing methods for valuing nonmarketed goods and services. With monetary values ready at hand, one can calculate the NPV of various management alternatives and allocate scarce resources to the highest-valued uses.

Payments for environmental (or *ecosystem*) *services* (PES) are receiving increasing interest around the world as financial incentives for nontimber values of forests. These programs rely on processes in which the beneficiaries of services pay for services provided by forests and other natural resources. Beneficiaries can be located far away from the location of the forests providing the services, can be direct users of these services, or can be governments that act on behalf of service users. These types of programs are attractive in settings where the landowners are poor or are marginalized, such as in developing countries (Engel et al., 2008). Global, national, regional, and local policy development processes can affect whether PES become self-sustaining markets. For example, conflicting direction may arise from two institutions operating at two distinct levels of social organization (e.g., a national organization setting national policy and a local community setting local policy); this is referred to as *vertical interplay*. In addition, conflicting direction may arise from two institutions operating at the same level of social organization (e.g., a national agricultural organization and a national forestry organization, each setting national policy); this is referred to as *horizontal interplay* (Corbera et al., 2009). Markets related to PES are currently evolving, and there are a number of challenges related to the efficiency at which services are provided and the equity in payments for services (Pascual et al., 2010).

Another service, sequestered forest carbon, is being traded within established forest carbon credit markets such as the California Air Resources Board. The additional carbon sequestered through afforestation projects and improved forest management projects can be traded in specialized markets that have been growing slowly during the early part of the 21st century. Incidentally, *improved forest management* projects are ones in which forests are managed specifically to increase carbon storage or reduce carbon losses through management activities (Diaz et al., 2011). There was a surge in the forest carbon market in 2010, for the most part owing to contracts arising from large Reduced Emissions from Deforestation and Forest Degradation (REDD) projects. Over half of the supply of carbon credits has arisen lately from Latin America. However, North America is a large source of both the supply and demand in the carbon market. Recent prices for forestry and land-use related carbon credits have been about US$3.20 per metric ton of carbon dioxide equivalent (tCO$_2$e) (Donofrio et al., 2019).

In an effort to better account for various goods and services generated by natural ecosystems, environmental economists have developed the concept of *total economic value* (TEV), which enables accounting for all the main outputs and their values, as well as other values generated by the mere existence of natural ecosystems (Freeman, 2003; Pearce and Turner, 1991). *Use values* constitute the first main component of TEV (Fig. 13.18). These values are generated by the active use of the environment, either direct or indirect. A direct use value involves outputs that are consumed directly (e.g., fodder, food, game, medicines, and timber). An indirect use value relates primarily to ecological services (e.g., carbon storage, flood control, and wildlife habitat). The last component of use value is an option value, which represents benefits associated with preserving forest resources for future direct or indirect uses, some of which may not even be known today.

Nonuse values, which are not related to any use, either current or future, constitute the second main component of TEV. Nonuse values include an existence value that represents the benefits of knowing that certain resources exist. For example, although a person may never travel

to the Grand Canyon or Yosemite Valley in the United States, they may feel good knowing that they exist, and thus place a value on their existence. Nonuse values also include a bequest value, which represents benefits to future generations. For example, a person may value the fact that areas of large California redwoods (*Sequoia sempervirens*) will be preserved for future generations to enjoy. Since these nonuse values tend to be intrinsic in nature, involve value judgments, and do not necessarily pertain to economic efficiency, there is ongoing debate regarding whether they should be subject to economic inquiry. If this point of view prevails, then we should only be concerned with use value.

Markets generally reveal only direct use values. When market values are not available, we are basically left with two choices: (1) values of nonmarketed goods and services can be derived either from observed market behavior (revealed preference methods); or (2) values of nonmarketed goods and services can be derived from responses to hypothetical questions (stated preference methods) (Freeman, 2003). Techniques based on observed market behavior work well when products from the activity have market values. These techniques generally rely on market prices for goods and services that are traded in commercial markets or an assessment of the contribution of ecosystem goods and services to the production of other marketed goods or services. For example, Peters et al. (1989) conducted an evaluation of an Amazonian rainforest and included in the analysis nonwood outputs used by the locals such as fruits, medicines, oils, and rubber. These nonwood goods could be brought to larger towns and markets to be traded, in which cases their prices were available and production costs could be estimated. They calculated the BLV of three management scenarios: (1) forest clearing followed by pulpwood plantation management; (2) forest clearing followed by pasture; and (3) sustainable selective tree and nonwood goods harvest. The third management alternative yielded the highest BLV, indicating that SFM that incorporates both wood and nonwood outputs is also an efficient economic outcome, and the conclusion drawn was used as an argument against forest clearing and land-use conversion. In cases when markets for ecosystem goods and services are not well defined, sometimes the value of these goods can be inferred from markets for related goods, called surrogate markets. In the valuation of resource values, labor and property markets are often used as surrogate markets.

The *Travel Cost Method* (TCM) is used to value nonmarketed resources using indirectly observed surrogates of market behaviors. The TCM measures the value of a recreational resource, such as a state or national park. This method assumes that travel expenses and time incurred by visitors represent the price of

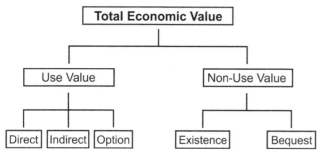

FIGURE 13.18 Components of the total economic value of a resource.

access to the site (Perman et al., 2003). One variation of this method, called the *zonal TCM*, groups visitors into zones with similar travel costs. Consider, for example, the Great Smoky Mountains National Park in the southern United States. Local visitors in neighboring counties would be grouped into Zone 1, visitors from nearby states would form Zone 2, and visitors from elsewhere in the United States and beyond would form Zone 3. To obtain this information, a simple license plate survey could be employed or the visitors could be asked for their postal codes when buying a ticket. As part of this effort, we would also obtain information about the number of visitors. Then, for each zone an estimate of the average travel distance and associated expenses can be determined. From this and other information, a demand function for the average visitor can be estimated, along with the total economic benefits of the particular recreational site. The TCM was used, for example, by Tobias and Mendelsohn (1991) to assess domestic ecotourism (recreation) values of the Monteverde Cloud Forest Biological Reserve in Costa Rica (Fig. 13.19). The reserve, located in rugged mountains on the Continental Divide, is well known for its foggy weather as well as for its diverse ecosystem. The estimated recreational value for the reserve ranged from US$2.4 million to US$2.9 million, depending on the assumptions used.

Another example of the surrogate market approach and revealed preference method is the *hedonic price method*. This approach is used to develop values for resource outputs that directly influence prices of marketed goods. Consider the task of determining housing prices. Imagine two homes with exactly the same amenities, except that they are located in two different areas of a city. More than likely, their prices would differ; these differences can arise from many sources, one of which is

the availability of local environmental goods and services. These may include the availability of, or proximity to, beautiful vistas, clean air, tree cover, or waterfront. Once information about amenities, site and neighborhood characteristics, and any other factors that may influence a home's price are available, we could then infer the impact of environmental attributes on house values. Applying the hedonic pricing method, Tyrväinen and Miettinen (2000) found that in Finland the presence of forest vistas increased housing prices by nearly 5%.

In some cases, however, no market evidence is available and stated preference methods must be relied on. The most widely used method in such circumstances is the *Contingent Valuation Method* (CVM). While it can be used to estimate the value of any resource, whether marketed or nonmarketed, CVM is mainly used to estimate nonuse values. The reason that CVM is termed a stated preference method is because it asks people directly to state how much they would be willing to pay for specific environmental output or how much compensation they would accept to forgo a specific environmental output. CVM relies on surveys to elicit individual valuation information. CVM is termed a contingent valuation because surveys ask respondents to value a particular resource based (or contingent) on a specific hypothetical scenario. Rubin et al. (1991) conducted a contingent valuation of the northern spotted owl and old-growth ecosystems in the Pacific Northwest region of the United States. Uses and values associated with the preservation of these resources included recreation as well as option, existence, and bequest values. The results indicated that survey respondents were willing to pay nearly US$50 per household per year. These results, adjusted for socioeconomic variables such as age, education and income, indicated a total annual willingness to pay nearly US$63 million in the state of Washington and US$1.5 billion nationwide.

While all valuation methods discussed here have their strengths and weaknesses, applications of the CVM have proven to be the most contentious because in this method people simply state their preferences as opposed to actually paying for the resource output in question. For example, in response to the Exxon Valdez oil spill in 1989, the CVM was extensively used in valuing environmental damage. The public's willingness to pay to avoid a future oil spill of this magnitude was estimated to be between US$2.8 billion and US$7 billion in 1990 (Carson et al., 2003), and these estimates (along with the methodology used for obtaining them) were severely contested. To assess the validity of the CVM, the U.S. National Oceanic and Atmospheric Administration appointed a panel of several eminent scientists to study the issue. The panel affirmed the validity of the CVM and also provided several recommendations regarding its usage (Arrow et al., 1993). The

FIGURE 13.19 Suspension bridge in the Monteverde Biological Reserve, Costa Rica. *Haakon S. Krohn through Wikimedia Commons.*
Image Link: https://commons.wikimedia.org/wiki/File:Monteverde_puente.jpg
License Link: https://creativecommons.org/licenses/by-sa/3.0/deed.en

proposed guidelines recommend, among others, conducting face-to-face surveys rather than over the telephone or by mail, developing accurate and explicit scenarios, reminding respondents that their willingness to pay for a resource output would leave less money for other purposes, and reminding respondents that substitutes may exist for the resource in question (Arrow et al., 1993). In a way, these guidelines advocated much more rigor in conducting contingent valuation studies. Since the Exxon Valdez oil spill and subsequent analyses, the CVM has been extensively used for valuing environmental issues (Schläpfer, 2008).

Depending on the circumstances, there are several methods from which to choose when valuing nonmarketed resource outputs. Although we have addressed only a few major methods in this chapter, a complete assessment of valuation methods can be found in Tietenberg (2006), Freeman (2003), Perman et al. (2003), and numerous other texts. All of these methods have their strengths and weaknesses but, when applied correctly, they will provide information that is useful for decision-making. Finally, as a word of caution, it should be noted that in addressing nonmarket benefits there may be a tendency to rely first on market values, especially when market-based surrogates exist, but to do so often obviates the necessity to estimate the value of nonmarket values.

Summary

Society today faces many substantial challenges regarding the use of forest and natural resources. Some nations must decide the manner in which to allocate scarce forest and natural resources for a wide range of uses. These choices are often difficult to make, but cannot be avoided, and often require real sacrifices. Difficulties exist because forest and natural resources are very diverse and fulfill many functions, resulting in a huge number of possible management alternatives. Choices involving the management of forests and natural resources cannot be avoided because resources are limited, and thus the full scope of humankind's wants and desires may not be addressed. Ultimately, sacrifices may be required because, in many cases, in order to secure necessary goods and services from a forest, communities are likely to lose the opportunity to enjoy other valuable goods or services.

Economics can assist forest and natural resource managers as they sort through and evaluate the numerous management alternatives that are available. Economic analyses can provide guidance for allocating available resources to the uses that will best satisfy society's needs and desires by allowing managers to compare the costs and benefits of the various alternatives. When benefits exceed costs, this is usually a good indication that a proposed course of action will increase the well-being of an organization, community, or country. Markets and governments both play important roles in ensuring that forest resources are allocated to the best uses. Markets work well in achieving resource efficiency by encouraging the highest-valued uses of forests, which helps eliminate waste. Governments often provide enabling policies and regulatory frameworks, and often address market failures when they occur. Through policies and law, governments can also ensure that important social objectives, such as equity and community stability, are addressed. However, the extent of government involvement in forest policies is likely to be a function of the economic, environmental, social, and political state (and history) of a given country.

Economics is a social science, and its vast collection of valuation and decision-making tools provides forest and natural resource managers with information that is useful for making management decisions. While economic analysis may not be able to address all of the natural resource management problems we face today, it certainly can assist in identifying efficient courses of action within many of the management situations that we face. While it is rare that forest and natural resource management decisions are based solely on economic criteria, economics has without doubt proven very useful for forest and natural resource decision-making processes and is likely to grow in importance in the future.

Questions

(1) *Scarcity.* Within a 15-minute period, develop an inventory of the natural resources that you rely on every day. Which of these resources would you consider to be scarce? Why? Summarize your findings in a brief memorandum to your course instructor.

(2) *Future value.* If you had the ability to place US$2000 in an investment today that generates a 4% annual rate of return using a compound interest process, how much money would you have 10 years from now if you kept

Questions (cont'd)

the investment for the entire decade, and the annual rate of return did not change?

(3) *Present value.* If you are told by a consulting forester that your stand of yellow-poplar (*Liriodendron tulipifera*) trees will be worth US$50,000 when it matures in seven years, what is the present value of this forest? In conducting this analysis, choose a discount rate that you feel is appropriate to the situation. Develop a short memorandum to illustrate the analysis and explain how you chose the discount rate that was used.

(4) *Internal rate of return.* You have an opportunity to place US$5000 in a forestry investment project that you believe will yield US$60,000 in 20 years. How fast will your money invested in this project grow? In other words, what is the rate of return you expect to achieve from this investment?

(5) *Total economic value.* Consider a forested area that you have visited recently or are familiar with. Make a list of all of the values that this forested area generates and classify them into the components of the TEV. Describe in a short PowerPoint presentation how you would measure these values.

(6) *Value judgments.* The famous economist Adam Smith once wrote that items or resources that have the highest utility to humans may frequently have little value in a market, yet items or resources that have the highest value in a market may have little utilitarian value to humans. This concept is mildly controversial because both types of resources may not be compared using common units of measure and perhaps because moral judgments apply. However, develop a short paragraph that describes one item or resource of each type and, within a small group of others, describe your findings.

References

Amacher, G.S., Ollikainen, M., Koskela, E., 2009. Economics of Forest Resources. The MIT Press, Cambridge, MA, 397 p.

Arrow, K., Solow, R., Portney, P.R., Leamer, E.E., Radner, R., Schuman, H., 1993. Report of the NOAA panel on contingent valuation. Federal Register 58 (10), 4601–4614.

Bettinger, P., Boston, K., Siry, J.P., Grebner, D.L., 2017. Forest Management and Planning, second ed. Academic Press, New York, 362 p.

Bigelow, B., Ewing, R.A., Kumar, V., 2015. Weyerhaeuser, North Carolina, United States of America. In: Siry, J.P., Bettinger, P., Merry, K., Grebner, D.L., Boston, K., Cieszewski, C. (Eds.), Forest Plans of North America. Academic Press, London, pp. 351–357.

Boxall, P.C., Murray, G., Unterschultz, J.R., 2003. Non-timber forest products from the Canadian boreal forest: an explanation of arboriginal opportunities. Journal of Forest Economics 9, 75–96.

Bullard, S.H., Gunger, J.E., Doolittle, M.L., Arano, K.G., 2002. Discount rates for nonindustrial private forest landowners in Mississippi: how high a hurdle rate? Southern Journal of Applied Forestry 26 (1), 26–31.

Carson, R.T., Mitchell, R.C., Hanemann, M., Kopp, R.J., Presser, S., Ruud, P.A., 2003. Contingent valuation and lost passive use: Damages from the Exxon Valdez oil spill. Environmental and Resource Economics 25, 257–286.

Corbera, E., González Soberanis, C., Brown, K., 2009. Institutional dimensions of payments for ecosystem services: an analysis of Mexico's carbon forestry programme. Ecological Economics 68, 743–761.

Crowley, T., Dhubhain, A.N., Moloney, R., 2001. The economic impact of forestry in the Ballyvourney area of County Cork, Ireland. Forest Policy and Economics 3, 31–43.

Cubbage, F.W., O'Laughlin, J., Bullock III, C.S., 1993. Forest Resource Policy. John Wiley & Sons, Inc., New York, 562 p.

Diaz, D., Hamilton, K., Johnson, E., 2011. State of the Forest Carbon Markets 2011, from Canopy to Currency. Ecosystem Marketplace, Washington, D.C.

Dong, Z., Tan, Y., Duan, J., Fu, X., Zhou, Q., Huang, X., Zhu, X., Zhao, J., 2011. Computing payment for ecosystem services in watersheds: an analysis of the middle route project of south-to-north water diversion in China. Journal of Environmental Sciences 23, 2005–2012.

Donofrio, S., Maguire, P., Merry, W., Zwick, S., 2019. Financing Emissions Reductions for the Future, State of the Voluntary Carbon Markets 2019. Forest Trends, Washington, D.C.

Duerr, W.A., 1993. Introduction to Forest Resource Economics. McGraw-Hill, Inc., New York, 485 p.

Engel, S., Pagiola, S., Wunder, S., 2008. Designing payments for environmental services in theory and practice: an overview of the issues. Ecological Economics 65, 663–674.

Faustmann, M., 1995. Calculation of the value which forest land and immature stands possess for forestry. Journal of Forest Economics 1 (1), 7–44.

Freeman III, A.M., 2003. The Measurement of Environmental and Resource Values. Resources for the Future Press, Washington, D.C., 491 p.

Guariguata, M.R., Cronkleton, P., Shanley, P., Taylor, P.L., 2008. The compatibility of timber and non-timber forest product extraction and management. Forest Ecology and Management 256, 1477–1481.

Hartman, R., 1976. The harvesting decision when the standing forest has value. Economic Inquiry 14 (1), 52–58.

Hepburn, C.J., Koundouri, P., 2007. Recent advances in discounting: implications for forest economics. Journal of Forest Economics 13, 169–189.

Klemperer, W.D., 1996. Forest Resource Economics and Finance. McGraw-Hill, Inc., New York, 551 p.

Loomis, J.B., 1993. Integrated Public Lands Management. Columbia University Press, New York, 474 p.

Lutz, J., 2011. IRR vs. NPV. Forest Research Notes 8 (1), 1–4.

McTague, J.P., Oppenheimer, M.J., 2015. Rayonier, Inc., southern United States of America. In: Siry, J.P., Bettinger, P., Merry, K.,

Grebner, D.L., Boston, K., Cieszewski, C. (Eds.), Forest Plans of North America. Academic Press, London, pp. 395–402.

Nijnik, M., van Kooten, G.C., 2000. Forestry in the Ukraine: the road ahead? Forest Policy and Economics 1, 139–151.

Office of Management and Budget, 2015. Guidelines and Discount Rates for Benefit-Cost Analysis of Federal Programs. Office of Management and Budget, Washington, D.C. OMB Circular No. A-94. 23 p.

Pascual, U., Muradian, R., Rodríguez, L.C., Duraiappah, A., 2010. Exploring the links between equity and efficiency in payments for environmental services: a conceptual approach. Ecological Economics 69, 1237–1244.

Pearce, D.W., Turner, R.K., 1991. Economics of Natural Resources and the Environment. John Hopkins University Press, Baltimore, MD, 378 p.

Perman, R., Ma, Y., McGilvray, J., Common, M., 2003. Natural Resource and Environmental Economics. Prentice Hall, Upper Saddle River, NJ, 699 p.

Peters, C.M., Gentry, A.H., Mendelsohn, R.O., 1989. Valuation of an Amazonian rainforest. Nature 339, 655–666.

Rideout, D.B., Hesseln, H., 2001. Principles of Forest and Environmental Economics. Resource & Environmental Management LLC, Fort Collins, CO, 297 p.

Rubin, J., Hefland, G., Loomis, J., 1991. A benefit-cost analysis of the northern spotted owl. Journal of Forestry 89 (12), 25–30.

Samuelson, P.A., 1976. Economics of forestry in an evolving society. Economic Inquiry 14 (4), 466–492.

Samuelson, P.A., Nordhaus, W.D., 1995. Microeconomics, fifteenth ed. McGraw-Hill, Inc., New York, 500 p.

Schläpfer, F., 2008. Contingent valuation: a new perspective. Ecological Economics 64, 729–740.

Tietenberg, T., 2006. Environmental and Natural Resource Economics. Pearson, Addison Wesley, New York, 655 p.

TimberMart-South, 2020. Market News Quarterly, 1st Quarter 2020, vol. 25. Norris Foundation, Athens, GA, p. 64. No. 1.

Tobias, D., Mendelsohn, R., 1991. Valuing ecotourism in a tropical rainforest reserve. Ambio 20 (2), 91–93.

Turner, R.K., Paavola, J., Cooper, P., Farber, S., Jessamy, V., Georgiou, S., 2003. Valuing nature: lessons learned and future research directions. Ecological Economics 46, 493–510.

Tyrväinen, L., Miettinen, A., 2000. Property prices and urban forest amenities. Journal of Environmental Economics and Management 39 (1), 205–223.

U.S. Department of State, 2012. Foreign Affairs Handbook, Volume 5 Handbook 5-Information Technology Systems Handbook. 5 FAH-5 H-620 Benefit-Cost Analysis (BCA) Process. U.S. Department of State, Washington, D.C. https://fam.state.gov/FAM/05FAH05/05FAH050620.html (accessed 27.02.20).

Wei, Y., White, R., Hu, K., Willett, I., 2010. Valuing the environmental externalities of oasis farming in Left Banner, Alxa, China. Ecological Economics 69, 2151–2157.

Widayati, A., Carslisle, B., 2012. Impacts of rattan cane harvesting on vegetation structure and tree diversity of Conservation Forest in Buton, Indonesia. Forest Ecology and Management 266, 206–215.

C H A P T E R

14

Forest disturbances and health

In addition to changes in forest character that are due to active management of the landscape by humans, forests undergo changes as a result of natural disturbances, such as wind, ice, insects, fungal pathogens, or lightning. In either event, the health of a forest is either improved or degraded, as its character has changed. Many forest disturbances occur with a frequency and intensity that are very difficult to estimate. For instance, the pattern of harvesting activities that occurred in the late 20th century in the northern United States is randomly located across the landscape (Kittredge et al., 2003). Imagine trying to determine when and where the next land-falling major hurricane will arrive along the southern coasts of the United States. Some natural disturbances are relatively foreseeable. For example, during extended periods of drought, one can reasonably assume that forest fire risk levels will be elevated due to increase in dead and dry fuels and an increase in the combustibility of the fuels. Using information regarding climate (i.e., wind), topography, and the forest resources present in an area, one may be able to assess the potential risk of forest and natural resources to fire. Although there are times when we may assume that the risk of losing resources may be low, having an understanding of the impacts and consequences of disturbances on forest resources is important, and, therefore, after reading this chapter, you should

• be aware of the various types of natural and anthropogenic disturbances that may affect forests and natural resources;
• understand the expected potential damages caused by different types of disturbances; and
• understand the range of management strategies one might employ in reacting to forest disturbances.

The variety of natural and anthropogenic disturbances that could act upon or occur within a given area can be substantial. In addition to providing some contemporary examples of forest disturbances, this chapter briefly describes the broad classes and sources of these disturbances, where appropriate.

14.1 What are forest disturbances?

A *forest disturbance* is an environmental alteration of the character of a wooded area through natural or anthropogenic means (Nieuwenhuis, 2010) and consists of an event that alters the existing vegetative structure or disrupts ecological processes. For example, certain areas of the world's forests can be exposed to diseases, insect infestations (Fig. 14.1), severe windstorms (tropical cyclones or hurricanes), and wildfires (natural and human-caused). Some coniferous forests of western North America, for example, are vulnerable to both wildfires and severe

FIGURE 14.1   Dead and dying pine trees on a hillside in the southern United States caused by the southern pine beetle. *Robert L. Anderson, U.S. Forest Service, through Bugwood.org.*

FIGURE 14.2   Western spruce budworm damage in an interior Douglas-fir (*Pseudotsuga menziesii*) forest in eastern Oregon, United States. *Dave Powell, U.S. Forest Service, through Bugwood.org.*

windstorms, and could be susceptible to western spruce budworm (*Choristoneura occidentalis*) infestations (Fig. 14.2) and perhaps to volcanic eruptions. Human-caused disturbances arguably can create some of the most important local impacts on forests. Collectively, human-caused disturbances across broad regions can change the character of the landscape. Many common forestry practices can be categorized as forest disturbances. Other human actions, such as the intentional lighting of forest fires by arsonists, can also lead to forest disturbances. As with human-caused disturbances, natural disturbances are major factors in shaping the development, structure, and function of forests. Some even argue human activities can be used to emulate the effects and benefits of natural disturbances (Attiwill, 1994). Many have used the analogy of logging as a replacement for fire.

The frequency, type, and severity of each disturbance can vary greatly, depending on the structure and location of the existing forested ecosystem as well as the

destructive force of the disturbance event. For example, the frequency of canopy disturbances in oak (*Quercus* spp.) forests in Alabama can be as wide as 20 to 50 years at the local level (Goode et al., 2020). With respect to extent and severity, forest damage caused by a tornado, as compared to forest damage caused by a hurricane, is likely to be much more narrow in scope and quite severe within the path of the storm. When considering the impact of wildfire on a forest, a low-level ground fire may kill most of the small plants in the forested understory, but this is much less harmful than a crown fire that may also kill the overstory trees. At one extreme, a volcanic eruption can cause widespread forest disturbance by killing not only plants and wildlife, but also the unfortunate people caught in the path of the plume cloud or lahar (mud and debris flow). Volcanic fires, ash, toxic fumes, and lava are indiscriminate, and burn, suffocate, and bury everything in their path. At the other extreme, some natural diseases spread very slowly, perhaps from tree to tree when environmental conditions favor expansion. Further, the introduction of invasive diseases can affect particular tree and plant species and cause changes not only to the forest structure but also to other ecological services forests provide, such as wildlife habitat. Not to be outdone, population explosions of insects, whether native or exotic, can slowly defoliate large forested areas, causing significant changes in plant and animal dynamics. This chapter focuses on disturbances imposed on forests by wind, fire, insects, diseases, ice, snow, floods, volcanoes, and humans. However, other disturbances to forests can be attributed to wildlife. For example, bark can be stripped from trees by deer (Yokoyama et al., 2001) and bear (Manning and Baltzer, 2011), and leaves can be consumed by deer (Kraft et al., 2004) and other animals. While not discussed in this chapter, these types of events can also impact the composition, structure, and reproductive potential of plants in the understory of forests.

14.2 Disturbances created by wind

Wind is the movement of air and gases in our atmosphere and occurs naturally at large and small scales. Wind is caused by the rotation of the planet as well as differential heating and cooling of the land. Wind can have an important cooling effect at night in communities situated near coastlines, as well as in communities located in rural or urban areas. For thousands of years, wind has served as a primary natural force for the movement of ships across freshwater and saltwater bodies. The evolution of sail technology on ships involved the construction of ever more complicated sail mechanisms designed to increase ship speed. During the pioneer days in the western United States, European descendants used wind power to pump water from underground aquifers. This practice is still used today in many locations around the world. Movement of wind within a forest governs the local microclimate and facilitates the movement of heat, gases, and water vapor to the atmosphere immediately above the forest canopy (Leahey and Hansen, 1987). Wind is also important for many natural processes. For instance, many plant species disseminate their seeds via wind processes. Some species have evolved wings on their seeds to enhance their dispersal. Some examples of tree seeds dispersed by wind include American elm (*Ulmus americana*), eucalyptus (*Eucalyptus* spp.), green ash (*Fraxinus pennsylvanica*), maple (*Acer* spp.), and red alder (*Alnus rubra*) (Burns and Honkala, 1990). Wind is also helpful for disseminating pollen for the sexual reproduction of numerous coniferous species (Fig. 14.3). It can also be beneficial for the generation of energy and for the disbursement of unpleasant or obnoxious odors.

Unfortunately, when severe, winds can be detrimental to both human communities and forested ecosystems. Often, when severe winds affect an area, the print, electronic, and televised media typically report the number of human casualties incurred and provide visual images of homes or buildings that have been damaged or destroyed (Fig. 14.4). However, severe windstorms can also result in changes to forest stand structure and composition (Brokaw and Walker, 1991), and damage to forests can be extensive. Much of the damage from wind-related events transitions forested areas naturally back to secondary succession processes. However, a distinct change in the density and character of a forest is likely to occur if the damage is not complete. In managed forests, the decision managers typically face is whether to allow the forest to recover without assistance or to apply management actions such as reforestation or salvage operations. The primary form of destructive wind events arises from tropical cyclones (or hurricanes), tornadoes, microbursts (wind gusts), and severe straight-line winds. Damage can escalate when combined with other atmospheric conditions such as ice that increases the weight on tree limbs. Severe winds can cause significant damage to forests, such as the pine forests surrounding the Gulf of Mexico or the boreal forests of Canada (Achim et al., 2005). These phenomena can cause excessive damage to forest ecosystems and significantly affect the numerous ecosystem services that they provide. Severe winds can also cause soil erosion, which can undermine the supportive aspect of ecosystem services. Evidence of large-scale wind erosion can be found in the Loess Hills of Mississippi and in some of the semiarid regions of Africa and Asia. Wind erosion generally reduces the productivity of a soil, which then diminishes the capacity of the soil to help produce healthy forested ecosystems.

FIGURE 14.3 Pollen from a pine tree in Pennsylvania, United States, being dispersed by wind. *Beatriz Moisset, through Wikimedia Commons. Image Link: https://commons.wikimedia.org/wiki/File:Pollen_from_pine_tree_2.jpg License Link: https://creativecommons.org/licenses/by-sa/3.0/deed.en*

FIGURE 14.4 Wind damage across residential property in coastal counties of Mississippi, United States, as a result of Hurricane Katrina. *Jon D. Prevost*

When subjected to wind, the tips of trees oscillate in an elliptical manner (Mergen, 1954), owing to asymmetrical root systems, crown abnormalities, friction, irregular tree stem shapes, and variations in wood elasticity (Mayer, 1987). In general, when subjected to severe winds, all sides of a tree stem and root system are stressed. The windward side of a tree is generally under tension and the leeward side is generally under compression. As a result of these actions, severe winds can lead to tree breakage (Fig. 14.5), permanent tree bending or lean (Fig. 14.6), or windthrow (Fig. 14.7). Broken trees are often killed unless some portion of the active crown remains on the bole that remains attached to the root system. Bent or leaning trees can

FIGURE 14.5 Tree breakage in a forest in southwestern France as a result of a windstorm in 2009. *Jean-Luc Peyron.*

FIGURE 14.6 Wind-bent trees in a forest in southwestern France as a result of a windstorm in 2009. *Jean-Luc Peyron.*

FIGURE 14.7 Windthrown trees in a spruce forest in Germany. *Walter J. Pilsak, through Wikimedia Commons.*
Image Link: https://commons.wikimedia.org/wiki/File:Windbruch-WJP-2.jpg
License Link: https://creativecommons.org/licenses/by-sa/3.0/deed.en

FIGURE 14.8 A tornado passing through forest and residential areas of Stoughton, Wisconsin, United States, in 2005. *Colin McDermott, U.S. National Oceanic and Atmospheric Administration, through Wikimedia Commons.*
Image Link: https://commons.wikimedia.org/wiki/File:Stoughton_Tornado.jpg
License Link: Public Domain

remain alive, but their form (and potential use) and the vertical structure of the forest are both severely altered. Windthrown trees usually perish, since most of the root system becomes exposed and the tree's connection to the soil resource is, for the most part, broken. The interlocking nature of the windthrown trees is a dangerous site as many are spring-loaded and can move dramatically during the clearing phase of a recovery operation.

A tornado (Fig. 14.8) is a rotating column of wind that is periodically formed in conjunction with thunderstorms. Although conditions can be favorable for the development of tornadoes when a warm weather front meets a cold weather front, scientists still have not found a definitive reason for their development (National Oceanic and Atmospheric Administration National Weather Service, 2011). Tornadoes can also form as a result of changes in temperature as supercells, or large rotating

thunderstorms, travel through an area. Tornadoes can be found on every continent in the world, but are most frequently found in North America, which hosts an area commonly known as *tornado alley*, a vast landscape located between the Rocky Mountains and the Appalachian Mountains. The states of Kansas, Oklahoma, and Texas have the greatest frequency of tornadoes, but powerful tornadoes can also periodically occur throughout the southern states (Francis et al., 2011). With the advent of YouTube and other forms of communication through the

FIGURE 14.9 Satellite view of Hurricane Katrina (2005) approaching the southern United States' coastline through the Gulf of Mexico. *Image courtesy of U.S. National Oceanic and Atmospheric Administration, through the IBTrACS website, World Data Center for Meteorology, Asheville, North Carolina.*

Internet, it is not difficult to locate homemade videos of tornadoes passing through a community and to witness forest damage occurring. When a tornado touches down in a forested landscape, it will often break the trunks of trees at some height, perhaps 3 or 6 meters (m; 10 or 20 feet [ft]), above the ground. This damage results in a loss of viable timber products, as well as a loss of wildlife habitat, and may eventually lead to disease and insect infestation problems.

Tropical cyclones, or hurricanes (Fig. 14.9), generally develop within 5 degrees of the equator from a combination of warm tropical oceans and atmospheric conditions. The specific number of events to expect each year and the approximate location of landfall is difficult to predict. Even the exact location of landfall of an approaching hurricane is difficult to pinpoint one or two days in advance (Broad et al., 2007). Although wind speeds within hurricanes are generally not as strong as those within tornadoes, hurricanes can affect much larger areas and cause more extensive damage. In some areas of the world, hurricanes may have replaced other processes such as wildfire (given fire suppression activities) as the main natural forest disturbance (Conner et al., 1989). Hurricane-force winds can accompany long-duration, straight-line windstorms such as the Great Coastal Gale

of 2007 that affected the Pacific Northwest coast of the United States. These types of storms are generally not cyclonic in nature, but they could contain rotating winds and may also produce heavy amounts of precipitation in the form of both rain and hail. The saturated soils are weaker further contributing to treefall. *Squall lines* or *gust fronts* (outflow boundaries) are often examples of the leading edges of severe straight-line wind events (Fig. 14.10) and may produce downbursts that can damage trees. The more common impact of hurricanes and straight-line winds on forested ecosystems is similar to that caused by tornadoes; i.e., trees are often uprooted (windthrown), broken, or bent. However, forest damage is much more extensive after a hurricane than after a tornado, and we typically see the most extensive damage along the right-hand side of the leading edge of a hurricane as it moves inland. Fortunately, a hurricane will lose energy as it travels inland, and this reduces its potential impact on forested landscapes located further inland (Bettinger et al., 2009).

Two approaches that have been followed to understand how forest and natural resource managers can prepare their forests for the potential damage caused by severe wind events are (1) surveys of managers who have dealt with the aftermath of severe wind

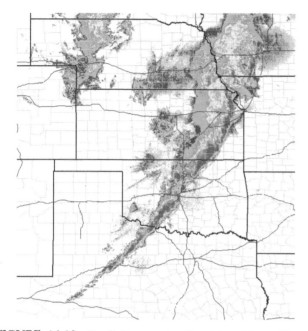

FIGURE 14.10 Squall line crossing the states of Iowa, Kansas, Missouri, Nebraska, Oklahoma, and Texas, United States, on June 5, 2008. *U.S. National Weather Service, through Wikimedia Commons.*
Image Link: https://commons.wikimedia.org/wiki/File:June_5_2008_squall_line.gif
License Link: Public Domain

events; and (2) site-specific research studies that have attempted to correlate damage with prestorm forest condition. Surveys by Merry et al. (2010) and Bettinger et al. (2010) have examined the perceptions of foresters in Mississippi in the wake of their experiences with Hurricane Katrina. Each has suggested that recent management activities and prior weather conditions (recent precipitation events) were the most important factors in determining a forest's resistance to wind damage. These perceptions are generally consistent with site-specific research studies; however, these generally place more emphasis on the amount of forest edge and the condition (i.e., age, diameter, height, and species) of trees. As you may gather, managing forested ecosystems to minimize wind damage is a difficult task. Some factors, such as recent weather activity, are unmanageable, while others, such as managing the order of activities, may be necessary to meet landowner goals. Further, actions such as reducing the amount of forest edge may increase a forest's resistance to extreme wind events; however, the loss or gain of forest edge habitat can affect wildlife species in different manners. Therefore, the potential trade-offs need to be assessed. Many researchers and foresters continue to evaluate the best course of action to select to address and react to extreme wind events. As noted in Chapter 3, choosing one course of action over another may lead to outcomes that are either consistent with, or in opposition to, a forest landowner's goals.

14.3 Disturbances created by fire

Fire is an event that produces heat, light, and smoke through the combustion of flammable materials such as wood, grass, paper, plastics, or metals. The combustion process requires sufficient quantities of flammable materials (fuel) and oxygen (O_2) to create the necessary chemical reaction to initiate a fire after the materials are exposed to an appropriate ambient temperature or external heat source. The fire triangle (Fig. 14.11) has often been used as a conceptual model for understanding the ingredients (oxygen, heat, and fuel) that are necessary to support a fire. Without a sufficient supply of any one of three ingredients, fire cannot begin. Fire has been very useful to humans and the development of our civilizations throughout the ages. Fire is often used to cook food and to boil water in order to kill various microorganisms that can adversely affect human health and can, in many cases, cause death. In addition, the heat derived from fire is often used to warm human dwellings during cold periods. The light provides a security from the dark. Heat from fire has also been used in various industries for purposes such as smelting (producing metal from ore) and the creation of tools and weapons. Fire is also an integral aspect of the social (Fig. 14.12) and religious activities of various cultures and has in many ways been tamed to allow us to take advantage of the combustion process. For example, when one starts the engine of an automobile to drive to work or school in the morning, the battery of the vehicle provides an initial spark that initiates the combustion of gasoline and releases energy, which is then used to propel the vehicle. This is an example of a highly controlled fire. In natural resource management, large controlled fires can also be used in a forest to burn and kill undesirable vegetation or to create a desirable habitat for certain wildlife species (e.g.,

FIGURE 14.11   The fire triangle. *Gustavb, through Wikimedia Commons.*
Image Link: https://commons.wikimedia.org/wiki/File:Fire_triangle.svg
License Link: https://creativecommons.org/licenses/by-sa/3.0/deed.en

FIGURE 14.13   Wildfire in the Bitterroot National Forest in 2000, Montana, United States. *John McColgan, through Wikimedia Commons. Image Link: https://commons.wikimedia.org/wiki/File:Deerfire_high_res_edit. jpg*
License Link: Public Domain

FIGURE 14.12 Midsummer festival bonfire near Mäntsälä, Finland. *Janne Karaste, through Wikimedia Commons.*
Image Link: https://commons.wikimedia.org/wiki/File:Midsummer_bonfire.jpg
License Link: https://creativecommons.org/licenses/by-sa/2.0/deed.en

bobwhite quail; *Colinus virginianus*). In Chapter 11, it was noted that prescribed fires are commonly used to control the understory vegetation of an existing forest or to control pioneer vegetation in a cleared area and prepare the area for the establishment of a new forest. The use of fire by natural resource managers can reasonably be argued to be one of the most important tools for maintaining or manipulating live or dead vegetation.

Unfortunately, as either a natural or anthropogenic process, fire can be very destructive to the resources we value. Fire can injure and kill people, as well as other living organisms, such as animals, birds, insects, fungi, and plants. Fire can also damage or destroy homes, offices, vehicles, automobiles, airplanes, ships, tractors, and a multitude of other tangible items. As we noted, humans can benefit by using a controlled fire, yet, when not adequately controlled, fire may destroy anyone or anything we deem important. An uncontrolled fire in the natural environment is otherwise known as a *wildfire* (Fig. 14.13). A wildfire can severely damage or completely destroy an ecosystem, thus denying animal and plant communities the plethora of ecosystem services that the forested landscape could have provided. The impact of wildfires on forests will vary depending on the intensity and extent of the fire, and whether the forested ecosystem is sensitive to fire, influenced by fire, or independent in any regard (Food and Agriculture Organization of the United Nations, 2005). Much of the damage from fire-related events transitions forested areas naturally back to secondary succession processes, where seed from neighboring unaffected plants (or from reserves in the ground) provide the pioneering growth of a new forest. Small fires could simply serve to adjust the species composition and density of understory trees, shrubs, and low-lying vegetation (herbs, forbs, etc.).

From a forest and natural resource perspective, the main causes of wildfires are primarily lightning and human activities. Thunderstorms are important sources of lightning, a discharge of electricity from the atmosphere to the ground. According to the National Oceanic and Atmospheric Administration National Weather Service (2018), there are more than 25 million lightning flashes in the United States each year. Lightning striking a tree or the ground can cause a wildfire that can spread across a forested landscape. Fires not immediately attended to, especially those located in remote areas, can burn unabated for many days. Other sources of lightning arise from volcanic eruptions and dust storms. Colliding dust and ash particles can generate electrical discharges that can be seen in the atmosphere (Witze, 2010).

In developing a basic understanding of wildland fire, concepts of fire behavior, fuels, and weather need to be addressed. Following the work of Pyne (1984), we will utilize his framework to discuss fire behavior, fuels, and the importance of weather. Fire behavior is dependent on the combustion of suitable fuels and an ample oxygen supply. The process of combustion consumes fuels until a wildfire is either extinguished by an external source (e.g., rainfall) or the fuel has been fully consumed. Once a fire begins, it can spread to new areas containing combustible fuel or it can increase in intensity within a specific area, given a sufficient supply of combustible fuel. Fires can act to dry other nearby combustible material, which can lead to the ignition of previously untouched fuels and support the spread of fire. Fires begin as points of ignition and then, depending on the wind, expand in either a circular or an elliptical manner. Within wildfires, flames may burn upward and inward due to convection, or flames may burn upward and outward given prevailing winds. The shape of a fire may evolve into an ellipse and the sides of a fire may progress at a different rate and burn at a different intensity than the leading edge due to the existence of different fuel loads, the direction of prevailing winds, or the local topography. Fires burning downwind (i.e., with the wind) are called heading fires. Fires burning upwind (i.e., against the wind) are called backing fires, and fires that burn between these zones are called flanking fires.

Fuel size, shape, and composition can greatly affect the rate of combustion and influence the spread of a wildfire. For example, dry fuels can burn quickly and therefore lead to an increase in fire intensity, while wet fuels can slow the spread of a fire. Fuel sources that contain pitch or other extractives can lead to hot, more explosive fires. Ambient winds can help push a wildfire toward unignited fuels. The heat generated by a wildfire can be carried by the wind and used to dry other wetter fuels to increase their combustibility. In addition, winds can carry smoldering particles from an area actively burning to areas not previously ignited, thus creating a spot fire. It is intuitively apparent that stronger and faster winds can increase the rate of spread of a wildfire. Another key factor that affects the spread of a wildfire is the local topography, which can either promote the rapid spread of a wildfire or diminish it. Elevation, aspect, and local terrain conditions can influence the microclimate around fuel sources, which may then influence the spread of a wildfire. For instance, wind acceleration across ridgelines and through saddles can hasten the pace of wildfire spread (Pyne, 1984).

There are three basic categories of wildfires: ground fires, surface fires, and crown fires. Ground fires (Fig. 14.14) are those found burning in fuels up to 15.24 centimeters (cm; 6 inches [in]) above the ground. Surface fires (Fig. 14.15) are those found burning in fuels up to 1.83 m (6 ft) above the ground, while crown fires (Fig. 14.16) are those found burning in fuels located more than 1.83 m (6 ft) above the ground, usually in the crowns of the trees (Gaylor, 1974). All of these types of wildfires are influenced by the available fuel loads and the weather conditions.

FIGURE 14.14 Ground backing fire in southwestern Georgia, United States, in 1971. *Dale Wade, Rx Fire Doctor, through Bugwood.org.*

FIGURE 14.15 A surface fire in Georgia, United States. *Matt Elliott.*

FIGURE 14.16 Crown fire on Mirror Plateau, Yellowstone National Park, Wyoming, United States, in 1988. *Jim Peaco, through Wikimedia Commons.*
Image Link: https://commons.wikimedia.org/wiki/File:Firestorm_Mirror_Plateu.jpg
License Link: Public Domain

Forest fuels play an important role in the spread of any forest wildfire, depending on whether they are ground or latter fuels, continuously spread across the landscape, compact, or moist (Gaylor, 1974). Ground fuels include low-lying vegetation, logs, leaves, grasses, fallen limbs, duff, peat, and soils that contain high organic content. Basically, ground fuels include living or dead vegetation that is either in contact with the soil or in very close proximity to it. Ladder fuels are those that are basically located above the ground, such as standing live and dead trees, branches, bushes, and mosses hanging from trees. The effect of these fuels on the intensity or behavior of a forest fire depends on their size and character. Fine fuels, such as grasses and twigs, burn quickly when dry, while newly fallen tree boles will burn more slowly (if at all) than similar debris that has been lying on the ground for some time. This suggests that another key attribute of forest fuels is their moisture content level. Many people experience problems with this fuel attribute when they attempt to start a campfire or a fire in the fireplace of their home. If wood is wet or has a high moisture content level, it will not burn. This concept can be extended to the forested landscape as well: wet fuel, or fuel with a high moisture content, generally will not burn. However, as air temperatures increase, the drying of fuels is accelerated, and the fire risk may change throughout the course of a single day. The length of time it takes for a particular fuel type to dry depends on its size and exposure to wind. Any fuel source exposed to wind dries more quickly than fuel sources that are protected by shade or other enclosures. The fuel size is directly related to the length of time it takes to dry. Fuel sizes are often categorized into *time lags* to represent the responsiveness of the fuel moisture conditions to changing weather conditions. For example, many small, fine ground fuels are categorized as 1-hour fuels, which may only require 1 hour to dry, while large downed logs may be categorized as 1000-hour fuels.

Fuel continuity, especially fine fuels, across a forested landscape can greatly affect the extent and spread speed of a wildfire (Gaylor, 1974). Fuel continuity can be described as uniform or patchy. Uniform fuels imply that fires have a steady source of combustible material, thus allowing them to rapidly move from one location to another. Patchy fuels imply that fires may spread to a point where there is no fuel available, such as bare ground or a rock outcrop. Typically, forest firefighters build firebreaks (fuel breaks) or fire lines, i.e., long strips of bare ground, to starve forest fires of new fuels. In creating these obstacles, it is assumed that a wildfire will not jump the fire line due to windblown sparks or ignited particulates. Another important aspect of forest fuels is the compactness of the resource (Gaylor, 1974). In compacted fuels, there is less oxygen, which can slow the rate of heat transfer and reduce the rate of fire spread. Loose piles of needles or dried leaves that are perhaps not compacted at all can enhance the rate of wildfire spread because these fuels have a sufficient level of oxygen around the fuel particles to facilitate ignition and combustion processes. Fuel load guides have been developed for many different forest types and, in addition to providing photographic examples of fuel-loading conditions (Fig. 14.17), may provide the necessary landscape-level parameters required by wildfire behavior models (e.g., Brose, 2008; Stebleton and Bunting, 2009).

Weather is the other key element in controlling or managing wildfires. As previously mentioned, air temperature and wind conditions can play an important role in drying an existing fuel source. Weather is dependent on the rotational movement of the Earth around its axis, as well as the transfer of heat between the equator and the polar regions (Gaylor, 1974). The circulatory pattern of air and the atmospheric pressure, which is

FIGURE 14.17 Ericaceous shrubs of fuel class 2 in the Delaware State Forest in Pennsylvania, United States. *From Brose (2008).*

FIGURE 14.18 Cumulus clouds in 1978 over Coconut Creek, Florida, United States. *Ralph F. Kresge, U.S. National Oceanic and Atmospheric Administration, through Wikimedia Commons.*
Image Link: https://commons.wikimedia.org/wiki/File:Cumulus20_-_NOAA.jpg
License Link: Public Domain

the weight of air mass pressing against the Earth, can influence the formation of wind patterns across a landscape. Winds can affect wildfires in a number of ways. They can

- dry fuels quickly;
- carry sparks to new fuel sources and hasten the spread of a wildfire;
- increase the oxygen supply to a fire;
- facilitate certain heat sources in igniting a fire; and
- influence the direction of fire spread.

Atmospheric stability and humidity are other factors that can affect the spread and control of wildfires. Unstable atmospheric conditions are those that have good visibility, have cumulus-type clouds (Fig. 14.18), and facilitate smoke rising to great elevations (Gaylor, 1974). Cumulus clouds may also indicate the potential for greater wind speeds and greater potential changes in wind direction at ground levels.

Invisible water vapor that is contained in the air is called humidity. A general lack of humidity in the atmosphere (i.e., dry air) can also produce conditions favorable for the spread of wildfires. Humidity can be measured in either absolute or relative terms. Absolute humidity is the weight of water within a given volume of air. Relative humidity is the amount of water in the air compared to the maximum amount of water that the air can carry, which will change with air temperature. As a rule, as air temperatures increase, the associated relative humidity (given the same amount of water in the air) decreases, because warmer air can hold more water than colder air. Thus, from morning to afternoon during a typical day, air temperatures increase and relative humidity decreases, leading to a greater fire risk.

Wildfires are natural events within a forested ecosystem, although they may also be initiated by humans, and the changes they make to a landscape and the resources located therein can be dramatic. If intense enough, wildfires can revert a landscape almost to its primary successional processes. While we frequently characterize wildfires as being detrimental to human values, they are often beneficial to and necessary for many ecological processes. Longleaf pine (*Pinus palustris*) forests, for example, have evolved with and adapted to wildfire in the southern United States. Frequent ground fires act to control deciduous tree competition, and the seedling (grass) stage of longleaf pine is highly resistant to the effects of ground fires. Lodgepole pine (*Pinus contorta*) in the western United States is an example of a tree species that can quickly reestablish a burned area because its cones have a level of protection to fire and are considered serotinous. Cones with this quality open and provide seed in response to an environmental cue, in this case heat from a fire. As a result, policies have been developed in various organizations around the world that allow wildfires to progress unimpeded through natural systems or are only designed to prevent wildfires from spreading outside protected areas (thus allowing them to burn within protected areas).

Effective fire control can require extensive planning, especially when managing a large forested landscape of 100,000 ha (nearly 250,000 ac) to 1 million ha (roughly 2.5 million ac) in size or more. In the United States, widespread fire suppression activities were employed throughout most of the 20th century on both public and private land; however, environmental policies have redirected efforts on some public lands and now allow fires to burn naturally in some areas. Since fire is a necessary part of the life cycle of some forested ecosystems, some public agencies allow wildfires to burn in protected or wilderness areas, although these are monitored and control is attempted only when they threaten human developments. This change in the manner in which wildfire is viewed involves a philosophical change from control measures to management measures. All federal and state agencies in the United States have a fire management division. While state agencies manage wildfires on state-owned land, they also respond to the prevention and control needs of private landowners. Other countries may also have developed a formal fire control or management agency. Canada, for example, has developed the Canadian Interagency Forest Fire Center that provides operational fire control services.

Organizations that manage large forested areas, such as the U.S. Forest Service, which manages the 78 million ha (nearly 193 million ac) of U.S. National Forests, generally have a hierarchical personnel structure trained in the various logistical and mechanical aspects of firefighting. This group continuously updates its skills in order to maintain proficiency within its workforce over time. Larger natural resource management organizations typically also maintain inventories of firefighting equipment that smaller organizations or individuals may not. This equipment includes fire trucks, water pumps, hoses, shovels, chainsaws, axes, rakes, radios, fire retardants, bulldozers, helicopters, and airplanes.

The two general approaches for combating a wildfire involve direct and indirect methods (Gaylor, 1974). Direct methods are common for combating surface and ground fires that spread through fuels such as leaves, needles, fallen branches, twigs, grasses, and duff. Direct firefighting requires working on the edge of fires, spraying water onto flames, smothering flames, or building fire lines to prevent the movement of flames. Fire lines may be constructed using hand tools or machines and may be irregular and quite long. While natural barriers (e.g., water bodies) may also be used, fire lines may also include barriers meant for other human purposes (e.g., roads). One advantage of a direct firefighting method is that, in general, a smaller area is burned (Gaylor, 1974). One disadvantage of a direct firefighting method, however, is that firefighters are exposed to heat, flames, and smoke. However, once burned, a land area can be used as a refuge for fleeing firefighting crews. Indirect methods for combating wildfires typically require building firebreaks some distance from the burning fire. This approach is commonly used when a wildfire is too hot and the risk of injury or loss of human life is too great. This method also commonly involves natural barriers such as roads, cliffs, and water bodies as firebreaks. One advantage of the indirect method of firefighting is that firefighters are working in a safer environment; another is that there is generally more time to build adequate firebreaks in accessible locations (Gaylor, 1974). In general, there is a better use of natural barriers with this method, and shorter fire lines are built. As a

FIGURE 14.19 Portion of the coastal coniferous forest consumed by the Tillamook Burn of October 1944, Tillamook County, Oregon, United States. *Russell Lee, through Wikimedia Commons.*
Image Link: https://commons.wikimedia.org/wiki/File:Tillamookburn.jpg
License Link: Public Domain

FIGURE 14.20 Firestorm approaches the Old Faithful complex in Yellowstone National Park, Wyoming, United States, in 1988. *Mongo, through Wikimedia Commons.*
Image Link: https://commons.wikimedia.org/wiki/File:Crown_fire_Old_Faithful.jpg
License Link: Public Domain

result, fire crews can expend more effort addressing hot spots along a fire's edge. Disadvantages of the indirect method of firefighting are that, in general, more land area is burned, fire crews can be outflanked, and changing weather conditions can negate much of the work previously accomplished.

In recent history, some notable wildfires have severely altered what we consider to be natural landscapes. In the United States, one such famous fire was the Tillamook Burn in the early part of the 20th century (Fig. 14.19). The Tillamook Burn was a series of fires that affected the same area at different times from the early 1930s to the early 1950s. The first fire began by accident on August 14, 1933, as a logging crew was shutting down activities for the day (Tillamook State Forest, 2008). A logging cable, hauling in the last log of the day, created a friction spark that ignited a fire that eventually burned about 97,125 ha (240,000 ac), of which nearly 81,000 ha (200,000 ac) burned in just one day. Subsequent fires in the same area in 1945 and 1951 extended the area of damage up to about 143,663 ha (355,000 ac). Although the damage caused by the forest fires was extensive, perhaps the most important result of the fire has been the renewal of the forest over the past 75 years. In the aftermath of the fires, an extensive reforestation effort was implemented to reclaim the burned areas. Nearly 72 million tree seedlings were planted from 1949 onward in the burned area (Tillamook State Forest, 2008). Today, this particular state forest hosts a lush forested landscape and is a popular outdoor recreation destination. It offers a broad array of hiking, horseback riding, and mountain biking trails, and campgrounds, as well as an extensive trail system dedicated for motorcycles and four-wheel-drive vehicles.

Another famous forest fire in the United States was the Yellowstone Fire of 1988 that affected one of the

United States' most popular national parks (Fig. 14.20). This fire developed from many smaller fires in the general Yellowstone area and combined into a large firestorm. Numerous factors contributed to the spread of this fire, such as the presence of dense forests composed of lodgepole pine and mixed forest age classes composed of both mature and dead overstory trees and young understory trees. In addition, unstable atmospheric and drought conditions led to favorable fire conditions. However, due to the patchy nature of the fuel source, the fire did not burn the entire national park. Approximately 323,750 ha (800,000 ac) were burned (Franke, 2000), while landmarks such as the Old Faithful Hotel were spared. Fortunately, since 1988, natural regeneration of trees in this area has been largely successful. The impact of the wildfire on recreation and leisure activities at this very popular park proved to be a catalyst for debates concerning the fire management policies of the United States.

14.4 Disturbances created by volcanic eruptions

A volcano is a rupture in the Earth's crust that allows hot ashes, gases, and magma (lava) to escape into the air and travel along the nearby land surface. Volcanoes are generally caused by the movement of convergent or divergent tectonic plates under the Earth's surface. When a volcano erupts, the natural environment (animals, insects, plants, etc.) can be decimated within some distance of the event and, in many cases, plant life reverts to primary or secondary successional processes. Volcanoes can create climatic conditions on at least a hemispheric scale that may affect the growth rates of trees (Salzer and Hughes, 2007). Volcanoes can conjure up in our minds various images of homes and businesses being burned and destroyed, as well as the tragic loss of life. The word *volcano* comes from the name of a small Italian island off the coast of Sicily in the Mediterranean Sea (Tilling, 1999). Nearby inhabitants believed that the volcano was the forge chimney of Vulcan, a blacksmith serving the Roman gods. In addition to humankind's constant exposure to these violent events, the modern film industry reminds us of their potential danger through movies such as *The Last Days of Pompeii*, *When Time Ran Out*, *Dante's Peak*, and *2012*. Despite the negative aspects of volcanoes portrayed by the film industry, volcanoes play an important process in land formation on Earth. Volcanoes create new land from the molten lava and ash that arise from eruptions. Over time, the deposited lava, ash, and rocky debris are weathered by the sun, water, and wind to form the basis of soil, which is critical for the primary succession of plant life and to the needs of other organisms on Earth.

Volcanoes are a type of mountain formed from the process of molten rock, commonly known as magma, being lifted from beneath the Earth's lithosphere, commonly known as the crust. The crust covers the liquid mantle, with a partially molten layer in between, known as the asthenosphere. Magma, ash, and rocky debris exit a volcano through a central vent or through fissures or vents at the sides or base of the volcano. The ejection of material onto the Earth's surface and into the atmosphere promotes the formation of new land but is also the cause of destruction of organisms that exist in the area near the eruption. The material emitted can be substantial. Forests buried 0.5 to 1 million years ago by volcanic activity have been found on Honshu Island in Japan (Noshiro et al., 2002). In 1749, excavators near Pompeii discovered that another Roman community had previously existed at the site but was buried beneath 5 to 6 m (16—19 ft) of ash and hardened lava (Özgenel, 2008). In the case of forested landscapes, plant and animal life may be either burned by hot volcanic gases and moving lava or buried by the ash deposit falling from the atmosphere (Fig. 14.21).

Volcanoes are generally caused by the movement of tectonic plates beneath the Earth's surface; therefore, the pattern of volcano locations is not random. Volcanoes can typically be found around the edges of tectonic plates that move distances as small as a few centimeters or inches per year. Tectonic plates are portions of the Earth's crust that sit on top of the moving molten magma found in the mantle, and these plates shift relative to each other over the Earth's surface. Tectonic plates covering the Earth include the African plate, the Antarctic plate, the Arabian plate, the Australian-Indian plate, the Caribbean plate, the Cocos plate, the Eurasian plate, the Nazca plate, the North American plate, the Philippine plate, the Pacific plate, and the South American plate (Press and Siever, 1982). When these plates move against each other, the lithosphere of one, which forms

FIGURE 14.21 Windthrown, broken, and burned trees in the Smith Creek Valley as a result of the Mount St. Helens eruption in 1980, Washington, United States. *Lyn Topinka, through Wikimedia Commons. Image Link: https://commons.wikimedia.org/wiki/File:MSH80_blowdown_smith_creek_09-24-80.jpg*
License Link: Public Domain

the crust, can slip beneath the other plate that it is pressing against. Earthquakes of various magnitudes are then felt but, as the crust of one plate slides under that of another, heat from the asthenosphere melts it, and this material can then be pushed upward and ejected from volcanic vents. The colliding plates also create mountains when the crust is crumpled and pushed up into mounds.

There are numerous types of volcanoes, including the cinder cone, the composite volcano, the shield volcano, and the lava dome (Tilling, 1999). A cinder cone volcano is formed from the accumulation of particles and congealed lava around a main vent which form an oval or circular cone shape similar to a bowl. These types of volcanoes are numerous in the western United States and other volcanic regions of the world. A composite volcano has the classic symmetrically shaped cone commonly seen in many movies (Fig. 14.22). These are formed from the ash, lava, cinders, and rocks that form layers down their steep slopes over time. Composite volcanoes can be found in North and South America, as well as in Asia. Shield volcanoes are formed from lava flows that progress across a landscape from a central vent and form a domed shaped structure with the profile of a shield. Lava can also seep from vents and fissures along the side of the central cones. The Hawaiian Islands are formed from these types of volcanoes. The lava flow from the shield volcanoes can form broad plateaus across the countryside. Lava domed volcanoes are relatively small and are characterized by viscous lava piling up around and over their vents. Over time, the vent expands and pops. Sometimes, these volcanoes are formed in conjunction with composite volcanoes. In other words, they can form on the side of a composite volcano or within its crater; thus, the source of magma

FIGURE 14.22 Volcán Arenal, Alajuela, Costa Rica. *Stacey Herrin.*

FIGURE 14.23 Mount St. Helens eruption May 18, 1980, Washington, United States. *U.S. Geological Survey, Austin Post and Carol Spears, through Wikimedia Commons.*
Image Link: https://commons.wikimedia.org/wiki/File:MSH80_eruption_mount_st_helens_05-18-80-dramatic-edit.jpg
License Link: Public Domain

for these two types of volcanoes is the same. Some notable lava domes can be found in the Cascade Range and the Sierra Nevada of Oregon and California.

Although there are many noteworthy volcanic eruptions in human history, one that occurred in modern times that significantly affected forested ecosystems was the eruption of Mount St. Helens on May 18, 1980 (Fig. 14.23). Mount St. Helens is a composite volcano located approximately 154 kilometers (km; 96 miles [mi]) south of Seattle, Washington, and 80 km (50 mi) northeast of Portland, Oregon. This violent explosion ejected large volumes of lava, gas, and ash from the north-facing slope of the volcano. The explosion flattened all forests and buildings over a 600-square-kilometer (km^2; 230-square-mile [mi^2]) area, and released approximately 2 million metric tons (2.2 million US short tons) of sulfur dioxide into the atmosphere (Gerlach and McGee, 1994). The explosion on the north face of the mountain caused a landslide debris avalanche (lahar) that was deemed the largest in recorded history (Tilling et al., 1990). The collapse of the north-facing slope triggered a mass movement of 3 million cubic meters (3.9 million cubic yards) of volcanic mud, ice, snow, and water down into two river basins, destroying roads, recreational sites, trails, highways, bridges, and logging camps (Tilling et al., 1990) (Fig. 14.24). The eruption produced an ash cloud that deposited ash 1036 km (644 mi) away in Edmonton, Alberta, Canada (Samuels, 2011). Many agricultural, forest, and range areas of central and eastern Washington also received a blanket of ash from this eruption (Fig. 14.25).

FIGURE 14.24 Mailboxes trapped in mudslide debris caused by the Mount St. Helens eruption. *Lyn Topinka, through Wikimedia Commons.*
Image Link: https://commons.wikimedia.org/wiki/File:After_Effects_(25727393810).jpg
License Link: Public Domain

The damage caused by the eruption of Mount St. Helens was extensive. In addition to the deaths of 57 people, almost 320 km (200 mi) of roads and railways were damaged, along with hundreds of homes and businesses and tens of thousands of hectares of old forests (Tilling et al., 1990). It has been suggested that over 9.4 million cubic meters (m^3; 4 billion board feet) of merchantable timber was destroyed or damaged, over

FIGURE 14.25　Ash from Mount St. Helens eruption deposited nearly 250 km (155 mi) away in Connell, Washington, United States. *Lyn Topinka, through Wikimedia Commons.*
Image Link: https://commons.wikimedia.org/wiki/File:MSH80_may18_ash_connell_washington.jpg
License Link: Public Domain

12 million salmon fingerlings perished in destroyed hatcheries, approximately 400,000 young and adult salmon perished in the turbine blades of hydroelectric generators due to mudflows and flooding, and over 7000 bear, deer, and elk were killed (Tilling et al., 1990). Although the damage from the volcanic eruption was extensive, the forests and environment around Mount St. Helens have begun to recover. In planted areas, thinning is now occurring. Salvage logging crews recovered approximately 25% of the damaged trees, the introduction of ash to local agricultural areas is expected to boost productivity, and the effect of ash that was transported into local water bodies appears to have been minor and temporary (Tilling et al., 1990). Since the eruption, the area has become a magnet for visitors interested in learning more about the effects of a volcanic eruption. Two years after the event, in 1982, the Mount St. Helens National Volcanic Monument was created (U.S. Geological Survey, 2020).

14.5 Disturbances created by ice and snow

Another type of disturbance that a forest may encounter is one that is caused by ice and snow. Water becomes frozen around 0 degrees Celsius (°C; 32 degrees Fahrenheit [°F]), and ice is water in its solid state. Snow is a granular form of small ice particles. We may typically think of ice in a positive manner as something we place in our glass on a warm evening to cool down tea or soda. In areas such as Minnesota during the winter, the formation of ice on a local pond or lake can allow seasonal ice-

skating opportunities or facilitate ice fishing. However, people often view ice or snow in a negative manner when it collects on roadways, since these events can cause a motorized vehicle to slide off a road and perhaps hit a tree. Further, in both southern and northern communities of the United States, ice is typically considered troublesome when it forms and hangs on power lines (Fig. 14.26). Ice storms generally occur when precipitation passes from a cold air layer through a warm air layer, and finally through a cold air layer located just above the ground. A thin layer of cold air just above the ground can supercool precipitation, which freezes when it lands on trees, buildings, or the ground. The weight of an ice sheet on power lines can lead to electrical outages that affect homes and businesses.

The build-up of ice and snow in a forest or natural resource setting can be viewed as either beneficial or harmful. From a beneficial perspective, water stored as ice or snow can become a suitable water supply when released through melting processes in the spring and summer months. Further, the layering of icicles on tree branches, on frozen waterfalls, and on cliff faces can be aesthetically pleasing to many people, as can a snow-covered landscape. From a harmful perspective, the formation and weight of ice and snow on trees can adversely affect forest structure and composition, although the impacts of individual storms are generally not consistent across a broad landscape (Irland, 2000). The main damage to trees caused by ice and snow is the breakage of limbs and tops due to the added weight of frozen water; however, only occasionally will trees fall under the weight of ice and snow. When combined with wind, it can be significant. For example, a 1998 ice storm in the northeastern United States and Canada damaged

FIGURE 14.26　Ice storm damage. *Brian0918, U.S. National Oceanic and Atmospheric Administration, through Wikimedia Commons.*
Image Link: https://commons.wikimedia.org/wiki/File:IceStormPowerLines.png
License Link: Public Domain

large expanses (millions of hectares) of forests in this manner. Many different factors have been suggested as influential in determining the impact of ice storms on forest structure, including forest density, recent management activity, topography, and tree species, but the relationship between these factors and the damage experienced is complex (Bragg et al., 2003). Much of the damage from ice and snow events creates changes in the structural characteristics of forests. If severe, individual tree death can occur, affecting the density and species composition of a forested area.

Another phenomenon associated with ice and snow is an avalanche. Avalanches can influence the structure and composition of nearby forests but their effects are limited to areas of around 100 ha (roughly 250 ac) (Fig. 14.27). Avalanches occur in areas with topographic conditions that favor their development and are among the most important natural disturbance processes in mountainous environments through their influence on forest dynamics and their effect on plant development and mortality processes (Viglietti et al., 2010). Where avalanches are common, forests are composed of trees that are relatively smaller and shorter than normal, with lower tree densities per unit area and with greater structural diversity. Forest conditions that reduce the likelihood of an avalanche include those with crown cover greater than 30% and those with fewer and smaller openings (Bebi et al., 2009).

FIGURE 14.27 Avalanche chute in the Glacier Peak Wilderness, Washington, United States. ©2006 Walter Siegmund, through Wikimedia Commons.
Image Link: https://commons.wikimedia.org/wiki/File:Avalanche_path_7271.JPG
License Link: https://creativecommons.org/licenses/by-sa/3.0/deed.en

14.6 Disturbances created by floods

Some forests of the world depend on flood events for renewal and for nutritional supplements. Mangrove forests, for instance, depend on regular tidal floods for certain stages of succession. Some types of forests, however, are less tolerant to flooding events. Pine forests along the southern United States coast, for example, are sensitive to changes in salt concentrations that may occur with tropical storm-related floods or storm surges. Floods may arise from lakes and rivers located inland and at higher elevations, or from seas and oceans located at lower elevations. Floods can be catastrophic and can promote moderate to severe changes in forest dynamics. Much of the damage from flooding events can transition a forest to secondary successional processes or, depending on the tolerance of trees to inundation or salination (if floods consisted of seawater), individual tree death can occur, affecting the density and species composition of a forested area. Evidence in one southeastern area of the South Island of New Zealand suggests that inland floods that occurred during the Middle Jurassic period buried at least 10 coniferous forests (Pole, 2001). More recently, Yin et al. (2009) observed moderate changes in the structure of forests following a flood in the Upper Mississippi River floodplain. In this case, deciduous tree species seemed to possess varying levels of disturbance tolerance and, therefore, exhibited different survival and succession strategies.

Tropical cyclones or hurricanes can generate storm surges along coastlines, causing extensive damage, as was the case with Hurricane Katrina in 2005 in coastal Alabama, Louisiana, and Mississippi. A storm surge consists of a long peak wave of seawater that may be as great as 10 m above normal water levels (Fritz et al., 2007). Forests located on coastline areas are vulnerable to flooding from storm surges. If trees are not uprooted from this flooding process or overly stressed due to excessive water loads, increased salt levels can result in severe forest damage. Excessive sodium levels result in increased levels of leaf and needle litterfall, along with decreased levels of water uptake through the tree root system (Blood et al., 1991). Some tree species of the southern United States that are relatively resistant to wind damage, such as baldcypress (*Taxodium distichum*), may be more sensitive to saltwater inundation (Barry et al., 1982). Some tree species, such as live oak (*Quercus virginiana*), are better adapted to handle saltwater intrusion than other native species (Fig. 14.28). Even a year after saltwater inundation, ammonia concentrations can remain high in standing water or groundwater, perhaps because the uptake of nitrogen by stressed and dying trees has become limited

FIGURE 14.28   Coastal oak trees in 2005 that survived a Gulf of Mexico storm surge during Hurricane Katrina, Mississippi, United States. *Jon D. Prevost.*

(Gardner et al., 1991). As if all of this were not bad enough, the increased amount of dead and dying vegetation can also influence the risk of insect and disease outbreaks and of wildfire occurrence and damage. However, the flooding associated with hurricanes can provide a beneficial phosphorus input for mangrove forests that positively influence their productivity (Castañeda-Moya et al., 2020).

14.7   Disturbances created by diseases

Although perhaps less dramatic in their appearance than other forest disturbances, forest diseases can impose a substantial disturbance on a forested ecosystem and the natural resources contained within. To simplify our discussion, we will focus on plant diseases, as their impact on forest structure and wildlife habitat affects ecological processes within a forested ecosystem. Before we begin, it is worthwhile to briefly acknowledge the roles that plant diseases have played in human society and the historical perceptions of these. For much of human history, plant diseases, as well as those involving other living organisms, were viewed as the divine intervention of God in response to the sins of humankind. This perspective existed from ancient times through the Middle Ages, until about the 18th century (Manion, 1981). Although some earlier observations on plants, such as those made by Theophrastus in the 4th century BC, had suggested different impacts of diseases on wild versus domestically cultivated trees, the taxonomic work of the botanist Carl Linnaeus in the mid-1700s significantly advanced the study of plant diseases, despite the continued view by some of divine intervention (Manion, 1981). This early taxonomic work was important for recognizing that environmental

FIGURE 14.29   Field elm (*Ulmus minor*) trees killed by Dutch elm disease in Prádena de la Sierra, Segovia, Central Spain. *Luis Fernández García, through Wikimedia Commons.*
Image Link: https://commons.wikimedia.org/wiki/File:Grafiosis.jpg
License Link: https://creativecommons.org/licenses/by-sa/2.1/es/deed.en

factors can play a significant role in the development and spread of plant diseases. The seriousness of plant diseases is illustrated by one of the most memorable plant disease outbreaks in the mid-1800s, potato blight, which led to the Irish Potato famine (Manion, 1981). Damage to potato crops caused by the plant pathogen *Phytophthora infestans* resulted in 1.5 million human deaths from starvation and other ailments, and mass emigration from Ireland (Bourke, 1964; Ristaino, 2002). During the latter part of the 1800s, a number of famous scientists, including Louis Pasteur, Anton de Bary, and Justus von Liebig, focused on the importance of fungi and bacteria in the development of diseases in different organisms. Since then, the study of plant diseases has increased in importance. Notable forest plant diseases, such as American chestnut blight and Dutch elm disease, have prompted nearly 100 years of research effort in an attempt to save once prominent tree species.

A disease is an entity that disturbs the normal function or condition of a living organism. In this chapter, the focus is on the disruption of the normal function or condition of plants, particularly trees. Diseases are caused by pathogens, and there are numerous types of pathogens that may be abiotic or biotic in nature. When forested landscapes, including urban forests, are disturbed by a disease, a decline in the provisioning service of the ecosystem is likely to occur. For instance, if a plant disease kills the dominant tree species in a forest (Fig. 14.29), the loss of tree cover may lead to a reduction in the provisioning and supporting ecosystem services. As an example, there once were about 53,000 Dutch elm (*Ulmus hollandica*) trees along streets and in parks in the state of New York; now there are less than 300 (Manion, 1981). The widespread decline of the American chestnut in the eastern United States because of chestnut blight is another example (Burke, 2011). Effects such as increased soil erosion and altered vertical structure of a forest can lead to shifts in the plant and wildlife species composition in an area, which may affect other provisional ecosystem services, including the type of recreational opportunity available.

Determining whether a tree or plant has a disease requires both education and experience. Damage from plant disease outbreaks can transition a forest to secondary successional processes if it significantly affects most trees in a given area. Depending on the tolerance of trees to disease outbreaks, individual tree death can occur, affecting the density and species composition of a forested area. As noted earlier, a plant disease disrupts the normal function or condition of a plant and is caused by a pathogen. Plant diseases are manifested by outward symptoms, as occurs in humans or other animals during sickness (Manion, 1981; Edmonds et al., 2011). However, because animals and plants are different, the symptoms and diseases of each are also different. For plants such as trees, symptoms may include a canker or gall on the side of a tree or on its leaves; reduced height or diameter growth of the tree; root rot or rootlet necrosis; vascular wilts; sap or heart rot; shoot blight; and chlorosis of the foliage (from insufficient chlorophyll) or other foliar distortions. Foresters and natural resource managers will sometimes see a single symptom or a combination of symptoms, depending on the disease.

Plants encounter three general types of diseases: biotic, abiotic, and decline diseases (Manion, 1981; Edmonds et al., 2011). Biotic plant diseases are a product of the interaction of plants, their environment, and the particular pathogen involved. Symptoms may include those that are plant-part specific, in which only part of a plant is infected, and the spread of the infected area is noticeable. Abiotic plant diseases are caused by agents such as temperature extremes, nutritional imbalances, soil-oxygen deficiencies, phytotoxic gases, and moisture stresses. Abiotic disease symptoms may or may not be plant-part specific and are usually uniform across the plant. Decline plant diseases are diseases that involve either abiotic or biotic factors in their development and are typically associated with blight or dieback. Decline plant disease symptoms affect some trees more heavily than others and can thus sometimes form a trail through a population consisting of both diseased and dead trees.

Abiotic vectors can cause plant disease; these include, among others, soil condition, winter damage, and air pollution (Manion, 1981). Soil conditions that can cause the occurrence of abiotic diseases include a deficiency in mineral nutrients, too little or too much moisture, excessive salt concentrations, and poor soil aeration. All healthy plants require macronutrients such as calcium, magnesium, nitrogen, phosphorus, potassium, and sulfur, and small quantities of micronutrients such as boron, copper, iron, manganese, molybdenum, and zinc. Plants will express symptoms of abiotic diseases when they grow in soils with insufficient amounts of these minerals. Soils also have capillary structures for holding moisture (Brady and Weil, 1999). Trees, other plants, and microorganisms can access water-pores where water tension is less than forces exerted by the rood; this is commonly from macrocapillary structures formed by the existence of organic matter in the soil. Therefore, a lack of organic matter in the soil can lead to lower soil moisture levels, thus causing water stress in plants and leading to disease development. Plant species that live in coastal environments are usually salt-tolerant (Edmonds et al., 2011). Trees that are planted in urban forests are typically not tolerant of salt in the soil but are often exposed to salt as a by-product of the runoff from deicing agents applied to roads and highways (Fig. 14.30). Salt inhibits uptake of potassium by plant roots, and the accumulation of salt in a soil can adversely affect the binding of other mineral ions to clay, thereby preventing the availability of nutrients to plant roots. Salt can also decrease the vapor pressure of soil, which can decrease a tree's ability to absorb water.

FIGURE 14.30 Salt damage to roadside trees. *U.S. Department of Agriculture, Forest Service—Northeastern Area Archive, through Bugwood. org.*

Ample soil aeration is critical for normal plant development and, in general, the proportion of oxygen and carbon dioxide in the soil is the same as that found in the atmosphere (Manion, 1981). Disruption of this ratio due to excessive water, soil compaction, cement from sidewalks and buildings, and increased microorganism activity can lead to a reduction in oxygen and thus a reduction in root growth. Water infiltration and soil-oxygen concentration can decline and soil carbon dioxide concentration can increase, due to processes such as soil compaction (Watson and Kelsey, 2006). A decline in root growth can affect a tree's ability to absorb water, acquire nutrients, and fend off pathogens. However, some tree species are adapted to a low soil-oxygen environment and, therefore, can persist in compacted, wet, or urban environments, although growth rates may be lower and tree form may be affected (Rodríguez-González et al., 2010).

Winter damage, another abiotic agent, affects plants in temperate and boreal regions through physiological chlorosis, desiccation, rapid temperature change, prolonged low temperature, and late spring and early fall frosts (Manion, 1981). This type of damage is usually not a problem for tree species that are adapted to ecosystems with near-freezing or freezing temperatures during winter months. Physiological chlorosis can occur in some coniferous species when seasons change, and this is typically expressed in the needles as a red-to-bronze color change. Desiccation (a state of extreme dryness) often occurs when nonnative species are planted in more northerly latitudes. Depending on the specific species, lower temperatures may limit water uptake leading to plant desiccation. Some tree species, such as jack pine (*Pinus banksiana*), red pine (*Pinus resinosa*), and Scotch pine (*Pinus sylvestris*), are desiccation resistant (Manion, 1981). Desiccation symptoms include foliage and stem color becoming yellowish-brown to reddish-brown (Fig. 14.31). Rapid temperature change is another form

of winter damage to both natives and plants that are outside their normal distribution range. A rapid lowering of air temperature can lead to intercellular ice formation, which disrupts cell membranes and denatures some proteins. Low air temperatures can also lead to reduced quantities of cell water leading to changes in cellular solute concentration, changes in cellular pH, and cell shrinkage (Manion, 1981). Frost cracking and stem splitting of the main tree bole of a tree can also be caused by low air temperature. Late spring frosts and early fall frosts affect plants when they are most vulnerable. In early fall, plants have not undergone the physiologic process of becoming cold-hardy for the winter season. In the spring, plants come out of their cold hardiness stage in order to take advantage of the warmer temperatures and begin a new growth season. If a late spring frost occurs, the cellular structure of the tree can be damaged. In this way, gardeners have lost many vegetable plants by planting them too early in the year.

Air pollution is an abiotic agent manifested by a range of pathogenic air pollutants, certain types of weather conditions, and acid precipitation (Manion, 1981). Plant pathogenic air pollutants are emitted from a wide range of sources such as emissions associated with the burning of fossil fuels and emissions from coal-burning electrical plants, chemical processing facilities, and metal ore smelting facilities, such as copper smelting (Smith, 1990). The toxicity of air pollutants to plants can vary by broad tree species groups, and the resistance and uptake of chemicals by trees can vary according to the time of day and other local climatic variables (e.g., humidity) (Tsai et al., 2010). Excessive concentrations of chemicals such as sulfur dioxide can cause necrosis (premature death of cells) and interveinal browning in deciduous trees (Smith, 1990). Further, ozone (O_3) can cause brown flecking on the upper surface of leaves, and fluoride can cause necrosis or marginal chlorosis (Manion, 1981). Excessive concentrations of sulfur dioxide or fluoride can cause tip burn in coniferous trees. Excessive ozone can also lead to increased tree transpiration rates (Wang et al., 2011) as well as tip burn, banding, and flecking on leaves and needles. Weather inversions, in which warmer air sits atop cooler air, can concentrate air pollutants in the lower atmosphere at levels that are not only hazardous to plants but also to humans. There are many types of weather inversions, including anticyclone, frontal, geographic, radiation, turbulence, and valley inversions. One main difference between the different types of inversions is whether their extent is local or regional. For example, in valley inversions, colder air sinks below warmer air, sometimes trapping harmful pollutants in the entire valley. Finally, acid precipitation is a condition in which dilute forms of chemicals such as sulfuric and nitric

FIGURE 14.31 Desiccation injury to tree tops above the level of a winter snow pack. *U.S. Department of Agriculture, Forest Service Archive, through Bugwood.org.*

FIGURE 14.32 Effects of acid rain in the Jizera Mountains of the Czech Republic. *Lovecz, through Wikimedia Commons.*
Image Link: https://commons.wikimedia.org/wiki/File:Acid_rain_woods1.JPG
License Link: Public Domain

FIGURE 14.33 Pine wilt nematode *(Bursaphelenchus xylophilus)* damage. *U.S. Department of Agriculture, Forest Service—North Central Research Station Archive, through Bugwood.org.*

acids leave the atmosphere through a rainfall event that is commonly known as *acid rain*. Acid rain can affect not only forests (Fig. 14.32) but also weather rates from geological materials, and aquatic systems by lowering their pH levels. As a result, plants can experience cuticle erosion and nutrients can be leached from the soil.

Biotic vectors are another class of pathogens that cause diseases in trees (Edmonds et al., 2011). For example, nematodes (round worms) are biotic vectors that can infect animals, small invertebrates, and plants. Some examples of plant parasitic nematodes include dagger nematodes *(Xiphinema* spp.), lesion nematodes *(Pratylenchus* spp.), and pine wood (or wilt) nematodes *(Bursaphelenchus xylophilus)*, root-knot nematodes *(Meloidogyne* spp.), and stubby-root nematodes *(Trichodorus christiei)*. Plant parasitic nematodes use fine, tube-like appendages called stylets to pierce a plant's cell wall and extract material from inside. Plant parasitic nematodes feed on plant roots and provide a vector for bacteria, fungi, and viruses to enter the plant. The life cycle of a nematode progresses from an egg through a larval stage to an adult plant-feeding stage. As an individual nematode can produce up to 500 eggs and its life cycle lasts about 1 month, more than one generation of nematodes can be produced in a single temperate forest growing season. Nematode symptoms in trees are similar to those expressed in a nutrient-deficient plants and include sparse yellow foliage (Fig. 14.33), reduced stem and root growth, premature leaf drop, abnormal wilting, root lesions, galls, excessive root branching, and an abnormal root-to-shoot ratio (Manion, 1981). Bacteria associated with nematodes can result in the release of toxins that can increase the risk of tree mortality, as in the case of pine wilt disease in Europe (Roriz et al., 2011). In addition to the use of breeding programs that boost

nematode resistance in plant species, control measures include adding chemical nematocides to the plant and sterilants to the soil.

Viruses are another set of biotic pathogens that cause tree diseases; they are considered to be an infectious agent composed of protein and nucleic acid (Edmonds et al., 2011). The simplest form of a virus has a short strand of ribonucleic acid (RNA). Some complex viruses consist of proteins and RNA while others contain DNA and have a structure that includes structural proteins as well as enzymes. In general, when viruses replicate within a plant the normal cellular structure is disrupted (Manion, 1981). Plant viruses are widespread, operate in association with their host (the plant), and can cause disease (Malmstrom et al., 2011). Trees can be infected by viruses through the soil and their root system (perhaps via nematodes) or through buds, leaves, or seeds (perhaps via sap-sucking insects such as aphids) (Nienhaus and Castello, 1989). Basically, a plant virus starves a cell and the organism it has infected, although the general effect on plants may be minor compared to the effects of bacteria and fungi pathogens. Typical symptoms expressed by viruses include chlorosis, dwarfing, foliage flecks, foliage ring spots, leaf curling, necrotic lesions, mottling, rosette formation, and witches broom. The primary control measure for viruses involves preventing their occurrence at the seedling stage, which could involve the use of suitable insecticides, hormones, and tissue culture techniques.

Bacteria are another important biotic pathogen that can cause tree diseases (Edmonds et al., 2011). Bacteria can be found in all types of habitat, can survive in extremely cold and extremely hot temperatures, and can occur at densities of millions per cubic centimeter of topsoil (Manion, 1981). Bacteria are unicellular prokaryotic microorganisms that may be spherical, spring-shaped, pleomorphic, mycelial, or rod-like, and can only be viewed through a microscope. Plant pathogenic bacteria use cellulose and

FIGURE 14.34   Slime flux associated with bacteria in an American elm (*Ulmus americana*). *Joseph O'Brien, U.S. Department of Agriculture, Forest Service, through Bugwood.org.*

pectinase enzymes to dissolve plant cell walls, while their inherent toxins disrupt the plasma membrane that controls the osmotic process of a cell. Bacteria usually cannot infect a plant on their own; they need an open wound (Fig. 14.34) caused by fungi, human action, insects, wind, or other mechanical means. Some bacteria can enter through the stomata of leaves. Bacteria reproduce through transverse binary fission, which involves duplication of their singular chromosome followed by separation into two juvenile cells. This replication process can lead to mutations over time. As with other types of biotic pathogens, different species of bacteria produce specific symptoms in specific plant parts. Common symptoms include formation of witch's broom, hypertrophy of stem tissue, increased permeability, leaf necrosis and maceration of tissues, phloem yellowing, and wilting. Primary measures for controlling bacterial infections in plants include addressing disease vectors such as insects and improving soil aeration. It is important to note that bacteria play an important role in breaking down organic matter in the soil and in supporting the natural carbon, nitrogen, and sulfur cycles that play critical roles in some of the basic ecosystem services provided by forested landscapes. Further, given recent attention to energy created through intensive forest biomass programs, fertilizers have been shown in at least one case (Cambours et al., 2006) to induce higher levels of bacterial attacks and subsequent health problems in trees.

The last key biotic pathogen we will discuss in this section is fungi (Edmonds et al., 2011). Fungi provide a number of beneficial functions to forests and humankind. In principle, fungi are active during the decay process of organic compounds. Some fungi are a source of food for humans and also for other living organisms. Viking warriors once ate *Amanita* fungi to induce their famed berserk rages in battle (Manion, 1981). Fungi can also help form symbiotic relationships that assist plants in their uptake of nitrogen when they are growing in nitrogen-poor environments. However, fungi play an important role in the transmission of plant diseases. There are approximately 100,000 species of fungi, including the general types of food, parasitic, saprophytic, and symbiotic fungi (Manion, 1981). Fungi usually require cellulose, starch, organic nitrogen, and water for growth and reproduce by producing spores in their macroscopic fruiting structures that are sometimes seen by naturalists and other outdoor enthusiasts as they walk through the woods (Fig. 14.35). The classification of fungi is a complex process that is dependent on many factors. As with the other biotic pathogens that have been mentioned, greater knowledge in this area requires the reader to study more advanced texts. The impact of fungi within a specific ecosystem may still be poorly understood and tree disease may not be attributed to these alone, when considered pathogens (Matusick et al., 2010). The life cycle of a plant disease involving fungi begins with the process of inoculation, during which fungal spores come into contact with a plant (Manion, 1981). Spores take up water and nutrients and then germinate into an actively growing organism. Infection occurs when the fungi derive nutrients from the host plant and subsequently proceed to invade the host more extensively.

FIGURE 14.35   *Meripilus giganteus* fungus on a beech tree, Edinburgh, Scotland. *M.J. Richardson, through Wikimedia Commons. Image Link: https://commons.wikimedia.org/wiki/File:Meripilus_giganteus_-_geograph.org.uk_-_510110.jpg License Link: https://creativecommons.org/licenses/by-sa/2.0/deed.en*

In contrast to pathogenic fungi, some fungi form a symbiotic relationship with their plant hosts. One important example is mycorrhizae or root fungi. Mycorrhizae are classified according to the manner in which they interact with the host plant. The three main types of mycorrhizae are endomycorrhizal, ectomycorrhizal, and ectendomycorrhizal. Endomycorrhizal fungi invade the cortical cells of a plant's roots and are commonly found in grasses, legumes, and some hardwoods. The most common type of mycorrhizae associated with pines and some deciduous tree species are the ectomycorrhizal fungi (Fig. 14.36) that form a mantle surrounding the roots of a plant without damaging the plant's cortical cells. All species of pines, beech, and birch as well as other tree species are colonized by ectomycorrhizal fungi and, in total, about 6000 species of these types of fungi have been classified thus far (Lilleskov et al., 2011). Ectendomycorrhizal fungi also form a mantle around the roots of a plant, but these invade the cortical cells and are typically found in nurseries and on species of hardwoods and pines. Mycorrhizae fungi increase the ability of a plant's root system to more efficiently absorb water and nutrients from the soil, thus allowing the fungus to survive by obtaining carbohydrates and nutrients from the plant. Mycorrhizal fungi have been extensively used in tree nurseries in an attempt to produce seedlings that have a better chance of survival on land with a low inherent fertility.

Although fungi have many beneficial attributes, some species of fungi can cause extensive damage to forest trees and plants, including those that cause canker disease, foliage disease, root disease (Matusick et al., 2010), rust disease, and wilt disease (Edmonds et al., 2011). Foliage disease symptoms caused by fungi include the complete necrosis and shriveling of leaves, which can cause a tree to shed leaves and produce a second leaf batch within the same growing season. Terminal shoot dieback and reduced growth caused by fungi can lead to the excessive branching that forms a witch's broom. A common fungal-based foliar disease on oaks is leaf blister (Fig. 14.37), caused by the *Taphrina* fungi (Spooner, 2007). Spores of this fungus overwinter on the bark of a tree and are washed by rainwater onto its leaves, causing deformed leaf growth and the development of blisters. Since the early 1990s, a fungus-like mold (*Phytophthora ramorum*) has had a significant effect on the leaves of deciduous trees in the western United States, and its preference for oak species (Brown and Allen-Diaz, 2009) has prompted the resulting disease to be named *sudden oak death*.

Rust diseases can be expressed by the presence of cankers or galls on a tree (Edmonds et al., 2011). Rust diseases can reduce the growth of plants and cause portions of the stem to die. For example, white pine blister rust (*Cronartium ribicola*) causes the tops and branches of white pine (*Pinus strobus*) to die back and can cause stem cankers (Lu et al., 2005). Another important fungal disease is known as fusiform rust (Fig. 14.38), which is caused by the *Cronartium quercuum* f.sp. *fusiforme* fungus. Fusiform rust is native to the southern United States and has become a problem for loblolly pine and slash pine (*Pinus elliottii*). The life cycle of this fungus requires the presence of various red oak (*Quercus*) tree species (Perkins and Matlack, 2002). Fungal spores infect newly formed pine growth, usually causing galls to form on both twigs and the stem of the tree. Fusiform rust reduces tree growth, affects stem form, and makes trees

FIGURE 14.36 Ectomycorrhizae (*Thelephora terrestris*) at the base of a seedling. *Robert L. Anderson, U.S. Department of Agriculture, Forest Service, through Bugwood.org.*

FIGURE 14.37 Leaf blister (*Taphrina caerulescens*) on a white oak *Quercus alba* leaf. *Joseph O'Brien, through Bugwood.org.*

FIGURE 14.38   Pine stand in the southern United States with stem cankers caused by *Cronartium quercuum* f.sp. *fusiforme*. *U.S. Department of Agriculture, Forest Service—Region 8—Southern Archive, through Bugwood.org.*

FIGURE 14.39   American chestnut tree afflicted with chestnut blight. *U.S. Department of Agriculture, Forest Service, through Wikimedia Commons.*
Image Link: https://commons.wikimedia.org/wiki/File:Chestnut_blight.jpg
License Link: Public Domain

more vulnerable to wind breakage. Canker disease symptoms include the callusing of annual cankers, large areas of dead bark, and bark discoloration in some species. An infamous canker disease in American history was the spread of the chestnut blight (Fig. 14.39), caused by the fungus *Cryphonectria parasitica* (Burke, 2011). This particular fungus affects the cambium of a tree, essentially killing the entire tree in a very short time. The fungus spreads by wind-aided spores, which enter a tree bole through wounds (or perhaps where the bark cracks) and begin to grow. As the fungus grows, oxalic acid is produced, thus lowering the pH of the infected tissue and effectively girdling the tree. As a result of the spread of this fungus, approximately 3.6 million ha (9 million ac) of formerly ubiquitous American chestnut forests were lost by about 1940 (Manion, 1981).

Vascular wilt diseases can be caused by either fungi or bacteria (Edmonds et al., 2011). When a wilt-disease pathogen invades a host, a plant's new vessels are plugged, thus preventing the movement of water from the plant's roots to the foliage, causing leaf wilt and ultimately causing the plant or tree to die. An infamous case of vascular wilt diseases involves Dutch elm disease, which occurs when the *Ophiostoma ulmi* fungus enters wounds in a tree caused by the European elm bark beetle (*Scolytus multistriatus*). Infection of the disease usually occurs in the spring when new spring wood vessels are formed. The damage caused by Dutch elm disease has been extensive in North America and Europe (Temple et al., 2006).

14.8   Disturbances created by insects

Forest insects play a number of critical roles in the function and operation of the many ecosystem services provided by forested landscapes. Several of these roles are beneficial and include those involving pollination processes and the generation of food and other products (Coulson and Witter, 1984). Many insects produce products that are coveted by humans, such as silk, which is produced by the silkworm (*Bombyx mori*), a type of caterpillar. The Indian lac (*Laccifer lacca*), a South Asian scale insect, secretes a resin material onto trees that can be used as shellac, an important ingredient in various inks, paints, sealants, and varnishes. Several human cultures enjoy insects in their daily diet as an important protein component. Further, some insects are an important food source for other insects and many species of reptiles, fish, amphibians, and birds, which acts as a means to lower the population of undesirable insects. Interestingly, approximately 50% of insects are herbivores and as such they serve to combat undesirable species of plants (Coulson and Witter, 1984). Insects also play an active role as scavengers by eating and breaking down dead and decaying plants and animals. Finally, because of their short lifespan and ability to produce several generations within a year, insects can be of great value to scientific research in the areas of genetics, population dynamics, and developmental biology, as well as being indicators of environmental change.

Unfortunately, insects have undesirable qualities as well. In the Democratic People's Republic of Korea (North Korea), an ongoing outbreak of pine moth (*Dendrolimus spectabilis*) has affected the growth and status of over 100,000 ha (247,000 ac) of Korean red pine (*Pinus densiflora*) (Food and Agriculture Organization of the United Nations, 2010). In North America, a defoliating insect, the gypsy moth (*Lymantria dispar*), accidentally introduced in 1869, has gradually expanded its range, and now occupies about a quarter of the larch (*Larix*), oak (*Quercus*), and poplar (*Populus*) forests on the continent (Liebhold et al., 2000). The Asian longhorned beetle (*Anoplophora glabripennis*), which feeds on the leaves, twigs, bark, phloem, and xylem of poplar and maple (*Acer*) trees (among others), is one of the major causes of forest damage in China and has recently (1996) been reported in the United States (MacLeod et al., 2002). Finally, the hemlock wooly adelgid (*Adelges tsugae*) feeds on the phloem sap of eastern hemlock (*Tsuga canadensis*) trees, leading to the loss of needles. This insect has helped reduce the abundance of these trees in eastern North America since the mid-20th century (Weckel et al., 2006). In summary, under the right conditions some insects can create widespread disturbances in forested ecosystems through the defoliation or death of trees (Edmonds et al., 2011).

When we think of harmful insects that affect human health, we typically think of insects such as mosquitoes that may carry malaria and dengue fever or of deer ticks (*Ixodes scapularis*) that may carry Lyme disease. We may also think of the harm that occurs to other animals, both domesticated and wild, from direct injuries such as bites or stings, or from transmitted diseases. Here, we will focus on insects that adversely affect growing plants, as well as the services provided by forested ecosystems. As with diseases, damage from insect outbreaks can transition a forest to secondary successional processes if most of the trees in a given area are significantly affected. Depending on the tolerance of trees to insect outbreaks, individual tree death can occur, which may reduce the density and species composition of a forested area.

Given the plethora of insect species that exist in the world, it is difficult to justifiably discuss their structure, function, and development in a small section of a book. Briefly, however, insects have heads that contain eyes, antennae, and mouth parts (Fig. 14.40). The shape of an insect's head is affected by its feeding mechanism. The compound eye structure can, in some species, have up to 30,000 sensory units, which can detect movement much faster than the human eye can (Coulson and Witter, 1984). Antennae are primarily sensory in nature and can be used for detecting features through touch as well as through odor, temperature, humidity, and sound. Mouth parts can range from those that chew to those that are specialized for piercing and sucking. Insects also have a thorax, legs, abdomen, exoskeleton, and in some cases, wings. The exoskeleton is the hard-outer shell (nonliving layer) of an insect, in contrast with the endoskeleton of humans and other mammals. As with mammalian organisms, insects have circulatory, digestive, endocrine, excretory, nervous, reproductive,

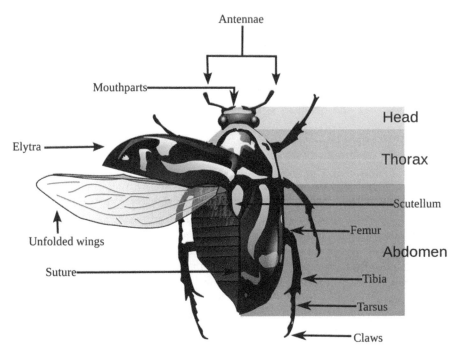

FIGURE 14.40 Morphology of a beetle. *Nicholas Caffarilla, through Wikimedia Commons.*
Image Link: https://commons.wikimedia.org/wiki/File:Fiddler_beetle_morphology_diagram.svg
License Link: https://creativecommons.org/licenses/by-sa/3.0/deed.en

and respiratory systems. Insects can secrete hormones internally to maintain the effective operation of their internal organs and can also secrete pheromones (externally) to influence the behavior of other individuals of the same species.

Generally, insects begin life as an egg, the form of which varies greatly by species. The eggs are sometimes found in clusters and at other times as individuals. Typically, insect eggs need to be fertilized by a male to reproduce male and female individuals, although some species such as parasitic wasps can produce individuals of one sex from unfertilized eggs (Coulson and Witter, 1984). After hatching, the immature insects grow and develop within the confines of their exoskeleton. When their existing exoskeletons become an impediment to their growth, then the insects molt. Molting is the process whereby an insect sheds its exterior shell and develops a new one. Some species molt many times. Other species go through a process of metamorphosis where the insect changes form during its development. A classic example of metamorphosis is when a caterpillar forms a cocoon from which it becomes a butterfly.

An insect's wings increase its ability to survive and disperse to new environments. An insect's small size facilitates faster maturity rates and reduces the quantity of resources it needs to survive. Further, the diverse feeding habits of insects allow them to utilize a variety of food sources. Some insects can enter a stage of quiescence, i.e., a simple form of dormancy, while others can enter a stage of diapause, or arrested development; such tactics are used to avoid adverse conditions such as persistent temperature extremes. Some also use rocks, leaf litter, soil, and live plant parts for shelter, while others develop protective color patterns to camouflage themselves from predators. Some insects employ a strategy of producing a large number of eggs, many of which can hatch and form a new generation despite losses from predators. Insects also have a variety of defensive mechanisms that can be behavioral, morphological, or chemical in nature.

The most extensive forest disturbance currently caused by insects may be the outbreak of mountain pine beetles (*Dendroctonus ponderosae*) (Fig. 14.41) in western North America, which has affected several million hectares of pine species (*Pinus albicaulis, Pinus contorta, Pinus monticola,* and *Pinus ponderosa*) (Food and Agriculture Organization of the United Nations, 2010). In Europe, the main forest disturbance agent is the European spruce bark beetle (*Ips typographus*) (Fahse and Heurich, 2011). Arguably, the most important insect related to forest disturbances in the southern United States is the southern pine beetle (*Dendroctonus frontalis*) (Payne, 2002). The southern pine beetle is one of 12 American species of *Dendroctonus* and can be found not only across the southern states but also in Mexico

FIGURE 14.41 Coniferous forest in Oregon, United States, damaged by a mountain pine beetle outbreak. *Dave Powell, U.S. Department of Agriculture, Forest Service, through Bugwood.org.*

and Central America. It is an aggressive bark beetle that, in large enough populations, can kill healthy trees and produce outbreaks lasting two to three years (Payne, 2002). Initially, *pioneer* beetles must discriminate between host trees by boring through the bark and feeding on the phloem (Coulson, 2002; Payne, 2002). Once inside the tree, the beetle lays eggs that produce the next generation, which can go on to produce another generation during a single growing season. This cycle can instigate a tree-wide attack that may be effective in overcoming or killing the host tree. The life stage of the southern pine beetle begins with an egg, and moves through the larva, pupa, and adult stages over 26 to 54 days. As a result, three generations of the beetle can be produced in one growing season. Female southern pine bark beetles are monogamous and mate when the resin flow stops in a tree. Females produce a distinctive S-shaped tunnel gallery (Fig. 14.42), active portions of which are kept free of debris by both male and female beetles. Forest practices such as thinning, prescribed burning, and replanting with tree species more resistant to the southern pine beetle have been promoted to influence beetle population dynamics (Rossi et al., 2011), although final harvests (clearcutting) have also been employed to stop the progress of beetle infestations. While a few direct measures for controlling southern pine beetles exist, techniques such as salvage and utilization of infested trees seem to be the best approaches (Billing, 2002). Other measures for directly combating southern pine beetles include the use of insecticides and the development of synthetic pheromones. Silvicultural guidelines, such as maintaining healthy stands, promoting individual tree resistance, removing high-risk trees, maintaining proper stand density, minimizing logging damage, favoring beetle-resistant species, regulating age classes by shortening the rotation age, and protecting the site, may also help (Belanger, 2002).

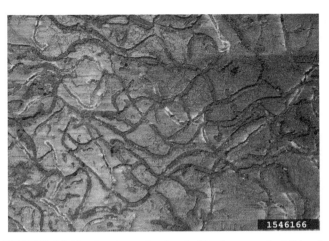

FIGURE 14.42 Southern pine beetle (*Dendroctonus frontalis*) gallery. *Roger Anderson, Duke University, through Bugwood.org.*

The behavior of insect-related disturbances such as those caused by the southern pine beetle can vary by season, with dispersing pioneers attacking potential hosts before slowing down during the winter months. Factors that draw pioneers to a stand of trees include forest metrics (e.g., average age, diameter, diseases, height, radial growth, species composition, tree density, and the presence of other insects), site factors (e.g., land form, site index, soil chemical properties, soil texture, and water regime), and weather-related variables (Hicks, 2002). For instance, air temperatures may directly affect insect survival by affecting developmental rates and progress through the life cycle (Hicks, 2002; Payne, 2002), and dense forests may be associated with low tree vigor, thus increasing vulnerability to infestation as chemicals to reduce infestation are a biologically expensive chemical for a tree to produce. For example, in coastal plain forests of the southern United States, southern pine beetle infestations often occur in overstocked pine forests

with slower growth rates. The presence of other diseases such as annosus root rot (caused by the *Heterobasidion annosum* fungus) may weaken a tree and make it vulnerable to insect infestation, although evidence to the contrary has also been noted (Erbilgin and Raffa, 2002). However, low tree vigor or low radial growth and stress from other forest disturbances or diseases might increase the risk of insect infestation.

14.9 Disturbances created by humans

Humans can disturb forests in a number of ways, from developmental actions to the harvesting of forest products. A transition from forest to nonforest use is perhaps the most important human impact on the land (Fig. 14.43). Manhattan Island and the surrounding areas of New York City are examples of locations that once supported northern hardwood forests, but that now support an array of human infrastructure mainly consisting of roads and buildings. In these areas, we would not expect to renew a forest within our lifetime; however, some beneficial aspects of forests can be obtained through urban forestry practices. In many of these areas, we would expect processes similar to primary succession to occur, depending on the level of soil removal or topographic change that had previously occurred. In areas where we expect no transition from forest to nonforest uses, forest harvesting (Fig. 14.44) is perhaps the most important anthropogenic disturbance, and here we would expect secondary succession processes to dominate. Periodic disturbances occurring across a landscape can exert a major influence on forest composition and forest dynamics. As an example of the extent of recent human activity, during a 16-year period at the end of the 20th century, over 25% of a 168,000 ha

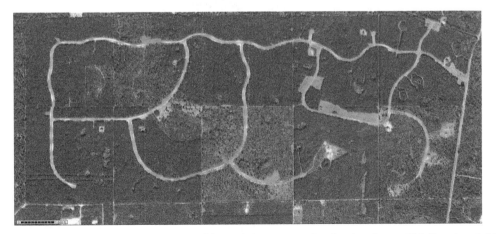

FIGURE 14.43 Forestland near Tallahassee, Florida, United States, being converted to developed uses. *U.S. Department of Agriculture, Natural Resources Conservation Service (2020).*

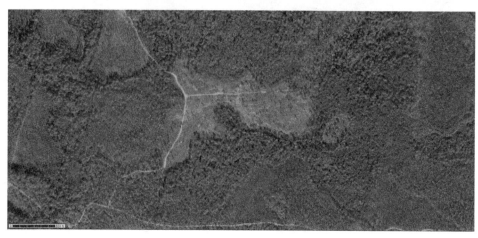

FIGURE 14.44 Forest harvesting disturbances near Tallahassee, Florida, United States. *U.S. Department of Agriculture, Natural Resources Conservation Service (2020).*

(415,137 ac) forested landscape in the northeastern United States was harvested in one manner or another, and the spatial pattern of harvesting operations seemed random with regard to most physical, biological, or cultural factors (Kittredge et al., 2003). Changes in the management of land have been postulated to facilitate the dispersal, movement, and occurrence of other disturbances. For example, changes in the management of southern United States forests over the 20th century have arguably increased the connectivity of landscape features and facilitated southern pine beetle outbreaks and the spread of fusiform rust (Perkins and Matlack, 2002). Further, forest harvesting activities have been noted to influence the risk of other disturbances occurring, such as landslides and wind damage (Tang et al., 1997).

Summary

Disturbances within a forested landscape can play a critical role in forest dynamics and the types of ecosystem services that can be provided. The capacity of forests to persist over time are based on both current physical conditions and historical conditions that change due to environmental disturbances (Iglesias and Whitlock, 2020). Studies regarding resistence and resilience to disturbance (e.g., Goode et al., 2020) are of great interest to researchers. The type, magnitude, and frequency of disturbance may lead to different successional stages. For instance, tornadoes can violently break individual trees, but the overall damage to a forested ecosystem is usually much less than damage caused by the passage of a strong hurricane. Forest fires, whether initiated by lightning strikes or carelessly thrown cigarettes, can cause extensive damage to forested ecosystems. Fortunately, many natural systems require periodic fire for the successful regeneration of certain forest vegetation. Further, volcanic eruptions can cause much more extensive damage than that caused by either wildfires or windstorms. Volcanic eruptions can level existing vegetation and bury it in molten lava or ash fallout. Despite the destruction that volcanic eruptions can bring to a forested landscape in the short term, over long periods volcanic activity can initiate the development of new soils and boost the productivity of current soils through the input of new nutrients to the ecosystem. Less dramatic, although still quite important, are disturbances caused by diseases and insects. Diseases can affect both plants and animals, but our focus in this book is on plant diseases. In the last century, diseases such as American chestnut blight have significantly transformed the composition of forests of the eastern United States. Disturbances created by forest insects can lead to widespread defoliated forests but can also play an important role in the transmission of biotic pathogens into forested vegetation. Unlike some of the other forest disturbances, insects are an important source of food for many bird and bat species, and they also assist in the breakdown of organic matter, an important step in cycling nitrogen, carbon, and sulfur through the ecosystem. Human-caused disturbances, either through development or harvesting activities, can significantly alter the ecological relationships from those governing a wilder environment.

We have discussed some of the major forest disturbances in this chapter. Other disturbances such as snow accumulation, grazing, and shifting cultivation may also be important in some areas of the world (Iqbal, 2019), and are not to be diminished in importance by their absence here. Although forest disturbances can greatly disrupt the ecosystem services that a forested landscape provides, benefits from these disturbances can also accrue. For example, flooding associated with hurricanes may provide valuable nutrient inputs to mangrove forest systems. However, forest and natural resource managers should prepare contingency plans for each of the potential disturbances that can influence forest dynamics of the land that they manage.

Questions

(1) *Wildfires.* Over the course of a week, monitor the news media, identify the number of forest fires that you learn of, and determine where they are occurring. Using a map of your area of the world, make a note of where the wildfires occur. Are the fires clustered or do they seem randomly located? Select one particular fire and find specific information about it through the Internet. Write a one- or two-page essay describing the event (when it began, how it began, how it has spread, where it is spreading to, etc.).

(2) *Windstorms.* Develop a one-page memorandum that describes the principal type of wind-related risk to the forests of your area of the world. If you perceive that the risk of wind damage is low, then describe why this is so. If the risk of wind damage accompanies or facilitates other types of disturbances, describe these interactions as well.

(3) *Volcanoes.* It has been suggested that volcanoes can affect forests hundreds of kilometers away from the direct impact zone. In one paragraph, describe the potential impacts to these forests. In a small group of four or five students, describe your impressions of the influence of volcanoes on forest development.

(4) *Floods.* Perform some investigative work on the impact of floods on the dynamics of forests in an area of the world of interest to you. In a short PowerPoint presentation, describe the types of floods that are common there and the main interactions of forests with floods. Further, describe your impression of the future vulnerability of these forests to flooding events.

(5) *Insects and disease.* Select one of the more extensive insect and disease problems of late, such as the emerald ash borer, the hemlock woolly adelgid, or annosus root rot. Write a one- to two-page fact sheet on how the insect or disease can affect forest ecosystem services. In this fact sheet, describe the current opinion of the threat and future impacts it may have if left unattended.

References

Achim, A., Ruel, J.-C., Gardiner, B.A., Laflamme, G., Meunier, S., 2005. Modelling the vulnerability of balsam fir forests to wind damage. Forest Ecology and Management 204, 35—50.

Attiwill, P.M., 1994. The disturbance of forest ecosystems: the ecological basis for conservative management. Forest Ecology and Management 63, 247—300.

Barry, P.J., Anderson, R.L., Swain Sr., K.M., 1982. How to Evaluate and Manage Storm-Damaged Forest Areas. U.S. Department of Agriculture, Forest Service, Southeastern Area, Forest Pest Management, Asheville, NC. Forestry Report SA-FR-20. 15 p.

Bebi, P., Kulakowski, D., Rixen, C., 2009. Snow avalanche disturbances in forest ecosystems—state of research and implications for management. Forest Ecology and Management 257, 1883—1892.

Belanger, R.P., 2002. Silvicultural guidelines for reducing losses to the southern pine beetle. In: Thatcher, R.C., Searcy, J.L., Coster, J.E., Hertel, G.D. (Eds.), The Southern Pine Beetle. U.S. Department of Agriculture, Forest Service, Expanded Southern Pine Beetle Research and Applications Program, Science and Education Administration, Pineville, LA. http://www.barkbeetles.org/spb/spbbook/Chapt9.html (accessed 26.03.20).

Bettinger, P., Merry, K.L., Grebner, D.L., 2010. Two views of the impact of hurricanes on the forests of the southern United States. Southeastern Geographer 50 (3), 291—304.

Bettinger, P., Merry, K.L., Hepinstall, J., 2009. Average tropical cyclone intensity along the Georgia, Alabama, Mississippi, and north Florida coasts. Southeastern Geographer 49 (1), 49—66.

Billing, R.F., 2002. Direct control. In: Thatcher, R.C., Searcy, J.L., Coster, J.E., Hertel, G.D. (Eds.), The Southern Pine Beetle. U.S. Department of Agriculture, Forest Service, Expanded Southern Pine Beetle Research and Applications Program, Science and Education Administration, Pineville, LA. http://www.barkbeetles.org/spb/spbbook/Chapt10.html (accessed 26.03.20).

Blood, E.R., Anderson, P., Smith, P.A., Nybro, C., Ginsberg, K.A., 1991. Effects of Hurricane Hugo on coastal soil solution chemistry in South Carolina. Biotropica 23, 348—355 (4a).

Bourke, P.M., 1964. Emergence of potato blight, 1843—46. Nature 203, 805—808.

Brady, N.C., Weil, R.R., 1999. The Nature and Properties of Soils, twelfth ed. Prentice Hall, Inc., Upper Saddle River, NJ, 881 p.

Bragg, D.C., Shelton, M.G., Zeide, B., 2003. Impacts and management implications of ice storms on forests in the southern United States. Forest Ecology and Management 186, 99—123.

Broad, K., Leiserowitz, A., Weinkle, J., Steketee, M., 2007. Misinterpretations of the "cone of uncertainty" in Florida during the 2004 hurricane season. Bulletin of the American Meteorological Society 88, 651—667.

Brokaw, N.V.L., Walker, L.R., 1991. Summary of the effects of Caribbean hurricanes on vegetation. Biotropica 23, 442—447 (4a).

Brose, P.H., 2008. Photo Guide for Estimating Fuel Loading and Fire Behavior in Mixed-Oak Forests of the Mid-Atlantic Region. U.S. Department of Agriculture, Forest Service, Northern Research Station, Newtown Square, PA. General Technical Report NRS-45. 104 p.

Brown, L.B., Allen-Diaz, B., 2009. Forest stand dynamics and sudden oak death: mortality in mixed-evergreen forests dominated by coast live oak. Forest Ecology and Management 257, 1271—1280.

Burke, K.L., 2011. The effects of logging and disease on American chestnut. Forest Ecology and Management 261, 1027—1033.

Burns, R.M., Honkala, B.H., 1990. Silvics of North America, vol. 2. Hardwoods. U.S. Department of Agriculture, Forest Service, Washington, D.C. Agriculture Handbook 654, 877 p.

Cambours, M.A., Heinsoo, K., Granhall, U., Nejad, P., 2006. Frost related dieback in Estonian energy plantations of willows in relation to fertilisation and pathogenic bacteria. Biomass and Bioenergy 30, 220—230.

Castañeda-Moya, E., Rivera-Monroy, V.H., Chambers, R.M., Zhao, X., Lamb-Wotton, L., Gorsky, A., Gaiser, E.E., Troxler, T.G., Kominoski, J.S., Hiatt, M., 2020. Hurricanes fertilize mangrove forests in the Gulf of Mexico (Florida Everglades, USA). Proceedings of the National Academy of Sciences United States of America 117 (9), 4831—4841.

Conner, W.H., Day Jr., J.W., Baumann, R.H., Randall, J.M., 1989. Influence of hurricanes on coastal ecosystems along the northern Gulf of Mexico. Wetlands Ecology and Management 1 (1), 45—56.

Coulson, R.N., 2002. Population dynamics. In: Thatcher, R.C., Searcy, J.L., Coster, J.E., Hertel, G.D. (Eds.), The Southern Pine Beetle. U.S. Department of Agriculture, Forest Service, Expanded Southern Pine Beetle Research and Applications Program, Science and Education Administration, Pineville, LA. http://www.barkbeetles.org/spb/spbbook/Chapt5.html (accessed 26.03.20).

Coulson, R.N., Witter, J.A., 1984. Forest Entomology: Ecology and Management. John Wiley & Sons, New York, 669 p.

Edmonds, R.L., Agee, J.K., Gara, R.I., 2011. Forest Health and Protection, second ed. Waveland Press, Inc., Long Grove, IL, 667 p.

Erbilgin, N., Raffa, K.F., 2002. Association of declining red pine stands with reduced populations of bark beetle predators, seasonal increases in root colonizing insects, and incidence of root pathogens. Forest Ecology and Management 164, 221—236.

Fahse, L., Heurich, M., 2011. Simulation and analysis of outbreaks of bark beetle infestations and their management at the stand level. Ecological Modelling 222, 1833—1846.

Food and Agriculture Organization of the United Nations, 2005. Global Forest Resources Assessment 2005. Food and Agriculture Organization of the United Nations, Rome, Italy. FAO Forestry Paper 147. 320 p.

Food and Agriculture Organization of the United Nations, 2010. Global Forest Resources Assessment 2010. Food and Agriculture Organization of the United Nations, Rome, Italy. FAO Forestry Paper 163. 340 p.

Francis, E., Hubbard, J., Tanglao, L., James, M.S., 2011. Storms, Tornadoes Leave Dozens Dead in Alabama, Mississippi, Georgia, and Tennessee. ABC News, New York. http://abcnews.go.com/US/massive-tornado-hits-alabama-storms-leave-16-dead/story?id=13465028 (accessed 26.03.20).

Franke, M.A., 2000. Yellowstone in the Afterglow, Lessons from the Fires. U.S. National Park Service, Yellowstone Center for Resources, Yellowstone National Park, Mammoth Hot Springs, WY. YCR—NR—2000—03.

Fritz, H.M., Blount, C., Sokoloski, R., Singleton, J., Fuggle, A., McAdoo, B.G., Moore, A., Grass, C., Tate, B., 2007. Hurricane Katrina storm surge distribution and field observations on the Mississippi Barrier Island. Estuarine, Coastal and Shelf Science 74, 12—20.

Gardner, L.R., Michener, W.K., Blood, E.R., Williams, T.M., Lipscomb, D.J., Jefferson, W.H., 1991. Ecological impact of Hurricane Hugo—salinization of a coastal forest. Journal of Coastal Research 8 (SI), 301—317.

Gaylor, H.P., 1974. Wildfires Prevention and Control. Robert J. Brady Company, Bowie, MD, 319 p.

Gerlach, T.M., McGee, K.A., 1994. Total sulfur dioxide emissions and pre-eruption vapor-saturated magma at Mount St. Helens, 1980-1988. Geophysical Research Letters 21, 2833—2836.

Goode, J.D., Barefoot, C.R., Hart, J.L., Dey, D.C., 2020. Disturbance history, species diversity, and structural complexity of a temperate deciduous forest. Journal of Forestry Research 31, 397—414.

Hicks, R.R., 2002. Climatic, site, and stand factors. In: Thatcher, R.C., Searcy, J.L., Coster, J.E., Hertel, G.D. (Eds.), The Southern Pine Beetle. U.S. Department of Agriculture, Forest Service, Expanded Southern Pine Beetle Research and Applications Program, Science and Education Administration, Pineville, LA. http://www.barkbeetles.org/spb/spbbook/Chapt4.html (accessed 26.03.20).

Iglesias, V., Whitlock, C., 2020. If the trees burn, is the forest lost? Past dynamics in temperate forests help inform management strategies. Philosophical Transactions of the Royal Society B 375. Article 20190115.

Iqbal, J., 2019. Forest disturbance and degradation in western Himalayan moist temperate forest of Pakistan. Asian Journal of Agriculture and Biology 7, 538—547.

Irland, L.C., 2000. Ice storms and forest impacts. The Science of the Total Environment 262, 231—242.

Kittredge Jr., D.B., Finley, A.O., Foster, D.R., 2003. Timber harvesting as ongoing disturbance in a landscape of diverse ownership. Forest Ecology and Management 180, 425—442.

Kraft, L.S., Crowe, T.R., Buckley, D.S., Nauertz, E.A., Zasada, J.C., 2004. Effects of harvesting and deer browsing on attributes of understory plants in northern hardwood forests, Upper Michigan, USA. Forest Ecology and Management 199, 219—230.

Leahey, D.M., Hansen, M.C., 1987. Observations of winds above and below a forest canopy located near a clearing. Atmospheric Environment 21, 1227—1229.

Liebhold, A., Elkinton, J., Williams, D., R.-Muzika, M., 2000. What causes outbreaks of the gypsy moth in North America? Population Ecology 42, 257—266.

Lilleskov, E.A., Hobbie, E.A., Horton, T.R., 2011. Conservation of ectomycorrhizal fungi: exploring the linkages between functional and taxonomic responses to anthropogenic N deposition. Fungal Ecology 4, 174—183.

Lu, P., Sinclair, R.W., Boult, T.J., Blake, S.G., 2005. Seedling survival of *Pinus strobus* and its interspecific hybrids after artificial inoculation of *Cronartium ribicola*. Forest Ecology and Management 214, 344—357.

MacLeod, A., Evans, H.F., Baker, H.A.R., 2002. An analysis of pest risk from an Asian longhorn beetle (*Anoplophora glabripennis*) to hardwood trees in the European community. Crop Protection 21, 635—645.

Malmstrom, C.M., Melcher, U., Bosque-Pérez, N.A., 2011. The expanding field of plant virus ecology: historical foundations, knowledge gaps, and research directions. Virus Research 159, 84—94.

Manion, P.D., 1981. Tree Disease Concepts. Prentice-Hall, Inc., Englewood Cliffs, NJ, 399 p.

Manning, J.L., Baltzer, J.L., 2011. Impacts of black bear baiting on Acadian forest dynamics—an indirect edge effect? Forest Ecology and Management 262, 838—844.

Matusick, G., Eckhardt, L.G., Somers, G.L., 2010. Susceptibility of longleaf pine roots to infection and damage by four root-inhabiting ophiostomatoid fungi. Forest Ecology and Management 260, 2189—2195.

Mayer, H., 1987. Wind-induced tree sways. Trees 1, 195—206.

Mergen, F., 1954. Mechanical aspects of wind-breakage and wind firmness. Journal of Forestry 52 (2), 119—125.

Merry, K.L., Bettinger, P., Grebner, D.L., Hepinstall, J., 2010. Perceptions of wind damage in Mississippi forests. Southern Journal of Applied Forestry 34 (3), 124–130.

Nienhaus, F., Castello, J.D., 1989. Viruses in forest trees. Annual Review of Phytopathology 27, 165–186.

Nieuwenhuis, M., 2010. Terminology of Forest Management, Terms and Definitions in English, second revised ed. International Union of Forest Research Organizations, Vienna, Austria. IUFRO World Series Volume 9-en.

National Oceanic and Atmospheric Administration National Weather Service, 2011. The Online Tornado Frequently Asked Questions. U.S. Department of Commerce, National Oceanic and Atmospheric Administration, National Weather Service, Silver Spring, MD. http://www. spc.ncep.noaa.gov/faq/tornado/ (accessed 26.03.20).

National Oceanic and Atmospheric Administration National Weather Service, 2018. Understanding Lightning Science. U.S. Department of Commerce, National Oceanic and Atmospheric Administration, National Weather Service, Silver Spring, MD. www.weather.gov/ safety/lightning-science-overview (accessed 31.03.20).

Noshiro, S., Suzuki, M., Tsuji, S., 2002. Three buried forests of the last glacial stage and middle holocene at ooyazawa on northern Honshu Island of Japan. Review of Palaeobotany and Palynology 122, 155–169.

Özgenel, L., 2008. A tale of two cities: in search for ancient Pompeii and Herculaneum. METU Journal of the Faculty of Architecture 25 (1), 1–25.

Payne, T.L., 2002. Life history and habits. In: Thatcher, R.C., Searcy, J.L., Coster, J.E., Hertel, G.D. (Eds.), The Southern Pine Beetle. U.S. Department of Agriculture, Forest Service, Expanded Southern Pine Beetle Research and Applications Program, Science and Education Administration, Pineville, LA. http://www.barkbeetles. org/spb/spbbook/Chapt2.html (accessed 26.03.20).

Perkins, T.E., Matlack, G.R., 2002. Human-generated pattern in commercial forests of southern Mississippi and consequences for the spread of pests and pathogens. Forest Ecology and Management 157, 143–154.

Pole, M., 2001. Repeated flood events and fossil forests at Curio Bay (middle Jurassic), New Zealand. Sedimentary Geology 144, 223–242.

Press, R., Siever, R., 1982. Earth, third ed. W.H. Freeman and Company, San Francisco, CA, 613 p.

Pyne, S.J., 1984. Introduction to Wildland Fire: Fire Management in the United States. John Wiley & Sons, Inc., New York, 455 p.

Ristaino, J.B., 2002. Tracking historic migrations of the Irish potato famine pathogen, Phytophthora infestans. Microbes and Infection 4, 1369–1377.

Rodríguez-González, P.M., Stella, J.C., Campelo, F., Ferreira, M.T., Albuquerque, A., 2010. Subsidy or stress? Tree structure and growth in wetland forests along a hydrological gradient in Southern Europe. Forest Ecology and Management 259, 2015–2025.

Roriz, M., Santos, C., Vasconcelos, M.W., 2011. Population dynamics of bacteria associated with different strains of the pine wood nematode Bursaphelenchus xylophilus after inoculation in maritime pine (Pinus pinaster). Experimental Parasitology 128, 357–364.

Rossi, F.J., Carter, D.R., Alavalapati, J.R.R., Nowak, J.T., 2011. Assessing landowner preferences for forest management practices to prevent the southern pine beetle: an attribute-based choice experiment approach. Forest Policy and Economics 13, 234–241.

Salzer, M.W., Hughes, M.K., 2007. Bristlecone pine tree rings and volcanic eruptions over the last 5000 years. Quaternary Research 67, 57–68.

Samuels, E.B., 2011. Mount St. Helens. Destination Guides, New York, NY. http://dguides.com/portland/features/day-trips-from-port land/mount-st-helens/ (accessed 26.03.20).

Smith, W.H., 1990. Air Pollution and Forests Interaction between Air Contaminants and Forest Ecosystems, second ed. Springer-Verlag, New York, 618 p.

Spooner, B., 2007. A Taphrina species on Quercus ilex in Britain. Field Mycology 8, 77–79.

Stebleton, A., Bunting, S., 2009. Guide for Quantifying Fuels in the Sagebrush Steppe and Juniper Woodlands of the Great Basin. U.S. Department of the Interior, Bureau of Land Management, Denver, CO. Technical Note 430. 18 p.

Tang, S.M., Franklin, J.F., Montgomery, D.R., 1997. Forest harvest patterns and landscape disturbance processes. Landscape Ecology 12, 349–363.

Temple, B., Pines, P.A., Hintz, W.E., 2006. A nine-year genetic survey of the causal agent of Dutch elm disease, Ophiostoma novo-ulmi in Winnipeg, Canada. Mycological Research 110, 594–600.

Tillamook State Forest, 2008. Welcome to Tillamook State Forest. Oregon Department of Forestry, Tillamook Forest Center, Tillamook, OR. http://www.tillamookforestcenter.com/resources/Tillamook WelcomeGuide.pdf (accessed 26.03.20).

Tilling, R.I., 1999. Volcanoes. U.S. Department of the Interior, Geological Survey, Reston, VA. http://pubs.usgs.gov/gip/volc/text. html (accessed 26.03.20).

Tilling, R.I., Topinka, L., Swanson, D.A., 1990. Report: Eruptions of Mount St. Helens: Past, Present, and Future. U.S. Department of the Interior, Geological Survey, Reston, VA. U.S. Geological Survey Special Interest Publication. http://vulcan.wr.usgs.gov/Volcanoes/ MSH/Publications/MSHPPF/MSH_past_present_futu re.html (accessed 26.03.20).

Tsai, J.-L., Chen, C.-L., Tsuang, B.-J., Kuo, P.-H., Tseng, K.-H., Hsu, T.-F., Sheu, B.-H., Liu, C.-P., Hsueh, M.-T., 2010. Observation of SO_2 dry deposition velocity at a high elevation flux tower over an evergreen broadleaf forest in Central Taiwan. Atmospheric Environment 44, 1011–1019.

U.S. Department of Agriculture, 2020. Natural Resources Conservation Service. Web Soil Survey. U.S. Department of Agriculture, Natural Resources Conservation Service, Washington, D.C. https:// websoilsurvey.sc.egov.usda.gov/App/HomePage.htm (accessed 01.04.20).

U.S. Geological Survey, 2020. Mount St. Helens National Volcanic Monument. U.S. Department of the Interior, Geological Survey, Reston, VA. www.usgs.gov/volcanoes/mount-st-helens/mount-st-helens-national-volcanic-monument?qt-science_support_page_ related_con=2#qt-science_support_page_related_con (accessed 31.03.20).

Viglietti, D., Letey, S., Motta, R., Maggioni, M., Freppaz, M., 2010. Snow avalanche release in forest ecosystems: a case study in the Aosta Valley Region (NW-Italy). Cold Regions Science and Technology 64, 167–173.

Wang, H., Ouyang, Z., Chen, W., Wang, X., Zheng, H., Ren, Y., 2011. Water, heat, and airborne pollutants effects on transpiration of urban trees. Environmental Pollution 159, 2127–2137.

Watson, G.W., Kelsey, P., 2006. The impact of soil compaction on soil aeration and fine root density of Quercus palustris. Urban Forestry and Urban Greening 4, 69–74.

Weckel, M., Tirpak, J.M., Nagy, C., Christie, R., 2006. Structural and compositional change in an old-growth eastern hemlock Tsuga canadensis forest, 1965–2004. Forest Ecology and Management 231, 114–118.

Witze, A., April 12, 2010. Why volcanic eruptions can spark lightning. Wired Science. http://www.wired.com/wiredscience/2010/04/ electric-dust-grains/ (accessed 26.03.20).

Yin, Y., Wu, Y., Bartell, S.M., Cosgriff, R., 2009. Patterns of forest succession and impacts of flood in the Upper Mississippi River floodplain ecosystem. Ecological Complexity 6, 463–472.

Yokoyama, S., Maeji, I., Ueda, T., Ando, M., Shibata, E., 2001. Impact of bark stripping by sika deer, Cervus nippon, on subalpine coniferous forests in central Japan. Forest Ecology and Management 140, 93–99.

Forest policies are tightly integrated with the management of natural resources, particularly when public or communal lands are being considered. In the United States, political controversy over the use and management of natural resources was instrumental in the development of public lands and has shaped the laws governing their administration and disposition since the mid-19th century (Muhn and Stuart, 1988). Policies arise from controversies, and because the array of controversies regarding the management of natural resources may change over time, real or perceived controversies will continue to shape the management of natural resources. Upon completion of this chapter, readers should be able to understand

- what constitutes a policy;
- the role of national level forest policies in the daily operational management of forests;
- the role of state and local policies in influencing forest management activities;
- the role of voluntary certification programs in forest management organizations and how they relate to forest policies;
- that some organizations impose policies of their own that influence forest management;

- the influence of the Kyoto Protocol on global forest management; and
- the concepts of carbon sequestration trade and of permanence, additionality, and leakage.

A forest landowner or a forest management organization is responsible for the plans, policies, and oversight of practices on the land they own or manage for others. Forests and natural resources may be used, conserved, developed, or protected to meet the objectives of the landowner. The actions employed should be designed in a manner consistent or compatible with economic, ecological, and social norms. In many cases, the actions employed must adhere to specific guidelines, as specified in various regulations or laws. Forest policies guide these actions.

15.1 What is a forest policy?

A *policy* is a set of general rules and acceptable protocols and procedures that should be followed. Often they are developed by a governmental agency, institution, group, or individual in order to achieve certain desirable outcomes (Schwarz et al., 1976). A *forest policy* is a purposeful course of action, or inaction, undertaken by an individual or an organization that is dealing with a concern regarding the use of forest resources (Cubbage et al., 1993). A forest policy can be as simple as *lock the gate after you leave a resource area* or as complicated as *maintaining suitable habitat for an endangered species*. Policies can be formally described through legislation, or informally implemented through budgeting processes (Cubbage and Newman, 2006). A policy may contain information regarding which actions that must be performed, when actions must be performed, why the actions must be performed, and to whom the policy applies. Policies developed by governmental entities and their associated agencies, a legislature, or courts of law are considered public policies. Forests provide an extremely broad range of products and fulfill numerous functions (Fig. 15.1), and because each forest cannot provide all of the products, services, and functions desired by each of the forest's potential stakeholders, conflicts can arise. Conflicts thus lead to policies regarding the management of the forest resources.

Policies are developed to address real or perceived problems in the management of lands or businesses. The policy development process usually consists of several stages that begin with the identification of the problem (Cubbage et al., 1993). Other key components of policies include the identification of the policy's authority and the determination of those required to follow and adhere to the policy. As suggested in Chapter 13, governments often form policies and regulatory frameworks to address market failures. Government policies are developed to ensure that important social objectives are addressed. A market failure represents the inability of a free market to allocate resources efficiently. With regard to public goods, when these are difficult to value or when there are externalities in production and consumption, they cause problems that prevent markets from working well and can result in market failures. Therefore, public policy development actions may be necessary based on market failures or other perceived social problems. Private organizations or individuals can also perceive management problems, which can lead to the development of policies that are meant to guide the actions of people working within their organization or working upon their land. Once alternative policies are assessed, a policy is adopted and implemented by the appropriate organization, group, or individual. Some policies may then be monitored and evaluated in order to improve the ongoing policy formulation process, often called adaptive management.

15.2 International forest policies

Chapter 1 provided a brief overview of forests in the current world political and environmental context. The 1992 Rio Earth Summit was discussed, which galvanized international cooperation on forest sustainability issues. In addition, some follow-up initiatives, such as the Montréal Process and the United Nations Forum on Forests, were discussed. These are all examples of international forest policies. Whether or not they are binding at the national level depends on the policy's wording and the responsibilities accepted by a national government through international agreements. This section provides an overview of other recent conventions, processes, and organizations that address forests and natural resources on the international stage.

Concerns regarding the possible damaging effects of global climate change led to the formation of the United Nations Framework Convention on Climate Change (UNFCCC) in 1994 (United Nations Framework Convention on Climate Change, 2020a). The purpose of this convention was to establish a framework for governments to use when addressing threats posed by global climate change. Signatory governments (i.e., those that agree in writing) are obliged to collect and share information on greenhouse gas emissions, develop approaches to reduce those emissions, and prepare adaptive strategies that address the negative impacts of climate change. Because this convention was meant to act as a framework for further work, specific action targets and achievement dates were not directly identified and, thus, the convention encouraged individual countries to reduce their greenhouse gas emissions through individual efforts (Holmgren, 2010). Specific requirements for the reduction of greenhouse gases were introduced through the Kyoto Protocol to the UNFCCC. These requirements essentially imposed mandatory greenhouse gas emission targets for industrialized countries.

FIGURE 15.1 View from a recreational trail below the summit of Polica, Poland, with a telecommunications tower in the distance. *Rafał Kozubek, through Wikimedia Commons.*
Image Link: https://commons.wikimedia.org/wiki/File:Okraglica_spod_Policy.jpg
License Link: https://creativecommons.org/licenses/by-sa/4.0/

The Kyoto Protocol, which was adopted in 1997 and enforced in 2005, regulates the emissions of six greenhouse gases: carbon dioxide (CO_2), methane (CH_4), nitrous oxide (NO_2), perfluorocarbons, hydrofluorocarbons, and sulfur hexafluoride (SF_6). The signatory industrialized nations (listed in Annex I of the Convention) are required to reduce their emissions by approximately 18% below their 1990 levels during the period 2013 through 2020 (United Nations Framework Convention on Climate Change, 2020b). The Protocol imposes a heavier burden on industrialized nations, recognizing that they may be primarily responsible for global climate change. While countries are expected to meet most of their emission reduction targets nationally, in an effort to reduce compliance costs and increase the efficiency of mitigation measures, the Protocol allows some flexibility by accepting three market-based mechanisms: emission trading (carbon markets); the clean development mechanism (CDM); and joint implementation (JI). In brief, the European Union Emission Trading Scheme (EU ETS) serves as a platform to implement emission trading and the CDM allows industrialized countries to develop emission-reducing projects in developing countries that are using the Protocol. The JI mechanism allows joint emission reduction projects to be developed between countries with reduction targets. If these two mechanisms (CDM and JI) yield emission reductions, they can be credited under the EU ETS.

Forests are relevant to this process because they store large amounts of carbon as biomass (Fig. 15.2) and in soils and, thus, they serve as sinks or storage areas of carbon. The Kyoto Protocol requires that countries report greenhouse gas balances associated within sectors related to forestry, other land uses, and land-use changes. Land-use changes, such as deforestation, will result in the release of carbon into the atmosphere, while changes such as afforestation will result in the capture of additional carbon. Through various methods of carbon accounting, it is now known that deforestation accounts for a substantial portion of global greenhouse gas emissions (Food and Agriculture Organization of the United Nations, 2010a,b). Concerns over land-use changes have led to a number of discussions regarding the policy instruments human society can develop to avoid deforestation and the resulting greenhouse gas emissions (e.g., Barua et al., 2012). Some of the resulting efforts have been collectively called Reduced Emissions from Deforestation and Forest Degradation (REDD) programs. Many of these programs are conducted as part of the United Nations REDD Program, as well as the World Bank's Carbon Partnership Facility (Holmgren, 2010). Further implications of the Kyoto Protocol for forestry and natural resource management are discussed in greater detail later in this chapter.

Enacted in 2011, the Convention on Biological Diversity is another important international agreement signed during the 1992 Rio Earth Summit (Convention on Biological Diversity, 2011). This agreement was designed to promote the conservation of the world's biodiversity, the sustainable use of its components, and the fair-sharing of benefits arising from the use of genetic resources. The convention requires its signatories to develop national plans to protect their biodiversity resources. The convention is supplemented by the Cartagena Protocol, which addresses safe handling of living modified organisms (commonly known as genetically modified organisms [GMOs]) through biotechnology to prevent negative impacts on both natural biodiversity and human health as a consequence of their introduction (Harrop and Pritchard, 2011). Another supplement to the Convention on Biological Diversity, the Nagoya Protocol, requires the fair and equitable sharing of benefits arising from biodiversity utilization, as well as access to genetic resources and relevant technologies. The Nagoya Protocol addresses equity issues and *biopiracy*, or the exploitation of natural resources in poor, yet biodiversity rich countries by developed countries, where benefits are not shared with the poor countries (Harrop and Pritchard, 2011). The Convention on Biological Diversity applies to forests worldwide, since there is much terrestrial biodiversity found in forested ecosystems. The provisions addressing various threats to biodiversity, for example, pertain to the development and deployment of genetically modified trees (which at this time is not a widespread concern), and perceived illegal logging, each of which may influence forest operations in various ways, depending on their context.

The United Nations Convention to Combat Desertification (UNCCD) (2020) was adopted to arrest land degradation in arid, semiarid, and dry areas of the world, mitigate the impact of droughts, and improve human living

FIGURE 15.2 Carbon being stored in a mixed woodland in Belhus Park, South Ockendon/Aveley, England, United Kingdom, part of the Thames Chase Community Forest. *Glyn Baker, through Wikimedia Commons. Image Link: https://commons.wikimedia.org/wiki/File:Carbon_Capture%5E_-_geograph.org.uk_-_1415890.jpg License Link: https://creativecommons.org/licenses/by-sa/2.0/deed.en*

FIGURE 15.3 Remnants of a forest in the process of conversion to an agricultural use in Zambia. *Colalife, through Wikimedia Commons. Image Link: https://commons.wikimedia.org/wiki/File:Citemene.JPG License Link: https://creativecommons.org/licenses/by-sa/3.0/deed.en*

conditions (Longjun, 2011). Desertification by its very nature is closely related to deforestation and is most severe on the continent of Africa (Fig. 15.3). The process suggested by the convention does not address the expansion of existing deserts; rather, it addresses the destruction of human-inhabited dry ecosystems and the rehabilitation of, among other ecosystems, natural desert woodlands (Longjun, 2011). Dry ecosystems cover as much as one-third of the world's land area and, as a result, they are very susceptible to overuse or misuse, which undermines the productive capacity of the land and could result in severe land degradation. Therefore, the UNCCD addresses the causes of deforestation in dry regions and attempts to develop tools to prevent deforestation through national and international initiatives.

The International Tropical Timber Agreement (ITTA) of 2006 is a commodity-centered agreement negotiated under the United Nations Conference on Trade and Development. The origin of the ITTA dates back to about 1976 when it arose from concerns over (1) the general fate of tropical forests, and (2) the notion that tropical timber trade is an important component of economic development in tropical countries. The ITTA agreement is unique because it supports trade but also emphasizes conservation and development. The reconciliation of these two seemingly incongruous issues is now the domain of the International Tropical Timber Organization (ITTO), which oversees the implementation of the agreement. The ITTO promotes international trade in tropical timber and promotes products that are legally harvested from sustainably managed forests (Palmberg and Esquinas-Alcazar, 1990). Tropical forests managed in such a manner can be viewed as tools toward achieving sustainable development, since well-managed forests can provide income and employment while protecting land from degradation and destruction (Blaser et al., 2011). As you may come to realize, the ITTA resembles, at least

to some degree, contemporary forest certification efforts. Indeed, the most recent version of the agreement (2006) focuses on tropical timber trade and sustainable forest management (SFM) and promotes timber trade that is consistent with efforts to improve forest management.

The Convention Concerning the Protection of the World Cultural and Natural Heritage has become an effective mechanism for protection of sites that are determined to be part of the world's heritage (Holmgren, 2010). In 2001, this convention developed a special program for forests under the name of the World Heritage Forest Program (United Nations Educational, Scientific and Cultural Organization, 2020a). A special agency of the United Nations, the United Nations Educational, Scientific and Cultural Organization, has developed a World Heritage list which contains land areas to be preserved due to their universal value. Forested areas on this list range from those as small as a few hectares (ha) to those as large as the almost 9 million ha (about 22.2 million acres [ac]) Lake Baikal area in the Russian Federation. Also on the list are 90 forested sites covering a total area of more than 62.7 million ha (about 154.9 million ac). Among the other sites on this list are the ancient beech forests of Germany, the Białowieża Forest in Poland—the home of the ancient wisent, or European bison (*Bison bonasus*), the Canadian Rocky Mountains parks (Fig. 15.4), the Central Amazon Conservation Complex in Brazil, the Greater Blue Mountains area in Australia, Kinabalu Park in Malaysia, Niokolo-Koba National Park in Senegal, and Yellowstone and Yosemite National Parks in the United States (United Nations Educational, Scientific and Cultural Organization, 2020b). These areas are of high ecological value and are protected by the individual countries in which they are located.

The Convention on International Trade in Endangered Species of Wild Fauna and Flora (CITES), also known as the Washington Convention, is an

FIGURE 15.4 View of Peyto Lake in Banff National Park, Alberta, Canada. *Tobias Alt, through Wikimedia Commons. Image Link: https://commons.wikimedia.org/wiki/File:Peyto_Lake-Banff_NP-Canada.jpg License Link: https://creativecommons.org/licenses/by-sa/4.0/*

international agreement that has the goal of preventing unsustainable exploitation of plant and animal species due to international trade (Holmgren, 2010). The CITES Convention was open for signing in 1973 and came into force in 1975. Wildlife trade, consisting of living organisms and products derived from them, is worth billions of dollars every year. The most visible efforts under CITES are those related to ivory and rhinoceros horn trade, in an effort to protect endangered elephants and rhinoceros in Africa and elsewhere (Bulte and van Kooten, 1999). It has been extended to many wood products, especially some tropical hardwoods that are often used in the construction of musical instruments. These and many other species reside in or are part of forested ecosystems. Today, more than 34,000 species receive some measure of protection under the CITES Convention (Convention on International Trade in Endangered Species of Wild Fauna and Flora, 2018).

15.3 National, state, and provincial policies

In both the United States and Canada, the authority to govern forests is divided among federal governments, states or provinces, and, in some cases, counties and local jurisdictions. While this section focuses on national (or federal) and state forest policies, county and local jurisdiction policies can also affect the implementation of forestry activities and be influenced by, for example, tree ordinances imposed by these governmental bodies. The U.S. federal government has a primary role in governing national forests and national parks, while in Canada much of the regulatory power rests with the provinces that own most of the public forestland (Kern et al., 1998). However, as noted in Chapter 2, the Canadian Forest Service is a research and policy organization of the government that is currently acting to promote both the sustainability of Canadian forests and the economic competitiveness of the Canadian forest sector. One clearly defined shift in public forest policy since the end of the 20th century, particularly notable in the United States, is represented by the change in emphasis from timber-oriented forest management in favor of a broad range of nontimber forest outputs.

In the early 20th century, federal action with regard to forest policy in the United States involved piecemeal approaches that addressed one problem after another. These involved the creation of the National Forest System, the public acquisition of private land, public cooperation with private landowners, and the development of comprehensive forest research programs (Clapp, 1941). Throughout the mid- to late-20th century, a number of important federal policies were enacted that influence the management of both public and private land in the United States, including the Multiple-Use Sustained

Yield Act, the Endangered Species Act, and the Clean Water Act. Federal forest policy development in the United States has been relatively static in the latter part of the 20th century and early 21st century, as limited government budgets, development interest groups, and market pressures have all contributed to incremental and limited expansion of congressional actions and agency regulations. During the same period, more policy development actions have been performed by state-level organizations, nongovernmental organizations (NGOs), and others involved in the development of international agreements (Cubbage and Newman, 2006). Some of the major national laws in the United States that affect the management of forests include the following:

- *Multiple-Use and Sustained Yield Act (1960).* Authorizes the Secretary of Agriculture to develop and administer the renewable resources on U.S. national forests for multiple uses and the sustained yield of products and services.
- *National Environmental Policy Act (1969).* Enacted to require the development of environmental impact statements outlining the actions that affect the quality of the environment.
- *Endangered Species Act (1973).* Designed to protect imperiled species (and the ecosystems on which they depend) from extinction as a consequence of human activities.
- *National Forest Management Act (1976).* Governs the management of U.S. national forests and provides guidance for the development of forest plans.
- *Clean Water Acts (1972, 1977, 1987).* A series of acts that address various aspects of water pollution, including the source and levels of contamination.
- *Farm Bill (periodic).* An agricultural and food policy tool that is used periodically to amend some permanent laws and suspend the provisions of others; used to bring about new policy provisions for limited periods.

Much of federal forest policy in the United States is directed toward the management of public (i.e., federal) lands. However, in the United States, federal water and air pollution laws, as well as endangered species protection laws, may influence and impact management activities on privately owned forestlands. State and provincial governments use both regulatory and nonregulatory approaches to manage and control activities within privately owned forests. Some states, primarily the western states, such as California, Oregon, and Washington, have broad forest laws that affect the management of private forests. Other states, such as Georgia, have little state-level legislation regarding private forest management. States can impose policies such as severance taxes on the value of timber harvested. West Virginia, for example, imposes a tax of 1.5% on

FIGURE 15.5 Construction of a firebreak around a loblolly pine (*Pinus taeda*) stand in the southern United States. *James H. Miller, United States Department of Agriculture Forest Service, through Bugwood.org.*

the gross value of timber harvested (West Virginia State Tax Department, 2018). States can use natural resource laws to regulate forest practices in an attempt to protect environmental quality and ensure sustainable management of forest resources, two real or perceived problem areas (Cubbage et al., 1993). States can also use nonregulatory approaches to induce certain behaviors on the part of private forest owners by offering technical assistance, fire protection assistance (Fig. 15.5), or subsidized tree seedlings. The federal government in the United States has also developed approaches such as these by funding forest assistance programs, often in cooperation with individual states.

In Central and South America, national forest policies have evolved over time to reduce the emphasis on wood product−centered forest management (Kern et al., 1998). Initially, public forest policies in many countries were focused on promoting the development of wood processing industries. Forest policies were initially aimed at generating income and increasing employment levels by utilizing forest resources, and conservation policies were developed later as national parks and forest reserves were established. As some Central and South American countries have tried to modernize their economies, they relied to a great extent on value-added forest production rather than simple resource extraction and exports of unprocessed logs. As noted in Chapter 2, countries such as Guatemala have enacted forest laws that require forest management plans and promote the concept of productive management of natural forests in order to both conserve biodiversity and improve living conditions for forest-dependent communities. Panama also enacted a Reforestation Incentive Law and has provided guidelines for forest management through other laws and decrees. Today, as a result of such policies, countries such as Brazil or Chile have well-developed forest products sectors comprised of intensively managed

plantation forests and advanced, globally competitive forest product manufacturing facilities. Therefore, whether reforestation is specifically regulated or not, the motivation to reforest may be guided by a perceived obligation to adhere to law, as in the cases of Brazil (Walters et al., 2005) and Panama (Simmons, 1997).

While the rights of forest-dependent peoples in Central and South America, including indigenous populations, were not recognized initially, over time some have regained their property rights to the forests. More recently, forest-dependent communities were recognized as important stakeholders, whose participation is necessary for successful public forest policies. An example of this process involves Mexico's *ejidos*, the owners of which over time received rights to harvest on their lands (Kern et al., 1998). This also has set the stage for a greater appreciation of community and private forestry as opposed to dominant public ownership and centralized decision-making. Following global trends, SFM has become the overriding concern of many public forest policies. This manifested itself through the recognition of the role forests have in fulfilling environmental functions, including the conservation of biological diversity (Stupak et al., 2011; Le et al., 2012). Many public policies now provide guidance in an attempt to ensure the achievement of SFM.

Public forest policies in Asia and the Pacific region have also undergone substantial change since the late 20th century. In countries such as China, where forests are mainly publicly owned, recent reforms assign forest property rights to local communities and even to individuals (Petry et al., 2010). In many cases, existing customary rights to forests are officially recognized (Kern et al., 1998). As with other world regions, the sustainability of forest management and the role of forests in protecting the environment and conserving biodiversity are increasingly important. Problems associated with forest degradation and deforestation in some countries are reflected in public policies, which in general serve to protect the existing forests as well as providing reforestation and afforestation incentives. For example, Japan's Basic Forest and Forestry Law promotes sustainable forestry and the development of multi functional forests.

Similar forest public policy changes have taken place in central and eastern Europe following the restoration of democracy to several countries in the region. Forest policies of the early to mid-20th century had a commodity production orientation, an objective that was eventually abandoned in favor of a more holistic view of the roles that forest resources fulfill. In the latter part of the 20th century, forest laws were revised and refocused on the achievement of broad forest objectives, which acknowledge the productive forest function but also encompass a range of environmental services (Cirelli et al., 2001). In addition, with the reemergence of private forestlands, consisting of lands once nationalized and then returned to their original owners, public policy also was developed to help new owners in managing their forests sustainably during the transition to market economies. In addition, to reflect democratic values, forest decision-making processes were open to public input. However, even in countries where land tenure continues to rest with the national government, such as Belarus, national strategies and forest codes guide the development of forests and natural resources (Gerasimov and Karjalainen, 2010). Further, a number of EU regulations directly or indirectly affect the management of forests (e.g., afforestation and protection activities) (Voitleithner, 2002).

In contrast, public forest policies in other parts of Europe are relatively mature, although they evolved to address new challenges and recognize the importance of SFM. Here, public policies underlie the importance of forest management planning and emphasize multiple uses, the conservation of biodiversity, and close-to-nature silvicultural operations (Cirelli et al., 2001). One example is the Everyman's Right (i.e., right of public access) in Finland, which allows the use of land owned by others for bicycling, hiking (Fig. 15.6), horseback riding, skiing, temporary camping purposes, and the extraction of some nontimber forest products such as picking wild berries (Parviainen et al., 2010). In many instances, public participation in policy-making processes is not only encouraged but is sometimes required. In addition, cooperation among individual landowners is encouraged to improve forest management and to deal with the impacts of excessive forest fragmentation.

FIGURE 15.6 Hiker walking along the Nordkalottleden Trail near Meekonjärvi Lake in Finland. *Matti Paavola, through Wikimedia Commons. Image Link: https://commons.wikimedia.org/wiki/File:Kalottireitti_Meekonjärvellä. JPG*
License Link: https://creativecommons.org/licenses/by-sa/3.0/deed.en

National level public forest policies in African countries tend to support management planning, including the recognition of social and ecological values, and are viewed as one of the tools necessary to help achieve SFM (Cirelli et al., 2001). Recent policy changes also reflect the requirement for forest-dependent communities to be part of the planning process. However, in some cases land tenure is unclear or unenforceable due to recent wars and civil unrest, thus preventing proactive management of forests (Sander et al., 2010). Some policies in these regions provide people with management and use rights, and with incentives devised to encourage the sustainable use of forests and to arrest forest degradation or deforestation. Public policies also have the object of enhancing forest conservation by establishing protected areas.

15.4 Organizational and individual policies

Many private and public organizations have developed policies of their own to address various aspects of forestry and natural resource management operations. These policies are often in addition to, or independent of, other policies or regulations developed by governmental bodies. The policies may, for example, pursue more advanced environmental protection measures as additional contributions to ecological diversity. As previously discussed, some organizations may even consider voluntary environmental protection measures and choose to certify the management of their forests, which requires adherence to a number of voluntary practices. Other organizational policies may address the flow of resource information within an organization (to ensure that information is effectively used and protected), safety concerns, and collaborative opportunities.

Individual private landowners have a multitude of objectives and constraints, many of which result in forest management policies. For example, a private landowner in central Alabama may desire to provide habitat for white-tailed deer (*Odocoileus virginianus*) and, thus, their inherent management policy may be to promote the development of mast-producing deciduous tree species. Landowners can also voluntarily enter into agreements with governments or NGOs in which they offer to manage their land in a certain manner in return for a reduction in property taxes or a fixed or regular payment of money (Reeves et al., 2018). The Forest Biodiversity Program in Finland is an example in which private landowners are paid to adapt their forest management practices in a manner that increases certain measures of biodiversity (Ministry of Agriculture and Forestry, 2008). In effect, the manner in which landowners manage their forests under the voluntary agreements become forest management policies.

15.5 Forest certification programs

Forest certification involves a set of forest policies that a landowner voluntarily enters into after adhering to a number of preacceptance (and, later, postacceptance, or audited) management criteria. Contemporary forest certification programs were initiated by environmental groups concerned with widespread deforestation and forest degradation, and most were begun after the Earth Summit in Rio de Janeiro in 1992 (the United Nations Conference on Environment and Development). From the Earth Summit came a nonbinding *Statement of Forest Principles*, consisting of 17 points providing the guidelines and means for protecting the world's forests. These guidelines were to be used by countries around the world in the development of regional and international criteria and indicators for measuring and monitoring successes in achieving SFM objectives (Siry et al., 2003). However, many environmental groups concerned about the rapid destruction of the world's forests considered the Earth Summit a failure because of the inability of stakeholders to agree on legally binding measures to protect the world's forests. Therefore, some forest certification programs were developed in response to this perceived failure. Forest certification programs are generally market-based initiatives aimed at improving the quality of forest management and promoting higher prices (price premiums) or better market access for wood products derived from sustainably managed forests. For a landowner, the promise of forest certification results in enhanced forest management and protection, and the generation of adequate financial returns from certified forests to sustain their existence. Certification programs were initially designed to improve the management of tropical forests, where deforestation and forest degradation were (and still are) of greatest concern. However, certification programs have now been designed for all forests of the world.

In general, forest certification programs consist of the following components:

(1) Standards (acceptable levels of forest management)
(2) Inspections and audits (assessments of the fulfillment of standards)
(3) Chain-of-custody (CoC) audits
(4) Certification (a labeling process that indicates that the three previous elements were met)

Typically, two types of certificates are awarded: (1) SFM certificates that verify forests are being managed sustainably; and (2) CoC certificates that verify forest products are made from wood harvested from certified forests.

While the largest forest certification programs in existence today originated after the Earth Summit, forest certification actually began with the American Tree Farm System in 1941. Contemporary forest certification began in 1993, with the Forest Stewardship Council (FSC). Other major programs include the Program for the Endorsement of Forest Certification (PEFC), the Canadian Standards Association (CSA) SFM program, and the Sustainable Forestry Initiative (SFI). The PEFC now functions as an umbrella program endorsing national forest certification systems, including North American programs such as the American Tree Farm System, the CSA SFM program, and the SFI. Unless stated otherwise, PEFC program information discussed in this chapter includes these North American programs.

15.5.1 Status of forest certification

As of February 2020, the total area certified by either the FSC or the PEFC programs amounted to 430 million ha (about 1.06 billion ac), which represents over 10% of the world's forest area. These certified forests supply about a quarter of the global industrial roundwood harvests, which are converted into lumber, paper, paperboard, and other forest products. Since forest products certification is much less common than forest management certification, only a fraction of products made from wood harvested in certified forests is recognized as such.

The PEFC is the largest forest certification program, as measured by the area of certified forests, which recently amounted to about 325 million ha (803 million ac). Currently, the program endorses 47 national certification programs (Programme for the Endorsement of Forest Certification, 2020) and is particularly popular in North America (the United States and Canada, under the SFI), Finland, and Norway. Forest area certified by the FSC amounts to another 143 million ha (353 million ac). While the FSC is present in 81 countries, much like the PEFC, most of its certified area is in North America (particularly in Canada) and the Russian Federation. Compared to the exponential growth of the 1990s, the rate of growth in certified forest area had slowed remarkably to about 7% per year as of 2011 (Kraxner, 2011). One of the main reasons is that the majority of sustainably managed forests, particularly in the Northern Hemisphere, have already been certified by one program or another.

Canada and the United States lead the world in forest certification, followed by the Russian Federation, Finland, and Sweden. Australia, Brazil, and Malaysia also have relatively large areas of forests certified. Other countries that have already certified or are in the process of certifying some of their forests include the Democratic Republic of Congo, Gabon, Chile, Uruguay, China, and South Africa. Other than the Russian Federation, further growth in forest certification in temperate and boreal regions is likely to be limited. In the United States, a substantial portion of forests are owned by smaller private owners, especially in the southeastern United States and this land tenure situation makes certification under one of the larger programs more difficult. Tropical regions still offer vast opportunities to expand forest certification, as only about 2% of their forest area is certified. However, many tropical forests are not managed sustainably, and deforestation and forest degradation are still widespread. These problems may hinder forest certification efforts. In addition, forest certification involves costs that may be too high for landowners or land management agencies in many poor countries.

CoC certification was introduced in 1997 and has been growing fairly rapidly since. CoC involves the identification of a product's origin (Fig. 15.7) and tracking the product and its derivatives through the system (Fig. 15.8). The number of CoC certificates issued by the two major certification schemes now exceeds 28,000, with most certificates issued in the United States, the United Kingdom, Germany, and France (Kraxner, 2011). This growth represents a sustained interest by certificate holders for demonstrating their commitment to SFM, as well as a desire to access certain markets in an effort to gain a competitive edge.

15.5.2 Major forest certification programs

For forest managers, certification may raise important issues because many commercial forests tend to be intensively managed, grown in plantations, and sometimes composed of exotic species. Such forestry approaches allow the achievement of high growth rates, yet some of the management practices can be highly controversial. One certification debate involves planted

FIGURE 15.7 FSC-certified pine logs from the Black Forest, Germany. *Gerhard Elsner, through Wikimedia Commons.*
Image Link: https://commons.wikimedia.org/wiki/File:Fsc-zertifizierung.jpg
License Link: https://creativecommons.org/licenses/by-sa/3.0/deed.en

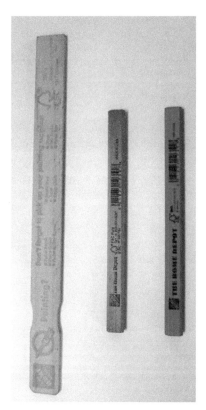

FIGURE 15.8 Some FSC-certified wood products. *Jacek P. Siry.*

forests, because of their perceived environmental shortcomings. As a result, some certification programs have been reviewing their protocols for certification of plantations. Interestingly, many certified forests include intensively managed plantations composed of exotic tree species, as is the case in Brazil, Chile, and several other countries. The certification of these and other intensively managed forests has helped us understand how they can contribute to achieving SFM goals. These types of issues may have significant consequences for forest and natural resource management; therefore, forest decision-makers should have a good understanding of the implications of certification under the various programs available today. The next few sections provide more detailed information concerning the three most important programs: the FSC, the SFI, and the PEFC.

15.5.2.1 Forest Stewardship Council

The FSC certification program is an independent, nongovernmental, nonprofit organization established to promote the sustainable management of the world's forests (Forest Stewardship Council, 2020). The organization's mission is to promote forest management that is environmentally appropriate, socially beneficial, and economically viable. At first glance of the mission, it is apparent that the overall forest certification goals are guided by environmental and social objectives. The program's principles and criteria (Table 15.1) describe how

TABLE 15.1 Principles associated with Forest Stewardship Council certification (Forest Stewardship Council, 2010).

Principle	Description
1	Comply with all applicable laws, FSC principles, and international treaties.
2	Obtain legally established, demonstrated and uncontested, clearly defined, long-term land tenure and use rights.
3	Recognize and respect indigenous peoples' rights to own, use, and manage their lands.
4	Maintain or enhance long-term social and economic well-being of forest workers and local communities.
5	Use efficiently the multiple products and services derived from the forest to ensure economic viability and environmental and social benefits.
6	Conserve biological diversity and maintain ecological function and integrity of the forest.
7	Develop an appropriate and continuously updated management plan, with clearly stated long-term objectives.
8	Monitor and assess activities to assess the condition of the forest, yields of forest products, the chain of custody of forest products, and the social and environmental impacts of management activities.
9	Maintain or enhance high conservation value forests of outstanding significance or critical importance.
10	Assure that plantations contribute to the reduction of pressure on, and promote the restoration and conservation of, natural forests.

forests are to be managed to meet the cultural, economic, environmental, and social needs of present and future generations.

The FSC standards require that a written forest management plan has been developed and implemented and is continuously updated and improved. Principle 10 addresses plantation forestry and requires that plantation design and layout should be done in such a way as to promote the protection of natural forests and not lead to increasing pressures on natural forests (Forest Stewardship Council, 2010). This principle advocates greater diversity of plantations, as measured by their age classes, size, spatial distribution, species, and structures. In the FSC program, native species have priority over exotic species, which can be used only in specific situations where their superiority is clearly demonstrated, and their management is closely monitored. In addition, part of the overall plantation of a forest should be devoted to the restoration of natural forest cover, as conversion of natural forest to plantations is not allowed. When certified by the FSC program, every effort is made to reduce and eliminate the use of

chemicals such as pesticides and fertilizers. Broad social and ecological assessments of the impacts of all plantation projects are also required. Interestingly, plantations established after 1994 are not eligible for certification, with the exception of cases in which the current landowner was not responsible for the conversion from native forests to plantation forests. Given the requirements and constraints of this specific certification program, organizations that pursue other forest certification programs may face the same challenging management issues.

15.5.2.2 Sustainable Forestry Initiative

The SFI program was initiated in 1994 by members of the American Forest and Paper Association, which was primarily composed of forest products companies. Members of the SFI program agreed to operate under a set of forestry principles demonstrating the use of SFM. This program was also initiated to demonstrate to the public that the forest industry can manage its forests in a sustainable manner. The SFI program was supplemented by a third-party verification process in 1998. Today, the SFI certification program is managed as an independent, nonprofit organization that maintains and oversees the forest certification standards, and the forest area in North America certified under the SFI program amounts to about 147 million ha (about 364 million ac) (Sustainable Forestry Initiative, 2019).

The SFI program promotes SFM through 15 objectives, which relate to aspects of sustainable forestry; forest productivity and health; the protection of water resources and biological diversity; the maintenance of aesthetic and recreation values; the protection of special sites; and fiber supply chain, research, training, and legal issues. These objectives are further refined into 37 performance measures and a number of indicators of performance. Program participants are required to have developed a written forest plan that, when implemented, aims to achieve goals related to the principles.

Participants in the SFI program must abide by its objectives (Table 15.2) in a manner according to their land ownership or wood-using status. SFI objectives 1–7 provide the means for evaluating compliance with the standards on forestland owned and leased by participants. This is achieved through forest management planning, protection, and maintenance of long-term forest productivity, protection of water resources, conservation of biodiversity, and of other environmental values and outputs, efficient use of forest resources, and other means. Objectives 8–15 address wood procurement systems through landowner outreach, recognition of indigenous peoples' rights, adherence to laws and best management practices, use of qualified logging professionals, and elimination of illegal logging. Further, they address legal

TABLE 15.2 Sustainable forestry initiative objectives for 2015–21 (Sustainable Forestry Initiative, 2015).

Objective	Description
1	Ensure forest management plans include long-term sustainable harvest levels
2	Maintain forest health and productivity
3	Protect and maintain water resources
4	Conserve biological diversity
5	Manage visual quality and recreational benefits
6	Protect special sites
7	Use efficiently fiber resources
8	Recognize and respect indigenous peoples' rights
9	Comply with laws and regulations
10	Invest in forestry research, science, and technology
11	Train and educate
12	Involve communities and landowners
13	Public land management responsibilities
14	Communications and public reporting
15	Management review and continuous improvement

TABLE 15.3 Programme for the Endorsement of Forest Certification (2018) criteria for sustainable forest management.

Criterion	Description
1	Maintenance and appropriate enhancement of forest resources and their contribution to the global carbon cycle
2	Maintenance of forest ecosystem health and vitality
3	Maintenance and encouragement of productive functions of forests
4	Maintenance, conservation, or appropriate enhancement of biological diversity in forest ecosystems
5	Maintenance and appropriate enhancement of protective functions in forest management
6	Maintenance of other socioeconomic functions and conditions

and regulatory compliance, and research, training and education, as well as community involvement, communication, management review, and continual improvement. Through adherence to these standards, participants are said to have demonstrated their commitment to sustainable management, the environment, and society, while recognizing the importance of commercially viable forests.

15.5.2.3 The Program for the Endorsement of Forest Certification

The PEFC program began in 1999 as an alternative to the Forest Stewardship Certification process. In addition to forest management certification, it also offers CoC certification and product ecolabeling. It is structured as an international, nonprofit NGO dedicated to the promotion of SFM. It is organized as an umbrella scheme, which endorses other national forest certification programs. For example, the American Tree Farm System is endorsed by the PEFC program. Each of the national programs then undergoes a rigorous third-party evaluation to ascertain its compliance with sustainability measures (Programme for the Endorsement of Forest Certification, 2020). The program thus uses a bottom-up process for developing national forest certification programs in order to ensure the rigor, robustness, affordability, and overall effectiveness deemed necessary.

Forest management standards, which apply to all forests regardless of their location, are summarized in six

criteria (Table 15.3), and progress in achieving SFM is measured by certification criteria developed for each standard (Programme for the Endorsement of Forest Certification, 2018). These criteria focus on the maintenance of forest resources and their productive and protective functions, and the fulfillment of important socioeconomic functions. The PEFC certification process offers group and regional certification status, making it a leader among small individual and community forest owners. The process prohibits forest conversion, use of the most hazardous chemicals, and, like the FSC program, the use of GMOs. Consultations with local stakeholders and regulatory compliance also receive a high priority in this program. Forest certification under this program suggests that biodiversity is maintained and enhanced, ecosystem services are sustained, worker's rights are protected, local employment is encouraged, indigenous rights are respected, and all relevant laws are followed.

15.5.3 Benefits and costs of forest certification

The costs of forest certification vary substantially because they involve document preparation and application, the certification process itself, and follow-up audits and recertification efforts. The more suited an organization and its information systems are to accommodate the certification process, the more efficient the process can become. Certification costs also include those associated with any management changes required by the certification system, such as reductions in harvests or more expensive logging operations. Certification costs can vary from US$5 to US$148 per ha (US$2 to US$60 per ac). The costs per unit area tend to be lower for larger forest estates, where fixed certification costs are applied to larger areas of land. One concern is that certification costs are primarily incurred by forest landowners with little ability to pass these costs onto wood products customers. However, there is some evidence that in certain cases pass-through costs are possible for solid wood products in well-developed markets (Kraxner, 2011).

Commonly recognized benefits of the certification of forestland include improved forest management, better documentation of an organization's operations, higher management credibility, market access, and, of course, price premiums for certified wood products. It has been reported that price premiums for certified wood products in the United States and Europe can reach 5% (Kraxner, 2011), depending on the market. It also has been noted that low-priced certified products (such as pencils or copy paper) tend to attract higher premiums than more expensive products. However, the current demand for certified wood products is still relatively low, although the demand may increase with the growth of green building programs and green procurement policies implemented by some governments. To date, most green building programs give credit for solely using FSC-certified forest products (Fig. 15.9). Wood product certification and labeling may be heavily influenced by developmental trends in Asia, where most of the new building construction projects are taking place.

One might expect that if the various forest certification systems could agree on mutual recognition, growth in the market share for certified wood products might be facilitated. However, mutual recognition is unlikely and there continues to be strong competition between the FSC and SFI programs in the United States, as well as international competition between the PEFC and FSC

programs. Unfortunately, Fernholz et al. (2010) indicate that consumers in North America have difficulty distinguishing between the various certification programs. The challenges involved in educating the public about the advantages and disadvantages of the programs make the achievement of true market success of certified forest products difficult.

Forest owners who desire to certify their forestry and manufacturing operations need to recognize the actions that may be necessary to meet certification requirements, and then to estimate the costs associated with these requirements, as well as the potential benefits of forest certification. While all major certification programs offer (or require) third-party audits, CoC certification, public reporting, stakeholder consultation, independent governance, and product labeling, there are some important differences related to the allowable management actions (Fernholz et al., 2010). In addition, the choice of a certification program may influence the marketing of wood products, due to consumer perception of particular certification programs. In today's management environment in the United States, forest certification is still a voluntary endeavor, and in some situations, it may be neither necessary nor required for the efficient and sustainable management of forest resources.

15.6 Trade issues

With the growth of economic globalization, the share of forest products traded internationally has been gradually increasing since the late 20th century. While in the 1980s only about 20% of the world's sawnwood (lumber), panel, paper and paperboard, and wood pulp output was traded internationally, this share has increased to approximately 30% in recent years (Fig. 15.10), and it can be expected to continue to rise. The rise in international trade in forest products should come as no surprise. It results from the abolition of barriers to trade and follows broad changes in world trade. Moreover, as long as certain conditions are met, international trade also makes perfect economic sense. On one hand, there are countries with very large forest resource endowments and limited domestic markets (such as the Russian Federation, Canada, or Brazil), while on the other hand there are rapidly growing economies that are experiencing domestic wood deficits (for example, China and India). Thus, forested countries benefit from utilizing their abundant forest resources, while countries with limited forest resources are able to acquire wood for a broad range of beneficial uses.

An examination of the types of forest products traded indicates that the share of unprocessed logs has been the lowest (Fig. 15.10). This trend is logical, as logs form a

FIGURE 15.9 Mixed-use loft style apartment building in Hollywood, California, United States, consisting of 100% FSC-certified wood products. *Calderoliver, through Wikimedia Commons.*
Image Link: https://commons.wikimedia.org/wiki/File:LawrenceScarpa_Cherokee_007780-2.jpg
License Link: https://creativecommons.org/licenses/by-sa/3.0/deed.en

bulky yet relatively low value commodity that does not usually justify shipping over long distances. The Russian Federation is the largest log exporter in the world, followed by the United States, New Zealand, Germany, and Malaysia (Food and Agriculture Organization of the United Nations, 2019a,b). As forest industries develop in resource-rich countries, logs of all quality will be traded internationally. The most widely traded forest products are sawnwood, panels, paper and paperboard, and wood pulp. Sawnwood exports are led by Canada, the Russian Federation, and Sweden. Much of the Russian log and sawnwood production is destined for China, currently the world's largest log importer and second largest sawnwood importer.

Wood pulp exports are dominated by Brazil, Canada, and the United States, which have well-developed pulp and paper industries (Food and Agriculture Organization of the United Nations, 2019a,b). The largest exporters of paper and paperboard products are Germany, the United States, Sweden, Finland, and China, which are established forest products suppliers that rely on temperate and boreal forest resources. The United States is by far the largest exporter of recycled paper, which is shipped primarily to China to feed what is today the largest pulp and paper industry sector in the world. China and Germany also lead the world in wood panel exports.

Forest products trade flows are influenced by forest resource endowments, the level of economic development, and supply and demand forces, as well as currency exchange rates. For example, the United States forestry trade balance appears to be inversely related to the strength of the dollar (Fig. 15.11). When the dollar is strong, the forestry trade balance is generally negative, for example, the United States imports more forest products than it exports. This is because forest products made

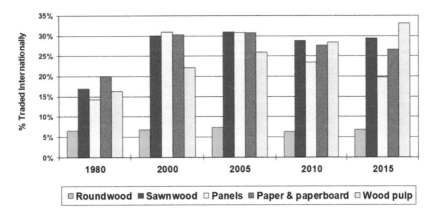

FIGURE 15.10   Changes in value-added products traded internationally. *Caulfield (2003) and Food and Agriculture Organization of the United Nations (2020).*

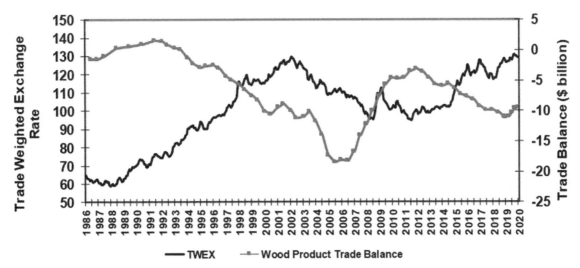

FIGURE 15.11   Is the forestry trade balance inversely related to strength of the dollar? *Caulfield (2003), Federal Reserve Bank (2020), and U.S. Foreign Agricultural Service (2020).*

in the United States are more expensive for foreign buyers to purchase when expressed in their own currencies. At the same time, foreign suppliers are very interested in selling products to the United States, as they then receive more money for their exports when the dollar is converted to their domestic currencies. This, in turn, has consequences for the forest industry in the United States in terms of wages, employment, the level of wood harvest, and income for forest landowners.

Currency exchange rates, along with other factors, are important for determining the global competitiveness of leading wood supply regions. Since the early 1990s, pulpwood prices (delivered to the gate of a processing facility) have continued to increase in Brazil, Chile, western Canada, and Sweden, which all rely heavily on export markets, while prices in the southern United States have remained relatively stable, thus making it one of the most competitive wood-growing regions in the world (Fig. 15.12). Similar trends were noted for sawtimber producers such as Brazil, New Zealand, and Sweden (Fig. 15.13). Currency exchange rates can strongly influence competitiveness and, consequently, trade flows, particularly in the short term.

Softwood lumber produced in Canada for export markets is primarily destined for the United States, where it accounts for as much as one-third of the lumber consumed. Such a substantial share of lumber consumption has resulted in a very long trade dispute, dating back to 1982. Lumber producers in the United States claim that lumber producers in Canada receive an illegal subsidy in the form of low fees for the right to harvest timber (stumpage fees) and are thus able to produce lumber more cheaply. As a result, Canadian lumber producers can sell large volumes of lumber in the United States at a lower cost than that of lumber produced in the United States. Part of the problem is that most of the forestland in Canada is owned by provincial governments, which determine the stumpage fees. Because they are set centrally and are lower than their complement in the United States, these fees lead to the claim of an illegal subsidy. In the United States, where most forestland is privately owned, stumpage fees are set

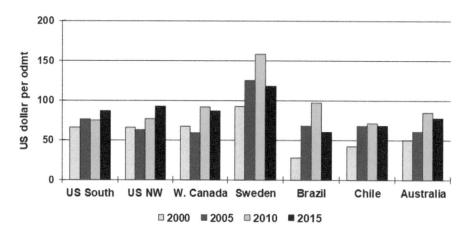

FIGURE 15.12   Conifer pulpwood delivered prices per oven-dry metric ton (odmt). NW, northwestern; W, western. *Ghasemi et al. (2020).*

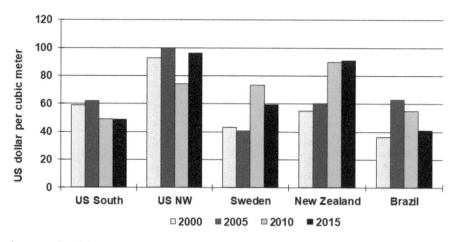

FIGURE 15.13   Conifer sawtimber delivered prices. NW, northwestern. *Ghasemi et al. (2020).*

through a competitive sale process. Under trade remedy laws in the United States, countervailing duty tariffs can be imposed on foreign goods benefiting from subsidies, which has been done in the past. While both governments have long tried to resolve this trade dispute, a fair compromise has proven elusive. Interestingly, while countervailing duties can benefit lumber producers in the United States, they can also harm millions of consumers in the United States who must pay higher lumber and, consequently, higher housing prices.

The relationship between forest products trade and forest sustainability has been a focal point of numerous environmental debates. While one fear has been that trade practices may lead to deforestation, trade practices have also been viewed as a tool that may help achieve SFM. After all, the principle on which forest certification programs were developed is that certified wood derived from sustainably managed forests should command higher prices and better market access across the world.

Illegal wood harvesting, however, is a persistent problem in much the world. Illegal harvesting can reduce biodiversity and other important environmental forest functions such as, for example, carbon storage, and lower the value for legally produced products. Illegal harvesting may consist of harvests in protected forest areas (where no harvesting is allowed), harvests in excess of legal and sustainable harvest levels, or harvests in violation of relevant international agreements such as the CITES. According to a study conducted by Seneca Creek Associates and Wood Resources International (2004), between 5% and 10% of the world's industrial roundwood output has been harvested illegally. It also appears that most illegally harvested wood is used domestically and that only a small fraction is traded internationally. Illegal harvesting is purported to reduce average world wood prices by as much as 16%. The cost of illegal harvest to producers based in the United States was estimated at US$460 million in terms of missed export opportunities and lower wood prices. The primary reasons behind illegal harvests are unclear or poorly enforced property rights, inadequate forest monitoring and law enforcement, and inefficient or corrupt policy institutions. One example of a governmental response to the problem of illegal logging is the EU's Forest Law Enforcement, Governance and Trade program. This program supports improved governance in timber producing countries, develops voluntary partnership agreements to prevent illegal wood from entering the EU, and introduces measures to reduce and eliminate the use of illegal wood within the EU.

International trade in forest products is increasingly important and seemingly beneficial. Some consumers and producers gain from trade and others lose from trade. As long as trade gains exceed losses, and as long as there are no externalities or unfair competition, trade should be beneficial, especially if the gainers would compensate the losers, which rarely occurs in real life. Trade can also have both positive and negative impacts on resource sustainability. It appears, however, that well-designed trade policies with appropriate safeguards in place do not need to degrade forest resources. On the contrary, in many instances trade can be used as a tool to promote SFM.

In 2011, about 37% of the fiber used to make paper products in the United States came from recycled sources (U.S. Environmental Protection Agency, 2016). Therefore, the recycling of wood-based products (paper, lumber, and even chemically treated wood) is an important aspect of international trade, for both the production and consumption of wood-based forest products. The EU, for example, developed directives for its member states that have increased recycling rates of products such as corrugated boxes to 50% or more and seek to reduce biodegradable landfill waste to as little as 35% of that produced (Monte et al., 2009), which may involve international trade. In the United States, nearly 69 million tons of paper and paperboard are produced annually, while about 43 million tons of paper was recycled in 2013, representing a recycling rate of about 63% (U.S. Environmental Protection Agency, 2016). In some areas of the world, the recycling of domestically produced or imported scrap paper may be the most efficient manner for paper production (Liang et al., 2012). Nearly 42% of the recycled paper in the United States was exported to overseas markets in 2011 and, although international trade that uses corrugated boxes tends to muddle the picture, about 89% of the corrugated boxes in the United States were said to have been recycled in 2013 (U.S. Environmental Protection Agency, 2016).

15.7 Forest carbon sequestration and markets

In an effort to reduce carbon dioxide emissions, those operations that emit too much carbon dioxide may choose to reduce the carbon intensity of their production processes by switching to renewable energy sources and nuclear power, improving energy efficiency, or sequestering more carbon through direct capture and storage or enhancement of natural sinks, such as forests. In general, each of these efforts involves an increase in the cost of doing business. To reduce the cost of carbon dioxide emission compliance, several emissions trading (cap-and-trade) programs have been developed. Under these programs, governments establish an overall emission limit (a *cap*) which is then allocated to individual emitters. In cases where emissions allowances are insufficient to cover an organization's carbon dioxide emissions, the organization can purchase emission allowances (*trade*) from others or purchase certified emission offsets such as those generated by forest carbon sequestration programs. Organizations in certain industries can lower their carbon emissions more easily and cheaply than can organizations in other industries, such as energy-intensive cement manufacturing. Therefore, some organizations may decide to purchase emission offsets rather than attempt to reduce their own emissions, which may turn out to be expensive and perhaps even not feasible given current technologies.

Forest carbon sequestration is one of the key approaches for reducing atmospheric carbon dioxide concentrations. Trees continuously remove carbon dioxide from the atmosphere through the photosynthesis process and store carbon in their biomass. This natural process is considered to be a safe, environmentally acceptable, and cost-effective way to capture and store substantial amounts of carbon dioxide. An intensively managed southern loblolly pine (*Pinus taeda*) plantation, for example, can sequester as many as 840 tons per ha carbon dioxide (340 metric tons per ac) during a single 25-year rotation (De La Torre and Smith, 2008). However, detailed analyses of each forest management situation should be conducted, as expectations of the cost-effectiveness of forest-based carbon sequestration processes may need to be adjusted (Smith and Applegate, 2004). Enkvist et al. (2007) estimated that activities such as protecting, planting, and replanting forests can constitute as much as 25% of the total global abatement potential. Further, the Food and Agriculture Organization of the United Nations (2010) estimated that forest ecosystems store more carbon dioxide than is present in the atmosphere. Recent research indicates that net annual forest carbon dioxide sequestration accounts for more than 10% of all anthropogenic emissions (Pan et al., 2011), which further underscores the important role that global forests have in regulating the atmosphere. Perhaps not surprisingly to a forester, young forests are particularly adept at rapidly sequestering atmospheric carbon (Pan et al., 2011).

Management practices, including afforestation, reforestation, and harvest, can substantially influence the carbon sequestration potential of forestland. It is obvious that forests sequester and store large amounts of carbon in the above-ground portion of a tree, but carbon is also stored below ground, in the root systems of trees (Fig. 15.14) and in the soil. Forest mitigation strategies may involve eliminating forestland conversions (especially in the case of tropical deforestation), postponing harvests, reducing burning, or increasing carbon uptake through intensified forest management and conversion of agricultural land to forestry (Alig, 2003). Appropriate management approaches and technologies are available to move these efforts forward. Like other basic materials, once carbon (in the form of wood) is removed from a forest, it may be stored in a forest product (chair, table, etc.)

FIGURE 15.14   Root systems of several trees, where soil has been washed away. *Edwin Klein, through Wikimedia Commons.*
Image Link: https://commons.wikimedia.org/wiki/File:Tree_Roots.JPG
License Link: Public Domain

for decades. One advantage of using wood as a basic material rather than other materials (steel, concrete, or plastics), beside the fact that forests are renewable, is that in most cases less fossil fuel is required to produce a wood product than a comparable steel, concrete, or plastic product. Finally, once wood products have surpassed their useful life, they may still provide biomass for energy-generation purposes.

The most important international agreement to date, the Kyoto Protocol, clearly recognized the role that forests and forest management play in reducing carbon dioxide emissions (United Nations Framework Convention on Climate Change, 1997). While the Kyoto Protocol allows carbon stored in forests to be used to meet certain emission reduction targets, it also places several restrictions on how this can be achieved. These restrictions are related to baseline, permanence, additionality, and leakage principles and have led to a complex set of rules for recognizing and rewarding the development of forest carbon sequestration projects. Unfortunately, only a few Kyoto Protocol-compliant forestry projects have been developed to date, and the leading mandatory emission reduction schemes exclude forestry projects from participating in any carbon trading.

The Kyoto Protocol states that only forests established after 1990 are eligible for carbon offsets. Several other trading schemes assume the same base date (*baseline*). This may represent a problem for regions in which a large amount of afforestation took place prior to 1990, as in the southern United States, or in New Zealand, effectively making projects in these areas ineligible for the development of carbon credits. The Kyoto Protocol also requires that forest carbon capture projects demonstrate additionality. A carbon emission reduction is *additional* only when it was developed exclusively for the purpose of climate change mitigation. Projects implemented under *business as usual* or as required by other laws and regulations are not considered additional. As one might guess, determining the usual management practices in the real world can be quite difficult.

The Kyoto Protocol further requires that carbon emission reductions are *permanent*. In other words, through carbon sequestration projects, carbon dioxide is permanently removed from the atmosphere. However, forest carbon sequestration, by its very nature, is a temporary endeavor. Although trees can store carbon for several decades or more, eventually they will die and the carbon will once again be released, some of it rapidly while others may last for decades. Harvested trees that become products used by humans may also eventually be discarded, and the carbon may effectively be released back into the environment. The lifespans of wood products, and their disposal and decomposition rate, have a large impact on the amount of carbon both sequestered and emitted (Lun et al., 2012). There probably is some value in the temporary storage of carbon in trees, as it at least provides more time for the development of alternative, permanent carbon emissions reduction technologies. Carbon stored in plants or underground is always at risk of leaking back into the atmosphere. Shifting activities (i.e., moving an activity that causes carbon emission rather than permanently avoiding the activity) and outsourcing services or commodities that were previously produced are examples of *leakage* in the amount of carbon intended to be stored in forest carbon sequestration projects (Aukland et al., 2003). Leakage can also occur at a much greater scale through the international trade of products (Ståhls et al., 2011). The impact of carbon dioxide leakage from carbon sequestered in forests will ultimately depend on the volume of carbon stored in these projects and the leakage rate (Teng and Tondeur, 2007). As a consequence of these issues, some doubt whether forestry-based sequestration projects are real or additional, although they may be as valid as some other energy projects for addressing carbon emissions reduction targets (Chomitz, 2002).

Other key questions related to the process of storing carbon in forests concern how carbon should be valued and how forest product harvesting and wood products manufacturing should be treated. Carbon can be stored in wood products for many years, but many carbon trading schemes do not consider forest harvesting or allow credit to be obtained for carbon stored in forest products. The resolution of these issues is critical for determining the role that managed forests may assume in climate change mitigation efforts. While managed forests will continue to sequester carbon and provide certain storage benefits, their true potential for increasing carbon sequestration above the current, natural (without extra management effort) levels may never be realized.

Carbon transactions have been conducted on both voluntary and regulated markets. The voluntary markets conduct carbon transactions that are not required by regulation. The Chicago Climate Exchange (CCX) was the world's largest voluntary cap-and-trade scheme for greenhouse gas emissions reductions until 2010, when it ceased regular trading operations due to a lack of activity in the United States. There are regulated markets that conduct carbon transactions arising from commitments under legally binding agreements, such as the Kyoto Protocol, as well as regional, national, and state-level programs. These include the EU ETS, the world's largest greenhouse gas emissions trading program, which covers 40% of the total carbon dioxide emissions in Europe (Kautto et al., 2012). Carbon transactions are measured and traded in standard units representing a quantity of metric tons of carbon dioxide equivalent (tCO_2e). To facilitate trade, carbon transactions need to be converted into common units with the same global warming potential, such as tCO_2e. These units are termed *carbon credits* and represent the right to emit one metric ton of carbon dioxide. In 2007, nearly 3 billion tCO_2e were traded globally, in both voluntary and regulated markets, reaching a market value of US$66.4 billion (Hamilton et al., 2008). While voluntary markets continue to develop, their share is still minuscule. Regulated markets, such as the EU ETS, accounted for 98% of the total trade volume and continue to dominate carbon markets.

Interestingly, forest carbon credits are only a very small part of the carbon trading market. While the EU ETS is the largest carbon trading program, it does not allow forest carbon credits to be traded. Therefore, forest-based carbon trade under various provisions of the Kyoto Protocol is effectively nonexistent. By 2008, only one forestry project was ratified (DANA Publishing, 2008). The share of forest carbon trade in the CCX was also modest before it folded in 2010. Examples of carbon trading programs that accept forest-based carbon include the Australian Emissions Reductions Fund, New Zealand Emission Trading System, Korea Emission Trading System, California Cap-and-trade, and the Regional Greenhouse Gas Initiative. Most of these programs are still in the early stages of development. Between January and March of 2018, forestry and land-use projects accounted for nearly 59% of carbon emission projects offsets (Hamrick and Gallant, 2018).

Although carbon sequestered in forests has been traded openly in various markets, carbon offset prices have been relatively low. In 2016, the average price in the voluntary over-the-counter markets was estimated at US$5.20 per tCO$_2$e (Hamrick and Gallant, 2018). Popular forestry projects included afforestation, reforestation, reduced emissions from deforestation, and improved forest management. The reasons for relatively low prices (compared with regulated markets) are numerous. For one, there are virtually hundreds of trading programs and numerous concerns regarding the quality of carbon offsets traded. In addition, due to the large number of forest project developers, there are significant differences between project development standards. In an effort to increase quality and consistency of forest carbon credits, voluntary markets rely increasingly on standards that focus broadly on the quality of carbon credits (measurement and monitoring) and cobenefits, such as socioeconomic and environmental benefits of forest carbon sequestration projects. Popular standards include the Climate Action Reserve protocols, the Climate, Community, and Biodiversity Standards, the Gold Standard, and the Voluntary Carbon Standard.

While it is apparent that forests, and sustainably managed forests in particular, can play an important role in climate change mitigation through increased uptake and storage of carbon dioxide, several important challenges must be addressed before forests can fulfill their potential. These challenges range from developing effective forest carbon sequestration rules to compliance requirements and market considerations. These arise from the nature of forest carbon sequestration projects and management characteristics of the forestry profession. The Kyoto Protocol and several other cap-and-trade systems require that forest carbon projects meet certain additionality, baseline, leakage, and permanence standards. It appears that the two critical standards are related to additionality and permanence, because they are directly related to the real environmental benefits that forest carbon sequestration can yield. The baseline requirement was developed to support additionality and, once additionality can be clearly demonstrated, baseline relevance will diminish. The same applies to the problem of leakage, which can be prevented by appropriate project design and verification.

As mentioned previously, forest carbon sequestration is a temporary endeavor. While carbon may be stored for decades in forests and wood products, it will eventually be released back into nature. Even when discarded in landfills or left idle outdoors, wood products will decay (Fig. 15.15). However, by using forest carbon sequestration we may gain time needed to develop effective permanent carbon storage approaches. Another option is to reforest areas that are in a low carbon landscape into a

FIGURE 15.15 Decaying pile of old logs, now supporting a thriving group of mushrooms. *Kate Jewell, through Wikimedia Commons. Image Link: https://commons.wikimedia.org/wiki/File:Toadstools_in_Plungar_Wood_-_geograph.org.uk_-_157168.jpg License Link: https://creativecommons.org/licenses/by-sa/2.0/deed.en*

higher carbon landscape that may result in more carbon stored in forests over time, but it requires an increase in the total amount of forested area. Consider, for example, Georgia Power, an organization that supplies electricity to citizens of the state of Georgia, mostly from coal-fired power plants. The company would like to improve its carbon efficiency, in part by switching to renewable sources of energy or to nuclear power. This approach requires several years before substantial reductions in carbon emissions can be achieved and, in the meantime, forestry projects can be used to meet some of these goals. However, a valuation system is needed to assess the benefits of temporary storage. A rental payment approach and carbon banking may provide a valuation solution for carbon emission reductions that are not permanent (Sedjo and Marland, 2003; Bigsby, 2009).

The second commitment period associated with the Kyoto Protocol is set to expire at the time of publication of this book. It appears that the Protocol has been at best only moderately successful and has failed to arrest the rise in carbon dioxide emissions (World Bank, 2010). What will happen next is uncertain. Therefore, the development of new rules for curbing carbon emissions may (or may not) represent an opportunity to develop international agreements that truly recognize the role of forests in global change mitigation and to propose rules that will help to mitigate global climate change. Further, some have suggested that changes in technology can lead to a carbon dioxide net balance in certain manufacturing processes, such as the production of paper (Farahani et al., 2004). These processes may involve the use of forest biomass as a substitute for fossil fuels. Therefore, the manner in which society addresses the rise in carbon dioxide and the associated challenges will certainly be of interest to forest and natural resource managers.

15.8 Renewable energy and forest resources

While carbon climate regulations have not been implemented in the United States, more progress has been made on the closely related issue of renewable energy in general, and woody biomass in particular. It is postulated that a greater reliance on bioenergy may help reduce dependence on energy imports, avoid disruptions associated with global energy supply shocks, and provide certain carbon benefits (Guo et al., 2007). Wood is a renewable source of energy that circulates carbon already present in the biosphere, as opposed to fossil fuels, which add more carbon to the biosphere with their combustion (Richter et al., 2009). The issue of purported carbon benefits associated with the use of biofuels is still far from settled, however, and the U.S. Environmental Protection Agency (EPA) continues to evaluate greenhouse gas emissions associated with the use of biofuels (Aguilar et al., 2011).

Wood can be used to generate heat and electricity (Fig. 15.16) and to produce liquid biofuels such as cellulosic ethanol or biodiesel. Wood combustion technologies can be considered as relatively well-developed, but liquid fuel technologies have been developed more recently and several challenges remain, as the presence of lignin makes some wood-to-liquid fuel conversion processes challenging. The success of the commercialization of biofuels depends on the capacity and usefulness of the existing forest harvesting and wood transport infrastructure, the development of cost-effective conversion technologies, the ability to deliver large volumes of biomass at competitive costs, and the ability to satisfy mandates that biomass is greenhouse

gas negative, i.e., that the use of biomass results in reduced greenhouse gas emissions. The demand for biofuels is driven by the price of fossil fuels (primary substitutes), conversion technologies, the cost of feedstock, and regulations that stimulate demand. Of particular interest are liquid biofuels that have the potential to replace at least some of the crude oil used to make transportation fuels. While, for example, cellulosic ethanol can be produced from corn and other energy crops such as grasses, wood appears to have several advantages over other sources of cellulosic biomass. For example, wood can be stored longer and cheaper, and wood has high bulk density which lowers effective transportation costs. Wood generally has a higher sugar content than other sources of cellulosic biomass, and its production requires less water and fertilizer. In addition, wood collection systems are relatively well established.

The role of public policies in the EU, the United States, or even China cannot be stressed enough in promoting renewable energy and, consequently, the use of woody biomass for energy purposes. The EU, for example, has developed a binding agreement whereby it expects renewable energy to provide 32% of the total energy it consumes by 2030 (Aguilar et al., 2018). France also has the ambitious goal of obtaining 23% of its energy from renewable sources by 2020 (Ministère de l'Écologie, de l'Énergy, du Développement durable et de la Mer, 2009). In 2017, renewable energy in the EU accounted for about 17% of the total energy consumed, and much of the renewable energy arises from woody biomass (European Commission, 2019). Achieving such impressive renewable energy targets may require massive volumes of wood that cannot be produced solely in Europe or China, thus making wood imports necessary. This, in turn, raises questions regarding potential wood sources, and subsequently the impacts of increased wood harvesting on forest sustainability in potential supply regions. In the United States, the renewable fuel standard in the Energy Bill requires a large amount of biomass, which will certainly increase demand for wood produced locally. In 2016, wood sources account for only about 2% of the annual energy consumption in the United States, and about 21% of the renewable energy consumed (U.S. Energy Information Administration, 2020a,b). To date, wood energy has been consumed primarily by industries such as pulp and paper manufacturers to generate heat and power for internal uses. In recent years, however, the United States government has introduced several incentive programs that promote the increased use of biofuels (Guo et al., 2007).

Wood pellets derived from forest biomass are becoming an important energy source in some areas of the world. While wood pellets have so far been mostly used for space and water heating, they are increasingly being used for power generation. The estimated wood

FIGURE 15.16 Bales of scrap wood that will be used to generate electricity at a power station in Scotland. *Richard Webb, through Wikimedia Commons.*
Image Link: https://commons.wikimedia.org/wiki/File:Baled_brashings_-_geograph.org.uk_-_1462955.jpg
License Link: https://creativecommons.org/licenses/by-sa/2.0/deed.en

pellet production was more than 33 million metric tons in 2017 (FAO, 2019a,b). Currently, the United States is the world's largest pellet producer. In response to European demand, several pellet projects are being developed, including those in Asia and the southern United States. However, relatively little southern United States forestland is certified because of fragmented holdings and private ownership, which may prove to be an obstacle for accessing European bioenergy markets that require certified wood products. Given the costs involved and the relatively low market price for wood pellet products, suppliers of wood pellets will need seaport access, a strong biomass capacity, and a high availability of logging and mill residues. In summary, wood energy markets are driven by renewable energy requirements in several regions of the world. As a result, the demand for wood pellets continues to grow.

Summary

Policies are the accepted guidelines for performing specific activities to achieve short-term or long-term goals. As such, they can have an important role in influencing how forests and natural resources are managed, whether developed at the international, national, state, or local level. Some international policies attempt to deal with worldwide issues such as climate change and illegal logging. National or state-level policies can be more regional in scope but can still have worldwide implications in today's global society. Trade issues can also exert pressure on a nation's ability to meet specific resource management goals and can affect domestic and international policies for sustainably managing forests and natural resources. A key point from this discussion is that forest policy development is an evolving process for developing guidelines and protocols for the management of forests and natural resources. Policies are often evaluated for their benefits and costs, and debates often arise on ways to improve their effectiveness. Since policies usually only arise after real or perceived problems are identified, and since natural resource management problems can vary considerably in size and scope, policies may be developed by different levels of government (federal, state, local, organizational, or individual) or may be directed toward specific land ownership classes (public or private) to achieve certain objectives.

Questions

(1) *National forest policy.* Select a national forest policy from the country in which you live. Develop a one-page summary of the important aspects of the policy. In a small group of other students, describe the intent of the policy you selected, who might be affected, and the advantages and disadvantages of its implementation.

(2) *State or local forest policy.* In a manner similar to Question 1, select a state, provincial, or local forest policy from the area in which you live. Develop a one-page summary of the important aspects of the policy. Again, in a small group of other students, describe the intent of the policy you selected, who might be affected, and the advantages and disadvantages of its implementation.

(3) *Forest certification.* Select a forest certification program. Develop a PowerPoint presentation that explains the goals and intents of two or three indicators of one of the diverse criteria. Further, provide pictures or diagrams that help support or describe how the indicators are measured.

(4) *Carbon credits.* There are at least four important aspects of a forest carbon sequestration project. Describe these in a short summary and present them in a memorandum to your instructor. Perform a search of the Internet for peer-reviewed journal articles or published government reports that provide information to support each aspect. Do not use the references we have provided in this book.

(5) *Kyoto Protocol.* Using the information provided in this chapter in conjunction with an Internet search, decide which of the various impacts and contributions of the Kyoto Protocol you feel is most important. In a short memorandum, describe your rationale for selecting the impact or contribution of the Kyoto Protocol.

(6) *Renewable energy.* Given the economic and energy issues of the early 21st century, develop a bulleted list of the influences of renewable energy issues on forest management. Insert this bulleted list into a one-page fact sheet (perhaps with accompanying pictures) that can be used to inform nonprofessionals.

(7) *Exploring your creative side.* With the concepts this chapter provided in mind (i.e., forest policies), and working within a group of four or five other students, brainstorm the design of an Internet application that could be used, through a phone or tablet, to provide people with information on these topics. Describe your ideas in a short PowerPoint presentation to your class.

References

Aguilar, F., Gaston, C., Hartkamp, R., Mabee, W., Skog, K., 2011. Wood energy markets, 2010–2011. In: Forest Products Annual Market Review 2010–2011. United Nations Economic Commission for Europe/Food and Agriculture Organization of the United Nations, Forestry and Timber Section. Geneva Timber and Forest Study Paper 27, Geneva, Switzerland, pp. 85–98.

Aguilar, F.X., Abt, K., Dmitriev, V., Glavonjić, B., Mabee, W., Sudekum, H., Vasilyev, O., 2018. Wood energy markets. In: UNECE/FAO Forest Products Annual Market Review, 2017-2018. United Nations Economic Commission for Europe/Food and Agriculture Organization of the United Nations, Forestry and Timber Section, Geneva, Switzerland, pp. 92–103.

Alig, R., 2003. U.S. landowner behavior, land use and land cover changes, and climate mitigation. Silva Fennica 37, 511–527.

Aukland, L., Costa, P.M., Brown, S., 2003. A conceptual framework and its application for addressing leakage: the case of avoided deforestation. Climate Policy 3, 123–136.

Barua, S.K., Uusivuori, J., Kuuluvainen, J., 2012. Impacts of carbon-based policy instruments and taxes on tropical deforestation. Ecological Economics 73, 211–219.

Bigsby, H., 2009. Carbon banking: creating flexibility for forest owners. Forest Ecology and Management 257, 378–383.

Blaser, J., Sarre, A., Poore, D., Johnson, S., 2011. Status of Tropical Forest Management 2011. International Tropical Timber Organization, Yokohama, Japan. Technical Series 38, 418.

Bulte, E.H., van Kooten, G.C., 1999. Economic efficiency, resource conservation and the ivory trade ban. Ecological Economics 28, 171–181.

Caulfield, J., 2003. Timber Value Trends: Shock and Awe. Florida Forests, vol. 7(2): 12-15. Florida Forestry Association, Tallahassee, FL.

Chomitz, K.M., 2002. Baseline, leakage and measurement issues: how do forestry and energy projects compare? Climate Policy 2, 35–49.

Cirelli, M.T., Smithüsen, F., Texier, J., Young, T., 2001. Trends in Forest Law in Europe and Africa. Food and Agriculture Organization of the United Nations, Rome, Italy. FAO Legislative Study 66, 147.

Clapp, E.H., 1941. Federal forest policies of the future. Journal of Forestry 39, 80–83.

Convention on Biological Diversity, 2011. Convention on Biological Diversity. Secretariat of the Convention on Biological Diversity, Montreal, Quebec. www.cbd.int (accessed 08.01.20).

Convention on International Trade in Endangered Species of Wild Fauna and Flora, 2018. The CITES Species. Convention on International Trade in Endangered Species of Wild Flora and Fauna, Geneva, Switzerland. https://www.cites.org/eng/disc/species.php (accessed 08.01.20).

Cubbage, F.W., Newman, D.H., 2006. Forest policy reformed: a United States perspective. Forest Policy and Economics 9, 261–273.

Cubbage, F.W., O'Laughlin, J., Bullock III, C.S., 1993. Forest Resource Policy. John Wiley & Sons, Inc., New York, 562 p.

DANA Publishing, 2008. The International Tree-Based Carbon Emissions Trading Industry. DANA Publishing, Rotorua, New Zealand.

De La Torre, R., Smith, M., 2008. Increasing carbon sequestration returns. Forest Landowner 67 (3), 5–10.

Enkvist, P., Naucler, T., Rosander, J., 2007. A cost curve for greenhouse gas reduction. The McKinsey Quarterly 2007 (1), 35–45.

European Commission, 2019. Report from the Commission to the European Parliament, the Council, the European Economic and Social Committee, and the Committee of the Regions, Renewable Energy Progress Report. European Commission, Brussels. COM (2019) 225 final.

Farahani, S., Worrell, E., Bryntse, G., 2004. CO$_2$—free paper? Resources. Conservation and Recycling 42, 317–336.

Federal Reserve Bank, 2020. Nominal broad dollar Index. http://www.federalreserve.gov/Releases/H10/Summary/indexb_m.txt (accessed 08.01.20).

Fernholz, K., Howe, J., Bratkovich, S., Bowyer, J., 2010. Forest Certification: A Status Report. Dovetail Partners, Inc., Minneapolis, MN.

Food and Agriculture Organization of the United Nations (FAO), 2010. Global Forest Resources Assessment 2010. Food and Agriculture Organization of the United Nations, Rome, Italy. FAO Forestry Paper 163.

Food and Agriculture Organization of the United Nations (FAO), 2019a. Forest Products 2017. FAO Statistics, Yearbook. Food and Agriculture Organization of the United Nations, Rome, Italy.

Food and Agriculture Organization of the United Nations (FAO), 2019b. FAOSTAT database. http://www.fao.org/faostat/en/#data/FO (accessed 08.01.20).

Food and Agriculture Organization of the United Nations (FAO), 2020. FAOSTAT Data. http://www.fao.org/faostat/en/#data/FO (accessed 12.03.20).

Forest Stewardship Council, 2010. FSC-US Forest Management Standard. Forest Stewardship Council, Minneapolis, MN, 109 p.

Forest Stewardship Council, 2020. Home Page. Forest Stewardship Council, Minneapolis, MN. https://us.fsc.org/en-us (accessed 13.02.20).

Gerasimov, Y., Karjalainen, T., 2010. Atlas of the Forest Sector in Belarus. Working Papers of the Finnish Forest Research Institute, No. 170. Finnish Forest Research Institute, Vantaa, Finland.

Ghasemi, S., Barynin, P., Malual, P., 2020. The World Timber Price Quarterly. RISI, Inc. Bedford, MA.

Guo, Z., Sun, C., Grebner, D.L., 2007. Utilization of forest derived biomass for energy production in the USA: status, challenges, and public policies. International Forestry Review 9, 748–758.

Hamilton, K., Sjardin, M., Marcello, T., Xu, G., 2008. Forging a Frontier: State of the Voluntary Carbon Markets 2008. Ecosystem Marketplace and New Carbon Finance, Washington, D.C.

Hamrick, K., Gallant, M., 2018. Fertile Ground: State of Forest Carbon Finance 2017. Forest Trends' Ecosystem Marketplace, Washington, D.C.

Harrop, S.R., Pritchard, D.J., 2011. A hard instrument goes soft: the implications of the Convention on Biological Diversity's current trajectory. Global Environmental Change 21, 474–480.

Holmgren, L., 2010. International Forest Policy—An Overview. Royal Swedish Academy of Agriculture and Forestry (KLSA). TIDSKRIFT nr 6 2010.

Kautto, N., Arasto, A., Sijm, J., Peck, P., 2012. Interaction of the EU ETS and national climate policy instruments—impact on biomass use. Biomass and Bioenergy 38, 117–127.

Kern, E., Rosenbaum, K.L., Silva, R.S., Young, R., 1998. Trends in Forestry Law in America and Asia. Food and Agriculture Organization of the United Nations, Rome, Italy. FAO Legislative Study 66.

Kraxner, F., 2011. Certified Forest Products Markets, 2010–2011. UNECE/FAO Forest Products Annual Market Review, 2010–2011. United Nations Economic Commission for Europe and FAO Forestry Department, Geneva, Switzerland. Geneva Timber and Forest Study Paper 27, ECE/TIM/SP/27.

Le, H.D., Smith, C., Herbohn, J., Harrison, S., 2012. More than just trees: assessing reforestation success in tropical developing countries. Journal of Rural Studies 28, 5–19.

Liang, S., Zhang, T., Zu, Y., 2012. Comparisons of four categories of waste recycling in China's paper industry based on physical input—output life-cycle assessment model. Waste Management 32, 603–612.

Longjun, C., 2011. UN convention to combat desertification. Encyclopedia of Environmental Health 504–517.

Lun, F., Li, W., Liu, Y., 2012. Complete forest carbon cycle and budget in China, 1999–2008. Forest Ecology and Management 264, 81–89.

Ministère de l'Écologie, de l'Énergy, du Développement durable et de la Mer, 2009. National Action Plan for the Promotion of Renewable Energies, 2009-2020. Ministère de l'Écologie, de l'Énergy, du Développement durable et de la Mer, Paris.

Ministry of Agriculture and Forestry, 2008. Government Resolution on the METSO Programme 2008–2016 (in Finnish). Ministry of Agriculture and Forestry, Helsinki, Finland.

Monte, M.C., Fuente, E., Blanco, A., Negro, C., 2009. Waste management from pulp and paper production in the European Union. Waste Management 29, 293–308.

Muhn, J., Stuart, H.R., 1988. Opportunity and Challenge, the Story of the BLM. U.S. Department of Interior. Bureau of Land Management, Washington, D.C.

Palmberg, C., Esquinas-Alcazar, J.T., 1990. The role of the United Nations agencies and other international organizations in the conservation of plant genetic resources. Forest Ecology and Management 35, 171–197.

Pan, Y., Birdsey, R.A., Fang, J., Houghton, R., Kauppi, P.E., Kurz, W.A., Phillips, O.L., Shvidenko, A., Lewis, S.L., Canadell, J.G., Ciais, P., Jackson, R.B., Pacala, S., McGuire, A.D., Piao, S., Rautiainen, A., Sitch, S., Hayes, D., 2011. A large and persistent carbon sink in the world's forests. Science 333, 988–993.

Parviainen, J., Västilä, S., Suominen, S., 2010. Finnish Forests and Forest Management. Finland's Forests in Changing Climate, Working Papers of the Finnish Forest Research Institute, No. 159. Finnish Forest Research Institute, Vantaa, Finland.

Petry, M., Lei, Z., Zhang, S., 2010. China—Peoples Republic of: Forest Products Annual Report 2010. U.S. Department of Agriculture, Foreign Agricultural Service, Washington, D.C. GAIN Report Number CH100042.

Programme for the Endorsement of Forest Certification, 2018. Sustainable Forest Management - Requirements. Programme for the Endorsement of Forest Certification Council, Geneva, Switzerland.

Programme for the Endorsement of Forest Certification, 2020. Facts and Figures. Programme for the Endorsement of forest certification, Geneva, Switzerland. www.pefc.org/discover-pefc/facts-and-figures (accessed 13.02.20).

Reeves, T., Mei, B., Bettinger, P., Siry, J., 2018. Review of the effects of conservation easements on surrounding property values. Journal of Forestry 116 (6), 555–562.

Richter, D.B., Jenkins, D.H., Karakash, J.T., Knight, J., McCreery, L.R., Nemestothy, K.P., 2009. Wood energy in America. Science 323, 1432–1433.

Sander, K., Peter, C., Gros, C., Huemmer, V., Sago, S., Kihulla, E., Daulinge, E., 2010. Enabling Reforms: A Stakeholder-Based Analysis of the Political Economy of Tanzania's Charcoal Sector and the Poverty and Social Impacts of Proposed Reforms. The World Bank, Washington, D.C. Report Number 55140.

Schwarz, C.F., Thor, E.C., Elsner, G.H., 1976. Wildland Planning Glossary. U.S. Department of Agriculture, Forest Service, Pacific Southwest Forest and Range Experiment Station, Berkeley, CA. General Technical Report PSW-13.

Sedjo, R., Marland, G., 2003. Inter-trading permanent emissions credits and rented temporary carbon emission offsets: some issues and alternatives. Climate Policy 3, 435–444.

Seneca Creek Associates and Wood Resources International, 2004. Illegal Logging and Global Wood Markets: The Competitive Impacts on the U.S. Wood Products Industry. A report prepared for the American Forest & Paper Association, Washington, D.C.

Simmons, C.S., 1997. Forest management practices in the Bayano region of Panama: cultural variations. World Development 25, 989–1000.

Siry, J.P., Cubbage, F.W., Ahmed, M.R., 2003. Sustainable forest management: global trends and opportunities. Forest Policy and Economics 7, 551–561.

Smith, J., Applegate, G., 2004. Could payments for forest carbon contribute to improved tropical forest management? Forest Policy and Economics 6, 153–167.

Ståhls, M., Saikku, L., Mattila, T., 2011. Impacts of international trade on carbon flows of forest industry in Finland. Journal of Cleaner Production 19, 1842–1848.

Stupak, I., Lattimore, B., Titus, B.D., Smith, C.T., 2011. Criteria and indicators for sustainable forest fuel production and harvesting: a review of current standards for sustainable forest management. Biomass and Bioenergy 35, 3287–3308.

Sustainable Forestry Initiative, 2015. SFI 2015-2019 Extended through December 2021 Forest Management Standard. Sustainable Forestry Initiative, Inc., Washington, D.C.

Sustainable Forestry Initiative, 2019. Forests of Opportunity, 2019 Progress Report. Sustainable Forestry Initiative, Inc., Washington, D.C.

Teng, F., Tondeur, D., 2007. Efficiency of carbon storage with leakage: physical and economical approaches. Energy 32, 540–548.

United Nations Convention to Combat Desertification, 2020. UNCCD, United Nations Convention to Combat Desertification. Secretariat of the United Nations Convention to Combat Desertification, Bonn, Germany. https://www.unccd.int/ (accessed 13.02.20).

United Nations Educational, Scientific and Cultural Organization, 2020a. World Heritage Forest Programme. United National Educational, Scientific and Cultural Organization, Paris, France. https://whc.unesco.org/en/forests/ (accessed 13.02.20).

United Nations Educational, Scientific and Cultural Organization, 2020b. World Heritage List Statistics. United National Educational, Scientific and Cultural Organization, Paris, France. https://whc.unesco.org/en/list/stat#d32 (accessed 29.02.20).

United Nations Framework Convention on Climate Change, 1997. Kyoto Protocol. United Nations Framework Convention on Climate Change, Bonn, Germany. https://unfccc.int/sites/default/files/resource/docs/cop3/l07a01.pdf#page=24a (accessed 13.02.20).

United Nations Framework Convention on Climate Change, 2020a. The United Nations Convention of Climate Change. Secretariat of the United Nations Framework Convention on Climate Change, Bonn, Germany. https://unfccc.int/ (accessed 13.02.20).

United Nations Framework Convention on Climate Change, 2020b. What is the Kyoto Protocol? Secretariat of the United Nations Framework Convention on Climate Change, Bonn, Germany. https://unfccc.int/kyoto_protocol (accessed 13.02.20).

U.S. Energy Information Administration, 2020a. Table 10.1. Renewable Energy Production and Consumption by Source. U.S. Energy Information Administration, Washington, D.C. www.eia.gov/totalenergy/data/monthly/pdf/sec10_3.pdf (accessed 03.03.20).

U.S. Energy Information Administration, 2020b. Table 2.1. Energy Consumption by Sector. U.S. Energy Information Administration, Washington, D.C. www.eia.gov/totalenergy/data/monthly/pdf/sec2_3.pdf (accessed 03.03.20).

U.S. Environmental Protection Agency, 2016. Wastes - Resource Conservation - Common Wastes & Materials - Paper Recycling. Frequent Questions. U.S. Environmental Protection Agency, Washington, D.C. https://archive.epa.gov/wastes/conserve/materials/paper/web/html/faqs.html (accessed 13.02.20).

U.S. Foreign Agricultural Service, 2020. Historical Foreign Agricultural Service U.S. Trade Online Database. U.S. Department of Agriculture, Foreign Agricultural Service, Washington, D.C. https://apps.fas.usda.gov/gats/default.aspx (accessed 14.03.20).

Voitleithner, J., 2002. The National Forest Programme in light of Austria's law and political culture. Forest Policy and Economics 4, 313–322.

Walters, B.B., Sabogal, C., Snook, L.K., de Almeida, E., 2005. Constraints and opportunities for better silvicultural practice in tropical forestry: an interdisciplinary approach. Forest Ecology and Management 209, 3–18.

West Virginia State Tax Department, 2018. Timber Severance Tax. West Virginia State Tax Department, Tax Services Division, Charleston, WV. Publication TSD-211.

World Bank, 2010. World Development Report 2010: Development and Climate Change. World Bank, Washington, D.C., 439 p.

CHAPTER

16

Urban forestry

Although many people have the impression that forestry and natural resource management activities are practiced solely within forests and wilderness areas, in fact these types of activities can also be practiced within cities, municipalities, or in any area that grows trees. Urban and suburban parks are examples of forests within areas developed by humans, although perhaps for different purposes than more remotely situated forest areas. Trees that line roadways and city streets can also be considered urban forests. The focus on trees that are established in developed areas makes the practice of urban forestry somewhat different from traditional forestry and natural resource management, and both arboricultural and silvicultural activities may be necessary during their lifespan. For example, while practices such as thinning and pruning that have been described earlier may be necessary in urban forest areas, other practices that address individual tree crown structure may also be necessary to address tree health and human safety concerns. The challenges posed by citizens who reside near urban forests also make work in this area unique, as significant existence and aesthetic values may be involved. This chapter provides an overview of forestry concepts that can be applied in an urban environment. Upon its completion, readers should be able to

- understand the concept of urban forestry;
- identify the benefits urban forests offer;

- understand the environmental conditions faced by urban forests; and
- understand the basic tools for managing urban forests.

The tending of plants within areas inhabited or frequently used by humans can involve planting, pruning, fertilization, pest control, and perhaps removal activities. *Arboriculture* is the management of individual trees or shrubs, usually within developed areas, that does not necessarily involve financial objectives. This practice may date back at least 4000 years to the tending of shade-intolerant oil palm (*Elaeis guineensis*) trees in western Africa (Logan and D'Andrea, 2012). In managing land to meet the needs of humans and other species today, arborists or tree wardens implement actions on individual trees that are necessary to maintain or enhance aesthetic values, property values, and infrastructural services (e.g., electricity). However, the use of management activities on urban trees to enhance ecosystem services (e.g., energy and climate management, water quality, biodiversity) is becoming increasingly important and is expected to become as significant as some traditional objectives (beautification or public services) (Young, 2010), even though the removal of dead trees or trees that present a hazard remains the top priority in some areas of the world (Rines et al., 2010).

Introduction to Forestry and Natural Resources, Second Edition
https://doi.org/10.1016/B978-0-12-819002-9.00016-X

16.1 What is urban forestry?

When tree management activities are performed within a developed area and are applied to groups of trees along streets, within parks (Fig. 16.1), and around buildings, these types of actions are collectively known as urban forestry practices and are implemented by either arborists or foresters. In states where both are licensed, there is some controversy of what work areas are exclusive to forestry and what areas are exclusive to the arborist. Many who work in the area find it best to obtain both credentials. *Urban forestry* is therefore the practice of forest and natural resource management situated within developed areas (e.g., cities, towns, and suburban areas) that involves the management of trees and other natural systems for the health and well-being of human society. A more formal definition, provided by the Society of American Foresters in its *Dictionary of Forestry*, suggests that urban forestry is "the art, science, and technology of managing trees and forest resources in and around urban community ecosystems for the physiological, sociological, economic, and aesthetic benefits trees provide society" (Helms, 1998). While urban forestry uses some of the same practices as general forestry, it involves challenges that are somewhat different. For example, urban forests are regularly subjected to the influences of modern human life; they may be under stress due to restricted growing areas and exposure to pollution, and therefore more susceptible to insect and disease problems than their more natural counterparts (Tait et al., 2009).

The concept of *urban forestry* can be viewed differently from one country to the next, and it is important to understand the subtle differences and similarities involved. Konijnendijk (2003), for example, summarizes a number of ways in which urban forests and urban forestry are viewed in Europe. Finland defines an urban forest as one that is located within or near an urban area, where the main function of the forest is to facilitate recreational activities. These areas consist mainly of natural forest vegetation and exclude areas such as built parks with grassy lawns. Although Germany has no specific term for urban forests or urban forestry, urban forests are generally considered built and are often located on former agricultural or derelict lands. Further, the forests are specifically designed and managed to support recreational activities (Fig. 16.2). Greece defines urban forests as urban green spaces situated in towns and along city streets, in city parks and gardens, and on land that surrounds cities and towns that includes trees. Iceland defines urban forestry as the planting of trees within the legal boundaries of urban areas for the purpose of providing amenity values to local citizens. These amenity values include aesthetic values, recreational opportunities, shelter, and the production of timber or other products, as long as removals do not detract from other amenity values. Ireland considers the concept of urban forests similar to the way the concept is viewed in North America. Italy uses the term *urban greenery* instead of *urban forestry* and defines urban greenery as a designated open space in an urban area containing vegetation that is regularly managed. In Lithuania, urban forests include forests, trees, and other green areas within towns, cities, and other communities. Forests and parks in urban areas of Slovenia that focus on environmental and social functions, rather than the production of wood-based products, are considered urban forests. The Netherlands uses the seemingly singular term *urban green* to refer to natural areas, urban woodlands, parks, public gardens, and roadside trees. Finally, the United Kingdom views the practice of urban forestry as a multidisciplinary set of activities that benefit humans, involving the planning, design, establishment, and management of forests, trees, and other woodlands and the associated natural resources contained within (Konijnendijk, 2003).

FIGURE 16.1 Forested park below the Palace of the Popes in Viterbo, Italy. *Donald L. Grebner.*

FIGURE 16.2 Walking path near the Arena Nürnberger Versicherung in Nürnberg, Germany. *Brenda F. Grebner.*

These somewhat varied viewpoints of what urban forests and urban forestry entail have a common focus: the management of artificial or natural forests that are located in urban communities. Urban forestry often seems like an odd profession to forestry and natural resource students because it generally refers to the management of trees and forests located in urban areas instead of rural areas or more remote wildlands. However, urban forestry plays an important role in influencing how human societies value forests and natural resources, and is one way in which human communities can integrate themselves into local and regional ecosystems. In fact, for many urban people, experiences within urban forests may be their only contact with nature. In Chapter 7, we discussed the importance of forest recreation to human society and, although much of the material was devoted to forest recreational opportunities supported by national forests and parks, municipal lands also supply a large array of outdoor opportunities for urban and suburban residents. However, as might be gathered in the following discussion, urban forestry involves aspects of organization and management that extend well beyond the provision of recreational opportunities to the public.

16.2 A brief history of urban forestry

Forests and their associated natural resources have played an important role in human civilization for thousands of years. Throughout this book, trees and forests have been discussed primarily from the viewpoint of natural or regenerated stands located within rural landscapes. However, ancient African civilizations are known to have tended trees in urban communities over 4000 years ago (Phillips, 1993; Logan and D'Andrea, 2012). The ancient Babylonians built the Hanging Gardens as a series of roof gardens organized on ziggurat (stepped pyramids) terraces (Tian and Jim, 2011) that were irrigated with water pumped from the Euphrates River. In urban gardens in Islamic, Spanish, and other Mediterranean communities, figs and plane trees were planted around homes (Phillips, 1993). Ancient civilizations such as the Phoenicians and the Chinese used trees for aesthetic purposes as well as for sacred groves, and the Greeks enclosed sports grounds in groves of plane trees (Koch, 2000).

The turbulence of the Dark Ages and the emergence of the Middle Ages had an important impact on urban trees. The Dark Ages occurred following the gradual societal collapse of the Roman Empire in the 5th century AD. The Middle Ages, which lasted for about 1000 years, ended in the 14th century with the beginning of the Renaissance movement (14th to 17th centuries). With the decline of the Roman Empire, the population of many urban communities also declined as people returned to the countryside (Koch, 2000). The reduced size of existing communities lessened the need for urban trees, but home gardens were still commonly found. Over time, walled cities grew in number, and greater human population density led to intense pressures on open spaces within these communities. During the Middle Ages in Europe, humankind began to make conscious decisions regarding population density and associated land use (Williams, 2000). Upper segments of society began to consider the aesthetic value of trees, initiating the transformation of urban gardens into modern public green zones and street trees (Fig. 16.3) (Koch, 2000). During the Middle Ages, some urban forest areas in Europe were significant economic drivers and were used as sources of firewood for sale to citizens in nearby villages (Szabó, 2010).

The Renaissance period in Italy and France during the 16th century promoted the development of urban forests through the development of *garden allées*, or tree-lined paths or access corridors connecting landscape gardens and possibly enclosed by walls (Koch, 2000). Over time, these allées were transformed into wooded boulevards that extended everywhere and became centers of recreational use. In addition, wall promenades came into being as more and more trees were planted near city walls (Fig. 16.4). This expanded the environmental benefits generated by the garden allée and boulevard trees and led to the development of waterside promenades and malls. The development of waterside promenades coincided with increased river commerce traffic and the emergence of urban development plans such as Amsterdam's Plan of the Tree Canals in 1615, which required the planting of elm trees along new canals in residential neighborhoods (Mumford, 1961). Malls were grassy, open areas lined with trees and commonly used for

FIGURE 16.3 Public garden near Buckingham Palace, London, England, United Kingdom. *Donald L. Grebner.*

FIGURE 16.4 Trees planted near an old fortification wall in Nürnberg, Germany. *Brenda F. Grebner.*

recreational purposes. These malls were used for many years until they were converted to other uses (Koch, 2000). The French monarchy created recreational zones near the edges of its cities when city walls were torn down (Lawrence, 1988). As previously stated, trees had been extensively planted alongside the original city defenses, and the conversion of old military structures into recreational areas became popular over time, but especially so after the Napoleonic Wars. As many urban centers were treeless, urban expansion and reorganization led to the planting of street trees in different urban designs from either along the sides of transportation corridors or through their center (Fig. 16.5). The advent of the sidewalk (or pavement) by the British moved street tree planting away from the road's edge, which improved tree survival and helped reduce soil compaction (Lawrence, 1988). During the Renaissance

period and for nearly 200 years afterward, coppice forests of urban woodlands in what is today the Czech Republic were managed on regular entry cycles even though income from woodlands (mainly through sales of firewood) became less economically significant (Szabó, 2010).

The start of the Industrial Revolution in the United Kingdom and the United States had an important impact on the development of their respective urban forests. The British tightly controlled land use and city size while developing greenbelts around urban areas; these were linked to natural and undeveloped areas (Koch, 2000). The American experience was different, given that early European settlers perceived there to be limitless expanses of forestlands, which prompted widespread deforestation for agricultural and developmental purposes. As with the British experience, the advent of the Industrial Revolution in America, around 1850, led to overcrowding, social disorder, and greater pollution in numerous communities and prompted local concern for the development of urban forests (Fig. 16.6). In the American experience, the park development movement was seen as a reaction to the tensions created through industrialization (Pipkin, 2005), and this led to an interest by city leaders in beautification projects, such as New York City's Central Park. Unfortunately, population shifts prompted by outmigration of city residents, the creation of the national highway system, changes in industries, the aging of structures, changes in neighborhood economic status, increases in vehicular traffic, and other avoidance phenomena have led to urban or suburban decline (Lucy and Phillips, 1997; Koch, 2000; Rosenthal, 2008).

The modern urban forestry discipline in the United States began during the 1960s and 1970s (Johnston, 1996). Although technical assistance, training, and

FIGURE 16.5 Tree-lined street in downtown Viterbo, Italy. *Donald L. Grebner.*

FIGURE 16.6 Urban forest park in downtown London, England, United Kingdom. *Donald L. Grebner.*

research on urban forestry were available at this time, the Cooperative Forest Management Act of 1972 (16 USC. § 564) specifically promoted the establishment of trees and shrubs within urban communities. Through this Act (since repealed and replaced by the Cooperative Forestry Assistance Act of 1978 [16 USC. § 2101−2105]), a variety of programs for large urban areas and small towns were established in California, Colorado, Florida, Georgia, Kansas, Maryland, and Missouri (Grey and Deneke, 1986). In addition, the Society of American Foresters created an Urban Forestry Working Group in 1972 as a mechanism for integrating the resource management of urban communities within the broader fields of forestry and natural resource management. During the 1970s and 1980s, national level funding for urban forestry activities in the United States fluctuated with economic and political conditions, but the 1990 Farm Bill solidified federal support for urban forestry by expanding the authority of the U.S. Forest Service to enable it to work with states on urban forest issues (Alvarez, 2001; Jones, 2011). Furthermore, the America the Beautiful Act of 1990 (16 USC. § 2101[note]) helped states hire urban forestry coordinators and establish state urban forestry advisory councils (Jones, 2011). Since this time, numerous communities around the United States have expanded their urban forestry programs (Fig. 16.7), which has been accompanied by growth in research areas that address urban forest issues. Extensive interest by both small and large communities in some states, such as Mississippi, for implementing urban and community forestry programs has recently been reported (Grado et al., 2006). Unfortunately, a lack of funding and limited information is still a barrier to the establishment of useful urban forestry programs.

Although some countries in Europe were involved in the early development of urban forestry concepts, urban forestry as a scientific discipline did not emerge until the 1990s (Konijnendijk, 2003). Interestingly, the growing demand for outdoor amenities by European citizens spurred greater interest in urban ecology and urban green planning during the 1970s and 1980s. Scientific interaction with North American urban forestry professionals resulted in technology transfer conferences associated with the World Urban Forum of the United Nations, the International Union of Forest Research Organizations (IUFRO), and the Dutch State Forest Research Institute.

While prevalent in North America and other areas of the world, modern urban forestry seems to have begun in Great Britain (Konijnendijk, 2003) with a national Community Forests program of tree planting and forest management for environmental, social, and economic development in 12 urban areas. Ireland and the Netherlands quickly followed Great Britain's example, and these early European efforts led to the development of urban forestry networks in Scandinavian and Baltic countries. Furthermore, the European Forest Institute implemented a number of urban forestry research projects during the mid-1990s. Other urban forestry initiatives have since been sponsored by the European Union through seminars and conferences. In addition, the European Urban Forestry Research and Information Center was formed in 2001, which serves as a vehicle for assisting urban and community forestry initiatives across the European continent. The greenways movement, which can involve urban forests or urban forestry principles, has also spread throughout Europe and North America since the start of the 21st century. A greenway (Fig. 16.8) is a route (path, pedestrian zone,

FIGURE 16.8 Portion of the Irwin Creek greenway in Charlotte, North Carolina, United States. *Mx. Granger, through Wikimedia Commons. Image Link: https://commons.wikimedia.org/wiki/File:Irwin_Creek_Greenway,_Charlotte,_North_Carolina.jpg License Link: https://creativecommons.org/publicdomain/zero/1.0/deed.en*

FIGURE 16.7 Shade trees near the edge of the Drill Field at Mississippi State University, Starkville, Mississippi, United States. *Donald L. Grebner.*

etc.) reserved exclusively for nonmotorized travel and designed to connect people to the landscape, enhance the quality of the environment, and preserve cultural heritage, among other purposes (Toccolini et al., 2006). The conversion of a railway to a recreational trail, through the Rails to Trails programs in North America and Europe, can be considered as greenway development (Houston and Zuñiga, 2019).

16.3 The benefits of urban forestry

Some of the benefits of practicing urban forestry are similar to those of managed or natural forests located in rural areas and remote wildlands. However, there are some important differences, and in this section we will discuss several key benefits of urban forests: climate amelioration, enhancements in the urban environment from engineering, architectural, social, and aesthetic perspectives, and enhancement of real estate and wildlife habitat values.

16.3.1 Climate amelioration

Climate amelioration occurs when humans modify the climate to make their life more comfortable. Modification can take the form of changing the air temperature or reducing wind movement (Grey and Deneke, 1978; Miller, 1988; Phillips, 1993; Randrup et al., 2005). The modification of air temperature around buildings through the use of trees or other plants is one way that humans can control an aspect of the climate within urban environments. This type of act has been practiced by humans within large urban areas and around homesteads in rural areas for thousands of years. On the Earth's surface, heat arises from numerous sources: it is produced by the sun as well as by living organisms. As discussed in Chapter 6, electromagnetic energy from the sun is absorbed during photosynthesis. However, dark-colored features on Earth absorb this energy in greater quantities than do light-colored features. For example, a black surface will absorb more electromagnetic energy from the sun than will a white surface. Lighter colors, including white, reflect more energy back into the atmosphere and away from the Earth. A key difference between urban and rural communities is the average air temperature. Large urban centers are generally 0.5–1.5 degrees Celsius (°C; 0.9–2.7°degrees Fahrenheit [°F]) warmer than rural areas (Federer, 1976) and they do not cool as rapidly at night. Although this is not a large difference, it is enough to impact a person's living environment at different times of the year. The main explanation for this temperature difference is the greater abundance and density per unit area of

buildings, roads, and other built objects that absorb or generate heat in urban areas. During the day, buildings of all sizes and materials absorb electromagnetic energy from the sun. Materials such as asphalt, concrete, glass, tar on roofs, steel, shingles, and slate are poor insulators and absorb and dissipate heat more readily than do soil or vegetation (Grey and Deneke, 1978). Heat is then usually transferred by convection, a method used to cook food in an oven, from these materials to the surrounding air.

The presence of trees and other plants can, through the generation of shade, reduce local air temperatures; thus, urban vegetation can be used as a means for addressing goals along these lines developed by a city or municipality (Nowak, 2006). Therefore, a key benefit of urban forestry is the ability of trees and plants to ameliorate air temperatures in urban environments (Grey and Deneke, 1978; Miller, 1988; Phillips, 1993; Randrup et al., 2005). Plants intercept light, absorb it from the sun, and either reflect the light back into the atmosphere or transmit it through the vegetation. Interestingly, the type of plant leaf (i.e., size, shape, and branching pattern) plays an important role in this process. As a result, many deciduous tree species may be very important in controlling heat levels within urban environments (Grey and Deneke, 1978). During the summer months, many people prefer the foliage of deciduous trees since they provide shade and reduce the temperature of the air underneath them (Fig. 16.9). Grey and Deneke (1978) refer to trees as nature's air conditioners because they can transpire up to 400 liters (L; 88 gallons [gal]) of water per day, which is equivalent to operating five average room air conditioners for 20 hours a day at 2.9 kWh per hour. During the winter months, deciduous trees lose their leaves, allowing

FIGURE 16.9 Shaded trees along walkway lower ambient temperatures underneath, Mississippi State University, Starkville, Mississippi, United States. *Donald L. Grebner.*

energy from the sun to penetrate the forest canopy and warm the ground. However, the lack of foliage on trees or plants can sometimes facilitate a decline in air temperature when winds are relatively strong.

Another aspect of heat transfer in an urban environment involves the time of day. During daytime hours, most materials that are directly contacted by sunlight absorb heat, but at night the process is reversed, and heat absorbed by these materials is emitted back into the atmosphere. The rate at which this occurs from various materials is important, as some materials are poor heat insulators and gain and dissipate heat quickly. Materials such as trees do not dissipate heat very quickly. During the day, trees provide a cooling effect (Fig. 16.10), while at night they provide a warming effect because of the slow rate at which they allow heat to be emitted back into the atmosphere. This warming effect can range from 6 to 8°C (10−15°F) (Federer, 1976). Air movement and wind protection are both important associated aspects of climate amelioration that are provided by trees in an urban environment. Wind can affect evaporative cooling processes, depending on the local environmental conditions. For example, on hot days, wind can be quite desirable; however, this is an illusionary effect and does not change the local air temperature. Unfortunately, wind can also displace the cool, moist air found under tree canopies with hotter, drier air from elsewhere.

Trees can also protect human communities from the effects of wind by disrupting wind direction and velocity, reducing radiant energy absorbed, reducing evapotranspiration, and creating sheltered areas (Baer, 1989; McPherson and Simpson, 2003). Trees planted to block direct contact with solar radiation during the hottest parts of a day, and those planted as wind barriers, can reduce temperature extremes; thus, this insulation layer may save energy and result in lower home heating and cooling bills (Perkins et al., 2004; Aboelata and Sodoudi, 2019). It has been suggested that about one-quarter of home heating costs can be saved during the winter with appropriately placed trees (Robinette, 1972). This can be especially desirable during periods of cold weather. In California, some suggest that strategic planting of shade trees on the east and west sides of buildings can reduce the peak load demand for energy by nearly 5% (McPherson and Simpson, 2003), while others suggest that urban tree planting programs may result in a 25% reduction in energy use for cooling and heating purposes (Akbari, 2002). Obtaining these effects often requires the presence of a tree species (such as conifers) with continuous foliage along the bole; this condition makes these species ideal for protecting human structures against windy conditions (Fig. 16.11). Dense rows of coniferous trees, for example, can disrupt wind flows near homes and create dead-air zones that act as a form of insulation (Grey and Deneke, 1978). Although we are currently discussing urban environments, the planting of trees around homes was quite common during the 1930s as a form of insulation. Tree planting programs in the Great Plains areas of the United States, such as those sponsored by the Civilian Conservation Corps, promoted the use of trees for this purpose (Baer, 1989).

FIGURE 16.10 Cooling effect of trees can protect sidewalks and automobiles. *Donald L. Grebner.*

FIGURE 16.11 Coniferous tree species are effective in sheltering residential buildings from weather conditions, Nürnberg, Germany. *Donald L. Grebner.*

Protecting structures or local environments from the effects of wind typically involves some form of a windbreak or shelterbelt as discussed in Chapter 11. A simple windbreak can consist of a single row of trees, which can act to reduce wind speeds and provide some insulation during the winter (Baer, 1989). Multiple rows can enhance this effect; however, it is difficult to establish multiple rows of trees in large urban areas where land area is expensive. There are two important issues to consider when using trees to protect against wind. First, when winds hit a line of trees, they will generally move up and over the top of the tree canopy (Baer, 1989); however, tree canopy characteristics and the resulting air flow permeability of forests can vary (Lee et al., 2010). A useful rule of thumb is that the average height of the tallest row of trees should represent the horizontal distance of protection behind the wind barrier. Second, when establishing wind barriers, trees should not be too densely planted. It is important to allow some permeability of wind into the windbreak to prevent a turbulence zone forming immediately behind the windbreak. Wind barriers that are nearly completely impermeable may be more vulnerable to damage, since the forces applied by the wind are not alleviated owing to the low permeability of the dense canopy.

In cold winter climates, tree barriers are also very effective in reducing the depth of snow and ice on roads (Baer, 1989) and sidewalks by trapping snowdrifts. This can effectively reduce the time and effort required to remove snow and ice from sidewalks, roads, and parking lots. Given the difficulties in today's global economies, this cost reduction in storm cleanup can be significant to local governments, private businesses, and private individuals. On another positive note, this function of windbreaks can also be beneficial for areas that need snow to accumulate, such as ski resorts and toboggan runs, and at areas requiring soil moisture replenishment (Grey and Deneke, 1978). Further, windbreaks have long been used to control soil erosion (Kučera et al., 2020).

As briefly discussed earlier, trees also play an important role in the hydrologic cycle. Trees and shrubs can intercept precipitation and reduce soil erosion processes; however, coniferous and deciduous species differ in their ability to do so. For example, some pine forests may allow only 60% of precipitation to reach the ground, while some deciduous forests may allow up to 80% (Grey and Deneke, 1978). In addition, the shade produced by trees not only decreases the air temperature underneath their canopy but also decreases water evaporation above and below ground. On the other hand, trees can also consume large quantities of water, depending upon the species. Eucalyptus, for example, is thought to be very efficient at capturing groundwater, while others, such as poplar (*Populus simonii*), may

exhibit low resistance to drought and therefore may not grow well when water levels are low (Liang et al., 2006).

16.3.2 Engineering benefits

Urban forestry plays an important role in addressing environmental engineering problems, including those related to erosion control, noise and air pollution abatement, wastewater management, watershed protection, and glare, reflection, and traffic control (Grey and Deneke, 1978; Miller, 1988). Although erosion control and watershed protection are important issues in rural areas, they are also a major concern in urban centers. Urban vegetation (trees and other plants) can be used to mitigate extreme stormwater runoff events in urban areas, and in many cases this role can justify the attention placed on the development and maintenance of urban forests (Sanders, 1986). Urban trees can affect stream flows as well, by their ability to intercept rainfall and affect soil infiltration rates of water. These acts can reduce surface water runoff, which can indirectly affect water quality by reducing the amount of sediment and urban pollutants entering a stream system (Nowak, 2006).

Erosion, however, can be caused by both wind and water, although wind has less of an effect than water does. Water-caused erosion is a particular concern in areas (Fig. 16.12) where bare soil is exposed to precipitation. Soil loss due to precipitation runoff can occur when rills (narrow incisions into the upper portion of a soil surface) or gullies (ditches resulting from significant erosion of a soil surface) are formed on the ground, when upper layers of soil are washed away in a large sheet, or when soil is lost in a mass, downhill movement known as a mudslide or landslide (Grey and Deneke, 1978). Landslides triggered by precipitation events are a perpetual

FIGURE 16.12 Urban erosion in Ankeny, Iowa. *Lynn Betts, U.S. Department of Agriculture, Natural Resources Conservation Service (http://luirig.altervista.org/pics/display.php?pos=137644).*

problem in many coastal areas of both developed and developing countries (Harp et al., 2009). Some methods for improving the hydrology of urban areas include perforating compacted areas to improve water absorption and infiltration, planting trees and shrubs to bind the soil, implementing terracing practices and contour planting, and applying mulch to construction sites (Grey and Deneke, 1978).

An important engineering issue that could be addressed by urban forests involves wastewater management (Miller, 1988). As the world population increases, the need to dispose of the daily wastewater produced will also increase, and urban forests have been used to absorb some of this. Several countries, including Egypt, Kuwait, Peru, and Yemen, use wastewater in their urban forests (Smit et al., 2001). Many countries around the world, including the United States, use this type of process for irrigating agricultural crops (Thebo et al., 2017), yet water supply and drainage systems are not perfect, thus the overland flow of wastewater can find its way into forested areas within and outside urban areas (Kaye et al., 2006). The effects are mixed, since nutrient levels can be limited in some forested ecosystems so adding nutrients to the system can potentially boost productivity; however, soils may retain heavy metals and high levels of phosphorous as a result (Brockway et al., 1986).

Another possible engineering use of an urban forest is for noise abatement purposes (Fig. 16.13). In general, taller trees and wider tree groups are most effective in reducing noise pollution (Fang and Ling, 2005). Noise pollution is composed of sounds, and sounds can be absorbed, deflected, reflected, or refracted. Reflected sounds bounce off objects, whereas absorbed sounds are trapped by objects. Refracted sounds are broken up and dissipated into the atmosphere, and deflected sounds bounce off objects and can be directed toward an area

of least concern (Grey and Deneke, 1978). Noise abatement is often necessary in areas where excessive noise occurs, such as along roadways. Given the high density of people living in urban areas and their associated transportation systems, noise often becomes a significant problem for local residents. Noise can be composed of different wavelengths of energy, and the manner in which it travels through the environment can be affected by many factors. For instance, the level of outdoor noise is a function of the source, the terrain, the surrounding vegetation, and atmospheric conditions such as wind speed and air temperature (Grey and Deneke, 1978; Miller, 1988). Typically, as noise emanates from a source it decreases in intensity the further away it travels. In other words, noise is loudest the closer one is to it. As noise travels, it spreads in a spherical pattern across the landscape away from its point of origin. Interestingly, urban forests that are closer than about eight average tree heights from a noise can reduce noise pollution more effectively than trees situated further away (Fang and Ling, 2005). Plant arrangement and density can moderate noise levels and combining plants with different noise attenuation characteristics can more effectively reduce unwanted sounds (Miller, 1988; Fan et al., 2010). Noise heard upwind may be reduced by 25 to 30 dB compared to noise downwind (Grey and Deneke, 1978). This effect occurs because noise is directed upward toward the atmosphere when it travels upwind and is directed downward to the ground when it travels downwind.

FIGURE 16.14 Public park serves as an air pollution filter in downtown Ljubljana, Slovenia. *Donald L. Grebner.*

FIGURE 16.13 Row of Leyland Cypress (*Cupressocyparis leylandii*) trees protect a residential house from traffic noise on a local road by acting as a sound barrier, Starkville, Mississippi, United States. *Donald L. Grebner.*

Urban forests can also contribute to reductions in air pollution and improvements in air quality (Escobedo et al., 2011). Trees found in urban settings play very important roles in cleansing pollutants from the air (Fig. 16.14). Whether trees are located in rows along a street, in parks, or in undeveloped areas, their leaves, branches, twigs, and boles can trap air pollutants. Typically, these trapped pollutants are washed into the ground during rainfall. Grey and Deneke (1978) refer to this process as airwashing. In addition, when trees are flowering, their floral scents can mask disagreeable odors. Tree and shrub vegetative surfaces, therefore, play an important role in the interception of particulate matter pollution (Tallis et al., 2011). The level of interception typically depends on particle shape and density, and the tree species employed. Plant vegetation can take up trace metal particles, such as lead, as well as radioactive particles, such as cesium isotopes (Smith, 1990). Trees, such as cottonwoods (*Populus* spp.), have been used for arsenic and trichloroethylene uptake from the soil (Cardellino, 2001; Rockwood et al., 2001; Wilde et al., 2003; U.S. Geological Survey, 2016).

Urban forests can also be used to facilitate carbon sequestering efforts. As discussed in Chapters 4 and 6, carbon sequestration occurs when trees accumulate carbon and release oxygen (O_2) essential for life. The conversion of forests to land uses such as roads, homes, office buildings, factories, shopping centers, sports stadiums, and airports releases carbon dioxide into the atmosphere, which contributes to climate change. Urban tree planting efforts, whether through green belts, parks, windbreaks, or shade trees around residential houses, can play an important role in the sequestration of carbon. One example is the nearly 55,000 tons of carbon that is expected to be sequestered through an urban tree program in Pretoria (Tshwane), South Africa (Stoffberg et al., 2010). The development and maintenance of urban forests, coupled with sustainable living concepts such as recycling and wind, solar, and other renewable energy technologies, can be used to address the growing concern of global climate change. Interestingly, in one analysis a park-like design of an urban forest seemed to be less effective for carbon sequestration purposes than a forest-like design due to emissions from construction and maintenance activities (Strohbach et al., 2012).

A final engineering use of urban forests can involve their ability to filter or block glare and reflected light (Beatty and Heckman, 1981; Smardon, 1988; Villalba et al., 2016). Materials such as light-colored concretes, glass, water, snow, and metals can reflect light in ways that cause difficulties involving increased heat and concentrated light energy. The strategic placement of trees can mitigate some of these problems. In addition to reducing glare imposed on homes, urban trees can reduce glare imposed on automobile drivers and pedestrians, which increases human health and safety conditions.

16.3.3 Architectural and aesthetic benefits

Trees and urban forests are also important for addressing numerous architectural issues commonly found in urban environments (Grey and Deneke, 1978; Miller, 1988; Phillips, 1993). These issues may include providing a sense of privacy in an area with many human inhabitants, providing space articulation, and providing aesthetically pleasing views or screening unsightly views (Fig. 16.15) such as junkyards or other undesirable features that can reduce neighborhood satisfaction (Ellis et al., 2006). Trees can be used to define space and can make larger areas appear smaller, while serving to draw attention to particular features of an urban landscape. In addition, the proper use of plants can make a place more inviting to visit. Even small urban forests can give those who visit it a sense of solitude, which could be important for their emotional, spiritual, and physical well-being. In addition, plants and trees can cause buildings to appear friendlier and to occupy more space than they actually do. In summary, urban trees can be used to address both aesthetic issues and utilitarian issues (noise or privacy) of humans (Crow et al., 2006).

16.3.4 Enhancement of real estate values

Another potential contribution of urban forests is the economic benefits accrued to residential homeowners

FIGURE 16.15   Two rows of young Leyland Cypress (*Cupressocyparis leylandii*) trees planted as a visual barrier in Starkville, Mississippi, United States. *Donald L. Grebner.*

and businesses, which can be both direct and indirect (Grey and Deneke, 1986; Miller, 1988). For homeowners, urban trees can offset heating and cooling costs while at the same time increasing the home's real estate value in the local market by increasing the home's street appeal. Work by Correll et al. (1978) showed that property values fell by US$13.78 per meter (m; US$4.20 per foot [ft]) the further a property was from a green belt. In Austin, Texas, Nicholls and Crompton (2005) showed that being near greenways may have significant positive impacts on a property's sale price. Another study (Carleyolsen et al., 2005) suggests that urban parks and trails not only generate US$13 million in total value-added income and 420 jobs in Jefferson County, Wisconsin, but the total market value of properties near park and state wildlife areas are likely to be enhanced from US$8 million to US$42 million due to their location. Business owners accrue some of the same benefits, as urban tree plantings can have a positive impact on their employees and can attract customers to their establishments (Dwyer et al., 1992). The value of benefits associated with beautification, privacy, and a sense of place and well-being provided by urban forests is difficult to determine; however, it might be reflected in the differences observed in the sales prices or property values of land with (or near) and without (or far from) urban forests or urban trees (McPherson and Simpson, 2002).

This discussion would not be complete without an acknowledgment of the costs associated with managing urban trees. In efforts to expand, improve, or simply maintain urban forests or urban trees, money and other resources are needed, and these may conceivably have been diverted from other potential uses (Dwyer et al., 1992). Costs of urban forest programs may include higher annual ground maintenance to dispose of fallen leaves, damaged branches, and tree trunks. Improper planting of trees can result in increased costs of maintenance and replacement (Dwyer et al., 1992). In addition, the roots from trees near sidewalks may disrupt the surface and damage concrete slabs, which would then require maintenance. Other maintenance costs also involve managing the presence of other undesirable tree species and possibly their seedlings (de Medeiros et al., 2019). Other direct costs may involve watering the trees and may also be incurred when tree limbs fall on power lines, homes, or automobiles. Further, costs can be incurred to remove and replace urban trees (Nowak and Aevermann, 2019). More indirect costs include an increased level of anxiety in people who harbor a fear of trees and associated wild environments (Dwyer et al., 1992), the potential for trees to exacerbate allergies, and their potential to drip sap on vehicles and other features of importance (Lohr et al., 2004).

16.3.5 Enhancement of wildlife habitat

Urban forests not only provide benefits to human homeowners but can also facilitate habitat requirements for a variety of wildlife species (Grado et al., 2008). Trees and shrubs located around homes and office buildings provide nesting environments for various avian species, as well as for insects and reptiles. The flowers and fruits produced act as food sources for many different species. In some cases, mammals, such as the white-tailed deer (*Odocoileus virginianus*), frequently enter residential areas to browse on the abundant plant life without fear of being hunted. The presence of these nuisance deer has spawned the development of a number of commercial deer repellents, as well as the development of capture and removal methods (Tregoning and Kays, 2003). Ultimately, a healthy urban forest can help protect urban streams and water bodies, promoting improved water quality and viable aquatic plant and fish populations. Some cities and private organizations are actively developing urban forest areas with the partial objective of developing wildlife habitat (Draus et al., 2019). However, urban forests can also invite unwanted wildlife, which can result in damage to plants and structures, threats to domestic pets, and the presence of wildlife droppings (defecation) (Dwyer et al., 1992).

16.3.6 Social benefits

The presence of urban forests can contribute to the quality of life of human civilizations and therefore transform an urban area into an enjoyable place to work, live, and recreate in (Dwyer et al., 1992); in general, urban forests can help people achieve a sense of calmness (Lohr et al., 2004). Urban forests can contribute to human health, a sense of community, and a sense of personal identity (Hull et al., 1994). One study conducted in Switzerland noted that leisure activities occurring in urban forests and public green spaces, and designed for children and youths, had the potential to facilitate social interaction between Swiss residents and young immigrant people, and could therefore play an important role in stimulating social interactions across cultures (Seeland et al., 2009). In Turkey, some kindergarten schools regularly take children to urban forest areas and, as they rise through the school system, these children seem to have improved learning skills (Kara et al., 2011). The emotional and symbolic experiences people enjoy within urban forests can be significant, and the enhancement of urban recreational opportunities may convince some people to spend their leisure time in close proximity to their homes. This type of decision can lead to a reduction in the use of fossil fuels (and perhaps air pollution) if the alternative were to travel to a more remote recreational area (Dwyer et al., 1992).

16.4 The environment of urban forestry

When the establishment of urban forests is being considered, it must be understood that the environment they will exist within what may be very different in character from natural wildland locations. While one can often find examples of large, old urban trees, their location (downtown vs. residential or park) and the stresses they face can influence their lifespan, which one meta-analysis of a set of urban forest studies suggests can be as low as 19 to 28 years for street trees (Roman and Scatena, 2011). When forests are converted to support a built structure, the conversion process can severely alter natural ecosystem processes. Several environmental issues can affect the development and maintenance of an urban forest, including space, soil, microclimate, pollution, and people (Grey and Deneke, 1978).

Space may appear to be an odd factor to consider in the cultivation of urban forests. However, it is relevant because it concerns the horizontal and vertical environment a tree or a plant needs to become established and to grow and maintain a healthy life cycle. An assessment of space requires that we must consider not only a tree's health but also how it will fit within the human structures already existing across an urban landscape. For instance, planting trees along a road or sidewalk (Fig. 16.16) offers many benefits, such as shade and wildlife habitat for birds. However, planting trees with too little ground area causes growth problems for the trees and, as they grow, their roots can buckle the nearby sidewalk or road surface (Fig. 16.17). When the vertical space requirements are being considered, the natural height that a tree or plant species will attain during its lifetime must be assessed. Planting a tree species that grows to 30 m or more (about 100 ft) directly beneath a power line can lead to a potential

FIGURE 16.17   Tree roots buckle sidewalk in Viterbo, Italy. *Donald L. Grebner.*

hazard. Another potential hazard can result from planting trees that are likely to become inordinately large and tall next to densely packed homes, owing to the risk of damage to the homes during inclement weather.

The conversion of a natural ecosystem to a type of built structure greatly impacts the supporting services that soils provide. When buildings are constructed, all of the existing vegetation is typically stripped from the ground and the topsoil is removed. The residual soils can also be compacted, and in this case preexisting drainage patterns may be altered. Parking lots and roads allow little or no absorption of rainwater into the soil horizon, leading to rapid stormwater runoff and erosion of exposed topsoil. The resulting unsuitable soil conditions may require some sort of amelioration prior to a tree planting effort. Therefore, certain tree species may be more vulnerable than others to windthrow or root damage when planted in unsuitable soil conditions, and root development can be enhanced by planting these species in engineered soils or soils that contain a mixture of crushed stone and clay (Bartens et al., 2010).

The type of microclimate surrounding trees is important to their survival. Urban environments are typically warmer during the summer and winter months than are rural landscapes. Warmer temperatures can expose trees to higher levels of evapotranspiration during the summer but can provide support against harsh conditions during the winter. It is therefore important to select the proper plant or tree species for planting in the various microclimates that may exist in urban areas.

Pollution is another common problem in urban forest environments and may appear in the form of air, soil, water, or light pollution. Urban forests suffer from a variety of air pollutants derived from automobiles, trucks, trains, local industries, and power generation facilities. The types of pollutants produced include nitrogen oxides,

FIGURE 16.16   Tree-lined bike paths and walkways provide bird and small animal habitat, Mississippi State University, Starkville, Mississippi, United States. *Donald L. Grebner.*

ammonia, sulfur dioxide, ozone (O₃), ethylene, chlorine, hydrogen chloride, fluorides, and particulates (Grey and Deneke, 1978). The impact of these chemicals on trees and plants varies depending on the density and frequency of chemical deposition as well as the physical characteristics of the plant species. Soil pollution can include the presence of oils, gasoline, salts, detergents, or other industrial chemicals, many of which can inhibit plant growth and are a hazard to humans. Water from rainstorms can leach harmful chemicals into the ground and once there they can be taken up by plant root systems. Roads that can become covered with snow and ice are commonly treated with salt products, which can accumulate over time and adversely impact plant growth. For these areas, an understanding of a plant species' level of tolerance to high salt concentrations can facilitate the successful development of an urban forest. Some tree species with good salt tolerance include green ash (*Fraxinus pennsylvanica*), paper birch (*Betula papyrifera*), and Siberian elm (*Ulmus pumila*) (Grey and Deneke, 1978).

Water pollution in an urban environment occurs when trash and unwanted waste is dumped directly into existing water bodies and when fuels and oils leak from watercraft and is also caused by stormwater runoff. Runoff can contain oil, gasoline, salt, and other chemicals that were lying on impermeable surfaces prior to a rainfall event. When these chemicals reach local bodies of water, they can adversely affect fish and insect populations, humans who swim or bathe in the contaminated waters, and those who eat contaminated fish from the waters. Light pollution can occur when plants are continuously exposed to an unwanted form of light, whether it is natural or artificial. Extensive light exposure can lead to a greater leaf area on a plant or tree, thus making the plant or tree more sensitive to air pollution and early winter frosts. Some plants with low light sensitivity are the European beech (*Fagus sylvatica*), ginkgo (*Ginkgo biloba*), and hornbeam (*Carpinus* spp.) (Cathey and Campbell, 1975).

Finally, the presence of people can affect the environment of urban forests in both a negative and a positive manner. Unfortunately, people can pollute soil and water resources by littering and discarding waste. However, people can also support the development and maintenance of urban forests by removing trash, perhaps as part of an Adopt-a-Highway program (Fig. 16.18) or a neighborhood cleanup day. People can further assist in the development and maintenance of an urban forest by planning, fertilizing, and otherwise tending to the trees.

16.5 The management of urban forests

The public perception of the value of urban trees is highly variable (Perkins, 2011). Until now, our

FIGURE 16.18 Forestry student picking up trash along the roadside on the outskirts of Starkville, Mississippi, United States. *Donald L. Grebner.*

discussion of urban forestry has not addressed the issue of urban forest ownership, although it is clear that the benefits of urban forests transcend ownership lines and therefore that people with no legal or financial interest in a forest can reap some of the benefits it provides. Urban forests can be managed in a number of ways, and the approach chosen will depend on factors including the objectives and goals of the landowner, the forest's condition, and the forest's location. Local governments often have jurisdiction for managing trees situated in areas adjacent to public streets and roadways and often along utility or power transmission lines. Local communities may own undeveloped forested areas or areas such as parks or public trails that are frequently visited. A community's governing authority often develops rules and regulations regarding the planting, protection, and management of trees, which are based on local, state, and federal laws, along with the wishes and desires of the local community. These may arise in the form of comprehensive plans, zoning ordinances, tree ordinances, subdivision regulations, and tree programs (Hill et al., 2010). Some communities also have urban tree boards that form decisions regarding how the community's forest resources will be managed.

The City of San Francisco might serve as an example of an organized city program. The Public Works Department of the city manages about 125,000 street trees through inspections and regular pruning activities. However, permits are needed by private landowners for pruning, planting, and removing these urban trees and these are approved by an arborculturist if they adhere to standards and best management practices (San Francisco Public Works, 2020). About another 575,000 urban trees are located on private lands and lands of other public agencies. An urban forest management plan (San Francisco Planning Department, 2014) outlines a number of strategies to grow, protect, manage, fund, and engage the public in caring for the urban forest. The plan suggests that the city will grow the urban forest by an additional 50,000 trees, addressing carbon sequestration, greenhouse gas, and social equitability concerns; however, the cost of implementing the plan requires an increase in funding to the Public Works Department, which poses a challenge for city managers and residents given other budgetary concerns such as education, health care, and public safety (Swae, 2015).

An important ownership group for perhaps significant areas of urban forests is private individual landowners. This group ranges from people who own a small plot of land to business owners who own 20 hectares (ha; about 50 acres [ac]) or more of land. Implementing urban forestry on these properties is highly dependent on the particular landowner's objectives for the property. A homeowner may be interested in providing bird boxes or birdseed in an effort to attract specific avian species for casual viewing at home (Fig. 16.19). A business owner may be interested in attracting customers by making their business environment more ecologically friendly. Other landowners may not prefer to have trees on their property, or trees of a specific species. Therefore, managing privately owned resources that may also provide public services can be a cumbersome and complex process. Some communities have addressed these issues by enacting public tree ordinances that dictate what actions can or

cannot be applied to land within urban forests or to trees on private properties. Most tree ordinances involve the maintenance of trees, a tree inventory, and rules regarding the preservation and potential removal of a certain number and type of tree species during developmental activities (Hill et al., 2010; Hilbert et al., 2019). While constricting at times, tree ordinances are often viewed as a necessary component of an urban forestry program (Rines et al., 2010). Some ordinances may have provisions related to private property care, penalties and fines for violations of rules, and regulations that focus on disease and abatement (Hill et al., 2010). These ordinances can involve actions beyond the management of trees, including the overall management of a property. For example, ordinances can require that runoff volume from developmental actions on private land not exceed the volume determined for the area prior to development (McPherson, 1992). However, some large cities may not have comprehensive ordinances and may therefore rely on voluntary compliance among landowners and developers to achieve urban greening goals (Jim, 2000).

The forestry practices commonly used in managing urban forests were discussed in Chapter 10. Some of these activities include planting, pruning, harvesting, or removing dead or live trees. Other practices used in managing urban forests may include controlling invasive plant populations and animal and insect species; establishing wildlife habitats such as bird and bat boxes (Fig. 16.20); maintaining trails; and attending to fire prevention activities. Implementing these practices requires a forester or

FIGURE 16.19   Eastern bluebird (*Sialia sialis*) box in front yard of residential house in Starkville, Mississippi, United States. *Donald L. Grebner.*

FIGURE 16.20   Bat box on coniferous tree outside Albuquerque, New Mexico, United States. *Donald L. Grebner.*

natural resource manager to consider the aesthetics, human use, local microclimatic environment, and locations of buildings, power lines, roads, and sidewalks.

Urban forests can be quite diverse in terms of tree species supported. For example, a recent survey of street trees in Melbourne (Australia) found that the nearly 1 million trees were represented by 1127 taxa, most of which were native species, yet few of which dominated (Table 16.1). In addition, a recent analysis of 10 cities in the Nordic (Denmark, Finland, Norway, and Sweden) region of Europe suggested that 16 tree genera comprised more than 92% of the total number of city trees (Table 16.2). However, an analysis of 12 cities in eastern North America suggests that tree diversity may be low; three of these cities are illustrated in Table 16.3. Due to their low level of diversity, these urban forests may be vulnerable to introduced insects such as the emerald ash borer (*Agrilus planipennis*) (Raupp et al., 2006). Many urban forest trees are planted by arborists and foresters. The selection of trees for urban forestry purposes can be based on criteria that involve the ability of tree species to withstand the expected stresses, as well as their ability to provide the desired benefits (Sjöman and Nielsen, 2010). Specific selection factors can include disease resistance, adaptability to restrictive soil restrictions, growth potential within crown space available, tolerance to expected pollution problems, growth form and resistance to breakage (limbs, particularly), and other aesthetic and social factors (Sæbø et al., 2003). Trees should be selected, and tree planting locations chosen, to avoid conflicts with power lines, sidewalks, and street lights. The amount of water necessary to maintain an urban tree and the ultimate cost-effectiveness of the tree

TABLE 16.2 Tree genera representing 2% or more of the street trees in 10 Nordic cities, around 2010.

Tree genera	General type of species	Proportion (%)
Tilia	Lime, linden, basswood	23.7
Acer	Maple	12.1
Sorbus	Whitebeam	11.6
Betula	Birch	10.9
Prunus	Cherry	3.9
Quercus	Oak	3.9
Ulmus	Elm	3.8
Populus	Poplar	3.4
Pinus	Pine	3.2
Fraxinus	Ash	2.8
Aesculus	Horse-chestnut	2.7
Salix	Willow	2.2
Crataegus	Hawthorn	2.1
Fagus	Beech	2.1
Platanus	Plane tree	2.1
Picea	Spruce	2.0

Sjöman et al. (2012).

planting endeavor should also be considered (McPherson and Simpson, 2003).

A series of guidelines for the effective management of urban forests was developed by Grado et al. (2008). Their work suggests that cities and municipalities should conduct a community-wide resource inventory. In addition, it is suggested that the conservation of local wooded areas is important, as well as the protection of urban and community streams and the development of strategies for controlling urban and community stormwater pollution. Local conditions at any point in time can facilitate a modification to an existing management strategy or tactic. As Grado et al. (2008) suggest, the first step in developing a sound urban forestry plan for any community is to assess the condition of the natural resources within a town or city. As discussed in Chapter 8, this requires that a community understands why an inventory of their natural resources is needed and the intended use of the information gathered. Most effective assessments start with a community map that includes building locations, parks, rail lines, roads, trails, water bodies, wooded areas, and the location of street trees. Unfortunately, many communities do not have a map or database of street tree locations. Obtaining this information on trees requires someone to traverse each street in a community and map the location, species, and other attributes of trees, such as height, diameter, and structural form. This

TABLE 16.1 Tree species representing 2% or more of the street trees in greater Melbourne, Australia, around the year 2000.

Common name	Scientific name	Proportion (%)
Queensland brush box	*Lophostemon confertus*	6.9
Snow-in-summer	*Melaleuca linariifolia*	5.2
Purple-leaf cherry plum	*Prunus cerasifera*	3.9
Prickly leaved paperbark	*Melaleuca styphelioides*	3.4
Willow bottlebrush	*Callistemon salignus*	3.0
London plane	*Platanus × acerifolia*	2.9
Yellow gum	*Eucalyptus leucoxylon*	2.3
Willow myrtle	*Agonis flexuosa*	2.2
Desert ash	*Fraxinus angustifolia*	2.2
Gum tree	*Eucalyptus* spp.	2.0

Frank et al. (2006).

TABLE 16.3   Tree genera representing 2% or more of the street trees in Kansas City, New York City, and Toronto, around 2000.

Genus	Common name	Kansas City proportion (%)	New York proportion (%)	Toronto proportion (%)
Acer	Maple	22	38	20
Quercus	Oak	14	10	16
Fraxinus	Ash	10	4	—
Ulmus	Elm	10	—	—
Pinus	Pine	3	—	—
Platanus	Sycamore	—	19	9
Gleditsia	Locust	—	7	3
Pyrus	Pear	—	7	9
Tilia	Basswood	—	6	4
Ginko	Ginko	—	3	2
Styphnolobium	Pagoda tree	—	2	—
Prunus	Cherry	—	—	8
Zelkova	Elm	—	—	8
Cornus	Dogwood	—	—	2

Raupp et al. (2006).

process is often costly, but recent advances in technology allow urban foresters to use small, portable global positioning systems that quickly record a street tree's location. There may be an opportunity to use online tools such as Google Earth to support data collection. Nevertheless, this process can also be expensive, which has prompted the use of other technologies to inventory urban natural resources. For example, Jones (2011) used Light Detecting and Ranging (LiDAR) data to assess the quantity of urban trees in Pass Christian and Hattiesburg, Mississippi. Although remotely sensed technologies such as LiDAR may initially be expensive to obtain, savings can be accrued through lower personnel costs, lower fuel consumption, and less depreciation of vehicles and other field equipment.

On a larger scale, when communities control areas of natural forests, a management structure is required to implement a series of best management practices (Grado et al., 2008). Guidelines for managing small, wooded areas are similar to those found for larger forested properties. The key is having defined goals and constraints for managing a property that reflect a community's desires. Once this is known, an urban forest management plan can be developed and used over a predefined period. For example, managing a community's wooded areas for fuelwood is different from managing it for recreational benefits. In addition, constraints will either permit or prohibit different types of management strategies. A lack of financial resources is a common problem for urban forest programs in many communities (Grado et al., 2006).

Another aspect of managing urban forests involves the protection of a community's streams and bodies of water (Fig. 16.21). An urban stream protection strategy could include many components (Grado et al., 2008). Water-based and land-use planning focuses on the control of impervious ground cover to ensure that water quality in urban streams and other bodies of water remains good. This process requires an assessment of existing conditions, a projection of future conditions, links to larger watershed planning methods, long-term

FIGURE 16.21   Urban stream in Dunbarton Oaks Park, Washington, D.C. *Smallbones, through Wikimedia Commons.*
Image Link: https://commons.wikimedia.org/wiki/File:Dunbarton_Park_Bridge_DC.jpg
License Link: https://creativecommons.org/publicdomain/zero/1.0/deed.en

monitoring for changes in water quality, and the protection of sensitive areas such as steep slopes, flood plains, wetlands, and shorelines, as well as mature forests located in critical areas. Strategies might involve reducing impervious cover by narrowing residential road widths, shortening road lengths, using shared parking and driveways, reducing cul-de-sac size, using vertical parking, and developing smaller parking stalls and buffers (Grado et al., 2008). Establishing a buffer system in a community could require identifying various land-use zones around a stream. A complete buffer system can reduce up to 75% of the sediment, 40% of the total nitrogen, 50% of the total phosphors, 60% of the trace metals, and 75% of the hydrocarbons in bodies of water. Maintaining a minimum buffer width, implementing a three-zone buffer system, using predevelopment vegetation as a target, using expanding and contracting buffers where necessary, incorporating provisions for buffer crossings, and other measures may be useful for both public and private urban forests (Grado et al., 2008).

Land development and construction projects are a major source of disturbance to urban forests. For preserving vegetation on site, buffer limits could be marked on construction plans, preconstruction buffers could be staked out to define disturbance areas or snow fencing, or signs could be used to prevent the entry of construction equipment (Grado et al., 2008). Specific recommendations for erosion control might also be defined, such as the need to spread mulch on an exposed site. Controlling stormwater runoff is very important in an urban forest environment, since this can cause soil erosion and carry pollutants into streams and other bodies of water and can overwhelm many municipal water treatment plants, too (Fig. 16.22). Other specific actions for controlling runoff include planting additional trees in suitable sites, improving the maintenance of existing trees, planting faster growing tree species and species that maximize interception, and matching tree species (based on their water requirements) to local precipitation levels (Grado et al., 2008).

The potential damage associated with natural disturbances can impact the way that urban forests are managed. In areas prone to natural disasters, such as windstorms, the ability of an urban forest to withstand natural disturbances can vary widely depending upon the type of disturbance that occurs. For example, aerial photos of the Mississippi coastline following the landfall of Hurricane Katrina in 2005 (Fig. 16.23) showed that the remaining trees were mainly live oaks (*Quercus virginiana*) and others that are relatively resistant to wind and saltwater damage. Live oak and sand oak (*Quercus geminata*) are both salt-tolerant trees and are generally resistant to wind breakage and windthrow. Other important salt-tolerant trees and plants in the southern United States include baldcypress (*Taxodium*

FIGURE 16.22 Storm water runoff, Rapid City, South Dakota. *Galen Hoogestraat, U.S. Geological Survey (2019).*

distichum), cabbage palm (*Sabal palmetto*), sweetgum (*Liquidambar styraciflua*), sycamore (*Platanus occidentalis*), and water oak (*Quercus nigra*).

Ice storms are another natural disturbance faced by urban foresters and natural resource managers. The frequency of these storms often delineates the northern ranges of many tree and shrub species. Tree species that naturally reside north of this zone survive snow and ice damage significantly better than those found south of this zone. Trees that are resistant to ice damage possess

FIGURE 16.23 Damage incurred in Long Beach, Mississippi, as a result of Hurricane Katrina. *U.S. Federal Emergency Management Agency and Mark Wolfe, through Wikimedia Commons.*
Image Link: https://commons.wikimedia.org/wiki/File:FEMA_-_18207_-_Photograph_by_Mark_Wolfe_taken_on_10-30-2005_in_Mississippi.jpg
License Link: Public Domain

attributes such as conical branching patterns, flexible but strongly attached branches, and lateral branches with reduced surface area. Some examples of these include eastern red cedar (*Juniperus virginiana*) and eastern hemlock (*Tsuga canadensis*). In general, deciduous tree species are more prone to ice damage than are coniferous tree species. Fire and drought are other natural disturbances that can adversely impact urban forests and are becoming more problematic as the number of residential homes constructed in forested areas increases. Drought conditions can also have an adverse impact on urban forests; therefore, planting drought-resistant plant and tree species can somewhat alleviate this problem.

Managing urban forests for a variety of objectives and situations is a complicated task. Many of the tools and practices used to do so are similar to those used in forest management situations conducted within rural landscapes. However, a key aspect of managing an urban forest may be to restore some of an area's basic ecosystem support services by minimizing impervious cover and reducing soil erosion and the toxins that enter local bodies of water. Understanding the most appropriate shrub or tree species to plant in a potentially stressful urban microenvironment can improve the health of an urban forest and reduce its vulnerability to damage arising from natural disturbances.

Summary

Trees and urban forests play an important role in how people manage the Earth's natural resources.

Urban forests have been part of human communities for thousands of years, but our modern interpretation of these did not begin until the 1960s. Although urban environments undergo dramatic changes from natural ecosystems to landscapes composed of concrete, asphalt, metals, and glass, some aspects of these environments can be managed, rehabilitated, or restored to provide a set of benefits that span all four categories of ecosystem services. Urban forests can ameliorate local temperatures, provide wildlife habitat, mask undesirable odors, abate unwanted noise, and provide spiritual and emotional comfort. Urban forests can also be used to sequester carbon, which can help to combat changes in the global climate. Urban forests are often located in more stressful growing environments than their rural cousins. Obstacles to urban forest management include human activities that result in increases in soil erosion and soil compaction, higher daily air temperatures, and drought conditions, which increase damage to tree limbs and trunks. Managing these types of forests successfully requires the integration of urban forest planning concepts at all levels of land development across a community, regardless of whether a small town or a large city with millions of inhabitants is being considered. Successful urban forest and natural resource management programs can lead to healthier urban forests that are more tolerant and resilient to natural disturbances such as hurricanes, tornadoes, wildfire, ice storms, insect and disease outbreaks, and drought.

Questions

(1) *Urban forests in your community.* For the local community where you currently live, obtain the answers to the following questions. You may need to visit your local town hall or city offices for this exercise. Provide the answers in a one-page memo addressed to your instructor.

   (a) Does your local community have urban forests? If so, how do you or would you use it?

   (b) Does your local community have an urban forestry program?

   (c) Who manages the urban forests?

   (d) Is the urban forestry work performed by town or city employees, or does your local community hire local businesses for the management and maintenance of the urban forests?

   (e) How many parks does your community maintain?

   (f) Does your local community have an active program for planting street trees or trees in open areas?

(2) *Residential property.* If you live in a house, a trailer, or an apartment complex, answer the following questions provided below, and describe the answers to these in a short PowerPoint presentation.

   (a) What percentage of the property is shaded?

   (b) How many trees does the property support?

   (c) Are the trees on the property pruned?

   (d) Are there vegetative sound or visual barriers on the property? If not, draw a simple map by hand and note on the map the locations that would be ideal for the barriers.

(3) *Benefits and disadvantages of urban forests.* In a brief one- or two-page memorandum, describe the benefits a community could derive from an urban forest and the

QUESTIONS (cont'd)

disadvantages of creating and maintaining an urban forest.

(4) *Planning an urban forest.* Imagine that a large city has purchased several parcels of land near the city center with the purpose of creating an urban forest or park. Imagine also that you are involved in the planning of

this urban forest. Develop 10 short, bulleted recommendations for consideration in the development of the urban forest. In a small group of four or five other students, describe and discuss your recommendations.

References

Aboelata, A., Sodoudi, S., 2019. Evaluating urban vegetation scenarios to mitigate urban heat island and reduce buildings' energy in dense built-up areas in Cairo. Building and Environment 166. Article 106407.

Akbari, H., 2002. Shade trees reduce building energy use and CO_2 emissions from power plants. Environmental Pollution 116, S119–S126.

Alvarez, M., 2001. A Brief History of the American Farm Bill. Society of American Foresters, Bethesda, MD.

Baer, N.W., 1989. Shelterbelts and windbreaks in the Great plains. Journal of Forestry 87 (4), 32–36.

Bartens, J., Wiseman, P.E., Smiley, E.T., 2010. Stability of landscape trees in engineered and conventional urban soil mixes. Urban Forestry and Urban Greening 9, 333–338.

Beatty, R.A., Heckman, C.T., 1981. Survey of urban tree programs in the United States. Urban Ecology 5, 81–102.

Brockway, D.G., Urie, D.H., Nguyen, P.V., Hart, J.B., 1986. Wastewater and sludge nutrient utilization in forest ecosystems. In: Cole, D.W., Henry, C.L., Nutter, W.L. (Eds.), The Forest Alternative for Treatment and Utilization of Municipal and Industrial Wastes. University of Washington Press, Seattle, WA, pp. 221–245.

Cardellino, R.W., 2001. Phytoremediation of Arsenic Contaminated Soils by Fast Growing Eastern Cottonwood Populus Deltoides (Bartr.) Clones. Master's Thesis. University of Florida, Gainesville, FL, 83 p.

Carleyolsen, S., Meyer, T., Rude, J., Scott, I., 2005. Measuring the Economic Impact and Value of Parks, Trails and Open Space in Jefferson County. Accounting for Current and Future Scenarios. Report Prepared for the Jefferson County Parks Department and Wisconsin Department of Natural Resources as Part of the Urban and Regional Planning Workshop. University of Wisconsin, Madison, WI.

Cathey, H.M., Campbell, L.E., 1975. Security lighting and its impact on the landscape. Journal of Arboriculture 1 (10), 181–187.

Correll, M.R., Lillydahl, J.H., Singell, L.D., 1978. The effects of greenbelts on residential property values: some findings on the political economy of open space. Land Economics 54, 207–217.

Crow, T., Brown, T., De Young, R., 2006. The Riverside and Berwyn experience: contrasts in landscape structure, perceptions of the urban landscape, and their effects on people. Landscape and Urban Planning 75, 282–299.

de Medeiros, P.I.S., de Souza Cabral, L.C., Carvalho, A.R., 2019. Cost to restore and conserve urban forest fragment. Urban Forestry and Urban Greening 46, 126465.

Draus, P., Lovall, S., Formby, T., Baldwin, L., Lowe-Anderson, W., 2019. A green space vision in Southeast Michigan's most heavily industrialized area. Urban Ecosystems 22, 91–102.

Dwyer, J.F., McPherson, E.G., Schroeder, H.W., Rowntree, R.A., 1992. Assessing the benefits and costs of the urban forest. Journal of Arboriculture 18, 227–234.

Ellis, C.D., Lee, S.-W., Kweon, B.-S., 2006. Retail land use, neighborhood satisfaction and the urban forest: an investigation into the moderating and mediating effects of trees and shrubs. Landscape and Urban Planning 74, 70–78.

Escobedo, F.J., Kroeger, T., Wagner, J.E., 2011. Urban forests and pollution mitigation: analyzing ecosystem services and disservices. Environmental Pollution 159, 2078–2087.

Fan, Y., Zhiyi, B., Zhujun, Z., Jiani, L., 2010. The investigation of noise attenuation by plants and the corresponding noise-reducing spectrum. Journal of Environmental Health 72 (8), 8–15.

Fang, C.F., Ling, D.L., 2005. Guidance for noise reduction provided by tree belts. Landscape and Urban Planning 71, 29–34.

Federer, C.A., 1976. Trees modify the urban microclimate. Journal of Arboriculture 2 (7), 121–127.

Frank, S., Waters, G., Beer, R., May, P., 2006. An analysis of the street tree population of Greater Melbourne at the beginning of the 21st century. Arboriculture and Urban Forestry 32, 155–163.

Grado, S.C., Grebner, D.L., Measells, M., Husak, A., 2006. Assessing the status, needs, and knowledge levels of Mississippi's governmental entities relative to urban forestry. Journal of Arboriculture and Urban Forestry 32 (1), 24–32.

Grado, S.C., Strong, S.S., Measells, M.K., 2008. Mississippi Urban and Community Forestry Management Manual, second ed. Forest and Wildlife Research Center, Mississippi State University, Starkville, MS. Publication FO 375. 218 p.

Grey, G.W., Deneke, F.J., 1978. Urban Forestry. John Wiley & Sons, Inc., New York, 279 p.

Grey, G.W., Deneke, F.J., 1986. Urban Forestry, second ed. John Wiley & Sons, Inc., New York, 299 p.

Harp, E.L., Reid, M.E., McKenna, J.P., Michael, J.A., 2009. Mapping of hazard from rainfall-triggered landslides in developing countries: examples from Honduras and Micronesia. Engineering Geology 104, 295–311.

Helms, J.A., 1998. The Dictionary of Forestry. Society of American Foresters, Bethesda, MD, 210 p.

Hilbert, D.R., Koeser, A.K., Roman, L.A., Hamilton, K., Landry, S.M., Hauer, R.J., Campanella, H., McLean, D., Andreu, M., Perez, H., 2019. Development practices and ordinances predict inter-city variation in Florida urban tree canopy coverage. Landscape and Urban Planning 190. Article 103603.

Hill, E., Dorfman, J.H., Kramer, E., 2010. Evaluating the impact of government land use policies on tree canopy coverage. Land Use Policy 27, 407–414.

Houston, D., Zuñiga, M.E., 2019. Put a park on it: how freeway caps are reconnecting and greening divided cities. Cities 85, 98–109.

Hull IV, R.B., Lam, M., Vigo, G., 1994. Place identity: symbols of self in the urban fabric. Landscape and Urban Planning 28, 109–120.

Jim, C.Y., 2000. The urban forestry programme in the heavily built-up milieu of Hong Kong. Cities 17, 271–283.

Johnston, M., 1996. A brief history of urban forestry in the United States. Arboricultural Journal 20, 257–278.

Jones, J.W., 2011. Using LiDAR Data with Remote Sensing and Geographical Information Systems (GIS) Technology to Assess Municipal Street Tree Inventories. PhD dissertation. Mississippi State University, Starkville, MS, 138 p.

Kara, B., Deniz, B., Kilicaslan, C., Polat, Z., 2011. Evaluation of Koçarlı Adnan Menderes urban forest in terms of the ecotourism. Procedia Social and Behavioral Sciences 19, 145–149.

Kaye, J.P., Groffman, P.M., Grimm, N.B., Baker, L.A., Pouyat, R.V., 2006. A distinct urban biogeochemistry? Trends in Ecology and Evolution 21, 192–199.

Koch, J., 2000. The origins of urban forestry. In: Kuser, J.E. (Ed.), Handbook of Urban and Community Forestry in the Northeast. Kluwer Academic, New York, pp. 1–10.

Konijnendijk, C.C., 2003. A decade of urban forestry in Europe. Forest Policy and Economics 5, 173–186.

Kučera, J., Podhrázská, J., Karásek, P., Papaj, V., 2020. The effect of windbreak parameters on the wind erosion risk assessment in agricultural landscape. Journal of Ecological Engineering 21, 150–156.

Lawrence, H.W., 1988. Origins of the tree-lined boulevard. Geographical Review 78, 355–374.

Lee, K.H., Ehsani, R., Castle, W.S., 2010. A laser scanning system for estimating wind velocity reduction through tree windbreaks. Computers and Electronics in Agriculture 73, 1–6.

Liang, Z.-S., Yang, J.W., Shao, H.-B., Han, R.-L., 2006. Investigation on water consumption characteristics and water use efficiency of poplar under soil water deficits on the Loess Plateau. Colloids and Surfaces B: Biointerfaces 53, 23–28.

Logan, A.L., D'Andrea, A.C., 2012. Oil palm, arboriculture, and changing subsistence practices during Kintampo times (3600–3200 BP, Ghana). Quaternary International 249, 63–71.

Lohr, V.I., Pearson-Mims, C.H., Tarnai, J., Dillman, D.A., 2004. How urban residents rate and rank the benefits and problems associated with trees in cities. Journal of Arboriculture 30, 28–35.

Lucy, W.H., Phillips, D.L., 1997. The post-suburban era comes to Richmond: city decline, suburban transition, and exurban growth. Landscape and Urban Planning 36, 259–275.

McPherson, E.G., 1992. Accounting for benefits and costs of urban greenspace. Landscape and Urban Planning 22, 41–51.

McPherson, E.G., Simpson, J.R., 2002. A comparison of municipal forest benefits and costs in Modesto and Santa Monica, California, USA. Urban Forestry and Urban Greening 1, 61–74.

McPherson, E.G., Simpson, J.R., 2003. Potential energy savings in buildings by an urban tree planting programme in California. Urban Forestry and Urban Greening 2, 73–86.

Miller, R.W., 1988. Urban Forestry: Planning and Managing Urban Greenspaces. Prentice Hall, Inc., Englewood Cliffs, NJ, 404 p.

Mumford, L., 1961. The City in History: Its Origins, its Transformations, and its Prospects. Harcourt, Brace, & World, Inc., New York, 657 p.

Nicholls, S., Crompton, J.L., 2005. The impact of greenways on property values: evidence from Austin, Texas. Journal of Leisure Research 37, 321–341.

Nowak, D.J., 2006. Institutionalizing urban forestry as a "biotechnology" to improve environmental quality. Urban Forestry and Urban Greening 5, 93–100.

Nowak, D.J., Aevermann, T., 2019. Tree compensation rates: compensating for the loss of future tree values. Urban Forestry and Urban Greening 41, 93–103.

Perkins, H.A., 2011. Gramsci in green: neoliberal hegemony through urban forestry and the potential for a political ecology of praxis. Geoforum 42, 558–566.

Perkins, H.A., Heynen, N., Wilson, J., 2004. Inequitable access to urban reforestation: the impact of urban political economy on housing tenure and urban forests. Cities 21, 291–299.

Phillips, L.E., 1993. Urban Trees: A Guide for Selection, Maintenance, and Master Planning. McGraw-Hill, Inc., New York, 273 p.

Pipkin, J.S., 2005. The moral high ground in Albany: rhetorics and practices of an 'Olmstedian' park, 1855–1875. Journal of Historical Geography 31, 666–687.

Randrup, T.B., Konijnendijk, C., Dobbertin, M.K., Pruller, R., 2005. The concept of urban forestry in Europe. In: Konijnendijk, C.C., Nilsson, K., Randrup, T.B., Schipperijn, J. (Eds.), Urban Forests and Trees. Springer-Verlag, Dordrecht, The Netherlands, pp. 10–21.

Raupp, M.J., Cumming, A.B., Raupp, E.C., 2006. Street tree diversity in eastern North America and its potential for tree loss to exotic borers. Arboriculture and Urban Forestry 32, 297–304.

Rines, D., Kane, B., Dennis, H., Ryan, P., Kittredge, D.B., 2010. Urban forestry priorities of Massachusetts (USA) tree wardens. Urban Forestry and Urban Greening 9, 295–301.

Robinette, G.O., 1972. Plants, People, and Environmental Quality. U.S. Department of the Interior, National Park Service, Washington, D.C., 129 p.

Rockwood, D.L., Ma, L.Q., Alker, G.R., Tu, C., Cardellino, R.W., 2001. Phytoremediation of Contaminated Sites Using Woody Biomass. Florida Center for Solid and Hazardous Waste Management. University of Florida, Gainesville, FL.

Roman, L.A., Scatena, F.N., 2011. Street tree survival rates: meta-analysis of previous studies and application to a field survey in Philadelphia, PA, USA. Urban Forestry and Urban Greening 10, 269–274.

Rosenthal, S.S., 2008. Old homes, externalities, and poor neighborhoods. A model of urban decline and renewal. Journal of Urban Economics 63, 816–840.

Sæbø, A., Benedikz, T., Randrup, T.B., 2003. Selection of trees for urban forestry in the Nordic countries. Urban Forestry and Urban Greening 2, 101–114.

Sanders, R.A., 1986. Urban vegetation impacts on the hydrology of Dayton, Ohio. Urban Ecology 9, 361–376.

San Francisco Planning Department, 2014. San Francisco Urban Forest Plan. Phase I: Street Trees. San Francisco Planning Department, San Francisco, CA.

San Francisco Public Works, 2020. Street Trees and Plants. San Francisco Public Works, San Francisco, CA. https://sfpublicworks.org/trees (accessed 02.04.20).

Seeland, K., Dübendorfer, S., Hansmann, R., 2009. Making friends in Zurich's urban forests and parks: the role of public green space for social inclusion of youths from different cultures. Forest Policy and Economics 11, 10–17.

Sjöman, H., Nielsen, A.B., 2010. Selecting trees for urban paved sites in Scandinavia—a review of information on stress tolerance and its relation to the requirements of tree planners. Urban Forestry and Urban Greening 9, 281–293.

Sjöman, H., Östberg, J., Bühler, O., 2012. Diversity and distribution of the urban tree population in ten major Nordic cities. Urban Forestry and Urban Greening 11, 31–39.

Smardon, R.C., 1988. Perception and aesthetics of the urban environment: review of the role of vegetation. Landscape and Urban Planning 15, 85–106.

Smit, J., Nasr, J., Ratta, A., 2001. Producing Food and Fuel in Urban Areas. Urban Agriculture Food, Jobs, and Sustainable Cities. The Urban Agriculture Network, Inc., Great Falls, VA. http://jacsmit.com/book/Chap05.pdf (accessed 02.04.20).

Smith, W.H., 1990. Air Pollution and Forests Interaction between Air Contaminants and Forest Ecosystems, second ed. Springer-Verlag, New York. 618 p.

Stoffberg, G.H., van Rooyan, M.W., van der Linde, M.J., Groeneveld, H.T., 2010. Carbon sequestration estimates of indigenous street trees in the City of Tshwane, South Africa. Urban Forestry and Urban Greening 9, 9–14.

Strohbach, M.W., Arnold, E., Haase, D., 2012. The carbon footprint of urban green space—a life cycle approach. Landscape and Urban Planning 104, 220–229.

Swae, J., 2015. City of San Francisco, California, United States of America. In: Siry, J.P., Bettinger, P., Merry, K., Grebner, D.L., Boston, K., Cieszewski, C. (Eds.), Forest Plans of North America. Academic Press, London, pp. 285–292.

Szabó, P., 2010. Driving forces of stability and change in woodland structure: a case-study from the Czech lowlands. Forest Ecology and Management 259, 650–656.

Tait, R.J., Allen, T.J., Sherkat, N., Bellett-Travers, M.D., 2009. An electronic tree inventory for arboriculture management. Knowledge-Based Systems 22, 552–556.

Tallis, M., Taylor, G., Sinnett, D., Freer-Smith, P., 2011. Estimating the removal of atmospheric particulate pollution by the urban tree canopy of London, under current and future environments. Landscape and Urban Planning 103, 129–138.

Thebo, A.L., Drechsel, P., Lambin, E.F., Nelson, K.L., 2017. A global, spatially-explicit assessment of irrigated croplands influenced by urban wastewater flows. Environmental Research Letters 12, Article 074008.

Tian, Y., Jim, C.Y., 2011. Factors influencing the spatial pattern of sky gardens in the compact city of Hong Kong. Landscape and Urban Planning 101, 299–309.

Toccolini, A., Fumagalli, N., Senes, G., 2006. Greenways planning in Italy: the Lambro River Valley Greenways System. Landscape and Urban Planning 76, 98–111.

Tregoning, D., Kays, J., 2003. Using Commercial Deer Repellents to Manage Deer Browsing in the Landscape. Maryland Cooperative Extension, University of Maryland, College Park, MD. Maryland Cooperative Extension Fact Sheet 810. https://extension.umd.edu/sites/extension.umd.edu/files/_docs/articles/FS810-A_UsingCommDeerReps.pdf (accessed 02.04.20).

U.S. Geological Survey, 2016. Can trees clean up ground water? Phytoremediation of trichloroethene-contaminated ground water at Air Force Plant 4, Fort Worth, Texas. U.S. Department of the Interior, U.S. Geological Survey, Washington, D.C.. http://toxics.usgs.gov/topics/rem_act/carswell.html (accessed 02.04.20).

U.S, Geological Survey, 2019. Surface Runoff and the Water Cycle. U.S. Department of the Interior, U.S. Geological Survey, Washington, D.C. https://www.usgs.gov/special-topic/water-science-school/science/surface-runoff-and-water-cycle?qt-science_center_objects=0#qt-science_center_objects (accessed 03.04.20).

Villalba, A.M., Pattini, A.E., Córica, M.L., 2016. Urban trees as sunlight control elements of vertical openings in front façades in sunny climates. Case Study: Morus alba on north façade. Indoor and Built Environment 25, 279–289.

Wilde, E.W., Brigmon, R.L., Berry, C.J., Altman, D.J., Rossabi, J., Looney, B.B., Harris, S.P., 2003. D-Area Drip Irrigation-Phytoremediation Project: SRTC Final Report. Westinghouse Savannah River Company, Savannah River Site, Aiken, SC. WSRC-TR-2,002–00,080.

Williams, M., 2000. Dark ages and dark areas: global deforestation in the deep past. Journal of Historical Geography 26, 28–46.

Young, R.F., 2010. Managing municipal green space for ecosystem services. Urban Forestry and Urban Greening 9, 313–321.

17

Ethics

The field of ethics is a branch of the philosophical sciences that concerns matters of right and wrong, i.e., good and bad, and thus of morality. Situations that require the consideration of professional ethics may arise through unforeseen business or policy decisions, events that tempt or influence decisions, or conflicting rules regarding the application of practices that are subject to debate (Irland, 1994a). An understanding of business and land ethics can be beneficial in making sound management decisions and helps people understand how they can contribute to society by using honest, productive work practices. Professional, ethical principles in the forestry and natural resource management fields are very closely associated with forest history, law, politics, and economics, and at times are directly related to the broader scope of land or environmental ethics (Klenk and Brown, 2007). Upon completion of this chapter, readers should have acquired

- an understanding of the scope of *ethics* and why it is important for both individuals and organizations;
- a feel for the language that might be contained within a published code of ethics of a forestry and natural resource organization; and

- an understanding of some of the important ethical issues posed by today's natural resource management environment.

In forestry and natural resource management, ethical practices are necessary in nearly every aspect of daily business, whether work is performed in the private sector or in the public sector. The use or nonuse of ethical practices illustrates a basic system of values held by an individual or an organization and can, in a sense, indicate how an individual or an organization wants to be treated by others. Ethics imply a sense of responsibility. Individuals and organizations are ultimately responsible for a broad set of values that exemplify justice (being conscientious and impartial, and acting in good faith), integrity (being honest and sincere), competence (being capable, reliable, and qualified), and utility (being responsible and improving the welfare of the public and others) (Raiborn and Payne, 1990).

17.1 Introduction

Professionals often find themselves in positions where they are required to resolve moral and ethical issues

(Pater and Van Gils, 2003), and perhaps the need for ethical behavior becomes evident in situations where the Golden Rule (i.e., *treat others as thou would like to be treated*) could be applied. In these situations, often a decision must be made that will affect (positively or negatively) another person, a resource, or the environment. In forestry and natural resource management settings, ethical behavior is often influenced by how we and others view the value of the outputs to be generated from a forested landscape. However, it could also be influenced by how we value our relationship with other professionals. These views can be in direct conflict with the views of others, who may place different values on the outputs of forests or on the relationships we have with others. In many cases one set of values can be neither inherently right nor wrong, which poses a problem for individuals whose extreme positions force people with differing viewpoints to feel as if their perspectives have been considered neither fairly nor honestly. As a result, an ethical decision faced by an individual can be prompted by social, cultural, environmental, organizational, or economic forces (Pater and Van Gils, 2003).

Professional ethics involve the implied or explicit duty an individual has to their client (e.g., an individual, an organization, or the public), the notion of fairness in competition and accounting, and the obligations forest managers have to their profession and other stakeholders (Coufal and Spuches, 1995). Environmental ethics involve the moral relationship between the status of the land (from the perspective of the human being) and the use of the land by humans and other living species. Patterson (1994) and Irland (1994b) proposed several questions foresters or natural resource managers might ask themselves when an ethical situation arises.

- What does your conscience say about the impending decision?
- What would you think if everyone were aware of the decision?
- What would you think if everyone acted in the same manner?
- What do you think the likely outcomes of the decision would look like?
- How would you feel if someone did the same thing to you?
- How would the decision ultimately affect your career or business?

In many instances, natural resource managers employed by large organizations have been severely scrutinized for the manner in which their forests and natural resources have been managed. As a result, many natural resource managers believe that their professional judgment has been undermined at one time or another. In small organizations and in entrepreneurial situations, scarce resources, high uncertainty and risk,

poorly defined responsibilities, and competitive pressures may lead to moral issues that involve choosing between the pursuit of self-interest and the maintenance of normative business ethics (Bryant, 2009).

This chapter focuses on a set of ethical considerations rooted in Western values. However, the spiritual practices and values of indigenous people around the world should also be recognized, since the views of indigenous people with regard to a land ethic may be distinctly different than those held by others (Groenfeldt, 2003). Many indigenous value systems reflect holistic, tradition-based management of the land. However, indigenous people such as the Māori in New Zealand also consider the trade-offs between culture, development, environmental protection, and profit (Loomis, 2000). Through the topics presented in this chapter, we hope to provide an overview of ethical considerations that are common for foresters and natural resource managers to prepare you for the challenges that may be encountered during your careers.

17.2 Philosophical approaches to land or conservation ethics

The field or concept of ethics in forestry and natural resource management is an evolving topical area that has been stimulated by the thoughts and actions of many individuals around the world. Within the United States, many noteworthy individuals have provided different perspectives on how forests and natural resources should be managed. At times, these perspectives have allowed individuals to assess their ethical standards for managing forests and natural resources through narrow prisms. Ultimately, the diversity of thought on these matters has provided a balance in the management of land for the wide variety of uses and products we desire. In this section, we briefly discuss the contributions of individuals including Henry David Thoreau, Bernhard Fernow, Gifford Pinchot, John Muir, Aldo Leopold, and Rachel Carson, as well as the Forest Service Employees for Environmental Ethics (FSEEE) organization. This is obviously not a complete list of the contributors to the evolution and science of land and professional ethics, since it only represents a few individuals from the United States. However, the discussion portrays the important roles their contributions have played and highlights the key turning points in society's ethical behavior toward the management of forests and natural resources in the United States.

Henry David Thoreau (1817–1862) is a famous philosopher and naturalist who lived in Concord, Massachusetts, during the 19th century and wrote extensively on subjects concerning natural history and botany. One of his more famous publications is the book *Walden*. He is

one of the founders of the American environmental movement owing to his eloquent prose that stressed the need for valuing and protecting untouched areas. Although today natural areas are changing rapidly due to developmental pressures, Thoreau also recognized similar changes occurring during his time (Primack et al., 2009).

Bernhard Fernow (1851–1923) was born in Posen in the Kingdom of Prussia (Fig. 17.1). After studying at the University of Königsberg and the Forest Academy at Münden, he immigrated to the United States and eventually became the chief of the United States Department of Agriculture (USDA) Division of Forestry (The Forest History Society, 2011a). Initially, he was the only person formally trained in forestry in the United States. Fernow was the primary promoter for the development of a national forest system in the United States and encouraged the implementation of what he viewed as scientific forestry, based on his earlier training in Prussia. Fernow eventually left the USDA Division of Forestry to become dean of the New York State College of Forestry at Cornell University in Ithaca, New York, the first professional forestry school in the United States. While there, Fernow witnessed a landscape that had been poorly managed and become a potential fire hazard. He attempted to implement a management system that required the conversion of the school's experimental forest to higher-valued trees, such as eastern white pine (*Pinus strobus*), using common

forestry and natural resource practices. Unfortunately, the scope of the project, poor public relations with neighboring landowners, and excessive smoke produced from the burning of slash and other debris caused quite a controversy and eventually led the New York state government to cut funding for and close the nascent forestry school in 1903. Fernow later became the founding dean of the Faculty of Forestry at the University of Toronto and worked there for the remainder of his career (Hosmer, 1923).

Fernow's legacy in the forestry and natural resources fields is immense even though he has been overshadowed by more famous individuals. Fernow had a utilitarian perspective on the ethics and use of forests and felt that forests and forest products play an important role in maintaining and supporting a viable human civilization (The Forest History Society, 2011a). His efforts laid the groundwork for the development of a United States national forest system that would be managed for future generations and not simply damaged through shifting logging practices. His approach to land ethics and the management of forests led to the development of the forestry curriculum at Cornell University and the University of Toronto and served as a model for similar professional forestry programs across North America (Adams, 1937; Miller and Lewis, 1999). However, his land ethic would later be challenged as societal values in North America evolved.

Gifford Pinchot (1865–1946) was born in Simsbury, Connecticut, United States (Fig. 17.2). He graduated from Yale University and attended the French National Forestry School in Nancy, France. Pinchot's family was involved in the lumber and land business and instilled in him the importance of the conservation ethic (The Forest History Society, 2011b). His family was instrumental in financing the development of Yale University's forestry graduate program. Pinchot became the chief of the USDA Division of Forestry following Fernow and, in 1905, became the first chief of the newly created United States Forest Service. Pinchot had a utilitarian perspective on the management of forests and natural resources but emphasized the development of new forestry practices adapted to North American forest conditions rather than the use of older European models. In 1905, he founded the Society of American Foresters, which brought credibility to the young, American forestry profession. This society requires that its members follow a code of ethics, which is outlined later in this chapter. Pinchot believed in the economically efficient management and protection of forest resources and felt that highly trained, professional foresters would help achieve that goal.

Pinchot's philosophical approach to the management of forests and natural resources was utilitarian in nature, as was the approach taken by Fernow. Pinchot's famous quote, "the greatest good for the greatest number in the long run," clearly echoes this philosophy (The Forest History Society, 2011b). He felt that forests should be sustainably managed to both be profitable and help human society. History has come to suggest that Pinchot was a conservationist, even though he placed a high value on the commercial use of forests. To the disdain of some, he viewed forests more as a crop than as a temple (Burns, 2009), but he also felt that timber companies of his time had narrow planning horizons and believed that the preservation of wilderness and scenery were of lesser importance. Pinchot was very successful in generating publicity for the Forest Service but was removed from his position over an article he helped create that was critical of the Secretary of the Interior's handling of an investigation into corruption involving public lands in Alaska (Ponder, 1990). Although Pinchot later became governor of Pennsylvania, his views on the use and management of public land and the policies he created while he was Chief of the Forest Service instigated considerable public debate with the well-known American naturalist, John Muir.

John Muir (1838–1914) was born in Dunbar, Scotland (Encyclopedia of World Biography, 2020a) (Fig. 17.3). His family immigrated to the United States in 1849 and settled in Wisconsin (Laurie, 1979). While young, he attended school, worked at various jobs, repaired machines, and learned about nature. Around 1868, he visited Yosemite and fell in love with its landscape. Over the years, he became a skilled guide and was respected for his knowledge of natural history. He

FIGURE 17.3 John Muir in 1907 (1838–1914). *Francis M. Fritz, through Wikimedia Commons.*
Image Link: https://commons.wikimedia.org/wiki/File:John_Muir_Cane.JPG
License Link: Public Domain

eventually began to write on natural history topics and his work was widely published. He viewed forested landscapes as a temple and, although raised a religious man, Muir felt that understanding God could also be achieved by studying nature. As a result, he placed great value on the spiritual and transcendental qualities of the forest.

Muir's legacy is extensive. He was the founder of the Sierra Club, which has since become an important environmental organization that supports the creation and maintenance of wilderness areas (Colby, 1967). Muir often lived within and played a pivotal role in the creation of Yosemite National Park. In the late 19th century, he was publicly critical of the uses of the land (for agriculture, logging, and tourism) that later became the park (Laurie, 1979). Muir was also very active in the Hetch Hetchy dam controversy, a political battle involving the damming of the Tuolumne River within Yosemite National Park. After the 1906 earthquake, public pressure mounted for a publicly owned water distribution system, as the privately owned water system had proven inadequate in responding to the disaster (Redmond, 2008). The Tuolumne River dam, called the O'Shaughnessy Dam (after the chief engineer), was located within Hetch Hetchy Valley, with the aim of supplying drinking water to the city of San Francisco and creating electricity for central California (Hamilton, 1971). Since the dam was to be located within a U.S. national park, an act of U.S. Congress (the Raker Act) was required to grant to the city and county of San Francisco certain rights of way within

FIGURE 17.4   Reservoir of water behind the O'Shaughnessy Dam in the Hetch Hetchy Valley, central California, United States. *Dan Lindsay, through Wikimedia Commons.*
Image Link: https://commons.wikimedia.org/wiki/File:HetchHetchyBW01.jpg
License Link: Public Domain

the public lands. After nearly a decade of analysis and heated debate, Muir and his colleagues (including the Sierra Club) lost this contentious battle in 1923 with the signing of the Raker Act. The dam was completed in 1923, although after construction it was argued (unsuccessfully) that the Raker Act was violated because the electricity was sold to a power company, who then sold the energy on to the public for a profit (Redmond, 2008). This series of events has often been used to provide examples of conflicts involving balancing the use of public land with the preservation of public land (Pincetl, 2006). Interestingly, discussions surrounding the removal of the dam have periodically taken place, although these raise a number of institutional and economic concerns related to water supply (Null and Lund, 2006), and political representatives of San Francisco have therefore opposed the idea. In addition, a program of controlled flooding of the valley below the dam to increase the release of water from the reservoir (Fig. 17.4) has been suggested as a means to restore hydrologic function along the Tuolumne River and provide recreational opportunities (Russo et al., 2012).

For a period, Muir and Pinchot were friends because of their mutual concern for the protection of forests and natural resources. Unfortunately, their relationship ended when Pinchot began supporting the grazing of sheep in forested areas, which was in direct conflict with Muir's preservationist philosophy and land ethic. Their philosophical conflicts split the then young conservation movement and created two distinctly different perspectives for managing forests and natural resources. These conflicts can still be observed in North America nearly a century later.

Aldo Leopold (1887–1948) was born in Burlington, Iowa, in the United States (The Wilderness Society, 2020) (Fig. 17.5). During his youth, he was actively engaged in outdoor activities in the woods. He was educated at the

FIGURE 17.5   Aldo Leopold in 1946 (1887–1948). *Howard Zahniser, U.S. Department of the Interior, Fish and Wildlife Service, through Wikimedia Commons.*
Image Link: https://commons.wikimedia.org/wiki/File:Aldo_Leopold,_1946_(cropped).jpg
License Link: Public Domain

Yale Sheffield Scientific School, now part of Yale University, and became a forester with the U.S. Forest Service in the Arizona and New Mexico territories of the United States. He eventually relocated to Wisconsin to work for the U.S. Forest Products Laboratory in Madison, and then served as a professor of game management at the University of Wisconsin. He purchased land in the central part of Wisconsin, which served as the setting for his famous book *Sand County Almanac*. Unfortunately, Leopold died of heart failure while fighting a brush fire along the Wisconsin River (Prince, 1995). His perspective on land ethics evolved over time, as he gradually came to believe that forest resources should be managed for more than traditional commodities and that sustaining the linkages between the levels of the food chain and the environment itself were a viable objective to achieve. His land ethic would later be considered an ecological approach to maintaining the diversity of all creatures in forested landscapes. Leopold rejected the purely utilitarian view of forest management, did not believe that private landowners should be free to use and dispose of land as they wished, and hoped that people would regard some natural resources as community goods (Prince, 1995). However, some argue that, in practice, Leopold was concerned with human economic exploitation of nature and maintaining ecological health, which when fully generalized becomes *sustainable development* (Callicott, 1992). Leopold's influence led to the foundation of the

FIGURE 17.6 Rachel Carson in 1944 (1907–1964). *U.S. Fish and Wildlife Service, through Wikimedia Commons.*
Image Link: https://commons.wikimedia.org/wiki/File:Rachel_Carson_w.jpg
License Link: Public Domain

Wilderness Society, which is dedicated to expanding and protecting American wilderness areas (The Wilderness Society, 2020). Two of his sons have had a significant impact on forestry. Starker was a prominent wildlife biologist and Luna a hydrologist-geomorphologist (Meagher, 2017; Hitch, 2020).

Rachel Carson (1907–1964) was born in Springdale, Pennsylvania, United States (Fig. 17.6), and grew up on a small farm (Encyclopedia of World Biography, 2020b). She attended the Pennsylvania College for Women and studied marine biology. After graduation, she began her professional career with the U.S. Bureau of Fisheries, which later became the U.S. Fish and Wildlife Service (Hynes, 1985). Her research responsibilities led to the publishing of numerous articles on the ocean's environment in newspapers and magazines. During the middle of her career, she became concerned with synthetic pesticides, and, during her investigations, learned about a conflict within the scientific community over the potential harm that could be caused by the presence of pesticides in the environment. At that time, some scientists did not want to believe that a problem existed, while others, such as Carson, disagreed. Her work prompted the evaluation of the interactions of pesticides with humans and provided evidence of potential links to cancer in people (Hynes, 1985). Her book *Silent Spring* argued that pesticides have a detrimental effect on the natural environment, as well as on the human body, and that their accumulation in the natural environment over time can have a lasting effect. Her work received fierce criticism and caused much controversy; the pesticide industry attempted to stop publication of *Silent*

Spring and one scientist suggested that her work would lead to famine and death due to the banning of pesticides for agricultural uses (Hynes, 1989). Shortly after the publication of *Silent Spring* a U.S. governmental committee recommended eliminating the use of persistent toxic pesticides (President's Science Advisory Committee Panel on the Use of Pesticides, 1963). While Carson may not be as well known as the individuals previously discussed, her work influenced protocols and practices for the sensible and careful (ethical) use of pesticides in the natural environment, led to the discontinued use of dichlorodiphenyltrichloroethane (DDT), and assisted in the birth of the environmental movement.

Organized groups can also be influential in developing or promoting ethical conduct among forestry and natural resource managers. As a group, FSEEE has the mission of protecting U.S. national forests and reforming the U.S. Forest Service (Forest Service Employees for Environmental Ethics, 2020). They are pursuing the goals associated with their mission by advocating the use of an environmental ethic, educating the public, and defending the rights of Forest Service employees who inform the public, or some other entity or authority, of alleged problems in the management of forests (i.e., whistleblowers). FSEEE believes that lands held in public trust should be valued for the benefits they might provide future generations. As such, and in pursuit of their mission, their objectives are to redefine timber and other commodity targets, to protect and promote the public's right to be informed, to champion the role of large intact ecosystems and natural disturbances, and to foster and develop a land management ethic (Forest Service Employees for Environmental Ethics, 2020). Although this is simply one organization that has expressed the need for environmental ethics, they have been actively collaborating with other environmental organizations such as Greenpeace, the National Audubon Society, the National Wildlife Federation, the Sierra Club, and the Wilderness Society.

A land or conservation ethic expands the traditional boundaries of ethical behavior to include issues related to the quality of soils, water, plants, and animals (Leopold, 1949). Therefore, a land or conservation ethic serves to guide people in their interaction with nature (Fig. 17.7). There are five distinct aspects of land ethics: ecological (Thompson, 2010), economic, egalitarian, libertarian, and utilitarian land ethics. An economic land ethic is one in which economic self-interest is considered to be the most important aspect of management and the plant or animal species that exist on the land and do not have an immediate economic value are disregarded. In the economic land ethic, value is only placed on those species that can provide a financial return to the landowner. A utilitarian land ethic is one that promotes the greatest good for the most people. Gifford Pinchot's famous statement, "the greatest good for the greatest number in the

FIGURE 17.7 Woman hiking with her dog in southern Finland.
Tuomas Työrinoja, through Wikimedia Commons.
Image Link: https://commons.wikimedia.org/wiki/File:Aarnikanmaki.jpg
License Link: https://creativecommons.org/licenses/by/3.0/deed.en

FIGURE 17.8 Farms along the Tambo River, Victoria, Australia.
John O'Neill, through Wikimedia Commons.
Image Link: https://commons.wikimedia.org/wiki/File:Farmland,-Bruthen,-Vic.jpg
License Link: https://creativecommons.org/licenses/by-sa/3.0/deed.en

long run," follows a utilitarian philosophy for managing forests and natural resources. Even though economic and utilitarian land ethics can often become intermingled, they are fundamentally different. A utilitarian land ethic is based on the moral justification that the conversion of a natural resource to a human need contributes to human health and welfare. The libertarian land ethic is one in which private property owners feel they have a natural right to do what they want with their forests and natural resources as long as their actions do not interfere with the freedoms and rights of their neighbors. The egalitarian land ethic suggests that a land management behavior is based on an opportunity right and should be employed to help those other than the landowner by protecting the productivity of forested landscapes for the wide range of products they provide. The ecological land ethic is based on stewardship and intrinsic values, even though numerous members of the food chain may not have an economic value. From this perspective, the management of forests and their natural resources is guided by a philosophy that respects the needs of humans and other species.

Differing perspectives with regard to land ethics can lead to conflict; thus, conflicts are an inevitable and unavoidable part of natural resource management, often involving the values and interests of stakeholders (Gritten et al., 2009). The following examples represent conflicts in natural resource management and are provided to illustrate the complex trade-offs that must often be addressed.

17.2.1 Management of forestland in Upper Lapland, Finland

Land use in Upper Lapland (Finland) is a contentious issue among reindeer herder cooperatives and the state forestry agency (Metsähallitus). The cooperatives feel that the management of forests by the state forestry agency is negatively affecting their ability to make a living from reindeer herding, and that additional constraints will further restrict their ability to do so. The state forestry agency is directed to foster employment in the forest products sector to operate in a profitable manner (as per legislative directive), and to operate in a sustainable manner (Gritten et al., 2009). Two moral issues are inherent in this debate: the ability of local (perhaps indigenous) people to continue their practice of reindeer herding, and the need by the state forestry agency to meet the social needs of local communities and adhere to the law. Therefore, the cooperatives and the state agency each see the actions of the other as a threat to the interests of their group. A resolution in favor of one group would ultimately be disadvantageous to the other.

17.2.2 Management of family farms in Victoria, Australia

The management of land for native biodiversity has become an important issue for the Australian government, and the maintenance and restoration of these values on private land are viewed as essential components of conservation programs (Farmar-Bowers and Lane, 2009). However, the goals of individual farmers generally involve their personal economic values rather than the maintenance of ecosystem services. In the case of Australian family farms in Victoria (Fig. 17.8), business decisions are typically justified in terms of economic rationale, and family decisions are justified in terms of fairness, protection, support, and agreement. Interestingly, the basic decisions concerning land use (where to farm or whether to farm) are often family decisions. How to accomplish the overall goal (to farm) and whether to purchase additional land are often then considered business decisions. However, decisions regarding the protection of native biodiversity are often initially made with the belief that they will be profitable in the long run; unfortunately, these decisions may later be rescinded when evidence indicates that they will not. If at this point a farmer wishes to continue the unprofitable practice, it becomes a family decision. The competition for family resources then becomes keen as business and family decisions conflict (Farmar-Bowers and Lane, 2009).

17.2.3 Bioenergy opportunities in sub-Saharan Africa

Containing over 30 of the poorest countries in the world, sub-Saharan Africa consists of a large number of rural people living in extreme poverty (Mangoyana, 2009). The production of crops for food is therefore paramount. However, a number of energy crops are now being promoted across the countries of sub-Saharan Africa in a quest to develop sustainable energy sources. These crops include cassava (*Manihot esculenta*), maize (*Zea mays*) (Fig. 17.9), sugarcane (*Saccharum* spp.), and sorghum (*Sorghum* spp.), which are also food staples. Some countries have even begun to use grants, subsidies, and tax exemptions to promote the production of biofuels from plants. These types of management systems are being recommended as ways through which governments cannot only meet energy needs but also reduce poverty, create employment opportunities, and

FIGURE 17.9 Maize (corn) field in South Africa. *Lotus Head, through Wikimedia Commons.*
Image Link: https://commons.wikimedia.org/wiki/File:Cornfield_in_South_Africa2.jpg
License Link: https://creativecommons.org/licenses/by-sa/3.0/deed.en

diversify rural economies. However, bioenergy land management systems may also act to divert food crops to energy uses, create competition for arable land, labor, and capital, and increase the price of food through increased demand for food crops. Further, bioenergy systems may ultimately involve the conversion of land from natural vegetative cover to crops that require intensive management (Mangoyana, 2009). Therefore, land-use changes induced by bioenergy concerns may significantly affect environmental sustainability, even though the goal (sustainable energy sources) seems necessary.

17.2.4 Clearcutting in the rainforest of British Columbia

Concern over clearcutting the forests of western British Columbia (Fig. 17.10), actions that were guided by forest management prescriptions designed under sustained yield of timber production principles, began in the early 1990s (Dempsey, 2011). The income and jobs created through this process were important objectives of the forest industry and the provincial government. However, the land involved was considered traditional territory of several indigenous tribes (First Nations) and these groups were not involved in the development of the land-use plans for the region. It is also argued that much of the benefits derived from the harvesting activities were accrued by people living outside the local communities. With the rise of these socioeconomic issues and citing forest fragmentation and a decline in forest health as ecological issues, environmental groups pressured the government and expressed the need for large reserves to counteract ecosystem decline. These groups also used end-market campaigns (i.e., boycotts of certain products) to successfully initiate a compromise agreement. Interestingly, in the aftermath of this agreement, many of the stakeholders were pleased. Environmental groups were content that some land was to be protected, forest products companies obtained a level of credibility and secured timber supplies, and some groups of indigenous peoples gained a voice in the management of forest areas. Not all stakeholders were pleased, however, as some were excluded from the development of the agreement, while others felt that the agreement had failed to protect enough grizzly bear (*Ursus arctos horribilis*) habitat (Dempsey, 2011).

17.3 Ethical conduct from a business perspective

In managing forests and natural resources, professionals can face many ethical issues related to their professional conduct, depending on their location and with whom they are working. Cultures and organizations

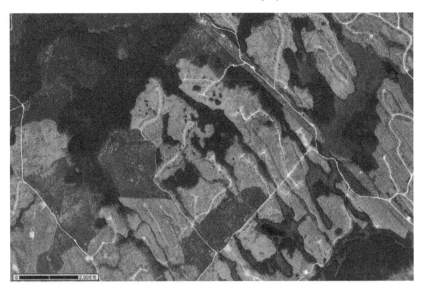

FIGURE 17.10 Clearcutting the forests of western British Columbia, Canada. *U.S. Department of Agriculture, Natural Resources Conservation Service (2020).*

often have different sets of values that affect the conduct of forest and natural resource management. Even within a single culture, multiple organizations may have developed separate codes of ethics. As we will illustrate later, following a code of conduct can help a professional avoid and prevent problems with the public, with clients, and with legal officials. This may become more important in the future, as in some professions there seems to be a changing relationship between employee and employer, in which some have suggested that a sense of loyalty to the employer (by the employee) is slowly being replaced by a sense of loyalty to oneself (Coughlan, 2001). Thus far, this chapter has focused on the general nature of land or conservation ethics and the behavior of individuals in the application of forest and natural resource management practices. However, professional business ethics are also paramount. Although there are numerous types of business-related ethical issues that forestry and natural resource management professionals may confront, a brief review of some noteworthy examples should help the reader to anticipate potential ethical problems.

One area of business ethical concern involves the manner in which an individual or an organization monitors the activities that they implement along with their revenues and costs. Bookkeeping involves methods that are used to monitor and track financial transactions, which may include the purchase of raw materials for manufacturing processes, expenditures for employees' wages, national and state tax payments, utility bill payments, the purchase of manufacturing equipment, and other such expenses. Revenues may include funds received from the sale of products, external grants received, the sale of stock, and rental income, to name a few. Some of these transactions may occur on a daily basis, while others may occur monthly or annually. Organizations typically have a formalized system for tracking and reporting this type of information using either a paper-based or electronic system. Commercial accounting software is widely available today. Bookkeeping allows organizations to understand their net revenues and determine whether a profit has been obtained through their business procedures. Government agencies also use bookkeeping methods to track the use of the funds they are allocated by their appropriate legislative bodies.

Accurate bookkeeping procedures reflect a state of management in which the entire financial process is transparent, thus allowing others to review the transactions that occur over a given period. Companies that employ people with responsibilities in these areas should encourage ethical behavior and financial executives should adhere to a specialized code of ethical conduct (Rezaee, 2004). One ethical issue concerning bookkeeping involves the reporting of accurate financial information. If an organization does not report all of the revenue it receives from the sales of a product or inflates their expenditures in order to reduce their perceived net revenue, these may be considered ethical issues because they are meant to deceive others. A cliché, often expressed in movies and on television, is that some organizations keep two sets of books: one book that is altered for tax officials and one real book for the business owners. This type of financial accounting logically causes some ethical concerns when viewed by those outside the organization, and, thus, independent and professional auditing should help control problems associated with this activity. Any organization can make honest mistakes when recording financial transactions; however, premeditation in this regard is a clear violation of ethical behavior. These types of transgressions reduce the quality of an organization's reputation and could affect local and regional governments, which may ultimately receive fewer funds for providing public services typically supported by government resources. Some countries have enacted legislation to specifically address accounting methods and auditing procedures. One such piece of legislation in the United States is the Sarbanes-Oxley Act of 2002 (Public Law 107–204, 116 Stat. 745–810), which emphasizes the accuracy and reliability of financial disclosures. In part, the law was developed to address certain assets or debts that are not reported on a company's financial balance sheet (i.e., *off-balance sheet* financial arrangements), some of which can be used to improve the financial position of an organization as viewed by investors. These financial arrangements include some leases, contingent obligations and guarantees, derivatives, and other contractual obligations (U.S. Securities and Exchange Commission, 2005).

Another ethical issue faced by some forest and natural resource managers involves bribery. Bribery is a form of corruption whereby an individual or organization provides money or a gift to a person with the implied intent that the person will alter his or her behavior or conduct upon its receipt. Bribery has been used to improve services obtained and, in social conditions where bribing is common, people who refuse the bribe are often given the worst service or may even be punished (Hunt and Laszlo, 2012). For instance, suppose a particular country has implemented a ban on the logging of a particular tree species or the hunting of a particular wildlife species. An unethical person, in an effort to obtain the banned commodity, may offer money to a local law enforcement official to ignore an illegal action that is likely to occur. Bribing is unfortunately an inherent part of logging operations in some tropical countries and may contribute to deforestation processes, as logging inspectors can be bribed by loggers to avoid fines associated with illegal logging (Amacher et al., 2012). Bribery can involve any type of gift given to change another's behavior. These gifts could include stock options, undocumented campaign funds, food, a favor, or any type of material item. Some organizations do not allow its employees to even accept hats of their contractors to avoid any appearance of wrongdoing.

Two other forms of bribery involve gift giving and kickbacks. As mentioned earlier, a gift is the transfer of an item to another without the expectation of a reciprocal transfer. Gift giving is a widely accepted tradition among many cultures around the world. People traditionally celebrate holidays by giving gifts to one another, and especially to children. Gift giving during a holiday season promotes sharing and goodwill among family members, friends, and others. However, gift giving becomes a form of bribery when preconditions are implicitly or explicitly made (rather than the gift being provided without the expectation of a reciprocal

transfer) and power may therefore be gained by the gift provider in matters related to the use of natural resources (Reed et al., 2009). In other words, one individual may present another with a gift but, perhaps without explicitly expressing it, would expect a certain behavior or act to occur as a result of the gift. Kickbacks are another form of bribery or corruption in which, during the normal course of employment of a person, a secret commission or profit is offered to that person without the knowledge of others. For example, perhaps a manufacturing organization desires a government contract to make airplanes or ships, and they offer influential governmental decision-makers money or another item of value to ensure they receive the contract. The money or other item of value received is the kickback. In a forest and natural resource management situation, a kickback could occur when a forest manager receives something of value from a logger in return for preferential timber harvesting rights or when a forest recreation manager receives something of value from an outdoor recreation organization for preferential access to a river. Robbins (2000) describes a case in India in which lower level natural resource management officials would allow local people to extract quantities of forest products above the amount allowed in exchange for kickbacks. This type of relationship, unfortunately, has become socially expected in some areas of the world.

Conflicts of interest are another set of important issues in the ethical behavior of foresters and natural resource managers. A conflict of interest occurs when a person or an organization that is entrusted with impartiality or a previous duty to one party develops multiple interests, at least one of which could potentially alter their behavior or conduct. For example, a conflict of interest could be the selection of a new employee from a list of three candidates, one of which is the decision-maker's child. As another example, a conflict of interest could occur when a teacher is instructing his or her own child in class and his or her desire for the child to do well influences him or her as he or she grades his or her child's examinations. A common remedy in this case would be to have someone without a potential conflict of interest grade the child's examinations. A conflict of interest could also involve a financial gain on the part of a decision-maker or his or her family and friends. In a forest and natural resource management setting, conflicts of interest can occur when decisions are not necessarily made in a rational or semirational manner. For example, imagine a public land manager who needs to employ a contractor to conduct management activities on a state forest. The manager may solicit bids for the work and then select a contractor who happens to be married to the manager's sister. A conflict of interest could also arise when organizations share sensitive information, and one of them (with a vested interest in the outcome) uses the information to develop a plan of action for all, as in the case of collaborative planning of transportation logistics in some forestry sectors (Audy et al., 2012). Fortunately, potential conflicts of interest can usually be identified and mitigated before any corruption or improper behavior occurs. The prevention of the appearance of conflicts of interest by foresters and natural resource managers can improve the perception of their ethical performance in the eyes of the public.

Another important issue to understand, and one that commonly occurs in the forest and natural resource management area, involves conflicts that occur through *moral absolutism*. Moral absolutism is a perspective shared by some individuals and organizations that certain actions are always either right or wrong, with no middle ground, regardless of context. This orientation establishes in people a set of imperatives for interpretation of issues and subsequent actions that are unconditional, overriding, and limiting to other actions (Walker, 2006). For instance, those adhering to moral absolutism would consider the act of lying to always be wrong, no matter what the situation. Therefore, regardless of the management style that you typically use, it would be considered unethical to express to a young employee that his or her performance was fair or okay when you or others think it was not, even if your intent was to avoid hurting the employee's feelings. Interestingly, in forest and natural resource management, the practice of clearcutting is viewed by many as being wrong and unethical, regardless of the regeneration requirements of shade-intolerant tree seedlings or the foraging requirements of some species of wildlife. Others may take the opposite view and suggest that clearcutting is the right approach for harvest and regeneration regardless of associated impacts (e.g., aesthetics). Thus, inherent value systems can sometimes help explain conflicts that arise regarding the management of forests and natural resources.

A final area of concern involves regulated and controlled antitrust and anticompetitive issues. Several United States laws (e.g., the Sherman Antitrust Act, the Clayton Antitrust Act, and the Hart-Scott-Rodino Antitrust Improvements Act) govern the behavior of individuals or organizations in areas considered to be anticompetitive. Other countries have developed similar legislation (e.g., the German Act Against Restraints of Competition and Canada's Competition Act) to govern anticompetitive behavior. These types of regulations were developed to encourage competition and protect consumers by discouraging actions that may lead to the exclusion of others or collusion within a group. Some examples of anticompetitive behavior include developing exclusive geographic territories, fixing prices, and rigging bids for contracts. In general, whether formal or informal, written or oral, discussions among competitors with respect to prices, costs, fees, terms and conditions of sales, or other transactional issues should be limited. For example, two prospective buyers of wood from a landowner should not discuss their timber valuations prior to bidding on the wood, as this may pose a disadvantage to the landowner. Nor should potential contractors discuss ways in which to increase the cost of management activities for landowners by limiting the services they offer in a specific geographic area. In addition, limiting the access of others to markets or excluding competitors from acquiring certain supplies from certain suppliers would also be considered anticompetitive behavior. In terms of forest certification programs, meeting the standards developed for entry into a program can allow an organization to legitimize their vision of the appropriate management of forest resources (McDermott, 2012). The exclusion of certain natural resource management organizations from either industry or environmentalist-sponsored certification programs when they are qualified and willing to participate may be considered anticompetitive, depending on the strictness of antitrust laws (van't Veld and Kotchen, 2011).

17.4 Example codes of ethics in natural resource management

Codes of professional ethics have been developed by many organizations in an effort to guide the behavior of employees (in the case of employer-employee relations) or associates (in the case of admission to a professional group) and to establish minimum standards of conduct for various situations that may arise during the normal course of business. Some of these relate to ethical considerations of the land, forests, and resources, while others relate to ethical considerations of daily business practices. The codes are generally developed to offer guidelines for decision-making processes when difficult issues arise to protect the personal integrity of the decision-maker and perceptions of their ability to handle their professional duties (Flanagan, 1994). External pressures may have provided the impetus to formalize ethical standards and thus to promote a sense of responsibility among those affected. It is suggested that in some cases codes of ethics may have been developed to ease societal concerns (Pater and Van Gils, 2003). However, codes of ethics can provide guidance to natural resource managers for effective communication and collaboration among colleagues and for effective development and maintenance of relations among employees, employers, clients, and the public (Lammi, 1994). Unfortunately, assessing the effectiveness of professional codes of ethics is limited and has mostly been performed within the United States (Pater and Van Gils, 2003). Nonetheless, there are numerous examples of codes of ethics that have been developed for natural resource professionals, and the samples provided here are meant to illustrate the depth at which ethical conduct in forest and natural resource management has been explored. Further, we provide a short discussion of the degree to which each of the codes reference four values of importance: justice, integrity, competence, and utility (Raiborn and Payne, 1990).

17.4.1 Society of American Foresters

As mentioned earlier, the Society of American Foresters is the national scientific and educational organization that represents the forestry profession in the United States. It was founded in 1900 by Gifford Pinchot. The society's mission is to

> advance sustainable management of forest resources through science, education, and technology; to enhance the competency of its members; to establish professional excellence; and to use our knowledge, skills, and conservation ethic to ensure the continued health, integrity, and use of forests to benefit society in perpetuity. *Society of American Foresters (2018)*

The Society of American Foresters has developed a code of conduct that members are required to follow. The Society of American Foresters' (2018) code of ethics, taken verbatim from their Internet site, is as follows:

Preamble: Service to society is the cornerstone of any profession. The profession of forestry serves society by fostering stewardship of the world's forests. Because forests provide valuable resources and perform critical ecological functions, they are vital to the well-being of both society and the biosphere. Members of the Society of American Foresters have a deep and enduring love for the land, and are inspired by the profession's historic traditions, such as Gifford Pinchot's utilitarianism and Aldo Leopold's ecological conscience. In their various roles as practitioners, teachers, researchers, advisers, and administrators, foresters seek to sustain and protect a variety of forest uses and attributes, such as aesthetic values, air and water quality, biodiversity, recreation, timber production, and wildlife habitat. The purpose of this Code of Ethics is to protect and serve society by inspiring, guiding, and governing members in the conduct of their professional lives. Compliance with the code demonstrates members' respect for the land and their commitment to the long-term management of ecosystems, and ensures just and honorable professional and human relationships, mutual confidence and respect, and competent service to society. On joining the Society of American Foresters, members assume a special responsibility to the profession and to society by promising to uphold and abide by the following:

Principles and Pledges:

1. Foresters have a responsibility to manage land for both current and future generations. We pledge to practice and advocate management that will maintain the long-term capacity of the land to provide the variety of materials, uses, and values desired by landowners and society.

2. Society must respect forest landowners' rights and correspondingly, landowners have a land stewardship responsibility to society. We pledge to practice and advocate forest management in accordance with landowner objectives and professional standards, and to advise landowners of the consequences of deviating from such standards.

3. Sound science is the foundation of the forestry profession. We pledge to strive for continuous improvement of our methods and our personal knowledge and skills; to perform only those services for which we are qualified; and in the biological, physical, and social sciences to use the most appropriate data, methods, and technology.

4. Public policy related to forests must be based on both scientific principles and societal values. We pledge to use our knowledge and skills to help formulate sound forest policies and laws; to challenge and correct untrue statements about forestry; and to foster dialogue among foresters, other professionals, landowners, and the public regarding forest policies.

5. Honest and open communication, coupled with respect for information given in confidence, is essential to good service. We pledge to always present, to the best of our ability, accurate and complete information; to indicate on whose behalf any public statements are made; to fully disclose and resolve any existing or potential conflicts of interest; and to keep proprietary information confidential unless the appropriate person authorizes its disclosure.

6. Professional and civic behavior must be based on honesty, fairness, goodwill, and respect for the law. We pledge to conduct ourselves in a civil and dignified manner; to respect the needs, contributions, and viewpoints of others; and to give due credit to others for their methods, ideas, or assistance.

There are a number of associated bylaws that specify the processes through which violation of the code may lead to reprimand, censure, or expulsion from the society. However, within just the preamble and the six principles, there are references to the utility value of a professional (responsibility to society, service, and stewardship) and a number of references to personal and organizational integrity (confidentiality, conflicts of interest, honesty, and high standards). Further, this portion of the Society of American Foresters code of ethics emphasizes aspects of justice (uphold laws) and competence (continuous education and appropriate behavior). The code of ethics is therefore fairly comprehensive in its treatment of the appropriate behavior of forestry professionals.

17.4.2 Institute of Foresters of Australia (IFA)

The Institute of Foresters of Australia (IFA) was established in 1935 and serves as the professional body for all forest management and conservation in Australia, including both public and private practi-

tioners and nonprofessional foresters (Institute of Foresters of Australia, 2019a). The Institute's main objectives are

> to advance and protect the cause of forestry; to promote professional standards and ethical practice among those engaged in forestry; to promote social intercourse between persons engaged in forestry; to publish and make educational, marketing and other materials available to those engaged in forestry; and to provide the services of the Institute to forestry organisations inside Australia and in overseas countries as the Board may deem appropriate. *Institute of Foresters of Australia (2019a)*

The IFA has a code of conduct that members are required to follow. The code includes seven principles that are supported by explanatory statements that elaborate the nature and application of the principles. These are not intended to address all of the potential issues associated with each principle, but rather to provide guidance to the members of the Institute when applying ethical principles during the conduct of their daily work activities. The Institute of Foresters of Australia (2019b) code of conduct, taken verbatim from their Internet site, is as follows:

The Institute of Foresters of Australia (IFA) Code of Conduct is the foundation upon which professional status is built and applies to all membership categories. All members of the IFA are required to adhere to, and promote this code, to uphold the integrity and reputation of the profession of forestry and to safeguard the public interest in matters of safety and health and otherwise.

The IFA's Code of Conduct is composed of 6 key standards. These standards also apply to registered professional foresters and consulting foresters. They also apply when members are practising overseas. The standards are supported by explanatory statements. These statements elaborate the nature and application of the principles but do not cover all issues associated with each principle; rather, they provide guidance for members of the IFA in applying ethical principles in the conduct of their profession. The sections also set out some behaviours and questions you could ask yourself when considering if your conduct meets professional standards.

Members of the IFA shall:

1. Act with Integrity
2. Always provide a high standard of service
3. Treat others with respect
4. Take responsibility
5. Act in a way that promotes trust in the profession
6. Have regard for sustainability throughout your work

Standard 1 — Act with integrity

Always put the interests of your clients or others to whom you have a professional responsibility above your own and respect their confidentiality. Always consider the wider interests of society when making your judgments. Always be honest. Be trustworthy in all that you do — do not deliberately mislead, whether by withholding or distorting information.

Standard 2 — Always provide a high standard of service

Be open and transparent in your dealings. Share the full facts with your clients, making things as plain and intelligible as possible. Know and act within your limitations. Be aware of the limits of your competence and do not be tempted to work beyond these. Only commit to what you can deliver. Be objective at all times. Give clear and appropriate advice. Never let sentiment or your own interests cloud your judgment. Maintain your professional competence in areas relevant to your work. Keep yourself informed of changes affecting the profession and broader developments relevant to your work and ensure your knowledge, skills and techniques are up to date. Apply this knowledge to the benefit of society.

Continued

Standard 3 — Treat others with respect

Treat everyone fairly with courtesy, politeness and respect and consider cultural sensitivities and business practices.

Standard 4 — Take responsibility

Be accountable for all your actions. Take full responsibility for your actions and do not blame others if things go wrong. Have the courage to make a stand. Be prepared to act if you suspect a risk to safety or malpractice of any sort.

Standard 5 — Act in a way that promotes trust in the profession

Set a good example. Remember that both your public and private behavior could affect your own reputation and that of the Institute and other members.

Standard 6 — Have regard for sustainability throughout your work

Practice your profession with due regard to sound ecological, social, economic and environmental principles to the advantage of present and future generations.

Within the preamble and the principles of the code of conduct developed by the Institute of Foresters of Australia there are a number of references to utility values of a professional and to personal and organizational integrity (honesty and confidentiality). The code of conduct also emphasizes aspects of justice and competence. Therefore, as with the previous example, this code of conduct is fairly comprehensive in its treatment of the appropriate behavior of forestry professionals.

17.4.3 Canadian Institute of Forestry

The Canadian Institute of Forestry (CIF/IFC) was established in 1908 and serves as Canada's national voice of forest practitioners and represents foresters, forest technologists and technicians, ecologists, biologists, geographers, educators, scientists, and others interested in forestry (Canadian Institute of Forestry, 2020). The Canadian Institute of Forestry (2014) code of ethics, taken verbatim from their Internet site, is as follows:

Purpose: The "Code of Ethics" is intended to guide the conduct of all members of the Canadian Institute of Forestry/Institut Forestier du Canada (CIF/IFC) in their business relationships with the public, their employers and/or clients, their employees and each other. While promoting just and honourable relationships, mutual confidence, respect and competent service, this Code is meant to encourage the highest possible standards of stewardship on forest lands, both public and private, under the care and management of CIF/IFC members. The "Guidelines and Standards" clarify the meaning and intent of the Code of Ethics by providing further guidance, examples and explanation. The Guidelines and Standards lay out in more detail the factors, issues and information that need to be considered in working toward achieving the high standards set forth in the Code of Ethics. It is recognized that aiming for the best possible conduct and highest possible standards is an ideal that cannot necessarily be legally enforced. This Code is intended to provide guidelines and examples whereby members are afforded instruction, assistance and suggestions should they encounter these situations. It is not intended to be a method to measure how far short of the ideal members might fall and how they should therefore be sanctioned.

Given the voluntary nature of CIF/IFC membership and the intent of this Code to offer guidance, disciplinary action against members, based on this Code of Ethics, is not intended. Nonetheless, it is a foundation of this organization that its members commit to the highest possible standards of behaviour and stewardship in their approach to the practice of forestry. It is therefore expected that all CIF/IFC members will fully consider the guidance offered in this Code of Ethics and will take the necessary steps to ensure proper conduct in themselves and others.

Factors Affecting CIF/IFC Members in the Practice of Forestry: CIF/IFC members represent a unique and, at the same time, very broad spectrum of expertise and interests in the practice of forestry. Membership is voluntary and includes woodlot owners, academics, scientists, politicians, forest technicians and technologists, foresters and those who are not directly involved in forest management. To describe this wide range of people and expertise, the term "forest practitioner" is used throughout this Code to reflect the diversity of CIF/IFC membership. This broad spectrum of interests and applications means that this Code of Ethics will offer more guidance to some members than to others.

Since the CIF/IFC is a national organization, issues, legislation and policy at the local, provincial, national and international levels affect its members. Such factors include the ethical requirements of provincial professional and technical associations along with the legal requirements of the various Acts, Regulations and policies of the jurisdiction in which CIF/IFC members work. Forestry related activities at the international level also have the potential to affect the way CIF/IFC members practice forestry both in Canada and abroad. The CIF/IFC Code of Ethics reflects the fact that many CIF/IFC members function within provincial jurisdictions and thus must be aware of and abide by the Codes of Ethics set out by provincial professional and technical bodies. As such, the CIF/IFC Code of Ethics was developed in concert with these existing Codes so as not to be in conflict with them.

A. Responsibility to Maintain the Public Good
   I. Forest Stewardship: To advocate, promote and practice the highest possible standards of forest stewardship, based on ecologically sound principles, which will maintain, protect and enhance the integrity, utility and value of the forest resource for the benefit of society, without compromising the opportunity for present and future generations to meet their objectives.
   II. Public Confidence: To inspire the public's confidence in the practice of forestry by maintaining high standards in one's conduct and daily work. To carry out such work in a spirit of integrity, honesty, fairness, good faith and courtesy.
   III. Public Understanding: To broaden the public's understanding of forests, the practice of forestry, the value of forestry to society and our commitment to the highest possible standards in the practice of forestry. The promotion of truthful and accurate statements on forestry matters should also be undertaken.
   IV. Public Welfare and Safety: To have proper regard, in all aspects of work, for the safety, health and welfare of the public and the potential impacts of forestry practices on public welfare.
B. Responsibility to the Profession
   I. Improve the Practice of Forestry: To work towards improving the standards, practices and policies that affect the stewardship of forest land.
   II. Ensure Competency: To undertake only such work as the person is competent to perform by virtue of their training and experience and to strive to improve the competence of all those practicing forestry.
C. Responsibility to the Employer/Client
   I. Consistent, Professional and Dedicated Service: To promote the best interests of an employer or client by consistently maintaining high standards of performance while acting in a conscientious, diligent and efficient manner.
   II. Consequences of Actions: To anticipate and advise employers or clients of the consequences of any contemplated policy, procedure or course of action which, based on professional judgment, is not consistent with the principles of sound forestry practice and best possible stewardship of forest land.
   III. Confidentiality: To hold as confidential and not to disclose information obtained as to the affairs, technical methods, practices and processes of the employer or client, unless released from this obligation by the employer or client, or except as required to do so by law.
   IV. Conflict of Interest: To ensure that activities related to all forestry undertakings do not conflict with the interests of their employer or client.
D. Responsibility to Other Professionals
   I. Fairness: To conduct oneself in a manner that demonstrates personal dignity and respect towards other forest practitioners, who are involved in the practice of forestry at all levels.
   II. Support: To provide advice, recognition, support and guidance to those practicing forestry in order to assist in furthering and enhancing their efforts and to ensure that the best possible practices and objectives are undertaken and recognized.
   III. Questionable Practice: To strive to avoid improper or questionable practices in their own work and in the work of others and to take steps as soon as possible to correct such practices and minimize their impacts on both the resource and the reputation of those engaged in the practice of forestry.

Within the code of ethics developed by the CIF, there are a number references to the utility values of a professional (stewardship) and to personal and organizational integrity (confidentiality, conflicts of interest, high standards, and honesty). This code of ethics emphasizes aspects of justice (respect and dignity for others) and competence (excellence in management). Therefore, as with the previous two examples, this code of ethics is also fairly comprehensive in its treatment of the appropriate behavior of forestry professionals.

17.4.4 Southern African Institute of Forestry

The Southern African Institute of Forestry (SAIF) is an association of forestry-related professionals designed to promote forestry, improve professional and technical aspects of forestry, enhance the public status of members while protecting members' interests, and represent the profession of forestry in Southern Africa (Southern African Institute of Forestry, 2018).

The Institute's mission is to

> assist members to achieve excellence in the practice of forestry, to promote growth and sustainability in the industry whilst being responsible as custodians of a sensitive environment. *Southern African Institute of Forestry (2018)*

The Southern African Institute of Forestry (2011) code of ethics, taken verbatim from their Internet site, is as follows:

1. Have the public interest at heart in all matters relevant to the forest and forest product industry.
2. Order conduct in such a manner as to uphold the dignity and reputation of the industry.
3. Strive to disseminate a true understanding of the forest and forest product industry.
4. Only issue statements on forestry policy and on technical matters while clearly indicating on whose authority they are acting.
5. When serving as an expert witness on forestry matters, base testimony on adequate knowledge of the subject matter.
6. Only undertake work for which education and experience does render them competent.
7. Clearly set out the possible consequences should professional judgement be overruled by a non-technical authority.
8. Only canvass professional employment or advertise consulting in an ethical manner.
9. Respect the professional reputation of any other practitioner in the industry.
10. Report convincing evidence of unprofessional conduct to the appropriate authority.

While brief, the code of ethics developed by the SAIF addresses all four of the values deemed important: justice (assess consequences), integrity (work with high standards), competence (only work in areas appropriately trained), and utility (duty to the public). While not as comprehensive as the others, it does outline the appropriate behavior of forestry professionals.

17.4.5 New Brunswick Forest Technicians Association

The New Brunswick Forest Technicians Association (NBFTA) was incorporated in 1996 (DeGrace, 1999) as an association of professionals that would not only address the image of forest technicians in New Brunswick but also promote the use of continuing education of members, and the protection and enhancement of ecosystems. NBFTA's objectives are

> to enhance the image of and encourage cooperation among forest technicians/technologists in New Brunswick; to provide one voice on forestry issues as they relate to forest technicians/technologists; to evaluate the qualifications (academic and experience) of graduate technicians/technologists who apply for membership in the Association and issue accreditation with respect thereto; to promote and assist in the continuing education of forest technicians/technologists and foster high standards; to promote the protection and enhancement of the New Brunswick biosphere; and to cooperate as much as possible with all associations with similar interests. *New Brunswick Forest Technicians Association (2020)*

The New Brunswick Forest Technicians Association (2019) code of ethics, taken verbatim from their Internet site, is as follows:

12.1 No Code of Ethics can prescribe appropriate conduct under all circumstances. Its purpose is to provide a general guide of ethical conduct principles to allow members to carry out their duties to the environment, Association members, the profession and the public. Members are bound by the law, acts, and regulations of the Province of New Brunswick and the rules and articles of the New Brunswick Forest Technicians Association Inc.

12.2 This code shall be interpreted by the following guidelines.

12.2-1 The Golden Rule ("Do unto others as you would have them do unto you").

12.2-2 Honorable judgment errors are not unethical and must be evaluated in that context rather than by strict adherence to the written Code of Ethics.

12.3 As a further guide, members are encouraged to ask themselves four critical questions when faced with ethical decisions:

12.3-1 Is the potential action legal in the criminal or civil sense?

12.3-2 Is the action within the guidelines of the NBFTA Code of Ethics?

12.3-3 Is the action balanced or will it heavily favor one party over another?

12.3-4 Am I comfortable with the action and willing to have it known by all?

12.4 Code of Ethics Articles

12.4-1 A member shall protect and promote safety, social, economic, and environmental interests.

12.4-2 A member shall only undertake work that he/she is competent to perform by virtue of training and experience.

12.4-3 A certified member shall maintain their technical competence through participation in Continuing Forestry Education (CFE).

12.4-4 A member shall recommend that other specialists be consulted on problems beyond his/her competence and shall cooperate with such specialists to the extent necessary.

12.4-5 A member shall regard the business of their employer as confidential unless released from this obligation.

12.4-6 A member, working simultaneously for clients whose interests in the work might be conflicting, shall notify all parties concerned.

12.4-7 A member shall be fair and honest in dealing with employer, client, subordinates, peers and members of the public.

12.4-8 A member shall be responsible to their employer or client and in the working relationship shall place their interests above everything except when those interests conflict with acceptable forestry practices.

12.4-9 A member, as a forest technician/technologist, shall not give opinions other than those based upon knowledge and experience.

12.4-10 A member shall not engage in any activity likely to result in an unfavorable reflection on the profession. This shall apply to conduct as a member of the profession and not as a citizen at large.

12.4-11 A member who engages in public discussion or controversy on forestry topics shall do so with dignity and honesty befitting his/her profession.

12.4-12 A member shall not condone untrue or misleading opinions concerning forestry.

12.4-13 A member shall not claim credit for facts or opinions, which are not his/her own.

12.4-14 A member shall not distort or withhold information for the purpose of supporting his/her opinions.

12.4-15 A member shall not make unsolicited or untrue comments to anyone about another forest technician/technologist's work or methods but shall disclose evidence of unethical actions to the Disciplinary Committee or the Appeal Board of the Disciplinary Committee of this Association.

12.4-16 A member shall not support the admission to this Association any person believed by them to be inadequately qualified or unworthy.

Within the objectives and the code of ethics developed by the NBFTA are a number of references to the justice values of a professional (upholding laws and regulations and treating others fairly) and there is a reference to the use of the Golden Rule. Further, there are a number of references to personal and organizational integrity (confidentiality and conflict of interest), utility (promote the protection of the biosphere), and competence (training, experience, and the pursuit of educational growth). While this code of ethics may be more specific with regard to certain points of discussion than some of the others, it is fairly comprehensive in its overall treatment of the appropriate behavior of forestry professionals.

This presentation of organizational codes of ethics by forest and natural resource organizations is, of course, neither exhaustive nor exclusive. However, in reviewing the different codes of ethics presented by these organizations, similarities can be seen. In their own manner, each of them conveys to their members that professional behavior includes being honest, fair, responsible, and, where appropriate, confidential. In addition, and where appropriate, professionals should acknowledge the public interest, identify consequences of one's actions, and use science, along with other considerations, to help guide decisions during the conduct of professional duties.

Summary

One of the reasons for delving into the issues concerning ethical conduct is to provide guidance regarding conduct, motivation, and choices made in the course of forest and natural resource management. Exploring the social norms for what is considered right or wrong may assist land managers as they think through the potential consequences of actions prior to implementing them. Ideally, in doing so a land manager would learn from both his or her accomplishments and his or her mistakes, and, through assistance and feedback, recognize his or her strengths and weaknesses and identify areas for improvement in professional dealings (Merchant, 1994). In addition, forestry and natural resource management professionals need to be aware of the various issues that can affect the public's perception of how forests and natural resources are managed. Treating others with fairness and honesty, as is implied in the Golden Rule, is perhaps the best course of action to take in many instances, although this should be consistent with the objectives of the landowner of the forest being managed. How a natural resource professional treats others and communicates their individual and organization's character to the public.

Questions

(1) *Privacy.* Using Internet resources, search for stories that relate potential ethical issues involving privacy. From your analysis, argue one way or another whether privacy really is an ethical issue. In doing so, describe why, or why not, privacy is important with regard to ethics. Further, develop a short list of other potential issues associated with privacy, and develop a one-page summary of your findings.

(2) *Sarbanes-Oxley Act of 2002.* Develop a short essay describing the purpose of this piece of United States legislation (Public Law 107–204, 116 Stat. 745–810). In your essay, describe how the Act relates to forest and natural resources. In addition, list some of the ethical issues you believe that the Act directly or indirectly addresses.

(3) *Conflict of interest.* Imagine that you are employed by a natural resource management organization in a region where management actions are intensely debated and hotly contested. Part of your job responsibilities entails developing environmental assessments of the potential impacts of management actions on water resources. Imagine now that your son has taken a summer job with a local timber company, and, further, that you regularly attend the same place of worship as the owner of another local timber company. In developing an environmental assessment for a proposed timber sale in which these two companies would ultimately be interested, describe the various conflicts of interest that could arise. In a small group of other students, describe how you would mitigate the potential conflicts of interest. Further, prepare a short, one-page summary of your thoughts on this matter.

(4) *Land ethics.* Of the five general types of land ethics we described in this chapter, decide which one closely emulates your value system at this stage in your life. Develop a one- to two-page summary of your thoughts on this subject and describe the circumstances under which you think your views might someday change.

(5) *Business ethics.* Imagine you are responsible for the management of a large public forest, and that you have direct influence on the awarding of contracts for timber management services. Now imagine that a local logger has invited you to a holiday party at his house, where common knowledge suggests that guests will receive elaborate gifts from the host. The opportunity is tempting. However, (A) what ethical issues should you consider, and (B) under what conditions, if any, would you attend the party? Develop a bulleted list of responses that address each of these two areas of concern.

(6) *Codes of ethics.* Locate a code of ethics for a natural resource management organization not presented in this chapter, perhaps the International Society of Arboriculture. Describe in a brief one-page report how the code of ethics addresses justice, integrity, competence, and utility values. Within a small group of colleagues, discuss your findings.

References

Adams, B., 1937. Cornell: an appreciation. Journal of Forestry 35, 649–653.

Amacher, G.S., Ollikainen, M., Koskela, E., 2012. Corruption and forest concessions. Journal of Environmental Economics and Management 63, 92–104.

Audy, J.-F., D'Amours, S., Rönnqvist, M., 2012. An empirical study on coalition formation and cost/savings allocation. International Journal of Production Economics 136, 13–27.

Bryant, P., 2009. Self-regulation and moral awareness among entrepreneurs. Journal of Business Venturing 24, 505–518.

Burns, K., 2009. Gifford Pinchot (1865–1946). WETA and the National Parks Film Project. LLC, Washington, D.C. http://www.pbs.org/nationalparks/people/historical/2/ (accessed 22.03.20).

Callicott, J.B., 1992. Principal traditions in American environmental ethics: a survey of moral values for framing an American ocean policy. Ocean & Coastal Management 17, 299–325.

Canadian Institute of Forestry, 2014. Code of Ethics. Canadian Institute of Forestry, Mattawa, ON. http://www.cif-ifc.org/wp-content/uploads/2014/05/Code_of_Ethics_English.pdf (accessed 22.03.20).

Canadian Institute of Forestry, 2020. About Us. Canadian Institute of Forestry, Mattawa, ON. https://www.cif-ifc.org/about/ (accessed 22.03.20).

Colby, W.E., 1967. The story of the Sierra Club. Sierra Club Bulletin. http://www.owensvalleyhistory.com/stories1/story_of_sierra_club.pdf (accessed 22.03.20).

Coufal, J.E., Spuches, C.M., 1995. Ethics in the forestry curriculum: a challenge for all foresters. Journal of Forestry 93 (9), 30–34.

Coughlan, R., 2001. An analysis of professional codes of ethics in the hospitality industry. Hospitality Management 20, 147–162.

DeGrace, B., 1999. From the President's desk. New Brunswick Forest Technicians Association, Fredericton, NB. The Azimuth 4 (3), 2–3.

Dempsey, J., 2011. The politics of nature in British Columbia's Great Bear Rainforest. Geoforum 42, 211–221.

Encyclopedia of World Biography, 2020a. John Muir Biography. Advameg, Inc., Flossmoor, IL. http://www.notablebiographies.com/Mo-Ni/Muir-John.html (accessed 22.03.20).

Encyclopedia of World Biography, 2020b. Rachel Carson Biography. Advameg, Inc., Flossmoor, IL. http://www.notablebiographies.com/Ca-Ch/Carson-Rachel.html (accessed 22.03.20).

Farmar-Bowers, Q., Lane, R., 2009. Understanding farmers' strategic decision-making processes and the implications for biodiversity conservation policy. Journal of Environmental Management 90, 1135–1144.

Flanagan, D.T., 1994. Legal considerations of professional ethics. In: Irland, L.C. (Ed.), Ethics in Forestry. Timber Press, Portland, OR, pp. 65–72.

Forest Service Employees for Environmental Ethics, 2020. Our Mission & Programs. Forest Service Employees for Environmental Ethics, Eugene, OR. https://www.fseee.org/about-us/ (accessed 22.03.20).

Gritten, D., Saastamoinen, O., Sajama, S., 2009. Ethical analysis: a structured approach to facilitate the resolution of forest conflicts. Forest Policy and Economics 11, 555–560.

Groenfeldt, D., 2003. The future of indigenous values: cultural relativism in the face of economic development. Futures 35, 917–929.

Hamilton, L.S., 1971. Concepts in planning for water resources development and conservation—the American experience. Biological Conservation 3, 107–112.

Hitch, G., 2020. Luna Leopold: A Visionary in Water Resource Management. Aldo Leopold Foundation, Baraboo, WI. https://www.aldoleopold.org/post/luna-leopold-visionary-water-resource-management/ (accessed 28.03.20).

Hosmer, R.S., 1923. Dr. Fernow's life work. Journal of Forestry 21, 320–323.

Hunt, J., Laszlo, S., 2012. Is bribery really regressive? Bribery's costs, benefits, and mechanisms. World Development 40, 355–372.

Hynes, H.P., 1985. Ellen Swallow, Lois Gibbs and Rachel Carson: catalysts of the American environmental movement. Women's Studies International Forum 8, 291–298.

Hynes, H.P., 1989. Silent spring: a feminist perspective. Environmental Impact Assessment Review 9, 3–7.

Institute of Foresters of Australia, 2019a. About the IFA. Institute of Foresters of Australia, Melbourne, Australia. https://www.forestry.org.au/Forestry/About_IFA/About_the_IFA___AFG/Forestry/About_the_IFA/About_the_IFA_AFG.aspx?hkey=6c52531a-405c-43b5-84db-c2aaf47fe18e (accessed 22.03.20).

Institute of Foresters of Australia, 2019b. Institute of Foresters of Australia - Code of Conduct. Institute of Foresters of Australia, Melbourne, Australia. https://www.forestry.org.au/Forestry/Membership/Code_of_Conduct/Forestry/Membership/Code_of_Conduct.aspx?hkey=358c8f4f-4cdd-411c-8a8b-5263af71cae0 (accessed 22.03.20).

Irland, L.C., 1994a. Introduction: developing ethical reflection. In: Irland, L.C. (Ed.), Ethics in Forestry. Timber Press, Portland, OR, pp. 15–18.

Irland, L.C., 1994b. Getting and keeping secrets: some ethical reflections for consultants. In: Irland, L.C. (Ed.), Ethics in Forestry. Timber Press, Portland, OR, pp. 105–110.

Klenk, N.L., Brown, P.G., 2007. What are forests for? The place of ethics in the forestry curriculum. Journal of Forestry 105 (2), 61–66.

Lammi, J.O., 1994. Professional ethics in forestry. In: Irland, L.C. (Ed.), Ethics in Forestry. Timber Press, Portland, OR, pp. 49–57.

Laurie, M., 1979. A history of aesthetic conservation in California. Landscape and Planning 6, 1–49.

Leopold, A., 1949. A Sand County Almanac. Oxford University Press, Oxford, United Kingdom, 226 p.

Loomis, T.M., 2000. Indigenous populations and sustainable development: building on indigenous approaches to holistic, self-determined development. World Development 28, 893–910.

Mangoyana, R.B., 2009. Bioenergy for sustainable development: an African context. Physics and Chemistry of the Earth 34, 59–64.

McDermott, C.L., 2012. Trust, legitimacy and power in forest certification: a case study of the FSC in British Columbia. Geoforum 43, 634–644.

Meagher, M., 2017. A. Starker Leopold 1913-1983. National Park Service, Washington, D.C. https://www.nps.gov/parkhistory/online_books/sontag/leopold.htm (accessed 28.03.20).

Merchant, R., 1994. Personal and professional ethics. In: Irland, L.C. (Ed.), Ethics in Forestry. Timber Press, Portland, OR, pp. 59–62.

Miller, C., Lewis, J.G., 1999. A contested past: forestry education in the United States, 1898–1998. Journal of Forestry 97 (9), 38–43.

New Brunswick Forest Technicians Association, 2019. Code of Ethics. New Brunswick Forest Technicians Association, Fredericton, NB. http://www.nbfta.org/code-of-ethics/ (accessed 22.03.20).

New Brunswick Forest Technicians Association, 2020. Constitution. New Brunswick Forest Technicians Association, Fredericton, NB. http://www.nbfta.org/constitution/ (accessed 22.03.20).

Null, S.E., Lund, J.R., 2006. Reassembling Hetch Hetchy: water supply without O'Shaughnessy Dam. Journal of the American Water Resources Association 42, 395–408.

Pater, A., Van Gils, A., 2003. Stimulating ethical decision-making in a business context: effects of ethical and professional codes. European Management Journal 21, 762–772.

Patterson, A.E., 1994. Ethics in forestry: four self-help questions. In: Irland, L.C. (Ed.), Ethics in Forestry. Timber Press, Portland, OR, pp. 45–47.

Pincetl, S., 2006. Conservation planning in the west, problems, new strategies and entrenched obstacles. Geoforum 37, 246–255.

Ponder, S., 1990. Progressive drive to shape public opinion, 1898–1913. Public Relations Review 16, 94–104.

President's Science Advisory Committee Panel on the Use of Pesticides, 1963. Chemical & Engineering News 41 (21), 102–115.

Primack, R.B., Miller-Rushing, A.J., Dharaneeswaran, K., 2009. Changes in the flora of Thoreau's Concord. Biological Conservation 142, 500–508.

Prince, H., 1995. A marshland chronicle, 1830–1960: from artificial drainage to outdoor recreation in central Wisconsin. Journal of Historical Geography 21, 3–22.

Raiborn, C.A., Payne, D., 1990. Corporate codes of conduct: a collective conscience and continuum. Journal of Business Ethics 9, 879–889.

Redmond, T., 2008. Power Struggle: How the Best-Laid Plans of John Edward Raker and the US Congress Were Scuttled by PG&E: A Chronology, 1848–1988. San Francisco Bay Guardian, San Francisco, CA.

Reed, M.S., Graves, A., Dandy, N., Posthumus, H., Hubacek, K., Morris, J., Prell, C., Quinn, C.H., Stringer, L.C., 2009. Who's in and why? A typology of stakeholder analysis methods for natural resource management. Journal of Environmental Management 90, 1933–1949.

Rezaee, Z., 2004. Corporate governance role in financial reporting. Research in Accounting Regulation 17, 107–149.

Robbins, P., 2000. The rotten institution: corruption in natural resource management. Political Geography 19, 423–443.

Russo, T.A., Fisher, A.T., Roche, J.W., 2012. Improving riparian wetland conditions based on infiltration and drainage behavior during and after controlled flooding. Journal of Hydrology 432−433, 98−111.

Society of American Foresters, 2018. SAF Code of Ethics. Society of American Foresters, Bethesda, MD. https://www.eforester.org/CodeofEthics.aspx (accessed 22.03.20).

Southern African Institute of Forestry, 2011. Code of Ethics. Menlopark, South Africa. http://www.saif.org.za/index.php?page=67 (accessed 03.10.11).

Southern African Institute of Forestry, 2018. About SAIF. Menlopark, South Africa. https://saif.org.za/about-us/ (accessed 22.03.20).

The Forest History Society, 2011a. Bernhard E. Fernow (1851−1923): 3rd Chief of the US Division of Forestry (1886−1898). Forest History Society, Durham, NC. https://foresthistory.org/research-explore/us-forest-service-history/people/chiefs/bernhard-e-fernow-1851-1923/ (accessed 22.03.20).

The Forest History Society, 2011b. Gifford Pinchot (1865−1946): 4th Chief of the US Division of Forestry (1898−1905), 1st Chief of the Forest Service, 1905−1910. Forest History Society, Durham, NC. https://foresthistory.org/research-explore/us-forest-service-history/people/chiefs/gifford-pinchot-1865-1946/ (accessed 22.03.20).

The Wilderness Society, 2020. Aldo Leopold. The Wilderness Society, Washington, D.C. https://www.wilderness.org/aldo-leopold (accessed 22.03.20).

Thompson, P.B., 2010. Land. In: Comstock, G.L. (Ed.), Life Science Ethics. Springer, New York, pp. 123−144.

U.S. Department of Agriculture, Natural Resources Conservation Service, 2020. Web Soil Survey. U.S. Department of Agriculture, Natural Resources Conservation Service, Washington, D.C. https://websoilsurvey.sc.egov.usda.gov/App/HomePage.htm (accessed 22.03.20).

U.S. Securities and Exchange Commission, 2005. Report and Recommendations Pursuant to Section 401(c) of the Sarbanes-Oxley Act of 2002 on Arrangements with Off-Balance Sheet Implications, Special Purpose Entities, and Transparency of Filings by Issuers. U.S. Securities and Exchange Commission, Office of the Chief Accountant, Office of Economic Analysis, Division of Corporation Finance, Washington, D.C.

van't Veld, K., Kotchen, M.J., 2011. Green clubs. Journal of Environmental Economics and Management 62, 309−322.

Walker, M.C., 2006. Morality, self-interest, and leaders in international affairs. The Leadership Quarterly 17, 138−145.

18

Forestry and natural resource management careers

Careers in forestry and natural resource management can be exciting and rewarding and can range from field-level management of resources to personnel management, accounting, auditing, and database management. You can have a variety of lifestyles from life in small rural communities or major urban centers. The Food and Agriculture Organization of the United Nations (2010) estimates that approximately 4 million people are employed full-time in forest management and conservation careers worldwide. The types of these careers vary from those with forest establishment roles to those with managerial and technical assistance roles. We approach the subject of careers in forestry and natural resource management by describing the various responsibilities some of these jobs encompass. Upon completion of this chapter, readers should

- know about the wide variety of jobs available to forestry and natural resource management professionals; and
- understand the types of skills needed to successfully obtain a particular position in natural resource management.

In addition to prior training and experience, potential employers may be looking for specific qualities in their pool of applicants. For most open positions in forestry and natural resource management, knowledge and experience with forest management activities is required. However, many entry-level positions now also require experience and training in technology such as the use of global positioning systems (GPS) and geographic information systems (GIS). Written and oral communication skills are very important for any professional to possess, since forestry and natural resource management professionals must effectively express their thoughts to individuals or groups and must adjust their language and terminology to ensure effective communication with their target audience. As an example, many recent job announcements express the following general traits desired of a new employee:

> In addition to minimum coursework in forestry and natural resource management topics, applicants should have a working knowledge of computer applications such as word processing programs and spreadsheets. Proficiency in written and verbal communication is also necessary. Individuals should be self-motivated with a strong work ethic and the ability to schedule, on their own, their projects and tasks. Individuals must also be able to communicate well with people.

As this suggests, job opportunities often require some demonstration of initiative or the ability to work independently or as part of a team to achieve goals and respond to the changing demands of a job. Further, employer expectations may include a set of more vague characteristics of a potential new employee that include demonstrating leadership, exhibiting a desired work ethic, and maintaining a positive attitude despite the weather, the steepness of the hillside or density of the brush you must walk through (Fig. 18.1). In the forestry and natural resource profession, it is essential that one can work effectively as part of a team and contribute to the mission of the organization. Leadership skills

are therefore important for providing direction to a forestry and natural resource management program. Further, many organizations expect an employee to be able to positively interact with other employees, landowners, contractors, consultants, the general public, and cooperating personnel from other organizations; thus, effective communication skills are valuable.

In pursuing a job opportunity, an applicant will be required to create a résumé, or short description of their qualifications. A number of résumé templates can be obtained by searching the Internet. The résumé will ideally describe a candidate's educational experience, work experience, awards and honors, and service to the profession and to society. Work experience can be presented in a reverse chronological order or organized by skills (Weinstein, 2012). In reading a résumé, an employer is likely to develop an impression of a person's ability and motivation, along with an idea of the steps a potential employee has taken to acquire both work-related and leadership skills. Further, an employer is likely to attempt to understand how trainable a potential employee is through their prior experiences. Ultimately, an employer will use a résumé to determine whether a potential employee has the competencies and traits required to be successful in the job being offered, and the résumé is usually used to screen those who have those competencies for a personal interview. The key behavioral aspects of people that employers are looking for relate to reasoning, problem-solving, communication, and negotiation skills, among others (Weinstein, 2012). To begin, a person should schedule an appointment with a career center (counselor or advisor) to discuss methods for locating potential career opportunities in the forestry and natural resource management fields. If work experience in forestry and natural resource management has yet to be acquired, an informal interview with a forestry or natural resource professional who works in one of the related fields would be highly beneficial. Further, acquiring volunteer experience would help a person determine whether a career in forestry and natural resource management is one that really appeals to him or her.

FIGURE 18.1   A positive attitude will go a long way in furthering a forestry and natural resource management career. *Jack Chappell.*

18.1 Forestry and natural resource management careers

As noted in the Introduction, a wide variety of forestry and natural resource management careers are available. We use the term *career* here loosely, since people often will move among different jobs as their working lives progress. Therefore, one should not necessarily assume that a career represents a single type of job that will be performed throughout their working life. The focus of the set of jobs listed below is on forestry and natural resource professionals. The number of these types of positions an organization employs depends on the size and mission of the organization. In many situations, organizations require that one person will be responsible for two or more of the jobs we describe. The set of jobs provided is not exhaustive. There are numerous other opportunities in natural resource management that can be pursued by motivated individuals. These include becoming an environmental educator, geneticist, tree nursery manager, and wood technologist, among many others. Their omission here does not detract from their value and service to society.

18.1.1 Forest ranger

While most of the other careers described later in this chapter generally require a four-year degree in the forestry or natural resource management fields, forest ranger positions generally require an associate's degree from a two-year community college or a technical school. The types of duties forest rangers (Fig. 18.2) typically perform include wildfire control, insect and disease control, forest inventory, tree planting inspection, site preparation work, and assessments of adherence to forest practices guidelines. In general, these positions involve firefighting and prescribed burning activities, and may require experience and training in the use of GIS or GPS technology.

FIGURE 18.2 Michael Starbuck, Forest Technician, United States Department of Agriculture, Forest Service, Mississippi, United States. *Michael Starbuck.*

18.1.2 Forest technician

Forest technicians perform a variety of activities, including reforestation, insect and disease control, stand examination, boundary line maintenance, and timber stand improvement activities. A four-year degree in the forestry or natural resource management fields or an associate's degree from a two-year course at a technical school will usually suffice in pursuit of these job opportunities. Forest technicians (Fig. 18.3) may also provide technical assistance for forestry cost-share programs, inspect logging jobs for best management practices or other rules, conduct prescribed burns, manage budgets, and manage facilities and equipment. Other duties may include maintaining databases, using GIS, and conducting forest inventories and associated analyses. Forest technicians should have practical knowledge of forestry methods and techniques and in many cases must possess the strength and endurance to work all day in rugged terrain under all types of weather conditions. Other than an appropriate educational background, there is often no prior experience necessary for these types of positions, leading them to serve as excellent opportunities to obtain entry-level forest management skills and experience.

18.1.3 Fire management officer and firefighter

Fire management officers and firefighters participate in prescribed and wildland fire management activities and in hazard fuels reduction projects such as vegetation cutting and slash piling activities using chainsaws and other equipment. Individuals employed in these types of positions (Fig. 18.4) may also be expected to perform and direct others in completing and processing resource

FIGURE 18.3 W. Christopher Kirk, Forest Technician, United States Department of Agriculture, Forest Service, Mississippi, United States. *Catherine Kirk.*

FIGURE 18.4 Mike Strange, Fire Engine Leader, Florida, United States. *Mike Strange.*

order requests for personnel, equipment, supplies, and aircraft needs for all types of incidents. They may also be required to coordinate with other dispatch organizations regarding the mobilization, reassignment, and demobilization of resources and to determine the status of available resources, while ensuring that this information is made available in a timely manner to those who need it. People in these positions may assist with management briefings and provide incident statistics and historical fire and weather analyses, as well as other data. They may be called upon to enter fire weather data into weather application software for the projection of fire behavior and to communicate the probabilities and forecasts obtained to field personnel and fire management staff.

Firefighters conduct wildland fire suppression operations and control activities, including fire suppression, preparedness, prevention, and monitoring. Other duties may involve supervising and directing engine operations, serving as the Initial Attack Incident Commander on wildland fires and prescribed burns, and providing fire and safety training and equipment support as needed. Other types of duties firefighters may be expected to perform include mapping fire perimeters and monitoring fire behavior and fire weather. These types of jobs may require the ability to operate two- and four-wheel-drive vehicles; to use computers for documentation, e-mail, evaluations, and program tracking; and to operate chainsaws. These positions generally require extensive travel during the typical wildfire season and may involve exposure to different regions, elevations, vegetation conditions, fuel types, and climates. Firefighters need to be in above-average physical condition, since their work requires prolonged standing, walking over uneven ground, repeated bending, reaching, and lifting, and carrying items that may weigh over 50 pounds. These activities are generally performed during emergency situations under adverse environmental conditions and over extended periods.

18.1.4 Industrial or operations forester

Industrial or operations foresters (Fig. 18.5) develop a variety of forest management plans, perform natural resource assessments, and implement management activities. These activities can include the full suite of management activities practiced within a district (e.g., site preparation, planting, thinning, or final harvest) or a single type of activity (e.g., planting) that is practiced over a broader area. Communication and negotiation skills are of value, since operations foresters plan activities and must work with contractors to ensure that the activities are conducted as they have been planned. As suggested, industrial or operations foresters also develop and administer contracts and provide compliance assessments for timber sales. Additional job responsibilities of these types of foresters may include tactical harvest scheduling, timber sale design, and preparation and execution of road maintenance or wood marketing program. Industrial or operations foresters may also be responsible for the logistical issues associated with wood deliveries to processing facilities. They are often responsible for meeting the variety of local, state and federal regulations that govern forestry operations.

FIGURE 18.5 Rachel Martin, Area Production Forester, Weyerhaeuser, Georgia, United States. *Rachel Martin.*

18.1.5 Consulting forester

Consulting foresters (Fig. 18.6) work with landowners to develop and implement management plans to meet a landowner's goals and objectives. To do so, they must plan and implement a wide variety of property management tasks that may include performing a timber inventory, laying out a timber sale and marking trees, administering a timber sale, supervising contractors, implementing silvicultural practices, supervising planting operations, and conducting financial analyses. Nowadays, to accomplish these tasks, the consulting forester probably needs experience with GIS and GPS and must have acceptable communication skills. A consulting forester is likely to work closely with each landowner client, prepare reports and billing statements, work with and manage company and subcontractor crews, perform site visits to estimate and audit forest activities, and perhaps work with public relations media. They commonly talk about two jobs, one is serving the client while the other is managing their consulting business.

18.1.6 Urban forester

Urban foresters are responsible for the development, implementation, and management of an urban forestry program for a state, city, county, or other municipality. Typical job responsibilities of urban foresters (Fig. 18.7) include analyzing and preparing elements of a tree management plan and implementing and enforcing city

FIGURE 18.7 Rachel Reyna, Program Leader, Department of Conservation and Natural Resources, Bureau of Forestry, Private Forestry and Urban Forestry Programs, Pennsylvania, United States. *Rachel Reyna.*

ordinances. Duties of urban foresters may also include tree planting, streetscape design, tree protection, tree survey and mapping activities, and the development of standards and protocols for tree placement. Communication skills are important, as urban foresters are typically called upon to present issues, problems, and changes related to urban forestry practices to various audiences such as city councils and community groups. Urban foresters may also be involved in supervising volunteers or groups involved in urban forestry projects, may serve as an information resource for the general public, and, therefore, may be involved in education and training programs. Further, urban foresters may be assigned the task of developing or managing databases related to urban trees or tree health, which requires knowledge and training in dendrology, entomology, GIS, and GPS, among other areas.

18.1.7 Utility forester

Utility foresters identify, inspect, and evaluate trees and brush along utility rights-of-way to maintain the safety along these lines. These professionals (Fig. 18.8) provide inspection results to clients, along with recommendations for further action. They plan, direct, and coordinate the activities of tree maintenance crews, and discuss and negotiate access issues. In essence, utility

FIGURE 18.6 Starling Childs, Forestry and Environmental Consultant, northern Connecticut, United States. *Starling Childs.*

FIGURE 18.8 Steve Hunt (middle) and Brian Reed (right), Utility Foresters with Georgia Transmission Corporation, United States, discuss operations with a contractor. *Brian Reed.*

foresters serve as a liaison between clients, tree maintenance crews, and the public. Some of the requirements of these positions include being able to identify local trees, having knowledge of tree growth rates, having the ability to work independently, being able to read and follow maps, and maintaining strong verbal and written communication skills. In seeking these types of positions, International Society of Arboriculture (ISA) certification and prior utility vegetation management experience is of value.

18.1.8 State and federal foresters

State and federal foresters (Fig. 18.9) work on state or federally managed lands and generally have duties that involve cruising and marking timber, delineating sale boundary lines, performing forest inventories, and administering timber sales. These duties may also include a variety of fire management and silvicultural activities related to reforestation, timber stand improvement, wildfire control, and fuels management goals. Some federal foresters (Fig. 18.10) may also have duties that relate to producing environmental assessment reports and environmental impact statements. Others may be charged with determining sale boundaries, designing harvest system layouts, determining acceptable marking and cutting methods, identifying the appropriate type of sampling system to use, and preparing logging plans. People in these positions may commonly participate on an interdisciplinary team to ensure that forest management activities meet legal requirements. In the United States, relevant laws might include the National Environmental Policy Act, the Endangered Species Act, the National Historic Preservation Act, and the National Forest Management Act.

FIGURE 18.9 Colleen Sloan, Federal Forester, United States Department of Agriculture, Forest Service, California, United States. *Colleen Sloan.*

FIGURE 18.10 Kimpton Cooper, District Ranger, United States Department of Agriculture, Forest Service, Texas, United States. *Kimpton Cooper.*

State or county foresters implement, coordinate, and administer forestry activities and programs with individual landowners, city or county governments or agencies, and conservation districts. People in these positions typically provide technical assistance, help write forest resource management plans, and provide educational information to ensure the use of effective forestry practices. Other duties of these foresters may include evaluating tree health, providing technical advice to local communities, assisting with special urban forestry field projects, assessing insect and disease management, and coordinating prescribed fire programs. Some state foresters (Fig. 18.11) are heavily involved in wildlife and nongame conservation activities. These foresters may have leadership roles in efforts to evaluate various management practices within the state and should generally be proficient with GIS and GPS technology. Duties might include conducting surveys of rare and endangered species, conservation planning activities, ecological restoration activities, and prescribed burns for habitat improvement.

Foresters working directly with military installations often become involved with a wide range of activities that involve timber management, soil and water conservation, insect and disease control, wildfire control, prescribed burning, and archeological and cultural site protection. In general, for these professionals the military mission is the highest priority; however, most military lands in the United States are managed on a multiple-use basis that includes forestry, fish and wildlife, outdoor recreation, and flood control concerns. In nearly every case, a state or federal forester will work with other professionals and specialists from other disciplines to plan and obtain the optimum mix of uses for the land.

FIGURE 18.11 Matt Elliott, Program Manager, Georgia Department of Natural Resources, Wildlife Resources Division, Nongame Conservation Section, Georgia, United States. *Joe Burnam.*

18.1.9 Procurement forester

Procurement foresters conduct fiber procurement activities in support of manufacturing facilities. In general, they are responsible for monitoring and scheduling fiber quality and flow to a facility. People in these positions are generally accountable for providing fiber to a manufacturing facility while meeting cost targets and, where appropriate, are responsible for supplier compliance with appropriate forest certification requirements. Procurement foresters ensure contractor compliance in daily operations, and tasks might include logging sale preparation, sale inspection, road layout, timber cruising, and landowner and logger relations. Procurement foresters may also be responsible for assuring that a business is compliant with the safety, environmental, and governmental regulations it must follow.

Procurement foresters commonly monitor the volume generated by and the performance of their wood suppliers. Since these people help to provide raw materials for manufacturing processes, mill processes and wood quality issues must be understood. They need a strong understanding of the conversion of logs to lumber. Timber sampling is a task that is required to properly appraise timber purchasing opportunities. In addition to having a formal degree in forestry, knowledge of wood procurement processes, timber markets, mill requirements and specifications, wood pricing, logging systems, and laws and guidelines for loggers are key requirements of these positions. A procurement forester should also have some training in finance and accounting concepts, sound negotiation skills, and refined communication skills.

18.1.10 Research forester

Research foresters investigate new techniques and methods to address questions related to forests and forest productivity. Some research foresters who work for private companies may also work with university cooperative programs to install, measure, and report the results of studies on silvicultural processes (e.g., fertilization or thinning), site preparation methods (e.g., herbaceous weed control), or planting stocks (e.g., genetically improved seedlings). These studies are often led and conducted solely by the company itself. Other research foresters may work for organizations, such as the U.S. Forest Service or universities, which address a broad range of forestry issues, including but not limited to economics, forest health, GPS, landscape fragmentation, and silvicultural practices. The duties of these positions generally require the collection of samples and measurements of relevant forest conditions and may also include statistical or spatial analyses. Research foresters must pay attention to detail, be able to record data accurately, and have good oral and written communication skills. Results of research findings may be presented to small groups of people at organizational meetings or to larger groups of people at regional, national, or international meetings. In addition, results of research studies are typically communicated through internal reports or peer-reviewed journal articles. Often, a graduate degree is required to directly obtain this type of a position immediately after the conclusion of a university education.

18.1.11 Inventory forester

Inventory foresters typically conduct or oversee fieldwork regarding the sampling of forest conditions and the compilation of sampling results for timberland appraisals. Inventory foresters need to be well organized, able to work alone or as part of a team, proficient in the use of data management software, and able to perform high-quality work, since the information developed often serves as the starting point for many other forest and natural resource management activities. Inventories are generally performed on property managed by an organization that employs the forester, and tasks often include timber cruising (sampling) processes, yet may also involve regeneration surveys, tree weight measurements, standing and felled-tree taper measurements, and merchantable product measurements. Inventory foresters may also perform log quality audits and may check-cruise (audit) other sampling efforts. Inventory analyses are generally performed to help improve the quality of forestry and natural resource management decision-making processes. They are increasing finding using LiDar as a data source for their analysis.

18.1.12 Forest analyst and planner

Forest analysts and planners generally need GIS, harvest scheduling, biometric, and database management skills to perform and coordinate planning efforts. People in these positions may often be asked to analyze potential activities for increasing efficiencies and improving the management procedures of an organization. And, these people are closely associated with strategic, tactical, and operational planning processes. A forest analyst or planner may work with others to assess management programs under consideration by an organization. Among other duties, forest analysts and planners (Fig. 18.12) maintain spatial and nonspatial databases (GIS and GPS), apply current and developing technologies to the databases, and manage the utilization and integration of new technology into an organization's daily activities. They may also coordinate the use of handheld

FIGURE 18.12 Erika Mavity, Forest Planner, United States Department of Agriculture, Forest Service, Georgia, United States. *Erika Mavity.*

data recorders, conduct forest inventory analyses, perform spatial analyses, or provide technical training opportunities for others. These foresters often perform miscellaneous administrative duties, such as report generation, written and oral responses to inquiries, and reviews of plans and documents prepared by others. Finally, forest analysts and planners are frequently asked to deliver presentations on activities within their area of responsibility and to participate in various internal and external committees. As forests are now considered an asset that many institutional investors use to diversify their portfolios, this has added new roles for the forest analyst and planner who is often required to meet the Security and Exchange Commission requirements on transparency of the investment's performance.

18.1.13 Forest engineer

Forest engineers perform tactical and operational planning activities, and harvest system selection and operational analyses in conjunction with forest harvesting activities. Road system design and cost analyses are two areas of responsibility for forest engineers (Fig. 18.13). In this regard, forest engineers provide road and harvest system designs for timber sales, which may include road layout, road construction, and bridge construction activities. Other duties may include the preparation of short- and long-term road management plans and the management of a comprehensive road system, with consideration of surfaces, culverts, and

FIGURE 18.13 Dana Mitchell, Research Engineer, United States Department of Agriculture Forest Service, Alabama, United States. *Wes Sprinkle.*

FIGURE 18.14 Zachary A. Parisa, International Development Forester and Entrepreneur, California, United States. *Zachary A. Parisa.*

roadside vegetation. Many are skilled in survey methods to collect the high-quality data necessary to design these structures. In addition, forest engineers are typically involved in contract preparation, inspection of construction and service contracts, the development of cooperative road maintenance agreements, and other related work. These types of jobs require knowledge of engineering principles, road design, and minor structure design, and other areas of coursework typically associated with the civil engineering field. In the western United States, there is often a separate license to credential forest engineers.

18.1.14 International development forester and natural resource manager

Many foresters and natural resource managers venture beyond their nation's borders to work in development areas for foreign countries (Fig. 18.14). These efforts are often made as part of afforestation or reforestation programs sponsored by a host country or an international organization. These positions may be traditional in a sense and involve working for a landowning forest products firm, or they may be more diverse, and involve developing new projects for a governmental or nongovernmental organization (NGO). In the United States, many young foresters and natural resource managers receive international work

experience through the U.S. Peace Corps, an independent governmental agency that has sponsored rural and urban development work in over 20 nations worldwide since its inception in the early 1960s. Other means for gaining international development work are through exchange programs or academic research projects that offer students or young professionals a chance to network with others and develop professional skills. Foresters and natural resource managers working in this arena commonly need to learn a language such as Spanish, Portuguese, Chinese, French, or Arabic.

18.1.15 Silviculturist

Silviculturists provide leadership, program management, and budgetary oversight of silvicultural programs that account for the growing of trees from seedling to a harvestable size. This leadership may include technical direction and oversight in all phases of a reforestation program, as well as of silviculture prescriptions to ensure that plans, treatment, and harvesting techniques are compatible with long-range land management plans. Silviculturists may also participate in the development, preparation, and review of environmental analysis reports and environmental impact statements, and provide assistance with sale preparation activities, fuel reduction projects, and prescribed burning activities (Fig. 18.15). Further, silviculturists may coordinate the field activities of tree improvement programs.

18.1.16 Ecologist

Ecologists generally must have experience working in forested ecosystems and in the collection and maintenance of ecological field data through the use of

FIGURE 18.16 Amy Castle Blaylock, Wildlife Bureau Director, Department of Wildlife, Fisheries, and Parks, Mississippi, United States. *Amy Castle Blaylock.*

FIGURE 18.15 Andrew McCarley, Silviculturist, United States Department of Agriculture Forest Service, eastern Tennessee, United States. *Andrew McCarley.*

standard forest ecosystem sampling methods. Therefore, a graduate degree is often required, as some familiarity with sample design and statistics is necessary. These positions can require one to understand and follow sampling protocols developed by others. Therefore, there is a need to follow procedures for data maintenance and delivery and to collect and maintain high-quality field notes. Ecologists must also be able to navigate through remote terrain to locate previously installed sample plots using a map, a compass, or GPS. Since these positions are closely associated with forest research positions, strong interpersonal, communication, and organizational skills are needed.

18.1.17 Wildlife biologist

Wildlife biologists work alongside other natural resource management professionals and may have a variety of responsibilities related to game and nongame management. For example, wildlife biologists (Fig. 18.16) may assist in the planning and implementation of wildlife censuses and surveys, and may be involved in recommending wildlife management practices. For some wildlife species, biologists may need to implement capture and monitoring efforts that include trapping, immobilizing, handling, and veterinary technician work (e.g., drug administration, blood collection and storage, and vital sign monitoring). Wildlife biologists may be involved in reviewing proposed projects, permits, and license applications in conjunction with biological assessments or in the development of

environmental impact statements. Along the way, biologists may use home range analysis, GIS, GPS, telemetry, and other techniques to analyze habitat quality and habitat use. In addition, biologists need good communication skills to provide reports, articles, instructions, and other technical documents to an organization or to the public.

18.1.18 Fisheries biologist

Fisheries biologists perform work related to the research, conservation, production, and management of fish or fishery resources. They may be involved in determining, establishing, and applying biological facts, methods, techniques, and procedures necessary for research, conservation, and management of fish and other aquatic animals. Some of the duties of these types of positions include field data collection, laboratory sample processing (Fig. 18.17), data management, and the preparation of reports for biological programs or for inclusion in procedures, processes, or services in support of a fisheries program. Fisheries biologists may also assist in the construction or placement of habitat improvement structures, fences, gabions, and other features within aquatic systems. People in these positions may also perform riparian studies on wetland areas and along stream corridors, which may include species inventory and habitat condition assessments. Some recent examples of the duties of fisheries biologists include the following:

- Bull trout population censuses that involve snorkeling and electroshocking activities.
- Studies of biological, physical, and chemical characteristics of a lake, including species of fish (angling and gill net collections), zooplankton, and

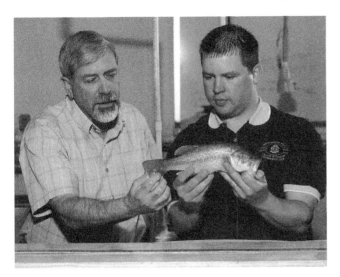

FIGURE 18.17 Michael Brown (left), professor of fish biology, South Dakota State University, South Dakota, United States. *Michael Brown.*

FIGURE 18.18 Todd Tietjen, Limnologist Project Manager, Southern Nevada Water Authority, Nevada, United States. *Todd Tietjen.*

phytoplankton; lake clarity and transparency; lake mixing; nutrient cycling; dissolved oxygen levels; and water quality.

18.1.19 Hydrologist and limnologist

Hydrologists and limnologists provide advice, training, direction, guidance, and leadership in the area of best hydrologic management practices (water movement) and their effects on water quality. In many cases, these duties involve protecting beneficial uses of water, such as recreation, habitat for fish and other aquatic organisms, and domestic water supplies. Their work may require water sampling activities, flow or discharge measurements, and chemical and biological analyses. Hydrologists may also conduct studies on the occurrence, distribution, quantity and quality of water within watersheds, and along these lines, some computer modeling may be performed. Limnologists (Fig. 18.18) focus on the physical, chemical, and biological processes in lakes and reservoirs that are influenced by the watershed, human activities, or changes within a water body. The types of jobs associated with this position require the analysis of field and laboratory data, preparation of written and oral reports, and the development of recommendations on hydrologic and water quality study findings into reports. Often, a graduate degree in these fields is required for positions that involve activities beyond the sampling of water resources.

18.1.20 Soil conservationist

Soil conservationists assist landowners and operators of agricultural, residential, recreational, and public lands in the planning, application, and maintenance of agronomic and engineering practices that relate to soil, water, air, plant, animal, and human resource issues. Along these lines, a soil conservationist (Fig. 18.19) may perform fieldwork that includes collecting resource data and interpreting the data to clearly define existing conditions and to propose practical and suitable alternatives for resources. Other duties include documenting decisions, providing technical assistance to implement a management plan, and addressing permit requirements. Individuals in these

FIGURE 18.19 Jessica Taylor, soil conservationist, Natural Resource Conservation Service, eastern Washington, United States. *Jessica Taylor.*

positions interact with landowners and operators, technical specialists, and the general public to obtain, clarify, or exchange information or facts needed to carry out cooperative conservation work. As suggested, a soil conservationist uses appropriate field tools to work with landowners and operators to assist in the determination of objectives, the identification of resource concerns, and the inventory of natural resources.

18.1.21 Technical services forester

Technical services foresters can be further classified as GIS managers or GPS specialists and may be involved in a range of geospatial and information technical support (Fig. 18.20). People in these positions typically support field personnel in the application of GIS and GPS for their needs, and the development and implementation of standards and protocols for issues regarding *corporate* (official) databases; thus, one primary focus of these positions may be database management. In this regard, technical services foresters help insure the integrity and accuracy of natural resource data through the development and administration of data collection procedures. They often acquire additional computer skills beyond that of the average forester.

18.1.22 Invasive species coordinator

Invasive species issues will vary by the region of the world in which a person is situated; however, a typical job in this area may include fieldwork for collecting information on, and implementing control of, invasive plants. People employed in these types of positions (Fig. 18.21) may be responsible for assisting with the implementation of control strategies, and an herbicide

FIGURE 18.21   Jon D. Prevost, Wildlife Forester/Invasive Species Coordinator, Delta Farm/Delta Wildlife, Mississippi, United States. *Jon D. Prevost.*

applicator's license may be required. Some of the duties of these positions may involve testing, monitoring, and evaluating invasive species control methods, planning and coordinating surveys for invasive plants, and using handheld computers and GPS to map invasive plants. People employed in these positions may assist with surveys of natural resources and damaged resources and work as part of a team in ecological restoration and site rehabilitation projects.

18.1.23 Forest recreation manager

Forest recreation managers plan and administer recreation program activities, which may include, among others, biking, hiking, water sports, or wilderness opportunities. These professionals (Fig. 18.22) may also provide technical assistance and advice on outdoor recreation and recreation resource issues, including policies, programs, studies, and permits. Recreation

FIGURE 18.20   Michael Torbett, GIS Specialist, Georgia Forestry Commission, Georgia, United States. *Michael Torbett*

FIGURE 18.22   Jason M. Hall, Forest Program Specialist, Department of Conservation and Natural Resources, Bureau of Forestry, Recreation Section, Pennsylvania, United States. *Photo courtesy of Jason M. Hall.*

managers may be involved in the processes that deal with easements, leases, licenses, and permits for a variety of uses, and may need to coordinate land-use authorizations. In terms of fieldwork, recreation managers may conduct inventories, develop and implement utilization surveys, and perform compliance inspections. As with other types of jobs, recreation managers are involved in producing an organization's budget proposals and with short- and long-term planning efforts. Recreation managers may also be involved in various aspects of a wilderness management program.

18.1.24 Park interpreter and guide/naturalist

Some of the more interesting careers in natural resource management involve direct contact with tourists and recreationists on a daily basis, as is the case with a park interpreter or guide. Many of these people are considered the frontline representatives of organizations that manage national, state, or local parks. They are often involved in guiding hikes or delivering presentations on the natural and managed ecology (and features) of a park, and therefore provide interpretive services. The development and delivery of programs aimed at educating visitors is important to this type of position, and research into natural phenomena may be required. Significant interaction with children may also be common. Depending on the location, a person working in this type of position may need to possess knowledge of local natural history, cultural history, and cultural values that are significant to the resources within and around the park. Some positions of this type may even require the development of press releases, visual displays, and grant applications, and the servicing of information centers. At times, these types of positions may require strenuous activity that involves standing, walking, and talking for extended periods through a typical workday, in addition to experiencing a high volume of personal contact with recreationists and tourists. Therefore, highly refined oral communication skills are valued along with interpersonal communication skills. Through directed and self-guided educational opportunities, interpreters and naturalists attempt to balance the need to protect natural resources with the need for people to obtain outdoor recreational experiences.

18.1.25 Forest entomologist

A forest entomologist performs a variety of field and laboratory tasks in support of pest management and forest health protection issues. These professionals (Fig. 18.23) provide forest and natural resource managers with professional advice and technical assistance that is useful for minimizing the damage caused by

FIGURE 18.23 Crew collecting data on ash tree decline and mortality due to the exotic insect emerald ash borer (*Agrilus planipennis*) in southeastern Michigan, United States. From left to right: Charles Flower, Stephanie Smith, Dr. Kamal Gandhi, and Danielle Lightle. *Dr. Kathleen Knight.*

forest insects. These types of jobs may involve planning, coordinating, and conducting pest management/forest health protection surveys and evaluations and providing technical assistance to other professionals or agencies. Forest entomologists analyze data and prepare comprehensive reports, make recommendations for achieving forest health objectives, and plan, organize, and conduct pilot projects, field tests, and demonstrations to determine the value of new or improved materials, strategies, or techniques. People in these positions may also coordinate with research scientists on new findings applicable to forest pest management/forest health protection.

18.1.26 Certification and sustainability auditor

As noted in Chapter 15, many land management organizations are pursuing certification of their land through the programs developed by the American Tree Farm System, the Forest Stewardship Council, the International Organization for Standardization, the Program for the Endorsement of Forest Certification, the Sustainable Forestry Initiative, and other programs. Experienced natural resource professionals (Fig. 18.24) are often called upon to review and audit the management plans and operational management of organizations seeking certification. This is often a role that consultants or industrial foresters may fill. Auditors document whether a landowner is in compliance with a set of certification standards through interviews and on-site observations. In conjunction with this process, an auditor may need to develop an action plan for a landowner to

FIGURE 18.24 Stephen C. Grado, SmartWood Certification Auditor, Rainforest Alliance, Northfield, Minnesota, United States. *Maralyn Renner.*

FIGURE 18.25 Thomas O'Shea, Assistant Director, Division of Fisheries and Wildlife, Massachusetts, United States. *Thomas O'Shea.*

outline ways in which they can address the difference between current management activity and certification standards. The reports and action plans developed through the auditing process are used to inform the certification organization of compliance levels and provide guidance in meeting certification standards.

18.1.27 General or regional manager

Usually through years of experience, general or regional managers have gained advanced knowledge of the concepts, principles, and practices of a wide variety of natural resource disciplines, such as forestry, range, and wildlife management; mineral management; land-use planning; and fire management and suppression. These positions require knowledge of an organization's missions and programs and of local government, state, and federal laws, policies, and procedures related to natural resource management. General and regional managers oversee budgeting activities and must possess advanced communication (both verbal and written) and negotiation skills for effective problem-solving and group decision-making activities. As suggested, these professionals (Figs. 18.25 and 18.26) need to be well versed in the areas of administration, mediation, and conflict management.

18.1.28 Educator

A professor of forestry or of another field in natural resource management now usually requires the completion of a doctoral degree (PhD). Many educators teaching four-year courses at colleges or universities (Figs. 18.27–18.30) are engaged in research, education, and outreach. Some educators are specifically involved in research, while the responsibilities of others may be confined to teaching courses that lead to an

FIGURE 18.26 Dan Hudnut, Vice President, Wagner Forest Management, Ltd., New Hampshire, United States. *Travis Howard.*

undergraduate or graduate degree. Some educators are involved solely in the outreach mission of a college or university system and conduct continuing education programs for professionals in their fields or more general programs for the public. Instructors of forestry or natural resource programs can often be hired with only a master's degree. These educators are generally employed to teach forestry and natural resources courses at 2-year college programs or at technical schools, and their main responsibilities involve teaching rather than research or outreach.

18.1.29 Law enforcement/conservation officers

Several career options within the natural resources arena involve the enforcement of laws, rules, and regulations. Conservation officers (Fig. 18.31) employed by

FIGURE 18.27 Samantha Langley, Vice Provost for Graduate Education, Research and Outreach, Northern Kentucky University, Kentucky, United States. *Samantha Langley.*

FIGURE 18.29 Paul Doruska, Professor, University of Wisconsin–Stevens Point, Wisconsin, United States. *Paul Doruska.*

FIGURE 18.28 Mark J. Ducey, Professor and Chair, Natural Resources and the Environment, University of New Hampshire, New Hampshire, United States. *Mark J. Ducey.*

FIGURE 18.30 Gustavo Perez-Verdin, Researcher, National Polytechnic Institute, CIIDIR, Durango, Mexico. *Gustavo Perez-Verdin.*

FIGURE 18.31 Wildlife Conservation Officers Travis McDonald (left) and Ryne Long are both sworn law enforcement officers that work for the Mississippi Department of Wildlife, Fisheries, and Parks, United States. *Travis McDonald.*

many state and federal wildlife agencies have various duties and responsibilities, including the enforcement of hunting-related laws and regulations. In addition to the enforcement of regulations, most wildlife law enforcement officers are also involved in many aspects of wildlife conservation, research, and public education programs. In addition to conservation officers, many state and federal agencies employ park rangers and game wardens to enforce park regulations. Other natural resource agencies, including the U.S. Forest Service, also employ law enforcement officers to enforce regulations and protect public interests on public lands. Some law enforcement opportunities are also available through private companies and organizations within the natural resource arena.

Summary

Where (geographically and thematically) a person begins his or her career and where he or she ends it are partly under his or her own control and partly dependent on market, political, and personal circumstances. Ideally, in preparation for a career as a specific type of forester or natural resource manager, a person would first try to determine the type of work that a job involves. Once ascertained, a motivated person can then develop the skills and acquire the tools necessary to market his- or herself to various natural resource management organizations. Common skills required by employers are the ability to communicate effectively in oral and written fashions. Employers are often looking for work-related and leadership experience in potential employees. Another skill that was alluded to, yet is necessary in nearly every job, is the ability to negotiate effectively. How well a person prepares for a career and how well he or she continues to improve his or her skills once his or her career begins are aspects of employment that are directly under his or her control.

Questions

(1) *Your dream career.* Ponder seriously for a few minutes the type of job you ideally want to perform during the early stages of your career. Develop a bulleted list of 10 items that you feel are necessary to prepare yourself for this type of opportunity. In a short, one-page memorandum addressed to your instructor, list these 10 items, and in a brief paragraph describe the actions and resources that may be necessary to address all 10 before you graduate from college.

(2) *Interview of a seasoned professional.* Locate a seasoned professional in your local area, and conduct a short interview concerning his or her career. In a short report, describe how he or she may have prepared his- or herself for his or her career. Further, describe his or her current job and the responsibilities that he or she has been given. Describe the types of positions he or she has held in the past. In a small group of students within your class, describe your case study of a seasoned professional.

(3) *Develop a résumé.* Using an Internet resource or the resources available at your school, develop a professional résumé. Gather some feedback by sharing the résumé with your instructors, close friends, and perhaps a professional working the natural resource management field. Then, develop a list of things you can do over the next year that will build upon or enhance your résumé.

References

Food and Agriculture Organization of the United Nations, 2010. Global Forest Resources Assessment 2010. Food and Agriculture Organization of the United Nations, Rome, Italy. FAO Forestry Paper 163.

Weinstein, D., 2012. The psychology of behaviorally focused résumés on applicant selection: are your hiring managers really hiring the "right" people for the "right" jobs? Business Horizons 55, 53—63.

Appendix A

Provided below is a list of the common and scientific names of nearly all of the living organisms found in the main text of this book. The species are listed alphabetically by scientific name (genus and species), and some of the common names are also provided.

Scientific Name	Common Name
Medicinal plant species	
Acacia farnesiana	Sweet acacia
Achillea millefolium	Common yarrow
Amaranthus spp.	Amaranth
Artocarpus incise	Breadfruit
Artocarpus altilis	Breadfruit
Asarum canadense	Canadian wild ginger
Atropa belladonna	Belladonna
Bixa orellana	Annatto or lipstick tree
Calligonum comosum	Abal
Cecropia peltata	Trumpet tree
Chondrodendron tomentosum	Curare or pareira
Eleutherococcus senticosus	Siberian ginseng
Gaylussacia spp.	Huckleberry
Glycyrrhiza glabra	Licorice
Hamamelis virginiana	Witch hazel
Hydrastis canadensis	Goldenseal
Lobelia inflate	Indian tobacco
Mirabilis jalapa	Clavillia or marvel of Peru
Panax quinquefolius	Ginseng
Peumus boldus	Boldo leaves
Pfaffia paniculata	Suma
Pinus strobus	White pine
Podophyllum peltatum	Mayapple
Quillaja saponaria	Soapbark tree
Rauwolfia serpentine	Serpent wood
Rubus spp.	Blackberry
Sanguinaria canadensis	Bloodroot
Sassafras albidum	Sassafras

Scientific Name	Common Name
Smilax spp.	Sarsaparilla
Theobroma cacao	Cocoa tree
Vaccinium spp.	Blueberry
Nontree species (berries, grasses, and mushrooms)	
Abrus precatorius	Jequirity or rosary pea
Achillea millefolium	Common yarrow
Adhatoda vasica	Malabar nut
Agropyron pungens	Sea couch grass
Alocasia macrorrhiza	Wild taro
Ammophila breviligulata	Marram grass or American beachgrass
Anaphalis margaritacea	Early everlasting
Andropogon virginicus	Broomsedge bluestem
Annona muricata	Soursop
Arceuthobium spp.	Dwarf mistletoe
Arctium lappa	Greater burdock
Areca catechu	Betel nut
Artemisia californica	Coastal sagebrush
Asarum canadense	Canadian wild ginger
Atropa belladonna	Belladonna
Bambusa spp.	Bamboo
Berberis spp.	Barberry
Bixa orellana	Annatto or lipstick tree
Bobgunnia madagascariensis	Mucherekese
Cananga odorata	Ylang-ylang
Carica papaya	Papaya
Caryota urens	Kitul palm tree or jaggery palm
Catharanthus roseus	Rosy periwinkle
Chondrodendron tomentosum	Curare or pareira
Cirsium arvense	Canada thistle
Cirsium spp.	Thistles
Cocos nucifera	Coconut
Colocasia esculenta	Sweet taro
Coriandrum sativum	Coriander

Scientific Name	Common Name	Scientific Name	Common Name
Cynodon dactylon	Bermuda grass	*Panax quinquefolius*	Ginseng
Cyperus rotundus	Nut-grass	*Paulownia* spp.	Paulownia
Cyrtosperma chamissonis	Swamp taro	*Phellinus* spp.	Abetasunk mushroom
Dichanthelium commutatum	Variable panicgrass	*Pennisetum purpureum*	Napier grass
Digitaria sanguinalis	Crabgrass	*Pfaffia paniculata*	Suma
Diodia teres	Buttonweed or poor joe	*Photinia fraseri*	Red robin photinia
Dioscorea spp.	Yams	*Photinia serrulata*	Chinese photinia
Elaeis spp.	Oil palm	*Piper methysticum*	Sakau
Encelia spp.	Brittlebush	*Pityopsis graminifolia*	Narrowleaf silkgrass
Epilobium angustifolium	Fireweed	*Plectranthus* spp.	Spur flowers
Euthamia tenuifolia	Goldenrod	*Polystichum munitum*	Western sword fern
Gaultheria shallon	Salal	*Poaceae* spp.	Wiregrass
Gaylussacia dumosa	Dwarf huckleberry	*Podophyllum peltatum*	Mayapple
Glycyrrhiza glabra	Licorice	*Polypremum procumbens*	Juniper leaf
Gnaphalium obtusifolium	Rabbit-tobacco	*Proustia pyrifolia*	Tola
Gnaphalium purpureum	Purple cudweed	*Pueraria montana* var. *lobata*	Kudzu
Hamamelis virginiana	Witch-hazel	*Quercus suber*	Cork oak
Haplopappus divaricatus	Yellow aster	*Rauwolfia serpentine*	Snakeroot or serpent wood
Heterotheca subaxillaris	Camphorweed		
Hibiscus tiliaceus	Hibiscus	*Rhynchospora* spp.	Beak-rush
Hydrastis canadensis	Goldenseal	*Rosa eglanteria*	Sweetbriar rose
Hypericum gentianoides	Pineweed or orange grass	*Rubus* spp.	Dewberry
Imperata cylindrica	Cogongrass	*Rubus fruticosus*	Blackberry
Justicia adhatoda	Malabar nut	*Rubus spectabilis*	Salmonberry
Lechea villosa	Pinweed	*Rubus strigosus*	Raspberry
Leptilon (or *Erigeron*) *canadensis*	Horseweed	*Rubus trivialis*	Southern dewberry
		Rumex acetosella	Sheep sorrel
Leucaena leucocephala	Leucaena	*Saccharum* spp.	Sugarcane
Leymus arenarius	Lyme grass or sand ryegrass	*Sabal palmetto*	Palm or cabbage palmetto
		Sanguinaria canadensis	Bloodroot
Linaria canadensis	Blue toadflax	*Sarracenia oreophila*	Green pitcher plant
Lobelia inflate	Indian tobacco	*Smilax regelii*	Sarsaparilla
Mahonia spp.	Mahonia	*Solanum* spp.	Nightshade
Manihot esculenta	Cassava	*Sorghum* spp.	Sorghum
Mirabilis jalapa	Clavillia or marvel of Peru	*Theobroma cacao*	Cocoa tree
Musa spp.	Bananas and plantains	*Tillandsia usneoides*	Spanish moss
Myrothamnus flabellifolius	Resurrection bush	*Typha* spp.	Cattails
Oenothera laciniata	Evening primrose	*Vaccinium* spp.	Blueberry
Oligostachyum lubricum	Bamboo	*Viscum album*	Mistletoe

Scientific Name	Common Name	Scientific Name	Common Name
Zea mays	Maize	*Chamaecyparis thyoides*	Southern Atlantic white cedar
Zingiber officinale	Ginger	*Cornus florida*	Dogwood
Native and exotic tree species		*Fagus grandifolia*	American beech
North America		*Fagus mexicana*	Mexican beech or haya
Abies balsamea	Balsam fir	*Fraxinus americana*	White ash
Abies fraseri	Fraser fir	*Fraxinus latifolia*	Oregon ash
Abies grandis	Grand fir	*Fraxinus pennsylvanica*	Green ash
Abies lasiocarpa	Rocky Mountain fir or subalpine fir	*Ilex opaca*	American holly
Abies procera	Noble fir	*Juglans cinerea*	Butternut
Acacia koa	Koa	*Juglans nigra*	Black walnut
Acer spp.	Maples	*Juniperus* spp.	Juniper
Acer circinatum	Vine maple	*Juniperus virginiana*	Eastern redcedar
Acer macrophyllum	Bigleaf maple	*Larix laricina*	Tamarack
Acer negundo	Boxelder	*Larix occidentalis*	Western larch
Acer rubrum	Red maple	*Leucaena leucocephala*	Leucaena or white leadtree
Acer saccharinum	Silver maple	*Liquidambar styraciflua*	Sweetgum
Acer saccharum	Sugar maple	*Liriodendron tulipifera*	Yellow-poplar or tulip tree
Alnus rubra	Red alder	*Lonchocarpus violaceus*	Balché tree
Arctostaphylos spp.	Manzanitas	*Maclura pomifera*	Osage orange
Betula spp.	Birch	*Magnolia grandiflora*	Southern magnolia
Betula alleghaniensis	Yellow birch	*Magnolia virginiana*	Sweetbay magnolia
Betula lenta	Sweet birch	*Manilkara zapota*	Sapodilla
Betula neoalaskana	Alaska birch or resin birch	*Metrosideros polymorpha*	'Ohi 'a lehua
Betula papyrifera	Paper birch	*Nyssa* spp.	Tupelo
Brosimum alicastrum	Breadnut	*Nyssa sylvatica*	Black gum
Calocedrus decurrens	Incense cedar	*Ostrya virginiana*	Eastern hophornbeam
Carya spp.	Hickories	*Picea glauca*	White spruce
Carya illinoinensis	Pecan tree	*Picea mariana*	Black spruce
Castanea dentata	American chestnut	*Picea pungens*	Blue spruce
Ceanothus spp.	Ceanothus	*Picea rubens*	Red spruce
Cedrela odorata	Cedro rojo or Spanish cedar	*Picea sitchensis*	Sitka spruce
Celtis occidentalis	Hackberry	*Pinus* spp.	Pines
Cercis canadensis	Eastern redbud	*Pinus banksiana*	Jack pine
Chamaecyparis spp.	Cedars	*Pinus contorta*	Lodgepole pine
Chamaecyparis lawsoniana	Port-Orford-cedar	*Pinus echinata*	Shortleaf pine
Chamaecyparis nootkatensis	Alaska yellow-cedar	*Pinus edulis*	Colorado pinyon

Scientific Name	Common Name	Scientific Name	Common Name
Pinus elliottii	Slash pine	*Sassafras albidum*	Sassafras
Pinus jeffreyi	Jeffrey pine	*Sequoia sempervirens*	Redwood
Pinus lambertiana	Sugar pine	*Swietenia macrophylla*	Caoba or mahogany
Pinus montícola	Western white pine	*Taxodium distichum* var. *distichum*	Baldcypress
Pinus palustris	Longleaf pine	*Taxodium distichum* var. *nutans*	Pondcypress
Pinus ponderosa	Ponderosa pine		
Pinus quadrifolia	Pinyon pine	*Taxus brevifolia*	Pacific yew
Pinus resinosa	Red pine	*Terminalia amazonia*	White olive
Pinus strobus	Eastern white pine	*Terminalia oblonga*	Guava or Peruvian almond
Pinus taeda	Loblolly pine		
Populus balsamifera	Balsam poplar	*Thuja occidentalis*	Northern white cedar or arborvitae
Populus deltoids	Eastern cottonwood		
Populus grandidentata	Bigtooth aspen	*Thuja plicata*	Western redcedar
Populus tremuloides	Quaking aspen or trembling aspen	*Tilia americana*	Basswood
		Tsuga canadensis	Eastern hemlock
Populus trichocarpa	Black cottonwood	*Tsuga heterophylla*	Western hemlock
Prunus emarginata	Bitter cherry	*Tsuga mertensiana*	Mountain hemlock
Prunus pensylvanica	Pin cherry	*Ulmus* spp.	Elms
Prunus serotina	Black cherry	*Ulmus alata*	Winged elm
Pseudotsuga menziesii	Douglas-fir	*Ulmus americana*	American elm
Psychotria hawaiiensis	Kopiko	**Central America**	
Psydrax odorata	Alahe'e	*Anacardium excelsum*	Wild cashew or espave
Pteridium aquilinum	Bracken	*Brosimum utile*	Cow tree or amapa
Quercus spp.	Oaks	*Calophyllum brasiliense*	Guanandi
Quercus alba	White oak	*Cavanillesia platanifolia*	Cuipo
Quercus falcate	Southern red oak	*Cecropia peltata*	Trumpet tree
Quercus geminate	Sand oak	*Cedrela odorata*	Spanish-cedar
Quercus incana	Bluejack oak	*Cordia alliodora*	Spanish elm or Ecuador laurel
Quercus laevis	Turkey oak		
Quercus margaretta	Sand post oak	*Dalbergia retusa*	Rosewood
Quercus nigra	Water oak	*Enterolobium cyclocarpum*	Guanacaste tree or ear tree
Quercus palustris	Pin oak	*Erisma uncinatum*	Cambara or cedrinho
Quercus robur	Pedunculate oak or English oak	*Erythrina poeppigiana*	Bucare ceibo
		Gliricidia sepium	Mata ratón or quickstick
Quercus velutina	Black oak	*Gmelina arborea*	White teak or beechwood
Quercus virginiana	Live oak	*Hymenaea courbaril*	Jatobá or guapinol or stinking toe
Robinia pseudoacacia	Black locust		
Salix spp.	Willows	*Inga laurina*	Ingá-branco
Salix barclayi	Barclay's willow	*Leucaena leucocephala*	Leucaena
Salix spp.	Willows	*Pachira quinata*	Pochote
Salix nigra	Black willow	*Parkia* spp.	Faveira or locust-bean

Scientific Name	Common Name	Scientific Name	Common Name
Pinus ayacahuite	Mexican white pine or pino blanco	*Pinus maximinoi*	Thinleaf pine
Pinus caribaea	Caribbean pine or pino costanero	*Pinus patula*	Patula pine
		Pinus radiata	Radiata pine
Pinus hartwegii	Hartweg's pine or pino de Montana	*Piptadenia macrocarpa*	Curupay
Pinus maximinoi	Thinleaf pine	*Prosopis pallida*	Huarango or mesquite or kiawe
Pinus oocarpa	Mexican yellow pine or Honduran yellow pine	*Quercus palustris*	Pin oak
Pinus pseudostrobus	Smooth-bark Mexican pine or pinabete	*Simarouba amara*	Marupa
		Swietenia macrophylla	Mara or mahogany
Pinus tecunumanii	Pino rojo or red pine	*Tectona* spp.	Teaks
Prioria copaifera	Cativo	**Europe**	
Swietenia macrophylla	Big leaf mahogany	*Abies alba*	European silver fir
Tabebuia rosea	Roble de sabana	*Acer opalus*	Italian maple
South America		*Aesculus hippocastanum*	Horse-chestnut or conker tree
Acacia mangium	Mangium	*Alnus glutinosa*	Europe black alder
Alnus acuminate	Andean alder or aliso	*Arctostaphylos* spp.	Manzanitas
Amburana cearensis	Ishpingo	*Betula pendula*	Silver birch or European white birch
Anadenanthera peregrine	Yopo	*Carpinus betulus*	European hornbeam
Araucaria araucana	Chilean pine or monkey puzzle tree	*Castanea sativa*	European chestnut
Aspidosperma macrocarpon	Pumaquiro	*Cedrus libani*	Lebanon cedar
Astronium urundeuva	Urundel	*Eucalyptus globulus*	Eucalyptus or blue gum
Bertholletia excelsa	Brazil nut	*Fagus sylvatica*	Dwarf beech or European beech
Bombacopsis quinata	Saqui-saqui	*Fraxinus* spp.	Ashes
Brosimum utile	Cow tree or amapa	*Ilex opaca*	American holly
Calophyllum brasiliense	Guanandi	*Larix* spp.	Larch
Cedrela odorata	Spanish-cedar	*Larix decidua*	European larch
Cinchona spp.	Cinchona tree	*Mangifera indica*	Indian mango
Erisma uncinatum	Cambara or cedrinho	*Picea abies*	Norway spruce
Eucalyptus globulus	Blue gum	*Picea sitchensis*	Sitka spruce
Hevea brasiliensis	Rubber tree	*Pinus halepensis*	Aleppo pine
Hura crepitans	Sandbox tree	*Pinus nigra*	Austrian or calabrian pine
Hymenaea courbaril	Jatobá or guapinol	*Pinus pinaster*	Maritime pine or mesogean
Juglans australis	Walnut	*Pinus pinea*	Stone pine
Lauraceae spp.	Laurels	*Pinus radiata*	Radiata pine or Monterey pine
Lonopterygium huasango	Hualtaco		
Meliaceae spp.	Mahogany	*Pinus sibirica*	Siberian pine
Nothofagus spp.	Beeches	*Pinus sylvestris*	Scot's pine
Parkia spp.	Faveira		

Scientific Name	Common Name
Populus nigra	European black poplar or Lombardy poplar
Pseudotsuga menziesii	Douglas-fir
Pteridium aquilinum	Bracken
Quercus petraea	Sessile oak
Quercus pyrenaica	Rebollo oak
Quercus robur	English oak or pedunculate oak
Quercus suber	Cork oak
Taxus brevifolia	Yew
Tilia spp.	Basswood
Ulmus spp.	Elms
Ulmus hollandica	Dutch elm
Asia	
Abies fargesii	Farges fir
Abies sibirica	Siberian fir
Aleurites fordii	Tung-oil tree
Alnus spp.	Alder
Arctostaphylos spp.	Manzanitas
Calocedrus formosana	Incense-cedar
Calocedrus macrolepis	Chinese incense-cedar
Cedrus deodara	Deodar cedar
Chamaecyparis spp.	Cedars
Chamaecyparis obtusa	Hinoki or Japanese cypress
Cinnamomum camphora	Camphor laurel
Cornus florida	Dogwood
Cryptomeria japonica	Sugi or Japanese cedar
Cunninghamia lanceolata	Chinese fir
Gonystylus bancanus	Ramin
Hevea brasiliensis	Rubberwood
Ilex opaca	American holly
Larix spp.	Larch
Larix olgensis	Olgan larch
Larix sibirica	Siberian larch
Lithocarpus spp.	Beech
Morinda citrifolia	Indian mulberry
Nerium indicum	Oleander
Phoebe nanmu	Nanmu
Picea asperata	Dragon spruce

Scientific Name	Common Name
Picea obovata	Siberian spruce
Pinus koraiensis	Korean pine or Chinese pinenut
Pinus merkusii	Sumatran pine
Pinus rigida	Pitch pine
Pinus strobus	White pine
Pinus yunnanensis	Yunnan pine
Populus spp.	Poplar
Populus adenopoda	Chinese aspen
Populus davidiana	Korean aspen
Pseudotsuga sinensis	Chinese Douglas-fir
Pterocarpus santalinus	Zitan or red sandalwood
Quercus acutissima	Sawtooth oak
Quercus mongolica	Mongolian oak
Robinia pseudoacacia	Black locust
Shorea albida	Light red meranti
Shorea almon	Philippine mahogany
Shorea robusta	Sal
Styrax benzoin	Kemenyan or styrax
Tectona grandis	Teak
Tilia spp.	Basswood
Viburnum odoratissimum	Arrowwood
Africa	
Acacia mearnsii	Black wattle
Acacia nilotica	Gum Arabic tree
Bombax flammeum	Gold Coast bombax
Calliandra calothyrous	Red calliandra
Casuarina equisetifolia	Filao tree or beach sheoak
Ceiba pentandra	Kapok or fromager
Chlorophora excelsa	Iroko or African teak
Diospyros spp.	Ebony or mabolo
Entandrophragma angolense	Tiama
Entandrophragma candollei	Kosipo
Entandrophragma cylindricum	Sapele
Entandrophragma utile	Sipo
Eucalyptus spp.	Eucalypts
Gambeya africana	Longhi or longui rouge
Ginkgo biloba	Ginkgo or maidenhair tree

Scientific Name	Common Name
Gossweilerodendron balsamiferum	Agba or tola
Grevillea robusta	Southern silky oak
Guarea cedrata	Bossé or guarea
Guibourtia pellegriniana	Bubinga
Hymenaea verrucosa	East African copal
Khaya ivorensis	Lagos mahogany
Lovoa trichilioides	African walnut or dibétou or African walnut
Mangifera indica	Indian mango
Mansonia altissima	Mansonia or ofun
Milicia excelsa	African teak
Millettia laurentii	Wengé
Mitragyna ciliata	Elolom or subaha
Nauclea diderrichii	Bilinga
Nesogordonia papaverifera	Danta or otutu
Ocotea bullata	Stinkwood
Olea capensis	Ironwood
Olinia ventosa	Hard pear
Pericopsis elata	Afrormosia
Pinus nigra	Austrian or Calabrian pine
Platylophus trifoliatus	White alder
Podocarpus latifolius	Yellowwood
Quercus suber	Cork oak
Sclerocarya birrea	Marula
Tectona grandis	Teak
Terminalia superba	Limba
Tieghemella heckelii	Makore
Triplochiton scleroxylon	Obeche or wawa
Uapaca spp.	Yeye

Oceania

Scientific Name	Common Name
Acacia spp.	Acacia
Acacia nilotica	Gum Arabic tree
Agathis australis	Kauri
Alectryon excelsus	Titoki
Araucaria hunsteinii	klinki
Beilschmiedia tawa	Tawa
Cordyline australis	Cabbage tree
Corynocarpus laevigatus	Karaka or New Zealand laurel

Scientific Name	Common Name
Dacrydium cupressinum	Rimu or red pine
Elaeocarpus dentatus	Hinau
Eucalyptus spp.	Eucalypts
Eucalyptus obliqua	Australian oak or messmate
Eucalyptus saligna	Saligna eucalyptus or Sydney bluegum
Griselinia littoralis	Kapuka or broadleaf
Intsia bijuga	Kwila or ifil
Juglans nigra	Black walnut
Leptospermum scoparium	Manuka or tea tree
Melaleucas spp.	Melaleucas
Meryta sinclairii	Puka
Metrosideros robusta	Northern rātā
Metrosideros umbellate	Southern rātā
Morinda citrifolia	Indian mulberry
Pinus radiata	Radiata pine or Monterey pine
Pometia pinnata	Taun
Pseudopanax crassifolius	Horoeka or lancewood
Psydrax odorata	Alahe'e
Quercus palustris	Pin oak
Sequoia sempervirens	Redwood
Toona ciliata	Australian red cedar
Ulmus spp.	Elms

Mammals

Scientific Name	Common Name
Ailuropoda melanoleuca	Giant panda
Alces alces	Moose
Antilocapra americana	Pronghorn antelope
Apodemus sylvaticus	Wood mouse
Arctonyx collaris	Hog badger
Bison bison athabascae	Wood bison
Blarina spp.	Short-tailed shrew
Blarina carolinensis	Southern short-tailed shrew
Canis latrans	Coyote
Canis lupus	Gray wolf
Castor canadensis	Beaver
Cervus canadensis	Elk
Dasycyon hagenbecki	Andean wolf
Didelphis virginiana	North American opossum

Scientific Name	Common Name	Scientific Name	Common Name
Elephas maximus maximus	Sri Lankan elephant	*Vombatus ursinus*	Common wombat
Eutamias sibiricus	Siberian chipmunk	*Vulpes vulpes*	Red foxes
Gulo gulo	Wolverines	**Birds**	
Herpestes vitticollis	Stripe-necked mongoose	*Aegolius funereus*	Tengmalm's owl or boreal owl
Lama guanicoe	Guanaco		
Lemur catta	Ring-tailed lemur	*Aegypius monachus*	Eurasian black vulture or cinereous vulture
Lepus mandschuricus	Manchurian hare		
Lontra canadensis	North American river otter	*Aix sponsa*	Wood duck
		Ammodramus savannarum	Grasshopper sparrow
Loxodonta africana cyclotis	Forest elephant	*Ara macao*	Scarlet macaw
Lycalopex culpaeus	Andean fox or culpeo	*Ardea herodias*	Great blue heron
Lycalopex griseus	South American gray fox or grey zorro	*Bonasa umbellus*	Ruffed grouse
		Branta canadensis	Canada goose
Macropus rufogriseus	Red-necked wallaby	*Bubo bubo*	Eurasian eagle owl
Mapudungun pudu	Pudu	*Buteo jamaicensis*	Red-tailed hawk
Marmota flaviventris	Yellow-bellied marmot	*Buteo lineatus*	Red-shouldered hawk
Marmota monax	Groundhog or woodchuck	*Cardinalis cardinalis*	Northern cardinal
		Carpodacus purpureus	Purple finch
Martes martes	European pine marten	*Catharus fuscescens*	Veery
Meles meles	European badger	*Colinus virginianus*	Northern bobwhite quail
Mephitis mephitis	Striped skunk	*Corvus corax*	Common raven or northern raven
Mustela spp.	Weasel		
Neotamias minimus	Western least chipmunk	*Corvus corone*	Carrion crow
Neovison vison	American mink	*Cygnus olor*	Mute swan
Ochotona princeps	American pika	*Ectopistes migratorius*	Passenger pigeon
Odocoileus hemionus	Mule deer	*Geococcyx californianus*	Roadrunner or greater roadrunner
Odocoileus virginianus	White-tailed deer		
Panthera pardus pardus	African leopard	*Geococcyx velox*	Roadrunner or lesser roadrunner
Panthera tigris tigris	Bengal tiger		
Peromyscus leucopus	White-footed mouse	*Grus japonensis*	Red-crowned crane
Procyon lotor	Raccoon	*Grus vipio*	White-naped crane
Puma concolor	Cougar or mountain lion	*Gymnogyps californianus*	California condor
Rangifer tarandus	Reindeer or caribou	*Haliaeetus leucocephalus*	Bald eagle
Rhinopithecus roxellana	Golden snub-nosed monkey	*Lagopus lagopus*	Willow ptarmigan
		Megalapteryx benhami	Flightless moa
Sciurus carolinensis	Eastern gray squirrel	*Melanerpes erythrocephalus*	Red-headed woodpecker
Sylvilagus floridanus	Eastern cottontail rabbit	*Meleagris gallopavo*	Wild turkey
Tamiasciurus hudsonicus	American red squirrel	*Melospiza melodia*	Song sparrow
Tapirus terrestris	South American tapir	*Micrastur ruficollis*	Barred forest-falcon
Ursus americanus	American black bear	*Pharomachrus mocinno*	Resplendent quetzal
Ursus arctos horribilis	Grizzly bear	*Phasianus colchicus*	Common pheasant

Scientific Name	Common Name	Scientific Name	Common Name
Picoides borealis	Red-cockaded woodpecker	*Litoria caerulea*	Australian green tree frog
Picoides villosus	Hairy woodpecker	*Litoria chloris*	Australian red-eyed tree frog
Poecile cinctus	Gray-headed chickadee	*Micrurus fulvius*	Eastern coral snake
Ramphastos sulfuratus	Keel-billed toucan	*Naja naja*	Indian cobra
Scolopax minor	American woodcock	*Oophaga pumilio*	Strawberry poison-dart frog
Sialia sialis	Eastern bluebird		
Stellula calliope	Calliope hummingbird	*Oplurus cuvieri*	Madagascan collard iguana or collared iguanid lizard
Strix occidentalis caurina	Northern spotted owl		
Sturnella magna	Eastern meadowlark	*Phrynosoma platyrhinos*	Horny toad or desert horned lizard
Turdus migratorius	American robin		
Tyto alba	Common barn owl	*Plethodon cinereus*	Red back salamander
Vultur gryphus	Andean condor	*Pseudotriton ruber*	Red salamander
Amphibians and reptiles		*Rana amurensis*	Siberian tree frog
Agkistrodon contortrix contortrix	Southern copperhead	*Sceloporus occidentalis*	Fence lizard
Amblyrhynchus cristatus	Marine iguana	*Terrapene carolina bauri*	Florida box turtle
Ambystoma opacum	Marbled salamander	*Thamnophis elegans terrestris*	Coast garter snake
Basiliscus basiliscus	Common basilisk	**Fish**	
Boa constrictor	Boa constrictor	*Micropterus salmoides*	Largemouth bass
Bufo bufo	Common toad	*Oncorhynchus kisutch*	Coho salmon
Bufo marinus	Cane toad	*Oncorhynchus mykiss*	Rainbow trout
Bufo periglenes	Golden toad	*Oncorhynchus nerka*	Sockeye salmon
Colostethus flotator	Rainforest rocket frog	*Oncorhynchus tshawytscha*	Chinook salmon
Corytophanes cristatus	Helmeted iguana	*Salmo salar*	Atlantic salmon
Crotalus adamanteus	Eastern diamondback	**Insects**	
Ctenosaura similis	Black spiny-tailed iguana	*Adelges tsugae*	Hemlock wooly adelgid
Dasypeltis fasciata	Central African egg-eating snake	*Anoplophora glabripennis*	Asian longhorn beetle
		Bombyx mori	Silkworm
Dendroaspis polylepis	Black mamba	*Bursaphelenchus xylophilus*	Pine wood or wilt nematode
Drymarchon couperi	Eastern indigo snake		
Enyalioides palpebralis	Horned wood lizard	*Choristoneura occidentalis*	Western spruce budworm
Gopherus polyphemus	Florida gopher tortoise	*Dendroctonus ponderosae*	Mountain pine beetle
Hyla arenicolor	Canyon tree frog	*Dendroctonus rufipennis*	Spruce beetle
Hynobius kimurae	Hida salamander	*Dendrolimus spectabilis*	Pine moth
Iguana iguana	Green iguana	*Ips typographus*	European bark beetle
Lachesis muta	South American bushmaster	*Ixodea scapularis*	Deer tick
		Laccifer lacca	Indian lac
Lampropeltis triangulum	Red milk snake or eastern milk snake	*Lymantria dispar*	Gypsy moth
		Malacosoma disstria	Forest tent caterpillar

Scientific Name	Common Name	Scientific Name	Common Name
Meloidogyne spp.	Root-knot nematode	**Pathogens/parasites/fungi**	
Pratylenchus spp.	Lesion nematode	*Amanita* spp.	Destroying angel or death cup
Trichodorus christiei	Stubby-root nematodes	*Cronartium quercuum* f. sp. *fusiforme*	Fusiform rust
Xiphinema spp.	Dagger nematode	*Heterobasidion annosum*	Annosus root rot
Diseases		*Phytophthora infestans*	Late blight
Cronartium ribicola	White pine blister rust	*Taphrina* spp.	Leaf curl
Cryphonectria parasitica	Chestnut blight		
Ophiostoma ulmi	Dutch elm disease		
Phytophthora ramorum	Sudden oak death		

Appendix B

Provided below is a partial list of the major forestry and natural resource management public or nongovernmental organizations from around the world. Our intent here is to provide an indication of the variety of organizations that have an interest in the management of forests and natural resources, and we acknowledge that there are many omissions. A complete enumeration of all agencies and organizations is beyond the scope of this book. However, we encourage readers to become aware of the organizations within their region that are influential in the forest and natural resource management issues.

Continent of Origin and Organization Name	Year Founded	Main Office Location	Internet Address in 2020
North America			
American Forestry Association	1875	Washington D.C., USA	www.americanforests.org
Bear Trust International	1999	Missoula, MT, USA	www.beartrust.org
Canadian Forest Service	1899	Ottawa, ON, Canada	www.cfs.nrcan.gc.ca
Canadian Institute of Forestry	1908	Mattawa, ON, Canada	www.cif-ifc.org
Canadian Parks and Wilderness Society	1963	Ottawa, ON, Canada	www.cpaws.org
Conservation International	1987	Arlington, VA, USA	www.conservation.org
Delta Waterfowl	1938	Bismarck, ND, USA	www.deltawaterfowl.org
Ducks Unlimited	1937	Memphis, TN, USA	www.ducks.org
Forest Products Association of Canada	1913	Ottawa, ON, Canada	www.fpac.ca
Forest Products Society	1945	Madison, WI, USA	www.forestprod.org
International Crane Foundation	1973	Baraboo, WI, USA	www.savingcranes.org
International Game Fish Association	1939	Dania Beach, FL, USA	www.igfa.org
International Society of Arboriculture	1924	Champaign, IL, USA	www.isa-arbor.com
International Wildlife Conservation Society	1968	Pacific Palisades, CA, USA	www.internationalwildlife.org
International Wildlife Rehabilitation Council	1972	Eugene, OR, USA	www.theiwrc.org
Mexico Secretariat of the Environment and Natural Resources	2000	Mexico City, Mexico	www.semarnat.gob.mx
Mexico National Commission for Knowledge and Use of Biodiversity	1992	Mexico City, Mexico	www.conabio.gob.mx/
Mexico National Institute for Agriculture and Forestry Research	—	Mexico City, Mexico	www.gob.mx/inifap
Mexico Natural Protected Areas and National Parks	2000	Mexico City, Mexico	www.gob.mx/conanp
National Wild Turkey Federation	1973	Edgefield, SC, USA	www.nwtf.org
National Wildlife Federation	1938	Reston, VA, USA	www.nwf.org

Continued

Continent of Origin and Organization Name	Year Founded	Main Office Location	Internet Address in 2020
Natural Resource Defense Council	1970	New York, NY, USA	www.nrdc.org
North American Forest Commission	1958	Washington, D.C., USA	www.fs.fed.us/global/nafc
North American Gamebird Association	1931	Cambridge, MD, USA	www.mynaga.org
North American Grouse Partnership	1999	Fruita, CO, USA	www.grousepartners.org
Pheasants Forever	1982	St Paul, MN, USA	www.pheasantsforever.org
Pronatura México	—	Mexico City, Mexico	www.pronatura.org.mx
Rocky Mountain Elk Foundation	1984	Missoula, MT, USA	www.rmef.org
Ruffed Grouse Society	1961	Coraopolis, PA, USA	www.ruffedgrousesociety.org
Society of American Foresters	1900	Bethesda, MD, USA	www.safnet.org
The Arbor Day Foundation	1972	Nebraska City, NE, USA	www.arborday.org
The Association of Fish and Wildlife Agencies	1902	Washington, D.C., USA	www.fishwildlife.org
The Nature Conservancy	1951	Arlington, VA, USA	www.nature.org
The Sierra Club	1892	San Francisco, CA, USA	www.sierraclub.org
The Society for Ecological Restoration	1988	Washington, D.C., USA	www.ser.org
The Society of Municipal Arborists	1964	Watkinsville, GA, USA	www.urban-forestry.com
The Wildlife Society	1937	Bethesda, MD, USA	www.wildlife.org
United Nations Division for Sustainable Development	1992	New York, NY, USA	www.un.org/esa/dsd
U.S. Bureau of Land Management	1946	Washington, D.C., USA	www.blm.gov
U.S. Bureau of Reclamation	1902	Washington, D.C., USA	www.usbr.gov
U.S. Department of the Interior	1849	Washington, D.C., USA	www.doi.gov
U.S. Fish and Wildlife Service	1956	Washington, D.C., USA	www.fws.gov
U.S. Forest Service	1905	Washington, D.C., USA	www.fs.fed.us
U.S. National Park Service	1916	Washington, D.C., USA	www.nps.gov
U.S. Natural Resource Damage Assessment and Restoration Program	—	Washington, D.C., USA	www.doi.gov/restoration
U.S. Natural Resources Conservation Service	1932	Washington, D.C., USA	www.nrcs.usda.gov
U.S. Office of Natural Resources Revenue	1982	Washington, D.C., USA	www.onrr.gov
Wild Sheep Foundation	1974	Cody, WY, USA	www.wildsheepfoundation.org
Wildlife Conservation Society	1895	Bronx, NY, USA	www.wcs.org
Wildlife Forever	1987	Brooklyn Center, MN, USA	www.wildlifeforever.org
Wildlife Management Institute	1911	Gardners, PA, USA	www.wildlifemanagementinstitute.org
Winrock International	1985	Morrilton, AR, USA	www.winrock.org
World Forestry Center	1964	Portland, OR, USA	www.worldforestry.org
World Wildlife Fund	1961	Washington, D.C., USA	www.worldwildlife.org

Central America

Belize Ministry of Natural Resources and Immigration	—	Belize City, Belize	http://mnra.gov.bz/natural-resources/

Continent of Origin and Organization Name	Year Founded	Main Office Location	Internet Address in 2020
Costa Rica Ministry of Environment, Energy, and Telecommunications	—	San José, Costa Rica	www.minae.go.cr
El Salvador Ministry of Environment and Natural Resources	—	San Salvador, El Salvador	www.marn.gob.sv
Guatemala Ministry of Environment and Natural Resources	2000	Guatemala City, Guatemala	www.marn.gob.gt
Honduras Ministry of Natural Resources, Environment and Mines	—	Tegucigalpa, Honduras	http://www.miambiente.gob.hn/
Nicaragua Ministry of the Environment and Natural Resources	—	Managua, Nicaragua	www.marena.gob.ni
Panama National Environmental Authority	1998	Panama City, Panama	www.miambiente.gob.pa
South America			
Argentina Ministry of Natural Resources and Sustainable Development	—	Buenos Aires, Argentina	www.ambiente.gob.ar
Bolivia Ministry of Hydrocarbons and Energy	1970	La Paz, Bolivia	www.hidrocarburos.gob.bo
Brazil Ministry of the Environment	1992	Brasilia, Brazil	www.mma.gov.br
Chilean Ministry for the Environment	—	Santiago, Chile	www.mma.gob.cl
Colombia Ministry of the Environment and Sustainable Development	2003	Bogota, Colombia	www.minambiente.gov.co
Ecuador Ministry of the Environment	1996	Cuenca, Ecuador	www.ambiente.gob.ec
Uruguay Ministry of the Environment	—	Montevideo, Uruguay	www.medioambiente.gub.uy
Venezuela Instituto Forestal Latinoamericano	1981	Mérida, Venezuela	www.forest.ula.ve
Europe			
Austria Federal Ministry of Agriculture, Forestry, Environment and Water	2000	Vienna, Austria	www.lebensministerium.at
Commonwealth Forestry Association	1921	Shropshire, United Kingdom	www.cfa-international.org
Czech Republic Ministry of the Environment	1990	Prague, Czech Republic	www.mzp.cz
Danish Environmental Protection Agency	—	Copenhagen, Denmark	www.mst.dk
Danish Ministry of Environment and Food	1971	Copenhagen, Denmark	https://en.mfvm.dk/the-ministry/
European Environment Agency	1990	Copenhagen, Denmark	www.eea.europa.eu
European Forest Institute	—	Joensuu, Finland	www.efi.int
European Forestry Commission	1947	Rome, Italy	www.fao.org/forestry/efc
Finland Ministry of Agriculture and Forestry	—	Helsinki, Finland	www.mmm.fi
Finnish Ministry of the Environment	—	Helsinki, Finland	www.environment.fi
Forestry Commission of Great Britain	—	—	www.forestry.gov.uk
Germany Friends of the Earth	1975	Berlin, Germany	www.bund.net
Germany Natural Parks		Bonn, Germany	www.naturparke.de

Continued

Continent of Origin and Organization Name	Year Founded	Main Office Location	Internet Address in 2020
Germany Federal Environment Agency	1974	Berlin, Germany	www.umweltbundesamt.de
Hessian Forest	—	Kassel, Germany	www.hessen-forst.de
International Institute for Environment and Development	1971	London, UK	www.iied.org
International Union for Conservation of Nature	1948	Gland, Switzerland	www.iucn.org
International Union of Forest Research Organizations	1892	Vienna, Austria	www.iufro.org
Ireland Department of Communications, Energy and Natural Resources	—	Dublin, Ireland	www.dcenr.gov.ie
Italian Ministry of Agriculture, Food, and Forestry Policies	1946	Rome, Italy	www.politicheagricole.it
Belarus Ministry of Natural Resources and Environmental Protection	1994	Minsk, Belarus	www.minpriroda.gov.by/en/
Nature and Biodiversity Conservation Union	—	Berlin, Germany	www.nabu.de
Norway Ministry of the Environment	1972	Oslo, Norway	www.regjeringen.no/en/dep/md.html?id=668
Poland Ministry of the Environment	—	Warsaw, Poland	www.mos.gov.pl
Portugal Ministry of Agriculture, Sea, Environment and Planning	—	Lisbon, Portugal	www.portugal.gov.pt
Russian Federal Forestry Agency	—	Moscow, Russia	www.rosleshoz.gov.ru
Slovenia Ministry of the Environment and Spatial Planning	—	Ljubljana, Slovenia	www.mop.gov.si
Spain Ministry of the Environment, Rural, and Marine	2008	Madrid, Spain	www.marm.es
Sweden Ministry of the Environment	—	Stockholm, Sweden	www.sweden.gov.se/sb/d/2066
Swedish Forest Agency	—	Jönköping, Sweden	www.skogsstyrelsen.se
The Ministry of the Environment and Natural Resources of the Russian Federation	—	Moscow, Russia	www.mnr.gov.ru
The Royal Society for the Protection of Birds	1889	Bedfordshire, UK	www.rspb.org.uk
The Slovak Republic Ministry of the Environment	2010	Bratislava, Slovak Republic	www.minzp.sk
The Wildlife Trust	1912	Nottinghamshire, UK	www.wildlifetrusts.org
United Kingdom Department for Environment, Food, and Rural Affairs	2001	London, UK	www.defra.gov.uk
United Kingdom Department of Energy and Climate Change	2008	London, UK	www.decc.gov.uk
Wildlife Britain	—	United Kingdom	www.wildlifebritain.com
Asia			
Bangladesh Ministry of Environment and Forest	1977	Dhaka, Bangladesh	www.moef.gov.bd

Continent of Origin and Organization Name	Year Founded	Main Office Location	Internet Address in 2020
Bhutan Ministry of Agriculture and Forest	2009	Thimphu, Bhutan	www.moaf.gov.bt
Cambodia Ministry of Environment	—	Phnom Penh, Cambodia	www.moe.gov.kh
Center for International Forestry Research	1971	Bogor, Indonesia	www.cifor.org
Culture and Environment Preservation Association	1995	Phnom Penh, Cambodia	www.cepa-cambodia.org
Indian Ministry of Environment	—	New Delhi, India	http://india.gov.in/sectors/environment/index.php
Japan Ministry of the Environment	—	Tokyo, Japan	www.env.go.jp/en
Malaysia Forest Department Sarawak	1919	Sarawak, Malaysia	www.forestry.sarawak.gov.my
Ministry of Natural Resources of the People's Republic of China	1998	Beijing, China	www.mnr.gov.cn
Mongolia Ministry of Nature, Environment and Tourism	—	Ulaanbaatar, Mongolia	www.mne.mn/mn
National Trust for Nature Conservation	1982	Khumaltar, Lalitpur, Nepal	www.ntnc.org.np
Pakistan Environmental Protection Agency	1997	Islamabad, Pakistan	www.environment.gov.pk
Philippines Department of Environment and Natural Resources	1997	Quezon City, Philippines	www.denr.gov.ph
Philippines Haribon Foundation	1972	Quezon City, Philippines	www.haribon.org.ph
Republic of Korea Ministry of Environment	1990	Gyeonggi-do, South Korea	http://eng.me.go.kr/
Singapore Ministry of Environment and Water Resources	2004	Singapore	www.mewr.gov.sg
Sri Lanka Ministry of Environment and Wildlife Resources	—	Battaramulla, Sri Lanka	www.msdw.gov.lk
Taiwan Environmental Protection Administration	1987	Taipei City, Taiwan	www.epa.gov.tw
Thailand Ministry of Natural Resources and Environment	1997	Bangkok, Thailand	www.mnre.go.th
The International Tropical Timber Organization	1986	Yokohama, Japan	www.itto.int
The Royal Society for the Conservation of Nature	1966	Jubeiha, Jordan	www.rscn.org.jo
Vietnam Ministry of Natural Resources and Environment	—	Hanoi, Vietnam	www.monre.gov.vn
Africa			
Republic of Congo Ministry of Forest Economy	—	Brazzaville, Congo	www.mefdd.cg
Democratic Republic of the Congo (DRC) Ministry of the Environment	—	Kinshasa, DRC	www.minenv.itgo.com
Forestry Research Institute of Nigeria	1954	Ibadan, Oyo State, Nigeria	www.frin.gov.ng

Continued

Continent of Origin and Organization Name	Year Founded	Main Office Location	Internet Address in 2020
Ghana Ministry of Lands, Forestry & Mines	—	Accra, Ghana	www.ghanaweb.com/GhanaHomePage/republic/ministry.profile.php?ID=46
Kenya Ministry of Environment and Forestry	—	Nairobi, Kenya	www.environment.go.ke
Mozambique International Organization for Conservation	1961	Maputo, Mozambique	www.wwf.org.mz
Namibia Ministry of the Environment and Tourism	1990	Windhoek, Namibia	www.met.gov.na
Nigeria Ministry of Environment	—	Abuja, Nigeria	www.nigeria.gov.ng
Republic of South Africa Department of Agriculture, Forestry, and Fisheries	—	Pretoria, South Africa	www.daff.gov.za
Republic of South Africa Department of Environmental Affairs	—	Pretoria, South Africa	www.environment.gov.za
Senegal Ministry of the Environment and Conservation	—	Dakar, Senegal	www.environnement.gouv.sn
Tanzania Ministry of Natural Resources and Tourism	—	Dar es Salaam, Tanzania	www.maliasili.go.tz
United Nations Environment Program	1972	Nairobi, Kenya	www.unep.org
World Agro-forestry Center	1978	Nairobi, Kenya	www.worldagroforestrycentre.org
Zimbabwe Ministry of Environment, Water and Climate	—	Harare, Zimbabwe	http://www.zarnet.ac.zw/evol/environ/
Oceania			
Australian Dept. of Sustainability, Environment, Water, Population and Communities	2010	Canberra, Australia	www.environment.gov.au
Australian National Environmental Protection Council	2001	Adelaide, Australia	www.nepc.gov.au
Fiji Department of Environment	—	Suva, Fiji	www.doefiji.wordpress.com
New Zealand Department of Conservation	—	Wellington, New Zealand	www.doc.govt.nz/parks-and-recreation
New Zealand Forest Owners Association	1926	Wellington, New Zealand	www.nzfoa.org.nz
New Zealand Institute of Forestry	1927	Wellington, New Zealand	www.nzif.org.nz
New Zealand Ministry of Primary Industries	—	Wellington, New Zealand	www.mpi.govt.nz
Papua New Guinea Conservation and Environment Protection Authority	1985	Boroko, Papua New Guinea	www.pngcepa.com
Royal Forest and Bird Protection Society of New Zealand	1923	Wellington, New Zealand	www.forestandbird.org.nz
Solomon Island Ministry of Environment, Climate Change, Disaster Management and Meteorology	2008	Honiara, Solomon Islands	www.mecdm.gov.sb

Glossary

Provided here is a glossary of terms used throughout this book. In cases where the definition provided is closely associated with a published source, the source is provided in numeric form in parentheses. The references section at the end of this Appendix provides a link of the numeric code to the original source of the definition.

Abiotic: The nonliving aspects of an ecosystem, including items such as rocks, minerals, compounds, soils, and water (1).

Abiotic disease: Any disease arising from a non-living source such as dust, temperature, sunlight, and moisture (1).

Aboriginal: The state of being the primary or initial species, or the type that originally resided in a given area (2).

Abut: The act of meeting along an outside edge or a cantilevering portion of another object. Alternatively, to end contact at a spot or to bend in order to receive support (2).

Accreditation: A system of review designed by an accrediting group that is utilized as a tool to assess and certify specific degree programs. Alternatively, it involves a review program that leads to the decision of whether or not organizations are certified by the forest certification system (1).

Acidity: The amount of acid within a given sample, or the condition of being overly acidic (2).

Acre: A unit of area, defined as covering 43,560 square feet, 4,840 square yards, or about 0.4 hectares (3).

Aerial fuel: Fuels that, either biotic or abiotic, have no direct contact with the ground but rather are supported upwards by something else. These fuels are typically thought of as being tree crowns (1).

Aerial logging: Any logging operation that utilizes flight as a means by which to move logs from the logging site to the landing (1).

Aesthetic: A philosophy that deals with beauty and taste, that focuses on the origin, creation, and the appreciation of art in any form. Alternatively, a personal opinion of the beauty of a piece of art or real estate, or an opinion of tastes by which a person determines the pleasing, or satisfying factors, of a certain work or object, according to their own senses specifically the sense of sight. Or, a feeling of satisfaction or pleasantry obtained from looking at an object or work (2).

Aesthetic landscape: Any landscape that, dependent upon taste, is pleasing to the senses especially that of sight. The feeling of aesthetics can be influenced by knowledge of potential economic gain. Often landscapes are designed in order to create an aesthetic feeling, or to increase revenue generation by increasing the aesthetics (3).

Aesthetically pleasing: A state of being appealing to one or more of the senses, particularly sight. What is considered aesthetically pleasing is very subjective and can vary greatly from person to person.

Afforestation: The planting of a forest on an area of land that has not had a forest recently existing on it (1).

Age: The average age of any aggregation of trees. For even-aged forests, the most common age is usually used represents the age of the entire stand. For trees, it is represented as the time that has passed since the seed was germinated, budded, planted, or reached breast height (4.5 feet above ground) (1).

Age class: A time or age period that is used to classify the trees that are located in a certain area (1).

Agrarian: A system that takes aspects of both horticulture and animal husbandry, and employs together under one functioning agricultural unit. Alternatively, anything having to do with agriculture, horticulture, farming, animal husbandry, or a combination of these (2,3).

Agrarian based community: A group of people in some type of social aggregate (village, city, town etc.) whose lifestyle, livelihood, and or economy is based on agricultural practices.

Agriculture: A system of growing or raising crops or livestock in order to produce food or revenue (2).

Agroforestry: An agricultural system that incorporates forestry, agriculture, and livestock into one system in order to receive the economical and natural benefits from all three, especially the ways in which the three benefit and complement one another (1).

All-aged stand: Any uneven-aged stand that contains trees from every or a significant amount of other age classes, including those that are merchantable. These stands typically are ideal as wildlife habitat and require selective harvesting operations to attain maximum revenue (1).

Allocation: The planning and distributing of one's fixed resources among uses, activities, or practices. Alternatively, the setting aside or reserving of a resource to put toward a specific use (1).

Allogenic: Anything that a system uses or has introduced that was not originally in that system or produced in that system (1).

Allowable cut: The number of trees that are able to be harvested and still retain the amount needed to have to stay in accordance with a sustained-yield forest plan (1).

Allowable cut effect (ACE): The harvesting of timber in anticipation of future availability of harvestable timber. This is a controversial tool employed to increase present timber output by using future growth throughout the rotation (1).

Alternate clear-strip felling: A felling operation in which strips of a predetermined width (spatially or by number of rows), are alternately clearcut or untouched. That is, every other row is clearcut and the rows in between are not harvested.

Amenity: The appealing non-physical characteristics that enable a place to be worth more monetarily such as pride of ownership, housing, location near schools or recreation (4).

Amphibian: An organism from the class Amphibia, characterized by being cold-blooded, vertebrates, having a gilled young stage, and an air-breathing mature stage. These are considered by some as a mixture between fish and reptiles in the evolutionary chain (2).

Angiosperm: A tree that produces fruit and or flowers, and whose seeds are coated. This group is typically thought of as common fruit trees (apple, cherry, etc.) and hardwoods (oaks, poplar, etc.). These can either be classified as Monocotyledonous or Dicotyledonous (1).

Animal stock: The collective group of one species of domesticated animals owned by a certain unit, i.e., farmer, town, company.

Anthropogenic: Any consequence or object of human cause, creation, or manipulation (1).

Apical meristem: The aggregation of cells at the tips of roots and shoots that will eventually become primary tissues by way of cell division and specialization (1).

461

Aquatic: A plant, animal, or action whose life cycle has a large emphasis on water, including those that grow or live in water for extended periods of time, or visit water on a regular basis in order to survive. The emphasis on water must exceed that of simple drinking or absorbing in small amounts strictly for the regulation of organs and cells (2).

Area-wide felling: see **Clearcut**.

Arid: An area of land that is exceedingly dry so much in that agricultural crops are difficult or impossible to be sustained. These are landscapes that are usually either excessively cold or overly hot (2).

Aridic: A term describing a climate or region that has no water availability for the residing vegetation for over half of the collective time that the temperature of the soil is over 5°C (41°F) at 50 cm (19.6 in) underneath the ground (1).

Aridity: The measure of how much the climate or region lacks in moisture content, this could be considered in air, soil, or both. Meteorologists consider aridity the opposite of humidity (3).

Artificial regeneration: Reforestation of a stand by means of direct planting of seedlings, seeds, or sprouts note. This can be done either by hand or by machine, but by no natural means.

Atmosphere: The layer of air covering the Earth, that is composed of 79% nitrogen, 20% oxygen, and 0.03% carbon dioxide that are essential for animal and plant development. The composition does not vary much throughout the atmosphere (3).

Autogenic: Anything that a system uses or employs which originally existed in that system, or was produced within that system (1).

Axiliary meristem: An aggregation of rapidly growing and reproducing cells at the tip of a stem or root of a plant.

Backing fire: A fire that is either advancing, or was initiated with the intent to advance in such a manner as to travel in an upwind direction, in the lack of wind, or downhill (1).

Bacteria: A microorganism whose composition is of a single cell, absent of chlorophyll, mitochondria, and a true nucleus, and that reproduces by way of fusion (1).

Basal area (BA): The cross-sectional area of a horizontal plane of an individual tree stem at breast height (1).

Base station: A fixed apparatus (antenna) that receives global navigation satellite system signals, calculates correctional values, and provides augmentation data to users of global navigation satellite system receivers.

Benefit: Anything good that comes out of an action. A possible output of every situation can be intentional or unintentional, and a benefit can arise from a situation that is deemed bad. What is beneficial or one person, group, species, or ecosystem may or may not be beneficial to another.

Best management practice(s) (BMP): A single practice or system of practices prepared by a state or specialized planning agency considered to be the best and most practical way in which point and nonpoint source pollution can be controlled at levels deemed acceptable (1).

Bio oil: Fuel that is extracted from woody or vegetative biomass as opposed to coal, oil, or natural gas.

Biodiversity: The variation of plant and animal species, genes, ecosystems, biomes, complexity and physical, topical, and geographical attributes within an ecosystem, community, region, or landscape (1).

Biofuel: A fuel that is derived from an organic or biological raw material (2).

Biomass: The summation of all the organic material in a tree, stand, or forest. Alternatively, the residual wood products left over, most often, from harvesting operations that may be used for the purpose of energy or biofuel production (1).

Biome: An area's ecosystem that has a defined aggregation of plant, animal, and microbial life as well as a topographical, and physical environment influencing and being influenced by the climate as well as soil conditions (1).

Biosphere: The section of the Earth that consists of everything from the surface of the land to the lower portion of the atmosphere where living creatures can and do live and survive. Alternatively, the section of the carbon cycle that includes living organisms and organic matter resultant from such that contain carbon (1).

Biotic disease: A plant disease that arises from a living source such as a virus, bacteria, or a poison or irritant from another organism.

Bird: An organism from the class Aves, these typically have such characteristics of being warm-blooded, vertebrates, and essentially entirely covered with some type of feathers and having modified forelimbs in the form of wings (2).

Blight: A disease or symptom that ultimately leads to the death of a plant, characterized by a swift discoloration and withering of the plant or plant parts before its demise (1).

Bole: The primary stem or trunk of a tree that is typically more or less perpendicular to the ground and supports all of the branches (1).

Bolt: A log or piece of pulpwood that is considered, by some standard, short (1).

Boreal: Having to do, or being associated with, the northern latitudes (1).

Boreal forest: Forests located in areas with relatively short, warm summers, and long cold winters.

Bottomland forest: A forest characterized by moist soils and large deciduous trees, typically found in swamps and bayous.

Brashing: A form of pruning in which a worker takes a tool such as a club and breaks off all the branches, whether live or dead, as far up as he can reach.

Breast height: A standard height at which tree diameter is measured for purposes of assessment and inventorying purposes, also this is where basal area is measured. In North America, this is 1.37 m (4.5 feet) above ground on the uphill side of a tree. In Europe, it is 1.3 m (4.25 feet) (1).

Brushing up: see **Brashing**.

Bucker: The job title of the person who is in charge of cutting (bucking) felled timber into logs or bolts in lengths specified by the market or mill requirements (1).

Bucking: The act of cutting felled timber into logs or bolts in lengths that are specified by the mill (1).

By-product: A good or externality that is produced in the making of another good, yet not a goal of production or even intentionally produced (5).

Cambial zone: An area of rapidly growing and dividing cells within the bole and stems of a tree.

Canopy: The foliage in a forest consisting of branches and leaves (1).

Capital: Products designed to be used in the production of additional goods. Alternatively, any resource that is owned and can be or is used in the production of a product, such as, time, land, money, etc. (5)

Carbon cycle: The scientific explanation of how carbon in all of its forms moves through the environment supplying the ecological needs (3).

Carbon dioxide (CO$_2$): A gas that is colorless, and relatively heavy, and mainly formed by animal respiration and the burning of animal or plant matter, although the gas itself is not flammable (2).

Carbon Sequestration: The storage of carbon in plant tissue, by means of plant growth (1).

Carriage: A wheeled vehicle used in harvesting operations, one that is mounted to either a cable or a track, and used to transport logs to a landing or through a saw in a milling operation (1).

Carrying capacity: The number of organisms that can reside in a defined area from one year to the next over a long period of time. This can be applied to wildlife, and livestock as well as recreational settings. This number will be highly dependent upon cover, food, water, and space, as well as the number of other species within an area (1).

Certify: To attest or guarantee that a seed, seedling, management operation, product, or professional is of high-quality and is competent to complete the goals that are expected of it within a reasonable amount of time (1).

Chain flail: A debarking or delimbing machine, used in harvesting operations, that employs revolving drum(s) with chains connected to it, to beat limbs from the boles of trees (1).

Chaining: A transport method where a chain is wrapped around logs in order to skid them to the landing from the site of the harvesting operation. Alternatively, a site preparation practice that is facilitated by dragging weighted chains along the ground to clear undesirable structures, such as small vegetation (1).

Channel: The radio frequency and circuitry necessary to track and tune a signal from a single global navigation satellite.

Check: The splitting of a log, timber, or board, along the surface, caused by drying.

Chemical thinning: The use of some type of herbicide to kill unwanted trees in a forest operation (1).

Chip-n-saw: A single production process that takes small logs and removes an outside layer producing chips rather than sawdust, and cants. Cants are then subsequently milled into lumber (1).

Chlorophyll: A composite material primarily composed of green pigments within the chloroplasts of a plant whose light-energy-transforming properties facilitates photosynthesis (1).

Chloroplasts: A structure containing chlorophyll that can be found in all plants and some algae species that, when exposed to light, is the location at which photosynthesis takes place (1).

Choker: A harvesting tool that consists of a piece wire rope or chain tied into a noose-type loop, which is then tightened around a log in order for it to be skidded to the landing (1).

Clean felling: see **Clearcut**.

Clearcut: A final harvest and site preparation method, where effectively all the mature trees, except those left for seeding purposes, are removed at the same time (1).

Climate: The typical weather or weather cycle of any given instance, or period of time, usually encompassing temperature, humidity, precipitation, wind, and atmospheric pressure. Often it is characterized by differing seasons throughout a 1 year cycle (1).

Climate amelioration: The improvement of climatic characteristics which can include air movement, temperature, and precipitation.

Climax: The highest possible stage of forest succession. This can vary from system to system, and is considered the greatest possibility of forest health and biodiversity without any resources being added. This is a very secure condition and will not easily be degraded by natural and small forces (1).

Closed-canopy: A canopy that essentially prevents any sunlight from reaching the forest floor, that is, the canopy is able to block 90—100% of the incoming sunlight.

Cloud forest: A forest of a tropical mountainous region that experiences excessive rainfall and virtually continuous condensation because of the cooling of very humid air rising up the mountains. These forests typically feature shorter more crooked trees and an abundance of mosses, ferns, lichens, and epiphytes growing on the forest floor and on the trees' trunks. Also these forests have the ability to produced atypically large plants where they are afforded ample growing space (1).

Coarse woody debris (CWD): Dead wood matter that is either on the forest floor or in a stream. The defining size of "coarse" varies between systems therefore is dependent upon where the CWD is located (1).

Coastal forest: A forest that is adjacent to, and that is highly affected and dependent upon an ocean or sea.

Commercial forest land (CFL): Land that is deemed by a certain perceived production rate, to be appropriate for the production of timber, and that is not by any regulation or policy prohibited from being used in this manner (1).

Commercial forestry: The practice of producing timber and other forest products in such a way that it generates revenue high enough to sustain the business (1).

Commercial thinning: A thinning operation that results in the generation of revenue high enough that it covers the expenditures of the thinning operation (1).

Commercial tree species: Tree species that are appropriate for use in producing industrial wood goods (1).

Commercial wood products: Any output from a processing facility that is composed of wood, and that would likely return revenue back to the producer.

Commodity: Uniform goods that are considered to be of some important use, that can be traded, and that are considered to arise typically from primary production (5).

Communal forest: A forest, typically in a developing nation, that is owned and managed by a small social aggregate whose residents share in the benefits, monetary or otherwise (1).

Communication skills: The ability to effectively and efficiently convey ideas, concepts, instructions, questions, or needs one has to a person or an audience. This can be in the form of spoken or written word, or a demonstration using a combination of the two.

Community forest: A forest that is, in a sense, owned or managed by a community that can equally share in its benefits (1).

Complete cutting: see **Clearcut**.

Complete felling: see **Clearcut**.

Composite: A product that is composed of at least two different substances, and that employs desirable physical qualities from both to create a product that has increased efficiency or function (2).

Compound interest: When dealing with lending or banking, this is interest added to the total principal amount and allowed to collect interest of its own (5).

Conifer: A tree that produces cones as opposed to flowers or fruits, typically the term conifer is used interchangeably with gymnosperm (1).

Conservation: The careful management, use, and consumption of a plant, animal, habitat, landscape, or other resource in such a way that it can continue to benefit and present and future consumers of this specific product (1).

Conservation ethic: A moral standard by which a conservationist of any fashion holds themselves up to, and one that can be applied to the quality, breadth, or extent by which they work to conserve the environment or a specific portion thereof.

Constraint: A condition that has to be satisfied for any set of activities to be feasible. Constraints can be natural (land limitations), technological (efficiency limitations), capital based (human capital limitation) or governmental (special taxes or fees). Constraints are typically thought of as the boundaries of economies and economic models. Constraints are basically the limits (quantity) or limitations (ability) of the resources that are used in production of some good (5).

Coordinated universal time (UTC): Time, kept by atomic clocks in global navigation satellites, that is a close approximation to Greenwich Mean Time, and is used as a standardized time.

Coppice: The act or resulting product of producing new saplings from older stumps or roots. Also, the act of cutting a tree, stump, or root in order to produce such products (1).

Coppice forest: A forest established entirely of the products of coppicing (1).

Corrugated board: A board, constructed of three sheets of heavy, pulp-based, paper material with two outer liner board layers and a middle corrugated (waved) layer. This product is typically used for packaging, shipping, and storing purposes.

Cost: The amount that one must pay for the varying inputs needed to produce a certain good or service (5).

Cost-benefit analysis: The practice of comparing the social costs and the social benefits in order to make a well-informed decision (5).

Cost-of-living index: The amount of expenditures required in order to keep an average household running and a normal, healthy, and comfortable level. The cost-of-living can vary greatly, even within states, and especially between nations. Some suggest that each family has a cost-of-living index based on what they perceive are necessities for daily life (5).

Cover: Any object or aggregation of objects that would provide protection or cloaking for an animal (1).

Crook: A sudden curve in a tree, log, or board (1).

Crown fire: A forest fire that burns within the tree tops or tall shrubbery of a forest with some level of independence from a surface fire (1).

Cruise, Cruising: The act of gathering information, or the information gathered, from a forest inspection. Usually, this is used to estimate the quality and quantity of the timber in a designated area, and involves some type of sampling scheme (1).

Cull: Any resource in that is removed from further development, or from a market, because it is deemed unfit or of undesirable quality for sale (1).

Culling: The act or practice of removing any and all items of a selection that are not up to a certain standard.

Cultural resource: An output provided due to the cultural traditions of a certain area, tribe, or country.

Cultural service: A forest or ecosystem service that enables a person, group, or community to perform some sort of cultural activity (i.e., traditional, spiritual, or religious).

Cut-to-length: A ground-based harvesting operation where trees are cut into specific lengths before being skidded or forwarded to a road or landing (1).

Cuticle: The thin, waterproof covering over the leaves and needles that provides protection, and helps control moisture content (1).

Cutter: A person whose job is to fell, limb, or buck trees (1).

Cutting: The harvest of a single tree or an entire forest. Alternatively, a piece of an existing plant removed in order to create a new plant out of the part removed (1).

Cutting cycle: The amount of time allowed to pass in between thinning entries in an uneven-aged stand (1).

Cytokinesis: A stage within the cycle of cell reproduction that is separate from mitosis or meiosis, and that results in the splitting of the cytoplasm.

Debarker: A machine that is used to remove bark from the bole of felled trees (1).

Deciduous: Trees or plants that have a typically uniform cycle of shedding their leaves after a period of time and growing them back again at the beginning of the next cycle (1).

Decision variable: Those variables within an experiment, model, or plan that can be controlled by the person conducting the research or planning exercise.

Deck: A pile of logs, or the location where the piles of logs are placed (1).

Decomposition: A relatively large-scale breakdown of organic matter into smaller particles that can be reused by plants to create new organic matter (1).

Defoliant: Any substance that causes a plant to lose its leaves when it comes into contact with the plant (1).

Defoliator: An organism or herbicide that causes a plant to lose its leaves (1).

Deforestation: The removal of forest from an area without the intention of regenerating or allowing regeneration of the forest (1).

Degradation: The process by which objects or land becomes more simplistic or eroded, or less useful or valuable (1).

Delimber: A machine that is used to remove the limbs from trees (1).

Demand inflation: Inflation that is induced because demand greatly exceeds supply (5).

Derelict land: An area that was at one point in the past used, but now has been abandoned (3).

Diameter at breast height (DBH, dbh): The diameter of a tree measured at **breast height** above the ground, on the uphill side of the tree (1).

Diameter inside bark (DIB): The diameter of only the wood portion of a stem, that is, the diameter of a stem minus the bark (1).

Diameter outside bark (DOB): The overall diameter of a stem, that is, the diameter of both the bark and the wood combined, or the DBH (1).

Differential correction: A data augmentation process commonly used to improve the quality of data by removing some of the errors introduced by atmospheric and other influences. This takes place after the data is collected and requires an additional set of data collected by a base station.

Diffuse porous: The characteristic of some hardwoods (e.g., yellow poplar, birch, and maple) in which the cells the in growth rings are all basically homogeneous in size and distribution. That is to say that when looking at a cross-section of these trees is it often difficult or impossible to differentiate the earlywood from the late-wood (1).

Digital orthophotograph: A digital or electronic aerial photograph or image that has been geometrically altered to account for factors such as terrain and camera tilt in order to provide a more perfect image of the landscape (1).

Dilution of precision (DOP): A measure of the strength of a satellite configuration in the sky, which has an impact on positional accuracy of global positioning systems when the strength is low.

Dioecious: A term used to describe those tree species where each individual plant exhibits either male or female sex organs, that is, they cannot have both.

Direct attack: Any forest fire extinguishing practice that is performed at an active face of a fire. Typically these practices include spraying with water, suffocating with a nonflammable material, or even physically separating the fire from any unburned fuels (1).

Discount rate: The rate used to account for the time value of money (5).

Discounting: An analysis that reduces future revenues and costs to today's values, in effect calculating the present values (5).

Disease: A detrimental change in the functioning of physiological processes (1).

Disposable income: Revenue of an individual or organization that can be spent on trivial items or activities that will not affect the proper management of the household or organization.

Disturbance: Any event or action that disrupts a system, cycle, or structure of an environment or ecosystem (1).

Dormancy: In animals, a state of very little or no movement or growth, sometimes simply a state of prolonged rest or sleep. In plants, it is a state of no growth for a prolonged period of time, this is sometimes an evolutionary trait that allows seeds to wait for more ideal germinating conditions before they try to establish themselves in soil. This is also sometimes considered a period of rest (1).

Double-grip harvester: A mechanical tree harvester that has a felling head and a separate delimber.

Downhill yarding system: A cable-based yarding system where the landing is located downhill of the site where the trees were felled.

Dry forest: Forests that typically experience long, dry periods, yet receive enough rainfall to support tree life.

Dynamics: The manner in which multiple units are intertwined and relate to one another. Alternatively, the study of how the introduction or removal of one individual affects all of the others that surrounded it.

Earlywood: The portion of the annual ring of a tree that was grown earlier in the season, is lighter in color, and is composed of less dense, large thin-walled cells (1).

Ecology: The study of how organisms and their environments interact and rely on one another, or the overall way in which an environment and its organisms relate with each other (2).

Ecology: The study of environments, which usually encompasses the study of all living and non-living aspects of the environment, how they interact, and the effects they have on each other and the environment as a whole.

Economic profit: The remaining funds after one subtracts total costs and opportunity costs from total revenues.

Economics: The study of the complex relationship between supply and demand of markets and how scarcity and utility affect these relationships. The information obtained and relationships analyzed should address three key questions: what should be produced, how it should be produced, and for whom should it be produced (5).

Ecosystem: The complex aggregation of all the living and nonliving objects and organisms in a defined space that interact, depend, or regulate each another in some way. There is no minimum or maximum size that defines an ecosystem (1).

Ecosystem resiliency: The ability of an ecosystem to recover from a type of degradation in the form of a natural disaster or intentional detrimental human action. This resiliency can be facilitated by species redundancy, immense genetic variation, and highly reproductive species.

Ecosystem services: Positive outcomes naturally provided by the environment. While these services occur naturally within the ecosystem, humans can affect them. Some ecosystem services work on a much larger scale than human technology can match, which makes these services valuable to society in many parts of the world.

Ecotourism: The visitation of natural habitats in such a way that minimizes human impacts on ecological attributes. Alternatively, the industry of supporting ecotourism by entire communities or countries (2).

Edaphic: A state of being connected to, caused by, or affected by a soil conditions (1).

Educator: A person who teaches others, or increases the level of their education (2).

Efficient: In seeking a goal, it is the use of the least amount of input, resources, or capital possible to obtain the desired end result. Alternatively, it describes the process of completely using all of the available resources within a given period of time in such a way that an objective is maximized or minimized.

Efficient resource allocation: A situation where all the resources available to a producer are used in a way that maximizes an objective, leaving no other way to improve the objective of the producer. In sum, they cannot make more of one unit of product or service without making less of another (5).

Endangered species: A species, plant or animal, that has a high risk of being completely eradicated from a major portion of its natural range (1).

Engineered wood: A larger product made from a combination of smaller pieces of wood, adhered together with glue to achieve a more ideal product, functionally or economically (1).

Environment: The culmination of all of the conditions within in a certain land area (1).

Environmental factor: Any variable mainly controlled by the natural world, such as topography, climate, or weather.

Ephemeral stream: The part of a stream system that only has water flow when sufficient precipitation has fallen recently. These types of streams receive little to no input from consistent water sources (1).

Epidermis: The outer layer of cells on the body of plants and animals that aides in protection of the organism (1).

Erosion: The degradation of a land surface, or the removal of layers of soil by means of water, wind, gravity, or other natural or human-caused sources (1).

Estimate: A guess or approximation based on little background knowledge or information (4).

Evapotranspiration: A change in the state of liquid water to that of water vapor (1).

Exploitation age: see **Felling age**.

Externality: A positive or negative result caused by one party yet affecting another (5).

Extinct species: A species that is thought to be completely eradicated from Earth, in that all known populations no longer exist. If only one or a few organisms of a species exist, yet are incapable of breeding, some consider the species to be extinct.

Extirpated species: A species that is extinct from certain parts of its natural range, or completely extinct from wild areas while some individuals exist in captivity.

Extraction: The removal of an object using some type of force, whether that be chemical or physical (1).

Extraction line: Also known as a skid road.

Extraction rack: see **Skid road**.

Extraction road: A typically semiprimitive road, developed within a forest, and used to move felled trees (or portions thereof) from a harvested area to a landing.

Factory: The place where a product is produced.

Faller: The person in a timber harvesting operation that is responsible for cutting down trees (1).

Fell-buncher: A machine that employs hydraulic arms to hold a tree while it is severed from its stump, then moves onto another tree while still holding the first and so on. This type of machine leaves piles of harvested trees on the ground in bunches for easier transport to a landing (1).

Felling: The act of cutting down trees (1).

Felling age: The age at which a tree was cut, or the average age that a forest was harvested.

Fiber product: Any non-solid wood product derived from wood and bark (1).

Field capacity: The amount of water that remains in the soil, measured in mass or volume, two to three-day subsequent of initial wetting, and after the normal drainage processes have occurred (1).

Final crop: The trees within a forest allocated to remain within the stand until the final harvest occurs (1).

Final cutting: The act of harvesting the remaining trees in a multientry thinning operation (1).

Final felling: see **Final cutting**.

Firebreak: A gap within a forest or stand, natural or man-made, that provides a break in fuels, and thus could be used to prevent the spread of forest fires (1).

Fish: An aquatic organism classified as Ichthyic that is characterized as having cold-blood, fins, a two-chambered heart, and gills (2).

Fish habitat: The aquatic and the closely surrounding terrestrial environment that provides the essential resources for a fish species during all of the stages of its life cycle (1).

Fishery: A place, natural or constructed, where fish that will eventually be harvested and processed for sale live. These fish, depending upon where they live and their perceived importance, may be fed, protected, or managed in order to obtain an objective.

Fixed cost: The costs that are accrued to an organization regardless of the level or lack of output (5).

Flanking fire: A fire that spreads perpendicular to the wind.

Foliage: The leaves, flowers, or branches of one or more plants, that may be utilized as ornamentation, food sources, habitat, or cover (2).

Food: Any beneficial material that provides nutrition or nourishment to the consumer. These typically consist of proteins, carbohydrates, fats, sugars, and oils (2).

Food patch: A natural or man-made area that contains some sort of food intended to provide nutrition to one or more wildlife species (1).

Forage: As a resource, this is any leafy or woody plant source used as a food source for wild or domesticated animals. As an act, this is the process of looking for, or eating some type of vegetation (1).

Forest: An ecosystem whose main vegetation consists of trees. Some definitions include minimum areas of land that contain trees of certain sizes (1).

Forest dependent communities: A group of people located in some type of social aggregate (village, city, town, etc.) whose lifestyle, livelihood, and or economy is based on the surrounding forests, and the management practices employed.

Forest dynamics: The manner in which all species of trees within a forest interact, including the effects of the removing or introducing other individuals or species.

Forest ecosystem: An ecosystem that is contained within, or depends heavily upon, a large aggregation of trees and related vegetation.

Forest entomology: The study of the interactions between insects and trees (1).

Forest fragmentation: The breakup of contiguous areas of forests, caused by human development, differentiation in forest ownership, and natural disturbances.

Forest fuel load: The sum (volume) of all of the consumable vegetative fuels within a forest.

Forest inventory: A measure of the diversity, population, distribution, and growth rate of trees within a forested area (1).

Forest litter: Limbs, leaves, and other woody debris that have fallen on the forest floor and have not yet fully decayed.

Forest management: The designation and application of management practices for forest stands in order to meet certain predetermined goals and objectives and thus to satisfy the demands of landowners or management organizations (1).

Forest plantation: An area of land planted with trees in rows.

Forest privatization: The act of converting forestland from public ownership to private ownership.

Forest product: Any commercial output that arises from a forest. The two types of forest products are commodities and non-timber forest products.

Forest ranger: An officer employed to manage and police a certain forest area, typically employed by a government to oversee a public forest (2).

Forest recreation: Any recreational activity that occurs in a forest, whether it be dependent or independent upon the actual existence of the forest.

Forest regeneration: The act of restoring a forest to a previous (younger) condition through artificial or natural means.

Forest reserve: An area of land that, through some form of legislation or regulation, can have timber harvested from it but cannot be converted into an area of nonforest use (1).

Forestation: Any act that results in the creation of a forest. The two common classifications are afforestation and reforestation.

Forester: A professional employed to manage a forested area (1).

Forestry: The act of employing the sciences, arts, and practices necessary to properly manage a forest and its associated natural resources in such a way to meet the goals and objectives of a landowner or land management organization (1).

Forwarder: A machine that moves trees and logs from a felling site to a landing (a primary transportation process), yet in a manner that they are entirely off of the ground (1).

Forwarding: The act of moving a log, from the felling site to a landing or collection site, with a forwarder. Also called pre-hauling (1).

Fragmentation: The breakup of contiguous areas of forests, caused by human development, differentiation in forest ownership, and natural disturbances.

Fuel compaction: The ratio of space taken up by a forest fuel in the total given area of space (1).

Fuel continuity: A level of consistency in a forest fuel may allow a fire to continue to spread. The two most common terms used to describe fuel continuity are uniform and patchy (1).

Fuel moisture content: The amount of water, typically expressed as a percentage, in a specific sample of forest fuels (1).

Fuelwood: Any wood source that is used, without alteration, as a type of fuel for heating, lighting, or cooking purposes. Often called firewood.

Fundamental niche: The service or place within an ecosystem that a species fulfills.

Fungi: An organism in the kingdom fungi characterized by the lack of chlorophyll, cell walls that are composed of cellulose and chitin, a vegetative body constructed of hyphae (referred to as a thallus), and the utilization of spores for reproduction purposes (1).

Future value: The projected value of a resource (e.g., money or timber) at some point in the foreseeable future.

Gap dynamics: The changes that occur in a forest when a portion of the tree canopy is removed through the harvest or death of one or more individual trees (1).

Gap-phase species: Those tree species that become established through seed under an existing forest canopy, and subsequently work their way into the canopy when opportunities become available.

Gate delimber: A sturdy gate that is fastened between two standing trees, and used to delimb trees by means of pushing or forcing them through it top-first (1).

Geographic information system (GIS): An organized collaboration of hardware, software, geographic and topological data, and personal knowledge that is brought together to manage, store, manipulate, and report geographic data. These systems are often used today to create maps.

Germination: The initiation of development of a seedling from a mature seed, spore, or pollen grain (1).

Global positioning system (GPS): An organized system of satellites, control stations, base stations, and hand-held receivers that provide, track, and store navigational coordinates based on an x, y, z coordinate system (1).

Goal: A predetermined end result that is worked toward over the course of a determined time period (1).

Goods and services: The outputs of a producer, including physical objects and physical jobs developed for a consumer (1).

Grapple: An articulated head, able to open and close (like tongs), and used to hold logs as they are being skidded to a landing or loaded onto a truck (1).

Grapple skidder: A skidder that employs a grapple during the primary transportation process (1).

Green energy: Energy that is created, generated, or used in some manner that has no significant negative effect on the environment.

Greenbelt: An area of land around or near a suburban or urban area, typically park-like, that remains undeveloped, mainly used for aesthetic purposes but also employed for recreational uses (1).

Greenhouse effect: The global warming caused by an accumulation of water vapor and carbon dioxide, among other gases (that are considered greenhouse gases), in the Earth's atmosphere (1).

Greenhouse gas: A gas, primarily water vapor, methane, chlorofluorocarbon, and CO_2, that absorbs radiated energy and emits it as heat (or thermal) energy (3).

Gross domestic product (GDP): The sum of production, of goods and services, of one country for one entire year (3).

Ground fire: A forest fire that occurs or spreads through the burning of fuel contained in the duff, roots, and organic matter of the soil layer (1).

Ground fuel: A fuel supply that is generally located underneath surface fuels (1).

Groundwater: The sum of water that is underneath the surface of the Earth, and not bound to any minerals. Underground streams are excluded (3).

Group selection harvest: A forest regeneration system where the canopy is opened through the removal of small groups of trees, which create gaps within which regeneration can occur.

Growth ring: The new tree growth from each growing cycle, consisting of early- and late-wood cells (1).

Growth-related respiration: The respiration of an organism that is above a rate needed to survive (maintenance-related respiration). This is respiration at such a rate as to facilitate large-scale cell growth and develop, as well as to allow normal metabolic functions to occur.

Guard cell: Tree cells, located on the sides of stomata, that regulate the intake and export of gases and water.

Gully: A canal created by intense water runoff (1).

Gymnosperm: Any organism of the group *Gymnospermae* characterized by being vascular, flowering, and producing a naked seed (a seed not sheltered by an ovary) (1).

Habitat: An area (horizontal or vertical or both), specific to each animal and plant species, that provides the proper climate, food, cover, space, and water necessary for the organism to live, grow, and reproduce without other natural limitations (1).

Hardwood: Generally speaking, a tree that is in the plant group Angiosperm, often a deciduous tree. The wood of these trees may or may not be considered physically hard (1).

Harvesting method: The manner in which trees are selected for removal and transport to a landing (1).

Heading fire: A forest fire that spreads into the wind.

Heartwood: The center part of the tree's stem that is used as a support system for the continued growth of the tree (1).

Heavy cooperage: Large wooden vats used to hold water or other liquids.

Hectare: A metric unit of land area defined as 10,000 m^2 (2.471 acres).

Herbicide: Any chemical or compound used to intentionally kill or reduce the growth of targeted plants (1).

Heterocellular: An organism, plant or animal, that is composed of two or more different types of cell.

High grading: A harvest process where the most valuable trees are removed, leaving trees and tree species deemed less valuable to continue to grow (1).

High pruning: The pruning of branches from live trees that are located above one's head.

High thinning: A thinning practice in which dominant, co-dominant, and suppressed tree classes are favored for removal.

High-lead logging: A logging system that employs a cable yarding process consisting of a metal rope passing through a block at the top of the head spar or a tower (1).

Homocellular: An organism, plant or animal, that is composed of only one type of cell.

Humidity: A measure of water vapor in the air (3).

Hurricane: A large rotating tropical storm (cyclone) that has a central powerful eye. These storms have the potential to cause great destruction and often are associated with very strong winds (perhaps tornadoes) and excessive rainfall.

Hydrocarbon: An organic compound that is composed only of hydrogen and carbon, typically present in natural or derived fuels (2).

Hydrologic cycle: The process by which water moves through the global environment. It begins as water is taken up into the atmosphere by means of evapotranspiration, sublimation, evaporation, and transpiration. Once in the atmosphere, water condenses into cloud coverage. Water is stored in the atmosphere until it returns to the ground in the form of precipitation. Precipitation can be intercepted, can become runoff, or can begin infiltrating and percolating into the soil. Once water is in the soil, it can be stored or begin to make its way back into larger water bodies to begin the cycle over again (1).

Hydrology: The study of water, encompassing flow processes, and effects on ecosystems and living organisms.

Incentives: Rewards or penalties that are used to influence a person or group of people to act in a manner which another set of people feels is best (5).

Indicator species: A species that represents the health of the environment by a direct or inverse relationship with its presence, population, health, or vitality (1).

Indigenous: Plants, animals, or societies that have inhabited an area for basically all of history, that is, they were not introduced by any outside means (1).

Indigenous people: The direct descendants of the inhabitants who lived in a given area when explorers or settlers of a different civilization or racial origin first arrived from a different part of the world (1).

Indirect attack: Any forest fire extinguishing practice that is performed away from an active face of a fire. Typically these practices include creating firebreaks, allowing the forest fire to spread to the control line, and back burning. These practices are typically applied to forest fires with a high heat content or a high rate of spread (1).

Induced features: Attributes of a landscape or ecosystem that did not naturally occur, but rather were introduced by human activity.

Industrial Revolution: A period of time during which a defined governed body experiences a relatively rapid change from an economy based mainly on agriculture to an economy based more on industrial processes. These revolutions are typically initiated by multiple inventions and innovations and are typically accompanied by urbanization around city centers.

Inflation: A continual rise in prices and salaries. Alternatively, a decrease in the value of money, or its purchasing power (5).

Inherent features: Attributes of a landscape or ecosystem that were not caused but an human actions, but rather naturally occur within that system.

Input: The resources and capital used to produce a product or output (1).

Insect: Any organism in the class Insecta, characterized by a three-segmented body (head, thorax, and abdomen), a pair of antennae, three pairs of legs, and sometimes up to two sets of wings (1).

Interest rate: The additional payment of debt accrued because of the lending or borrowing of money, usually represented as a percent (5).

Intermediate service: A service that must be performed by the environment in order for another environmental service to be performed, these are not considered environmental services themselves.

Interpretation: Any recreational activity with an educational function that is intended to explain or relate something through the use of models, personal experience, or illustrations, as opposed to lectures (1).

Interpretive areas: A section within a park or forest that is used to better explain facts or concepts about the natural world, using physical examples instead of simply words.

Invasive species: A species currently existing within an area, yet did not originate there, especially those species that cause problems for pre-existing or native species.

Knot: A part of a limb that is contained inside lumber. Sound knots were branches that were alive at the time of overgrowth of woody material, while loose knots are branches that were dead at the time of overgrowth of woody material (1).

Ladder fuels: Any fuel that extends from the ground surface to the canopy, presenting a means by which a forest fire can spread vertically (1).

Landing: A clearing where logs are brought to store, sort, limb, buck, or pile prior to secondary transport (1).

Lateral meristem: Those meristems that run vertically along the tree stem or root system which have cells that divide and expand laterally (1).

Latewood: The darker portion of the annual ring that was produced later in the growing season, and that was caused by smaller, thick-walled, harder, densely packed cells (1).

Leaf primordial: An aggregation of cells that will eventually form into a leaf, usually collected in buds along the stem of a plant.

Legume: Any member of the *Leguminasae* (or *Fabacea*) family that are characterized by being dicotyledonous and fruit bearing, and that possess bacteria on root nodules which fix nitrogen, a necessary element for all plant growth. Common legumes are plants such as beans, peas, and clovers, that provide high-quality forage for wildlife (2).

Limbing: The practice of removing limbs from merchantable sections of a tree bole in order to prepare the bole for future processing.

Live skyline: A skyline system that has the ability to move up and down during the yarding process in order to allow continued harvesting (1).

Log: A section of a felled tree stem that has been prepared for milling, and that is at least 1.8 meters (6 feet) long (2).

Logistics: The management or supervision of the particulars of a process or company (2).

Long run: In economics, the period of time that is so great that in fact no variable is fixed.

Low pruning: The removal of limbs that are generally below head level.

Low thinning: The harvesting of trees that are only as tall as the lower portion of the canopy, in order to improve the growing conditions for trees that are taller (1).

Lumber: A solid wood product that was sawn and milled to specific dimensions, to facilitate the development of furniture, homes, and other features (1).

Maintenance-related respiration: Respiration performed by an organism that supplies that organism with enough oxygen simply to maintain its current state, that is, it does not supply enough oxygen to facilitate any growth.

Mammal: Any organism in the class Mammalia, characterized by being highly evolved, vertebrate, warm-blooded, essentially completely covered with hair, and having females that are capable of lactation for the feeding of their young (2).

Management: The processes involved in forming decisions or courses of action regarding the use of resources, including the decision to do nothing at all.

Mangrove forest: A form of tidal forest found in the brackish wetlands between land and sea, in river deltas, and along sheltered coastlines.

Manual felling: The felling of timber by means of any hand-held equipment (axes, cross-cut saws, chainsaws. etc.).

Marginal: The positive or negative change per unit change in any variable (5).

Marginal benefit: The increase in benefit per unit of an increase in an activity (5).

Marginal cost: The increase in cost per unit of an increase in an activity (5).

Market access: The ability to enter and sell within a market (5).

Mask: When dealing with global satellite navigation systems, an electronic screening practice that ignores low-quality signals.

Maximum Sustained Yield (MSY): The maximum amount of timber (measured by number or trees, basal area, or weight) that can be harvested from an area in a uniform interval that will not deplete future harvests.

Measurable trade-off: Any trade-off that has a definite physical or fiscal value (e.g., salary or a product).

Mediterranean forest: Forests characterized by having limited annual precipitation events, that are concentrated during the winter months.

Merchantable: A forest, stem, or tree that is of adequate size or grade and would be able to be sold (1).

Merchantable height: The length from the cutting height (stump) up to where the stem reaches a diameter too small for typical wood product production (1).

Merchantable top diameter: The diameter that a stem reaches where everything above it is considered non-merchantable (1).

Mill: A place where logs are taken to in order to have then sawn, planed, and processed into dimensional lumber or other mass-produced wood products.

Mitosis: The division of a nucleus that produces two identical nuclei, both with a full set of chromosomes. This is the cell division all cells, except gametes, undergo (1).

Monoecious: A term describing tree species where each individual organism can exhibit both male and female sex organs.

Monsoon forest: A forest system that will experience a relatively long dry season, during which trees will most likely shed their leaves, followed by a rather short season of extensive and intense rainfall during which trees will re-grow their leaves (1).

Montane: Involving or pertaining to mountains or the climates, ecosystems, or species thereof (1).

Montane forest: Situated along mountain ranges, and can be characterized as boreal, temperate, or tropical, depending on the climate of the region.

Multipath: A problem with data collected with global satellite navigation systems in which the signal did not take a direct route from the satellite to the receiver, but instead reflected off of some surface such as the ground, a body of water, a building, or a tree.

Multiple-use forestry: Management practices or plans that accommodate more than one forest objective to be obtained (1).

National historic park: An area, owned by a federal government, set aside for the public because of some historical importance (1).

National monument: An area that is owned by a federal government and defined in policy, set aside for the public because of the archaeological, scientific, historic, or aesthetic interest as it pertains to that nation (1).

National park: An area, owned by a federal government and defined in policy, with the intention of conserving the aesthetic qualities, plant and animal life, and any items or areas that hold natural or historical significance within its boundaries for the public to be able to experience infinitely (1).

Native forest: A forest that is the same in the basic make up, form, and function presently as it was when the first non-native people discovered it. Therefore it is relatively free of large-scale human interaction and disturbance.

Native Species: A species that has inhabited an area since the beginning of time or as far back as historically recorded. Also, any species that inhabited an area at the time of settlement from an outside culture (1).

Natural disturbance: A form of disruption, degradation, or destruction of an area, cycle, landscape, or ecosystem that is not caused by man but rather some type of natural occurrence.

Natural resource: Any item that does not have to be produced by man but rather can be found naturally on, in, or above the Earth, that might be consumed by man in order to meet some type of need or desire (3).

Natural resource depletion: The consumption of a limited natural resource to the point that it is almost used completely up and therefore must the replaced by some other means, conserved, or renewed.

Natural resource management: The allocation of resources in such a way that they are sustainable for current and future use. Alternatively, the introduction of practices and activities in a manner that is intended to improve natural resources or for consumption, conservation, and preservation.

Naval stores: Items, materials, or substances that were typically used in the production and maintenance of wooden ships (pitch, tar, spirits, turpentine, rosin, etc.) (1).

Negative externality: A cost that is caused by one group, person, or activity yet imposed on another (5).

Nematode: An organism from the phylum Nematoda that is characterized by being simple, thin, unsegmented, round in cross-section, and sometimes parasitic (a worm) (1).

Nesting: The act of building or inhabiting a nest (2).

Net exporter: A nation where the total sum of all exported items exceeds that of which they import.

Net importer: A nation where the total sum of all imported items exceeds that of which they export.

Net primary production: The amount of material that was produced, minus the material that was consumed during the process of production.

Niche: The specific function that an organism or species performs within an ecosystem (1).

Noise abatement: The softening, ending, or reduction of noise that is usually caused by an intentional act of humans.

Nonexcludable: Any resource whose owner cannot in anyway impede a person from, or charge a fee for partaking in it.

Nonprofit organization: Any organization that does not have stock for sale as a way for people to invest in the company and does not divide up surplus income among owners or high ranking company workers.

Nonrenewable resource: A resource that is limited, and after being consumed has no way of being regenerated (3).

Nonrival: Any resource where consumption in any amount by one person or organization does not and cannot impede the consumption of another in either quantity or quality.

Nonsolid products: Any forest product that is not composed of a single forest input, but rather is made of multiple forest and nonforest products.

Nontemporary trade-off: Any trade-off that has a long term or eternal benefit or detriment.

Nontimber forest product: A product derived from a forest, including wildlife and vegetation, that is not a solid or engineered wood product (1).

Normative economics: The theory of how economics and economies work. This information is not based on actual economic data (5).

Objective: A desired end result that is meant to be obtained within a certain date or time (1).

Ocular estimation: An estimation based on a person's judgment according their eyesight, and supported by a sufficient knowledge in the particular field.

Old-growth: Vegetation, typically trees or an entire forest, that has developed to a late successional stage (1).

Opportunity cost: The goods, services, or revenue that would have been produced or received by performing certain practices, and that in essence are given up by deciding to employ other practices (5).

Organic matter: Matter or material that once arose from living organisms, in natural resource management this usually refers to matter on top of a soil, yet below forest litter, and that is decomposing.

Organism: A complicated living being or structure that is made up of differing, but interacting, organs and organ systems (2).

Oriented-strand board (OSB): A type of engineered wood that is essentially a sheet of multiple layers of flakes or strands of wood that are oriented at 90° angles of one another, and are glued, heated, and compressed (1).

Outputs: The end products, whether they be a service or a good, of any man-made or natural process.

Overbrowsing: The continual use of vegetation as a food source by ungulates to the point that it is detrimental or destroying to the regeneration of that vegetation (1).

Overconsumption: The use of a resource or input to the point at which it is detrimental. In renewable resources this makes it difficult to regenerate the resources; in nonrenewable resources this simply uses the resource to the point it cannot be replaced in a timely manner, efficiently, or at all.

Overgrazing: The repetitive consumption of rangeland vegetation in so much that it degrades the vegetation, regeneration capabilities of the vegetation, or the range itself (1).

Overhunting: The hunting and harvesting a specific population or species to the point of detriment. The amount of harvesting is so much that it has affected the ability of a population to reproduce enough young to sustain the usual carrying capacity.

Ozone: A molecule of oxygen containing three oxygen atoms, which acts as a helpful atmospheric gas by absorbing solar ultraviolet radiation that is detrimental to human health (3).

Paper: A thin sheet of dry plant fiber most often used for writing and packaging (2).

Parasite: An organism that inhabits the interior or exterior or another organism and benefits from it while the host organism suffers (1).

Park-like: A value statement describing a forest area that contains trees but very little shrubbery or forest litter and the ground, much like that of a residential lawn.

Partial cutting: The harvesting of only a portion of a forest without the intention of using the remaining timber as a means of regeneration (1).

Particle board: A relatively thick sheet of wood particles (typically flakes or sawdust) that are bound together through processes including gluing, heating, and compressing.

Particulates: Particles that are so small that they are easily suspended in water and air.

Pathogen: A parasite that has the direct ability to cause a disease (1).

Pecuniary externalities: Those externalities that do not directly affect other parties but rather affect them indirectly through the market (5).

Pedology: The study of soils and how they can be characterized, developed, and distributed (3).

Perennial stream: A stream that actively flows water all year, given normal climatic conditions (1).

Petroleum: A complicated compound of hydrocarbons and other trace substances that make up a variety of highly flammable liquids ranging in appearance from clear to black. Petroleum can be found in many forms such as gasoline and naphtha (2).

pH: A unit of measure for acidity (1).

Phloem: The layer of woody cells used to transport food within a plant from the leaves to the stem and root system, the actively growing phloem is located right underneath the bark (1).

Photosynthesis: A process by which plants make their own food, using sunlight, water, and carbon dioxide to create carbohydrates (1).

Pioneer species: A plant that, by natural means on a bare site, is one of the first species to inhabit the area (1).

Pith: The innermost core of a tree's stem, branches, and roots. The pith is created in the first year and is composed of mostly soft tissue (1).

Plantation: A stand of trees planted by artificial means, generally in rows, and that has the primary purpose of producing revenue (1).

Planting density: An economic, operational, and silviculturistic compromise between the ideal amount of spacing required for the optimum outcome of each respective field. Such a compromise should bring about the best mixture of spacing to be able to allow sufficient growth space, permit the use of operational practices, as well as the ability to maximized revenue but efficiently using all the resources provided.

Plywood: An engineered wood product that is formed by taking multiple layers of veneers and gluing them together in such a way that the grain of each veneer is perpendicular to the one(s) it is touching (1).

Pole: A tree in a middle stage of development that is no longer a sapling but not yet a mature tree, the determining size of poles varies greatly by species and by region. Alternatively, a mature tree that can be used as a telephone or utility pole (1).

Pollination: The movement of pollen (sperm) to the female part of the plant that holds the eggs (1).

Population: A group of organisms of the same species that all inhabit a defined geographic location (1).

Positional dilution of precision (PDOP): A measure of the mean dilution of precision value of a global navigation satellite system that takes into account both vertical and horizontal aspects of dilution of precision.

Positive economics: How economies and economic process actually work in real-world situations, the study of such economics requires much research and data (5).

Positive externality: A benefit that is unintentionally caused by one party but received without solicitation by another (5).

Potential vegetation: The sum of the vegetation that could develop given that all stages of succession were accomplished with the present state of the site (1).

Pre-commercial thinning (PCT): The felling of trees that are not harvested for direct economic return but rather to allow more desirable trees to develop more efficiently so that they will return greater revenue (1).

Precipitation: All forms of water that fall from the atmosphere back to the Earth's surface (1).

Prehauling: Moving logs or whole trees from a felling site to a landing.

Prescribed burn: The intentional burning of land in order to make the environment more ideal for new plant growth, wildlife habitat, or operational activities (1).

Present value: The value of a good or capital today, adjusted for inflation or deflation for a specific time in the future.

Preservation: The protection of something in its natural state, allowing no consumption even that which might be sustainable (3).

Preserve: An area set aside, protected from any harvesting activities, neither forest nor wildlife, considered a safe haven for wildlife (1).

Primary growth: Plant growth that is essentially vertical, and originates from an apical meristem (1).

Primary production: The human-caused production of natural resources (agricultural crops, planted or managed forests, etc.).

Primary sector: The portion of the economy that deals with the production and sale of natural resources. This includes agriculture, forestry, fisheries, wildlife, mineral extraction, as well as other operations (5).

Primary succession: The sequence of growth that begins with bare rock and climaxes with an old growth forest. Stages vary in length and difficulty to accomplish but in general become easier and shorter as the progress in succession.

Primary transport: see **Prehauling**.

Private ownership: The possession of a resource by a nonpublic entity that controls it in such a way that it, unless non-rival and non-excludable, is not allowed to be freely consumed by the public.

Private sector: The portion of the economy that is owned by nonpublic entities rather than any form of governmental body (3).

Processor: A machine employed during a harvesting operation that has the capability to perform more than one task, these tasks could include limbing, bucking, debarking, chipping, or measuring (1).

Product: The ultimate goal of production, or the final result of the process (4).

Production period: The overall period required to produce a certain product, that is, from the time the inputs are added and the product is ready for market.

Professional ethics: The guidelines by which people carry themselves as professionals.

Profit: Any revenue that exceeds the costs of production incurred by a producer. Accounting profit is revenue minus costs, and economic profit is revenue minus costs and other opportunity costs (5).

Progressive clearcut system: A harvesting system in which the trees are felled in a succession of strips, at times these are cut going toward usual wind flow to decrease losses due to windthrow (1).

Progressive shelterwood system: The practice of cutting trees along the edges of regeneration areas in order to advance the spread of the regeneration process.

Provisioning service: An ecosystem service that produces food, water, and other resources.

Pruning: The cutting away of live or dead branches from a tree bole in order to improve some aspect of the tree, these improvements may be timber productivity, aesthetics, or health (1).

Public land: An area, defined by policy that is owned and managed by a local, state, or national government (1).

Public ownership: The ownership of land, organizations, companies, or anything else by a local, state, or national government. Although the ownership of these entities is considered public they may not be freely or fully accessible by the general public (5).

Public sector: The portion of a government economy that is owned and controlled by the governing body (3).

Pulpwood: Any wood product that is processed into chips, flakes, wood flour, or other small particles which are then used for the production of wood pulp or paper (1).

Purchasing power: The actual amount of goods or services that can be bought with a certain unit of money (5).

Rainforest: A forest that is characterized by elevated temperature, high rainfall amounts (>60 inches or >1524 mm), and high humidity year-round (1).

Rangeland: Land that is generally open, with very few trees or large brush thickets, and typically covered with forage that is an essential aspect of habitat for many animals.

Rational: (economics) An assumption based on the premise that each person will act within their own best interest, regardless of others, even to the point of selfishness. Further, a person would not knowingly do something that was detrimental to their personal economic situation. This assumption is necessary for some economic models and for the development of supply and demand curves in order to best represent the market that they represent.

Rational: (management) A course of action that is used to assess alternatives to a plan or problem. It involves assessing all of the alternatives to a plan or problem and selecting the one with the greatest fitness or contribution to society or an organization.

Raw material: A crude or slightly processed substance that can be further manufactured into something different and perhaps more useful to society (2).

Ream: An amount of paper typically considered to be about 500 sheets (2).

Recreation: An activity that is enjoyed by an individual during their free time, or time separate from their normal work environment, and that is usually performed willingly by the participant (1).

Recreation facility: Any constructed feature of a recreational area that enhances the recreational experience. It can be as simple as a developed walking path through the woods (1).

Recreation resource management: The management of a recreational area and the surrounding resources that employs practices to allocate resources in the best manner to achieve the greatest recreational opportunities or user enjoyment (1).

Recreation site: An area of land or water that possesses characteristics that cause it to be used for recreational activities (1).

Recreation visitor day: Generally, 12 hours of recreational use incurred by a single person, or some combination of hours and people that are equivalent.

Recreational carrying capacity: The maximum number of people that can be accommodated by a recreational facility before social benefit begins to decline.

Recreational opportunity: The potential that an area possesses to host certain types of recreational activities.

Recycling: The act of taking a previously used material and either refashioning or reprocessing it so as to minimize waste (3).

Reforestation: The regeneration of previously forested land with trees, either naturally or artificially (1).

Refuge: An asylum or defense from threats or perils, or a location that provides such (2).

Refugium: An area that supports wildlife species that have had their habitat reduced (1).

Regeneration: The reestablishment of trees in a previously forested area, either by natural or artificial means (1).

Regolith: Rock material that has been broken off of a main rock by weathering, yet remains placed atop its the source. Regolith is different from soil in that it has not yet been combined with organic matter (3).

Regulating service: Benefits obtained from forests through the regulation of ecosystem processes, including the regulation of climate, water, and other processes.

Release activity: The freeing of young trees from the pressure of overstory competition, or the prevention of the development of undesirable plant species, or the removal of brush, weeds, grass, or other plant competition from the area around small trees.

Release treatment: Actions that are meant to act as release activities.

Renewable resource: Any resource that has the ability to be recreated or reformed over time, and whose current consumption, in general, does not affect future consumption (1).

Reptile: An organism from the class Reptilia, characterized as a vertebrate that breathes air, has an ossified skeleton with one occipital condyle, has a distinct quadrate bone, has ribs attached to a sternum, and has a body usually sheltered by either scales or bony plates (2).

Reservation: An area of federally owned land that is set aside for indigenous people, or alternatively a place where hunting is prohibited, often to provide a safe breeding area (2).

Resilience: The ability of a population or ecosystem to recover from some sort of disturbance (1).

Respiration: The process of breaking down carbon-containing energy sources, as well as facilitate competitive reactions during photosynthesis that result in the discharge of CO_2 into the atmosphere by plants (1).

Return interval: In a multiple-entry forest management plan, this is the amount of time allowed to pass before another thinning or harvesting entry is allowed.

Revenue: Money received from the sale of a good or service, or in the case of a government, money received from taxation (5).

Rills: Small crevasses on the Earth's surface that are about 5 to 2000 mm (0.2–79 inches) in width, and typically caused by heavy rainfall (3).

Ring porous: A characteristic of hardwood (e.g., ash and oak) tree species where the earlywood consists of well-defined, thin-walled cells and the late-wood consists of well-defined, thick-walled cells, making differentiation of each easy by the naked eye (1).

Riparian: A state of being associated with, or being located near a stream, wetland, or other body of water (1).

Riparian zone: The terrestrial area that immediately surrounds a lake, river, or stream (1).

Risk: A notion that the consequences of an action are not entirely known, especially when dealing with potentially negative consequences (5).

Root cap: A grouping of plant cells that create a protective covering over root meristems (1).

Root meristem: A collection of cells that divide to form different tissue of a plant root system.

Root suckers: A plant sprout that originates from a place in its root system.

Rooting zone: The area around a plant that is required for root development throughout the life of the plant.

Rotation: The period of time that is allowed to pass between regeneration and final harvest operations. this term is typically applied to even-aged forests, and the period of time can be based on many factors (1).

Rotation age: see **Felling age**.

Rotation age: The age of a forest at which nearly all of the trees are harvested.

Roughage: Plant matter that is composed mainly of fiber and trace amounts of digestible nutrients. It is typically bulky or coarse, and either dry or green (1).

Roundwood: A part of a tree that is perceivably very round in cross-section, and used in further manufacturing processes (1).

Row thinning: A thinning operation where every x number of rows in a plantation is harvested.

Row thinning with selection: A thinning operation where every x number of rows in a plantation is harvested, then a selection practice is implemented on the trees between those rows.

Running skyline: A cable-based harvesting system that utilizes at least two moving lines suspended in the air to transport logs from the location where trees were felled to a landing (1).

Runoff: The excess precipitation (water) that did not infiltrate the soil, and subsequently was allowed to flow to another area over land or by means of stream channels (1).

Rural—urban fringe: A general term for human development that has occurred near, or abuts, forested or other natural areas.

Salinity: The amount or concentration of salt that is within a given sample of soil or water, generally.

Salvage cutting: The harvesting of dead, dying, or damaged trees in order to capture what would be lost if the trees were not harvested (1).

Sample: A portion of a population that is examined on the premise that it is representative of the whole. Alternatively, the act of selecting and measuring certain individuals out of a population for examination (1).

Sample plot: A small location within a system (forest or otherwise) that is considered to be a good representation of the system as a whole and is therefore utilized for inventory and research purposes (1).

Sampling schemes: The patterned or randomized methods in which samples (i.e., trees) are selected to represent a larger population (i.e., forest).

Sanctuary: An area that protects wildlife from human pressure, and in some instances protects from predators and parasites (1).

Sanitation cutting: The felling of a tree(s) in a forest in order to protect the other individuals within the system, particularly from diseases or insects (1).

Sapwood: The layers of wood outside the heartwood, that consist of living cells whose purpose is to allow the osmosis of water through the tree (1).

Savannas: Grassy ecosystems that may contain a variable density forest where trees are widely spaced, thus the tree canopy does not completely close. Savannas can be found in any biome, and represent a transition between closed canopy forests and prairies or deserts.

Sawtimber: Trees or logs that are large enough (diameter and height), and of high enough quality to be appropriate for milling into lumber (1).

Scarcity: The characteristic of being in some way limited, either by speed of production or simply by known existence on the Earth (5).

Scenic river: A river or portion thereof, that has been designated by legislation as such, or as being without natural or artificial dams or reservoirs, and having a very basic shoreline and watershed that has been essentially undeveloped (1).

Secondary growth: The relatively horizontal enlargement of a tree's stem, branches, and roots for which the cambium is responsible (1).

Secondary succession: The successional stage that often occurs after a disturbance, natural or human-caused, such as a tornado or final harvest. Secondary succession differs from primary succession in that primary succession begins with bare material (rock) while secondary succession begins with at least bare soil. One can consider secondary succession as starting later in the successional process than primary succession.

Secondary transportation: The transportation of logs or other woody products, from a landing to a facility where they will be processed into usable wood products.

Sediment: Small particles of mineral and organic matter that are suspended and transported from one site to another by some force such as gravity, water (liquid or solid), or air. This term can also be applied to the accumulation of these individual particles in one area (1).

Sedimentation: The process by which sediment is created and accumulated in one place (1).

Seed-tree system: A forest regeneration system in which all the trees in a stand are harvested except for a select few left in strategic locations in order to naturally regenerate the area. Typically after the area has been successfully regenerated, another harvest entry is implemented to remove the seed trees (1).

Seed-tree: A tree that has been excluded from a harvesting operation in order to provide a natural means (seed) of regeneration (1).

Seepage: Water coming up from below the surface of the ground over an expanded area. Alternatively, the movement of water downward, naturally through the soil (1).

Selection thinning: A harvesting operation in which taller, more mature trees are harvested in order to aide in the growth of those that make up the lower canopy (1).

Selective cutting: A thinning operation that does not remove all of the trees in a forest stand (1).

Selective thinning: The removal of inferior trees in all tree size classes to promote the continued development of superior trees.

Self-supporting practice: A management practice that returns revenue in such a proportion that covers at least the cost of implementing the practice, that is, it generates enough income to pay for itself.

Seral stage: Any individual one of the many transitional stages a forest will experience during succession, from bare ground to climax condition (1).

Sere: A specific stage within forest succession (1).

Shade intolerant: Having the ability to live, grow, and compete for other resources, yet only in conditions with minimal amounts of shade (1).

Shade tolerant: Having the ability to live, grow, and compete for other resources in conditions with moderate to heavy amounts of shade (1).

Shake: A thin piece of wood used as a roofing material. Alternatively, a crevasse in a log or bole of a tree that runs alongside a growth ring for a considerable distance, or a crack in lumber running along the grain but not splitting all the way to the other side, caused by stress (1).

Shelterbelt: A section of trees or shrubbery that is managed in order to modify winds, and sometimes temperatures, in such a way as to protect something of importance (agricultural fields or buildings) (1).

Shelterwood: A harvesting system in which most of the trees are felled but some are left to provide protection for the new forest by way of either shade or wind protection (1).

Shoot apical meristem: The aggregation of cells at the tip of a shoot that divide to facilitate the growth of the root system.

Short rotation: A rotation system that is based on short harvesting intervals, these systems are typically employed on stands being harvested for pulpwood.

Short run: In economics, a length of proceeding time in which at least one variable cannot be changed making it a fixed variable, therefore limiting accessible capital of the producer.

Shovel logging: A skidding operation employing swing machines (loaders) to move felled timber from one pile to another in the direction in which the skid is moving (1).

Signal-to-noise ratio (SNR): A measure of the high-quality information content of a global navigation satellite system (e.g., GPS) signal relative to the lower-quality noise associated with the signal. If the SNR is 20 there is a 20:1 ratio of high-quality to lower-quality information.

Silviculture: The branch of forestry that deals with planting, managing, and maturing of forests in such a way that the landowner's objectives are met in a sustainable manner (1).

Silvopastoralism: A branch of agroforestry that focuses on trees and livestock, that is, both timber and animals are grown and managed under the same system (1).

Single-grip harvester: A harvester that has one articulated arm for felling and bucking a tree bole into specific lengths in one fluid motion.

Site: A defined area where a forest stand will be planted or where one currently grows (1).

Site class: A measurement of the quality a site possesses, often expressed in dominant tree height, or potential annual growth (1).

Site preparation: Any of the numerous practices that can be implemented in order to facilitate tree survival or germination in artificial or natural forest regeneration (1).

Skid road: A path created through the forest and used for skidding logs or whole trees. Also called a skid trail (1).

Skidder: Any of many machines used to transport logs from the felling site to a landing (primary transportation) (1).

Skidding: The movement of logs from a felling site to a landing (primary transportation) (1).

Skyline: A harvesting system component that consists of a cable pulled tight between two locations, that is used as a track for a yarding carriage (1).

Slack cooperage: Wooden barrels that are not tightly constructed or bound. Typically used to hold larger items that would not easily or at all pass through small cracks, such as nails.

Slackline system: A cable-based harvesting system that employs a moving skyline and a carriage attached to the mainline and haulback line (1).

Slide-boom delimber: A machine used to delimb trees at a landing.

Snag: A dead and un-merchantable, yet standing tree that is regarded as a good source of wildlife habitat. Alternatively, a log, stump, or entire tree that is wholly or partially underneath the surface of a body of water, or the cantilevering stob of a broken or trimmed branch (1).

Social cost: The sum of external (air, noise, water, and light pollution) and internal (actual cost to the owners) costs of an organization (5).

Socioeconomic outcomes: End results of an action, situation, or natural occurrence that affect the social and economic state of the people in the region in which it occurred.

Softwood: Any tree in the group *Gymnosperm* or the xylem thereof, basically consisting of longitudinal tracheids with small traces of rays and resin ducts. Typically coniferous tree species are called softwoods, regardless of how soft or hard the actual wood is (1).

Soil moisture deficit: The potential amount of water that can be added to a sample before it reaches 100% saturation.

Soil moisture tension: A measure of difficulty to move or remove moisture from the soil.

Space: The coverage of an area, which is typically shown by way of some measurement of length (ft^2, m^3, etc.) or area (hectares, acres) (3).

Spatial analysis: The evaluation of land using observation, modeling, and analysis to determine the potential and appropriateness of the land for different uses. Alternatively, geographic analyses for specific requests of the landowner or manager (1).

Spatial data: Specific measurements illustrating the site, form, position, contour, correlation, and relationship of geographic features, often represented by means of mapping (1).

Special interest species: A species whose population is in some way threatened, and that only inhabits an area during certain parts of the year (i.e., winter) or their life cycle (i.e., breeding).

Species: A group of animals that possess certain defining characteristics, have similar form, have the same morphology, and can interbreed and produce successful offspring (1).

Species redundancy: The environmental phenomena where there are more than one species that perform the same service for the ecosystem. This allows one species to become extinct or move out of the area, and as long as the other species' population can expand to a size at which they will be able to make up for the loss of the other species and the ecosystem can continue to function as normal.

Spot fire: A fire that is initiated outside of a main fire area by means of sparks or embers carried by the wind (1).

Sprout: A shoot that is growing out of the base of a plant (1).

Standing line: A cable, use in a cable-based harvesting system, that does not move during the operation, that is, it is secured at each end (1).

Steppe forest: Transition zones between boreal and temperate biomes.

Stomata: Tiny openings in the bottom of leaves that are the way by which gas and water are exchanged from the plant to the surrounding air (1).

Stratification: The dividing of a population, based upon certain characteristics or variables, into different strata. Alternatively, the silviculture practice of exposing a seed to ideal germination conditions in order to cause it to forgo dormancy and germinate (1).

Stroke-boom delimber: A machine used to delimb trees at a landing.

Sun-synchronous satellite: A satellite that travels around the Earth in a fixed relationship with the sun (3).

Sunk cost: A cost that is already incurred and cannot be retrieved (5).

Supporting service: Ecosystem services that are necessary for the production of all other ecosystem services.

Surface fire: A fire that occurs at ground level of a forest and generally only consumes the surface fuels (forest litter, coarse woody debris, and low-growing vegetation) (1).

Surface fuels: Any source of combustible material that is located on or near the forest floor, typically forest litter, coarse woody debris, and low-growing vegetation (1).

Sustainability: The ability of an object, population, or area to continue to function at its current or healthy level by natural means or through human intervention (1).

Sustainable forest management (SFM): The management, stewardship, preservation, and conservation of a forest or forestland in such a way that current use or harvesting operations do not adversely affect the ability of future generations to benefit likewise (1).

Sustained yield: The amount of timber that can be harvested at a consistent and continual rate, based on the current intensity of management practices. Alternatively, the state of a management system that allows continual harvesting of a certain number of trees in regular intervals, without extreme adjustment to current management practices (1).

Swamp: An area of land that is essentially always covered by water and that is inhabited with numerous trees, shrubs, and woody vegetation (1).

Sweep: A measure of the variance of a tree's bole from an ideal straight state, typically represented by degrees (1).

Systematic sample: A sampling scheme whose individuals are selected by some type of linear, geometric, or other nonrandom selection process, although the starting individual may be chosen at random (1).

Systematic thinning: A thinning operation in which every x number of rows is harvested, and a selective thinning is implemented within the remaining rows.

Tactical harvest schedule: A forest plan that is 2–20 years in length, in general, and accounts for spatial issues in the management of resources.

Temperate: A region, or something relating to such a region, characterized by having an average temperature of $\pm 50°F$ ($10°C$) for about two to four months of the year (1).

Temperate forest: A forest which is characterized by a well-defined mild winter and an average temperature range of -30 to 30 degrees Celsius (°C) (-22–86 degrees Fahrenheit (°F)). Rainfall is spread basically evenly throughout the year, and soils typically exhibit higher nutrient concentration and decomposition rates than those of a boreal forest due to the more extreme summer temperatures. There are two types of temperate forests, temperate coniferous and temperate deciduous. There are many subdivisions of these two types and are classified by changes in topography and climate.

Temporary trade-off: Any trade-off that can or will be recovered within a short period of time.

Terrestrial: Of or relating to dry land, or of a species that is dependent on dry land for the essential requirements of its life cycle (2).

Thinning entry: The act of entering a forested area and removing a portion of the trees for various reasons (i.e., to generate revenue, to improve health, etc.).

Thinning from above: The removal of the lower-quality trees in the uppermost crown classes in order to benefit those of higher quality in the same or lower crown classes (1).

Thinning from below: The removal of suppressed or intermediate crown class trees in order to reduce competition for those trees in the uppermost crown classes (1).

Thinning intensity: A combination of the number of trees removed and the entry interval (1).

Thinning interval: The period of time that is allowed to pass before another harvest entry occurs (1).

Thinning proportional to the diameter of distribution: The removal of trees of sizes proportional to the diameter distribution prior to thinning.

Thinning regime: A term that represents the basic sum of all thinning activities, including type, grade, entry interval, and often including the beginning and ending year (1).

Threatened species: Any species, plant or animal, that is at risk of becoming endangered throughout all or most of its natural range within the near future (1).

Tiaga: see **Boreal**.

Tight cooperage: Wooden barrels that are tightly constructed or bound. Typically these can hold things that will easily pass through the slightest cracks, such as wine.

Timber: A stand of merchantable forest crop, or alternatively wood that is used in the production of solid wood products (1).

Timber marking: The indication, by some means (flagging, painting, etc.), of single trees that need to have some type of forest practice implemented upon them (cutting, pruning, etc.) (1).

Timber sampling: The practice of gathering information about a stand (i.e., measurements or core samples) in order to gauge growth rates, habitat quality, or merchantable volume.

Timberbelt: A shelterbelt that is not only used from protection but also revenue generation through timber production (1).

Timberland: Forestland that is composed of merchantable trees (timber) (2).

Time preference: An economic assumption that a consumer would rather have goods or services in the present than sometime in the future, especially without compensation (5).

Topographical map: A map that shows the contour, to scale, of the geographic features of an area (3).

Topography: The art or practice of precisely documenting the geographic features and contour of an area on a two-dimensional, planar surface area. Alternatively, the physical characteristics, natural or man-made, of an area, including the grade and extent of the ground slope (2).

Tornado: A high energy terrestrial storm or cyclone that arises from high energy storm cells.

Tourism: The act of traveling to a place outside of where one resides in order to experience something new or different (2).

Tracheid: A wood cell, shaped much like an elongated football, that after maturation becomes a conducting cell (1).

Trade-off: The act of giving up one object, resource, or benefit in exchange for another. Alternatively it is often used to simply refer to what has been given up (5).

Tree: A plant characterized by a woody, well-defined stem or stems, as being perennial, and typically having a large, defined crown (1).

Tropical: Regions that are characterized by such climatic conditions as high temperatures, high humidity and rainfall, and rarely a light frost (1).

Tropical rainforest: A forest with generally two seasons, but one of them is not cold. Rainforests are generally found at lower latitudes around the equator, and trees may be required to endure both rainy and dry seasons. In general, the availability of daylight is very consistent, and variations in air temperature are minimal year-round.

Turbulence: The flow of wind or liquid which displays unpredictable fluctuations in velocity (1).

Turgor: The pressure of water within cells of a plant (1).

Uncertainty: An awareness of a lack of information or knowledge regarding a present or future situation, scenario, or possibility (5).

Understory: The plant life in a forest that resides and lives in the vertical space between the forest floor and the lower reaches of a tree canopy (1).

Ungulate: An animal that has hoofed feet (2).

Unlimited: Not bounded by any exception, constraint, or boundary (2).

Unlimited resource: A resource that literally, virtually, or conceivably cannot be completely consumed because of its vastness, regenerative characteristics, or ability to cycle through the environment.

Unreserved forest land: Forests that have no prohibiting regulations placed upon them to prevent actions, such as harvests, from occurring (1).

Urban forest: A forest that is located within or near city limits, and one that is used for, among other things, recreation, picnicking, walking, wildlife-watching, and other purposes.

Urban forestry: The planning and implementation of forest management plans for urban forests to serve an urban community and provide multiple benefits to their surroundings (1).

Urban hinterland: see **Rural—urban fringe**.

Urbanization: The movement in a human population from a rural setting to a more urbanized setting, caused or catalyzed by a perceived increase in jobs or wages in the urban areas (3).

Utility: The fitness, satisfaction, worth, or usability of a resource, providing an indication of the value of the resource (1).

Value statement: In real estate, any term usually used that is used to help describe an attribute of the land in question, which can have an effect land value. In other areas, any statement made by a person that is rooted in their value system.

Variable cost: A cost incurred that is directly dependent on the level of production (5).

Vascular cambium: A vertical ring of cells that is the secondary site of gas conduction in stems and roots (1).

Vegetation: Living plant material, or the sum all of plant material in a certain area (2).

Veneer: A solid wood product that is made by using a large blade or saw to cut, slice, or saw relatively thin sheet of wood from logs (1).

Vertical aerial photograph: An aerial photograph that was taken with the lens essentially perpendicular and the film essentially horizontal with the ground (1).

Virus: A parasite that is capable of producing a disease. These are characterized by being too small to be seen with an optical microscope, and as being composed of a nucleic acid surrounded by a protein coat, and only being capable of reproducing in living tissue (1).

Weeding: The practice of removing undesirable vegetation (weeds) in order to prevent competition for saplings, this practice is not employed once a forest has reached sapling stage (1).

Wet forest: A forest that receives moderate to excessive rainfall throughout the year and that typically does not experience long dry periods.

Wetland: The intermediary area between an aquatic and terrestrial ecosystem that is almost completely saturated for extended periods of time, allowing hydrophilic plants and animals to live there (1).

Wild river: A river or a portion thereof that is without human development, consisting of primitive shoreline and watershed, and difficult to access other than by means of a trail (1).

Wilderness: An area that has had no human development within its boundaries, or an area distinguished as such by legislation (1).

Wilderness area: An area that is protected from human development, even in its smallest form, by some aggregation or administration, these areas are typically large governmentally owned areas intended for the aesthetic qualities (2).

Wildfire: A fire that was either initiated or allowed to progress without intention, or a fire that has no organization or purpose, particularly those out of direct and total control (1).

Wildland: Land that has not been developed for any human use. Alternatively, land that, for a significant period of time, has not been used or developed for human activity (1).

Wildland urban interface: see **Rural—urban fringe**.

Wildlife: Any animal not successfully domesticated (1).

Wildlife corridor: A section of land that connects two or more separated habitat areas. These lands are typically narrow and highly used by animals traveling from one area to another (1).

Wildlife habitat: An area of land that meets the nesting, roosting, and foraging needs of a species in an acceptable manner.

Wildlife tree: A tree that is either planted, managed, or naturally occurring, and in some way benefits wildlife, especially in the form or habitat (1).

Wind: The natural or artificial movement of air, mainly horizontal, occasionally vertical (2).

Windbreak: A line of trees or shrubbery either planted or managed in such a way as to protect a building or crops, or alter climate or wind (1).

Wood: Any wooded area. Alternatively, the matter that is produced within the limbs and stems of all trees and woody vegetation (1).

Wood fiber: Individual fibers that compose the wood of a tree.

Wood product: Any product that is composed of wood tissue.

Wood residues: Any woody substance left over a harvesting or milling operation is complete.

Woods road: A road in a forest that has either a native surface or a rocked surface.

Woody biomass: Woody material that can be used to develop energy.

Xeric forest: Forests that are situated in an arid environment, such as those that occur in northern Mexico and the southwestern United States.

Xylem: Plant tissue that is responsible for the conduction of water (tracheids and vessels), structural support (tracheids and fibers), and food transport and storage (parenchyma) (1).

Yard: A place where logs are taken to be sorted, to be prepared to move again, or both (1).

Yarder: A cable-based harvesting machine or system that includes a tower and power-operated winches for moving logs from where trees were felled to a landing (1).

Yarding: The primary transport of harvested timber (logs, trees) from the location where trees were felled to a landing. This term is particularly applicable when the transportation is carried out by way of cable, balloon, or helicopter-based logging systems (1).

References

(1) Helms, J.A. (Ed.), 1998. The Dictionary of Forestry. The Society of American Foresters, Bethesda, MD, 210 p.

(2) Merriam-Webster, Inc., 2011. Merriam-Webster Dictionary Online. Merriam-Webster, Inc., Springfield, MA.

(3) Mayhew, S., 1997. A Dictionary of Geography. Oxford University Press, Oxford, UK, 460 p.

(4) Friedman, J.P., 1994. Dictionary of Business Terms. Barron's Educational Series, Inc., Hauppauge, NY, 692 p.

(5) Black, J., 1997. A Dictionary of Economics. Oxford University Press, Oxford, UK, 512 p.

Index

Note: Page numbers followed by "f" indicate figures and "t" indicate tables.